Student Solutions Manual

Chemistry
The Molecular Science

FOURTH EDITION

John W. Moore
University of Wisconsin, Madison

Conrad L. Stanitski
Franklin and Marshall College

Peter C. Jurs
Pennsylvania State University

Prepared by

Judy L. Ozment
Pennsylvania State University

Australia • Brazil • Japan • Korea • Mexico • Singapore • Spain • United Kingdom • United States

For product information and technology assistance, contact us at **Cengage Learning Customer & Sales Support, 1-800-354-9706**

For permission to use material from this text or product, submit all requests online at **www.cengage.com/permissions**
Further permissions questions can be emailed to **permissionrequest@cengage.com**

ISBN-13: 978-1-4390-4963-1
ISBN-10: 1-4390-4963-7

Brooks/Cole
20 Davis Drive
Belmont, CA 94002-3098
USA

Cengage Learning is a leading provider of customized learning solutions with office locations around the globe, including Singapore, the United Kingdom, Australia, Mexico, Brazil, and Japan. Locate your local office at: **www.cengage.com/global**

Cengage Learning products are represented in Canada by Nelson Education, Ltd.

To learn more about Brooks/Cole, visit **www.cengage.com/brookscole**

Purchase any of our products at your local college store or at our preferred online store **www.cengagebrain.com**

Printed in the United States of America
1 2 3 4 5 6 7 14 13 12 11 10

Table of Contents

Solutions for Blue-Numbered Questions for Review and Thought

Introduction

This solutions manual was written specifically to accompany the fourth edition of the textbook *Chemistry The Molecular Science*, by Moore, Stanitski, and Jurs. It presents detailed solutions for the some of the Questions for Review and Thought at the end of each chapter. If a question's number is bold and blue, then its solution will be in this book.

Using this Book

Many of these solutions are presented using the same format described in Chapter 1 and Appendix A. The *Strategy and Explanation* section usually uses three of the four-stage process: analyze the problem, plan a solution, and execute the plan, then the *Reasonable Answer Check* shows a check of the answer to see if it is reasonable. Following these stages should help to make methods more readily applicable to similar problems, or the same kind problem in a different context. Some of the solutions are patterned after the methods shown in Problem-Solving Examples throughout the text.

It is important to try to answer a question for yourself, before looking at the solutions book. When you find it necessary to use this book, try first to use it to get hints or directions by reading the first part of the *Strategy and Explanation* section to see if that information clarifies how you analyze the question. If you find that your interpretation or evaluation of the problem is routinely incorrect or incomplete, then you might seek help from your instructor, a teaching assistant, a learning center staff person, or a tutor in reading word problems for comprehension. Sometimes, the best help for some problems can be gained from math tutors, since they often have experience helping students specifically with reading word problems.

If you find that you usually analyze the problem in a similar fashion as described here, but still need help understanding how to plan a solution, then read how the plan is developed. This will give general step-by-step instructions for how the problem is solved. For numerical answers, it is at this point where you should try to estimate what answer you anticipate. Do you expect it to be a large or small number? What units and sign do you expect it to have? What significant figures will it have? The more you are able to frame an expectation for the answer, the less likely you are to make mistakes along the way.

Step-by-step solutions are shown after the plan is described. If you can analyze the questions, plan solutions, and execute the plans before actually looking here, you will start gaining confidence in your ability to learn how to answer the questions on your own.

Once you have the answer to a question, take a moment to think about whether it makes sense. Often, reflection will help you confirm the correctness of an answer, or expose its flaws. For example, if you just determined that an atom of gold weighs four times more than the mass of the entire earth, the *Reasonable Answer Check* stage of the problem-solving method might help you see that you made an error. It is wise to go back to the question and read it again, then ask yourself: Does this result answer this question? Is the result the right size and sign? Is it what you expected? It is important to check the units and the significant figures at this point, also.

A true sign that you are learning how to do chemistry problems is when you find yourself relying on this solutions book less and less. Set goals for yourself whenever you use this book to limit how often you consult it and how extensively. It is easy to such books as a crutch, and crutches keep you from walking on your own.

There are many ways to solve problems. Often times, the right answer can be derived in several ways that are equally valid. Your instructors may have different ways of describing how to work some of the problems solved here. All good methods have three things in common: (1) They always give the right answer. (2) They demonstrate how the answer was achieved. (3) And, they make sense. If any one of these three is missing, the method is flawed.

Cautions

Resist the temptation to just read the solutions in this book without doing any work on your own. There are two very obvious reasons for this: (1) Recognition is easier than recall. It is far easier to look over a solution done correctly and believe that you understand it, than it is to look at the same question followed by a blank space (as would happen on a test, for example) and recall how to do it. (2) Most teachers will not let you use this book when they are testing you; hence, these solutions will not be there when you need to know how to do things.

All teachers using the textbook *Chemistry – The Molecular Science* also have a copy of the solutions in this book. Resist the temptation to copy work from this book into your assignments and call it your work. Besides being unethical, such practice defeats the purpose of an assignment designed to have you practice problem-solving. The instructor is asking you to do your own work, and to show what you have learned about the subject.

Learn How to Learn

Finally, keep in mind that you are learning basic chemistry and introductory physical science as a building block for other things. Those things are much more complicated. Always try to learn a subject with maximum flexibility. Whenever possible, look for how a solution can be generalized. Relying on rote memorization and narrowly-defined systems used to solve very specific types of problems may work to get you through this course, but it can be detrimental to your learning science or learning how to apply science to more complicated things, including life, the universe and everything. Sometimes, you will have to give up on preconceived notions to be able to learn more.

> "The man who grasps principles can successfully select his own methods. The man who tries methods, ignoring principles, is sure to have trouble."
> – Ralph Waldo Emerson

> "Imagination is more important than knowledge."
> – Albert Einstein

> "I know I have not found the answers to all of my questions. The answers I have found only serve to raise a whole set of new questions. In some ways I am as confused as ever, but I believe that I am confused on a higher level and about more important things."
> – unknown

Acknowledgements

I want to especially thank Karen Pesis, from the American River College in California, for her patient and expert help in checking the accuracy of these solutions. I greatly appreciate the assistance of Stephen Meckler for his assistance in checking several key chapters in this edition. I greatly appreciate the assistance and support of all three of the book authors, Dr. Peter Jurs, Dr. Conrad Stantiski, and especially Dr. John Moore. I also want to thank Leslie Kinsland again, since much of what she helped me learn on the very first solutions manual I wrote is still a part of these newer ventures.

Lastly, I gratefully acknowledge the fantastic support of my family, especially Karen and Emma Pesis, Lynda Webb, Susan Thompson, and Loretta Ozment, and good friends, especially Sara Shriner, Paul Bomboy, Lori Campbell, Lynne O'Cain, Jared Lipton, Carolyn Weber, and Alex Thomson.

Chapter 1: The Nature of Chemistry

Solutions for Blue-Numbered Questions for Review and Thought

Topical Questions

How Science is Done (Section 1.3)

11. *Answer:* **(a) quantitative (b) qualitative (c) qualitative (d) quantitative and qualitative (e) qualitative**

Strategy and Explanation:

(a) The temperature at which an element melts (29.8 °C) is **quantitative** information. Information about what element it is (gallium) is **qualitative**.

(b) The color of a compound (blue) and information about specific elements it contains (cobalt and chlorine) are both **qualitative**.

(c) The fact that a metal conducts electricity and information about what element it is (aluminum) are **qualitative**.

(d) The temperature at which a compound boils (79 °C) is **quantitative** information. Information about what compound it is (ethanol) is **qualitative**.

(e) The details of the appearance of the crystals of a compound (shiny, plate-like, and yellow) and information about specific elements it contains (lead and sulfur) are both **qualitative**.

13. *Answer:* **(a) qualitative (b) quantitative (c) quantitative and qualitative (d) qualitative**

Strategy and Explanation:

(a) The details of the appearance of a substance (silvery-white) and information about the specific element it contains (sodium) are both **qualitative**.

(b) The temperature at which a solid melts (660 °C) is **quantitative** information. Information about what element it is (aluminum) is **qualitative**.

(c) The mass percentage of an element in the human body (about 23%) is **quantitative** information. Information about what element it is (carbon) is **qualitative**.

(d) The allotropic forms of an element (graphite, diamond, and fullerenes) and information about what element it is (carbon) are both **qualitative**.

Identifying Matter: Physical Properties (Section 1.4)

15. *Answer/Explanation:* Bromine is a reddish-brown liquid. Sulfur is a chalky yellowish solid. They appear to have no property in common. The physical phase, shape, color, and appearance are different.

17. *Answer:* **The liquid will boil because your body temperature of 37°C is above that boiling point of 20°C.**

Strategy and Explanation: Many Americans only remember the human body temperature in the Fahrenheit scale. That is 98.6 °F. If that is the case, we can quickly apply the °F to °C conversion equation, so we can compare it to the boiling point.

$$^{\circ}\text{C} = \frac{5}{9} \times \left(^{\circ}\text{F} - 32\right) = \frac{5}{9} \times \left(98.6 - 32\right) = 37.0^{\circ}\text{C}$$

If the liquid boils at a temperature of 20 °C and your hand is 37 °C, the liquid will boil when exposed to the heat energy emitted by your hand when you hold the sample.

19. *Answer:* **(a) 20 °C (b) 100 °C (c) 60 °C (d) 20 °F**

Strategy and Explanation: We use these questions about temperature using different scales to help us become familiar with the Celsius scale. So, it is important to try estimating the answer before using equations. For a while it is useful to have a few connection points to a familiar scale. For those reasons, let's make some key connections between the Celsius scale and the Fahrenheit scale

Compare two thermometers side-by-side, one calibrated with a Celsius scale and one calibrated with a Fahrenheit scale:

100 °C — 212 °F

37 °C — 98.6 °F

0 °C — 32 °F

(a) 20 °C (above freezing) is higher than 20 °F (below freezing).

(b) 100 °C (at boiling) is higher than 180 °F (below boiling)

(c) 100 °F is close to body temperature, which is around 40 °C. Therefore 60 °C is higher than 100 °F.

(d) – 12 °C and 20 °F are both below freezing, so we will use the conversion equation to calculate this one:

$$^{\circ}C = \frac{5}{9}\times\left({}^{\circ}F - 32\right) = \frac{5}{9}\times\left(20\ {}^{\circ}F - 32\right) = -\ 6.7\ {}^{\circ}C$$

The temperature of – 6.7 °C is warmer than – 12 °C, so 20 °F is a higher temperature than – 12 °C.

21. *Answer:* **copper**

Strategy and Explanation: We have the mass of the metal and some volume information. We need to determine the density. Use the initial and final volumes to find the volume of the metal piece, then use the mass and the volume to get the density. The metal piece displaces the water when it sinks, making the volume level in the graduated cylinder rise. The difference between the starting volume and the final volume, must be the volume of the metal piece:

$$V_{metal} = V_{final} - V_{initial} = (37.2\ mL) - (25.4\ mL) = 11.8\ mL$$

$$d = \frac{m}{V} = \frac{105.5\ g}{11.8\ mL} = 8{,}94\ \frac{g}{mL}$$

According to Table 1.1, this is very close to the density of copper (d = 8.93 g/mL).

✓ *Reasonable Answer Check:* The metal piece sinks, so the density of the metal piece needed to be higher than water. (Table 1.1 gives water density as 0.998 g/mL.)

23. *Answer:* **aluminum**

Strategy and Explanation: We have the three linear dimensions of a regularly shaped piece of metal.

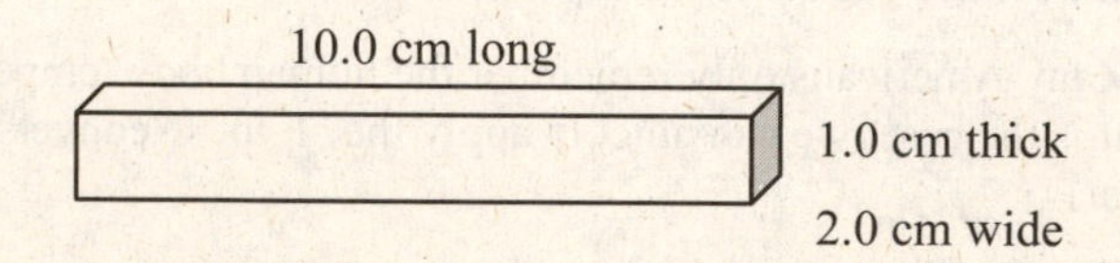

We also have its mass. We have a table of densities (Table 1.1). We need to determine the identity of the metal.

Use the three linear dimensions to find the volume of the metal piece. Use the volume and the mass to find the density. Use the table of densities to find the identity of the metal.

$$V = (\text{thickness}) \times (\text{width}) \times (\text{length}) = (1.0\ cm) \times (2.0\ cm) \times (10.0\ cm) = 20.\ cm^3$$

Using dimensional analysis, find the volume in mL: $20.\ \text{cm}^3 \times \dfrac{1\ \text{mL}}{1\ \text{cm}^3} = 20.\ \text{mL}$

Find the density: $d = \dfrac{m}{V} = \dfrac{54.0\ \text{g}}{20.\ \text{mL}} = 2.7\ \dfrac{\text{g}}{\text{mL}}$

According to Table 1.1, the density that most closely matches this one is **aluminum** (d = 2.70 g/mL).

✓ *Reasonable Answer Check:* An object whose mass is larger than its volume will have a density larger than one. The mass of this object is between two and three times larger than the volume.

25. *Answer:* **3.9 × 10³ g**

Strategy and Explanation: We have the three linear dimensions of a regularly shaped sodium chloride crystal:

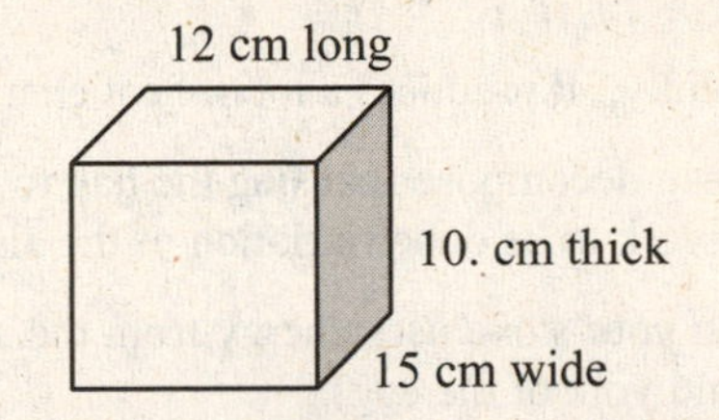

We have a table of densities (Table 1.1). We need to determine the mass of the crystal.

Use the three linear dimensions to find the volume of the crystal. Use the volume and density (Table 1.1) to find the mass.

$$V = (\text{thickness}) \times (\text{width}) \times (\text{length}) = (10.\ \text{cm}) \times (15\ \text{cm}) \times (12\ \text{cm}) = 1800\ \text{cm}^3 = 1.8 \times 10^3\ \text{cm}^3$$

Using dimensional analysis, find the volume in mL:

$$1.8 \times 10^3\ \text{cm}^3 \times \frac{1\ \text{mL}}{1\ \text{cm}^3} = 1.8 \times 10^3\ \text{mL}$$

Find the mass using the density: $1.8 \times 10^3\ \text{mL} \times \dfrac{2.16\ \text{g}}{1\ \text{mL}} = 3.9 \times 10^3\ \text{g}$

Notice: We're carrying two significant figures since the length data was only that precise.

✓ *Reasonable Answer Check:* This crystal is pretty big. So, while the mass calculated is a large number, the volume is still about half the mass, in keeping with a density around two.

Chemical Changes and Chemical Properties (Section 1.5)

27. *Answer:* **(a) physical (b) chemical (c) chemical (d) physical**

Strategy and Explanation:

(a) The normal color of bromine is a **physical** property. Determining the color of a substance does not change its chemical form.

(b) The fact that iron can be transformed into rust is a **chemical** property. Iron, originally in the elemental metallic state, is incorporated into a compound, rust, when it is observed to undergo this transformation.

(c) The fact that dynamite can explode is a **chemical** property. The dynamite is chemically changed when it is observed to explode.

(d) Observing the shininess of aluminum does not change it, so this is a **physical** property. Melting aluminum does not change it to a different substance, though it does change its physical state. It is still aluminum, so melting at 387 °C is a **physical** property of aluminum.

29. *Answer:* **(a) chemical (b) chemical (c) physical**

Strategy and Explanation:

(a) Bleaching clothes from purple to pink is a **chemical** change. The purple substance in the clothing reacts with the bleach to make a pink substance. The purple color cannot be brought back nor can the bleach.

(b) The burning of fuel in the space shuttle (hydrogen and oxygen) to form water and energy is a **chemical** change. The two elements react to form a compound.

(c) The ice cube melting in the lemonade is a **physical** change. The H_2O molecules do not change to a different form in the physical state change.

31. *Answer:* **(a) forcing a chemical reaction to occur (b) causing work to be done (c) causing work to be done (d) forcing a chemical reaction to occur**

Strategy and Explanation:

(a) The conversion of excess food into fat molecules is the body's way of storing energy for doing work later. So, this represents the outside source of energy (from the food we eat) is forcing a chemical reaction to occur (the production of fat).

(b) Sodium reacts with water rather violently. It produces a lot of heat energy and causes work to be done.

(c) Sodium azide in an automobile's airbag decomposes causing the bag to inflate. This uses a chemical reaction to release energy and cause work to be done (inflation of the air bag).

(d) The process of hard-boiling an egg on your stove uses energy from the stove to cause a chemical reaction to occur (the coagulation of the white and yolk of the egg).

Classifying Matter: Substances and Mixtures (Section 1.6)

33. *Answer/Explanation:* It is clear by visual inspection that the mixture is non-uniform (**heterogeneous**) at the macroscopic level. Iron could be separated from sand **using a magnet**, since iron is attracted to magnets and the sand is not.

35. *Answer:* **(a) homogenous (b) heterogeneous (c) heterogeneous (d) heterogeneous**

Strategy and Explanation: The terminology "heterogeneous" and "homogeneous" is somewhat subjective. In Section 1.6, the terms are described. A heterogeneous mixture is described as one whose uneven texture can be seen without magnification or with a microscope. A homogeneous mixture is defined as one that is completely uniform, wherein no amount of optical magnification will reveal variations in the mixture's properties. Notice the "gap" between these two terms. The question: "how close do I look?" comes to mind. The bottom line is this: These terms were designed to HELP us classify things, not to create trick questions. If you explain your answer with a valid defense, the answer ought to be right; however, don't go out of your way to imagine unusual circumstances that make the substance more difficult to classify. Think about what you would SEE. Identify whether what you see has variations, then make a case for the proper term.

(a) Vodka is classified as a **homogenous** mixture. It is a clear, colorless solution of alcohol, water, and probably some other minor ingredients.

(b) Blood appears smooth in texture and to our eyes most likely appears to be homogeneous. Upon closer examination it is found to have various particles within the liquid, and might, for that reason, be called **heterogeneous**.

(c) Cowhide is **heterogeneous**. Even folks who have never seen a cowhide might be able to imagine that brown cows, white cows, black cows and spotted cows probably have different coloration to their hide. There are probably pores where the hair grows out, making it rough in texture. Unless presented with a sample that visually changed our perception of what constitutes a cowhide, it is safe to call it heterogeneous.

(d) Bread is **heterogeneous**. The crust is a different color. Some parts of the bread have bigger bubbles than other parts. Some breads have whole grains in them. Some breads are composed of different colors (like the rye/white swirl breads). In general, most samples of bread have identifiable regions that are different from other regions.

Classifying Matter: Elements and Compounds (Section 1.7)

37. *Answer/Explanation:* Sometimes, it is necessary to try some tests to see if different parts of the mixture respond differently to physical separation techniques. It may require some experimentation, such as testing whether

samples of the pure substances dissolve in water or are attracted to a magnet. Such information may also be available on the internet.

(a) Table salt dissolved in water can be separated by **evaporating the water**, which would leave the dry salt.

(b) Testing shows that iron filings are attracted to a magnet, but magnesium pieces are not. **Using a magnet**, the iron filings can be lifted out of the mixture, leaving behind the magnesium pieces.

(c) This mixture will have shiny silver metal pieces in a white crystalline powder, so the first thing we could try is use tweezers or forceps to pick out the shiny metal zinc pieces from the white sugar crystals. Solubility testing shows that sucrose dissolves in water, but magnesium does not. Therefore, put the mixture in water, **dissolve the sucrose, separate the solution from the solid magnesium** using a filter or a sieve, then **evaporate the water**.

39. *Answer/Explanation:*

(a) A blue powder turns white and loses mass. The loss of mass is most likely due to the creation of a gaseous product. That suggests that the original material was a **compound that decomposed** into the white substance (a compound or an element) and a gas (a compound or an element).

(b) If three different gases were formed, that suggests that the original material was a **compound that decomposed** into three compounds or elements.

41. *Answer:* **(a) heterogeneous mixture (b) pure compound (c) heterogeneous mixture (d) homogeneous mixture**

Strategy and Explanation:

(a) A piece of newspaper is a **heterogeneous mixture**. Paper and ink are distributed in a non-uniform fashion at the macroscopic level.

(b) Solid granulated sugar is a **pure compound**. Sugar is made of two or more elements.

(c) Fresh squeezed orange juice is a **heterogeneous mixture**. The presence of unfiltered pulp makes part of the mixture solid and part of the mixture liquid.

(d) Gold jewelry is a **homogeneous mixture**, most of the time. Most jewelry is less than 24 carat—pure gold. If it is 18 carat, 14 carat, 10 carat gold, that means that other metals are mixed in with the gold (usually to make it more durable and cheaper). The combining of metals in a fashion that prevents us from seeing variations in the texture or properties of the metal qualifies it as a homogeneous mixture.

43. *Answer:* **(a) heterogeneous mixture (b) pure compound (c) element (d) homogeneous mixture**

Strategy and Explanation:

(a) Chunky peanut butter is definitely a **heterogeneous mixture**. The uncrushed peanut chunks do not have the same properties as the smooth, sweetened part of the mixture.

(b) Distilled water is a **pure compound**. The distillation process removes other minerals and substances from water, leaving it just water.

(c) Platinum is an **element** with the symbol "Pt".

(d) Air is usually considered to be a **homogeneous mixture**. Now, sometimes air has enough variable properties to qualify as heterogeneous, such as near the tailpipe of a diesel truck. However, most of the time, the gases in a sample of air are sufficiently well mixed such that there is no visible difference in the properties of various regions of that air sample.

45. *Answer:* **(a) No (b) Maybe**

Strategy and Explanation:

(a) The black substance was both the source of the element that contributes to the red-orange substance and the source of the oxygen in the water.

(b) The red-orange substance may be a combination of two or more elements including possibly hydrogen or oxygen, or it maybe an elemental substance, since the water produced could account for the fate of the hydrogen and the oxygen.

Nanoscale Theories and Models (Section 1.8)

47. *Answer:* **macroscopic world; a parallelpiped shape; the atom crystal arrangement is parallel piped shape.**

Strategy and Explanation: The crystal of halite pictured is in the macroscopic world. Its shape is cubic or parallelepiped. It could be expected that the arrangement of particles (atoms and/or ions) in the nanoscale world are also arranged in a cubic or parallelepiped fashion.

49. *Answer/Explanation:* Using Figure 1.14 to help define scales. The bacterium is in the **microscale world**.

51. *Answer/Explanation:* Using Figure 1.14 to help define scales. The images of electrons from the scanning tunneling microscope are at the **nanoscale**.

53. *Answer/Explanation:* When we open a can of a soft drink, the carbon dioxide gas expands rapidly as it rushes out of the can. At the nanoscale, this can be explained as large number of **carbon dioxide molecules crowded into the unopened can**. When the can is opened, the molecules that were about to hit the surface where the hole was made continue forward through the hole. A large number of the carbon dioxide particles that were contained within the can very quickly **escape through the** same **hole**.

55. *Answer/Explanation:* The atoms in the solid sucrose molecules start off at a relatively low energy and compose a rather complex molecule. A significant amount of heat energy must be added to increase the motion of these atoms so that they are able to break free of the bonds that hold them together in the sugar molecule and to interact with each other to make the "caramelization" products.

57. *Answer:* **(a) 3.275×10^4 m (b) 3.42×10^4 nm (c) 1.21×10^{-3} μm**

Strategy and Explanation: Use metric conversion factors:

(a) $$32.75\ \text{km} \times \frac{1000\ \text{m}}{1\ \text{km}} = 3.275 \times 10^4\ \text{meters}$$

(b) $$0.0342\ \text{mm} \times \frac{10^{-3}\ \text{m}}{1\ \text{mm}} \times \frac{1\ \text{nm}}{10^{-9}\ \text{m}} = 3.42 \times 10^4\ \text{nanometers}$$

(c) $$1.21 \times 10^{-12}\ \text{km} \times \frac{10^3\ \text{m}}{1\ \text{km}} \times \frac{1\ \mu\text{m}}{10^{-6}\ \text{m}} = 1.21 \times 10^{-3}\ \text{micrometers}$$

The Atomic Theory (Section 1.9)

59. *Answer/Explanation:* Conservation of mass is easy to see from the point of view of atomic theory. A chemical change is described as the rearrangement of atoms. Because the atoms in the starting materials must all be accounted for in the substances produced, and because the mass of each atom does not change, there would be no change in the mass.

61. *Answer/Explanation:* The law of multiple proportions can be explained using atomic theory. Consider two compounds that both contain the same two elements. The proportion of the two elements in these two compounds must be different. Compounds are composed of a specific number and type of atoms. If you pick samples of each of these compounds such that they contain the same number of atoms of the first element, then you'll find that the ratio of atoms on the other element is a small integer. Consider a concrete example. Fe_2O_3 and Fe_3O_4. Compare a sample containing three Fe_2O_3 to a sample containing two Fe_3O_4. These samples both have six Fe atoms. The first sample has nine oxygen atoms and the second sample has 8 oxygen atoms. Since all oxygen atoms weigh the same, the ratio of mass will also be the ratio of atoms; in this example, the ratio of mass and the ratio of atoms is $\frac{9}{8}$.

63. *Answer/Explanation:* The law of multiple proportions says that if two compounds contain the same elements and samples of those two compounds both contain the same mass of one element, then the ratio of the masses of the other elements will be a small whole number.

The Chemical Elements (Section 1.10)

65. *Answer/Explanation:* Many responses are equally valid here. Below are a few common examples. These lists are not comprehensive; many other answers are also right. The periodic table on the inside cover of your text is

color coded to indicate metals, non-metals and metalloids.

(a) Common metallic elements: iron, Fe; gold, Au; lead, Pb; copper, Cu; aluminum, Al

(b) Common non-metallic elements: carbon, C; hydrogen, H; oxygen, O; nitrogen, N

(c) Metalloids: boron, B; silicon, Si; germanium, Ge; arsenic, As; antimony, Sb; tellurium, Te

(d) Elements that are diatomic molecules: nitrogen, N_2; oxygen, O_2; hydrogen, H_2; fluorine, F_2; chlorine, Cl_2; bromine, Br_2; iodine, I_2

Communicating Chemistry: Symbolism (Section 1.11)

67. *Answer/Explanation:* Formula for each substance and nanoscale picture:

(a) Water H_2O

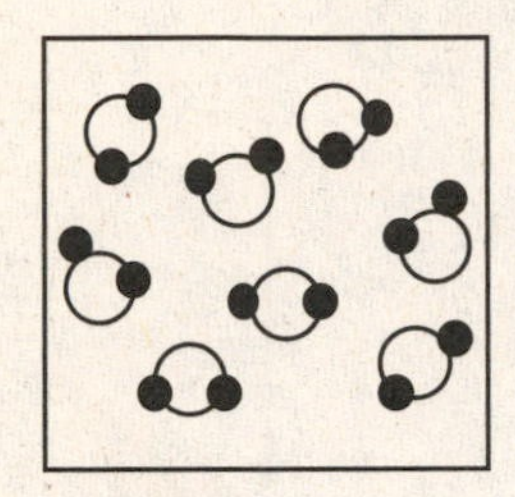

(b) Nitrogen N_2

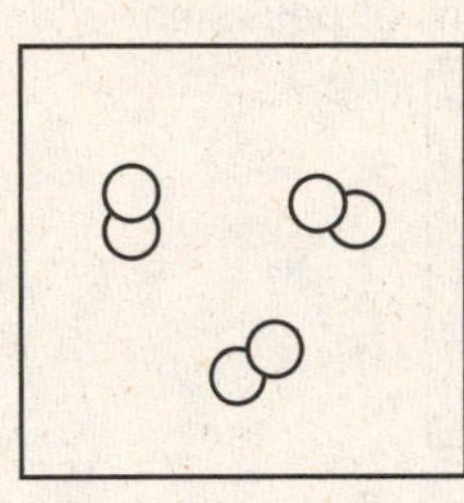

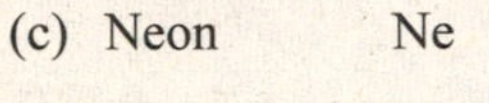

(c) Neon Ne

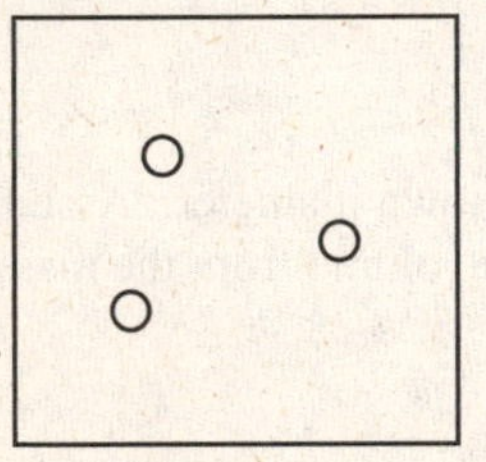

(d) Chlorine Cl_2

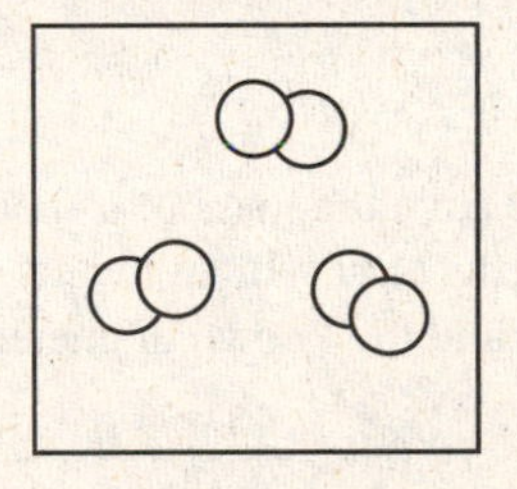

69. *Answer/Explanation:* $2\ H_2\ (g) + O_2\ (g) \longrightarrow 2\ H_2O\ (g)$

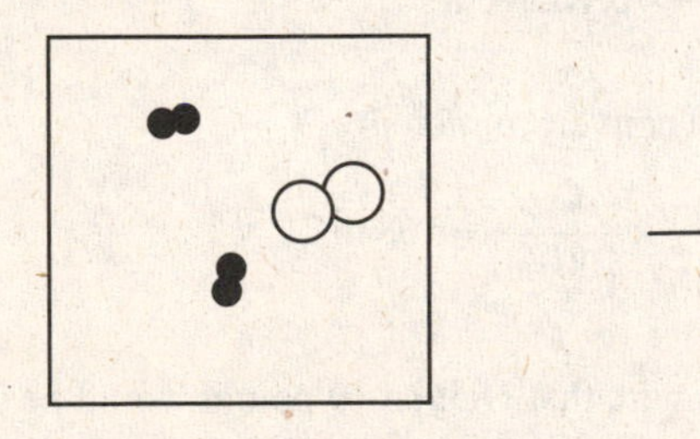

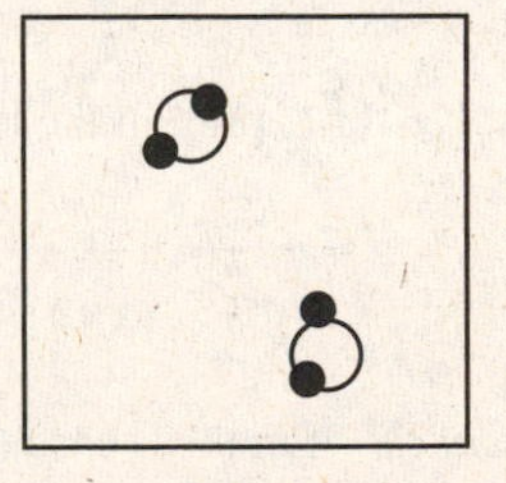

71. *Answer/Explanation:* $I_2\ (s) \longrightarrow I_2\ (g)$

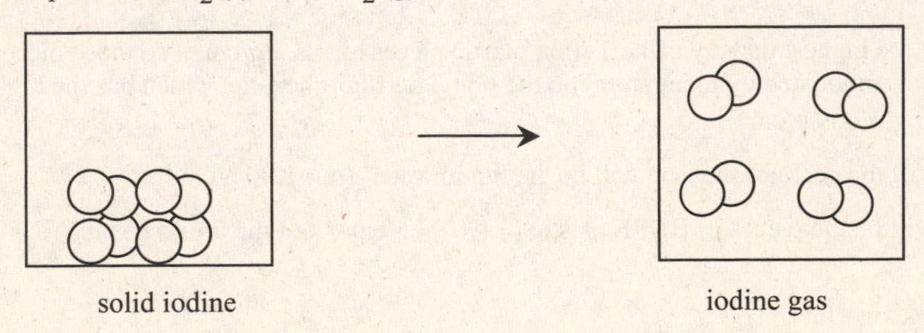

solid iodine

iodine gas

General Questions

73. *Answer/Explanation:*

(a) The mass of the compound (1.456 grams) is quantitative and relates to a physical property. The color (white), the fact that it reacts with a dye, and the color change in the dye (red to colorless) are all qualitative. The colors are related to physical properties. The reaction with the dye is related to a chemical property.

(b) The mass of the metal (0.6 grams) is quantitative and related to a physical property. The identity of the metal (lithium) and the identities of the chemicals it reacts with and produces (water, lithium hydroxide, and hydrogen) are all qualitative information. The fact that a chemical reaction occurs when the metal is added to water is qualitative information and related to a chemical property.

75. *Answer/Explanation:* If the density of solid calcium is almost twice that of solid potassium, but their masses are approximately the same size, then the volume must account for the difference. This suggests that the atoms of calcium are smaller than the atoms of potassium:

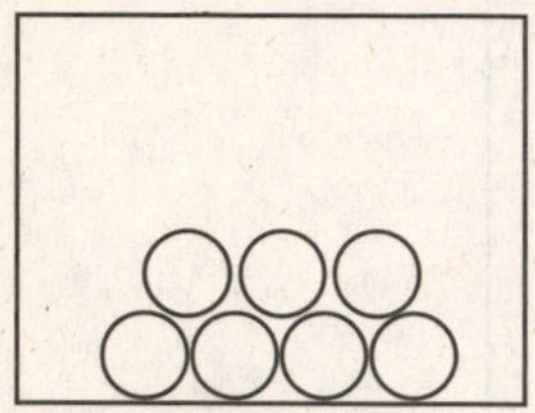

solid calcium
smaller atoms
closer packed
smaller volume

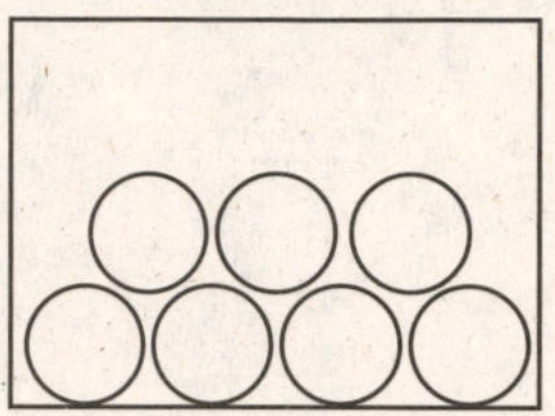

solid potassium
larger atoms
less closely packed
larger volume

77. *Answer:* **508 m**

Strategy and Explanation: We have the mass of a spool of aluminum wire with known diameter. Assuming the wire is a cylinder, find the length (ℓ) of wire in meters. Use the density to find the volume from the mass, then use the given volume equation and the known diameter to find the length.

$$10.0\text{ lb} \times \frac{453.59\text{ g}}{1\text{ lb}} \times \frac{1\text{ mL}}{2.70\text{ g}} \times \frac{1\text{ cm}^3}{1\text{ mL}} = 1680\text{ cm}^3$$

The radius is half the diameter. Determine the radius in centimeters:

$$R = \frac{1}{2} \times (0.0808\text{ in}) \times \frac{2.54\text{ cm}}{1\text{ in}} = 0.1026\text{ cm}$$

Rearrange $V = \pi r^2 \ell$ to solve for ℓ, plug in the known values and convert to meters:

$$\ell = \frac{V}{\pi r^2} = \frac{1680\text{ cm}^3}{(3.14159) \times (0.1026\text{ cm})^2} \times \frac{1\text{ m}}{100\text{ cm}} = 508\text{ m}$$

✓ *Reasonable Answer Check:* It makes sense that a quantity of wire that weighs 10 pound would be several hundred feet long.

79. *Answer/Explanation:* The highest density materials will sink to the bottom, with increasingly less dense materials floating on top.

The solid material with the highest density is the Teflon plastic pieces ($d = 2.3\text{ g/cm}^3$), so those pieces will be found at the bottom of the graduated cylinder sitting in the liquid perfluorohexane, which has the highest density of the liquids ($d = 1.669\text{ g/cm}^3$).

Floating on the surface of the perfluorohexane will be the liquid water ($d = 1.00\text{ g/cm}^3$).

Floating on the water will be the pieces of HDPE plastic ($d = 0.97\text{ g/cm}^3$) and the liquid hexane ($d = 0.766\text{ g/cm}^3$).

The following diagram shows these layers:

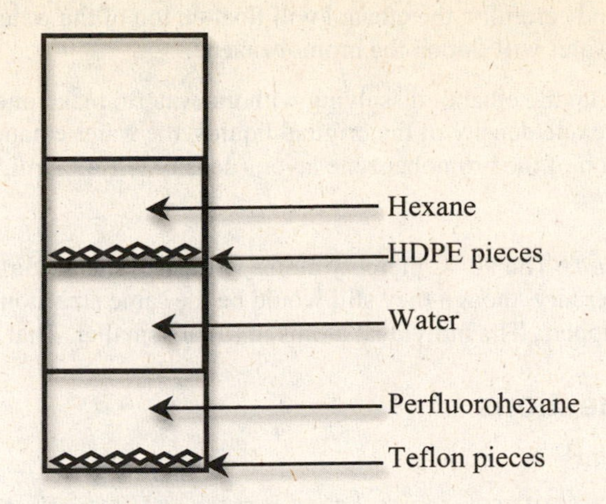

81. *Answer/Explanation:* Different allotropes are different at the nanoscale by how the atoms are organized and bonded together.

Applying Concepts

83. *Answer:* **(a) bromobenzene sample (b) gold sample (c) lead sample**

Strategy and Explanation:

(a) (Table 1.1) density of butane = 0.579 g/mL; density of bromobenzene = 1.49 g/mL 1 mL butane weighs less than 1 mL of bromobenzene so, 20 mL butane weighs less than 20 mL of bromobenzene. The bromobenzene sample has a larger mass.

(b) (Table 1.1) density of benzene = 0.880 g/mL; density of gold = 19.32 g/mL There are 0.880 grams of benzene in 1 mL of benzene, so there are 8.80 grams of benzene in 10 mL of benzene. Since 1.0 mL of gold has a mass of 19.32 grams that means the gold sample has a larger mass.

(c) (Table 1.1) density of copper = 8.93 g/mL; density of lead = 11.34 g/mL Any volume of lead has a larger mass than the same volume of copper. That means the lead sample has a larger mass.

85. (a) *Answer:* **2.7×10^2 mL ice**

Strategy and Explanation: Use the volume of the bottle and the densities of water and ice to determine the volume of ice formed from a fixed amount of water.

Use the volume of the bottle and the density of water to determine the mass of water frozen, then calculate the volume of the ice.

At 25 °C, density of water is 0.997 g/mL.

At 0 °C, density of ice = 0.917 g/mL

(b) If the bottle is made of flexible plastic, it might be deformed and bulging if not cracked and leaking ice. If the bottle is made of glass and the top came off, there would be ice (approximately 20 mL of it) oozing out of the top. Worst case scenario: if the bottle was glass and the top did not come off, it would be broken.

$$250 \text{ mL water} \times \frac{0.997 \text{ g } H_2O(\ell)}{1 \text{ mL water}} \times \frac{1 \text{ g } H_2O(s)}{1 \text{ g } H_2O(\ell)} \times \frac{1 \text{ mL ice}}{0.917 \text{ g } H_2O(s)} = 2.7 \times 10^2 \text{ mL ice}$$

✓ *Reasonable Answer Check:* The density of water is larger than the density of ice. It makes sense that the volume of the ice produced is larger than the volume of water.

87. *Answer/Explanation:* All substances with a density less than that of mercury's will float. The only substance in Table 1.1 with a density greater than that of mercury is **gold**. Prospectors in California used to use liquid mercury to separate and clean gold from rocks and dirt. The small particles of dirt and rocks floated in the mercury and could be skimmed off. The mercury then could be poured carefully off, leaving the gold behind.

89. *Answer/Explanation:*

(a) (Table 1.1) density of water = 0.998 g/mL, density of bromobenzene = 1.49 g/mL. Since water does not dissolve in bromobenzene, the lower density water will be the top layer of the immiscible layers.

(b) If poured slowly and carefully, the ethanol will float on top of the water and slowly dissolve in the water. Both ethanol and water will flot on the bromobenzene.

(c) Stirring will speed up the ethanol dissolving with the waterto make one phase. Assuming the new mixture has the average density of the original liquids, the water/ethanol layer (average density is 0.894 g/mL) will sit on top of the bromobenzene layer. (density is 1.49 g/mL)

91. *Answer:* **Drawing (b)**

Strategy and Explanation: The 90 °C mercury atoms will be a little bit further apart and moving somewhat more than the 10 °C mercury, though they still would be the same size atoms. The individual atoms in (c) are bigger. That doesn't happen. The individual atoms in (d) are smaller. That doesn't happen, either.

More Challenging Questions

93. *Answer:* $\mathbf{6.02 \times 10^{-29}\ m^3}$

Strategy and Explanation: We have the length of the edge of a cube. Find the volume of the cube in m^3:

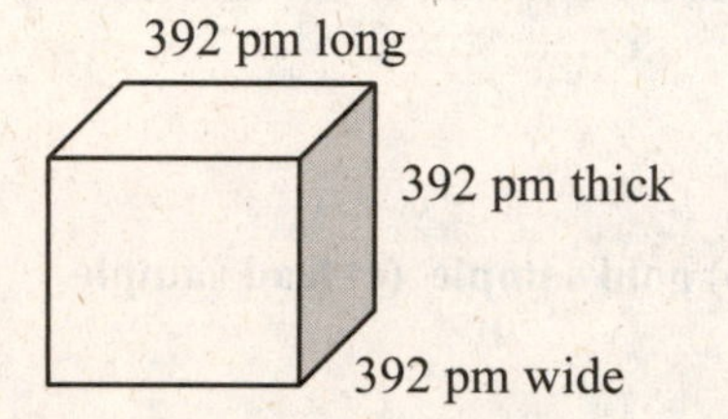

Use the linear dimensions to find the volume, then convert the volume to m^3 using metric conversions.

$$V = (\text{thickness}) \times (\text{width}) \times (\text{length}) = (392\ \text{pm}) \times (392\ \text{pm}) \times (392\ \text{pm}) = 6.02 \times 10^7\ \text{pm}^3$$

Using dimensional analysis, find the volume in m^3:

$$6.02 \times 10^7\ \text{pm}^3 \times \left(\frac{10^{-12}\ \text{m}}{1\ \text{pm}}\right)^3 = 6.02 \times 10^{-29}\ \text{m}^3$$

✓ *Reasonable Answer Check:* The cube is from the nanoscale, so it makes sense that it would be a very small volume using a macroscale unit of measure.

95. *Answer/Explanation:* Look at the periodic table, given.

(a) A colorless gas is a non-metal. Those gases are found in the **lavender area**.

(b) A solid that is ductile and malleable are metals. Solids with these characteristics are found in the **gray and blue areas**.

(c) Non-metals and metalloids are poor electrical conductors. Solids with this characteristic are found in the **orange and lavender areas**.

97. *Answer/Explanation:* Se and S have the greatest similarities in physical and chemical properties because they are both in the same periodic group (Group 6A).

99. A substance that can be broken down is not an element. A series of tests will result in a confirmation with one positive test. To prove that something is an element requires a battery of tests that all have negative results. A hypothesis that the substance is an element and cannot be broken down is more difficult to prove. (Section 1.4)

101. *Answer:* **(a) nickel, lead and magnesium (b) titanium**

Strategy and Explanation:

(a) According to the table of densities, a metal will float if the density is lower. That means that nickel, lead and magnesium will float on liquid mercury.

(b) The more different the densities, the smaller the fraction of the floating element will be below the surface. That means that titanium will float highest on mercury.

103. *Answer:* **(a) 32.1 g sulfur (b) 29.8 g zinc sulfide**

Strategy and Explanation: Given the mass of one reactant and the mass of product of their combination, find the mass the second reactant in the product, then determine the mass of product that could be formed from a different mass of reactant.

(a) The product is composed of two elements. The difference between the given masses must be the mass of the second element.

$$97.5 \text{ g zinc sulfide} - 65.4 \text{ g zinc} = 32.1 \text{ g sulfur}$$

(b) The elements combine in fixed ratios, so we can set up a zinc-to-product ratio to determine how the mass of product changes with a different mass of zinc.

$$\frac{32.1 \text{ g zinc}}{65.4 \text{ g zinc sulfide}} = \frac{20.0 \text{ g zinc}}{\text{x g zinc sulfide}}$$

$$\text{x} = 29.8 \text{ g zinc sulfide}$$

✓ *Reasonable Answer Check:* The sum of the reactant elements is the mass of the product, according to the conservation of mass. A smaller mass of one element will produce a smaller mass of product.

105. *Answer:* **No, the samples contain variable percentages of iron**

Strategy and Explanation: Given the mass of various portion of a sample known to contain only iron and sulfur and the mass of iron in each portion, determine if the sample is a compound of iron and sulfur. Explain.

If the sample is a compound, the elements will be combined in fixed proportion, so we can calculate the percentage of iron in each portion and compare them.

$$\%\text{ iron portion } 1 = \frac{0.964 \text{ g iron}}{1.518 \text{ g portion}} \times 100\,\% = 63.5\,\%$$

$$\%\text{ iron portion } 2 = \frac{1.203 \text{ g iron}}{2.056 \text{ g portion}} \times 100\,\% = 58.51\,\%$$

$$\%\text{ iron portion } 3 = \frac{1.290 \text{ g iron}}{1.873 \text{ g portion}} \times 100\,\% = 69.87\,\%$$

The percentage of iron changes from portion to portion, so the sample is not composed of a single compound of iron and sulfur.

✓ *Reasonable Answer Check:* The variable iron content proves that the sample cannot be a single compound.

107. *Answer:* **3.1 L**

Strategy and Explanation: Given the mass of one substance, the volume and density of a solution, and the mass of the solution after a reaction produces a gas with known density that escapes, determine the volume of the gas produced.

Calculate the total mass of the original mixture before the reaction occurs by adding the masses of the calcium carbonate and the hydrochloric acid solution:

$$\text{Total mass} = 12.6 \text{ g calcium carbonate} + 63.0 \text{ mL solution} \times \frac{1.096 \text{ g}}{1 \text{ mL}} = 81.6 \text{ g before reaction}$$

Use the conservation of mass to calculate the mass of escaped gas. Assuming that nothing else escaped the solution (such as water in the form of steam), the difference between the mass of the mixture before the reaction and the mass after must be the mass of the escaped gas:

$$\text{Mass gas} = 81.6 \text{ g before reaction} - 76.1 \text{ g after reaction} = 5.5 \text{ g gas}$$

Use the density of the gas to determine the volume from this mass:

$$5.5 \text{ g} \times \frac{1 \text{ L}}{1.798 \text{ g}} = 3.1 \text{ L}$$

✓ *Reasonable Answer Check:* It makes sense that the solution's mass is smaller than the original mixture's mass because a gas escaped. The low density of the gas produces a large volume of gas.

109. *Answer/Explanation:* To do the experiment, obtain four or more lemons. Do nothing prior to juicing to one of them; that serves as the "control" case. Perform the designated tasks to the remaining lemons, then juice all three lemons recording the results: juice volume and/or noticeable ease of squeeze or ease of juice delivery. Repeat several times to achieve better reliability. Hypothesis: disrupting "juice sacks" inside the pulp of the lemon will release juice.

Chapter 2: Atoms and Elements

Solutions for Blue-Numbered Questions for Review and Thought

Topical Questions

Units and Unit Conversions (Section 2.3)

7. *Answer:* **40,000 cm**

   Strategy and Explanation: If the nucleus is scaled to a diameter of 4 cm, determine the diameter of the atom.

   Find the accepted relationship between the size of the nucleus and the size of the atom. Use size relationships to get the diameter of the "artificially large" atom.

   The atom is about 10,000 times bigger than the nucleus.

$$10{,}000 \times 4 \text{ cm} = 40{,}000 \text{ cm}$$

   ✓ *Reasonable Answer Check:* A much larger nucleus means a much larger atom with a large atomic diameter.

9. *Answer:* **614 cm; 242 in; 20.1 ft**

   Strategy and Explanation: Given the pole vault record height in meters, use conversion factors to change the units to centimeters, feet, and inches.

   Use the metric relationship between m and cm to convert m into cm. Then, use the metric relationship between cm and inches to convert cm into inches. Then, use the relationship between inches and feet to convert from inches to feet.

   Notice: If any one of these questions were asked separately, we would start with the given information and apply the all the appropriate conversion factors in sequence. For example:

$$6.14 \text{ m} \times \frac{100 \text{ cm}}{1 \text{ m}} \times \frac{1 \text{ in}}{2.54 \text{ cm}} \times \frac{1 \text{ft}}{12 \text{ in}} = 20.1 \text{ ft}$$

   Only because we need to report the answers in each of the three units will we stop at each point and write down the intermediate answer, then use the result to begin the next calculation:

$$6.14 \text{ m} \times \frac{100 \text{ cm}}{1 \text{ m}} = 614 \text{ cm}$$

$$614 \text{ cm} \times \frac{1 \text{ in}}{2.54 \text{ cm}} = 242 \text{ in}$$

$$242 \text{ in} \times \frac{1 \text{ ft}}{12 \text{ in}} = 20.1 \text{ ft}$$

   ✓ *Reasonable Answer Check:* The number of centimeters should be larger than the number of meters, since centimeter is a smaller unit of measure. The number of inches should be smaller than the number of centimeters, since inch is a larger unit of measure. The number of feet should be smaller than the number of inches, since foot is a larger unit of measure. Notice: Some people may interpret this question as needing only two answers: one in cm (641 cm) and one in feet and inches (20 ft 2 in).

11. *Answer:* **76.2 kg**

   Strategy and Explanation: Given the mass in pounds, determine the mass in kilograms.

   Use conversion factor between pounds and grams, then grams and kilograms to change the units pounds to grams.

$$168 \text{ pounds} \times \frac{453.59 \text{ g}}{1 \text{ pound}} \times \frac{1 \text{ kg}}{1000 \text{ g}} = 76.2 \text{ kg}$$

   Notice: The original information has only three significant figures, so the result of these multiplications and divisions must also have three significant figures. (See Section 2.4)

   ✓ *Reasonable Answer Check:* Pounds are smaller than kilograms, so the number of kilograms should be smaller than the number of pounds.

13. *Answer:* **2.0×10^3 cm^3, 2.0 L**

Strategy and Explanation: Given displacement volume of 120. in^3, determine the volume in cm^3 and liters.

Use conversion factor between inches and centimeters to change the units from cubic inches to cubic centimeters. Then use conversion factors between cubic centimeters and milliliters, then milliliters and liters to change the units from cubic centimeters to liters.

NOTICE: Each of the "inch" units in "cubic inches" must be converted to centimeters. That is why the length conversion factor must cubed to obtain the volume conversion factor:

$$120.\ \text{in}^3 \times \frac{2.54\ \text{cm}}{1\ \text{in}} \times \frac{2.54\ \text{cm}}{1\ \text{in}} \times \frac{2.54\ \text{cm}}{1\ \text{in}} = 120.\ \text{in}^3 \times \left(\frac{2.54\,\text{cm}}{1\,\text{in}}\right)^3 = 1.97 \times 10^3\ \text{cm}^3$$

$$1.97 \times 10^3\ \text{cm}^3 \times \frac{1\ \text{mL}}{1\ \text{cm}^3} \times \frac{1\ \text{L}}{1000\ \text{mL}} = 1.97\ \text{L}$$

Notice: The metric definition of "milli-" applied to liters is: 10^{-3} L = 1 mL. The 1:1000 ratio used above is a very common variant that appropriately indicates a larger number (1000) of small things (mL) equal to a smaller number (1) of large things (L). Of course, it is perfectly acceptable, but not as common in practice, to use the metric definition directly as the conversion factor to get the same answer:

$$1.97 \times 10^3\ \text{cm}^3 \times \frac{1\ \text{mL}}{1\ \text{cm}^3} \times \frac{10^{-3}\ \text{L}}{1\ \text{mL}} = 1.97\ \text{L}$$

✓ *Reasonable Answer Check:* Centimeters are smaller than inches, so a cubic centimeter is much smaller than a cubic inch, so the number of cubic centimeters should be larger than the number of cubic inches. A liter is larger than a cubic centimeter, so the number of liters should be smaller than the number of cubic inches.

15. *Answer:* **1550 in^2**

Strategy and Explanation: Given one square meter, determine the number of square inches.

Use metric conversion between meters and centimeters, then the relationship between centimeters and inches.

$$1\ \text{m}^2 \times \left(\frac{100\ \text{cm}}{1\ \text{m}}\right)^2 \times \left(\frac{1\ \text{in}}{2.54\ \text{cm}}\right)^2 = 1550\ \text{in}^2$$

Notice: The significant figures are somewhat ambiguous, since the word "one" might be interpreted as 1, which has one significant figure. If that is the case, the answer would be 2000 in^2.

✓ *Reasonable Answer Check:* A meter is larger than an inch, so the number of square inches should be larger than the number of square meters.

17. *Answer:* **2.8×10^9 m^3**

Strategy and Explanation: Given a volume in cubic miles, determine the number of cubic meters.

Use the relationship between miles and meters.

$$0.67\ \text{miles}^3 \times \left(\frac{1000\ \text{m}}{0.62137\ \text{mile}}\right)^3 = 2.8 \times 10^9\ \text{m}^3$$

✓ *Reasonable Answer Check:* A meter is smaller than a mile, so the number of cubic meters should be larger than the number of cubic miles.

Significant Figures (Section 2.4)

22. *Answer:* **(a) four (b) three (c) four (d) four (e) three**

Strategy and Explanation: Given several measured quantities, determine the number of significant figures.

Use rules given in Section 2.4, summarized here: All non-zeros are significant. Zeros that precede (sit to the left of) non-zeros are never significant (e.g., 0.003). Zeros trapped between non-zeros are always significant (e.g., 3.003). Zeros that follow (sit to the right of non-zeros are (a) significant if a decimal point is explicitly given (e.g., 3300) OR (b) not significant, if a decimal point is not specified (e.g., 3300.).

(a) 1374 kg has **four** significant figures (The 1, 3, 7, and 4 digits are each significant.)

(b) 0.00348 s has **three** significant figures (The 3, 4, and 8 digits are each significant. The zeros are all before the first non-zero-digit 3 and therefore they are not significant.)

(c) 5.619 mm has **four** significant figures (The 5, 6, 1, and 9 digits are each significant.)

(d) 2.475×10^{-3} cm has **four** significant figures (The 2, 4, 7, and 5 digits are each significant.)

(e) 33.1 mL has **three** significant figures (The 3, 3, and 1 digits are each significant.)

✓ *Reasonable Answer Check:* Only one answer had zeros in it, and those (in (b)) were to the left of the first non-zero digit, so none of the zeros here were significant.

24. *Answer:* **(a) 4.33×10^{-4} (b) 4.47×10^{1} (c) 2.25×10^{1} (d) 8.84×10^{-3}**

Strategy and Explanation: Given several measured quantities, round them to three significant figures and write them in scientific notation. Use rules for rounding given in Section 2.4. If the last digit is below 5, then rounding does not change the digit before it. If the last digit is above a five, the digit before it is made one larger. If the last digit is exactly five, round the digit before it to an even number, up if odd and down if even. After rounding, adjust the appearance of the number to scientific notation (i.e., to a number between 1 and 9.999… that is multiplied by ten to a whole-number power.)

(a) 0.0004332 has four significant figures. To round it to three significant figures, we need to remove the fourth significant figure, the 2. Since 2 is below 5, the result is 0.000433. Putting this value in scientific notation, we get **4.33×10^{-4}**.

(b) 44.7337 has six significant figures. To round it to three significant figures, we need to remove the fourth, fifth and sixth significant figures. The removal of the last digit 7 rounds the 3 next to it up to a 4, but the removal of 4 doesn't change the 3 next to it, and the removal of 3 doesn't change the 7 next to it, the result is 44.7. Putting this value in scientific notation, we get **4.47×10^{1}**.

(c) 22.4555 has six significant figures. To round it to three significant figures, we need to remove the fourth, fifth and sixth significant figures. Since the removal of the last 5 rounds the 5 next to it up to a 6, and the removal of 6 rounds the 5 next to it up to a 6, and the removal of 6 rounds the 4 next to it to a 5, the result is 22.5. Putting this value in scientific notation, we get **2.25×10^{1}**.

(d) 0.0088418 has five significant figures. To round it to three significant figures, we need to remove the fourth and fifth significant figures. Since the removal of the last 8 rounds the 1 next to it up to a 2, and the removal of 2 doesn't change the 4 next to it, the result is 0.00884. Putting this value in scientific notation, we get **8.84×10^{-3}**.

✓ *Reasonable Answer Check:* Small numbers are still small, large numbers are still large, what remains after the rounding is the larger part of the value. Each answer has three significant figures and each number is represented in proper scientific notation.

26. *Answer:* **(a) 1.9 g/mL (b) 218.4 cm^3 (c) 0.0217 (d) 5.21×10^{-5}**

Strategy and Explanation: Given some numbers combined using calculations, determine the result with proper significant figures. Perform the mathematical steps according to order of operations, applying the proper significant figures (addition and subtraction retains the least number of decimal places in the result; multiplication and division retain the least number of significant figures in the result). *Notice: if operations are combined that use different rules, it is important to stop and determine the intermediate result any time the rule switches.*

(a) $$\frac{4.850\ \text{g} - 2.34\ \text{g}}{1.3\ \text{mL}}$$

The numerator uses the subtraction rule. The first number has three decimal places (the 8, the 5, and the 0 are all decimal places -- digit that follow the decimal point to the right) and the second number has two decimal places (the 3 and the 4 are both decimal places), so the result of the subtraction has two decimal places.

$$\frac{2.51\ \text{g}}{1.3\ \text{mL}}$$

The ratio uses the division rule. The numerator has three significant figures and the denominator has two significant figures, so the answer will have two significant figures. Therefore, we get **1.9 g/mL**.

(b) $$V = \pi r^3 = (3.1415926) \times (4.112\ \text{cm})^3$$

This whole calculation uses the multiplication rule, with four significant figures, limited by the measurement of r. The value of π should be carried to more than four significant figures, such as 3.14159... The answer comes out **218.4 cm³**.

(c) $$(4.66 \times 10^{-3}) \times 343.2$$

This calculation uses the multiplication rule. The first number, 4.66×10^{-3}, has three significant figures and the second number, 343.2, has four significant figures, so the answer has three significant figures **0.0217**.

(d) $$\frac{0.003400}{65.2}$$

This calculation uses the division rule. The numerator has four significant figures and the denominator has three significant figures, so the answer has three significant figures 0.0000521 or $\mathbf{5.21 \times 10^{-5}}$.

✓ *Reasonable Answer Check:* The proper significant figures rules were used. The size and units of the answers are appropriate.

Mass Spectrometry (Section 2.5)

28. *Answer/Explanation:* In Section 2.5, page 56, a "Tools of Chemistry" box explains the Mass Spectrometer. The species that is moving through a mass spectrometer during its operation are **ions** (usually +1 cations) that have been formed from the sample molecules by a bombarding electron beam.

30. *Answer/Explanation:* In Section 2.5, page 56, a "Tools of Chemistry" box explains the Mass Spectrometer. The ions in the mass spectrometer are separated from each other **using a magnetic field**, which causes the ions to have different paths through the instrument.

Isotopes (Sections 2.5 and 2.6)

35. *Answer:* **number of neutrons**

Strategy and Explanation: Uranium-235 differs from uranium-237 in terms of the number of neutrons in the atoms.

37. *Answer:* **27 protons, 27 electrons, and 33 neutrons**

Strategy and Explanation: Given the identity of an element (cobalt) and the atom's mass number (60), find the number of electrons, protons, and neutrons in the atom.

Look up the symbol for cobalt and find that symbol on the periodic table. The periodic table gives the atomic number. The atomic number is the number of protons. The number of electrons is equal to the number of protons since the atom has no charge. The number of neutrons is the difference between the mass number and the atomic number.

The element technetium has the symbol Co. On the periodic table, we find it listed with the atomic number 27. So, the atom has 27 protons, 27 electrons and (60 – 27 =) 33 neutrons.

✓ *Reasonable Answer Check:* The number protons and electrons must be the same (27=27). The sum of the protons and neutrons is the mass number (27 + 33 = 60). This is correct.

39. *Answer:* **78.92 amu/atom**

Strategy and Explanation: Given the average atomic weight of an element and the percentage abundance of one isotope, determine the atomic weight of the only other isotope.

Using the fact that the sum of the percents must be 100%, determine the percent abundance of the second isotope. Knowing that the weighted average of the isotope masses must be equal to the reported atomic weight, set up a relationship between the known atomic mass and the various isotope masses using a variable to describing the second isotope's atomic weight.

We are told that natural bromine is 49.31% ^{81}Br and that there are only two isotopes. To calculate the percentage abundance of the other isotope, subtract from 100%:

$$100.0\% - 49.31\% = 50.69\%$$

These percentages tell us that every 10000 atoms of bromine contains 4931 atoms of the ^{81}Br isotope and 5069 atoms of the other bromine isotope *(limited to 4 sig figs)*. The atomic weight for Br is given as

79.904 amu/atom. Table 2.3 in Section 2.6 gives the isotopic mass of ^{81}Br isotope as 80.916289 amu/atom. Let X be the atomic mass of the other isotope of bromine.

$$\frac{4931 \text{ atoms } ^{81}\text{Br}}{10000 \text{ Br atoms}} \times \left(\frac{80.916289 \text{ amu}}{1 \text{ atom } ^{81}\text{Br}}\right) + \frac{5069 \text{ atoms other isotope}}{10000 \text{ Br atoms}} \times \left(X \frac{\text{amu}}{\text{atom}}\right) = 79.904 \frac{\text{amu}}{\text{Br atom}}$$

Solve for X

$$39.90 + 0.5069\,X = 79.904$$

$$X = 78.92 \text{ amu/atom} \quad \textit{(limited to 4 sig figs)}$$

✓ *Reasonable Answer Check:* Table 2.3 in Section 2.6 gives the atomic weight of ^{79}Br to be 78.918336, which is the same as that given in Table 2.3 for ^{79}Br, within the permissible significant figures.

41. *Answer:* **(a) 9 (b) 48 (c) 70**

Strategy and Explanation: Given the identity of an element and the number of neutrons in the atom, determine the mass number of the atom.

Look up the symbol for the element and find that symbol on the periodic table. The periodic table gives the atomic number, which represents the number of protons. Add the number of neutrons to the number of protons to get the mass number.

(a) Beryllium has the symbol Be. On the periodic table, we find it listed with the atomic number 4. The given number of neutrons is 5. So, (4 + 5 =) **9** is the mass number for this Be atom.

(b) Titanium has the symbol Ti. On the periodic table, we find it listed with the atomic number 22. The given number of neutrons is 26. So, (22 + 26 =) **48** is the mass number for this Ti atom.

(c) Gallium has the symbol Ga. On the periodic table, we find it listed with the atomic number 31. The given number of neutrons is 39. So, (31 + 39 =) **70** is the mass number for this Ga atom.

✓ *Reasonable Answer Check:* Mass number should be close to (but not exactly the same as) the atomic weight given on the periodic table. Beryllium's atomic weight (9.0122) is close to its mass number of 9, titanium's (47.88) is close to 48, and gallium's (69.72) is close to 70.

43. *Answer:* **(a) $^{23}_{11}$Na (b) $^{39}_{18}$Ar (c) $^{69}_{31}$Ga**

Strategy and Explanation: Given the identity of an element and the number of neutrons in the atom, determine the atomic symbol $^{A}_{Z}$X.

Look up the symbol for the element and find that symbol on the periodic table. The periodic table gives the atomic number (Z), which represents the number of protons. Add the number of neutrons to the number of protons to get the mass number (A).

(a) The element sodium has the symbol Na. On the periodic table, we find it listed with the atomic number 11. The given number of neutrons is 12. So, (11 + 12 =) 23 is the mass number for this sodium atom. Its atomic symbol looks like this: **$^{23}_{11}$Na**.

(b) The element argon has the symbol Ar. On the periodic table, we find it listed with the atomic number 18. The given number of neutrons is 21. So, (18 + 21 =) 39 is the mass number for this argon atom. Its atomic symbol looks like this: **$^{39}_{18}$Ar**.

(c) The element gallium has the symbol Ga. On the periodic table, we find it listed with the atomic number 31. The given number of neutrons is 38. So, (31 + 38 =) 69 is the mass number for this gallium atom. Its atomic symbol looks like this: **$^{69}_{31}$Ga**.

✓ *Reasonable Answer Check:* Mass number should be close to (but not exactly the same as) the atomic weight also given on the periodic table. Sodium's atomic weight (22.99) is close to the 23 mass number. Argon's atomic weight (39.95) is close to the 39 mass number. Gallium's atomic weight (69.72) is close to the 69 mass number. These numbers seem reasonable.

45. *Answer:* **(a) 20 e^-, 20 p^+, 20 n^o (b) 50 e^-, 50 p^+, 69 n^o (c) 94 e^-, 94 p^+, 150 n^o**

Strategy and Explanation: Given the atomic symbol $^{A}_{Z}$X of the isotope, determine the number of electrons, protons, and neutrons. The atomic number (Z) represents the number of protons. In neutral atoms, the number of electrons is equal to the number of protons. To get the number of neutrons, subtract the number of protons from the mass number (A).

(a) The isotope given is ${}^{40}_{20}Ca$. That means A = 40 and Z = 20. So, the number of protons is 20, the number of electrons is 20, and the number of neutrons is (40 – 20 =) 20.

(b) The isotope given is ${}^{119}_{50}Sn$. That means A = 119 and Z = 50. So, the number of protons is 50, the number of electrons is 50, and the number of neutrons is (119 – 50 =) 69.

(c) The isotope given is ${}^{244}_{94}Pu$. That means A = 244 and Z = 94. So, the number of protons is 94, the number of electrons is 94, and the number of neutrons is (244 – 94 =) 150.

✓ *Reasonable Answer Check:* The number of protons and electrons must be equal in neutral atoms. The mass number must be the sum of the protons and neutrons.

47. *Answer:*

Z	A	Number of Neutrons	Element
35	81	46	Br
46	108	62	Pd
77	192	115	Ir
63	151	88	Eu

Strategy and Explanation: Fill in an incomplete table with Z, A, number of neutrons and element identity.

The atomic number (Z) represents the number of protons. The mass number (A) is the number of neutrons and protons. The element's identity can be determined using the periodic table by looking up the atomic number and getting the symbol.

Z	A	Number of Neutrons	Element
35	81	(a)	(b)
(c)	(d)	62	Pd
77	(e)	115	(f)
(g)	151	(h)	Eu

(a) Number of neutrons = A – Z = 81 – 35 = 46

(b) Z = 35. Look up element #35 on periodic table: Element = Br

(c) Look up Pd on the periodic table: Z = 46

(d) A = Z + number of neutrons = 46 + 62 = 108

(e) A = Z + number of neutrons = 77 + 115 = 192

(f) Z = 77. Look up element #77 on periodic table: Element = Ir

(g) Look up Eu on the periodic table: Z = 63

(h) Number of neutrons = A – Z = 151 – 63 = 88

✓ *Reasonable Answer Check:* The atomic number and the symbol must match what is shown on the periodic table. The mass number must be the sum of the atomic number and the number of neutrons.

49. *Answer:* $\mathbf{{}^{18}_{9}X}$**,** $\mathbf{{}^{20}_{9}X}$**, and** $\mathbf{{}^{15}_{9}X}$

Strategy and Explanation: The atomic symbol general form is ${}^{A}_{Z}X$, where A is the mass number and Z is the atomic number. Isotopes of the same element will have the same Z, but A will be different. So, $\mathbf{{}^{18}_{9}X}$**,** $\mathbf{{}^{20}_{9}X}$**, and** $\mathbf{{}^{15}_{9}X}$ are all isotopes of the same element with an atomic number of 9.

Percent (Section 2.6)

51. *Answer:* **80.1% silver and 19.9% copper**

Strategy and Explanation: Given a 17.6-gram bracelet that contains 14.1 grams of silver and the rest copper, determine the percentage silver and the percentage copper.

Determine the percentage of silver by dividing the mass of silver by the total mass and multiplying by 100%. Since the metal is made up of only silver and copper, determine the percentage of copper by subtracting the percentage of silver from 100%.

$$\frac{14.1 \text{ g silver}}{17.6 \text{ g bracelet}} \times 100\% = 80.1\% \text{ silver}$$

$$100\%\ \text{total} - 80.1\%\ \text{silver} = 19.9\%\ \text{copper}$$

Notice: When you are using grams of different substances, be careful to carry enough information in the units so you don't confuse one mass with another.

✓ *Reasonable Answer Check:* A significant majority of the metal in the bracelet is silver, so it makes sense that the percentage of silver is larger than the percentage of copper.

53. *Answer:* **245 g sulfuric acid**

Strategy and Explanation: Given the volume of a battery acid sample, the density and the percentage of sulfuric acid in the battery acid by mass, determine the mass of acid in the battery.

Always start with the sample. Use the density of the solution to create a conversion factor between milliliters and grams, so you can determine the mass of the battery acid in grams. Then use the mass percentage of sulfuric acid in the battery acid as a conversion factor to determine the mass of sulfuric acid in the sample.

Every 1.000 mL of battery acid solution weighs 1.285 grams. Every 100.00 g of battery acid solution contains 38.08 grams of sulfuric acid.

$$500.\ \text{mL solution} \times \frac{1.285\ \text{g solution}}{1.000\ \text{mL solution}} \times \frac{38.08\ \text{g sulfuric acid}}{100.00\ \text{g solution}} = 245\ \text{g sulfuric acid}$$

Notice: When you are using grams of different substances, be careful to carry enough information in the units so you don't confuse one mass with another.

✓ *Reasonable Answer Check:* Only about a third of the sample is sulfuric acid, so the mass of sulfuric acid should be smaller than the volume of the solution.

55. *Answer:* **0.93% Na**

Strategy and Explanation: Given the mass of a sample of cereal and the number of milligrams of sodium in that sample, determine the percentage of sodium in the cereal.

Determine the mass of sodium in the same units as the sample mass. Then determine the percentage sodium by dividing the mass of sodium by the sample mass and multiplying by 100%.

$$280.\ \text{mg Na} \times \frac{1\ \text{g Na}}{1000\ \text{mg Na}} = 0.280\ \text{g Na}$$

$$\frac{0.280\ \text{g Na}}{30.\ \text{g cereal}} \times 100\% = 0.93\%\ \text{Na in cereal}$$

✓ *Reasonable Answer Check:* A very small amount of sodium is present in the cereal, so it makes sense that the percentage mass is so low.

Atomic Weight (Section 2.6)

57. *Answer:* **6.941 amu/atom**

Strategy and Explanation: Using the exact mass and the percent abundance of several isotopes of an element, determine the atomic weight.

Calculate the weighted average of the isotope masses.

Every 1000 atoms of lithium contains 75.00 atoms of the ^{6}Li isotope and 925.0 atoms of the ^{7}Li isotope.

$$\frac{75.00\ \text{atoms}\ ^{6}\text{Li}}{1000\ \text{Li atoms}} \times \left(\frac{6.015121\ \text{amu}}{1\ \text{atom}\ ^{6}\text{Li}}\right) + \frac{925.0\ \text{atoms}\ ^{7}\text{Li}}{1000\ \text{Li atoms}} \times \left(\frac{7.016003\ \text{amu}}{1\ \text{atom}\ ^{7}\text{Li}}\right) = 6.941\ \text{amu/Li atom}$$

✓ *Reasonable Answer Check:* The periodic table value for atomic weight is the same as calculated here.

59. *Answer:* **60.12% ^{69}Ga, 39.87% ^{71}Ga**

Strategy and Explanation: Using the exact mass of several isotopes and the atomic weight, determine the abundance of the isotopes.

Establish variables describing the isotope percentages. Set up two relationships between these variables. The sum of the percents must be 100%, and the weighted average of the isotope masses must be the reported atomic mass. X% ^{69}Ga and Y% ^{71}Ga. This means: Every 100 atoms of gallium contains X atoms of the ^{69}Ga isotope and Y atoms of the ^{71}Ga isotope.

$$\frac{\text{X atoms } ^{69}\text{Ga}}{100 \text{ Ga atoms}} \times \left(\frac{68.9257 \text{ amu}}{1 \text{ atom } ^{69}\text{Ga}}\right) + \frac{\text{Y atoms } ^{71}\text{Ga}}{100 \text{ Ga atoms}} \times \left(\frac{70.9249 \text{ amu}}{1 \text{ atom } ^{71}\text{Ga}}\right) = 69.723 \frac{\text{amu}}{\text{Ga atom}}$$

And, X + Y = 100%. We now have two equations and two unknowns, so we can solve for X and Y algebraically. Solve the first equation for Y: Y = 100 – X. Plug that in for Y in the second equation. Then solve for X:

$$\frac{X}{100} \times (68.9257) + \frac{100-X}{100} \times (70.9249) = 69.723$$

$$0.689257X + 70.9249 - 0.709249X = 69.723$$

$$70.9249 - 69.723 = 0.709249X - 0.689257X = (0.709249 - 0.689257)X$$

$$1.202 = (0.019992)X$$

$$X = 60.12, \text{ so there is } 60.12\% \ ^{69}\text{Ga}$$

Now, plug the value of X in the first equation to get Y.

$$Y = 100 - X = 100 - 60.12 = 39.88, \text{ so there is } 39.88\% \ ^{69}\text{Ga}$$

Therefore the abundances for these isotopes are: 60.12% ^{69}Ga and 39.88% ^{71}Ga

✓ *Reasonable Answer Check:* The periodic table value for the atomic weight is closer to 68.9257 than it is to 70.9249, so it makes sense that the percentage of ^{69}Ga is larger than ^{71}Ga. The sum of the two percentages is 100.00%.

61. *Answer:* ^{7}**Li**

Strategy and Explanation: The two isotopes of lithium are ^{6}Li and ^{7}Li. The mass of ^{6}Li is close to 6 amu and the mass of ^{7}Li is close to 7 amu. Because lithium's atomic weight (6.941 amu) is much closer to 7 amu than to 6 amu, the isotopic ^{7}**Li is more abundant** than the isotope ^{6}Li.

The Mole, Molar Mass, and Problem Solving (Sections 2.7 and 2.8)

63. *Answer/Explanation:* A few common counting units are: pair (2), dozen (12), six-pack (6), gross (144), hundred (100), million (1,000,000), billion (1,000,000,000), etc.

66. *Answer:* **(a) 27 g B (b) 0.48 g O_2 (c) 6.98 × 10^{-2} g Fe (d) 2.61 × 10^{3} g H**

Strategy and Explanation: Determine mass in grams from given quantity in moles.

Look up the elements on the periodic table to get the atomic weight (with at least four significant figures). If necessary, calculate the molecular weight. Use that number for the molar mass (with units of grams per mole) as a conversion factor between moles and grams.

Notice: Whenever you use physical constants that you look up, it is important to carry more significant figures than the rest of the measured numbers, to prevent causing inappropriate round-off errors.

(a) Boron (B) has atomic number 5 on the periodic table. Its atomic weight is 10.811, so the molar mass is 10.811 g/mol.

$$2.5 \text{ mol B} \times \frac{10.811 \text{ g B}}{1 \text{ mol B}} = 27 \text{ g B}$$

(b) O_2 (diatomic molecular oxygen) is made with two atoms of element with atomic number 8 on the periodic table. Its atomic weight is 15.9994; therefore, the molecular weight of O_2 is 2×15.9994=31.9988, and the molar mass is 31.9988 g/mol.

$$0.015 \text{ mol } O_2 \times \frac{31.9988 \text{ g } O_2}{1 \text{ mol } O_2} = 0.48 \text{ g } O_2$$

(c) Iron (Fe) has atomic number 26 on the periodic table. Its atomic weight is 55.845, so the molar mass is 55.845 g/mol.

$$1.25 \times 10^{-3} \text{ mol Fe} \times \frac{55.845 \text{ g Fe}}{1 \text{ mol Fe}} = 6.98 \times 10^{-2} \text{ g Fe}$$

(d) Helium (He) has atomic number 2 on the periodic table. Its atomic weight is 4.0026, so the molar mass is 4.0026 g/mol.

$$653 \text{ mol He} \times \frac{4.0026 \text{ g He}}{1 \text{ mol He}} = 2.61 \times 10^{3} \text{ g He}$$

✓ *Reasonable Answer Check:* The moles units cancel when the factor is multiplied, leaving grams.

68. *Answer:* **(a) 1.9998 mol Cu (b) 0.499 mol Ca (c) 0.6208 mol Al (d) 3.1 × 10^{-4} mol K (e) 2.1 × 10^{-5} mol Am**

Strategy and Explanation: Determine the quantity in moles from given mass in grams.

Look up the elements on the periodic table to get the atomic weight. Use that number for the molar mass (with units of grams per mole) as a conversion factor between grams and moles.

Notice: Whenever you use physical constants that you look up, it is important to carry more significant figures than the rest of the measured numbers, to prevent causing inappropriate round-off errors.

(a) Copper (Cu) has atomic number 29 on the periodic table. Its atomic weight is 63.546, so the molar mass is 63.546 g/mol.

$$127.08 \text{ g Cu} \times \frac{1 \text{ mol Cu}}{63.546 \text{ g Cu}} = 1.9998 \text{ mol Cu}$$

(b) Calcium (Ca) has atomic number 20 on the periodic table. Its atomic weight is 40.078, so the molar mass is 40.078 g/mol.

$$20.0 \text{ g Ca} \times \frac{1 \text{ mol Ca}}{40.078 \text{ g Ca}} = 0.499 \text{ mol Ca}$$

(c) Aluminum (Al) has atomic number 13 on the periodic table. Its atomic weight is 26.9815, so the molar mass is 26.9815 g/mol.

$$16.75 \text{ g Al} \times \frac{1 \text{ mol Al}}{26.9815 \text{ g Al}} = 0.6208 \text{ mol Al}$$

(d) Potassium (K) has atomic number 19 on the periodic table. Its atomic weight is 39.0983, so the molar mass is 39.0983 g/mol.

$$0.012 \text{ g K} \times \frac{1 \text{ mol K}}{39.0983 \text{ g K}} = 3.1 \times 10^{-4} \text{ mol K}$$

(e) Radioactive americium (Am) has atomic number 95 on the periodic table. The atomic weight given on the periodic table is the weight of its most stable isotope 243, so the molar mass is 243 g/mol.

Convert milligrams into grams, first.

$$5.0 \text{ mg Am} \times \frac{1 \text{ g Am}}{1000 \text{ mg Am}} \times \frac{1 \text{ mol Am}}{243 \text{ g Am}} = 2.1 \times 10^{-5} \text{ mol Am}$$

✓ *Reasonable Answer Check:* Notice that grams units cancel when the factor is multiplied, leaving moles.

70. *Answer:* **2.19 mol Na**

Strategy and Explanation: Determine the quantity in moles from the given mass in grams.

Look up the Na on the periodic table to get the atomic weight. Use that number for the molar mass (with units of grams per mole) as a conversion factor between grams and moles.

Sodium (Na) has atomic number 11 on the periodic table. Its atomic weight is 22.9898, so the molar mass is 22.9898 g/mol.

$$50.4 \text{ g Na} \times \frac{1 \text{ mol Na}}{22.9898 \text{ g Na}} = 2.19 \text{ mol Na}$$

✓ *Reasonable Answer Check:* Notice that grams units cancel when the factor is multiplied, leaving just moles. The resulting number representing moles should be smaller than number representing the mass.

72. *Answer:* **9.42 × 10^{-5} mol Kr**

Strategy and Explanation: Determine the quantity in moles from given mass in grams.

Look up krypton on the periodic table to get the atomic weight. Use that number for the molar mass (with units of grams per mole) as a conversion factor between grams and moles.

Krypton (Kr) has atomic number 36 on the periodic table. Its atomic weight is 83.798, so the molar mass is 83.798 g/mol.

$$0.00789 \text{ g Kr} \times \frac{1 \text{ mol Kr}}{83.798 \text{ g Kr}} = 9.42 \times 10^{-5} \text{ mol Kr}$$

✓ *Reasonable Answer Check:* Notice that grams units cancel when the factor is multiplied, leaving just moles.

74. *Answer:* **4.131 × 10^{23} Cr atoms**

Strategy and Explanation: Given a chromium sample with known mass, determine the number of atoms in it.

Start with the mass. Use the molar mass of chromium as a conversion factor between grams and moles. Then use Avogadro's number as a conversion factor between moles of chromium atoms and the actual number of chromium atoms.

$$35.67 \text{ g Cr} \times \frac{1 \text{ mol Cr atoms}}{51.996 \text{ g Cr}} \times \frac{6.0221 \times 10^{23} \text{ Cr atoms}}{1 \text{ mol Cr atoms}} = 4.131 \times 10^{23} \text{ Cr atoms}$$

✓ *Reasonable Answer Check:* A sample of chromium that a person can see and hold is macroscopic. It will contain a very large number of atoms.

76. *Answer:* **1.055 × 10^{-22} g Cu**

Strategy and Explanation: Given the number of copper atoms, determine the mass.

Start with the quantity. Use Avogadro's number as a conversion factor between the number of copper atoms and moles of copper atoms. Then use the molar mass of copper as a conversion factor between moles and grams.

$$1 \text{ Cu atom} \times \frac{1 \text{ mol Cu atoms}}{6.022 \times 10^{23} \text{ Cu atoms}} \times \frac{63.546 \text{ g Cu}}{1 \text{ mol Cu atoms}} = 1.055 \times 10^{-22} \text{ g Cu}$$

✓ *Reasonable Answer Check:* An atom of copper is NOT very large. It will have a very small mass.

The Periodic Table (Section 2.9)

78. *Answer/Explanation:* A group on the periodic table is the collection of elements that share the same vertical column, whereas a period on the periodic table is the collection of elements that share the same horizontal row.

83. *Answer/Explanation:* Common transition elements are: iron, copper, chromium (Period 4). Halogens are the collection of elements in Group 7A. Common halogens are: fluorine and chlorine.

Alkali metals are the collection of metallic elements in Group 1A (Notice: metallic means excluding hydrogen). A common alkali metal is sodium.

85. *Answer/Explanation:* There are five elements in Group 4A of the periodic table. They are non-metal: carbon (C), metalloids: silicon (Si) and germanium (Ge), and metals: tin (Sn) and lead (Pb).

87. *Answer:* **(a) I (b) In (c) Ir (d) Fe**

Strategy and Explanation: Look up the four elements on Figure 2.6 in Section 2.9.

(a) I, iodine, is the halogen (because it is in Group 7A)

(b) In, indium, is a main group metal (because it is a metal found in an A group—Group 3A)

(c) Ir, iridium, is a transition metal (colored blue on Figure 2.6) in period 6. The period number 6 is given to the far left of the row in the periodic table next to the element Cs.

(d) Fe, iron, is a transition metal (colored blue on Figure 2.6) in period 4. The period number 4 is given to the far left of the row in the periodic table next to the element K.

89. *Answer/Explanation:* In the current periodic table, the period with the most known elements is the **sixth period**. It has a full transition metal series of ten elements, as well as the lanthanide series of 14 elements. The seventh period will have as many elements as the sixth period, if and when the elements above atomic number 112 are ever isolated.

91. *Answer:* **(a) Mg (b) Na (c) C (d) S (e) I (f) Mg (g) Kr (h) O (i) Ge** *Notice: There are multiple answers to (a), (b) and (i) in this Question. The ones given here are only examples.*

Strategy and Explanation: Use the periodic table and information given in Section 2.9.

(a) An element in Group 2A is magnesium (Mg).

(b) An element in the third period is sodium (Na).

(c) The element in the second period of Group 4A is carbon (C).

(d) The element in the third period in Group 6A is sulfur (S).

(e) The halogen in the fifth period is iodine (I).

(f) The alkaline earth element in the third period is magnesium (Mg).

(g) The noble gas element in the fourth period is krypton (Kr).

(h) The non-metal in Group 6A and the second period is oxygen (O).

(i) A metalloid in the fourth period is germanium (Ge).

93. *Answer:* **(a) iron or magnesium (b) hydrogen (c) silicon (d) iron (e) chlorine**

Strategy and Explanation: Use the chart showing the plot of relative abundance of the first 36 elements, the periodic table and information given in Section 2.9.

(a) The most abundant metal (of the first 36 elements) has atomic number 26. That is iron. The element that comes in a very close second has atomic number of 12. That is magnesium.

(b) The most abundant nonmetal (of the first 36 elements) has atomic number 1. That is hydrogen.

(c) The most abundant metalloid (of the first 36 elements) has atomic number 14. That is silicon.

(d) The most abundant transition element (of the first 36 elements) has atomic number 26. That is iron.

(e) Among the first 36 elements, three halogens are considered: fluorine (9), chlorine (17), and bromine (35). Of these three, the most abundant is chlorine.

General Questions

97. *Answer:* **(a) 0.197 nm (b) 197 pm**

Strategy and Explanation: A distance is given in angstroms (Å), which are defined. Determine the distance in nanometers and picometers.

Use the given relationship between angstroms and meters as a conversion factor to get from angstroms to meters. Then use the metric relationships between meters and the other two units to find the distance in nanometers and picometers.

$$1.97\ \text{Å} \times \frac{1 \times 10^{-10}\ \text{m}}{1\ \text{Å}} \times \frac{1\ \text{nm}}{1 \times 10^{-9}\ \text{m}} = 0.197\ \text{nm}$$

$$1.97\ \text{Å} \times \frac{1 \times 10^{-10}\ \text{m}}{1\ \text{Å}} \times \frac{1\ \text{pm}}{1 \times 10^{-12}\ \text{m}} = 197\ \text{pm}$$

✓ *Reasonable Answer Check:* The unit nanometer is larger than an angstrom, so the distance in nanometers should be a smaller number. The unit picometer is smaller than an angstrom, so the distance in picometers should be a larger number.

99. *Answer:* **(a) 0.178 nm^3 (b) 1.78 × 10^{-22} cm^3**

Strategy and Explanation: The edge length of a cube is given in nanometers. Determine the volume of the cube in cubic nanometers and in cubic centimeters.

Cube the edge length in nanometers to get the volume of the cube in cubic nanometers. Use metric relationships to convert nanometers into meters, then meters into centimeters. Cube the edge length in centimeters to get the volume of the cube in cubic centimeters.

$$V = (\text{edge length})^3 = (0.563\ \text{nm})^3 = 0.178\ \text{nm}^3$$

$$0.563\ \text{nm} \times \frac{1 \times 10^{-9}\ \text{m}}{1\ \text{nm}} \times \frac{100\ \text{cm}}{1\ \text{m}} = 5.63 \times 10^{-8}\ \text{cm}$$

$$V = (\text{edge length})^3 = (5.63 \times 10^{-8}\ \text{cm})^3 = 1.78 \times 10^{-22}\ \text{cm}^3$$

✓ *Reasonable Answer Check:* Cubing fractional quantities makes the number smaller. The unit centimeter is larger than a nanometer, so the volume in cubic centimeter should be a very small number.

102. *Answer:* **(a) 1.67 × 10^{21} molecules (b) 1.67 × 10^8 km**

Strategy and Explanation: Given the volume of one drop water and the size of a water molecule. Determine how many water molecules are in the drop and the distance spanned if the molecules were laid end to end.

(a) Find moles of water in 0.0500 mL, using the density and the molar mass, then calculate number of molecules.

$$2(1.0079\ \text{g H}) + 15.9994\ \text{g O} = 18.0152\ \text{g}\ H_2O$$

$$0.0500\ \text{mL}\ H_2O \times \frac{1.00\ \text{g}\ H_2O}{1\ \text{mL}\ H_2O} \times \frac{1000\ \text{g}}{1\ \text{kg}} \times \frac{1\ \text{mol}\ H_2O}{18.0152\ \text{g}\ H_2O} = 0.00278\ \text{mol}\ H_2O$$

$$0.0278\ \text{mol}\ H_2O \times \frac{6.022 \times 10^{23}\ \text{molecules}\ H_2O}{1\ \text{mol}\ H_2O} = 1.67 \times 10^{21}\ \text{molecules}\ H_2O$$

(b) Calculate the length of a chain of 1.67 × 10^{21} water molecules using the given size and metric conversions.

$$1.67 \times 10^{21}\ \text{molecules}\ H_2O \times \frac{100.\ \text{pm}}{1\ \text{molecule}\ H_2O} \times \frac{1 \times 10^{-12}\ \text{m}}{1\ \text{pm}} \times \frac{1\ \text{km}}{1000\ \text{m}} = 1.67 \times 10^{8}\ \text{km}$$

✓ *Reasonable Answer Check:* Molecules are nanoscale particles so it makes sense that there are a very large number of them in a drop of water. The size of one water molecule is also very small, so it makes sense that the distance spanned when these small water molecules are laid out end-to-end is also very large.

104. *Answer:* **89 tons/yr**

Strategy and Explanation: Start with the sample. Given the number of people in the city, use the volume of water each person needs per day, then calculate everyone's water needs for the day. Using the number of days in a year calculate the total volume of water used per year. Then using the mass of one gallon of water as a conversion factor, determine the grams of water. Convert the grams to tons. Then, using the fluoride concentration as a conversion factor, determine the number of tons of fluoride, and using the mass percentage of fluoride in sodium fluoride to determine the number of tons.

$$150{,}000\ \text{people} \times \frac{175\ \text{gal water / person}}{1\ \text{day}} \times \frac{365\ \text{days}}{1\ \text{year}} \times \frac{8.34\ \text{lb water}}{1\ \text{gal water}} \times \frac{1\ \text{ton water}}{2000\ \text{lb water}}$$

$$\times \frac{1\ \text{ton fluoride}}{1{,}000{,}000\ \text{tons water}} \times \frac{100\ \text{tons sodium fluoride}}{45.0\ \text{tons flouride}} = 89\ \frac{\text{tons sodium fluoride}}{\text{year}}$$

✓ *Reasonable Answer Check:* The significant figures are limited to two by the 150,000 figure. The mass units are appropriately labeled. The units cancel appropriately to give tons per year. This is a large number of people using a large amount of water so the large quantity of sodium fluoride makes sense.

106. *Answer/Explanation:* Potassium's atomic weight is 39.0983. The isotopes that contribute most to this mass are ^{39}K and ^{41}K, since the question tells us that ^{40}K has a very low abundance. Since the atomic mass is closer to 39 than 41, that confirms that the **^{39}K isotope** is more abundant.

109. *Answer/Explanation:*

(a) **Ti** is the symbol for titanium. It has atomic number = **22** and atomic weight = **47.88**.

(b) Titanium is in **Group 4B** and **Period 4**. The other elements in its group are **zirconium** (Zr), **hafnium** (Hf), and **rutherfordium** (Rf).

(c) Titanium is **light-weight and strong**, making it a good choice for something that needs to be sturdy and small.

(d) According to the New College Edition of the American Heritage Dictionary of the English Language (©1978): Titanium: A **strong, low-density, highly corrosion resistant**, lustrous white metallic element that **occurs widely** in igneous rocks and is used to alloy aircraft metals for **low weight, strength, and high-temperature stability**. (page 1348)

111. *Answer:* **0.038 mol**

Strategy and Explanation: Given the carat mass of a diamond and the relationship between carat and milligrams, determine how many moles of carbon are in the diamond.

Always start with the sample. Diamond is an allotropic form of pure carbon. Given the carats of the diamond, use the relationship between carats and milligrams as a conversion factor to determine milligrams of carbon. Then using metric relationships to determine grams of carbon, and the molar mass of carbon to determine the moles of carbon.

$$2.3 \text{ carats C} \times \frac{200. \text{ mg C}}{1 \text{ carat C}} \times \frac{1 \text{ g C}}{1000 \text{ mg C}} \times \frac{1 \text{ mol C}}{12.01 \text{ g C}} = 0.038 \text{ mol C}$$

✓ *Reasonable Answer Check:* A carat is 0.2 grams, so 2.3 carats is less than a half a gram of carbon. Since 12 grams of carbon represents a mole, half a gram should be a few hundredths of a mole.

113. *Answer:* **$5,700**

Strategy and Explanation: Given the moles of gold you want to buy, the relationship between troy ounces and grams, and the price of a troy ounce of gold, determine the amount of money you must spend.

The sample is the 1.00 mole of gold. Use the molar mass of gold to determine the grams of gold and the given relationship between grams and troy ounces to determine the troy ounces of gold, then use the price per troy ounce to determine how much that would cost.

$$1.00 \text{ mol Au} \times \frac{196.9666 \text{ g Au}}{1 \text{ mol Au}} \times \frac{1 \text{ troy ounce Au}}{31.1 \text{ g Au}} \times \frac{\$900.00}{1 \text{ troy ounce Au}} = \$5,700$$

✓ *Reasonable Answer Check:* The number of moles has three significant figures, so the answer must be reported with three significant figures. The units cancel properly. A mole of gold is pretty heavy, so this seems like a reasonable amount of money to spend.

115. *Answer:* **3.4 mol, 2.0 × 10^{24} atoms**

Strategy and Explanation: Given the dimensions of a piece of copper wire and the density of copper, determine the moles of copper and the number of atoms of copper in the wire.

Use the metric and English length relationships to convert the wire's dimensions to centimeters. Then use those dimensions to find its volume. Then use the density of copper to determine the mass of the wire. Then use the molar mass of copper to determine the moles of copper and Avogadro's number to determine the actual number of copper atoms.

$$\text{wire length in centimeters} = L = 25 \text{ ft long} \times \frac{12 \text{ in}}{1 \text{ ft}} \times \frac{2.54 \text{ cm}}{1 \text{ in}} = 7.6 \times 10^2 \text{ cm long}$$

$$\text{wire diameter in centimeters} = d = 2.0 \text{ mm diameter} \times \frac{1 \text{ m}}{1000 \text{ mm}} \times \frac{100 \text{ cm}}{1 \text{ m}} = 0.20 \text{ cm}$$

$$\text{wire radius in centimeters} = r = \frac{d}{2} = \frac{0.20 \text{ cm}}{2} = 0.10 \text{ cm}$$

$$\text{cylindrical wire's volume} = V = A \times L = (\pi r^2) \times L$$

$$V = (3.14159) \times (0.10 \text{ cm})^2 \times (7.6 \times 10^2 \text{ cm}) = 24 \text{ cm}^3$$

$$24 \text{ cm}^3 \text{ Cu} \times \frac{8.92 \text{ g Cu}}{1 \text{ cm}^3 \text{ Cu}} \times \frac{1 \text{ mol Cu atoms}}{63.546 \text{ g Cu}} = 3.4 \text{ mol Cu atoms}$$

$$3.4 \text{ mol Cu atoms} \times \frac{6.022 \times 10^{23} \text{ Cu atoms}}{1 \text{ mol Cu atoms}} = 2.0 \times 10^{24} \text{ Cu atoms}$$

✓ *Reasonable Answer Check:* The length and diameter of the wire both have two significant figures, so the answer must be reported with two significant figures. The units cancel properly. The number of atoms in a wire you can see and hold is conveniently represented in the quantity unit moles. The actual number of atoms is very large.

Applying Concepts

116. *Answer:* **(a) not possible (b) possible (c) not possible (d) not possible (e) possible (f) not possible**

Strategy and Explanation: Check to see if the following statements are true: The atomic number must be the same as the number of protons and the mass number must be the sum of the number of protons and neutrons.

(a) These values are **not possible**, since the atomic number (42) is not the same as the number of protons (19), and the sum of protons and neutrons (19+23=42) is not the same as the mass number (19).

(b) These values are **possible**. The atomic number (92) is the same as the number of protons (92), and the sum of protons and neutrons (92+143=235) is the same as the mass number (235).

(c) These values are **not possible**. The sum of protons and neutrons (131+79=210) is not the same as the mass number (53).

(d) These values are **not possible**. The sum of protons and neutrons (15+15=30) is not the same as the mass number (32).

(e) These values are **possible**. The atomic number (7) is the same as the number of protons (7), and the sum of protons and neutrons (7+7=14) is not the same as the mass number (14).

(f) These values are **not possible**. The sum of protons and neutrons (18+40=58) is not the same as the mass number (40).

118. *Answer:* **(a) same (b) second (c) same (d) same (e) same (f) second (g) same (h) first (i) second (j) first**

Strategy and Explanation: A particle is an atom or a molecule. The unit mole is a convenient way of describing a large quantity of particles. It is also important to keep in mind that 1 mol of particles contains 6.022×10^{23} particles and the molar mass describes the mass of 1 mol of particles.

(a) A sample containing 1 mol of Cl has the **same** number of particles as a sample containing 1 mol Cl_2, since each sample contains 1 mole of particles. (The masses of the samples are different, the numbers of atoms contained in the samples are different; however, those statements do not answer this question.)

(b) The **second** sample containing 1 mol of O_2 contains 6.022×10^{23} molecules of O_2. The 1-mol sample has more particles than the first sample containing just 1 molecule of O_2.

(c) Each sample contains only one particle. These samples have the **same** number of particles.

(d) A sample containing 6.022×10^{23} molecules of F_2 contains 1 mol of F_2 molecules. This sample has the **same** number of particles as 6.022×10^{23} molecules of F_2.

(e) The molar mass of Ne is 20.18 g/mol, so a sample of 20.2 grams of neon contains 1 mol of neon. Notice that, with the degree of certainty limited to one decimal place, the slightly smaller molar mass is indistinguishable from the sample mass. This 20.2-gram sample has the **same** number of particles as the sample with 1 mol of neon.

(f) The molar mass of bromine (Br_2) is (79.9 g/mol Br)×(2 mol Br) = 159.8 g/mol, so the **second** sample of 159.8 grams of bromine contains 1 mol bromine which equals 6.022×10^{23} molecules. This 159.8-gram sample has more particles than a sample containing just 1 molecule of Br_2.

(g) The molar mass of Ag is 107.9 g/mol, so a sample of 107.9 grams of Ag contains 1 mol of Ag. The molar mass of Li is 6.9 g/mol, so a sample of 6.9 grams of Li contains 1 mol of Li. These samples have the **same** number of particles, since each sample contains 1 mole of particles.

(h) The molar mass of Co is 58.9 g/mol, so the first sample of 58.9 grams of Co contains 1 mol of Co. The molar mass of Cu is 63.55 g/mol, so the second sample of 58.9 grams of Cu contains less than 1 mol of Cu. The 58.9-g **first** sample of Co has more particles than the second 58.9-g sample of Cu.

(i) The second sample containing 6.022×10^{23} atoms of Ca contains 1 mol of Ca atoms. The molar mass of Ca is 40.1 g/mol, so the first sample of 1 gram of Co contains less than 1 mol of Co. The 6.022×10^{23}-atoms **second** sample has more particles than the second 1-gram sample.

(j) A chlorine molecule contains two chlorine atoms. Since chlorine atoms all weigh the same, a chlorine molecule weighs twice as much as a chlorine atom. If the two samples of chlorine both weigh the same, the **first** sample of atomic chlorine (Cl) sample will have more particles in it than in the secomd sample of molecular sample (Cl_2).

121. *Answer:* **(a) K (b) Ar (c) Cu (d) Ge (e) H (f) Al (g) O (h) Ca (i) Br (j) P**

Strategy and Explanations:

(a) K is an alkali metal. (Group 1A)

(b) Ar is a nobel gas. (Group 8A)

(c) Cu is a transition metal. (Group 1B)

(d) Ge is a metalliod. (Group 4A)

(e) H is a group 1 nonmetal.

(f) Al is a metal that forms a 3+ ion. (Group 3A)

(g) O is a nonmetal that forms a 2– ion. (Group 6A)

(h) Ca is an alkaline earth metal. (Group 2A)

(i) Br is a halogen. (Group 7A)

(j) P is a nonmetal that is a solid. (Group 5A)

123. *Answer:* **(a) Se (b) ^{39}K (c) ^{79}Br (d) ^{20}Ne**

Strategy and Explanations:

(a) O is a nonmetal in Group 6A. The atomic number is the same as the number of protons. The number of protons is the same as the number of electrons in an uncharged atom. The element in Group 6A that has 34 electrons is Se.

(b) The alkali metal group is Group 1A. To get the number of neutrons, subtract the atomic number (Z) from the mass number (A). The element in Group 1A that has A – Z = 39 – 19 = 20 is ^{39}K.

(c) A halogen is an element in Group 7A. The atomic number is the number of protons, and the sum of the protons and the neutrons is the mass number. The element in Group 7A with Z = 35 and A = (35 + 44 =) 79 is ^{79}Br.

(d) A noble gas is an element in Group 8A. The atomic number is the number of protons, and the sum of the protons and the neutrons is the mass number. The element in Group 8A with Z = 10 and A = (10 + 10 =) 20 is ^{20}Ne.

More Challenging Questions

124. (a) *Answer:* **(a) 270 mL (b) No**

Strategy and Explanation: Given the volume of a bottle containing water, the density of water and ice (at different temperatures), determine the volume of ice.

Always start with the sample. Convert the volume of water to grams using the density of water. The entire mass of water freezes at the new temperature. Use the density of water as a conversion factor to determine the volume of ice.

$$250 \text{ mL water} \times \frac{0.997 \text{ g water}}{1 \text{ mL water}} \times \frac{1 \text{ g ice}}{1 \text{ g water}} \times \frac{1 \text{ mL ice}}{0.917 \text{ g ice}} = 272 \text{ mL ice} \approx 270 \text{ mL ice}$$

✓ *Reasonable Answer Check:* The density of ice is lower than the density of water, so the volume should be greater once the water freezes.

(b) The ice could not be contained in the bottle. The volume of ice exceeds the capacity of the bottle.

127. *Answer:* **(a) ^{79}Br–^{79}Br, ^{79}Br–^{81}Br, ^{81}Br–^{81}Br (b) 78.918 g/mol, 80.916 g/mol (c) 79.90 g/mol (d) 51.1% ^{79}Br, 48.9% ^{81}Br**

Strategy and Explanation: Using the mass of several diatomic molecules of bromine that differ by iostopic composition and the relative heights of their spectral peaks on a mass spectrum, determine the identity of isotopes in the molecules, the mass of each isotope of bromine, the average atomic mass of bromine, and the abundance of the isotopes.

Identify which of the two isotopes contribute to each of the mass spectrum peaks. Then use that information to find that atomic mass of each isotope. Use the relative peak heights and the isotope masses to find the average molar mass of Br_2 and the average atomic mass of Br, then establish variables describing the isotope percentages. Set up two relationships between these variables. The sum of the percents must be 100%, and the weighted average of the molecular masses must be the reported atomic mass.

(a) The peak representing the diatomic molecule with the lightest mass must be composed of two atoms of the lightest isotopes, ^{79}Br–^{79}Br. The peak representing the diatomic molecule with the largest mass must be composed of two atoms of the heavier isotopes, ^{81}Br–^{81}Br. The middle peak must represent molecules made with one of each, ^{79}Br–^{81}Br.

(b) The mass of the molecule composed of two lightweight atoms is 157.836 g/mol, so each atom must weigh 0.5×(157.836 g/mol) = 78.918 g/mol. The mass of the molecule composed of two heavy weight atoms is 161.832 g/mol, so each atom must weigh 0.5×(161.832 g/mol) = 80.916 g/mol.

(c) The average mass of the molecules would be the weighted average of these three variants:

$$\frac{25.54\text{ atoms }^{79}\text{Br}_2}{100\text{ Br}_2\text{ molecules}}\times\left(\frac{157.836\text{ amu}}{1\text{ atom }^{79}\text{Br}_2}\right)+\frac{49.99\text{ atoms }^{79}\text{Br}^{81}\text{Br}}{100\text{ Br}_2\text{ molecules}}\times\left(\frac{159.834\text{ amu}}{1\text{ atom }^{79}\text{Br}^{81}\text{Br}}\right)+$$

$$\frac{24.46\text{ atoms }^{81}\text{Br}_2}{100\text{ Br}_2\text{ molecules}}\times\left(\frac{161.832\text{ amu}}{1\text{ atom }^{81}\text{Br}_2}\right)=159.80\frac{\text{amu}}{\text{Br}_2\text{ molecules}}$$

The diatomic molecule mass gets divided by two, to determine the atomic mass: (159.80 g/mol)/2 = 79.90 g/mol.

(d) The X% ^{79}Br and Y% ^{81}Br, This means:

Every 100 atoms of silver contains X atoms of the ^{79}Br isotope.

Every 100 atoms of silver contains Y atoms of the ^{81}Br isotope.

$$X+Y=100\%$$

$$\frac{X\text{ atoms }^{79}\text{Br}}{100\text{ Br atoms}}\times\left(\frac{78.918\text{ amu}}{1\text{ atom }^{79}\text{Br}}\right)+\frac{Y\text{ atoms }^{91}\text{Br}}{100\text{ Ag atoms}}\times\left(\frac{80.916\text{ amu}}{1\text{ atom }^{91}\text{Br}}\right)=79.90\frac{\text{amu}}{\text{Br atom}}$$

We now have two equations and two unknowns, so we can solve for X and Y algebraically. Solve the first equation for Y: Y = 100 – X. Plug that in for Y in the second equation. Then solve for X:

$$\frac{X}{100}\times(78.918)+\frac{100-X}{100}\times(80.916)=79.90$$

$$0.78918\text{ X}+80.916-0.80916\text{ X}=79.90$$

$$80.916-79.90=0.80916\text{ X}-0.78918\text{ X}$$

$$1.02=(0.01998)\text{X}$$

$$\text{X}=51.1$$

Now, plug the value of X in the first equation to get Y. Y = 100 – X = 100 – 51.1 = 48.9

Therefore the abundance for these isotopes are: 51.1% ^{79}Br and 48.9% ^{81}Br.

✓ *Reasonable Answer Check:* These isotopes are about equally abundant as indicated by the near 25:50:25 ratio of the isotopes in the three mass spectrum peaks.

Chapter 3: Chemical Compounds
Solutions for Blue-Numbered Questions for Review and Thought

Topical Questions

Molecular and Structural Formulas (Section 3.1)

9. *Answer/Explanation:*

(a) Bromine trifluoride is $\mathbf{BrF_3}$.
(b) Xenon difluoride $\mathbf{XeF_2}$.
(c) Diphosphorus tetrafluoride is $\mathbf{P_2F_4}$.
(d) Pentadecane is $\mathbf{C_{15}H_{32}}$.
(e) Hydrazine is $\mathbf{N_2H_4}$.

11. *Answer/Explanation:*

	Molecular Formula	Condensed Formula	Structural Formula
butanol	$C_4H_{10}O$	$CH_3CH_2CH_2CH_2OH$	H–C(H)(H)–C(H)(H)–C(H)(H)–C(H)(H)–OH
pentanol	$C_5H_{12}O$	$CH_3CH_2CH_2CH_2CH_2OH$	H–C(H)(H)–C(H)(H)–C(H)(H)–C(H)(H)–C(H)(H)–OH

14. *Answer/Explanation:*

(a) Benzene Molecular Formula: $\mathbf{C_6H_6}$
(b) Vitamin C Molecular Formula: $\mathbf{C_6H_8O_6}$

16. *Answer:* **(a) 1 Ca, 2 C, 4 O (b) 8 C, 8 H (c) 2 N, 8 H, 1 S, 4 O (d) 1 Pt, 2 N, 6 H, 2 Cl (e) 4 K, 1 Fe, 6 C, 6 N**

Strategy and Explanation: Keep in mind that atoms found inside parentheses that are followed by a subscript get multiplied by that subscript.

(a) CaC_2O_4 contains one atom of calcium, two atoms of carbon, and four atoms of oxygen.
(b) $C_6H_5CHCH_2$ contains eight atoms of carbon and eight atoms of hydrogen.
(c) $(NH_4)_2SO_4$ contains two (1 × 2) atoms of nitrogen, eight (4 × 2) atoms of hydrogen, one atom of sulfur, and four atoms of oxygen.
(d) $Pt(NH_3)_2Cl_2$ contains one atom of platinum, two (1 × 2) atoms of nitrogen, six (3 × 2) atoms of hydrogen, and two atoms of chlorine.
(e) $K_4Fe(CN)_6$ contains four atoms of potassium, one atom of iron, six (1 × 6) atoms of carbon, and six (1 × 6) atoms of nitrogen.

Binary Inorganic Compounds (Section 3.2)

18. *Answer/Explanation:* A general rule for naming binary compounds is to name the first element then take the first part of the name of the second element and add the ending -ide. Prefixes given in Table 3.2 are used to designate the number of a particular kind of atom, such as mono- for one, di- for two, tri- for three, etc.

(a) SO_2 is **sulfur dioxide.**
(b) CCl_4 is **carbon tetrachloride.**
(c) P_4S_{10} is **tetraphosphorus decasulfide.**
(d) SF_4 is **sulfur tetrafluoride.**

20. *Answer/Explanation:* A general rule for naming binary compounds is to name the first element then take the first part of the name of the second element and add the ending -ide. Prefixes given in Table 3.2 are used to designate the number of a particular kind of atom, such as mono- for one, di- for two, tri- for three, etc.

(a) HBr is **hydrogen bromide**.

(b) ClF_3 is **chlorine trifluoride**.

(c) Cl_2O_7 is **dichlorine heptaoxide**.

(d) BI_3 is **boron triiodide**.

Hydrocarbons (Sections 3.3)

22. *Answer/Explanation:* Carbon makes four bonds. A carbon atom in an alkane chain is bonded to at least one other C atom, so that leaves up to three remaining bonds that may each be to an H atom. So, in a noncyclic alkane other than methane, the maximum number of hydrogen atoms that can be bonded to one carbon atom is **three**.

24. *Answer/Explanation:* The general formula for alkanes is C_nH_{2n+2}. Heptane has seven C atoms, so it's formula is C_7H_{16}. Therefore, heptane has **16** hydrogen atoms.

Constitutional Isomers (Section 3.4)

26. *Answer/Explanation:*

(a) Two molecules that are constitutional isomers have the same formula (i.e., on the molecular level, these molecules have **the same number of atoms of each kind**).

(b) Two molecules that are constitutional isomers of each other have their atoms in **different bonding arrangements**.

Predicting Ion Charges (Section 3.5)

28. *Answer:* **(a) Li^+ (b) Sr^{2+} (c) Al^{3+} (d) Ca^{2+} (e) Zn^{2+}**

Strategy and Explanation: A general rule for the charge on a metal cation: the group number represents the number of electrons lost. Hence, the group number will be the cation's positive charge.

(a) Lithium (Group 1A) Li^+

(b) Strontium (Group 2A) Sr^{2+}

(c) Aluminum (Group 3A) Al^{3+}

(d) Calcium (Group 2A) Ca^{2+}

(e) Zinc (Group 2B) Zn^{2+}

30. *Answer:* **Ba: 2+, Br: –1**

Strategy and Explanation: A general rule for the charge on a metal cation is as follows: the group number represents the number of electrons lost. Hence, the group number will be the cation's positive charge. For nonmetal elements in Groups 5A-7A, the electrons gained by an atom to form a stable anion are calculated using the formula: 8 – (group number). That means the (group number) – 8 is the negative charge of the anion.

Barium is in Group 2A so barium ions would have +2 charge. Bromine is in Group 7A, so bromide ions will have a –1 charge, calculated from 7– 8.

32. *Answer:* **(a) +2 (b) +2 (c) +2 or +3 (d) +3**

Strategy and Explanation: A general rule for the charge on a metal cation: the group number represents the number of electrons lost. Hence, the group number will be the cation's positive charge. Transition metals often have a +2 charge. Some have +3 and +1 charged ions, as well.

(a) magnesium (Group 2A) has a +2 charge. Mg^{2+}

(b) zinc (Group 8B) has a +2 charge. Zn^{2+}

(c) iron (a transition metal) has a +2 or +3 charge. Fe^{2+} or Fe^{3+}

(d) gallium (Group 3A) has a +3 charge. Ga^{3+}

34. *Answer:* **CoO, Co_2O_3**

Strategy and Explanation: Cobalt ions are Co^{2+} and Co^{3+}. Oxide ion is O^{2-}. The two compounds containing cobalt and oxide are made from the neutral combination of the charged ions:

one Co^{2+} and one O^{2-} [net charge = +2 + (–2) = 0] CoO

two Co^{3+} and three O^{2-} [net charge = 2 × (+3) + 3 × (–2) = 0] Co_2O_3

36. *Answer:* **(c) and (d) are correct formulas. (a) $AlCl_3$, (b) NaF**

Strategy and Explanation:

(a) Aluminum ion (from Group 3A) is Al^{3+}. Chloride ion (from Group 7A) is Cl^-.

AlCl is **not** a neutral combination of these two ions. The correct formula would be $AlCl_3$.

[net charge = +3 + 3 × (–1) = 0]

(b) Sodium ion (Group 1A) is Na^+. Fluoride ion (from Group 7A) is F^-.

NaF_2 is **not** a neutral combination of these two ions. The correct formula would be NaF.

[net charge = +1 + (–1) = 0]

(c) Gallium ion (from Group 3A) is Ga^{3+}. Oxide ion (from Group 6A) is O^{2-}.

Ga_2O_3 **is** the correct neutral combination of these two ions.

[net charge = 2 × (+3) + 3 × (–2) = 0]

(d) Magnesium ion (from Group 2A) is Mg^{2+}. Sulfide ion (from Group 6A) is S^{2-}.

MgS **is** the correct neutral combination of these two ions.

[net charge = +2 + (–2) = 0]

38. *Answer:* **Mn**

Strategy and Explanation: The atomic number of an atom is represented by the number of protons. The net charge of an ion shows the charge excess (if negative) or charge deficiency (if positive).

net charge = (number of protons) – (number of electrons)

number of protons = (number of electrons) + (net charge)

Given an ion with 23 electrons and a net ionic charge of +2, the number of protons can be calculated:

23 + 2 = 25. According to the periodic table, the element manganese has an atomic number of 25, so X = Mn.

Polyatomic Ions (Section 3.5)

40. *Answer:* **(a) Pb^{2+} and NO_3^-; 1 and 2, respectively (b) Ni^{2+} and CO_3^{2-}; 1 and 1, respectively (c) NH_4^+ and PO_4^{3-}; 3 and 1, respectively (d) K^+ and SO_4^{2-}; 2 and 1, respectively**

Strategy and Explanation:

(a) $Pb(NO_3)_2$ has one ion of lead(II) (Pb^{2+}) and two ions of nitrate (NO_3^-).

(b) $NiCO_3$ has one ion of nickel(II) (Ni^{2+}) and one ion of carbonate (CO_3^{2-}).

(c) $(NH_4)_3PO_4$ has three ions of ammonium (NH_4^+) and one ion of phosphate (PO_4^{3-}).

(d) K_2SO_4 has two ions of potassium (K^+) and one ion of sulfate (SO_4^{2-}).

42. *Answer:* **$BaSO_4$, barium ion, 2+, sulfate, 2–; $Mg(NO_3)_2$, magnesium ion, 2+, nitrate, 1–; $NaCH_3CO_2$, sodium ion, 1+, acetate, 1–**

Strategy and Explanation: Barium sulfate is $BaSO_4$. It contains barium ion (Ba^{2+}), with a 2+ electrical charge, and sulfate ion (SO_4^{2-}), with a 2– electrical charge. Magnesium nitrate is $Mg(NO_3)_2$. It contains magnesium ion (Mg^{2+}), with a 2+ electrical charge, and nitrate ions (NO_3^-), with a 1– electrical charge. Sodium acetate is $NaCH_3CO_2$ or $NaCH_3COO$. It contains sodium ion (Na^+), with a 1+ electrical charge, and acetate ion (CH_3COO^-), with a 1– electrical charge. (Notice: Occasionally the Na^+ is written on the other end of the acetate formula like this CH_3COONa. It is done that way because the negative charge on acetate is on one of the oxygen atoms, so that's where the Na^+ cation will be attracted.)

44. *Answer:* **(a) $Ni(NO_3)_2$ (b) $NaHCO_3$ (c) $LiClO$ (d) $Mg(ClO_3)_2$ (e) $CaSO_3$**

Strategy and Explanation:

(a) Nickel(II) ion is Ni^{2+}. Nitrate ion is NO_3^-. Use one Ni^{2+} and two NO_3^- to make neutral $Ni(NO_3)_2$.

(b) Sodium ion is Na^+. Bicarbonate ion is HCO_3^-. Use one Na^+ and one HCO_3^- to make neutral $NaHCO_3$.

(c) Lithium ion is Li^+. Hypochlorite ion is ClO^-. Use one Li^+ and one ClO^- to make neutral $LiClO$.

(d) Magnesium ion is Mg^{2+}. Chlorate ion is ClO_3^-. Use one Mg^{2+} and two ClO_3^- to make neutral $Mg(ClO_3)_2$.

(e) Calcium ion is Mg^{2+}. Sulfite ion is SO_3^{2-}. Use one Mg^{2+} and one SO_3^{2-} to make neutral $CaSO_3$.

Ionic Compounds (Sections 3.5, 3.6, 3.7)

46. *Answer:* **(b), (c), and (e) are ionic.**

Strategy and Explanation: To tell if a compound is ionic or not, look for metals and nonmetals together, or common cations and anions. If a compound contains only nonmetals or metalloids and nonmetals, it is probably not ionic.

(a) CF_4 contains only nonmetals. Not ionic.

(b) $SrBr_2$ has a metal and nonmetal together. Ionic.

(c) $Co(NO_3)_3$ has a metal and nonmetals together. Ionic.

(d) SiO_2 contains a metalloid and a nonmetal. Not ionic.

(e) KCN has a metal and a common diatomic ion (CN^-) together. Ionic.

(f) SCl_2 contains only nonmetals. Not ionic.

48. *Answer:* **Only (e) is ionic, with metal and nonmetal combined. (a)-(d) are composed of only nonmetals.**

Strategy and Explanation: To tell if a compound is ionic or not, look for metals and nonmetals together, or common cations and anions. If a compound contains only nonmetals or metalloids and nonmetals, it is probably not ionic.

(a) SF_6 contains only nonmetals. Not ionic.

(b) CH_4 contains only nonmetals. Not ionic.

(c) H_2O_2 contains only nonmetals. Not ionic.

(d) NH_3 contains only nonmetals. Not ionic.

(e) CaO has a metal and a nonmetal together. Ionic.

50. *Answer:* **(a) $(NH_4)_2CO_3$ (b) CaI_2 (c) $CuBr_2$ (d) $AlPO_4$**

Strategy and Explanation: Make neutral combinations with the common ions involved.

(a) Ammonium (NH_4^+) and carbonate (CO_3^{2-}) must be combined 2:1, to make $(NH_4)_2CO_3$.

(b) Calcium (Ca^{2+}) and iodide (I^-) must be combined 1:2, to make CaI_2.

(c) Copper(II) (Cu^{2+}) and bromide (Br^-) must be combined 1:2, to make $CuBr_2$.

(d) Aluminum (Al^{3+}) and phosphate (PO_4^{3-}) must be combined 1:1, to make $AlPO_4$.

52. *Answer:* **(a) potassium sulfide (b) nickel(II) sulfate (c) ammonium phosphate (d) aluminum hydroxide (e) cobalt(III) sulfate**

Strategy and Explanation: Give the name of the cation then the name of the anion.

(a) K_2S contains cation K^+ called potassium and anion S^{2-} called sulfide, so it is potassium sulfide.

(b) $NiSO_4$ contains cation Ni^{2+} called nickel(II) and anion SO_4^{2-} called sulfate, so it is nickel(II) sulfate.

(c) $(NH_4)_3PO_4$ contains cation NH_4^+ called ammonium and anion PO_4^{3-} called phosphate, so it is ammonium phosphate.

(d) $Al(OH)_3$ contains cation Al^{3+} called aluminum and anion OH^- called hydroxide, so it is aluminum hydroxide.

(e) $Co_2(SO_4)_3$ contains cation Co^{3+} called cobalt(III) and anion SO_4^{2-} called sulfate, so it is cobalt(III) sulfate.

54. *Answer:* **MgO; MgO has higher ionic charges and smaller ion sizes than NaCl**

Strategy and Explanation: Magnesium oxide is MgO, and it is composed of Mg^{2+} ions and O^{2-} ions. The relatively high melting temperature of MgO compared to NaCl (composed of Na^+ ions and Cl^- ions) is probably due to the larger ionic charges and smaller sizes of the ions. The large opposite charges sitting close together have very strong attractive forces between the ions. Melting requires that these attractive forces be overcome.

Electrolytes (Section 3.7)

56. *Answer:* **Conducts electricity in water; check electrical conductivity; examples: NaCl and $C_{12}H_{22}O_{11}$ (Other answers are possible.)**

Strategy and Explanation: An electrolyte is a compound that conducts electricity when dissolved in water because it dissociates into ions. Electricity is conducted when ions are present in the solution. We can differentiate a strong electrolye from a nonelectrolyte by checking whether the solution conducts electricity. When a strong electrolyte (such as NaCl) dissolves in water, it will ionize completely (producing Na^+ and Cl^-). When a nonelectrolyte (such as table sugar, sucrose, $C_{12}H_{22}O_{11}$) dissolves in water it does not ionizes (In the solution is found only the molecular form, $C_{12}H_{22}O_{11}$).

58. *Answer/Explanation:* "Molecular compounds are generally non-electrolytes." This general trend is sensible, since the molecular compounds are generally not ionic compounds, and therefore would not ionize in water.

60. *Answer:* **(a) K^+ and OH^- (b) K^+ and SO_4^{2-} (c) Na^+ and NO_3^- (d) NH_4^+ and Cl^-**

Strategy and Explanation: Identify the common ions present in the compounds.

(a) The ions present in a solution of KOH are K^+ and OH^-.

(b) The ions present in a solution of K_2SO_4 are K^+ and SO_4^{2-}.

(c) The ions present in a solution of $NaNO_3$ are Na^+ and NO_3^-.

(d) The ions present in a solution of NH_4Cl are NH_4^+ and Cl^-.

62. *Answer:* **(a) Box 2 (b) Box 3 (c) Box 1**

Strategy and Explanation: Identify the common ions present in the compounds.

(a) The ions present in a solution of $MgCl_2$ are Mg^{2+} and 2 Cl^-, in a 1:2 ratio. **Box 2** has eight 1– charges and four 2+ charges, giving a 1:2 ratio.

(b) The ions present in a solution of K_2SO_4 are 2 K^+ and SO_4^{2-}, in a 2:1 ratio. **Box 3** has eight 1+ charges and four 2– charges, giving a 2:1 ratio.

(c) The ions present in a solution of NH_4Cl are NH_4^+ and Cl^-, in a 1:1 ratio. **Box 1** has five 1+ charges and five 1– charges, giving a 1:1 ratio.

64. *Answer:* **(a) and (d)**

Strategy and Explanation: Determine if the compound is an electrolyte, by determining if it is ionic.

(a) NaCl is an ionic compound and will ionize to form Na^+ and Cl^-. When NaCl is dissolved in water, the resulting solution will conduct electricity.

(b) $CH_3CH_2CH_3$ is an organic hydrocarbon compound and will not ionize. When $CH_3CH_2CH_3$ is dissolved in water, the resulting solution will not conduct electricity.

(c) CH_3OH is an organic compound and will not ionize. When CH_3OH is dissolved in water, the resulting solution will not conduct electricity.

(d) $Ca(NO_3)_2$ is an ionic compound and will ionize to form Ca^{2+} and NO_3^-. When $Ca(NO_3)_2$ is dissolved in water, the resulting solution will conduct electricity.

Moles of Compounds (Section 3.8)

66. *Answer:*

	CH_3OH	Carbon	Hydrogen	Oxygen
No. of moles	**1 mol**	**1 mol**	**4 mol**	**1 mol**
No. of molecules or atoms	**6.022×10^{23} molecules**	**6.022×10^{23} atoms**	**2.409×10^{24} atoms**	**6.022×10^{23} atoms**
Molar mass (grams per mol methanol)	**32.0417**	**12.0107**	**4.0316**	**15.9994**

Strategy and Explanation: Consider a sample of 1 mol of methanol. The formula gives the mole ratio of each atom in one mole of methanol. Each mole is 6.022×10^{23} molecules. The molar masses can be determined by looking up each element on the period table, finding the atomic weight (which is also the grams/mol), and multiplying by the number of moles (see the solution to Question 58 for more details).

68. *Answer:* **(a) 159.688 g/mol (b) 67.806 g/mol (c) 44.0128 g/mol (d) 197.905 g/mol (e) 176.1238 g/mol**

Strategy and Explanation: To calculate the molar mass of a compound, we perform a series of steps: First, look up the atomic mass of each element on the periodic table or in a table of atomic masses to identify the molar mass of the element. Second, determine the number of moles of atoms in one mole of the compound by examining the formula. Third, multiply the number of moles of the element by the molar mass of the element. Last, add all individual masses.

(a) Iron(III) oxide is Fe_2O_3. 1 mol of Fe_2O_3 contains 2 mol of Fe and 3 mol of O.

$$\left(\frac{2\text{ mol Fe}}{1\text{ mol Fe}_2\text{O}_3}\times\frac{55.845\text{ g}}{1\text{ mol Fe}}\right)+\left(\frac{3\text{ mol O}}{1\text{ mol Fe}_2\text{O}_3}\times\frac{15.9994\text{ g}}{1\text{ mol O}}\right)=159.688\ \frac{\text{g}}{\text{mol Fe}_2\text{O}_3}$$

Once we get used to what the numbers represent, this calculation may be abbreviated as follows:

$$2\times(55.845\text{ g Fe})+3\times(15.9994\text{ g O})=159.688\text{ g/mol Fe}_2\text{O}_3$$

(b) Boron trifluoride is BF_3. 1 mol of BF_3 contains 1 mol of B and 3 mol of F.

$$\left(\frac{1\text{ mol B}}{1\text{ mol BF}_3}\times\frac{10.811\text{ g}}{1\text{ mol B}}\right)+\left(\frac{3\text{ mol F}}{1\text{ mol BF}_3}\times\frac{18.9984\text{ g}}{1\text{ mol F}}\right)=67.806\ \frac{\text{g}}{\text{mol BF}_3}$$

This calculation may be abbreviated as follows:

$$10.811\text{ g B}+3\times(18.9984\text{ g F})=67.806\text{ g/mol BF}_3$$

(c) Dinitrogen oxide is N_2O. 1 mol of N_2O contains 2 mol of N and 1 mol of O. The abbreviated calculation for the molar mass looks like this:

$$2\times(14.0067\text{ g N})+15.9994\text{ g O}=44.0128\text{ g/mol N}_2\text{O}$$

(d) Manganese(II) chloride tetrahydrate is $MnCl_2{\cdot}4\,H_2O$. 1 mol of $MnCl_2{\cdot}4\,H_2O$ compound has 1 mol of $MnCl_2$ with 4 mol of H_2O molecules. So, it contains 1 mol Mn, 2 mol Cl, (4 × 2 =) 8 mol H and (4 × 1 =) 4 mol O. The abbreviated calculation for the molar mass looks like this:

$$54.938\text{ g Mn}+2\times(35.453\text{ g Cl})+8\times(1.0079\text{ g H})+4\times(15.9994\text{ g O})=197.905\text{ g/mol MnCl}_2{\cdot}4\text{ H}_2\text{O}$$

(e) 1 mol of $C_6H_8O_6$ compound contains 6 mol of C, 8 mol of H, and 6 mol of O. The abbreviated calculation for the molar mass looks like this:

$$6\times(12.0107\text{ g C})+8\times(1.0079\text{ g H})+6\times(15.9994\text{ g O})=176.1238\text{ g/mol C}_6\text{H}_8\text{O}_6$$

70. *Answer:* **(a) 0.0312 mol (b) 0.0101 mol (c) 0.0125 mol (d) 0.00406 mol (e) 0.00599 mol**

Strategy and Explanation: Determine the number of moles in a given mass of a compound.

Use the formula and the periodic table to calculate the molar mass for the compound, then use the molar mass as a conversion factor between grams and moles.

(a) Molar mass CH_3OH = (12.0107 g) + 4 × (1.0079 g) + (15.9994 g) = 32.0417 g/mol

$$1.00 \text{ g } CH_3OH \times \frac{1 \text{ mol } CH_3OH}{32.0417 \text{ g } CH_3OH} = 0.0312 \text{ mol } CH_3OH$$

(b) Molar mass Cl_2CO = 2 × (35.453 g) + (12.0107 g) + (15.9994 g) = 98.916 g/mol

$$1.00 \text{ g } Cl_2CO \times \frac{1 \text{ mol } Cl_2CO}{98.916 \text{ g } Cl_2CO} = 0.0101 \text{ mol } Cl_2CO$$

(c) Ammonium nitrate is NH_4NO_3. Molar mass NH_4NO_3 = 2 × (14.0067 g) + 4 × (1.0079 g) + 3 × (15.9994 g) = 80.043 g/mol

$$1.00 \text{ g } NH_4NO_3 \times \frac{1 \text{ mol } NH_4NO_3}{80.043 \text{ g } NH_4NO_3} = 0.0125 \text{ mol } NH_4NO_3$$

(d) Magnesium sulfate heptahydrate is $MgSO_4{\cdot}7\ H_2O$. Molar mass $MgSO_4{\cdot}7\ H_2O$

= (24.305 g) + (32.065 g) + 11 × (15.9994 g) + 14 × (1.0079 g) = 246.474 g/mol

$$1.00 \text{ g } MgSO_4 \cdot 7H_2O \times \frac{1 \text{ mol } MgSO_4 \cdot 7H_2O}{246.474 \text{ g } MgSO_4 \cdot 7H_2O} = 0.00406 \text{ mol } MgSO_4 \cdot 7H_2O$$

(e) Silver acetate is $AgC_2H_3O_2$. Molar mass $AgC_2H_3O_2$

= (107.8682 g) + 2 × (12.0107 g) + 3 × (1.0079 g) + 2 × (15.9994 g) = 166.9121 g/mol

$$1.00 \text{ g } AgC_2H_3O_2 \times \frac{1 \text{ mol } AgC_2H_3O_2}{166.9121 \text{ g } AgC_2H_3O_2} = 0.00599 \text{ mol } AgC_2H_3O_2$$

✓ *Reasonable Answer Check:* The quantity in moles is always going to be smaller than the mass in grams.

72. *Answer:* **(a) 179.855 g/mol (b) 36.0 g (c) 0.0259 mol**

Strategy and Explanation: Determine the molar mass of a compound and then determine the mass of a given number of moles and the number of moles in a given mass.

Use the formula and the periodic table to calculate the molar mass for the compound, then use the molar mass as a conversion factor between grams and moles.

(a) Molar mass $Fe(NO_3)_2$ = 55.845 g + 2 × (14.0067 g) + 6 × (15.9994 g) = 179.855 g/mol

(b) $$0.200 \text{ mol } Fe(NO_3)_2 \times \frac{179.857 \text{ g } Fe(NO_3)_2}{1 \text{ mol } Fe(NO_3)_2} = 36.0 \text{ g } Fe(NO_3)_2$$

(c) $$4.66 \text{ g } Fe(NO_3)_2 \times \frac{1 \text{ mol } Fe(NO_3)_2}{179.855 \text{ g } Fe(NO_3)_2} = 0.0259 \text{ mol } Fe(NO_3)_2$$

✓ *Reasonable Answer Check:* The quantity in moles is always going to be smaller than the mass in grams.

74. *Answer:* **(a) 151.1622 g/mol (b) 0.0352 mol (c) 25.1 g**

Strategy and Explanation: Determine the molar mass of a compound and then determine the mass of a given number of moles and the number of moles in a given mass.

Use the formula and the periodic table to calculate the molar mass for the compound, then use the molar mass as a conversion factor between grams and moles.

(a) Molar mass $C_8H_9O_2N$ = 8 × (12.0107 g) + 9 × (1.0079 g) + 2 × (15.9994 g)

+ (14.0067 g) = 151.1622 g/mol

(b) $$5.32 \text{ g } C_8H_9O_2N \times \frac{1 \text{ mol } C_8H_9O_2N}{151.1622 \text{ g } C_8H_9O_2N} = 0.0352 \text{ mol } C_8H_9O_2N$$

(c) $$0.166 \text{ mol } C_8H_9O_2N \times \frac{151.1622 \text{ g } C_8H_9O_2N}{1 \text{ mol } C_8H_9O_2N} = 25.1 \text{ g } C_8H_9O_2N$$

✓ *Reasonable Answer Check:* The quantity in moles is always going to be smaller than the mass in grams.

76. *Answer:* **1.80 × 10^{-3} mol, 1.09 × 10^{21} molecules**

Strategy and Explanation: Given the mass and formula of a compound, determine the number of moles and molecules in that sample.

Use the formula and the periodic table to calculate the molar mass for the compound. Use the molar mass as a conversion factor between grams and moles, then use Avogadro's number to determine the actual number of molecules.

$$\text{Molar mass } C_9H_8O_4 = 9 \times (12.0107 \text{ g}) + 8 \times (1.0079 \text{ g}) + 4 \times (15.9994 \text{ g}) = 180.15710 \text{ g/mol}$$

$$325 \text{ mg } C_9H_8O_4 \times \frac{1 \text{ g}}{1000 \text{ mg}} \times \frac{1 \text{ mol } C_9H_8O_4}{180.15710 \text{ g } C_9H_8O_4} = 1.80 \times 10^{-3} \text{ mol}$$

$$1.80 \times 10^{-3} \text{ mol} \times \frac{6.022 \times 10^{23} \text{ molecules}}{1 \text{ mol}} = 1.09 \times 10^{21} \text{ molecules}$$

✓ *Reasonable Answer Check:* The number of molecules in a macroscopic sample will be huge.

78. *Answer:* **(a) 0.400 mol (b) 0.250 mol (c) 0.628 mol**

Strategy and Explanation: Given the mass of a substance, determine the number of moles.

Use the formula and the periodic table to calculate the molar mass for the compound, then use the molar mass as a conversion factor between grams and moles.

(a) Molar mass $H_2SO_4 = 2 \times (1.0079 \text{ g}) + (32.065 \text{ g}) + 4 \times (15.9994 \text{ g}) = 98.078 \text{ g/mol}$

$$39.2 \text{ g } H_2SO_4 \times \frac{1 \text{ mol } H_2SO_4}{98.078 \text{ g } H_2SO_4} = 0.400 \text{ mol } H_2SO_4$$

(b) Molar mass $O_2 = 2 \times (15.9994 \text{ g}) = 31.9988 \text{ g/mol}$

$$8.00 \text{ g } O_2 \times \frac{1 \text{ mol } O_2}{31.9988 \text{ g } O_2} = 0.250 \text{ mol } O_2$$

(c) Molar mass $NH_3 = 14.0067 \text{ g} + 3 \times (1.0079 \text{ g}) = 17.0304 \text{ g/mol}$

$$10.7 \text{ g } NH_3 \times \frac{1 \text{ mol } NH_3}{17.0304 \text{ g } NH_3} = 0.628 \text{ mol } NH_3$$

✓ *Reasonable Answer Check:* The quantity in moles is always going to be smaller than the mass in grams.

80. *Answer:* **2.7 × 10^{23} atoms**

Strategy and Explanation: Given the mass and formula of a compound, determine the number of atoms of an element in that sample.

Use the formula and the periodic table to calculate the molar mass for the compound. Use the molar mass as a conversion factor between grams and moles, use the formula stoichiometry to determine the number of moles of the element in the compound, then use Avogadro's number to determine the actual number of atoms.

$$\text{Molar mass } Cu(NO_3)_2 = 63.546 \text{ g} + 2 \times (14.0067 \text{ g}) + 6 \times (15.9994 \text{ g}) = 187.556 \text{ g/mol}$$

$$14.0 \text{ g } Cu(NO_3)_2 \times \frac{1 \text{ mol } Cu(NO_3)_2}{187.556 \text{ g } Cu(NO_3)_2} \times \frac{6 \text{ mol O}}{1 \text{ mol } Cu(NO_3)_2} \times \frac{6.022 \times 10^{23} \text{ atoms of O}}{1 \text{ mol O}}$$

$$= 2.70 \times 10^{23} \text{ atoms of O}$$

✓ *Reasonable Answer Check:* The number of atoms in a macroscopic sample will be huge.

82. *Answer:* **1.2 × 10^{24} molecules**

Strategy and Explanation: Given the mass of a compound, determine the number of molecules.

Use the formula and the periodic table to calculate the molar mass for the compound. Use the molar mass as a conversion factor between grams and moles, then use Avogadro's number to determine the actual number of molecules.

$$\text{Molar mass } CCl_2F_2 = (12.0107 \text{ g}) + 2 \times (35.453 \text{ g}) + 2 \times (18.9984 \text{ g}) = 120.914 \text{ g/mol}$$

$$250\text{ g }CCl_2F_2 \times \frac{1\text{ mol }CCl_2F_2}{120.914\text{ g }CCl_2F_2} \times \frac{6.022\times10^{23}\text{ molecules }CCl_2F_2}{1\text{ mol }CCl_2F_2} = 1.2\times10^{24}\text{ molecules }CCl_2F_2$$

✓ *Reasonable Answer Check:* The number of molecules for a macroscopic sample will be huge.

84. *Answer:* **(a) 0.250 mol CF_3CH_2F (b) 6.02 × 10^{23} F atoms**

Strategy and Explanation: Given the mass of a compound, determine the number of moles of that compound, and the number of atoms of one of the elements.

Use the formula and the periodic table to calculate the molar mass for the compound. Use the molar mass as a conversion factor between grams and moles, then use Avogadro's number to determine number of molecules, then use the formula stoichiometry to determine number of atoms of each type.

Molar mass CF_3CH_2F = 2 × (12.0107 g) + 4 × (18.9984 g) + 2 × (1.0079 g) = 102.0308 g/mol

(a)
$$25.5\text{ g }CF_3CH_2F \times \frac{1\text{ mol }CF_3CH_2F}{102.0308\text{ g }CF_3CH_2F} = 0.250\text{ mol }CF_3CH_2F$$

(b) Stoichiometry of the chemical formula: Each mol of CF_3CH_2F contains 4 mol of F atoms.

$$0.250\text{ mol }CF_3CH_2F \times \frac{4\text{ mol F}}{1\text{ mol }CF_3CH_2F} \times \frac{6.022\times10^{23}\text{ F atoms}}{1\text{ mol F}} = 6.02\times10^{23}\text{ F atoms}$$

✓ *Reasonable Answer Check:* The number of atoms in a macroscopic sample will be huge. The mass of a substance will always be larger than the number of moles.

Percent Composition (Section 3.9)

87. *Answer:* **(a) 239.3 g/mol PbS, 86.60% Pb, 13.40% S (b) 30.0688 g/mol C_2H_6, 79.8881% C, 20.1119% H (c) 60.0518 g/mol CH_3CO_2H, 40.0011% C, 6.7135% H, 53.2854% O (d) 80.0432 g/mol NH_4NO_3, 34.9979% C, 5.0368% H, 59.9654% O**

Strategy and Explanation: Given the formula of a compound, determine the molar mass, and the mass percent of each element.

Calculate the mass of each element in one mole of compound, while calculating the molar mass of the compound. Divide the calculated mass of the element by the molar mass of the compound and multiply by 100% to get mass percent. To get the last element's mass percent, subtract the other percentages from 100%.

(a)
Mass of Pb per mole of PbS = 207.2 g Pb

Mass of S per mole of PbS = 32.065 g S

Molar mass PbS = (207.2 g) + (32.065 g) = 239.3 g/mol PbS

$$\%\text{ Pb} = \frac{\text{mass of Pb per mol PbS}}{\text{mass of PbS per mol PbS}} \times 100\% = \frac{207.2\text{ g Pb}}{239.3\text{ g PbS}} \times 100\% = 86.60\%\text{ Pb in PbS}$$

% S = 100% – 86.60% Pb = 13.40% S in PbS

(b)
Mass of C per mole of C_2H_6 = 2 × (12.0107 g) = 24.0214 g C

Mass of H per mole of C_2H_6 = 6 × (1.0079 g) = 6.0474 g H

Molar mass C_2H_6 = (24.0214 g) + (6.0474 g) = 30.0688 g/mol C_2H_6

$$\%\text{ C} = \frac{\text{mass of C / mol }C_2H_6}{\text{mass of }C_2H_6\text{ / mol }C_2H_6} \times 100\% = \frac{24.0214\text{ g C}}{30.0688\text{ g }C_2H_6} \times 100\% = 79.8881\%\text{ C in }C_2H_6$$

% H = 100% – 79.8881% C = 20.1119% H in C_2H_6

(c)
Mass of C per mole of CH_3CO_2H = 2 × (12.0107 g) = 24.0214 g C

Mass of H per mole of CH_3CO_2H = 4 × (1.0079 g) = 4.0316 g H

Mass of O per mole of CH_3CO_2H = 2 × (15.9994 g) = 31.9988 g O

$$\text{Molar mass } CH_3CO_2H = (24.0214 \text{ g}) + (4.0316 \text{ g}) + (31.9988 \text{ g}) = 60.0518 \text{ g/mol } CH_3CO_2H$$

$$\%C = \frac{\text{mass of C/mol } CH_3CO_2H}{\text{mass of } CH_3CO_2H\text{/mol } CH_3CO_2H} \times 100\%$$

$$= \frac{24.0214 \text{ g C}}{60.0518 \text{ g } CH_3CO_2H} \times 100\% = 40.0011\% \text{ C in } CH_3CO_2H$$

$$\%\ H = \frac{\text{mass of H/mol } CH_3CO_2H}{\text{mass of } CH_3CO_2H\text{/ mol } CH_3CO_2H} \times 100\%$$

$$= \frac{4.0316 \text{ g H}}{60.0518 \text{ g } CH_3CO_2H} \times 100\% = 6.7135\% \text{ H in } CH_3CO_2H$$

$$\%\ O = 100\% - 40.0011\%\ C - 6.7135\%\ H = 53.2854\%\ O \text{ in } CH_3CO_2H$$

(d)

$$\text{Mass of N per mole of } NH_4NO_3 = 2 \times (14.0067 \text{ g}) = 28.0134 \text{ g N}$$

$$\text{Mass of H per mole of } NH_4NO_3 = 4 \times (1.0079 \text{ g}) = 4.0316 \text{ g H}$$

$$\text{Mass of O per mole of } NH_4NO_3 = 3 \times (15.9994 \text{ g}) = 47.9982 \text{ g O}$$

$$\text{Molar mass } NH_4NO_3 = (28.0134 \text{ g}) + (4.0316 \text{ g}) + (47.9982 \text{ g}) = 80.0432 \text{ g/mol } NH_4NO_3$$

$$\%\ N = \frac{\text{mass of N/mol } NH_4NO_3}{\text{mass of } NH_4NO_3\text{/mol } NH_4NO_3} \times 100\% = \frac{28.0134 \text{ g N}}{80.0432 \text{ g } NH_4NO_3} \times 100\%$$

$$= 34.9979\% \text{ N in } NH_4NO_3$$

$$\%\ H = \frac{\text{mass of H/mol } NH_4NO_3}{\text{mass of } NH_4NO_3\text{/mol } NH_4NO_3} \times 100\% = \frac{4.0316 \text{ g H}}{80.0432 \text{ g } NH_4NO_3} \times 100\%$$

$$= 5.0368\% \text{ H in } NH_4NO_3$$

$$\%\ O = 100\% - 34.9979\%\ C - 5.0368\%\ H = 59.9654\%\ O \text{ in } NH_4NO_3$$

Notice: When masses of different things are used in the same question, make sure your units clearly specify what each mass refers to.

✓ *Reasonable Answer Check:* Calculating the last element's mass percent using the formula gives the same answer as subtracting the other percentages from 100%.

89. *Answer:* **58.0% M in MO**

Strategy and Explanation: Given the mass percent of one compound, M_2O, containing one known element, O, and one unknown element, M, calculate the percent by mass of another compound, MO.

Choose a convenient sample mass of M_2O, such as 100.0 g. Find the mass of M and O in the sample, using the given mass percent. Using the molar mass of oxygen as a conversion factor, determine the number of moles of oxygen, then using the formula stoichiometry of M_2O as a conversion factor determine the number of moles of M. Find the molar mass of M by dividing the mass of M by the moles of M. Use the molar mass of M, and the formula stoichiometry of MO, to determine the mass percent of M in MO.

73.4% M in M_2O means that 100.0 grams of M_2O contains 73.4 grams of M.

$$\text{Mass of O} = 100.0 \text{ g } M_2O - 73.4 \text{ g M} = 26.6 \text{ g O}$$

Formula stoichiometry: 1 mol of M_2O contains 2 mol M and 1 mol O.

$$26.6 \text{ g O} \times \frac{1 \text{ mol O}}{15.9994 \text{ g O}} \times \frac{2 \text{ mol M}}{1 \text{ mol O}} = 3.33 \text{ mol M}$$

$$\text{Molar mass of M} = \frac{\text{mass of M in sample}}{\text{mol of M in sample}} = \frac{73.4 \text{ g M}}{3.33 \text{ mol M}} = 22.1 \frac{\text{g}}{\text{mol}}$$

$$\text{Molar mass of MO} = 22.1 \text{ g} + 15.9994 \text{ g} = 38.07 \text{ g/mol}$$

$$\% M = \frac{\text{mass of M / mol MO}}{\text{mass of MO / mol MO}} \times 100\% = \frac{22.1 \text{ g M}}{38.07 \text{ g MO}} \times 100\% = 58.0\% \text{ M in MO}$$

✓ *Reasonable Answer Check:* It makes sense that the compound with more atoms of M has a higher mass percent of M. The closest element to M's atomic mass (22.1) is sodium (atomic mass = 22.99). If M is sodium, the two compounds would probably be sodium oxide (Na_2O) and sodium peroxide (Na_2O_2, a compound made up of two Na^+ ions and one O_2^{2-} ion. The simple ratio of Na and O atoms in this compound is 1:1).

91. *Answer:* **245.745 g/mol, 25.858% Cu, 22.7992% N, 5.74197% H, 13.048% S, 32.5528% O**

Strategy and Explanation: Given the formula of a compound, determine the molar mass, and the mass percent of each element

Calculate the mass of each element in one mole of compound, while calculating the molar mass of the compound (see full method on Question 68). Divide the calculated mass of the element by the molar mass of the compound and multiply by 100% to get mass percent.

Mass of Cu per mole of $Cu(NH_3)_4SO_4{\cdot}H_2O$ = 63.546 g Cu

Mass of N per mole of $Cu(NH_3)_4SO_4{\cdot}H_2O$ = 4 × (14.0067) = 56.0268 g N

Mass of H per mole of $Cu(NH_3)_4SO_4{\cdot}H_2O$ = (3 × 4 + 2) × (1.0079 g) = 14 × (1.0079 g)= 14.1106 g H

Mass of S per mole of $Cu(NH_3)_4SO_4{\cdot}H_2O$ = 32.065 g S

Mass of O per mole of $Cu(NH_3)_4SO_4{\cdot}H_2O$ = (4 + 1) × (15.9994 g) = 79.9970 g O

Molar mass $Cu(NH_3)_4SO_4{\cdot}H_2O$ = (63.546 g) + (56.0268 g) + (14.1106 g) + (32.065 g) + (79.9970 g) = 245.745 g/mol $Cu(NH_3)_4SO_4{\cdot}H_2O$

$$\% \text{ Cu} = \frac{\text{mass of Cu / mol Co(NH}_3)_4\text{SO}_4 \cdot \text{H}_2\text{O}}{\text{mass of Co(NH}_3)_4\text{SO}_4 \cdot \text{H}_2\text{O / mol Co(NH}_3)_4\text{SO}_4 \cdot \text{H}_2\text{O}} \times 100\%$$

$$= \frac{63.546 \text{ g Cu}}{245.745 \text{ g Co(NH}_3)_4\text{SO}_4 \cdot \text{H}_2\text{O}} \times 100\% = 25.858\% \text{ Cu in Co(NH}_3)_4\text{SO}_4 \cdot \text{H}_2\text{O}$$

$$\% \text{ N} = \frac{\text{mass of N / mol Co(NH}_3)_4\text{SO}_4 \cdot \text{H}_2\text{O}}{\text{mass of Co(NH}_3)_4\text{SO}_4 \cdot \text{H}_2\text{O / mol Co(NH}_3)_4\text{SO}_4 \cdot \text{H}_2\text{O}} \times 100\%$$

$$= \frac{56.0268 \text{ g N}}{245.745 \text{ g Co(NH}_3)_4\text{SO}_4 \cdot \text{H}_2\text{O}} \times 100\% = 22.7992\% \text{ N in Co(NH}_3)_4\text{SO}_4 \cdot \text{H}_2\text{O}$$

$$\% \text{H} = \frac{\text{mass of H / mol Co(NH}_3)_4\text{SO}_4 \cdot \text{H}_2\text{O}}{\text{mass of Co(NH}_3)_4\text{SO}_4 \cdot \text{H}_2\text{O / mol Co(NH}_3)_4\text{SO}_4 \cdot \text{H}_2\text{O}} \times 100\%$$

$$= \frac{14.1106 \text{ g H}}{245.745 \text{ g Co(NH}_3)_4\text{SO}_4 \cdot \text{H}_2\text{O}} \times 100\% = 5.74197\% \text{ H in Co(NH}_3)_4\text{SO}_4 \cdot \text{H}_2\text{O}$$

$$\% \text{ S} = \frac{\text{mass of S / mol Co(NH}_3)_4\text{SO}_4 \cdot \text{H}_2\text{O}}{\text{mass of Co(NH}_3)_4\text{SO}_4 \cdot \text{H}_2\text{O / mol Co(NH}_3)_4\text{SO}_4 \cdot \text{H}_2\text{O}} \times 100\%$$

$$= \frac{32.065 \text{ g S}}{245.745 \text{ g Co(NH}_3)_4\text{SO}_4 \cdot \text{H}_2\text{O}} \times 100\% = 13.048\% \text{ S in Co(NH}_3)_4\text{SO}_4 \cdot \text{H}_2\text{O}$$

$$\% \text{O} = \frac{\text{mass of O / mol Co(NH}_3)_4\text{SO}_4 \cdot \text{H}_2\text{O}}{\text{mass of Co(NH}_3)_4\text{SO}_4 \cdot \text{H}_2\text{O / mol Co(NH}_3)_4\text{SO}_4 \cdot \text{H}_2\text{O}} \times 100\%$$

$$= \frac{79.9970 \text{ g O}}{245.745 \text{ g Co(NH}_3)_4\text{SO}_4 \cdot \text{H}_2\text{O}} \times 100\% = 32.5528\% \text{ O in Co(NH}_3)_4\text{SO}_4 \cdot \text{H}_2\text{O}$$

✓ *Reasonable Answer Check:* The sum of the percentages add up to 100%.

93. *Answer:* **One**

Strategy and Explanation: Given the molar mass of a compound with an unknown formula and the mass percent of an element in that compound, determine the number of atoms of that element in the compound.

Using a convenient sample size of compound, determine the mass of the element in that compound. Convert both masses to moles, then determine the mole ratio.

In 100.000 grams of the compound carbonic anhydrase (c.a.) there are 0.218 g Zn.

$$100.000 \text{ g c.a.} \times \frac{1 \text{ mole c.a.}}{3.00 \times 10^4 \text{ g c.a.}} = 0.00333 \text{ mole c.a.}$$

$$0.218 \text{ g Zn} \times \frac{1 \text{ mole Zn}}{65.409 \text{ g Zn}} = 0.00333 \text{ mole Zn}$$

0.00333 mole Zn : 0.00333 mole c.a.

1 mole Zn : 1 mole c.a.

One Zn atom in every molecule of c.a.

✓ *Reasonable Answer Check:* The contribution of the zinc mass to the mass of the molecule is very small, so it makes sense that the number of atoms of zinc is very small in this large molecule.

95. *Answer:* **(a) –2 (b) Al_2X_3 (c) Se**

Strategy and Explanation:

(a) A group 6A element is likely to have a –2 charge, since: anion charge = (group number) – 8 = 6 – 8 = –2

(b) Aluminum ion, Al^{3+} combines with X^{2-} to form Al_2X_3.

(c) Given the percent by mass of an element in a compound and the compound's formula including an unknown element, determine the identity of the unknown element.

Choose a convenient sample of Al_2X_3, such as 100.00 g. Using the percent by mass, determine the number of grams of Al and X in the sample. Use the molar mass of Al as a conversion factor to get the moles of Al. Use the formula stoichiometry as a conversion factor to get the moles of X. Determine the molar mass by dividing the grams of X in the sample, by the moles of X in the sample. Using the periodic table, determine which Group 6A element has a molar mass nearest this value.

The compound is 18.55% Al by mass. This means that 100.00 g of Al_2X_3 contains 18.55 grams Al and the rest of the mass is from X.

Mass of X in sample = 100.00 g Al_2X_3 – 18.55 g Al = 81.45 g X

Formula stoichiometry: 1 mol of Al_2X_3 contains 2 mol of Al atoms.

$$18.55 \text{ g Al} \times \frac{1 \text{ mol Al}}{26.9815 \text{ g Al}} \times \frac{3 \text{ mol X}}{2 \text{ mol Al}} = 1.031 \text{ mol X}$$

$$\text{Molar Mass of X} = \frac{\text{mass of X in sample}}{\text{moles of X in sample}} = \frac{81.45 \text{ g X}}{1.031 \text{ mol X}} = 78.98 \text{ g/mol}$$

The periodic table indicates that X = Se (Z = 34, with atomic weight = 78.96 g/mol).

✓ *Reasonable Answer Check:* The molar mass calculated is close to that of the Group 6A element, Se. Ions of the elements in Group 6A will typically have 2– charge, and the formula would be Al_2Se_3.

Empirical and Molecular Formulas (Section 3.10)

98. *Answer/Explanation:* **An empirical formula shows simplest whole-number ratio of the elements, e.g., CH_3, a molecular formula shows actual number of atoms of each element, e.g., C_2H_6.**

Strategy and Explanation: An empirical formula shows the simplest whole number ratio of the elements in a compound. The molecular formula gives the actual number of atoms of each element in one formula unit of the compound. For ethane, C_2H_6 is the molecular formula. CH_3 is the empirical formula, since 6 is exactly divisible by 2 to give the smallest whole number ratio of 1:3.

100. *Answer:* **$C_4H_4O_4$**

Strategy and Explanation: Given the empirical formula of a compound and the molar mass, determine the molecular formula.

Find the mass of 1 mol of the empirical formula. Divide the molar mass of the compound by the empirical mass to get a whole number. Multiply all the subscripts in the empirical formula by the whole number.

The empirical formula is CHO, the molecular formula is $(CHO)_n$.

Mass of 1 mol of CHO = 12.0107 g + 1.0079 g + 15.9994 g = 29.0180 g/mol CHO

$$n = \frac{\text{mass of 1 mol of molecule}}{\text{mass of 1 mol of CHO}} = \frac{116.4 \text{ g}}{29.0180 \text{ g}} = 4.011 \approx 4$$

Molecular Formula is $(CHO)_4 = C_4H_4O_4$

✓ *Reasonable Answer Check:* The molar mass is about 4 times larger than the mass of one mole of the empirical formula, so the molecular formula $C_4H_4O_4$ makes sense.

102. *Answer:* $\mathbf{C_2H_3OF_3}$

Strategy and Explanation: Given the molar mass of a compound and the percents by mass of elements in a compound, find the molecular formula.

Choose a convenient sample of $C_xH_yO_zF_w$, such as 100.00 g. Using the percent by mass, determine the number of grams of each element in the sample. Use the molar mass of each element to get the moles of that element. Set up a mole ratio to determine the values of x, y, z, and w. Use those integers as subscripts in the empirical formula. Next, calculate the molar mass of the empirical formula, and divide the given molar mass of the compound into this value to determine the number of empirical formulas in the molecular formula and use that to write the formula of the molecule.

Exactly 100.0 g of $C_xH_yO_zF_w$ contains 24.0 grams C, 3.0 grams H, 16.0 grams O, and 57.0 grams F.

$$24.0 \text{ g C} \times \frac{1 \text{ mol C}}{12.0107 \text{ g C}} = 2.00 \text{ mol C} \qquad 3.0 \text{ g H} \times \frac{1 \text{ mol H}}{1.0079 \text{ g H}} = 3.0 \text{ mol H}$$

$$16.0 \text{ g O} \times \frac{1 \text{ mol O}}{15.9994 \text{ g O}} = 1.00 \text{ mol O} \qquad 57.0 \text{ g F} \times \frac{1 \text{ mol F}}{18.9984 \text{ g F}} = 3.00 \text{ mol F}$$

Set up ratio and simplify: 2.00 mol C : 3.0 mol H : 1.00 mol O : 3.00 mol F

simplify and write the empirical formula: 2 C : 3 H : 1 O : 3 F $\quad C_2H_3OF_3$

Mass of 1 mol of $C_2H_3OF_3$ = 2 × (12.0107 g) + 3 × (1.0079 g) + 15.9994 g + 3 × (18.9984 g)
= 100.0397 g / mol $C_2H_3OF_3$

$$n = \frac{\text{mass of 1 mol of molecule}}{\text{mass of 1 mol of } C_2H_3OF_3} = \frac{100.0 \text{ g}}{100.0397 \text{ g}} = 1.000 \approx 1$$

Molecular Formula is $C_2H_3OF_3$

✓ *Reasonable Answer Check:* The empirical mass is very close to the given molecular mass.

104. *Answer:* $\mathbf{KNO_3}$

Strategy and Explanation: Given the percents by mass of elements in a compound, find the empirical formula.

Choose a convenient sample of $K_xN_yO_z$, such as 100.00 g. Using the percent by mass, determine the number of grams of each element in the sample. Use the molar mass of each element to get the moles of that element. Set up a mole ratio to determine the values of x, y, z, and w. Use those integers as subscripts in the empirical formula.

Exactly 100.0 g of $K_xN_yO_z$ contains 38.67 grams K, 13.85 grams N, and 47.47 grams O.

$$38.67 \text{ g K} \times \frac{1 \text{ mol K}}{39.0983 \text{ g K}} = 0.9890 \text{ mol K} \qquad 13.85 \text{ g N} \times \frac{1 \text{ mol N}}{14.0067 \text{ g N}} = 0.9888 \text{ mol N}$$

$$47.47 \text{ g O} \times \frac{1 \text{ mol O}}{15.9994 \text{ g O}} = 2.967 \text{ mol O}$$

Set up ratio and simplify: 0.9890 mol K : 0.9888 mol N : 2.967 mol O

simplify the ratio 1 K : 1 N : 3 O and write the empirical formula: KNO_3

✓ *Reasonable Answer Check:* The formula matches a common ionic compound, potassium nitrate.

106. *Answer:* $\mathbf{B_5H_7}$

Strategy and Explanation: Given the percent by mass of one element in a compound containing two elements with unknown subscripts and a list of possible empirical formulas, determine the empirical formula.

Choose a convenient sample of B_xH_y, such as 100.0 g. Using the percent by mass, determine the number of grams of B and H in the sample. Use the molar masses of B and H as conversion factors to get the moles of B and H. Set up a mole ratio to determine the y/x ratio. Compare to the list given to select the empirical formula that has the closest mole ratio.

The compound is 88.5% B by mass. This means that 100.0 g of B_xH_y contains 88.5 grams B and the rest of the mass is from H.

$$\text{Mass of H in sample} = 100.00 \text{ g } B_xH_y - 88.5 \text{ g B} = 11.5 \text{ g H}$$

$$88.5 \text{ g B} \times \frac{1 \text{ mol B}}{10.811 \text{ g B}} = 8.19 \text{ mol B} \qquad 11.5 \text{ g H} \times \frac{1 \text{ mol H}}{1.0079 \text{ g H}} = 11.41 \text{ mol H}$$

$$\text{Mole ratio} = \frac{\text{moles of H in sample}}{\text{moles of B in sample}} = \frac{11.41 \text{ mol H}}{8.19 \text{ mol B}} = 1.39 \approx \frac{7 \text{ mol H}}{5 \text{ mol B}}$$

Therefore, the formula is B_5H_7.

✓ *Reasonable Answer Check:* The other possible formulas' mole ratios were 2.5, 2.2, and 2.0, so the 1.4 ratio (from 7/5) was the closest to 1.39.

108. *Answer:* **(a)** $\mathbf{C_3H_4}$ **(b)** $\mathbf{C_9H_{12}}$

Strategy and Explanation: Given the percent by mass of one element in a compound containing two elements with unknown subscripts and the molar mass, determine the empirical formula and the molecular formula.

Choose a convenient sample of C_xH_y, such as 100.00 g. Using the percent by mass, determine the number of grams of C and H in the sample. Use the molar masses of C and H as conversion factors to get the moles of C and H. Set up a mole ratio to determine the whole number y/x ratio for the empirical formula. Find the mass of 1 mol of the empirical formula. Divide the molar mass of the compound by the calculated empirical mass to get a whole number. Multiply all the subscripts in the empirical formula by this whole number.

(a) The compound is 89.94% C by mass. This means that 100.00 g of C_xH_y contains 89.94 grams C and the rest of the mass is from H.

Mass of H in sample = 100.00 g C_xH_y – 89.94 g C = 10.06 g H

Find moles of C and H in the sample:

$$89.94 \text{ g C} \times \frac{1 \text{ mol C}}{12.0107 \text{ gC}} = 7.488 \text{ mol C} \qquad 10.06 \text{ g H} \times \frac{1 \text{ mol H}}{1.0079 \text{ g H}} = 9.981 \text{ mol H}$$

$$\text{Mole ratio} = \frac{9.981 \text{ mol H}}{7.488 \text{ mol C}} = 1.333 = \frac{4 \text{ mol H}}{3 \text{ mol C}}$$

The empirical formula is C_3H_4, so the molecular formula is $(C_3H_4)_n$.

(b) Mass of 1 mol $C_3H_4 = 3 \times (12.0107 \text{ g}) + 4 \times (1.0079 \text{ g}) = 40.0637 \text{ g / mol}$

$$n = \frac{\text{mass of 1 mol of molecule}}{\text{mass of 1 mol of } C_3H_4} = \frac{120.2 \text{ g}}{40.0637 \text{ g}} = 3.000 \approx 3$$

$$\text{Molecular Formula is } (C_3H_4)_3 = C_9H_{12}$$

✓ *Reasonable Answer Check:* The mole ratio of C and H in the sample is very close to $\frac{4}{3}$, so the empirical formula makes sense. The molar mass is 3 times larger than the mass of one mole of the empirical formula, so the molecular formula C_9H_{12} makes sense.

110. *Answer:* **(a) C_5H_7N (b) $C_{10}H_{14}N_2$**

Strategy and Explanation: Given the percent by mass of all the elements in a compound and the molar mass, determine the empirical formula and the molecular formula.

Choose a convenient sample of $C_xH_yN_z$, such as 100.00 g. Using the percent by mass, determine the number of grams of C, H, and N in the sample. Use the molar masses of C, H, and N to get the moles of C, H and N. Set up a mole ratio to determine the whole number x:y:z ratio for the empirical formula. Find the mass of 1 mol of the empirical formula. Divide the molar mass of the compound by the calculated empirical mass to get a whole number. Multiply all the subscripts in the empirical formula by this whole number.

(a) The compound is 74.0% C, 8.65% H, and 17.35% N by mass. This means that 100.00 g of $C_xH_yN_z$ contains 74.0 grams C, 8.65 grams H, and 17.35 grams N.

Find moles of C, H, and N in the sample:

$$74.0 \text{ g C} \times \frac{1 \text{ mol C}}{12.0107 \text{ g C}} = 6.16 \text{ mol C} \qquad 8.65 \text{ g H} \times \frac{1 \text{ mol H}}{1.0079 \text{ g H}} = 8.58 \text{ mol H}$$

$$17.35 \text{ g N} \times \frac{1 \text{ mol N}}{14.0067 \text{ g N}} = 1.239 \text{ mol N}$$

Mole ratio: 6.16 mol C : 8.58 mol H : 1.239 mol N

Divide each term by the smallest number of moles, 1.239 mol

Atom ratio: 5 C : 7 H : 1 N

(b) The molecular formula is $(C_5H_7N)_n$.

Mass of 1 mol $C_5H_7N = 5 \times (12.0107 \text{ g}) + 7 \times (1.0079 \text{ g}) + 14.0067 \text{ g} = 81.1155 \text{ g / mol}$

$$n = \frac{\text{mass of 1 mol of molecule}}{\text{mass of 1 mol of } C_5H_7N} = \frac{162 \text{ g}}{81.1155 \text{ g}} = 2.00 \approx 2$$

Molecular Formula is $(C_5H_7N)_2 = C_{10}H_{14}N_2$

✓ *Reasonable Answer Check:* The mole ratios of C, H, and N in the sample are close to integer values, so the empirical formula makes sense. The molar mass is twice the mass of one mole of the empirical formula, so the molecular formula $C_{10}H_{14}N_2$ makes sense.

112. *Answer:* **$C_5H_{14}N_2$**

Strategy and Explanation: Given the percent by mass of all the elements in a compound and the molar mass, determine the molecular formula.

Choose a convenient sample of $C_xH_yN_z$, such as 100.00 g. Using the percent by mass, determine the number of grams of C, H, and N in the sample. Use the molar masses of C, H, and N as conversion factors to get the moles of C, H and N. Set up a mole ratio to determine the whole number x:y:z ratio for the empirical formula. Find the mass of 1 mol of the empirical formula. Divide the molar mass of the compound by the calculated empirical mass to get a whole number. Multiply all the subscripts in the empirical formula by this whole number.

The compound is 58.77% C, 13.81% H, and 27.42% N by mass. This means that 100.00 g of $C_xH_yN_z$ contains 58.77 grams C, 13.81 grams H, and 27.42 grams N.

Find moles of C, H, and N in the sample:

$$58.77 \text{ g C} \times \frac{1 \text{ mol C}}{12.0107 \text{ g C}} = 4.893 \text{ mol C} \qquad 13.81 \text{ g H} \times \frac{1 \text{ mol H}}{1.0079 \text{ g H}} = 13.70 \text{ mol H}$$

$$27.42 \text{ g N} \times \frac{1 \text{ mol N}}{14.0067 \text{ g N}} = 1.958 \text{ mol N}$$

Mole ratio: 4.893 mol C : 13.70 mol H : 1.958 mol N

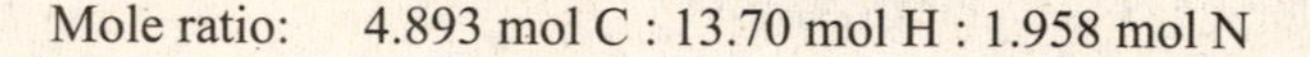

Divide each term by the smallest number of moles, 1.958 mol, to get atom ratio:

Atom ratio: 2.5 C : 7 H : 1 N

Multiply each term by 2 to get whole number atom ratio:

Whole number atom ratio: 5 C : 14 H : 2 N

The empirical formula is $C_5H_{14}N_2$, so the molecular formula is $(C_5H_{14}N_2)_n$.

Mass of 1 mol $C_5H_{14}N_2$ = 5 × (12.0107 g) + 14 × (1.0079 g) + 2 × (14.0067 g) = 102.1775 g / mol

$$n = \frac{\text{mass of 1 mol of molecule}}{\text{mass of 1 mol of } C_5H_{14}N_2} = \frac{102.2 \text{ g}}{102.1775 \text{ g}} = 1.000 \approx 1$$

Molecular Formula is $(C_5H_{14}N_2)_1 = C_5H_{14}N_2$

✓ *Reasonable Answer Check:* The mole ratios of C, H, and N in the sample are close to integer values, so the empirical formula makes sense. The molar mass is almost the same as the mass of one mole of the empirical formula, so the molecular formula $C_5H_{14}N_2$ makes sense.

114. *Answer:* **five**

Strategy and Explanation: The mass of a sample of a hydrate compound is given. The formula of the hydrate is known except for the amount of water in it. All the water is dried out of the sample using high temperature, leaving a mass of completely dehydrated salt. Determine out the number of water molecules in the formula of the hydrate compound.

Use the molar mass of the dehydrated compound as a conversion factor to convert the mass of the dehydrated compound into moles. Since the only thing lost was water, a mole relationship can be established between the dehydrated and hydrated compound. In addition, the difference between the mass of the hydrated compound and the mass of the dehydrated compound gives the amount of water lost by the sample. Convert that water into moles, using the molar mass of water. Divide the moles of water by the moles of hydrate, to determine how many moles of water are in one mole of hydrate compound.

5.172 g of hydrated compound, $CuSO_4{\cdot}xH_2O$, is dehydrated to make 3.306 g of $CuSO_4$.

Molar Mass $CuSO_4$ = 63.546 g + 32.065 g + 4 × (15.9994 g) = 159.609 g/mol

Molar Mass H_2O = 2 × (1.0079 g) + 2 × (15.9994 g) = 18.0152 g/mol

Find moles of $CuSO_4{\cdot}xH_2O$ in the sample:

$$3.306 \text{ g } CuSO_4 \times \frac{1 \text{ mol } CuSO_4}{159.609 \text{ g } CuSO_4} \times \frac{1 \text{ mol } CuSO_4 \cdot xH_2O}{1 \text{ mol } CuSO_4} = 2.071 \times 10^{-2} \text{ mol } CuSO_4{\cdot}xH_2O$$

Find mass of water lost by the sample: (5.172 g $CuSO_4{\cdot}xH_2O$) – (3.306 g $CuSO_4$) = 1.866 g H_2O

Find moles of H_2O lost from sample: $1.866 \text{ g } H_2O \times \frac{1 \text{ mol } H_2O}{18.0152 \text{ g } H_2O} = 0.1036 \text{ mol } H_2O$

$$\text{Mole ratio} = \frac{\text{mol } H_2O \text{ from sample}}{\text{mol hydrate is sample}} = \frac{0.1036 \text{ mol } H_2O}{2.071 \times 10^{-2} \text{ mol hydrate}} = 5.002 \approx 5$$

The proper formula for the hydrate is $CuSO_4{\cdot}5H_2O$ and there are five molecules of water in one formula unit.

✓ *Reasonable Answer Check:* The mole fraction is very close to the whole number 5.

General Questions

122. *Answer:* **(a) $C_7H_5N_3O_6$ (b) $C_3H_7NO_3$**

Strategy and Explanation:

(a) Trinitrotoluene, TNT, has seven C atoms, six in the ring and one more in the $-CH_3$ at the top of the structure. It has five H atoms, two on ring carbons and three more in the $-CH_3$ at the top of the structure. It has three N atoms (one in each of the three $-NO_2$ groups attached to the ring carbons). It also has six O atoms (two in each of the three $-NO_2$ groups attached to the ring carbons). So, the molecular formula is $C_7H_5N_3O_6$.

(b) Serine has three carbon atoms. It has three hydrogen atoms attached to carbons, two H atoms in the two $-OH$ groups, and two H atoms in the $-NH_2$ groups. It has one N atom. It has two O atoms in the $-OH$ groups and one more in the =O. So, the molecular formula is $C_3H_7NO_3$.

125. *Answer:* **(a) (i) chlorine tribromide (ii) nitrogen trichloride (iii) calcium sulfate (iv) heptane (v) xenon tetrafluoride (vi) oxygen difluoride (vii) sodium iodide (viii) aluminum sulfide (ix) phosphorus pentachloride (x) potassium phosphate (b) (iii), (vii), (viii), and (x)**

Strategy and Explanation:

(i) binary molecular compound (both nonmetals; not ionic): chlorine tribromide

(ii) binary molecular compound (both nonmetals; not ionic): nitrogen trichloride

(iii) ionic compound (common cation, Ca^{2+}, and anion, SO_4^{2-}): calcium sulfate

(iv) organic compound (alkane): heptane

(v) binary molecular compound (both nonmetals; not ionic): xenon tetrafluoride

(vi) binary molecular compound (both nonmetals; not ionic): oxygen difluoride

(vii) ionic compound (common cation, Na^+, and anion, I^-): sodium iodide

(viii) ionic compound (common cation, Al^{3+}, and anion, S^{2-}): aluminum sulfide

(ix) binary molecular compound (both nonmetals; not ionic): phosphorus pentachloride

(x) ionic compound (common cation, K^+, and anion, PO_4^{3-}): potassium phosphate

127. *Answer:* **(a) 2.39 mol (b) 21.7 cm³**

Strategy and Explanation: Given the mass of a block of platinum in troy ounces and the density of platinum, determine the number of moles of metal and the volume of the block in cubic centimeters.

Use the given relationship between troy ounces and grams as a conversion factor to determine the mass of the block in grams. (a) Use the periodic table to get the molar mass of platinum (Pt). Use the molar mass as a conversion factor to get the number of moles of Pt in the block. (b) Use the density as a conversion factor to get the block's volume in cubic centimeters.

(a) $$15.0 \text{ troy oz Pt} \times \frac{31.1 \text{ g Pt}}{1 \text{ troy oz Pt}} \times \frac{1 \text{ mol Pt}}{195.078 \text{ g Pt}} = 2.39 \text{ mol Pt}$$

(b) $$15.0 \text{ troy oz Pt} \times \frac{31.1 \text{ g Pt}}{1 \text{ troy oz Pt}} \times \frac{1 \text{ cm}^3 \text{ Pt}}{21.45 \text{ g Pt}} = 21.7 \text{ cm}^3 \text{ Pt}$$

✓ *Reasonable Answer Check:* The macroscopic sample has a convenient number of moles. The volume should be approximately $\frac{3}{2}$ times the troy ounces.

129. *Answer:* **(a) CO_2F_2 (b) CO_2F_2**

Strategy and Explanation: Given the percent by mass of all the elements in a compound and the molar mass, determine the empirical formula and the molecular formula.

Choose a convenient sample of $C_xO_yF_z$, such as 100.0 g. Using the percent by mass, determine the number of grams of C, O, and F in the sample. Use the molar masses of C, O, and F as conversion factors to get the moles of C, O and F. Set up a mole ratio to determine the whole number x:y:z ratio for the empirical formula. Find the mass of 1 mol of the empirical formula. Divide the molar mass of the compound by the calculated empirical mass to get a whole number. Multiply all the subscripts in the empirical formula by this whole number.

(a) The compound is 14.6% C, 39.0% O, and 46.3% F by mass. This means that 100.00 g of $C_xO_yF_z$ contains 14.6 grams C, 39.0 grams O, and 46.3 grams F.

Find moles of C, O, and F in the sample:

$$14.6 \text{ g C} \times \frac{1 \text{ mol C}}{12.0107 \text{ g C}} = 1.22 \text{ mol C} \qquad 39.0 \text{ g O} \times \frac{1 \text{ mol O}}{15.9994 \text{ g O}} = 2.44 \text{ mol O}$$

$$46.3 \text{ g F} \times \frac{1 \text{ mol F}}{18.9984 \text{ g F}} = 2.44 \text{ mol F}$$

Mole ratio: 1.22 mol C : 2.44 mol O : 2.44 mol F

Divide each term by the smallest number of moles, 1.22 mol

Atom ratio: 1 C : 2 O : 2 F

The empirical formula is CO_2F_2

(b) The molecular formula is $(CO_2F_2)_n$.

Mass of 1 mol CO_2F_2 = 12.0107 g + 2 × (15.9994 g) + 2 × (18.9984 g) = 82.0063 g/mol

$$n = \frac{\text{mass of 1 mol of molecule}}{\text{mass of 1 mol of } CO_2F_2} = \frac{82 \text{ g}}{82.0063 \text{ g}} = 1.00 \approx 1$$

The molecular formula is $(CO_2F_2)_1 = CO_2F_2$

✓ *Reasonable Answer Check:* The simplest ratios of C, O, and F in the sample end up very close to whole number values, so the empirical formula makes sense. The molar mass is the same as the mass of one mole of the empirical formula, so the molecular formula CO_2F_2 makes sense.

131. *Answer:* $\mathbf{MnC_9H_7O_3}$

Strategy and Explanation: Given the percent by mass of all the elements in a compound, determine the empirical formula.

Choose a convenient sample of $Mn_vC_xH_yO_z$, such as 100.0 g. Using the percent by mass, determine the number of grams of C, H, O, and Mn in the sample. Use the molar masses of C, H, O, and Mn as conversion factors to get the moles of C, H, O, and Mn. Set up a mole ratio to determine the whole number x:y:z:v ratio for the empirical formula.

The compound is 49.5% C, 3.2% H, 22.0% O, and 25.2% Mn by mass. This means that 100.0 g of $Mn_vC_xH_yO_z$ contains 49.5 grams C, 3.2 grams H, 22.0 grams O, and 25.2 grams Mn.

Find moles of C, H, O, and Mn in the sample:

$$49.5 \text{ g C} \times \frac{1 \text{ mol C}}{12.0107 \text{ g C}} = 4.12 \text{ mol C}$$

$$3.2 \text{ g H} \times \frac{1 \text{ mol H}}{1.0079 \text{ g H}} = 3.2 \text{ mol H}$$

$$22.0 \text{ g O} \times \frac{1 \text{ mol O}}{15.9994 \text{ g O}} = 1.38 \text{ mol O}$$

$$25.2 \text{ g Mn} \times \frac{1 \text{ mol Mn}}{54.938 \text{ g Mn}} = 0.459 \text{ mol Mn}$$

Mole ratio: 0.459 mol Mn : 4.12 mol C : 3.2 mol H : 1.38 mol O

Divide each term by the smallest number of moles, 0.459 mol, and round to whole numbers:

1 Mn : 9 C : 7 H : 3 O

The empirical formula is $MnC_9H_7O_3$

✓ *Reasonable Answer Check:* The simplest ratios of Mn, C, H, and O in the sample end up very close to whole number values, so the empirical formula makes sense.

133. *Answer:* **(a) 1.66×10^{-3} mol (b) 0.346 g**

Strategy and Explanation: Given the number of tablets consumed, the mass of a compound in each tablet, and the formula of the compound, determine the moles of the compound consumed and the mass of one element consumed.

Always start with the sample – in this case, the number of tablets consumed. We assume that $C_7H_5BiO_4$ is the "active ingredient". Use the mass of $C_7H_5BiO_4$ per tablet to determine the mass of $C_7H_5BiO_4$ in the sample. Then use the molar mass of $C_7H_5BiO_4$ to determine the moles of $C_7H_5BiO_4$ in the sample. Then use the formula stoichiometry to get the moles of Bi in the sample. Then use the molar mass of Bi to get the grams of Bi in the sample.

The sample is composed of two tablets of Pepto-Bismol. Find grams of $C_7H_5BiO_4$:

$$2 \text{ tablets} \times \frac{300.\text{ mg } C_7H_5BiO_4}{1 \text{ tablet}} \times \frac{1 \text{ g } C_7H_5BiO_4}{1000 \text{ mg } C_7H_5BiO_4} = 0.600 \text{ g } C_7H_5BiO_4$$

Molar mass of $C_7H_5BiO_4$

$$= 7 \times (12.0107 \text{ g}) + 5 \times (1.0079 \text{ g}) + 208.9804 \text{ g} + 4 \times (15.9994 \text{ g}) = 362.0924 \text{ g/mol}$$

(a) Find moles of $C_7H_5BiO_4$ in the sample

$$0.600 \text{ g } C_7H_5BiO_4 \times \frac{1 \text{ mol } C_7H_5BiO_4}{362.0924 \text{ g } C_7H_5BiO_4} = 1.66 \times 10^{-3} \text{ mol } C_7H_5BiO_4$$

(b) Find grams of Bi in the sample

$$1.66 \times 10^{-3} \text{ mol } C_7H_5BiO_4 \times \frac{1 \text{ mol Bi}}{1 \text{ mol } C_7H_5BiO_4} \times \frac{208.9804 \text{ g Bi}}{1 \text{ mol Bi}} = 0.346 \text{ g Bi}$$

✓ *Reasonable Answer Check:* (a) The sample is somewhere between macroscale and microscale, so it makes sense that the number of moles is somewhat small. (b) The bismuth is almost 60% of the $C_7H_5BiO_4$ compound mass so it makes sense that the number of grams of Bi in the sample is about 60% of the mass of the compound in the sample.

135. *Answer:* **2.14×10^5 g**

Strategy and Explanation: Given the mass of a sample of an ore, the mass percent of titanium in the ore, the fact that the ore contains a compound ilmenite, and the formula of ilmenite ($FeTiO_3$), determine the mass of $FeTiO_3$ in the sample of ore.

Alway start with the sample – in this case, 1.00 metric ton of ore. Use the given relationship and metric conversions to determine the sample mass in grams. Then use the mass percent of Ti as a conversion factor to determine the mass of Ti in the sample. Then use the molar mass of Ti to find the moles of Ti in the sample. Then use the formula stoichiometry of ilmenite to determine the moles of ilmenite, $FeTiO_3$. Then use the molar mass of $FeTiO_3$ to get the grams of $FeTiO_3$ in the sample.

The ore is 6.75% Ti. That means, 100.00 grams of ore contains 6.75 grams Ti.

Formula Stoichiometry: 1 mol Ti is contained in 1 mol $FeTiO_3$.

Molar mass of $FeTiO_3$ = 55.845 g + 47.867 g + 3 × (15.9994 g) = 151.710 g/mol

$$1.00 \text{ metric ton ore} \times \frac{1000 \text{ kg ore}}{1 \text{ metric ton ore}} \times \frac{1000 \text{ g ore}}{1 \text{ kg ore}} \times \frac{6.75 \text{ g Ti}}{100 \text{ g ore}} \times \frac{1 \text{ mol Ti}}{47.867 \text{ g Ti}}$$

$$\times \frac{1 \text{ mol } FeTiO_3}{1 \text{ mol Ti}} \times \frac{151.720 \text{ g } FeTiO_3}{1 \text{ mol } FeTiO_3} = 2.14 \times 10^5 \text{ g } FeTiO_3$$

✓ *Reasonable Answer Check:* Ti represents about one third of the mass of $FeTiO_3$. This mass of $FeTiO_3$ in the ore is about three times more than the% Ti in the ore.

138. (a) A crystal of sodium chloride has alternating lattice of Na^+ and Cl^- ions.

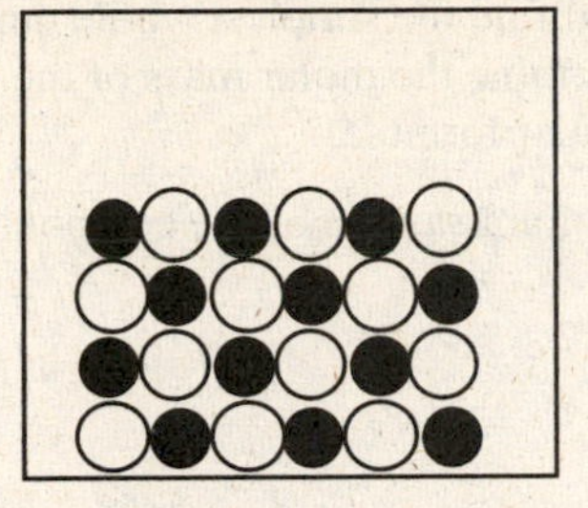

(b) The sodium chloride after it is melted has paired ions randomly distributed.

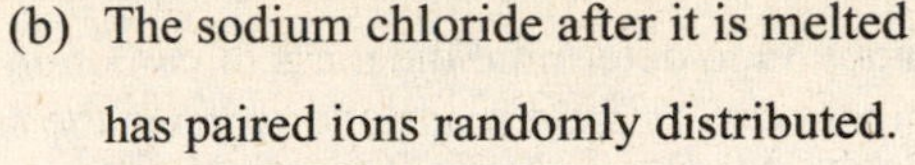

Applying Concepts

140. *Answer:* **(a) three (b) see identical pairs, below.**

Strategy and Explanation: There are three isomers given for C_3H_8O. Each of these isomers is represented by two of the structures. The following pairs are identical:

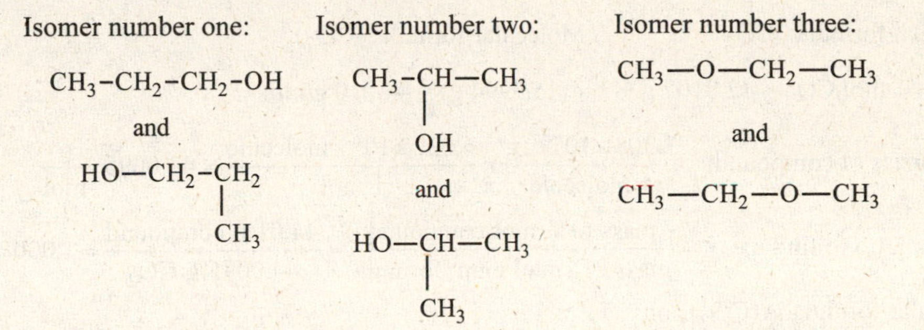

The first pair has an OH bonded to and end carbon. The second pair has an OH bonded to the middle carbon. The third pair has an O bonded between two carbon atoms.

142. *Answer:* **Tl_2CO_3, Tl_2SO_4**

Strategy and Explanation: Thallium nitrate is $TlNO_3$. Since NO_3^- has a –1 charge, that means that thallium ion has a +1 charge, and is represented by Tl^+. The carbonate compound containing thallium will be a combination of Tl^+ and CO_3^{2-}, and the compound's formula will look like this: Tl_2CO_3. The sulfate compound containing thallium will be a combination of Tl^+ and SO_4^{2-}, and the compound's formula will look like this: Tl_2SO_4.

144. *Answer:* **(a) perbromate, bromate, bromite, hypobromite (b) selenate, selenite**

Strategy and Explanation:

(a) Based on the guidelines for naming oxyanions in a series, relate the chloro-oxyanions to the bromo-oxyanions, since chlorine and bromine are in the same group (Group 7A).

BrO_4^- is perbromate (like ClO_4^- is perchlorate) BrO_3^- is bromate (like ClO_3^- is chlorate)

BrO_2^- is bromite (like ClO_2^- is chlorite) BrO^- is hypobromite (like ClO^- is hypochlorite)

(b) Based on the guidelines for naming oxyanions in a series, relate the sulfur-oxyanions to the selenium-oxyanions.

SeO_4^{2-} is selenate (like SO_4^{2-} is sulfate) SeO_3^{2-} is selenite (like SO_3^{2-} is sulfite)

More Challenging Questions

147. *Answer:* **CO_2**

Strategy and Explanation: Given the mass of one molecule of a compound, the mass percentage of carbon, and the fact that the rest of the compound is oxygen, determine the identity of the compound.

Pick a convenient sample size, such as 100.0 g of compound, C_xO_y. Use the percent by mass of carbon in the compound to determine the mass of carbon in the sample. Subtract the mass of carbon from the mass of the compound to find the mass of oxygen in the sample. Then use the molar masses of C and O to find the moles of C and moles of O in the sample. Then set up a mole ratio and find the simplest whole number relationship between the atoms. Use the mass of one molecule to determine the molar mass of the compound. Use the molar mass and the empirical formula to determine the molecular formula.

The compound is 27.3% C and the rest is O. That means that a 100.0-gram sample of the compound will contain 27.3 grams of carbon. The rest is oxygen:

$$100.0 \text{ g } C_xO_y - 27.3 \text{ g C} = 72.7 \text{ g O}$$

Determine the moles of C and O in the sample.

$$27.3 \text{ g C} \times \frac{1 \text{ mol C}}{12.0107 \text{ g C}} = 2.27 \text{ mol C} \qquad 72.7 \text{ g O} \times \frac{1 \text{ mol O}}{15.9994 \text{ g O}} = 4.54 \text{ mol O}$$

Set up mole ratio: 2.27 mol C : 4.54 mol O

Divide all the numbers in the ratio by the smallest number of moles, 2.27 mol, and round to whole numbers: 1 C : 2 O

Empirical formula: CO_2 Molecular formula: $(CO_2)_n$

Mass of 1 mol CO_2 = 12.0107 g + 2 × (15.9994 g) = 44.010 g/mol

$$\text{Molar mass of compound: } \frac{7.308 \times 10^{-23} \text{ g}}{1 \text{ molecule}} \times \frac{6.022 \times 10^{23} \text{ molecule}}{1 \text{ mol}} = 44.0095 \frac{\text{g}}{\text{mol}}$$

$$\text{Set up a ratio to find n: } n = \frac{\text{mass of 1 mol compound}}{\text{mass of 1 mol emp. formula}} = \frac{44.01 \text{ g compound}}{44.0095 \text{ g } CO_2} = 1.000 \cong 1$$

Molecular formula is $(CO_2)_1$ or CO_2

✓ *Reasonable Answer Check:* The mole ratio is very close to a whole number, and we recognize this compound as carbon dioxide, a common compound of carbon and oxygen.

149. *Answer:* **$MgSO_3$, magnesium sulfite**

Strategy and Explanation: Given the mass percentage of two elements in a compound, and the fact that the rest of the compound is a third element, determine the empirical formula and name of the compound. Pick a convenient sample size, such as 100.0 g of compound, $Mg_xS_yO_z$. Use the percent by mass of Mg and O in the compound to determine the mass of S in the sample. Subtract the mass of Mg and O from the mass of the compound to find the mass of S in the sample. Then use the molar masses of Mg, S, and O to find the moles of Mg, S and O in the sample. Then set up a mole ratio and find the simplest whole number relationship between the atoms.

Exactly 100.0 grams of the compound will contain 23.3 grams of Mg and 46.0 grams of O. The rest is chlorine:

$$100.0 \text{ g } Mg_xS_yO_z - 23.3 \text{ g Mg} - 46.0 \text{ g O} = 30.7 \text{ g S}$$

Determine the moles of Mg, S and O in the sample.

$$23.3 \text{ g Mg} \times \frac{1 \text{ mol Mg}}{24.3050 \text{ g Mg}} = 0.959 \text{ mol Mg} \qquad 30.7 \text{ g S} \times \frac{1 \text{ mol S}}{32.065 \text{ g S}} = 0.957 \text{ mol S}$$

$$46.0 \text{ g O} \times \frac{1 \text{ mol O}}{15.9994 \text{ g O}} = 2.88 \text{ mol O}$$

Set up mole ratio: 0.959 mol Mg : 0.957 mol S : 2.88 mol O

Divide all the numbers in the ratio by the smallest number of moles, 0.957 mol, to get whole numbers:

1 Mg : 1 S : 3 O

Empirical formula: $MgSO_3$ This is the ionic compound called magnesium sulfite.

✓ *Reasonable Answer Check:* The mole ratio is very close to a whole number, and we recognize this compound as carbon dioxide, a common compound of carbon and oxygen.

150. *Answer:* **(a) 0.0130 mol Ni (b) NiF_2 (c) nickel(II) fluoride**

Strategy and Explanation: Given the dimensions of a sample of nickel foil, the density of nickel metal, and the mass of a nickel fluoride compound (Ni_xF_y) produced using that foil as a source of nickel, determine the moles of nickel, and the formula of the nickel fluoride compound.

Always start with the sample – in this case, the dimensions of the Ni foil sample. Use metric conversion factors and the relationship between length, width, and height and volume to determine the sample's volume. Then use the density to determine the mass of Ni in the sample. Subtract the mass of Ni from the mass of the compound produced to find the mass of F in the sample. Then use the molar masses of Ni and F to find the moles of Ni and moles of F in the sample. Then set up a mole ratio and find the simplest whole number relationship between the atoms.

The foil is 0.550 mm thick, 1.25 cm long and 1.25 cm wide.

0.550mm thick → 1.25 cm long 1.25 cm wide

$$V = (\text{thickness}) \times (\text{length}) \times (\text{width})$$

$$V = \left(0.550\ \text{mm} \times \frac{1\ \text{m}}{1000\ \text{mm}} \times \frac{100\ \text{cm}}{1\ \text{m}}\right) \times (1.25\ \text{cm}) \times (1.25\ \text{cm}) = 0.0859\ \text{cm}^3$$

(a) $$0.0859\ \text{cm}^3\ \text{Ni} \times \frac{8.908\ \text{g Ni}}{1\ \text{cm}^3\ \text{Ni}} \times \frac{1\ \text{mol Ni}}{58.69\ \text{g Ni}} = 0.0130\ \text{mol Ni}$$

(b) $$0.0859\ \text{cm}^3\ \text{Ni} \times \frac{8.908\ \text{g Ni}}{1\ \text{cm}^3\ \text{Ni}} = 0.765\ \text{g Ni}$$

Mass of sulfur in the sample = Ni_xS_y compound mass – nickel mass

$$1.261\ \text{g}\ Ni_xS_y - 0.765\ \text{g Ni} = 0.496\ \text{g F}$$

Find moles of F in sample: $0.496\ \text{g F} \times \frac{1\ \text{mol F}}{18.9984\ \text{g F}} = 0.0261\ \text{mol F}$

Set up mole ratio: 0.0130 mol Ni : 0.0261 mol F

Divide all the numbers in the ratio by the smallest number of moles, 0.0130 mol, and round to whole numbers: 1 Ni : 2 F. That gives an empirical formula of NiF_2

(c) NiF_2 is called nickel(II) fluoride. (It looks composed of nickel(II) ion, Ni^{2+}, and fluoride ion, F^-)

✓ *Reasonable Answer Check:* The mole ratio was very close to a whole number ratio, so the empirical formula of NiF_2 makes sense. One of the common ionic forms of nickel is nickel(II) ion, so this also confirms that the result is sensible.

153. *Answer:* **5.0 lb N, 4.4 lb P, and 4.2 lb K**

Strategy and Explanation: Given the percent by mass of some elements and compounds in a fertilizer, determine the mass (in pounds, lb) of three elements in a given mass of the fertilizer.

Use the given mass and the definition of percent (parts per hundred) to determine the mass of each compound in the sample. Then determine the mass of each element in one mole of each of the compounds (as is done when determining mass percent of the element), and use that ratio to determine the mass of the elements in the sample.

Assuming at least two significant figures precision, the fertilizer is 5.0% N, 10.% P_2O_5, and 5.0% K_2O. That means, a sample of 100. lb of the fertilizer mixture, has 5.0 lb N, 10. lb P_2O_5, and 5.0 lb K_2O.

Mass of P in one mole of P_2O_5 = 2 × (30.9738 g) = 61.9476 g P

Molar mass P_2O_5 = 61.9476 g + 5 × (15.9994 g) = 141.9446 g P_2O_5

$$10.\text{ lb } P_2O_5 \times \frac{453.59237\text{ g}}{1\text{ lb}} \times \frac{61.9476\text{ g P}}{141.9446\text{ g } P_2O_5} \times \frac{1\text{ lb}}{453.59237\text{ g}} = 4.4\text{ lb P}$$

Mass of K in one mole of K_2O = 2 × (39.0983 g) = 78.1966 g K

Molar mass K_2O =78.1966 g + 15.9994 g = 94.196 g K_2O

$$5.0\text{ lb } K_2O \times \frac{453.59237\text{ g}}{1\text{ lb}} \times \frac{78.1966\text{ g K}}{94.196\text{ g } K_2O} \times \frac{1\text{ lb}}{453.59237\text{ g}} = 4.2\text{ lb K}$$

So, 100. lb of the fertilizer contains 5.0 lb N, 4.4 lb P, and 4.2 lb K.

✓ *Reasonable Answer Check:* It makes sense that the percentage of P_2O_5 had to be higher to get approximately the same mass of P, since that compound has a significant mass of oxygen.

155. *Answer:* **75.0% sulfate**

Strategy and Explanation: Given a mixture of two sulfate compounds and the mass percent of one of the two compounds in the mixture, determine the mass percent of sulfate in the mixture.

Using a 100.00 gram sample of the mixture, determine the mass percent of the second compound, then calculate the mass of sulfate from each compound in the sample. Add these two masses and determine the mass percent of sulfate in the 100.00 g sample.

100.0 g sample – 32.0 g $MgSO_4$ = 68.0 g $(NH_4)_2SO_4$

Molar mass $MgSO_4$ = 24.3050 g1 × (32.065 g) + 4 × (15.9994 g) = 120.368 g/mol

Molar mass $(NH_4)_2SO_4$ = 2 × (14.0067 g) + 8 × (1.0079 g) + (32.065 g) + 4 × (15.9994 g) = 132.139 g/mol

Molar mass SO_4^{2-} = 1 × (32.065 g) + 4 × (15.9994 g) = 96.063 g/mol

From $MgSO_4$

$$\text{mass } SO_4^{2-} = 32.0\text{ g } MgSO_4 \times \frac{1\text{ mol } MgSO_4}{120.368\text{ g } MgSO_4} \times \frac{1\text{ mol } SO_4^{2-}}{1\text{ mol } MgSO_4} \times \frac{96.063\text{ g } SO_4^{2-}}{1\text{ mol } SO_4^{2-}} = 25.54\text{ g } SO_4^{2-}$$

From $(NH_4)_2SO_4$

$$\text{mass } SO_4^{2-} = 68.0\text{ g } (NH_4)_2SO_4 \times \frac{1\text{ mol } (NH_4)_2SO_4}{132.1392\text{ g } (NH_4)_2SO_4} \times \frac{1\text{ mol } SO_4^{2-}}{1\text{ mol } (NH_4)_2SO_4} \times \frac{96.063\text{ g } SO_4^{2-}}{1\text{ mol } SO_4^{2-}}$$

$$= 49.43\text{ g } SO_4^{2-}$$

Total mass SO_4^{2-} = 25.54 g + 49.43 g = 74.97 g (round to one decimal place)

$$\%\, SO_4^{2-} = \frac{74.97\text{ g } SO_4^{2-}}{100.0\text{ g sample}} \times 100\% = 75.0\%\ SO_4^{2-} \text{ (3 sig figs)}$$

✓ *Reasonable Answer Check:* The other atoms in these compounds are lightweight compared to sulfate, so it makes sense that a majority of the mass percent is sulfate.

Chapter 4: Quantities of Reactants and Products
Solutions for Blue-Numbered Questions for Review and Thought

Topical Questions

Stoichiometry (Section 4.1)

10. *Answer:*

	NH_3	O_2	NO	H_2O
No. molecules	4	5	4	6
No. atoms	16	10	8	18
No. moles of molecules	4	5	4	6
Mass (g)	68.1216	159.9940	120.0244	108.0912
Total mass of reactants (g)	228.1156		–	
Total mass of products (g)	–		228.1156	

Strategy and Explanation: Given the equation for the reaction, find the related molecules, atoms, moles, masses of each of the reactants and products, and total masses of the reactants and the products.

The stoichiometric coefficients (the numbers in front of the molecules' formulas) in the balanced equation can be interpreted as the related number of molecules or the related number of moles of molecules. The molar mass of each molecule is multiplied by the number of moles of that molecule represented in the equation to find its mass in grams. Add the mass of NH_3 and O_2 for the total reactant mass. Add the mass of NO and H_2O for the total product mass.

$$4\ NH_3(g) + 5\ O_2(g) \longrightarrow 4\ NO(g) + 6\ H_2O(g)$$

The stoichiometric coefficients are 4 for NH_3, 5 for O_2, 4 for NO and 6 for H_2O; these numbers represent relative numbers of molecules and relative numbers of moles of molecules.

	NH_3	O_2	NO	H_2O
No. molecules	4	5	4	6
No. atoms	4 × (1 N+ 3 H) = 16	5 × (2 O) = 10	5 × (1 N + 1 O) = 8	6 × (2 H + 1 O) = 18
No. moles of molecules	4	5	4	6
Mass	4 mol × (1×14.0067g/mol + 3×1.0079g/mol) = 68.1216 g	5 mol × (2×15.9994g/mol) = 159.9940 g	4 mol × (1×14.0067g/mol + 1×15.9994g/mol) = 120.0244 g	6 mol × (2×1.0079 g/mol + 1×15.9994g/mol) = 108.0912 g
Total mass of reactants	68.1216 g + 159.9940 g = 228.1156 g		–	
Total mass of products	–		120.0244 g + 108.0912 g =228.1156 g	

✓ *Reasonable Answer Check:* A properly balanced equation has the same numbers of atoms of each type (4 N, 12 H, and 10 O) in the products and reactants, so the totals on each side should be equal, also (16+10=8+18). The law of conservation of mass says that mass of the reactants must equal the mass of the products, 228.1156g.

12. *Answer:* **(a) 1.00 g (b) 2 for Mg, 1 for O_2, and 2 for MgO (c) 50 atoms of Mg**

Strategy and Explanation: Given the balanced equation for a reaction, identify the stoichiometric coefficients in this equation, and relate the quantity of products to reactants and vice versa.

(a) The law of conservation of mass says that the total mass of the reactants is the same as the total mass of the products. (b) The stoichiometric coefficients are the numbers in front of each formula in the equation. (c) The stoichiometric coefficients can be interpreted as the related number of reactants. Use the formula stoichiometry of O_2 to find the number of molecules that reacted. Use equation stoichiometry to find out how many atoms of Mg reacted with that many O_2 molecules.

First, balance the equation: $2\ Mg(s) + O_2(g) \longrightarrow 2\ MgO(g)$

(a) The total mass of product, MgO, is 1.00 g. The total mass of the reactants (mass of Mg plus mass of O_2) that reacted must also be 1.00 g, due to the conservation of mass.

(b) The stoichiometric coefficients are 2 for Mg, 1 for O_2, and 2 for MgO.

(c) 50 atoms of oxygen make up 25 molecules of O_2, since there are two O atoms in each molecule. Since the stoichiometry is 1:1, that means 50 atoms of Mg were needed to react with this much O_2.

✓ *Reasonable Answer Check:* Though it is an unnecessary calculation, we can check the answer to part (a) by calculating the masses of Mg and O_2 that reacted to make 1.00 gram of MgO,

$$1.00\text{ g MgO} \times \frac{1\text{ mol MgO}}{40.3044\text{ g MgO}} \times \frac{1\text{ mol Mg}}{1\text{ mol MgO}} \times \frac{24.3050\text{ g Mg}}{1\text{ mol Mg}} = 0.603\text{ g Mg}$$

$$1.00\text{ g MgO} \times \frac{1\text{ mol MgO}}{40.3044\text{ g MgO}} \times \frac{1\text{ mol O}_2}{2\text{ mol MgO}} \times \frac{31.9988\text{ g O}_2}{1\text{ mol O}_2} = 0.397\text{ g O}_2$$

Adding them up reproduces the product mass: 0.603 g + 0.397 g = 1.000 g.

14. The following is the balanced equation for the combination reaction. $\mathbf{4\ Fe(s) + 3\ O_2(g) \longrightarrow 2\ Fe_2O_3(s)}$

✓ *Reasonable Answer Check:* 4 Fe and 6 O atoms on each side.

16. *Answer:* **Equation (b)**

Strategy and Explanation: In the first box, three A_2 molecules are combined with six B atoms. After the reaction occurs, the box contains only six AB molecules. That means symbolically: $3\ A_2 + 6\ B \longrightarrow 6\ AB$

We might be tempted to choose (a) after this observation. However, stoichiometric coefficients are supposed to be the smallest whole numbers relating the reactants and products and we find that all of these coefficients are divisible by 3. Therefore, the equation given in (b), $A_2 + 2\ B \longrightarrow 2\ AB$ is the best description of the proper stoichiometry of this reaction.

18. $2 \times 4\ A_2 + 2 \times 3\ B \longrightarrow 2 \times 1\ B_3A_8$

$8\ A_2 + 6\ B \longrightarrow 2\ B_3A_8$

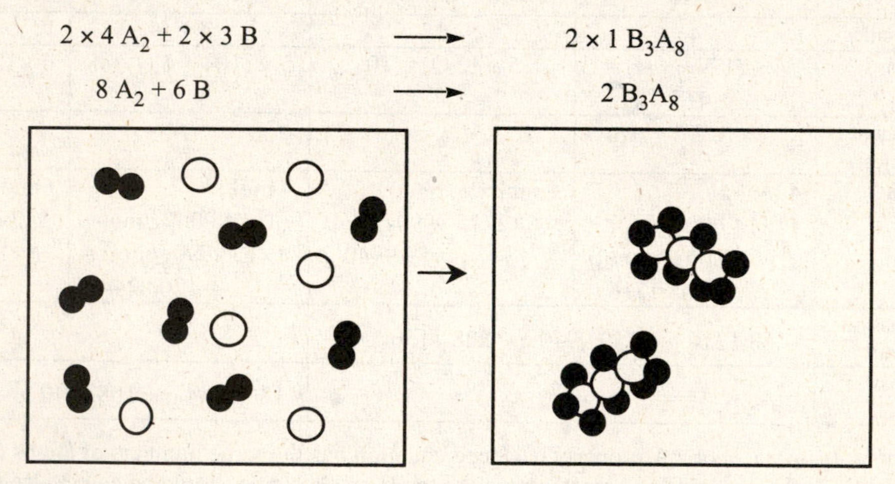

20. *Answer:* **(a) Box (a)** **(b) Box (c)**

Strategy and Explanation:

(a) The reactants must all be diatomic molecules. The stoichiometry of the balanced equation indicates that the relative quantity must be in a ratio of 1: 2. **Box (a)** has nine diatomic molecules. Three of them are blue, X_2, and six of them are pink, Y_2, for a ratio of 1:2.

(b) The products must be triatomic molecules, with the formula XY_2, so each molecule must be composed of one blue (X) atom and two pink (Y), atoms. **Box (c)** has six molecules of XY_2 so this box represents the products.

22. *Answer:* $X_2 + Y_2 \longrightarrow XY_2 + X$

Strategy and Explanation: In the first box, there are six diatomic molecules made with pink (X) atoms and six diatomic molecules made with purple (Y) atoms, so the reaction between X_2, and Y_2 is shown with a 1:1 stoichiometry. The products are composed of six pink (X) atoms and six triatomic molecules made up of two purple atoms and one pink atom (XY_2), also in a 1:1 stoichiometry.

$$6\,X_2 + 6\,Y_2 \longrightarrow 6\,XY_2 + 6\,X$$

Divide by six to get the smallest integer coefficient:

$$X_2 + Y_2 \longrightarrow XY_2 + X$$

Classification of Chemical Reactions (Section 4.2)

24. *Answer:* **(a) combination (b) decomposition (c) exchange (d) displacement**

Strategy and Explanation: Use the patterns of chemical reactions described in Section 4.2 to determine the type of reaction identified in each equation. The four major types of reactions (combination, decomposition, displacement, and exchange) can be identified by these four general equations:

Combination: $X + Y \longrightarrow XY$ Decomposition: $XY \longrightarrow X + Y$

Displacement: $A + BX \longrightarrow B + AX$ Exchange: $AX + BY \longrightarrow AY + BX$

(a) Two substances (Cu and O_2) combine to make a single product (CuO) in this equation, so it is a "Combination" reaction.

(b) Two substances (N_2O and H_2O) are produces when the reaction (NH_4NO_3) decomposes in this equation, so it is a "Decomposition" reaction.

(c) Two compounds ($BaCl_2$ and K_2SO_4) exchange ion partners to form two new compounds ($BaSO_4$ and KCl) in this equation, so it is an "Exchange" reaction.

(d) An element (Mg) replaces another element (H_2) in a compound in this equation, so it is a "Displacement" reaction.

✓ *Reasonable Answer Check:* All of these reactions follows one of the predicted patterns.

26. *Answer:* **(a) decomposition (b) displacement (c) combination (d) exchange**

Strategy and Explanation: Use the patterns of chemical reactions described in Section 4.2 as summarized in the solution to Question 24.

(a) Two substances (PbO and CO_2) are produces when the reactant ($PbCO_3$) decomposes in this equation, so it is a "Decomposition" reaction.

(b) An element (Cu) replaces another element (H) in a compound in this equation, so it is a "Displacement"

(c) Two substances (Zn and O_2) combine to make a single product (ZnO) in this equation, so it is a "Combination" reaction.

(d) Two compounds ($BaCl_2$ and K_2SO_4) exchange ion partners to form new compounds ($BaSO_4$ and KCl) in this equation, so it is an "Exchange" reaction.

✓ *Reasonable Answer Check:* All of these reactions follows one of the predicted patterns.

Balancing Equations (Section 4.3)

28. *Answer:* **(a) $2\,C_4H_{10}(g) + 13\,O_2(g) \longrightarrow 8\,CO_2(g) + 10\,H_2O(g)$**

(b) $C_6H_{12}O_6(s) + 6\,O_2(g) \longrightarrow 6\,CO_2(g) + 6\,H_2O(g)$

(c) $2\,C_4H_8O(\ell) + 11\,O_2(g) \longrightarrow 8\,CO_2(g) + 8\,H_2O(g)$

Strategy and Explanation: Write the balanced combustion equations.

The products of combustion depend on the elements present in the compound being combusted:

Compound Contains:	C	H
Combustion Product is:	CO_2	H_2O

The compound and O_2 are the reactant. The products are CO_2 and H_2O for these combustion reactions. Balance the equations, starting with C, then H, then O. If the coefficient of O_2 ends up a fraction, multiply everything in the reaction by two, to make all coefficients whole numbers.

(a) Unbalanced: $? \, C_4H_{10}(g) + ? \, O_2(g) \longrightarrow ? \, CO_2(g) + ? \, H_2O(g)$

Balance C: $C_4H_{10}(g) + ? \, O_2(g) \longrightarrow 4\, CO_2(g) + ? \, H_2O(g)$ 4 C's

Balance H: $C_4H_{10}(g) + ? \, O_2(g) \longrightarrow 4\, CO_2(g) + 5\, H_2O(g)$ 10 H's

Balance O: $C_4H_{10}(g) + \frac{13}{2}\, O_2(g) \longrightarrow 4\, CO_2(g) + 5\, H_2O(g)$ 13 O's

Multiply by 2: $2\, C_4H_{10}(g) + 13\, O_2(g) \longrightarrow 8\, CO_2(g) + 10\, H_2O(g)$

(b) Unbalanced: $? \, C_6H_{12}O_6(s) + ? \, O_2(g) \longrightarrow ? \, CO_2(g) + ? \, H_2O(g)$

Balance C: $C_6H_{12}O_6(s) + ? \, O_2(g) \longrightarrow 6\, CO_2(g) + ? \, H_2O(g)$ 6 C's

Balance H: $C_6H_{12}O_6(s) + ? \, O_2(g) \longrightarrow 6\, CO_2(g) + 6\, H_2O(g)$ 12 H's

Balance O: $C_6H_{12}O_6(s) + 6\, O_2(g) \longrightarrow 6\, CO_2(g) + 6\, H_2O(g)$ 18 O's

(c) Unbalanced: $? \, C_4H_8O(\ell) + ? \, O_2(g) \longrightarrow ? \, CO_2(g) + ? \, H_2O(g)$

Balance C: $C_4H_8O(\ell) + ? \, O_2(g) \longrightarrow 4\, CO_2(g) + ? \, H_2O(g)$ 4 C's

Balance H: $C_4H_8O(\ell) + ? \, O_2(g) \longrightarrow 4\, CO_2(g) + 4\, H_2O(g)$ 8 H's

Balance O: $C_4H_8O(\ell) + \frac{11}{2}\, O_2(g) \longrightarrow 4\, CO_2(g) + 4\, H_2O(g)$ 12 O's

Multiply by 2: $2\, C_4H_8O(\ell) + 11\, O_2(g) \longrightarrow 8\, CO_2(g) + 8\, H_2O(g)$

✓ *Reasonable Answer Check:* (a) 8 C, 20 H, & 26 O (b) 6 C, 12 H, & 18 O (c) 8 C, 16 H, & 24 O

30. *Answer:* **(a) $2\, Mg(s) + O_2(g) \longrightarrow 2\, MgO(s)$, magnesium oxide (b) $2\, Ca(s) + O_2(g) \longrightarrow 2\, CaO(s)$, calcium oxide (c) $4\, In(s) + 3\, O_2(g) \longrightarrow 2\, In_2O_3(s)$, indium(III) oxide**

Strategy and Explanation: Balance the equation for the reaction between a metal and oxygen.

The products are metal oxides. Write the product in the form of a neutral metal oxide using the periodic table to predict the metal ion's charge. Balance the equations. Notice that the oxide anion has a –2 charge. Name the ionic compound product.

(a) Magnesium is in Group 2A. Its cation will have +2 charge.

$2\, Mg(s) + O_2(g) \longrightarrow 2\, MgO(s)$ The product is called magnesium oxide.

(b) Calcium is in Group 2A. Its cation will have +2 charge.

$2\, Ca(s) + O_2(g) \longrightarrow 2\, CaO(s)$ The product is called calcium oxide.

(c) Indium is in group 3A. Its cation will have +3 charge.

$4\, In(s) + 3\, O_2(g) \longrightarrow 2\, In_2O_3(s)$ The product is called indium(III) oxide.

✓ *Reasonable Answer Check:* (a) 2 Mg & 2 O (b) 2 Ca & 2 O (c) 4 In & 6 O

32. *Answer:* **(a) $2\, K(s) + Cl_2(g) \longrightarrow 2\, KCl(s)$, potassium chloride (b) $Mg(s) + Br_2(\ell) \longrightarrow MgBr_2(s)$, magnesium bromide (c) $2\, Al(s) + 3\, F_2(g) \longrightarrow 2\, AlF_3(s)$, aluminum fluoride**

Strategy and Explanation: Balance the equation for the reaction between a metal and a halogen.

Write the product formula in the form of a neutral halide salt. Balance the equations. Notice that the halogens form anions with a –1 charge. Name the products as ionic compounds.

(a) Potassium is in group 1A. Its cation will have +1 charge.

$2\ K(s) + Cl_2(g) \longrightarrow 2\ KCl(s)$ The product is named potassium chloride.

(b) Magnesium is in group 2A. Its cation will have +2 charge.

$Mg(s) + Br_2(\ell) \longrightarrow MgBr_2(s)$ The product is named magnesium bromide.

(c) Aluminum is in group 3A. Its cation will have +3 charge.

$2\ Al(s) + 3\ F_2(g) \longrightarrow 2\ AlF_3(s)$ The product is named aluminum fluoride.

✓ *Reasonable Answer Check:* (a) 2 K & 2 Cl (b) 1 Mg & 2 Br (c) 2 Al & 6 F

34. *Answer:* **(a) $4\ Al(s) + 3\ O_2(g) \longrightarrow 2\ Al_2O_3(s)$ (b) $N_2(g) + 3\ H_2(g) \longrightarrow 2\ NH_3(g)$ (c) $2\ C_6H_6(\ell) + 15\ O_2(g) \longrightarrow 6\ H_2O(\ell) + 12\ CO_2(g)$**

Strategy and Explanation: Balance the given equations.

When balancing equations, select a specific order in which the elements are balanced. Select first the atoms that are only in one product and one reactant, since they will be easier. Select next those elements that are in more than one reactant or product (such as H or O) and then those which are present in elemental form in either the reactants or products. Be systematic. If any of the coefficients end up fractional, multiply every coefficient by the same constant to eliminate the fraction.

(a) $?\ Al(s) + ?\ O_2(g) \longrightarrow ?\ Al_2O_3(s)$ Select order: Al then O

$2\ Al(s) + ?\ O_2(g) \longrightarrow 1\ Al_2O_3(s)$ 2 Al

$2\ Al(s) + \frac{3}{2}\ O_2(g) \longrightarrow 1\ Al_2O_3(s)$ 3 O

$4\ Al(s) + 3\ O_2(g) \longrightarrow 2\ Al_2O_3(s)$

(b) $?\ N_2(g) + ?\ H_2(g) \longrightarrow ?\ NH_3(g)$ Select order: H then N

$?\ N_2(g) + 3\ H_2(g) \longrightarrow 2\ NH_3(g)$ 6 H

$1\ N_2(g) + 3\ H_2(g) \longrightarrow 2\ NH_3(g)$ 2 N

(c) $?\ C_6H_6(\ell) + ?\ O_2(g) \longrightarrow ?\ H_2O(\ell) + ?\ CO_2(g)$ Select order: C, H then O

$C_6H_6(\ell) + ?\ O_2(g) \longrightarrow ?\ H_2O(\ell) + 6\ CO_2(g)$ 6 C

$C_6H_6(\ell) + ?\ O_2(g) \longrightarrow 3\ H_2O(\ell) + 6\ CO_2(g)$ 6 H

$C_6H_6(\ell) + \frac{15}{2}\ O_2(g) \longrightarrow 3\ H_2O(\ell) + 6\ CO_2(g)$ 15 O

$2\ C_6H_6(\ell) + 15\ O_2(g) \longrightarrow 6\ H_2O(\ell) + 12\ CO_2(g)$

✓ *Reasonable Answer Check:* (a) 4 Al & 6 O (b) 2 N & 6 H (c) 12 C, 12 H & 30 O

36. *Answer:* **(a) $UO_2(s) + 4\ HF(\ell) \longrightarrow UF_4(s) + 2\ H_2O(\ell)$ (b) $B_2O_3(s) + 6\ HF(\ell) \longrightarrow 2\ BF_3(s) + 3\ H_2O(\ell)$ (c) $BF_3(g) + 3\ H_2O(\ell) \longrightarrow 3\ HF(\ell) + H_3BO_3(s)$**

Strategy and Explanation: Follow the method described in the solution for Question 34.

(a) $?\ UO_2(s) + ?\ HF(\ell) \longrightarrow ?\ UF_4(s) + ?\ H_2O(\ell)$ Select order: U, F, H, O

$UO_2(s) + ?\ HF(\ell) \longrightarrow UF_4(s) + ?\ H_2O(\ell)$ 1 U

$UO_2(s) + 4\ HF(\ell) \longrightarrow UF_4(s) + ?\ H_2O(\ell)$ 4 F

$UO_2(s) + 4\ HF(\ell) \longrightarrow UF_4(s) + 2\ H_2O(\ell)$ 4 H and 2 O

(b) $?\ B_2O_3(s) + ?\ HF(\ell) \longrightarrow ?\ BF_3(s) + ?\ H_2O(\ell)$ Select order: B, F, H, O

$B_2O_3(s) + ?\ HF(\ell) \longrightarrow 2\ BF_3(s) + ?\ H_2O(\ell)$ 2 B

$B_2O_3(s) + 6\ HF(\ell) \longrightarrow 2\ BF_3(s) + ?\ H_2O(\ell)$ 6 F

$B_2O_3(s) + 6\ HF(\ell) \longrightarrow 2\ BF_3(s) + 3\ H_2O(\ell)$ 6 H and 3 O

(c) $? BF_3(g) + ? H_2O(\ell) \longrightarrow ? HF(g) + ? H_3BO_3(s)$ Select order: B, F, H, O

$BF_3(g) + ? H_2O(\ell) \longrightarrow ? HF(\ell) + H_3BO_3(s)$ 1 B

$BF_3(g) + ? H_2O(\ell) \longrightarrow 3 HF(\ell) + H_3BO_3(s)$ 3 F

$BF_3(g) + 3 H_2O(\ell) \longrightarrow 3 HF(\ell) + H_3BO_3(s)$ 6 H and 3 O

✓ *Reasonable Answer Check:* (a) 1 U, 2 O, 4 H & 4 F (b) 2 B, 3 O, 6 H & 6 F (c) 1 B, 3 F, 6 H & 3 O

38. *Answer:* **(a) $H_2NCl(aq) + 2 NH_3(g) \longrightarrow NH_4Cl(aq) + N_2H_4(aq)$**

(b) $(CH_3)_2N_2H_2(\ell) + 2 N_2O_4(g) \longrightarrow 3 N_2(g) + 4 H_2O(g) + 2 CO_2(g)$

(c) $CaC_2(s) + 2 H_2O(\ell) \longrightarrow Ca(OH)_2(s) + C_2H_2(g)$

Strategy and Explanation: Follow the method described in the solution for Question 34.

(a) $? H_2NCl(aq) + ? NH_3(g) \longrightarrow ? NH_4Cl(aq) + ? N_2H_4(aq)$ Select order: Cl, N, H

$H_2NCl(aq) + ? NH_3(g) \longrightarrow NH_4Cl(aq) + ? N_2H_4(aq)$ 1 Cl

$H_2NCl(aq) + 2 NH_3(g) \longrightarrow NH_4Cl(aq) + N_2H_4(aq)$ 3 N and 8 H

(b) $? (CH_3)_2N_2H_2(\ell) + ? N_2O_4(g) \longrightarrow ? N_2(g) + ? H_2O(\ell) + ? CO_2(g)$ Select order: H, C, O, N

$(CH_3)_2N_2H_2(\ell) + ? N_2O_4(g) \longrightarrow ? N_2(g) + 4 H_2O(g) + ? CO_2(g)$ 8 H

$(CH_3)_2N_2H_2(\ell) + ? N_2O_4(g) \longrightarrow ? N_2(g) + 4 H_2O(g) + 2 CO_2(g)$ 2 C

$(CH_3)_2N_2H_2(\ell) + 2 N_2O_4(g) \longrightarrow ? N_2(g) + 4 H_2O(g) + 2 CO_2(g)$ 8 O

$(CH_3)_2N_2H_2(\ell) + 2 N_2O_4(g) \longrightarrow 3 N_2(g) + 4 H_2O(g) + 2 CO_2(g)$ 6 N

(c) $? CaC_2(s) + ? H_2O(\ell) \longrightarrow ? Ca(OH)_2(s) + ? C_2H_2(g)$ Select order: Ca, C, O, H

$CaC_2(s) + ? H_2O(\ell) \longrightarrow Ca(OH)_2(s) + ? C_2H_2(g)$ 1 Ca

$CaC_2(s) + ? H_2O(\ell) \longrightarrow Ca(OH)_2(s) + C_2H_2(g)$ 2 C

$CaC_2(s) + 2 H_2O(\ell) \longrightarrow Ca(OH)_2(s) + C_2H_2(g)$ 2 O and 4 H

✓ *Reasonable Answer Check:* (a) 8 H, 3 N & 1 Cl (b) 2 C, 8 H, 6 N & 8 O (c) 1 Ca, 2 C, 4 H & 2 O

40. *Answer:* **(a) $C_6H_{12}O_6 + 6 O_2 \longrightarrow 6 CO_2 + 6 H_2O$ (b) $C_5H_{12} + 8 O_2 \longrightarrow 5 CO_2 + 6 H_2O$**

(c) $2 C_7H_{14}O_2 + 19 O_2 \longrightarrow 14 CO_2 + 14 H_2O$ (d) $C_2H_4O_2 + 2 O_2 \longrightarrow 2 CO_2 + 2 H_2O$

Strategy and Explanation: Balance the given combustion equations.

When balancing combustion equations, select a C, H, O as the order in which the elements are balanced. Be systematic. If the coefficient of O_2 ends up a fraction, multiply every coefficient by 2 to make all of the coefficients whole numbers.

(a) $? C_6H_{12}O_6 + ? O_2 \longrightarrow ? CO_2 + ? H_2O$

$C_6H_{12}O_6 + ? O_2 \longrightarrow 6 CO_2 + ? H_2O$ 6 C

$C_6H_{12}O_6 + ? O_2 \longrightarrow 6 CO_2 + 6 H_2O$ 12 H

$C_6H_{12}O_6 + 6 O_2 \longrightarrow 6 CO_2 + 6 H_2O$ 18 O

(b) $? C_5H_{12} + ? O_2 \longrightarrow ? CO_2 + ? H_2O$

$C_5H_{12} + ? O_2 \longrightarrow 5 CO_2 + ? H_2O$ 5 C

$C_5H_{12} + ? O_2 \longrightarrow 5 CO_2 + 6 H_2O$ 12 H

$C_5H_{12} + 8 O_2 \longrightarrow 5 CO_2 + 6 H_2O$ 16 O

(c) $? C_7H_{14} + ? O_2 \longrightarrow ? CO_2 + ? H_2O$

$C_7H_{14}O_2 + ? O_2 \longrightarrow 7 CO_2 + ? H_2O$ 7 C

$C_7H_{14}O_2 + ? O_2 \longrightarrow 7 CO_2 + 7 H_2O$ 14 H

$C_7H_{14}O_2 + \frac{19}{2} O_2 \longrightarrow 7 CO_2 + 7 H_2O$ 21 O

$2 C_7H_{14}O_2 + 19 O_2 \longrightarrow 14 CO_2 + 14 H_2O$

(d) $? C_2H_4O_2 + ? O_2 \longrightarrow ? CO_2 + ? H_2O$

$C_2H_4O_2 + ? O_2 \longrightarrow 2 CO_2 + ? H_2O$ 2 C

$C_2H_4O_2 + ? O_2 \longrightarrow 2 CO_2 + 2 H_2O$ 4 H

$C_2H_4O_2 + 2 O_2 \longrightarrow 2 CO_2 + 2 H_2O$ 6 O

✓ *Reasonable Answer Check:* (a) 6 C, 12 H & 18 O, (b) 5 C, 12 H & 16 O, (c) 14 C, 28 H, 42 O (d) 2 C, 4 H, 6 O

The Mole and Chemical Reactions (Section 4.4)

42. *Answer:* **50.0 mol HCl**

Strategy and Explanation: Given a balanced chemical equation and the number of moles of a product formed, determine the moles of reactant that was needed.

Use the stoichiometry of the balanced equation as a conversion factor to convert the moles of product to moles of reactant.

The balanced equation says: 4 mol HCl are needed to make 1 mol Cl_2.

$$12.5 \text{ mol } Cl_2 \times \frac{4 \text{ mol HCl}}{1 \text{ mol } Cl_2} = 50.0 \text{ mol HCl}$$

✓ *Reasonable Answer Check:* More HCl molecules are needed than Cl_2 molecules are formed. It makes sense that moles of HCl are greater.

44. *Answer:* **12.8 g**

Strategy and Explanation: Given a balanced chemical equation and the number of grams of a product, determine the mass of the reactant that are needed to form it.

Use the molar mass of the product to find the moles of that substance. Then use the stoichiometry of the balanced equation as a conversion factor to relate the moles of product to moles to that of reactant, then use the molar mass of the reactant to find the grams.

$$5.00 \text{ g } O_2 \times \frac{1 \text{ mol } O_2}{31.9988 \text{ g } O_2} = 0.156 \text{ mol } O_2$$

The balanced equation says: 3 mol O_2 is produced from 2 mol $KClO_3$.

$$0.156 \text{ mol } O_2 \times \frac{2 \text{ mol } KClO_3}{3 \text{ mol } O_2} = 0.104 \text{ mol } KClO_4$$

$$0.104 \text{ mol } KClO_4 \times \frac{122.550 \text{ g } KClO_4}{1 \text{ mol } KClO_4} = 12.8 \text{ g } KClO_4$$

✓ *Reasonable Answer Check:* The mass of $KClO_4$ is larger than the mass of the oxygen produced. This makes sense because it contains the elements potassium and chlorine, also.

46. *Answer:* **1.1 mol O_2, 35 g O_2, 1.0 × 10^2 g NO_2**

Strategy and Explanation: Given a balanced chemical equation and the number of moles of a reactant, determine the moles and grams of another reactant needed and the grams of product produced.

Use the stoichiometry of the balanced equation as a conversion factor to convert the moles of the one reactant to moles of the other reactant. Use the molar mass to convert from moles to grams. Use the stoichiometry of the equation to find the moles of product formed, then use the molar mass of the product to find the grams.

The balanced equation says: 2 mol NO react with 1 mol O_2.

$$2.2 \text{ mol NO} \times \frac{1 \text{ mol } O_2}{2 \text{ mol NO}} = 1.1 \text{ mol } O_2$$

$$1.1 \text{ mol } O_2 \times \frac{31.9988 \text{ g } O_2}{1 \text{ mol } O_2} = 35 \text{ g } O_2$$

The balanced equation says: 2 mol NO produces 2 mol NO_2.

$$2.2 \text{ mol NO} \times \frac{2 \text{ mol } NO_2}{2 \text{ mol NO}} \times \frac{46.0055 \text{ g } NO_2}{1 \text{ mol } NO_2} = 1.0 \times 10^2 \text{ g } NO_2 \text{produced}$$

✓ *Reasonable Answer Check:* Fewer moles of O_2 are needed than moles of NO. The mass of NO used is 2.2 mol × (30.0061 g/mol) = 66 g. The sum of the reactant masses (66 g + 35 g) is equal to the products mass (1.0 × 10^2 g), within known significant figures.

48. *Answer:* **(a) 12.7 g Cl_2 (b) 0.179 mol $FeCl_2$, 22.7 g $FeCl_2$ expected**

Strategy and Explanation: Given a balanced chemical equation and the number of grams of a reactant, determine the grams of another reactant needed and the moles and grams of product produced.

In (a), use the molar mass of the first reactant to find the moles of that substance. Then use the stoichiometry of the balanced equation as a conversion factor to convert the moles of the one reactant to moles of the other reactant. Use the molar mass to convert from moles to grams. For (b), use the stoichiometry of the equation to find moles of product formed, then use the molar mass of the product to find grams.

(a) The balanced equation says: 1 mol Fe reacts with 1 mol Cl_2.

$$10.0 \text{ g Fe} \times \frac{1 \text{ mol Fe}}{55.845 \text{ g Fe}} = 0.179 \text{ mol Fe}$$

$$0.179 \text{ mol Fe} \times \frac{1 \text{ mol } Cl_2}{1 \text{ mol Fe}} \times \frac{70.906 \text{ g } Cl_2}{1 \text{ mol } Cl_2} = 12.7 \text{ g } Cl_2$$

(b) The balanced equation says: 1 mol Fe produces 1 mol $FeCl_2$.

$$0.179 \text{ mol Fe} \times \frac{1 \text{ mol } FeCl_2}{1 \text{ mol Fe}} = 0.179 \text{ mol } FeCl_2 \text{ expected}$$

$$0.179 \text{ mol } FeCl_2 \times \frac{126.751 \text{ g } FeCl_2}{1 \text{ mol } FeCl_2} = 22.7 \text{ g } FeCl_2 \text{ expected}$$

✓ *Reasonable Answer Check:* The sum of the masses of the reactants must add up to the total mass of the product. 10.0 g + 12.7 g = 22.7 g. This is the same product mass as that calculated above.

50. *Answer:* The complete table looks like this:

$(NH_4)_2PtCl_6$	Pt	HCl
12.35 g	**5.428** g	**5.410** g
0.02782 mol	**0.02782** mol	**0.1484** mol

Strategy and Explanation: Given a balanced chemical equation and the number of grams of a reactant, determine the moles of the reactant and the mass and moles of two products.

Use the molar mass of the reactant to find the moles of that substance. Then use the stoichiometry of the balanced equation as a conversion factor to convert the moles of reactant to moles of each product formed, then use the molar mass of each product to find the grams.

Molar mass of $(NH_4)_2PtCl_6$ =

$$2 \times [1 \times (14.0067 \text{ g}) + 4 \times (1.0079 \text{ g})] + (195.078 \text{ g}) + 6 \times (35.453 \text{ g}) = 443.873 \text{ g/mol}$$

$$12.35 \text{ g } (NH_4)_2PtCl_6 \times \frac{1 \text{ mol } (NH_4)_2PtCl_6}{443.873 \text{ g } (NH_4)_2PtCl_6} = 2.782 \times 10^{-2} \text{ mol } (NH_4)_2PtCl_6$$

The balanced equation says: 3 mol $(NH_4)_2PtCl_6$ react to form 3 mol Pt.

$$2.782 \times 10^{-2} \text{ mol } (NH_4)_2PtCl_6 \times \frac{1 \text{ mol Pt}}{1 \text{ mol } (NH_4)_2PtCl_6} = 2.782 \times 10^{-2} \text{ mol Pt}$$

$$2.782 \times 10^{-2} \text{ mol Pt} \times \frac{195.078 \text{ g Pt}}{1 \text{ mol Pt}} = 5.428 \text{ g Pt}$$

The balanced equation says: 3 mol $(NH_4)_2PtCl_6$ react to form 16 mol HCl.

$$2.782 \times 10^{-2} \text{ mol } (NH_4)_2PtCl_6 \times \frac{16 \text{ mol HCl}}{3 \text{ mol } (NH_4)_2PtCl_6} = 0.1484 \text{ mol HCl}$$

$$0.1484 \text{ mol HCl} \times \frac{36.461 \text{ mol HCl}}{1 \text{ mol HCl}} = 5.410 \text{ g HCl}$$

✓ *Reasonable Answer Check:* The moles are smaller than the grams. This is appropriate. The moles of HCl are larger than the moles of $(NH_4)_2PtCl_6$ and Pt.

52. *Answer:* **(a) 0.148 mol (b) 5.89 g TiO_2 and 10.8 g HCl**

Strategy and Explanation: Given a balanced chemical equation and the number of grams of a reactant, determine the moles of another reactant that are needed and the masses of the two products expected.

Use the molar mass of the reactant to find the moles of that substance. Then use the stoichiometry of the balanced equation as a conversion factor to convert the moles of one reactant to moles of the other reactant. Use the stoichiometry to determine the moles of each of the products produced, then use the molar mass of each product to find the grams.

(a) The balanced equation says: 1 mol $TiCl_4$ reacts with 2 mol H_2O.

$$14.0 \text{ g } TiCl_4 \times \frac{1 \text{ mol } TiCl_4}{189.688 \text{ g } TiCl_4} = 0.0738 \text{ mol } TiCl_4$$

$$0.0738 \text{ mol } TiCl_4 \times \frac{2 \text{ mol } H_2O}{1 \text{ mol } TiCl_4} = 0.148 \text{ mol } H_2O$$

(b) The balanced equation says: 1 mol $TiCl_4$ produces 1 mol TiO_2.

$$0.0738 \text{ mol } TiCl_4 \times \frac{1 \text{ mol } TiO_2}{1 \text{ mol } TiCl_4} \times \frac{79.866 \text{ g } TiO_2}{1 \text{ mol } TiO_2} = 5.89 \text{ g } TiO_2$$

The balanced equation says: 1 mol $TiCl_4$ produces 4 mol HCl.

$$0.0738 \text{ mol } TiCl_4 \times \frac{4 \text{ mol HCl}}{1 \text{ mol } TiCl_4} \times \frac{36.461 \text{ g HCl}}{1 \text{ mol HCl}} = 10.8 \text{ g HCl}$$

✓ *Reasonable Answer Check:* The sum of the masses of the reactants (14.0 g + 2.67 g = 16.7 g) adds up the mass of the product (5.90 g + 10.8 g = 16.7 g).

54. *Answer:* **2.0 mol, 36.0304 g**

Strategy and Explanation: Given the identity of a reactant undergoing a named reaction and the number of moles of a reactant, determine the number of moles and the mass of one product formed.

Balanced the equation for the reaction. Using the moles of reactant and the stoichiometry of the balanced equation, determine the moles of the product produced. Then use the molar mass of the product to find the grams.

Combustion of the hydrocarbon, C_3H_8, involves its reaction O_2 to produce CO_2 and H_2O.

$$C_3H_8 + 5\ O_2 \longrightarrow 3\ CO_2 + 4\ H_2O$$

The balanced equation says: 4 mol H_2O is produced when 5 mol O_2 react.

$$2.5 \text{ mol } O_2 \times \frac{4 \text{ mol } H_2O}{5 \text{ mol } O_2} = 2.0 \text{ mol } H_2O$$

$$2.0 \text{ mol } H_2O \times \frac{18.0152 \text{ g } H_2O}{1 \text{ mol } H_2O} = 36.0304 \text{ g } H_2O$$

✓ *Reasonable Answer Check:* The moles of H_2O are is smaller than the moles of O_2. The number representing the mass of a molecule will always be larger than the number of moles.

56. *Answer:* **0.699 g Ga and 0.751 g As**

Strategy and Explanation: Given the formula of a compound and its mass, determine the mass of the elements required to make it.

Use the molar mass of the compound to find the moles of that substance. Then use the stoichiometry of the balanced equation to determine the moles of each element needed. Then use the molar masses of these elements to find the grams.

The formula gives: 1 mol GaAs is produced from 1 mol Ga and 1 mol As.

$$1.45 \text{ g GaAs} \times \frac{1 \text{ mol GaAs}}{144.6446 \text{ g GaAs}} \times \frac{1 \text{ mol Ga}}{1 \text{ mol GaAs}} \times \frac{69.723 \text{ g Ga}}{1 \text{ mol Ga}} = 0.699 \text{ g Ga}$$

$$1.45 \text{ g GaAs} \times \frac{1 \text{ mol GaAs}}{144.6446 \text{ g GaAs}} \times \frac{1 \text{ mol As}}{1 \text{ mol GaAs}} \times \frac{74.9215 \text{ g As}}{1 \text{ mol As}} = 0.751 \text{ g As}$$

✓ *Reasonable Answer Check:* The sum of the reactant masses (0.699 g + 0.751 g = 1.450 g) is the same as the total mass of the compound.

58. *Answer:* **(a)** $\mathbf{4\ Fe(s) + 3\ O_2(g) \longrightarrow 2\ Fe_2O_3(s)}$ **(b) 7.98 g (c) 2.40 g**

Strategy and Explanation: Given the products and reactants of a reaction and the mass of one reactant, balance the chemical equation, determine the mass of the product produced, and determine the mass of the other reactant.

Balance the equation from the given formulas of the reactants and product. Use the molar mass of the reactant to find the moles of that substance. Then use the stoichiometry of the equation to determine the moles of the product produced and the moles of the other reactant required. Then use the molar mass of each to find the grams.

(a) Balance O atoms, then Fe atoms: $4\ Fe(s) + 3\ O_2(g) \longrightarrow 2\ Fe_2O_3(s)$

(b) The balanced equation says: 4 mol Fe produces 2 mol Fe_2O_3.

$$5.58 \text{ g Fe} \times \frac{1 \text{ mol Fe}}{55.845 \text{ g Fe}} \times \frac{2 \text{ mol } Fe_2O_3}{4 \text{ mol Fe}} \times \frac{159.688 \text{ g } Fe_2O_3}{1 \text{ mol } Fe_2O_3} = 7.98 \text{ g } Fe_2O_3$$

(c) The balanced equation says: 4 mol Fe reacts with 3 mol O_2.

$$5.58 \text{ g Fe} \times \frac{1 \text{ mol Fe}}{55.845 \text{ g Fe}} \times \frac{3 \text{ mol } O_2}{4 \text{ mol Fe}} \times \frac{31.9988 \text{ g } O_2}{1 \text{ mol } O_2} = 2.40 \text{ g } O_2$$

✓ *Reasonable Answer Check:* The sum of the reactant masses is: 5.58 g + 2.40 g = 7.98 g. This is the same as the total mass of the product compound.

60. *Answer:* **(a)** $\mathbf{CCl_2F_2 + 2\ Na_2C_2O_4 \longrightarrow C + 4\ CO_2 + 2\ NaCl + 2\ NaF}$ **(b) 170. g** $\mathbf{Na_2C_2O_4}$ **(c) 112 g** $\mathbf{CO_2}$

Strategy and Explanation: Given the reactants and products of a reaction and the mass of a reactant, balance the chemical equation, determine the mass of another reactant, and determine the mass of one product produced.

Given the formulas of the reactants and products, balance the equation, selecting appropriate order for the systematic balancing of all the atoms. Then use the molar mass of the reactant to find the moles of that substance. Then use the stoichiometry of the equation to determine the moles of the other reactant and product produced. Then use the molar mass of each to find the grams.

(a) $? CCl_2F_2 + ? Na_2C_2O_4 \longrightarrow ? C + ? CO_2 + ? NaCl + ? NaF$ Select order: Cl, F, Na, O, C

$CCl_2F_2 + ? Na_2C_2O_4 \longrightarrow ? C + ? CO_2 + 2 NaCl + ? NaF$ 2 Cl

$CCl_2F_2 + ? Na_2C_2O_4 \longrightarrow ? C + ? CO_2 + 2 NaCl + 2 NaF$ 2 F

$CCl_2F_2 + 2 Na_2C_2O_4 \longrightarrow ? C + ? CO_2 + 2 NaCl + 2 NaF$ 4 Na

$CCl_2F_2 + 2 Na_2C_2O_4 \longrightarrow ? C + 4 CO_2 + 2 NaCl + 2 NaF$ 8 O

$CCl_2F_2 + 2 Na_2C_2O_4 \longrightarrow C + 4 CO_2 + 2 NaCl + 2 NaF$

(b) The balanced equation says: 1 mol CCl_2F_2 requires 2 mol $Na_2C_2O_4$.

$$76.8 \text{ g } CCl_2F_2 \times \frac{1 \text{ mol } CCl_2F_2}{120.914 \text{ g } CCl_2F_2} = 0.635 \text{ mol } CCl_2F_2$$

$$0.635 \text{ mol } CCl_2F_2 \times \frac{2 \text{ mol } Na_2C_2O_4}{1 \text{ mol } CCl_2F_2} \times \frac{133.9986 \text{ g } Na_2C_2O_4}{1 \text{ mol } Na_2C_2O_4} = 170. \text{ g } Na_2C_2O_4$$

(c) The balanced equation says: 1 mol CCl_2F_2 produces 4 mol CO_2.

$$0.635 \text{ mol } CCl_2F_2 \times \frac{4 \text{ mol } CO_2}{1 \text{ mol } CCl_2F_2} \times \frac{44.0095 \text{ g } CO_2}{1 \text{ mol } CO_2} = 112 \text{ g } CO_2$$

✓ *Reasonable Answer Check:* (a) Check the number of atom of each type in the reactants and products: 5 C, 2 Cl, 2 F, 4 Na, 8 O. The equation is properly balanced. (b) The masses of $Na_2C_2O_4$ and CO_2 are both larger than the mass of CCl_2F_2.

62. *Answer:* **(a) 699 g (b) 526 g**

Strategy and Explanation: Given a balanced chemical equation and the mass of a reactant in kilograms, determine the mass of one of the products produced and the mass of another reactant required.

Use metric conversion factors to convert kilograms to grams. Use the molar mass of the reactant to find the moles of that substance. Then use the stoichiometry of the equation to determine the moles of the product produced and the other reactant required. Then use molar masses to find the grams.

$$1.00 \text{ kg } Fe_2O_3 \times \frac{1000 \text{ g } Fe_2O_3}{1 \text{ kg } Fe_2O_3} \times \frac{1 \text{ mol } Fe_2O_3}{159.688 \text{ g } Fe_2O_3} = 6.26 \text{ mol } Fe_2O_3$$

(a) The balanced equation says: 1 mol Fe_2O_3 produces 1 mol Fe.

$$6.26 \text{ mol } Fe_2O_3 \times \frac{2 \text{ mol Fe}}{1 \text{ mol } Fe_2O_3} \times \frac{55.845 \text{ g Fe}}{1 \text{ mol Fe}} = 699 \text{ g Fe}$$

(b) The balanced equation says: 1 mol Fe_2O_3 requires 3 mol CO.

$$6.26 \text{ mol } Fe_2O_3 \times \frac{3 \text{ mol CO}}{1 \text{ mol } Fe_2O_3} \times \frac{28.0101 \text{ g CO}}{1 \text{ mol CO}} = 526 \text{ g CO}$$

✓ *Reasonable Answer Check:* The mass of iron produced is less than the mass of iron(III) oxide it was produced from. The mass of CO used is less than the mass of iron(III) oxide even with a larger stoichiometric coefficient because CO contains lighter-weight atoms.

Limiting Reactant (Section 4.5)

64. *Answer:* **$BaCl_2$, 1.12081 g $BaSO_4$**

Strategy and Explanation: Given a balanced chemical equation and the masses of both reactants, determine the limiting reactant and the mass of the product produced.

Here, we use a slight variation of what the text calls "the mole method." We will calculate a directly comparable quantity, the moles of the desired product. Use the molar mass of the reactants to find the moles of the reactant substances. Then use the stoichiometry of the equation to determine the moles of the product produced in each case. Identify the limiting reactant from the reactant that produces the least number of products. Then, use the moles of product produced from the limiting reactant and the molar mass of the product to find the grams.

Exactly 1 gram of each reactant is present initially.

The balanced equation says: 1 mol Na_2SO_4 produces 1 mol $BaSO_4$.

$$1 \text{ g } Na_2SO_4 \times \frac{1 \text{ mol } Na_2SO_4}{142.042 \text{ g } Na_2SO_4} \times \frac{1 \text{ mol } BaSO_4}{1 \text{ mol } Na_2SO_4} = 0.00704016 \text{ mol } BaSO_4$$

The balanced equation says: 1 mol $BaCl_2$ produces 1 mol $BaSO_4$.

$$1 \text{ g } BaCl_2 \times \frac{1 \text{ mol } BaCl_2}{208.233 \text{ g } BaCl_2} \times \frac{1 \text{ mol } BaSO_4}{1 \text{ mol } BaCl_2} = 0.00480231 \text{ mol } BaSO_4$$

The number of $BaSO_4$ moles produced from $BaCl_2$ is smaller (0.00480231 mol < 0.00704016 mol), so $BaCl_2$ is the limiting reactant.

Find the mass of 0.00480231 mol $BaSO_4$:

$$0.00480231 \text{ mol } BaSO_4 \times \frac{233.390 \text{ g } BaSO_4}{1 \text{ mol } BaSO_4} = 1.12081 \text{ g } BaSO_4$$

✓ *Reasonable Answer Check:* There are fewer moles of $BaSO_4$ present than $BaCl_2$, and the equation needs the same amount of $BaSO_4$ as $BaCl_2$, so it makes sense that $BaCl_2$ is the limiting reactant.

66. *Answer:* **(a) Cl_2 is limiting. (b) 5.08 g Al_2Cl_6 (c) 1.67 g Al unreacted**

Strategy and Explanation: Given a balanced chemical equation and the masses of both reactants, determine the limiting reactant, the mass of the product produced, and the mass remaining of the excess reactant when the reaction is complete.

Here, we use a slight variation of what the text calls "the mole method." We will calculate a directly comparable quantity, the moles of product. (a) Use the molar mass of the reactants to find the moles of the reactant substances. Then use the stoichiometry of the equation to determine the moles of the product produced in each case. Identify the limiting reactant from the reactant that produces the least number of products. (b) Use the moles of product produced from the limiting reactant and the molar mass of the product to find the grams. (c) From the limiting reactant quantity, determine the moles of the other reactant needed for complete reaction. Convert that number to grams using the molar mass. Then subtract the quantity used from the initial mass given to get the mass of excess reactant.

(a) The balanced equation says: 2 mol Al produces 1 mol Al_2Cl_6.

$$2.70 \text{ g Al} \times \frac{1 \text{ mol Al}}{26.9815 \text{ g Al}} \times \frac{1 \text{ mol } Al_2Cl_6}{2 \text{ mol Al}} = 0.0500 \text{ mol } Al_2Cl_6$$

The balanced equation says: 3 mol Cl_2 produces 1 mol Al_2Cl_6.

$$4.05 \text{ g } Cl_2 \times \frac{1 \text{ mol } Cl_2}{70.906 \text{ g } Cl_2} \times \frac{1 \text{ mol } Al_2Cl_6}{3 \text{ mol } Cl_2} = 0.0190 \text{ mol } Al_2Cl_6$$

The number of Al_2Cl_6 moles produced from Cl_2 is smaller (0.0190 mol < 0.0500 mol), so Cl_2 is the limiting reactant and Al is the excess reactant.

(b) Find the mass of 0.0190 mol Al_2Cl_6:

$$0.0190 \text{ mol } Al_2Cl_6 \times \frac{266.682 \text{ g } Al_2Cl_6}{1 \text{ mol } Al_2Cl_6} = 5.08 \text{ g } Al_2Cl_6$$

(c) The balanced equation says: 3 mol Cl_2 react with 2 mol Al.

$$4.05 \text{ g } Cl_2 \times \frac{1 \text{ mol } Cl_2}{70.906 \text{ g } Cl_2} \times \frac{2 \text{ mol Al}}{3 \text{ mol } Cl_2} \times \frac{26.9815 \text{ g Al}}{1 \text{ mol Al}} = 1.03 \text{ g Al}$$

2.70 g Al initial – 1.03 g Al used up = 1.67 g Al remains unreacted

✓ *Reasonable Answer Check:* There are fewer moles of Cl_2 present than Al, and the equation needs more Cl_2 than Al, so it makes sense that Cl_2 is the limiting reactant. In addition, the calculation in (c) proved that the initial mass of Al was larger than required to react with all of the Cl_2. The sum of the masses of the reactants that reacted (4.05 g + 1.03 g = 5.08 g) equals the mass of the product produced (5.08 g).

68. *Answer:* **(a) CO (b) 1.3 g H_2 (b) 85.2 g**

Strategy and Explanation: Given a balanced chemical equation and the masses of both reactants, determine the limiting reactant, the mass of the product produced, and the mass remaining of the excess reactant when the reaction is complete.

To solve (a): use the molar mass of the reactants to find the moles of the reactant substances. Then use the stoichiometry of the equation to determine the moles of the product produced in each case. Identify the limiting reactant from the reactant that produces the least number of products. To solve (b): from the limiting reactant quantity determine the moles of the other reactant needed for complete reaction. Convert that number to mass using the molar mass. Then subtract the quantity used from the initial mass given, to get the mass of excess reactant. (c) Use the moles of product produced from the limiting reactant and the molar mass of the product to find the grams.

(a) The balanced equation says: 1 mol CO produces 1 mol CH_3OH.

$$74.5 \text{ g CO} \times \frac{1 \text{ mol CO}}{28.0101 \text{ g CO}} \times \frac{1 \text{ mol } CH_3OH}{1 \text{ mol CO}} = 2.66 \text{ mol } CH_3OH$$

The balanced equation says: 2 mol H_2 produces 1 mol CH_3OH.

$$12.0 \text{ g } H_2 \times \frac{1 \text{ mol } H_2}{2.0158 \text{ g } H_2} \times \frac{1 \text{ mol } CH_3OH}{2 \text{ mol } H_2} = 2.98 \text{ mol } CH_3OH$$

The number of CH_3OH moles produced from CO is smaller (2.66 mol < 2.98 mol), so CO is the limiting reactant and H_2 is the excess reactant.

(b) The balanced equation says: 2 mol H_2 react with 1 mol CO.

$$74.5 \text{ g CO} \times \frac{1 \text{ mol CO}}{28.0101 \text{ g CO}} \times \frac{2 \text{ mol } H_2}{1 \text{ mol CO}} \times \frac{2.0158 \text{ g } H_2}{1 \text{ mol } H_2} = 10.7 \text{ g } H_2$$

12.0 g H_2 initial – 10.7 g H_2 used up = 1.3 g H_2 remains unreacted

(c) Find the mass of 2.66 mol CH_3OH:

$$2.66 \text{ mol } CH_3OH \times \frac{32.0417 \text{ g } CH_3OH}{1 \text{ mol } CH_3OH} = 85.2 \text{ g } CH_3OH$$

✓ *Reasonable Answer Check:* The calculation in (b) proved that the initial mass of H_2 was larger than required to react with all the CO. The sum of the masses of the reactants that reacted (74.5 g + 10.7 g = 85.2 g) equal the mass of the product produced (85.2 g).

70. *Answer:* **0 mol CaO, 0.19 mol NH_4Cl, 2.00 mol H_2O, 4.00 mol NH_3, 2.00 mol $CaCl_2$**

Strategy and Explanation: Given a balanced chemical equation and the masses of both reactants, determine the moles of reactants and products present when the reaction is finished.

Use the molar mass of the reactants to find the moles of the reactant substances present initially. Then use the stoichiometry of the equation to determine the moles of one of the products. Identify the limiting reactant from the reactant that produces the least number of products. From the limiting reactant quantity, and using the stoichiometry of the balanced equation, determine the moles of the other reactant and products.

The balanced equation says: 1 mol CaO produces 1 mol H_2O.

$$112 \text{ g CaO} \times \frac{1 \text{ mol CaO}}{56.077 \text{ g CaO}} \times \frac{1 \text{ mol } H_2O}{1 \text{ mol CaO}} = 2.00 \text{ mol } H_2O$$

The balanced equation says: 2 mol NH_4Cl produces 1 mol H_2O.

$$224 \text{ g } NH_4Cl \times \frac{1 \text{ mol } NH_4Cl}{53.4913 \text{ g } NH_4Cl} \times \frac{1 \text{ mol } H_2O}{2 \text{ mol } NH_4Cl} = 2.09 \text{ mol } H_2O$$

The number of H_2O moles produced from CaO is smaller (2.00 mol < 2.09 mol), so CaO is the limiting reactant and NH_4Cl is the excess reactant.

The balanced equation says: 1 mol CaO produces 2 mol NH_3.

$$112 \text{ g CaO} \times \frac{1 \text{ mol CaO}}{56.077 \text{ g CaO}} \times \frac{2 \text{ mol } NH_3}{1 \text{ mol CaO}} = 4.00 \text{ mol } NH_3$$

The balanced equation says: 1 mol CaO produces 1 mol $CaCl_2$.

$$112 \text{ g CaO} \times \frac{1 \text{ mol CaO}}{56.077 \text{ g CaO}} \times \frac{1 \text{ mol } CaCl_2}{1 \text{ mol CaO}} = 2.00 \text{ mol } CaCl_2$$

The balanced equation says: 1 mol CaO reacts with 2 mol NH_4Cl.

$$112 \text{ g CaO} \times \frac{1 \text{ mol CaO}}{56.077 \text{ g CaO}} \times \frac{2 \text{ mol } NH_4Cl}{1 \text{ mol CaO}} = 4.00 \text{ mol } NH_4Cl \text{ reacted}$$

$$224 \text{ g } NH_4Cl \times \frac{1 \text{ mol } NH_4Cl}{53.4913 \text{ g } NH_4Cl} = 4.19 \text{ mol } NH_4Cl \text{ present initially}$$

$$4.19 \text{ mol } NH_4Cl \text{ initial} - 4.00 \text{ mol } NH_4Cl \text{ reacted} = 0.19 \text{ mol } NH_4Cl \text{ left}$$

✓ *Reasonable Answer Check:* The excess moles of NH_4Cl prove that the right limiting reactant was determined, since there are zero moles of CaO left. The moles of products are stoichiometrically appropriate multiples of 2.00 moles, as they should be.

72. *Answer:* **1.40 kg Fe**

Strategy and Explanation: Given a balanced chemical equation and the masses of both reactants, determine the maximum amount of product that can be produced.

Use a metric conversion and the molar mass of the reactants to find the moles of the reactant substances present initially. Then use the stoichiometry of the equation to determine the moles of one of the products. Identify the limiting reactant from the reactant that produces the least number of products. From the limiting reactant quantity, determine the moles of the product, then use the molar mass of the product to get the grams and a metric conversion to get kilograms.

The balanced equation says: 1 mol Fe_2O_3 produces 2 mol Fe.

$$2.00 \text{ kg } Fe_2O_3 \times \frac{1000 \text{ g } Fe_2O_3}{1 \text{ kg } Fe_2O_3} \times \frac{1 \text{ mol } Fe_2O_3}{159.6882 \text{ g } Fe_2O_3} \times \frac{2 \text{ mol Fe}}{1 \text{ mol } Fe_2O_3} = 25.0 \text{ mol Fe}$$

The balanced equation says: 3 mol CO produces 2 mol Fe.

$$2.00 \text{ kg CO} \times \frac{1000 \text{ g CO}}{1 \text{ kg CO}} \times \frac{1 \text{ mol CO}}{28.0101 \text{ g CO}} \times \frac{2 \text{ mol Fe}}{3 \text{ mol CO}} = 47.6 \text{ mol Fe}$$

The number of Fe moles produced from Fe_2O_3 is smaller (25.0 mol < 47.6 mol), so Fe_2O_3 is the limiting reactant and CO is the excess reactant.

Find the mass of the Fe from the limiting reactant:

$$25.0 \text{ mol Fe} \times \frac{55.845 \text{ g Fe}}{1 \text{ mol Fe}} \times \frac{1 \text{ kg Fe}}{1000 \text{ g Fe}} = 1.40 \text{ kg Fe}$$

✓ *Reasonable Answer Check:* The reactants are present in the same mass quantities, but Fe_2O_3 has a larger molar mass, so it makes sense that it is the limiting reactant. The mass of iron should be smaller than the mass of iron (III) oxide.

74. *Answer:* Graph A

Strategy and Explanation: Graph A is the best representation of the formation of SO_3 in this Question because the mass of the product SO_3 increase (starting from zero) as S is added to the O_2 After a certain amount of S is added all the O_2 will run out, at which time the mass of SO_3 will remain unchanged. This is represented by a line that starts at zero, then increases linearly until a certain point, then becomes a flat straight line after that point.

75. *Answer:* **(a) CH_4 (b) 200 g (c) 700 g**

Strategy and Explanation: Given a balanced chemical equation and the masses of both reactants, determine the limiting reactant, the mass of the product produced, and the mass remaining of the excess reactant when the reaction is complete.

For part (a): Use the molar mass of the reactants to find the moles of the reactant substances. Then use the stoichiometry of the equation to determine the moles of the product produced in each case. Identify the limiting reactant from the reactant that produces the least number of products. For part (b): Use the moles of product produced from the limiting reactant and the molar mass of the product to find the grams. For part (c): From the quantity of limiting reactant, determine the moles of the other reactant needed for complete reaction. Convert that number to grams using the molar mass. Then subtract the quantity used from the initial mass given to get the mass of excess reactant.

Notice that 500 grams has 1 significant figure and 1300 grams has 2 significant figures.

(a) The balanced equation says: 1 mol CH_4 produces 3 mol H_2.

$$500 \text{ g } CH_4 \times \frac{1 \text{ mol } CH_4}{16.0423 \text{ g } CH_4} \times \frac{3 \text{ mol } H_2}{1 \text{ mol } CH_4} = 90 \text{ mol } H_2$$

The balanced equation says: 1 mol H_2O produces 3 mol H_2.

$$1300 \text{ g } H_2O \times \frac{1 \text{ mol } H_2O}{18.0152 \text{ g } H_2O} \times \frac{3 \text{ mol } H_2}{1 \text{ mol } H_2O} = 220 \text{ mol } H_2$$

The number of H_2 moles produced from CH_4 is smaller (90 mol < 220 mol), so CH_4 is the limiting reactant and H_2O is the excess reactant.

(b) Find the mass H_2:

$$90 \text{ mol } H_2 \times \frac{2.0158 \text{ g } H_2}{1 \text{ mol } H_2} = 200 \text{ g } H_2$$

(c)

$$500 \text{ g } CH_4 \times \frac{1 \text{ mol } CH_4}{16.0423 \text{ g } CH_4} \times \frac{1 \text{ mol } H_2O}{1 \text{ mol } CH_4} \times \frac{18.0152 \text{ g } H_2O}{1 \text{ mol } H_2O} = 600 \text{ g } H_2O$$

1300 g H_2O initial – 600 g H_2O used up = 700 g H_2O remains unreacted

✓ *Reasonable Answer Check:* The small number of significant figures makes the uncertainty in this calculation somewhat high. However, the general expectations are still met. There is a larger mass of H_2O present, and the molar masses of the reactants are about the same size. Since the equation indicates that equal moles of H_2O and CH_4 react, it makes sense that CH_4 is the limiting reactant. In addition, the calculation in (c) proves that the initial mass of H_2O was larger than required to react with all the CH_4.

Percent Yield (Section 4.6)

76. *Answer:* **699 g, 93.5%**

Strategy and Explanation: Given the balanced chemical equation, the mass of the limiting reactant and the actual yield, determine the theoretical yield and the percent yield.

First, calculate the theoretical yield by determining the maximum mass of product that could have been made from the given quantity of reactant: Take the mass of the limiting reactant and convert it to moles. Then use stoichiometry to find the moles of product. Then convert to grams using molar mass. Take the given actual yield and divide by the calculated theoretical yield and multiply by 100% to get percent yield.

The limiting reactant is Fe_2O_3. From its mass, find the maximum grams of Fe that could be made. The mole ratio comes from the balanced equation.

$$1.00 \text{ kg } Fe_2O_3 \times \frac{1000 \text{ g } Fe_2O_3}{1 \text{ kg } Fe_2O_3} \times \frac{1 \text{ mol } Fe_2O_3}{159.688 \text{ g } Fe_2O_3} \times \frac{2 \text{ mol Fe}}{1 \text{ mol } Fe_2O_3} \times \frac{55.845 \text{ g Fe}}{1 \text{ mol Fe}} = 699 \text{ g Fe}$$

The given mass of Fe is the actual yield. Use these two masses to calculate the percent yield.

$$\frac{654 \text{ g Fe actual}}{699 \text{ g Fe theoretical}} \times 100\% = 93.5\% \text{ yield}$$

✓ *Reasonable Answer Check:* Close to the maximum quantity of iron was produced, so it makes sense that the percent yield is over ninety percent.

77. *Answer:* **86.3 g**

Strategy and Explanation: Given a balanced chemical equation, the volume of a liquid limiting reactant, and the density of the liquid, determine the maximum theoretical yield of a product.

Use the density of the reactant to find the grams of the reactant. Then use the molar mass of the reactant to find the moles of reactant. Then use the stoichiometry of the equation to determine moles of product. Then use the molar mass of the product to get the mass of the product.

$$25.0 \text{ mL } Br_2 \times \frac{3.1023 \text{ g } Br_2}{1 \text{ mL } Br_2} \times \frac{1 \text{ mol } Br_2}{159.808 \text{ g } Br_2} \times \frac{1 \text{ mol } Al_2Br_6}{3 \text{ mol } Br_2} \times \frac{533.387 \text{ g } Al_2Br_6}{1 \text{ mol } Al_2Br_6} = 86.3 \text{ g } Al_2Br_6$$

✓ *Reasonable Answer Check:* The units all cancel, and the numerical multipliers are larger than the dividers.

79. *Answer:* **56.0%**

Strategy and Explanation: Given the theoretical yield and the actual yield, determine the percent yield.

Divide the actual yield by the theoretical yield and multiply by 100% to get percent yield. *(Note that the balanced equation is given, too, but you don't need to use it to answer this question.)*

$$\frac{36.7 \text{ g CaO actual}}{65.5 \text{ g CaO theoretical}} \times 100\% = 56.0\% \text{ yield}$$

✓ *Reasonable Answer Check:* A little more than half the maximum quantity of quicklime was produced, so it makes sense that the percent yield is a little more than 50%.

81. *Answer:* **8.8%**

Strategy and Explanation: Given the balanced chemical equation, the mass of the limiting reactant and the actual yield, determine the percent yield.

First, we need to calculate the theoretical yield. We get this by determining the maximum amount of product that could have been made from the given quantities of reactant: Take the mass of the limiting reactant and convert it to moles. Then use stoichiometry to find the moles of product. Then convert to grams using molar mass. Take the actual yield and divide by the calculated theoretical yield and multiply by 100% to get percent yield.

The limiting reactant is H_2. From its mass, find the maximum grams of CH_3OH that could be made. The mole ratio comes from the balanced equation.

$$5.0\times10^3 \text{ g } H_2 \times \frac{1 \text{ mol } H_2}{2.0158 \text{ g } H_2} \times \frac{1 \text{ mol } CH_3OH}{2 \text{ mol } H_2} \times \frac{32.0417 \text{ g } CH_3OH}{1 \text{ mol } CH_3OH} = 4.0\times10^4 \text{ g } CH_3OH$$

The given mass of CH_3OH is the actual yield. Use these two masses to calculate the percent yield.

$$\frac{3.5\times10^3 \text{ g } CH_3OH \text{ actual}}{4.0\times10^4 \text{ g } CH_3OH \text{ theoretical}} \times 100\% = 8.8\% \text{ yield}$$

✓ *Reasonable Answer Check:* The actual mass produced is roughly a factor of ten less than the maximum quantity of ethanol produced, so it makes sense that the percent yield is around 10%.

83. *Answer:* **5.3 g SCl_2**

Strategy and Explanation: Given the balanced chemical equation, the desired mass of the product, and the percent yield, determine the amount of limiting reactant that must be used.

Interpret the percent yield as the relationship between the actual grams and theoretical grams. Use that relationship to find the theoretical yield mass of the product. Then use stoichiometry to find the moles of limiting reactant. Then, using molar mass, convert to actual grams of reactant needed.

The 51% percent yield tells us the following:

To make 51 grams of S_2Cl_2, we need to have enough limiting reactant to make 1.19 grams of S_2Cl_2.

$$1.19 \text{ g } S_2Cl_2 \text{ actual} \times \frac{100. \text{ g } S_2Cl_2}{51 \text{ g } S_2Cl_2 \text{ actual}} = 2.3 \text{ g } S_2Cl_2$$

Use this mass of product to determine what mass of limiting reactant to use.

$$2.3 \text{ g } S_2Cl_2 \times \frac{1 \text{ mol } S_2Cl_2}{135.036 \text{ g } S_2Cl_2} \times \frac{3 \text{ mol } SCl_2}{1 \text{ mol } S_2Cl_2} \times \frac{102.971 \text{ g } SCl_2}{1 \text{ mol } S_2Cl_2} = 5.3 \text{ g } SCl_2$$

✓ *Reasonable Answer Check:* The yield suggests that we need to try to make about twice as much. The stoichiometry is 3:1, so it makes sense that a larger mass of SCl_2 is needed than the mass of S_2Cl_2 formed.

Empirical Formulas (Section 4.7)

86. *Answer:* **SO_3**

Strategy and Explanation: Given percent mass of elements in a compound, determine the empirical formula.

Choose a convenient sample mass of S_xO_y, such as 100.0 g. Find the mass of S and O in the sample, using the given mass percent. Use the molar mass of the elements to find their moles, then use a whole-number mole ratio to determine the empirical formula.

100.0 g of .S_xO_y contains 40.0 g S and 60.0 g O.

$$40.0 \text{ g S} \times \frac{1 \text{ mol S}}{32.065 \text{ g S}} = 1.25 \text{ mol S} \qquad 60.0 \text{ g O} \times \frac{1 \text{ mol O}}{15.9994 \text{ g O}} = 3.75 \text{ mol O}$$

Set up mole ratio and simplify by dividing by the smallest number of moles:

1.25 mol S : 3.75 mol O

1 S : 3 O

Use the whole number ratio for the subscripts in the formula. The empirical formula is SO_3.

✓ *Reasonable Answer Check:* The percent by mass of oxygen in the compound is somewhat greater than the percent by mass of sulfur. Since sulfur atoms weigh more than oxygen atoms, it makes sense that there are several O atom per S atom present in this formula. SO_3 is a common oxide of sulfur.

88. *Answer:* **CH**

Strategy and Explanation: Given the mass of a compound, the identity of the elements in the compound, and the identity and masses of all the products produced, determine the empirical formula.

Use the molar mass of the products to find their moles and use the stoichiometry of their formulas to determine the moles of the elements in the compound. Then use a whole-number mole ratio to determine the empirical formula.

The compound is a hydrocarbon and contains only C and H: C_iH_j. Its combustion produced H_2O and CO_2. Use molar mass and formula stoichiometry to determine the moles of C and H.

$$1.481 \text{ g } CO_2 \times \frac{1 \text{ mol } CO_2}{44.0095 \text{ g } CO_2} \times \frac{1 \text{ mol C}}{1 \text{ mol } CO_2} = 0.03365 \text{ mol C}$$

$$0.303 \text{ g } H_2O \times \frac{1 \text{ mol } H_2O}{18.0152 \text{ g } H_2O} \times \frac{2 \text{ mol H}}{1 \text{ mol } H_2O} = 0.0336 \text{ mol H}$$

Set up mole ratio and simplify by dividing by the smallest number of moles:

$$0.03365 \text{ mol C} : 0.0336 \text{ mol H}$$

$$1 \text{ C} : 1 \text{ H}$$

Use the whole number ratio for the subscripts in the formula. The empirical formula is CH.

✓ *Reasonable Answer Check:* These are not necessary calculations, but we can calculate the masses of each element, C and H.

$$0.03365 \text{ mol C} \times \frac{12.0107 \text{ g C}}{1 \text{ mol C}} = 0.4041 \text{ g C} \qquad 0.0336 \text{ mol H} \times \frac{1.0079 \text{ g H}}{1 \text{ mol H}} = 0.0339 \text{ g H}$$

The sum of the masses (0.4041 g + 0.0339 g = 0.4380 g) adds up to the mass of the compound (0.438 g).

90. *Answer:* $\mathbf{C_3H_6O_2}$

Strategy and Explanation: Given the mass of a compound, the identity of the elements in the compound, and the identity and masses of all the products produced, determine the empirical formula.

Combustion uses oxygen. When the compound also contains oxygen, determine the amount of oxygen after the other elements. Use the molar mass of the products to find their moles and use the stoichiometry of their formulas to determine the moles of the elements that are not oxygen in the compound. Find the masses of those elements, and subtract them from the total mass of the compound to get the mass of oxygen in the compound. Then use the molar mass to calculate the moles of oxygen in the compound. Then use a whole-number mole ratio to determine the empirical formula.

The compound contains C, H, and O: $C_iH_jO_k$. Its combustion produced H_2O and CO_2. Use molar mass and formula stoichiometry to determine the moles of C and H. Use the whole number ratio for the subscripts in the formula.

$$0.421 \text{ g } CO_2 \times \frac{1 \text{ mol } CO_2}{44.0095 \text{ g } CO_2} \times \frac{1 \text{ mol C}}{1 \text{ mol } CO_2} = 9.56 \times 10^{-3} \text{ mol C}$$

$$0.172 \text{ g } H_2O \times \frac{1 \text{ mol } H_2O}{18.0152 \text{ g } H_2O} \times \frac{2 \text{ mol H}}{1 \text{ mol } H_2O} = 1.91 \times 10^{-2} \text{ mol H}$$

Calculate the masses of C and H.

$$9.56 \times 10^{-3} \text{ mol C} \times \frac{12.0107 \text{ g C}}{1 \text{ mol C}} = 0.115 \text{ g C} \qquad 0.0191 \text{ mol H} \times \frac{1.0079 \text{ g H}}{1 \text{ mol H}} = 0.0192 \text{ g H}$$

Calculate the masses of O by subtracting the masses of C and H from the given total compound mass.

$$0.236 \text{ g } C_iH_jO_k - 0.115 \text{ g C} - 0.0192 \text{ g H} = 0.102 \text{ g O}$$

Calculate the moles of O. $\quad 0.102 \text{ g O} \times \frac{1 \text{ mol O}}{15.9994 \text{ g O}} = 6.37 \times 10^{-3} \text{ mol O}$

Set up mole ratio and simplify by dividing by the smallest number of moles:

$$9.56 \times 10^{-3} \text{ mol C} : 1.91 \times 10^{-2} \text{ mol H} : 6.37 \times 10^{-3} \text{ mol O}$$

$$1.5 \text{ C} : 3 \text{ H} : 1 \text{ O}$$

Multiply by 2, to get a whole number ratio: 3 C : 6 H: 2 O

Use the whole number ratio for the subscripts in the formula.

The empirical formula is $C_3H_6O_2$.

✓ *Reasonable Answer Check:* The mole ratio came out very close to whole number values.

94. *Answer:* **(a) $C_9H_{11}NO_4$ (b) $C_9H_{11}NO_4$**

Strategy and Explanation: Given the percent mass of elements in an organic compound and the compounds molar mass, determine the empirical formula and the molecular formula.

Choose a convenient sample mass of product, $C_xH_yN_zO_w$, such as 100.00 g. Find the mass of C and H in the sample, using the given mass percent, then subtract those masses from the total sample mass to get the mass of O. Use the molar mass of the elements to find their moles, then use a whole-number mole ratio to determine the empirical formula. For combustion, use the formula of the organic compound and O_2 as reactants and CO_2 and H_2O as products, then balance the equation.

(a) 100.00 g of .$C_xH_yN_zO_w$ contains 54.82 g C, 7.10 g N, and 32.46 g O.

$$100.00 \text{ g of .}C_xH_yN_zO_w - 54.82 \text{ g C} - 7.10 \text{ g N} - 32.46 \text{ g O} = 5.62 \text{ g H}$$

$$54.82 \text{ g C} \times \frac{1 \text{ mol C}}{12.0107 \text{ g C}} = 4.564 \text{ mol C} \qquad 7.10 \text{ g N} \times \frac{1 \text{ mol N}}{14.0067 \text{ g N}} = 0.507 \text{ mol N}$$

$$32.46 \text{ g O} \times \frac{1 \text{ mol O}}{15.9994 \text{ g O}} = 2.029 \text{ mol O} \qquad 5.62 \text{ g H} \times \frac{1 \text{ mol H}}{1.0079 \text{ g H}} = 5.58 \text{ mol H}$$

Set up mole ratio and simplify by dividing by the smallest number of moles:

4.564 mol C : 5.58 mol H : 0.507 mol N : 2.029 mol O

9 C : 11 H : 1 N : 4 O

Use the whole number ratio for the subscripts in the formula. The empirical formula is $C_9H_{11}NO_4$.

(b) The molar mass of the empirical formula = 197.1875 g/mol $C_9H_{11}NO_4$

$$\frac{197.19 \text{ g/mol compound}}{197.1875 \text{ g/mol emp. formula}} = 1 \text{ emp. formula/compound}$$

The molecular formula is $C_9H_{11}NO_4$.

✓ *Reasonable Answer Check:* The mole ratio is quite close to whole number values. The empirical formula is very close to the molecular formula.

General Questions

95. *Answer:* **21.6 g N_2**

Strategy and Explanation: Given the unbalanced chemical equation and the mass of the limiting reactant, determine the mass of a product that can be isolated.

Balance the equation. Use the molar mass of the reactant to find moles of reactant, then use the stoichiometry of the equation to determine moles of product, then use the molar mass of the product to get the mass of the product.

Select a balance order: H, O, N, Cu

$$2 \, NH_3 + 3 \, CuO \longrightarrow N_2 + 3 \, Cu + 3 \, H_2O$$

Check the balance: 2 N, 6 H, 3 Cu, 3 O

$$26.3 \text{ g } NH_3 \times \frac{1 \text{ mol } NH_3}{17.0304 \text{ g } NH_3} \times \frac{1 \text{ mol } N_2}{2 \text{ mol } NH_3} \times \frac{28.0134 \text{ g } N_2}{1 \text{ mol } N_2} = 21.6 \text{ g } N_2$$

✓ *Reasonable Answer Check:* It makes sense that the initial mass of NH_3 is nearly the same as the mass of N_2 created, since the NH_3 molar mass is about half that of N_2 and the stoichiometry is 2:1.

97. *Answer:* **Element (b)**

Strategy and Explanation: Given the mass of the only reactant, the formula of the reactant with one element unknown, a balanced equation showing its decomposition, and the mass and identity of one of the products, determine the identity of the unknown element from a given list.

Use the molar mass of the product to find moles of product, then use the stoichiometry of the equation to determine moles of reactant. Then divide the grams of reactant by the calculated moles of reactant to determine the molar mass of the reactant. Subtract the molar masses of the known elements in the compound to determine the molar mass of the unknown element. Compare the molar masses of the elements on the list and find the one that most closely matches it.

Use mass of CO_2 to find moles of MCO_3

$$0.376 \text{ g } CO_2 \times \frac{1 \text{ mol } CO_2}{44.0095 \text{ g } CO_2} \times \frac{1 \text{ mol } MCO_3}{1 \text{ mol } CO_2} = 8.54 \times 10^{-3} \text{ mol } MCO_3$$

Divide given mass by moles: $\dfrac{1.056 \text{ g } MCO_3}{8.54 \times 10^{-3} \text{ mol } MCO_3} = 124 \dfrac{\text{g}}{\text{mol}} = \text{molar mass of } MCO_3$

Subtract the molar mass of C and three times the molar mass of O from this molar mass to find the molar mass of M:

$$\frac{124 \text{ g } MCO_3}{\text{mol } MCO_3} - \frac{12.0107 \text{ g C}}{\text{mol } MCO_3} - \frac{3 \times 15.9994 \text{ g O}}{\text{mol } MCO_3} = 64 \text{ g M}$$

The element in the given list with the closest molar mass to 64 g/mol is (b) Cu.

✓ *Reasonable Answer Check:* The molar mass of Cu (63.5 g/mol) is 64. None of the others in the list (Ni at 58.7 g/mol, Zn at 65.4 g/mol, or Ba at 137.2 g/mol) are this close, though within ±1 g/mol Zinc almost qualifies. If we carry more decimal places than strictly allowed by the rules of significant figures, the molar mass of MCO_3 is 123.65 g/mol and the molar mass of M is 63.6 g/mol. It might make us feel better picking Cu, although the additional significant figures are unknown with the limited data provided.

99. *Answer:* **12.5 g $Pt(NH_3)_2Cl_2$**

Strategy and Explanation: Given the balanced chemical equation and the mass of one reactant and the moles of the other reactant, determine the maximum theoretical mass of a product.

Find the limiting reactant by determining moles of a product made from each reactant. Then using the molar mass of the product, determine the mass from the moles produced by the limiting reactant.

$$15.5 \text{ g } (NH_4)_2PtCl_4 \times \frac{1 \text{ mol } (NH_4)_2PtCl_4}{372.967 \text{ g } (NH_4)_2PtCl_4} \times \frac{1 \text{ mol } Pt(NH_3)_2Cl_2}{1 \text{ mol } (NH_4)_2PtCl_4} = 0.0416 \text{ mol } Pt(NH_3)_2Cl_2$$

$$0.15 \text{ mol } NH_3 \times \frac{1 \text{ mol } Pt(NH_4)_2Cl_2}{2 \text{ mol } NH_3} = 0.075 \text{ mol } Pt(NH_4)_2Cl_2$$

$(NH_4)_2PtCl_4$ is the limiting reactant, since fewer moles of $Pt(NH_3)_2Cl_2$ are produced from the $(NH_4)_2PtCl_4$ reactant (0.0416 mol) than produced from the NH_3 reactant (0.075 mol). Therefore, use 0.0416 mol $Pt(NH_3)_2Cl_2$ and the molar mass to determine the grams of $Pt(NH_3)_2Cl_2$ produced:

$$0.0416 \text{ mol } Pt(NH_3)_2Cl_2 \times \frac{300.0448 \text{ g } Pt(NH_3)_2Cl_2}{1 \text{ mol } Pt(NH_3)_2Cl_2} = 12.5 \text{ g } Pt(NH_3)_2Cl_2$$

✓ *Reasonable Answer Check:* It is satisfying that $(NH_4)_2PtCl_4$ is the limiting reactant, because that reactant contains the expensive platinum metal and the experimenter would want to make sure all of that got used up. Since the molar mass of the product is somewhat smaller than the molar mass of the limiting reactant and their stoichiometry is 1:1, it makes sense that a slightly smaller mass of product is formed in this reaction.

101. *Answer:* **SiH_4**

Strategy and Explanation: Given the mass of a compound, the identity of the elements in the compound, and the identity and masses of all the products produced, determine the empirical formula.

Use the molar mass of the products to find their moles and use the stoichiometry of their formulas to determine the moles of the elements in the compound. Then use a whole-number mole ratio to determine the empirical formula.

The compound is Si_xO_y. Its combustion produced SiO_2 and H_2O. Use molar mass and formula stoichiometry to determine the moles of Si and H.

$$11.64 \text{ g } SiO_2 \times \frac{1 \text{ mol } SiO_2}{60.0843 \text{ g } SiO_2} \times \frac{1 \text{ mol Si}}{1 \text{ mol } SiO_2} = 0.1937 \text{ mol Si}$$

$$6.980 \text{ g } H_2O \times \frac{1 \text{ mol } H_2O}{18.0152 \text{ g } H_2O} \times \frac{2 \text{ mol H}}{1 \text{ mol } H_2O} = 0.7749 \text{ mol H}$$

Set up mole ratio and simplify by dividing by the smallest number of moles: 0.1937 mol Si : 0.7749 mol H

$$1 \text{ Si} : 4.000 \text{ H}$$

Use the whole number ratio for the subscripts in the formula. The empirical formula is SiH_4.

✓ *Reasonable Answer Check:* These are not necessary calculations, but we can calculate the masses of each element, Si and H.

$$0.1937 \text{ mol Si} \times \frac{28.0855 \text{ g Si}}{1 \text{ mol Si}} = 5.440 \text{ g Si} \qquad 0.7749 \text{ mol H} \times \frac{1.0079 \text{ g H}}{1 \text{ mol H}} = 0.7810 \text{ g H}$$

The sum of these masses (5.440 g + 0.7810 g = 6.221 g) add up to the mass of the original compound (6.22 g), to the given significant figures.

103. *Answer:* **KOH, KOH**

Strategy and Explanation: Given equal moles of all the reactants, determine the limiting reactant. Given equal masses of all the reactants, determine the limiting reactant.

When the same moles of every reactant is present, the reactant with the largest stoichiometric coefficient is going to be the limiting reactant. In the second part, use the molar mass of each reactant to find their moles, then use the stoichiometry of the reaction to determine the moles of a product formed from each reactant. The reactant that produces the least amount of product is the limiting reactant.

First question: 5 mol of each reactant is present. The stoichiometric coefficients are: 4 for KOH, 2 for MnO_2, 1 for O_2 and 1 for Cl_2. The reactant with the largest stoichiometric coefficient is KOH. So, KOH is the limiting reactant.

Second question: 5 grams of each reactant is present. Determine the moles of KCl they each form:

$$5 \text{ g KOH} \times \frac{1 \text{ mol KOH}}{56.1056 \text{ g KOH}} \times \frac{2 \text{ mol KCl}}{4 \text{ mol KOH}} = 0.04 \text{ mol KCl}$$

$$5 \text{ g } MnO_2 \times \frac{1 \text{ mol } MnO_2}{86.9368 \text{ g } MnO_2} \times \frac{2 \text{ mol KCl}}{2 \text{ mol } MnO_2} = 0.06 \text{ mol KCl}$$

$$5 \text{ g } O_2 \times \frac{1 \text{ mol } O_2}{31.9988 \text{ g } O_2} \times \frac{2 \text{ mol KCl}}{1 \text{ mol } O_2} = 0.3 \text{ mol KCl}$$

$$5 \text{ g } Cl_2 \times \frac{1 \text{ mol } Cl_2}{70.906 \text{ g } Cl_2} \times \frac{2 \text{ mol KCl}}{1 \text{ mol } Cl_2} = 0.1 \text{ mol KCl}$$

Only 0.04 mol of HCl is produced from KOH, so the KOH is the limiting reactant here, too.

✓ *Reasonable Answer Check:* The first question is easy. The reaction says that more moles of KOH are needed that any other reactant, so when equal moles of reactants are present, it runs out first. The differences in molar mass are insufficient to keep the large stoichiometric coefficient for KOH from making it the limiting reactant when equal masses of the reactants are present.

Applying Concepts

105. Two butane molecules react with 13 diatomic oxygen molecules to produce eight carbon dioxide molecules and ten water molecules.

Two mol of gaseous butane molecules react with 13 mol of gaseous diatomic oxygen molecules to produce eight mol of gaseous carbon dioxide molecules and ten mol of liquid water molecules.

107. *Answer:* **A_3B**

Strategy and Explanation: Balance the equation for this reaction: $4\ A_2 + AB_3 \longrightarrow 3\ \_\_\_\_$

Reactant number of A atoms = 4 × 2 + 1 = 9 Product number of A atoms = 9 = 3 × 3

Reactant number of B atoms = 3 Product number of A atoms = 3 = 3 × 1

So, the product molecule has 3 A atoms and 1 B atom. Its formula is A_3B.

✓ *Reasonable Answer Check:* This product balances the equation: $4\ A_2 + AB_3 \longrightarrow 3\ A_3B$

109. *Answer:* **Ag^+, Cu^{2+}, and NO_3^-**

Strategy and Explanation: Given the moles of all the reactants, determine the limiting reactant and the ions present at the end of the reaction.

Use the stoichiometry of the equation to determine the moles of a product formed from each reactant. The reactant that produces the least amount of product is the limiting reactant. The excess reactant and the products are present at the end of the reaction. From that information, determine what ions are in the solution after the reaction is complete.

$$1.5 \text{ mol Cu} \times \frac{1 \text{ mol Cu(NO}_3)_2}{1 \text{ mol Cu}} = 1.5 \text{ mol Cu(NO}_3)_2$$

$$4.0 \text{ mol AgNO}_3 \times \frac{1 \text{ mol Cu(NO}_3)_2}{2 \text{ mol AgNO}_3} = 2.0 \text{ mol Cu(NO}_3)_2$$

Only 1.5 mol $Cu(NO_3)_2$ can be formed, so the limiting reactant is Cu and the excess reactant is $AgNO_3$.

That means the solution contains: Ag^+ ions, Cu^{2+} ions, and NO_3^- ions after the reaction is over. (The question only asked what ions are present not how many.)

If we want to know how many, we need to determine the moles of $AgNO_3$ actually reacted:

$$1.5 \text{ mol Cu} \times \frac{2 \text{ mol AgNO}_3}{1 \text{ mol Cu}} = 3.0 \text{ mol AgNO}_3 \text{ reacted}$$

The mol of $AgNO_3$ left = 4.0 mol $AgNO_3$ initial – 3.0 $AgNO_3$ mol reacted = 1.0 mol $AgNO_3$ left

Quantitatively, the solution has 1.0 mol Ag^+ ions, 1.5 mol Cu^{2+} ions, and (1.0 mol + 2 × 1.5 mol =) 4.0 mol NO_3^- ions.

✓ *Reasonable Answer Check:* The excess reactant is an ionic compound and one of the products is an ionic compound, so some of the ions present at the beginning are still present when the Cu runs out. The quantities of Ag^+ ions have dropped, since some of the silver atoms are now part of the Ag solid product. The quantity of NO_3^- ions doesn't change, since neither N nor O are part of the solid product formed.

112. *Answer:* **Equation (b)**

Strategy and Explanation: Four XY_3 molecules are made from two diatomic X_2 molecules and six diatomic Y_2 molecules. So, symbolically, the reaction is $2\ X_2 + 6\ Y_2 \longrightarrow 4\ XY_3$, and the stoichiometric equation representing that reaction is (b) $X_2 + 3\ Y_2 \longrightarrow 2\ XY_3$.

114. *Answer:* $A_2 + 4\ BC_2 \longrightarrow 4\ C_2 + 2\ AB_2$

Strategy and Explanation: Three A_2 molecules react with 12 BC_2 molecules to make 12 diatomic C_2 molecules and six AB_2 molecules. So, symbolically, the reaction is $3\ A_2 + 12\ BC_2 \longrightarrow 12\ C_2 + 6\ AB_2$ Dividing each coefficient by 3, gives the smallest integer coefficients: $A_2 + 4\ BC_2 \longrightarrow 4\ C_2 + 2\ AB_2$.

116. *Answer:* **When the metal mass is less than 1.2 g, the metal is the limiting reactant. When the metal mass is greater than 1.2 g, the bromine is the limiting reactant.**

Strategy and Explanation: When masses smaller than 1.2 gram of the metal are added, the metal is the limiting reactant, so the mass of the compound produced is directly proportional to the mass of metal present (shown by the straight line rising up to the right on the graph). More metal makes more products when the mass of metal is less than 1.2 gram.

When the mass larger than 1.2 g of metal are added, the bromine is the limiting reactant, so the particular mass of the metal is independent of how much compound is made. Since the amount of bromine is held constant, the mass of compound formed is also a constant (shown by a horizontal line on the graph).

More Challenging Questions

118. *Answer:* $H_2(g) + 3\ Fe_2O_3(s) \longrightarrow H_2O(\ell) + 2\ Fe_3O_4(s)$

Strategy and Explanation: Given the formulas of the reactants and one product and the percent mass of elements in the second product, determine the balanced chemical equation.

Choose a convenient sample mass of product, Fe_xO_y, such as 100.0 g. Find the mass of Fe and O in the sample, using the given mass percent. Use the molar mass of the elements to find their moles, then use a whole-number mole ratio to determine the empirical formula. Using the formulas of the reactants and products balance the equation.

100.0 g of .Fe_xO_y contains 72.3 g Fe and 27.7 g O.

$$72.3\ \text{g Fe} \times \frac{1\ \text{mol Fe}}{55.845\ \text{g Fe}} = 1.29\ \text{mol Fe} \qquad 27.7\ \text{g O} \times \frac{1\ \text{mol O}}{15.9994\ \text{g O}} = 1.73\ \text{mol O}$$

Set up mole ratio and simplify by dividing by the smallest number of moles:

1.29 mol Fe : 1.73 mol O

1 Fe : 1.34 O

Multiply by 3 to get whole numbers: 3 Fe : 4 O

Use the whole number ratio for the subscripts in the formula. The empirical formula is Fe_3O_4.

$?\ H_2(g) + ?\ Fe_2O_3(s) \longrightarrow ?\ H_2O(\ell) + ?\ Fe_3O_4(s)$	Select order: Fe, O, H
$?\ H_2(g) + 3\ Fe_2O_3(s) \longrightarrow ?\ H_2O(\ell) + 2\ Fe_3O_4(s)$	6 Fe
$?\ H_2(g) + 3\ Fe_2O_3(s) \longrightarrow 1\ H_2O(\ell) + 2\ Fe_3O_4(s)$	9 O
$1\ H_2(g) + 3\ Fe_2O_3(s) \longrightarrow 1\ H_2O(\ell) + 2\ Fe_3O_4(s)$	2 H

✓ *Reasonable Answer Check:* Fe_3O_4 is a common oxide of iron. 6 Fe, 9 O and 2 H on each side.

120. *Answer:* **(a) 2/1 (b) C**

Strategy and Explanation: Given the formula of two binary oxide compounds both containing an unknown element and an unknown amount of oxygen, and given the mass of the both elements in a given sample of each compound, we need to determine the ratio of oxygen atoms in the two formulas. Given the number of oxygen atoms in one formula, determine the identity of the unknown element.

We have incomplete information in this problem, so we'll let the ratio help us eliminate the need to know information common between the two substances. Start by converting the mass of O in each sample to moles of O. We then use the mass of A and its molar mass to determine moles of A, then use the formula stoichiometry to determine the moles of AO_x. Dividing the moles of O by the moles of AO_x tells us the value of x. Follow the same procedure to get y. We don't know molar mass of A, but it will cancel in the ratio.

(a) Find moles of O: $3.2\text{ g O} \times \frac{1\text{ mol O}}{15.9994\text{ g O}} = 0.20\text{ mol O}$

Find moles of AO_x, using M_A for the molar mass of unknown element A:

$$1.2\text{ g A} \times \frac{1\text{ mol A}}{M_A\text{ g A}} \times \frac{1\text{ mol }AO_x}{1\text{ mol A}} = \frac{1.2}{M_A}\text{ mol }AO_x$$

Find moles of AO_y in a similar fashion:

$$2.4\text{ g A} \times \frac{1\text{ mol A}}{M_A\text{ g A}} \times \frac{1\text{ mol }AO_y}{1\text{ mol A}} = \frac{2.4}{M_A}\text{ mol }AO_y$$

Now set up the ratio to get $\frac{x}{y}$: $\frac{\text{mol O}/\text{mol }AO_x}{\text{mol O}/\text{mol }AO_y}$

$$\frac{0.20\text{ mol O}\Big/\frac{1.2}{M_A}\text{ mol }AO_x}{0.20\text{ mol O}\Big/\frac{2.4}{M_A}\text{ mol }AO_y} = \frac{\frac{2\text{ mol O}}{1\text{ mol }AO_x}}{\frac{1\text{ mol O}}{1\text{ mol }AO_y}} \qquad \frac{x}{y} = \frac{2}{1}$$

(b) If x = 2, we can use the moles of O to tell use how many moles of A are present in the AO_x sample.

$$0.20\text{ mol O} \times \frac{1\text{ mol A}}{2\text{ mol O}} = 0.10\text{ mol A}$$

Divide mass of A in the sample by moles of A in the sample to get molar mass of A:

$$\frac{1.2\text{ g A}}{0.10\text{ mol A}} = 12\text{ g/mol}$$ The element with this molar mass is carbon, C.

✓ *Reasonable Answer Check:* (a) The law of multiple proportions assures us that the ratio of elements in binary compounds is a small whole number, so the ratio looks reaonable. There are probably many other legitimate ways to solve this problem. (b) The molar mass of the unknown element was very close to the molar mass of carbon, and both CO and CO_2 are known compounds, so this result is reasonable, also.

122. *Answer:* **44.9 amu**

Strategy and Explanation: Given the mass of a sample of a compound X_2S_3 that is then roasted to form a given mass of X_2O_3, determine the atomic mass of X.

Because all the X atoms from X_2S_3 end up in X_2O_3, then it must be true that each sample must contains the same number of moles of X. Because each compound has the same number of X atoms, it must be true that each sample must contain the same number of moles of compound. Using a variable, x, for the molar mass of X, write an expression that shows the calculation of moles for each compound and equate these two expressions. Solve for x.

Let x = molar mass of X and y = molar mass of Y

$$\text{mol of }X_2Y_3 = \text{mass of }X_2Y_3 \times \frac{1\text{ mol }X_2Y_3}{(2x + 3y)}$$

$$10.00\text{ g }X_2S_3 \times \frac{1\text{ mol }X_2S_3}{[2x + 3(32.065)\text{ g S}]} = 7.410\text{ g }X_2O_3 \times \frac{1\text{ mol }X_2O_3}{[2x + 3(15.9994)\text{ g O}]}$$

$$10.00 \times \frac{1}{(2x + 96.195)} = 7.410 \times \frac{1}{(2x + 47.9982)}$$

$$10.00\,(2x + 47.9982) = 7.410\,(2x + 96.195)$$

$$20.00\,x + 479.982 = 14.82\,x + 712.805$$

$$20.00\,x - 14.82\,x = 712.805 - 479.982$$

$$5.18\,x = 232.813$$

$$x = 44.9 \text{ g/mol}$$

$$\text{Atomic mass} = 44.9 \text{ amu}$$

✓ *Reasonable Answer Check:* The calculated atomic mass is very close to that of Sc, which does form 3+ ions and would combine with 2- ions like O^{2-} and S^{2-} to form Sc_2S_3 and Sc_2O_3.

124. *Answer:* **0 g $AgNO_3$, 9.82 g Na_2CO_3, 6.79 g Ag_2CO_3, 4.19 g $NaNO_3$**

Strategy and Explanation: Given the names of products and reactants of a chemical reaction, and the masses of the two reactants, determine the masses of all the products and reactants after the reaction is complete.

Determine the formulas of the reactants and products and set up and balance the chemical equation. Use the molar mass of the reactants to find the moles of the reactant substances. Then use the stoichiometry of the equation to determine the moles of the product produced in each case. Identify the limiting reactant from the reactant that produces the least number of products. Use the moles of product produced from the limiting reactant, the molar masses of the products to find the grams of products. Use the moles of product produced, determine the moles of the other reactant needed for complete reaction. Convert that number to grams using the molar mass. Then subtract the quantity used from the initial mass given to get the mass of excess reactant.

Silver nitrate is $AgNO_3$. Sodium carbonate is Na_2CO_3. Silver carbonate is Ag_2CO_3. Sodium nitrate is $NaNO_3$.

$? \, AgNO_3 + ? \, Na_2CO_3 \longrightarrow ? \, Ag_2CO_3 + ? \, NaNO_3$ Order: Ag, Na, N, C, O

$2 \, AgNO_3 + ? \, Na_2CO_3 \longrightarrow 1 \, Ag_2CO_3 + ? \, NaNO_3$ 2 Ag

$2 \, AgNO_3 + 1 \, Na_2CO_3 \longrightarrow 1 \, Ag_2CO_3 + 2 \, NaNO_3$ 2 Na, 2 N, 1 C, 9 O

The balanced equation says: 1 mol Na_2CO_3 produces 1 mol Ag_2CO_3.

$$12.43 \text{ g } Na_2CO_3 \times \frac{1 \text{ mol } Na_2CO_3}{105.9885 \text{ g } Na_2CO_3} \times \frac{1 \text{ mol } Ag_2CO_3}{1 \text{ mol } Na_2CO_3} = 0.1173 \text{ mol } Ag_2CO_3$$

The balanced equation says: 2 mol $AgNO_3$ produces 1 mol Ag_2CO_3.

$$8.37 \text{ g } AgNO_3 \times \frac{1 \text{ mol } AgNO_3}{169.8731 \text{ g } AgNO_3} \times \frac{1 \text{ mol } Ag_2CO_3}{2 \text{ mol } AgNO_3} = 0.0246 \text{ mol } Ag_2CO_3$$

The number of Ag_2CO_3 moles produced from $AgNO_3$ is smaller ($0.0246 \text{ mol} < 0.1173 \text{ mol}$), so $AgNO_3$ is the limiting reactant and Na_2CO_3 is the excess reactant. Therefore, at the end of the reaction, the mass of $AgNO_3$ present is zero grams.

Find the mass Ag_2CO_3: $$0.0246 \text{ mol } Ag_2CO_3 \times \frac{275.7453 \text{ g } Ag_2CO_3}{1 \text{ mol } Ag_2CO_3} = 6.79 \text{ g } Ag_2CO_3$$

Find the mass $NaNO_3$: $$0.0246 \text{ mol } Ag_2CO_3 \times \frac{2 \text{ mol } NaNO_3}{1 \text{ mol } Ag_2CO_3} \times \frac{84.9947 \text{ g } NaNO_3}{1 \text{ mol } NaNO_3} = 4.19 \text{ g } NaNO_3$$

Find the mass Na_2CO_3 used up, then subtract from initial for mass unreacted:

$$0.0246 \text{ mol } Ag_2CO_3 \times \frac{1 \text{ mol } Na_2CO_3}{1 \text{ mol } Ag_2CO_3} \times \frac{105.9885 \text{ g } Na_2CO_3}{1 \text{ mol } Na_2CO_3} = 2.61 \text{ g } Na_2CO_3 \text{ used up}$$

$$12.43 \text{ g } Na_2CO_3 \text{ initial} - 2.61 \text{ g } Na_2CO_3 \text{ used up} = 9.82 \text{ g } Na_2CO_3 \text{ remains unreacted}$$

✓ *Reasonable Answer Check:* The total mass before the reaction = 12.43 g + 8.37 g = 20.80 g is equal to the total mass after the reaction = 6.79 g + 4.19 g + 9.82 g = 20.80 g, in accordance with the conservation of mass.

126. *Answer:* **99.7% CH_3OH, 0.3% C_2H_5OH**

Strategy and Explanation: Given the mass of a sample of a liquid containing unknown amounts of two organic compounds and given the mass of one of the products of a chemical reaction, determine the composition of the liquid.

First, balance the combustion equation for ethyl alcohol and methyl alcohol. Set two variables X = the mass in grams of methyl alcohol and Y = the mass in grams of ethyl alcohol. Then establish two equations, one describing the total mass of the sample related to the two components and one describing the moles of carbon dioxide produced when burning the sample. Algebraically solve for X and Y, then use those masses and the total sample mass to find the percentage by mass of the two compounds.

Unbalanced: $? \ C_2H_5OH(\ell) + ? \ O_2(g) \longrightarrow ? \ CO_2(g) + ? \ H_2O(\ell)$ Order: C, H, then O

$1 \ C_2H_5OH(\ell) + ? \ O_2(g) \longrightarrow 2 \ CO_2(g) + ? \ H_2O(\ell)$ 2 C's

$1 \ C_2H_5OH(\ell) + ? \ O_2(g) \longrightarrow 2 \ CO_2(g) + 3 \ H_2O(\ell)$ 6 H's

$1 \ C_2H_5OH(\ell) + 6 \ O_2(g) \longrightarrow 2 \ CO_2(g) + 3 \ H_2O(\ell)$ 7 O's

Unbalanced: $? \ CH_3OH(\ell) + ? \ O_2(g) \longrightarrow ? \ CO_2(g) + ? \ H_2O(\ell)$ Order: C, H, then O

$1 \ CH_3OH(\ell) + ? \ O_2(g) \longrightarrow 1 \ CO_2(g) + ? \ H_2O(\ell)$ 1 C's

$1 \ CH_3OH(\ell) + ? \ O_2(g) \longrightarrow 1 \ CO_2(g) + 2 \ H_2O(\ell)$ 4 H's

$1 \ CH_3OH(\ell) + \frac{3}{2} \ O_2(g) \longrightarrow 1 \ CO_2(g) + 2 \ H_2O(\ell)$ 4 O's

Let X = grams of ethyl alcohol and Y = grams of methyl alcohol

Total mass of sample = 0.280 g = X + Y

Moles of carbon dioxide produced when burned $= 0.385 \text{ g } CO_2 \times \frac{1 \text{ mol } CO_2}{44.0095 \text{ g } CO_2} = 0.00875 \text{ mol } CO_2$

$$0.00875 \text{ mol } CO_2 = X \text{ g } C_2H_5OH \times \frac{1 \text{ mol } C_2H_5OH}{46.0682 \text{ g } C_2H_5OH} \times \frac{2 \text{ mol } CO_2}{1 \text{ mol } C_2H_5OH}$$

$$+ Y \text{ g } CH_3OH \times \frac{1 \text{ mol } CH_3OH}{32.0417 \text{ g } CH_3OH} \times \frac{1 \text{ mol } CO_2}{1 \text{ mol } CH_3OH}$$

$0.280 = X + Y$ and $0.00875 = 0.0434139 \ X + 0.0312093 \ Y$

Solve for X:

$$0.00875 = 0.0434139 \ X + 0.0312093(0.280 - X)$$

$$0.00875 = 0.0434139 \ X + 0.000874 - 0.0312093 \ X$$

$$0.00875 - 0.000874 = (0.0434139 - 0.0312093) \ X$$

$$0.00001 = 0.0122046 \ X$$

$$X = 0.0008 \text{ g ethyl alcohol}$$

$$Y = 0.280 - X = 0.280 - 0.0008 = 0.279 \text{ g methyl alcohol}$$

$$\frac{0.279 \text{ g } CH_3OH}{0.280 \text{ g liquid}} \times 100\ \% = 99.7\ \%\ CH_3OH \qquad \frac{0.001 \text{ g } C_2H_5OH}{0.280 \text{ g liquid}} \times 100\ \% = 0.3\ \%\ C_2H_5OH$$

✓ *Reasonable Answer Check:* Assuming the liquid is pure methyl alcohol gives the following mass of CO_2:

$$0.280 \text{ g } CH_3OH \times \frac{1 \text{ mol } CH_3OH}{32.0417 \text{ g } CH_3OH} \times \frac{1 \text{ mol } CO_2}{1 \text{ mol } CH_3OH} \times \frac{44.0095 \text{ g } CO_2}{1 \text{ mol } CO_2} = 0.385 \text{ g } CO_2$$

So it makes sense that the liquid is almost one hundred percent methanol.

Chapter 5: Chemical Reactions

Solutions for Blue-Numbered Questions for Review and Thought

Topical Questions

Solubility (Section 5.1)

11. *Answer:* **(a) soluble, Fe^{2+} and ClO_4^- (b) soluble, Na^+ and SO_4^{2-} (c) soluble, K^+ and Br^- (d) soluble, Na^+ and CO_3^{2-}**

Strategy and Explanation: Given a compound's formula, determine whether it is water-soluble.

Identify the cation and the anion in the salt, then use Table 5.1 on page 163.

Notice: Any rule that applies is sufficient to determine the compound's solubility. For example, if Rule 1 applies, there is no need to look for other rules related to the anion.

Notice: The Question asks which ions are present, not how many. When answering this Question, numbers that were part of the compound's formula stoichiometry are not included.

(a) $Fe(ClO_4)_2$ contains iron(II) ion and perchlorate ion

Rule 6: "All perchlorates are soluble." $Fe(ClO_4)_2$ is soluble.

$Fe(ClO_4)_2$ ionizes to form Fe^{2+} and ClO_4^-.

(b) Na_2SO_4 contains sodium ion (Group 1A) and sulfate ion

Rule 1: "All ... sodium ... salts are soluble." Na_2SO_4 is soluble.

Na_2SO_4 ionizes to form Na^+ and SO_4^{2-}.

(c) KBr contains potassium ion (Group 1A) and bromide ion

Rule 1: "All ... potassium ... salts are soluble." KBr is soluble.

KBr ionizes to form K^+ and Br^-.

(d) Na_2CO_3 contains sodium ion (Group 1A) and carbonate ion

Rule 1: "All ... sodium ... salts are soluble." Na_2CO_3 is soluble.

Na_2CO_3 ionizes to form Na^+ and CO_3^{2-}.

13. *Answer:* **(a) soluble, K^+ and HPO_4^{2-} (b) soluble, Na^+ and ClO^- (c) soluble, Mg^{2+} and Cl^- (d) soluble, Ca^{2+} and OH^- (e) soluble, Al^{3+} and Br^-**

Strategy and Explanation: Given a compound's formula, determine whether it is water-soluble.

Identify the cation and the anion in the salt, then use Table 5.1 on page 163.

Notice: Any rule that applies is sufficient to determine the compound's solubility. For example, if Rule 1 applies, there is no need to look for other rules related to the anion.

Notice: The Question asks which ions are present, not how many. When answering this Question, numbers that were part of the compound's formula stoichiometry are not included.

(a) Potassium hydrogen phosphate is K_2HPO_4.

Rule 1: "All ... group 1A ... salts are soluble." Potassium is in group 1A, so potassium hydrogen phosphate is soluble. It ionizes to form potassium ion, K^+, and hydrogen phosphate ion, HPO_4^{2-}.

(b) Sodium hypochlorite is NaOCl.

Rule 1: "All ... group 1A ... salts are soluble." Sodium is in group 1A, so sodium hypochlorite is soluble.

It ionizes to form sodium ion, Na^+, and hypochlorite ion, ClO^-.

(c) Magnesium chloride is $MgCl_2$.

Rule 3: "All common... chlorides ... are soluble, except AgCl, Hg_2Cl_2, and $PbCl_2$..." Magnesium chloride is soluble.

It ionizes to form magnesium ion, Mg^{2+}, and chloride ion, Cl^-.

(d) Calcium hydroxide is $Ca(OH)_2$.

Rule 10: " ... $Ca(OH)_2$ is slightly soluble." That means it dissolves only slightly, but the ions that will be formed are calcium ion, Ca^{2+}, and hydroxide ion, OH^-.

(e) Aluminum bromide is $AlBr_3$.

Rule 3: "All common... bromides ... are soluble, except AgBr, Hg_2Br_2, and $PbBr_2$..." Aluminum bromide is soluble. It ionizes to form aluminum ion, Al^{3+}, and bromide ion, Br^-.

15. *Answer:* **Box (b)**

Strategy and Explanation: Calcium chloride has a 1:2 ion stoichiometry in the formula: $CaCl_2$ due to the differences in their charges. So, locate the box with twice as many pink atoms (Cl^-) as blue atoms (Ca^{2+}). **Box (b)** shows 16 pink atoms and eight blue atoms.

17. *Answer:* **Box (b)**

Strategy and Explanation: Calcium phosphate has a 3:2 ion stoichiometry in the formula: $Ca_3(PO_4)_2$ due to the differences in their charges. So, locate the box with two pink atoms (PO_4^{3-}) for every three blue atoms (Ca^{2+}). **Box (b)** shows 12 pink atoms and 18 blue atoms giving a 2:3 ratio.

Exchange Reactions (Sections 5.1 & 5.2)

19. Nitric acid is HNO_3 and calcium hydroxide is $Ca(OH)_2$. Every OH^- ion reacts with one H^+ ion to make one water molecule, so we will need two nitric acid molecules to neutralize the hydroxide in calcium hydroxide.

$$\mathbf{2\ HNO_3(aq) + Ca(OH)_2(aq) \longrightarrow 2\ H_2O(\ell) + Ca(NO_3)_2(aq)}$$

21. *Answer:* **(a) $MnCl_2(aq) + Na_2S(aq) \longrightarrow MnS(s) + 2\ NaCl(aq)$ (b) no precipitate ("NP")**
(c) no precipiate ("NP") (d) $Hg(NO_3)_2(aq) + Na_2S(aq) \longrightarrow HgS(s) + 2\ NaNO_3(aq)$
(e) $Pb(NO_3)_2(aq) + 2\ HCl(aq) \longrightarrow PbCl_2(s) + 2\ HNO_3(aq)$
(f) $BaCl_2(aq) + H_2SO_4(aq) \longrightarrow BaSO_4(s) + 2\ HCl(aq)$

Strategy and Explanation: Given the reactants' formulas, determine if a precipitation reaction occurs. If so, write the balanced equation for the reaction. Create appropriate products of the possible exchange reaction, then check Table 5.1 on page 163 for insoluble products that would form a precipitate, then balance the equation.

Notice: Common acids (listed in Table 5.2 on page 170) are soluble.

(a) Possible exchange products are: MnS (Rule 13) Insoluble solid precipitate

NaCl (Rule 1) Soluble

$$MnCl_2(aq) + Na_2S(aq) \longrightarrow MnS(s) + 2\ NaCl(aq)$$

(b) Possible exchange products are: $Cu(NO_3)_2$ (Rule 2) Soluble

H_2SO_4 (Table 5.2) Soluble strong electrolyte

$$HNO_3(aq) + CuSO_4(aq) \longrightarrow \text{“NP.”}$$

(c) Possible exchange products are: H_2O Molecular liquid; no significant ionization

$NaClO_4$ (Rule 1) Soluble

$$NaOH(aq) + HClO_4(aq) \longrightarrow \text{“NP.”}$$

(d) Possible exchange products are: HgS (Rule 13) Insoluble solid precipitate

$NaNO_3$ (Rule 1) Soluble

$$Hg(NO_3)_2(aq) + Na_2S(aq) \longrightarrow HgS(s) + 2\ NaNO_3(aq)$$

(e) Possible exchange products are: $PbCl_2$ (Rule 3 exception) Insoluble solid precipitate

HNO_3 (Table 5.2) Soluble strong electrolyte

$$Pb(NO_3)_2(aq) + 2\ HCl(aq) \longrightarrow PbCl_2(s) + 2\ HNO_3(aq)$$

(f) Possible exchange products are: $BaSO_4$ (Rule 4 exception) Insoluble solid precipitate

HCl (Table 5.2) Soluble strong electrolyte

$$BaCl_2(aq) + H_2SO_4(aq) \longrightarrow BaSO_4(s) + 2\ HCl(aq)$$

✓ *Reasonable Answer Check:* All equations with insoluble products are identified as precipitation reactions. Those equations with all products soluble are identified with “NP.” Notice that, in both of the “NP” cases above, a reaction **does** occur; they are, however, not precipitation reactions.

23. *Answer:* **(a) CuS insoluble; $Cu^{2+} + H_2S(aq) \longrightarrow CuS(s) + 2\ H^+$; spectator ion is Cl^-. (b) $CaCO_3$ insoluble; $Ca^{2+} + CO_3^{2-} \longrightarrow CaCO_3(s)$; spectator ions are K^+ and Cl^-. (c) AgI insoluble; $Ag^+ + I^- \longrightarrow AgI(s)$; spectator ions are Na^+ and NO_3^-.**

Strategy and Explanation: Given overall chemical equations, identify the water-insoluble products, write the net ionic equations, and determine the spectator ions.

Use the solubility rules to determine the actual physical state of the products. Identify the water-insoluble product. Use the solubility rules to determine the actual physical state of the reactants. Write the complete ionic equation by writing ionized forms for any soluble ionic compound and all strong acids and bases. Ions that remain completely unchanged in the complete ionic equation (same physical phase and same ionized form), i.e., any ions that are the same on the product side as on the reactant side, are eliminated to produce the net ionic equation. The ions eliminated are the spectator ions.

Notice: All soluble ionic compounds and strong acids and bases are ionized in the complete ionic equation. All solids, weak acids and bases, and molecular compounds remain un-ionized.

All ions have aqueous phase in the equations below.

(a) CuS is insoluble (Table 5.1, Rule 13), and HCl is a common strong acid (Table 5.2).

The water-insoluble product is CuS.

$CuCl_2$ is soluble (Table 5.1, Rule 3). H_2S is not a common strong acid (Table 5.2), so assume it is weak.

$$Cu^{2+} + 2\ Cl^- + H_2S(aq) \longrightarrow CuS(s) + 2\ H^+ + 2\ Cl^- \qquad \text{complete ionic eq.}$$

Eliminate Cl^- ions on each side:

$$Cu^{2+} + H_2S(aq) \longrightarrow CuS(s) + 2\ H^+ \qquad \text{net ionic equation}$$

The spectator ion is Cl^-.

(b) KCl is soluble (Table 5.1, Rule 1). $CaCO_3$ is insoluble (Table 5.1, Rule 9).

The water-insoluble product is $CaCO_3$.

$CaCl_2$ is soluble (Table 5.1, Rule 3). K_2CO_3 is soluble (Table 5.1, Rule 1).

$$Ca^{2+} + 2\ Cl^- + 2\ K^+ + CO_3^{2-} \longrightarrow CaCO_3(s) + 2\ K^+ + 2\ Cl^- \qquad \text{complete ionic eq.}$$

Eliminate K^+ and Cl^-

$$Ca^{2+} + CO_3^{2-} \longrightarrow CaCO_3(s) \quad \text{net ionic equation}$$

The spectator ions are K^+ and Cl^-.

(c) AgI is insoluble (Table 5.1, Rule 3). $NaNO_3$ is soluble (Table 5.1, Rule 1).

The water-soluble product is AgI.

$AgNO_3$ is soluble (Table 5.1, Rule 2). NaI is soluble (Table 5.1, Rule 1).

$$Ag^+ + NO_3^- + Na^+ + I^- \longrightarrow AgI(s) + Na^+ + NO_3^- \quad \text{complete ionic equation}$$

Eliminate Na^+ and NO_3^-.

$$Ag^+ + I^- \longrightarrow AgI(s) \quad \text{net ionic equation}$$

The spectator ions are Na^+ and NO_3^-.

✓ *Reasonable Answer Check:* All spectator ions have been identified and eliminated to make the net ionic equation. The atoms are all balanced and the charges are all balanced in each equation.

25. *Answer:* **Complete ionic: $2\ K^+ + CO_3^{2-} + Cu^{2+} + 2\ NO_3^- \longrightarrow CuCO_3(s) + 2\ K^+ + 2\ NO_3^-$;**

net ionic: $CO_3^{2-} + Cu^{2+} \longrightarrow CuCO_3(s)$; precipitate is copper(II) carbonate.

Strategy and Explanation: Use the method described in the solution to Question 23. Given the names of two reactants combined to form an unidentified precipitate, write complete and the net ionic equations for this reaction and name the precipitate.

Determine the formulas of the reactants, then determine the resulting products by rearranging the ions in the reactants to form new compounds. Balance the complete equation. Ionize the reactants and use the solubility rules to determine the actual physical state of the products. Ions that remain completely unchanged (same physical phase and same ionized form) are the spectator ions. Any ions that are the same on the product side as on the reactant side are eliminated to produce the net ionic equation. The insoluble product compound should then be named as the precipitate.

Notice: All soluble ionic compounds and strong acids and bases are ionized in the complete ionic equation. All solids, weak acids and bases, and molecular compounds remain un-ionized.

All ions have aqueous phase in the equations below.

Potassium carbonate is K_2CO_3 and copper(II) nitrate is $Cu(NO_3)_2$.

The cross combinations are copper (II) carbonate, $CuCO_3$, which is insoluble (according to rule 9: All carbonates are insoluble…) and potassium nitrate, KNO_3, which is soluble (according to rule 1: All Group 1A… ion compounds are soluble.)

balanced overall equation: $K_2CO_3\ (aq) + Cu(NO_3)_2(aq) \longrightarrow CuCO_3(s) + 2\ KNO_3(aq)$

$$2\ K^+ + CO_3^{2-} + Cu^{2+} + 2\ NO_3^- \longrightarrow CuCO_3(s) + 2\ K^+ + 2\ NO_3^- \quad \text{complete ionic equation}$$

Eliminate spectator ions, K^+, potassium ions and NO_3^-, nitrate ions.

$$CO_3^{2-} + Cu^{2+} \longrightarrow CuCO_3(s) \quad \text{net ionic equation}$$

The precipitate is copper(II) carbonate.

✓ *Reasonable Answer Check:* One insoluble compound was formed, as indicated in the question. All spectator ions have been identified and eliminated to make the net ionic equation. The atoms are all balanced and the charges are all balanced in each equation.

27. *Answer:* **(a) $Zn(s) + 2\ HCl(aq) \longrightarrow H_2(g) + ZnCl_2(aq)$; complete ionic: $Zn(s) + 2\ H^+ + 2\ Cl^- \longrightarrow H_2\ (g) + Zn^{2+} + 2\ Cl^-$; net ionic: $Zn(s) + 2\ H^+ \longrightarrow H_2(g) + Zn^{2+}$ (b) $Mg(OH)_2(s) + 2\ HCl(aq) \longrightarrow MgCl_2(aq) + 2\ H_2O(\ell)$; complete ionic: $Mg(OH)_2(s) + 2\ H^+ + 2\ Cl^- \longrightarrow Mg^{2+} + 2\ Cl^- + 2\ H_2O(\ell)$; net ionic: $Mg(OH)_2(s) + 2\ H^+ \longrightarrow Mg^{2+} + 2\ H_2O(\ell)$ (c) $2\ HNO_3(aq) + CaCO_3(s)$**

$\longrightarrow$ $Ca(NO_3)_2(aq) + H_2O(\ell) + CO_2(g)$; complete ionic: $2\ H^+ + 2\ NO_3^- + CaCO_3(s) \longrightarrow Ca^{2+} + 2\ NO_3^- + H_2O(\ell) + CO_2(g)$; net ionic: $2\ H^+ + CaCO_3(s) \longrightarrow Ca^{2+} + H_2O(\ell) + CO_2(g)$ (d) $4\ HCl(aq) + MnO_2(s) \longrightarrow MnCl_2(aq) + Cl_2(g) + 2\ H_2O(\ell)$; complete ionic: $4\ H^+ + 4\ Cl^- + MnO_2(s) \longrightarrow Mn^{2+} + 2\ Cl^- + Cl_2(g) + 2\ H_2O(\ell)$; net ionic: $4\ H^+ + 2\ Cl^- + MnO_2(s) \longrightarrow Mn^{2+} + Cl_2(g) + 2\ H_2O(\ell)$

Strategy and Explanation: Use the method described in the solution to Question 23. Given overall chemical equations, balance them, then write complete and net ionic equations.

Use standard balancing techniques to balance the overall equation. Use the solubility rules to determine the actual physical state of the reactants and products. Write complete ionic equations using that information. Ions that remain completely unchanged (same physical phase and same ionized form) are the spectator ions. Any ions that are the same on the product side as on the reactant side are eliminated to produce the net ionic equation.

Notice: Just like with balancing elements, we can balance ions as units: Select a balancing order and follow it.

Notice: All soluble ionic compounds and strong acids and bases are ionized in the complete ionic equation. All solids, weak acids and bases, and molecular compounds remain un-ionized.

All ions have aqueous phase in the equations below.

(a) $Zn(s) + 2\ HCl(aq) \longrightarrow H_2(g) + ZnCl_2(aq)$

$Zn(s) + 2\ H^+ + 2\ Cl^- \longrightarrow H_2\ (g) + Zn^{2+} + 2\ Cl^-$ complete ionic equation

Eliminate spectator ion, Cl^-. $Zn(s) + 2\ H^+ \longrightarrow H_2(g) + Zn^{2+}$ net ionic equation

(b) $Mg(OH)_2(s) + 2\ HCl(aq) \longrightarrow MgCl_2(aq) + 2\ H_2O(\ell)$

$Mg(OH)_2(s) + 2\ H^+ + 2\ Cl^- \longrightarrow Mg^{2+} + 2\ Cl^- + 2\ H_2O(\ell)$complete ionic equation

Eliminate spectator ion, Cl^-. $Mg(OH)_2(s) + 2\ H^+ \longrightarrow Mg^{2+} + 2\ H_2O(\ell)$ net ionic equation

(c) $2\ HNO_3(aq) + CaCO_3(s) \longrightarrow Ca(NO_3)_2(aq) + H_2O(\ell) + CO_2(g)$

$2\ H^+ + 2\ NO_3^- + CaCO_3(s) \longrightarrow Ca^{2+} + 2\ NO_3^- + H_2O(\ell) + CO_2(g)$ complete ionic equation

Eliminate spectator ion, NO_3^-. $2\ H^+ + CaCO_3(s) \longrightarrow Ca^{2+} + H_2O(\ell) + CO_2(g)$ net ionic equation

(d) $4\ HCl(aq) + MnO_2(s) \longrightarrow MnCl_2(aq) + Cl_2(g) + 2\ H_2O(\ell)$

$4\ H^+ + 4\ Cl^- + MnO_2(s) \longrightarrow Mn^{2+} + 2\ Cl^- + Cl_2(g) + 2\ H_2O(\ell)$ complete ionic equation

Eliminate spectator ions, two of the four Cl^-.

$4\ H^+ + 2\ Cl^- + MnO_2(s) \longrightarrow Mn^{2+} + Cl_2(g) + 2\ H_2O(\ell)$ net ionic equation

✓ *Reasonable Answer Check:* All of the equations are balanced. The net ionic equations have no spectator ions. The atoms are all balanced and the charges are all balanced in each equation.

29. *Answer:* **(a) $Ca(OH)_2(s) + 2\ HNO_3(aq) \longrightarrow Ca(NO_3)_2(aq) + 2\ H_2O(\ell)$; complete ionic: $Ca(OH)_2(s) + 2\ H^+ + 2\ NO_3^- \longrightarrow Ca^{2+} + 2\ NO_3^- + 2\ H_2O(\ell)$; net ionic: $Ca(OH)_2(s) + 2\ H^+ \longrightarrow Ca^{2+} + 2\ H_2O(\ell)$ (b) $BaCl_2(aq) + Na_2CO_3(aq) \longrightarrow BaCO_3(s) + 2\ NaCl(aq)$; complete ionic: $Ba^{2+} + 2\ Cl^- + 2\ Na^+ + CO_3^{2-} \longrightarrow BaCO_3(s) + 2\ Na^+ + 2\ Cl^-$; net ionic: $Ba^{2+} + CO_3^{2-} \longrightarrow BaCO_3(s)$ (c) $2\ Na_3PO_4(aq) + 3\ Ni(NO_3)_2(aq) \longrightarrow Ni_3(PO_4)_2(s) + 6\ NaNO_3(aq)$; complete ionic: $6\ Na^+ + 2\ PO_4^{3-} + 3\ Ni^{2+} + 6\ NO_3^- \longrightarrow Ni_3(PO_4)_2(s) + 6\ Na^+ + 6\ NO_3^-$; net ionic: $2\ PO_4^{3-} + 3\ Ni^{2+} \longrightarrow Ni_3(PO_4)_2(s)$**

Strategy and Explanation: Given overall chemical equations, balance them, then write complete and net ionic equations. Use the method described in the solution to Question 23. All ions have aqueous phase in the equations below.

(a) $Ca(OH)_2$ is slightly soluble (Table 5.1, Rule 10). HNO_3 is a strong acid (Table 5.2). $Ca(NO_3)_2$ is soluble (Table 5.1, Rule 2). H_2O is a liquid molecular compound.

$$Ca(OH)_2(s) + 2\ HNO_3(aq) \longrightarrow Ca(NO_3)_2(aq) + 2\ H_2O(\ell) \quad \text{balanced}$$

$$Ca(OH)_2(s) + 2\ H^+ + 2\ NO_3^- \longrightarrow Ca^{2+} + 2\ NO_3^- + 2\ H_2O(\ell) \quad \text{complete ionic equation}$$

Eliminate spectator ion, NO_3^-. $Ca(OH)_2(s) + 2\ H^+ \longrightarrow Ca^{2+} + 2\ H_2O(\ell)$ net ionic equation

(b) $BaCl_2$ is soluble (Table 5.1, Rule 3). Na_2CO_3 is soluble (Table 5.1, Rule 1).

$BaCO_3$ is insoluble (Table 5.1, Rule 9), NaCl is soluble (Table 5.1, Rule 1).

$$BaCl_2(aq) + Na_2CO_3(aq) \longrightarrow BaCO_3(s) + 2\ NaCl(aq) \quad \text{balanced}$$

$$Ba^{2+} + 2\ Cl^- + 2\ Na^+ + CO_3^{2-} \longrightarrow BaCO_3(s) + 2\ Na^+ + 2\ Cl^- \quad \text{complete ionic equation}$$

Eliminate spectator ions, Cl^- and Na^+. $Ba^{2+} + CO_3^{2-} \longrightarrow BaCO_3(s)$ net ionic equation

(c) Na_3PO_4 is soluble (Table 5.1, Rule 1). $Ni(NO_3)_2$ is soluble (Table 5.1, Rule 2). $Ni_3(PO_4)_2$ is insoluble (Table 5.1, Rule 8). $NaNO_3$ is soluble (Table 5.1, Rule 1).

$$Na_3PO_4(aq) + Ni(NO_3)_2(aq) \longrightarrow Ni_3(PO_4)_2(s) + NaNO_3(aq) \quad \text{unbalanced}$$

Select balancing order: Ni^{2+}, then PO_4^{3-}, then NO_3^-, and Na^+

$$2\ Na_3PO_4(aq) + 3\ Ni(NO_3)_2(aq) \longrightarrow Ni_3(PO_4)_2(s) + 6\ NaNO_3(aq) \quad \text{balanced}$$

$$6\ Na^+ + 2\ PO_4^{3-} + 3\ Ni^{2+} + 6\ NO_3^- \longrightarrow Ni_3(PO_4)_2(s) + 6\ Na^+ + 6\ NO_3^- \quad \text{complete ionic equation}$$

Eliminate spectator ions, NO_3^- and Na^+. $2\ PO_4^{3-} + 3\ Ni^{2+} \longrightarrow Ni_3(PO_4)_2(s)$ net ionic equation

✓ *Reasonable Answer Check:* All of the balanced equations are balanced. The net ionic equations have no spectator ions. The atoms are all balanced and the charges are all balanced in each equation.

31. *Answer:* $\mathbf{Ba(OH)_2(aq) + 2\ HNO_3(aq) \longrightarrow Ba(NO_3)_2(aq) + 2\ H_2O(\ell)}$

Strategy and Explanation: Given the names of the reactants and one product of a reaction, balance the equation that occurs.

Use the solubility rules and acid and base identities (Table 5.1 on page 163 and Table 5.2 on page 170) to determine solubility of salts and strength of acids and bases to determine the actual physical state of the reactants and exchange products. Use standard balancing techniques to balance the overall equation.

Barium hydroxide is $Ba(OH)_2$, a soluble hydroxide compound (Table 5.1, Rule 10). Nitric acid is HNO_3, a strong acid (Table 5.2). Each OH^- ion in the solution reacts with one H^+ ion to make one H_2O molecule. The ionic compound produced during this neutralization is $Ba(NO_3)_2$. It is soluble (Table 5.1, Rule 10).

$$Ba(OH)_2(aq) + 2\ HNO_3(aq) \longrightarrow Ba(NO_3)_2(aq) + 2\ H_2O(\ell) \quad \text{balanced}$$

✓ *Reasonable Answer Check:* The equation is balanced. This acid-base neutralization reaction produces liquid water.

33. *Answer:* $\mathbf{CdCl_2(aq) + 2\ NaOH(aq) \longrightarrow Cd(OH)_2(s) + 2\ NaCl(aq)}$**; complete ionic:** $\mathbf{Cd^{2+} + 2\ Cl^- + 2\ Na^+ + 2\ OH^- \longrightarrow Cd(OH)_2(s) + 2\ Na^+ + 2\ Cl^-}$**; net ionic:** $\mathbf{Cd^{2+} + 2\ OH^- \longrightarrow Cd(OH)_2(s)}$

Strategy and Explanation: Given an overall chemical equation for a precipitation, balance it, then write complete and net ionic equations. Use the method described in the solution to Question 23. All ions have aqueous phase in the equations below.

$CdCl_2$ is soluble (Table 5.1, Rule 3). NaOH is soluble (Table 5.1, Rule 1). $Cd(OH)_2$ is insoluble (Table 5.1, Rule 10). NaCl is soluble (Table 5.1, Rule 1)

$$CdCl_2(aq) + 2\ NaOH(aq) \longrightarrow Cd(OH)_2(s) + 2\ NaCl(aq) \quad \text{balanced overall equation}$$

$$Cd^{2+} + 2\ Cl^- + 2\ Na^+ + 2\ OH^- \longrightarrow Cd(OH)_2(s) + 2\ Na^+ + 2\ Cl^- \quad \text{complete ionic equation}$$

Eliminate spectator ions, Cl^- and Na^+. $Cd^{2+} + 2\ OH^- \longrightarrow Cd(OH)_2(s)$ net ionic equation

✓ *Reasonable Answer Check:* The atoms and the charges in each equation are balanced. This precipitation reaction produces an insoluble solid.

35. *Answer:* **$Pb(NO_3)_2(aq) + 2\ KCl(aq) \longrightarrow PbCl_2(s) + 2\ KNO_3(aq)$; reactants: lead(II) nitrate and potassium chloride; products: lead(II) chloride and potassium nitrate**

Strategy and Explanation: Given the reactants of a precipitation reaction, balance the equation.

Use the method described in the solution to Question 33.

Lead(II) nitrate is $Pb(NO_3)_2$, a soluble compound (Table 5.1, Rule 2). Potassium chloride is KCl, a soluble compound (Table 5.1, Rule 1). The products of the reaction would be lead(II) chloride, $PbCl_2$, an insoluble compound (Table 5.1, Rule 3) and potassium nitrate, KNO_3, a soluble compound (Table 5.1, Rule 1).

$$Pb(NO_3)_2(aq) + 2\ KCl(aq) \longrightarrow PbCl_2(s) + 2\ KNO_3(aq) \quad \text{balanced overall equation}$$

✓ *Reasonable Answer Check:* The equation is balanced. This precipitation reaction produces an insoluble solid.

38. *Answer:* **(a) base, strong, K^+ and OH^- (b) base, strong, Mg^{2+} and OH^- (c) acid, weak, H^+ and ClO^- (d) acid, strong, H^+ and Br^- (e) base, strong, Li^+ and OH^- (f) acid, weak; H^+, HSO_3^-, and SO_3^{2-}**

Strategy and Explanation: Given some chemical formulas, identify if they are acids or bases, identify whether they are weak or strong, and determine what ions produce with they dissolve in water.

Acids produce H^+ in aqueous solution. Bases produce OH^- in aqueous solution. In several of these cases, Table 5.1 on page 163 or Table 5.2 on page 170 can be used to determine whether a large or small amount of ions are produced, by determining if the compound is soluble and/or weak or strong. In some cases, however, it is not possible to look up that information in Chapter 5. However, it is almost always true that, if an acid is **not** one of the common strong acids listed in Table 5.2, it is a **weak** acid. That is the case with the weak acids in this Question. If the acid or base is strong, it will ionize. If the acid or base is weak it will not ionize to a great extent, remaining primarily in the molecular form.

(a) KOH is a strong base. (Given in Table 5.2), producing K^+ and OH^- ions when dissolved in water.

(b) $Mg(OH)_2$ is an insoluble ionic compound (Table 5.1, Rule 10) so few ions are produced in water, though the $Mg(OH)_2$ that dissolves does ionize completely. So, practically, it may be considered weak, because the OH^- ion concentration will never be very large. Technically, it may be considered to be strong, because all the dissolved $Mg(OH)_2$ is ionized, producing Mg^{2+} and OH^- ions when dissolved in water.

(c) HClO is a weak acid. (It's not listed as a common strong acid in Table 5.2.). It will not ionize very much, remaining mostly in the HClO(aq) form. The small amount that does ionize will form H^+ and ClO^- ions.

(d) HBr is a strong acid (Given in Table 5.2), producing H^+ and Br^- ions when dissolved in water.

(e) LiOH is a strong base (Given in Table 5.2), producing Li^+ and OH^- ions when dissolved in water.

(f) H_2SO_3 is a weak acid (It's not listed as a common strong acid in Table 5.2.). It will not ionize very much, remaining mostly in the $H_2SO_3(aq)$ form. The small amount that does ionize will form H^+, HSO_3^-, and SO_3^{2-} ions.

✓ *Reasonable Answer Check:* All of the ions produced are common ions, most of them are found in Figure 3.2 and Table 3.7. Most of these are given specifically in Table 5.2. If you want to look far into your chemistry future, Table 16.2 on page 770 can also be used to confirm that the two weak acids given here in (c) and (f) are indeed weak.

40. *Answer:* **(a) HNO_2; NaOH; complete ionic: $HNO_2(aq) + Na^+ + OH^- \longrightarrow H_2O(\ell) + Na^+ + NO_2^-$; net ionic: $HNO_2(aq) + OH^- \longrightarrow H_2O(\ell) + NO_2^-$ (b) H_2SO_4; $Ca(OH)_2$; complete ionic & net ionic: $H^+ + HSO_4^- + Ca(OH)_2(s) \longrightarrow 2\ H_2O(\ell) + CaSO_4(s)$ (c) HI; NaOH; complete ionic: $H^+ + I^- + Na^+ + OH^- \longrightarrow H_2O(\ell) + Na^+ + I^-$; net ionic: $H^+ + OH^- \longrightarrow H_2O(\ell)$ (d) H_3PO_4; $Mg(OH)_2$; complete ionic & net ionic: $2\ H_3PO_4(aq) + 3\ Mg(OH)_2(s) \longrightarrow 6\ H_2O(\ell) + Mg_3(PO_4)_2(s)$**

Strategy and Explanation: Given some chemical formulas for salts, identify the acids and bases that would react to form them, then write the overall neutralization reaction both in complete and net ionic form.

The formation of a salt using neutralization comes from reacting an acid containing the salt's anion and a base containing the salt's cation. Set up the complete equation, then use the method described in previous questions to find the complete and net ionic equations.

(a) $NaNO_2$ is formed from the neutralization of HNO_2 (to supply the nitrite anion) and NaOH (to supply the sodium cation). HNO_2 is a weak acid. NaOH is a strong base. $NaNO_2$ is a soluble compound.

$HNO_2(aq) + NaOH(aq) \longrightarrow H_2O(\ell) + NaNO_2(aq)$ balanced overall equation

$HNO_2(aq) + Na^+ + OH^- \longrightarrow H_2O(\ell) + Na^+ + NO_2^-$ complete ionic equation

Eliminate spectator ion, Na^+. $HNO_2(aq) + OH^- \longrightarrow H_2O(\ell) + NO_2^-$ net ionic equation

(b) $CaSO_4$ is formed from the neutralization of H_2SO_4 (to supply the sulfate anion) and $Ca(OH)_2$ (to supply the calcium cation). The first ionization of H_2SO_4 is strong, but HSO_4^- is a weak acid. The reactant $Ca(OH)_2$ is only slightly soluble. The product, $CaSO_4$, is insoluble.

$H_2SO_4(aq) + Ca(OH)_2(s) \longrightarrow 2\ H_2O(\ell) + CaSO_4(s)$ balanced overall equation

$H^+ + HSO_4^- + Ca(OH)_2(s) \longrightarrow 2\ H_2O(\ell) + CaSO_4(s)$ complete ionic and net ionic equations

(c) NaI is formed from the neutralization of HI (to supply the iodide anion) and NaOH (to supply the sodium cation). HI is a strong acid and NaOH is a strong base. The product, NaI, is soluble.

$HI(aq) + NaOH(aq) \longrightarrow H_2O(\ell) + NaI(aq)$ balanced overall equation

$H^+ + I^- + Na^+ + OH^- \longrightarrow H_2O(\ell) + Na^+ + I^-$ complete ionic equation

Eliminate spectator ions, Na^+ and I^-. $H^+ + OH^- \longrightarrow H_2O(\ell)$ net ionic equation

(d) $Mg_3(PO_4)_2$ is formed from the neutralization of H_3PO_4 (to supply the phosphate anion) and $Mg(OH)_2$ (to supply the magnesium cation). H_3PO_4 is a weak acid. The reactant $Mg(OH)_2$ is insoluble. The product, $Mg_3(PO_4)_2$, is insoluble.

$$2\ H_3PO_4(aq) + 3\ Mg(OH)_2(s) \longrightarrow 6\ H_2O(\ell) + Mg_3(PO_4)_2(s)$$

balanced overall, complete ionic, and net ionic equations

✓ *Reasonable Answer Check:* These acids and bases undergo neutralization to produce the appropriate salt and water. The net ionic equation does not always includes the whole salt, when one or both of the ions of the salt are found to be spectator ions.

42. *Answer:* **(a) precipitation reaction; products are NaCl and MnS; $MnCl_2(aq) + Na_2S(aq) \longrightarrow 2\ NaCl(aq) + MnS(s)$ (b) precipitation reaction; products are NaCl and $ZnCO_3$; $Na_2CO_3(aq) + ZnCl_2(aq) \longrightarrow 2\ NaCl(aq) + ZnCO_3(s)$ (c) gas-forming reaction; products are $KClO_4$, H_2O and CO_2; $K_2CO_3(aq) + 2\ HClO_4(aq) \longrightarrow 2\ KClO_4(aq) + H_2O(\ell) + CO_2(g)$**

Strategy and Explanation: Given the reactants of reactions, classify the reaction that occurs, identify the products, and balance the equations.

To classify these reactions, determine the exchange products and check their solubility and/or strength using Tables 5.1 on page 163 and 5.2 on page 170. Remember that carbonate compounds reacting with acids produce CO_2 gas.

(a) When $MnCl_2$ reacts with Na_2S, the exchange products are NaCl and MnS. Checking their solubility, we find that NaCl is soluble, but MnS is insoluble. That makes this a **precipitation reaction.**

$MnCl_2(aq) + Na_2S(aq) \longrightarrow 2\ NaCl(aq) + MnS(s)$ balanced overall equation

(b) When Na_2CO_3 reacts with $ZnCl_2$, the exchange products are NaCl and $ZnCO_3$. Checking the solubility, we see that NaCl is soluble, but $ZnCO_3$ is insoluble, so this is a **precipitation reaction**.

$Na_2CO_3(aq) + ZnCl_2(aq) \longrightarrow 2\ NaCl(aq) + ZnCO_3(s)$ balanced overall equation

(c) When K_2CO_3 reacts with $HClO_4$, the exchange products are $KClO_4$ and H_2CO_3, which decomposes into liquid H_2O and CO_2 gas. That makes this a **gas-forming reaction**. $KClO_4$ is soluble.

$K_2CO_3(aq) + 2\ HClO_4(aq) \longrightarrow 2\ KClO_4(aq) + H_2O(\ell) + CO_2(g)$ balanced overall equation

✓ *Reasonable Answer Check:* Precipitation reactions form insoluble compounds and gas-forming reactions produce gases. The atoms are all balanced and the charges are all balanced in each equation.

Oxidation-Reduction Reactions (Sections 5.3 & 5.4)

44. *Answer:* **(a) Ox. # O = –2, Ox. # S = +6 (b) Ox. # O = –2, Ox. # H = +1, Ox. # N = +5 (c) Ox. # K = +1, Ox. # O = –2, Ox. # Mn = +7 (d) Ox. # O = –2, Ox. # H = +1 (e) Ox. # Li = +1, Ox. # O = –2, Ox. # H = +1 (f) Ox. # Cl is –1, Ox. # H = +1, Ox. # C = 0**

Strategy and Explanation: Given several formulas of compounds, determine the oxidation numbers of the atoms in each of them.

For this Question, the rules spelled out in Section 5.4 on page 183 are used extensively. Several elements in compounds have predictable oxidation numbers (Rules 1 to 3). The oxidation numbers of the remaining atom(s) can be determined using the sum rule (Rule 4). The term "oxidation number" is abbreviated below as "Ox. #".

(a) SO_3 contains oxygen. Rule 3 gives us Ox. # O = –2. This is a molecule, so the sum of the oxidation numbers is zero.

$$0 = 1 \times (\text{Ox. \# S}) + 3 \times (-2)$$

Therefore, Ox. # S = +6.

(b) HNO_3 contains oxygen and hydrogen. Rule 3 gives us Ox. # O = –2 and Ox. # H = +1. We use the sum rule to find the Ox. # N. This is a molecule, so the sum of the oxidation numbers is zero.

$$0 = 1 \times (+1) + 1 \times (\text{Ox. \# N}) + 3 \times (-2)$$

Therefore, Ox. # N = +5.

(c) $KMnO_4$ contains a monatomic cation, K^+, and a polyatomic anion, MnO_4^-. Rule 2 gives us Ox. # K = +1. Rule 3 gives us Ox. # O = –2. We use the sum rule to find the Ox. # Mn. For a polyatomic anion, the sum of the oxidation numbers is equal to the ion's charge of 1–.

$$-1 = 1 \times (\text{Ox. \# Mn}) + 4 \times (-2)$$

Therefore, Ox. # Mn = +7.

(d) H_2O contains oxygen and hydrogen. Rule 3 gives us Ox. # O = –2 and Ox. # H = +1.

(e) LiOH contains a monatomic cation, Li^+, and a diatomic anion, OH^-. Rule 2 gives us Ox. # Li = +1. Rule 3 gives us Ox. # O = –2 and Ox. # H = +1.

(f) CH_2Cl_2 contains carbon, hydrogen, and chlorine. Since chlorine makes a –1 ion when part of ionic compounds, we will assume its oxidation number is –1 (as was described in Section 5.4 on page 183 for the compound PCl_3). Rule 3 gives us Ox. # H = +1. We use the sum rule to find the Ox. # C. This is a molecule, so the sum of the oxidation numbers is zero.

$$0 = (\text{Ox. \# C}) + 2 \times (+1) + 2 \times (-1)$$

Therefore, Ox. # C = 0.

✓ *Reasonable Answer Check:* The non-metal elements farther to the right on the periodic table have consistently more negative oxidation numbers. The metallic elements and nonmetals farther to the left on the periodic table have more positive oxidation numbers.

46. *Answer:* **(a) Ox. # O = –2, Ox. # S = +6 (b) Ox. # O = –2, Ox. # N = +5 (c) Ox. # O = –2, Ox. # Mn = +7 (d) Ox. # O = –2, Ox. # H = +1, Ox. # Cr = +3 (e) Ox. # O = –2, Ox. # H = +1, Ox. # P = +5 (f) Ox. # O = –2, Ox. # S = +2**

Strategy and Explanation: Given several formulas of ions, determine the oxidation numbers of the atoms in each one.

For this Question, the rules spelled out in Section 5.4 on page 183 are used extensively. Several elements in compounds have predictable oxidation numbers (Rules 1-3). The oxidation numbers of the remaining atom(s) can be determined using the sum rule (Rule 4). The term "oxidation number" is abbreviated below as "Ox. #".

(a) SO_4^{2-}. Rule 3 gives us Ox. # O = –2. We use the sum rule to find the Ox. # S. This is a polyatomic anion, so the sum of the oxidation numbers is equal to the ion's charge of 2–.

$$-2 = 1 \times (\text{Ox. \# S}) + 4 \times (-2)$$

Therefore, Ox. # S = +6.

(b) NO_3^- contains oxygen. Rule 3 gives us Ox. # O = –2. We use the sum rule to find the Ox. # N. This is a polyatomic anion, so the sum of the oxidation numbers is equal to the ion's charge of 1–:

$$-1 = 1 \times (\text{Ox. \# N}) + 3 \times (-2)$$

Therefore, Ox. # N = +5.

(c) MnO_4^- contains oxygen. Rule 3 gives us Ox. # O = –2. We use the sum rule to find the Ox. # Mn. For a polyatomic anion, the sum of the oxidation numbers is equal to the ion's charge of 1–.

$$-1 = 1 \times (\text{Ox. \# Mn}) + 4 \times (-2)$$

Therefore, Ox. # Mn = +7.

(d) $Cr(OH)_4^-$ contains oxygen and hydrogen. Rule 3 gives us Ox. # O = –2 and Ox. # H = +1. We use the sum rule to find the Ox. # Cr. For a polyatomic anion, the sum of the oxidation numbers is equal to the ion's charge of 1–.

$$-1 = 1 \times (\text{Ox. \# Cr}) + 4 \times [1 \times (-2) + 1 \times (+1)]$$

Therefore, Ox. # Cr = +3.

(e) $H_2PO_4^-$ contains oxygen and hydrogen. Rule 3 gives us Ox. # O = –2 and Ox. # H = +1. We use the sum rule to find the Ox. # P. For a polyatomic anion, the sum of the oxidation numbers is equal to the ion's charge of 1–.

$$-1 = 2 \times (+1) + 1 \times (\text{Ox. \# P}) + 4 \times (-2)$$

Therefore, Ox. # P = +5.

(f) $S_2O_3^{2-}$ contains oxygen. Rule 3 gives us Ox. # O = –2. We use the sum rule to find the Ox. # S. For a polyatomic anion, the sum of the oxidation numbers is equal to the ion's charge of 2–.

$$-2 = 2 \times (\text{Ox. \# S}) + 3 \times (-2)$$

Therefore, 2 × (Ox. # S) = +4 and Ox. # S = +2.

✓ *Reasonable Answer Check:* The non-metal elements farther to the right on the periodic table have consistently more negative oxidation numbers. The metallic elements and nonmetals farther to the left on the periodic table have more positive oxidation numbers.

48. *Answer:* **(a) –1 (b) +1 (c) +3 (d) +5 (e) +7**

Strategy and Explanation: Given several formulas, determine the oxidation state of the Cl atoms in each one.

Use the rules spelled out in Section 5.4 on page 183. Use Rule 3 to determine the oxidation numbers of chlorine, when possible, and other elements, as needed.

The term "oxidation number" is abbreviated below as "Ox. #". Rule 3 gives us Ox. # O = –2 and Ox. # H = +1. This information is used in several parts of this question.

(a) HCl is a neutral compound, but. Rule 3 tells us that us Ox. # Cl = –1. We can check this by noting that Rule 3 also tells us that us Ox. # H = +1, and the sum rule confirms that the neutral charge (0) is the sum of the two oxidation numbers: 0 = –1 + 1.

(b) HClO is a neutral compound, so the sum of the oxidation numbers is equal to zero.

$$0 = 1 \times (+1) + 1 \times (\text{Ox. \# Cl}) + 1 \times (-2)$$

$$0 = +1 + \text{Ox. \# Cl} - 2 \qquad \text{Ox. \# Cl} = +1.$$

(c) $HClO_2$ is a neutral compound, so the sum of the oxidation numbers is equal to zero.

$$0 = 1 \times (+1) + 1 \times (\text{Ox. \# Cl}) + 2 \times (-2)$$

$$0 = +1 + \text{Ox. \# Cl} - 4 \qquad \text{Ox. \# Cl} = +3.$$

(d) $HClO_3$ is a neutral compound, so the sum of the oxidation numbers is equal to zero.

$$0 = 1 \times (+1) + 1 \times (\text{Ox. \# Cl}) + 3 \times (-2)$$

$$0 = +1 + \text{Ox. \# Cl} - 6 \qquad \text{Ox. \# Cl} = +5.$$

(e) $HClO_4$ is a neutral compound, so the sum of the oxidation numbers is equal to zero.

$$0 = 1 \times (+1) + 1 \times (\text{Ox. \# Cl}) + 4 \times (-2)$$

$$0 = +1 + \text{Ox. \# Cl} - 8 \qquad \text{Ox. \# Cl} = +7.$$

✓ *Reasonable Answer Check:* Because the oxygen atom carries a negative oxidation number, the chlorine atoms must ends up with positive oxidation numbers in all but (a).

50. *Answer:* **(a) –2 (b) 0 (c) +2 (d) +4 (e) +6**

Strategy and Explanation: Given several formulas, determine the oxidation state of the S atoms in each one.

For this Question, the rules spelled out in Section 5.4 on page 183 are used extensively. Several elements in compounds have predictable oxidation numbers (Rules 1-3). The oxidation number (which identifies the oxidation state) of S can be determined using the sum rule (Rule 4).

The term "oxidation number" is abbreviated below as "Ox. #".

(a) H_2S, a binary acid compound. Rule 3 gives us Ox. # H = +1. We use the sum rule to find the Ox. # S. This is a neutral acid compound, so the sum of the oxidation numbers is equal to zero.

$$0 = 2 \times (+1) + 1 \times (\text{Ox. \# S})$$

Therefore, Ox. # S = –2. Sulfur is in the –2 oxidation state.

(b) S_8, an elemental form of sulfur. Rule 1 gives us Ox. # S =0. Sulfur is in the zero oxidation state.

(c) SCl_2 is a binary compound containing the halogen chlorine. Rule 3 gives us Ox. # Cl = –1. We use the sum rule to find the Ox. # S. For a compound, the sum of the oxidation numbers is equal to zero.

$$0 = 1 \times (\text{Ox. \# S}) + 2 \times (-1)$$

Therefore, Ox. # S = +2. Sulfur is in the +2 oxidation state.

(d) SO_3^{2-} is an ion that contains oxygen. Rule 3 gives us Ox. # O = –2. We use the sum rule to find the Ox. # S. For a polyatomic anion, the sum of the oxidation numbers is equal to the ion's charge of 2–.

$$-2 = 1 \times (\text{Ox. \# S}) + 3 \times (-2)$$

Therefore, Ox. # S = +4. Sulfur is in the +4 oxidation state.

(e) K_2SO_4 contains potassium ion, K^+, and sulfate ion, SO_4^{2-}, which contains oxygen. Rule 2 gives us the Ox. # K = +1. Rule 3 gives us Ox. # O = –2. We use the sum rule to find the Ox. # S. For a compound, the sum of the oxidation numbers is equal to zero.

$$0 = 2 \times (+1) + (\text{Ox. \# S}) + 4 \times (-2)$$

Therefore, Ox. # S = +6. Sulfur is in the +6 oxidation state.

✓ *Reasonable Answer Check:* When sulfur is bonded to non-metal elements (such as O or Cl) farther to the right on the periodic table it ends up in a positive oxidation state. When it is bonded to hydrogen (farther to the left), it is in a negative oxidation state. When it is bonded to itself, the oxidation state is zero.

52. *Answer:* **Only reaction (b) is an oxidation-reduction reaction; oxidation numbers change. Reaction (a) is a precipitation reaction; reaction (c) is an acid-base neutralization reaction.**

Strategy and Explanation: Given several reactions, determine if they are oxidation-reduction reactions and classify the remaining reactions.

To decide if a reaction is an oxidation-reduction reaction, we need to see if any of the elements change oxidation state. Oxidation-reduction reactions are ones in which the atoms have different oxidation states before and after the reaction. If no change in oxidation state is observed, then it is not an oxidation-reduction reaction.

(a) The ionic compounds representing reactants and products in this reaction all contain Cd^{2+}, Cl^-, Na^+, and S^{2-} ions. Therefore, this is **not an oxidation-reduction reaction**. The formation of insoluble CdS classifies this reaction as a **precipitation reaction.**

(b) The elements Ca and O_2, both with zero oxidation states, are combined into an ionic compound, CaO, with Ca^{2+} and O^{2-} ions. That means the oxidation numbers did change and this is an **oxidation-reduction reaction**.

(c) The ionic compounds representing reactants contain Ca^{2+}, OH^-, H^+, and Cl^- ions. The product ionic compound contains Ca^{2+} and Cl^-. The other product, water, has O in the –2 oxidation state and H in the +1 oxidation state. Therefore, this is **not an oxidation-reduction reaction**. The formation of water from the reaction of OH^- and H^+ classifies this reaction as an **acid-base neutralization** reaction.

✓ *Reasonable Answer Check:* The reactions that are not oxidation-reduction reactions are easily classified as one of the other reactions we have studied in this chapter.

55. *Answer:* **Substances (b), (c), and (d)**

Strategy and Explanation: (b) O_2, (c) HNO_3 and (d) MnO_4^- are common oxidizing agents. They are all good at oxidizing other chemicals because they are readily reduced.

57. *Answer:* **(a) CO_2 or CO (b) PCl_3 or PCl_5 (c) $TiCl_2$ or $TiCl_4$ (d) Mg_3N_2 (e) Fe_2O_3 (f) NO_2**

Strategy and Explanation: Given reactants of oxidation-reduction reactions, identify the combination reactions' products.

A combination reaction is the production of a product by combining two other chemicals. In some cases, more than one product might form. In those cases, all possibilities have been included.

(a) $C(s) + O_2(g) \longrightarrow CO_2(g)$

$2\ C(s) + O_2(g) \longrightarrow 2\ CO(g)$

(b) $P_4(s) + 6\ Cl_2(g) \longrightarrow 4\ PCl_3(g)$

$P_4(s) + 10\ Cl_2(g) \longrightarrow 4\ PCl_5(g)$

(c) $Ti(s) + Cl_2(g) \longrightarrow TiCl_2(s)$

$Ti(s) + 2\ Cl_2(g) \longrightarrow TiCl_4(s)$

(d) $3\ Mg(s) + N_2(g) \longrightarrow Mg_3N_2(s)$

(e) $4\ FeO(s) + O_2(g) \longrightarrow 2\ Fe_2O_3(s)$

(f) $2\ NO + O_2(g) \longrightarrow 2\ NO_2(g)$

✓ *Reasonable Answer Check:* The single product formed is a combination of the two substances given.

59. *Answer:* **F_2 is the best oxidizing agent; I^- anion is the best reducing agent.**

Strategy and Explanation: F_2 is the strongest oxidizing agent, since it is most easily reduced. I^- anion is the strongest reducing agent, since it is the most easily oxidized.

61. *Answer:* **(b) Br_2 is reduced; NaI is oxidized; Br_2 is the oxidizing agent; NaI is the reducing agent. (c) F_2 is reduced; NaCl is oxidized; F_2 is the oxidizing agent; NaCl is the reducing agent. (d) Cl_2 is reduced; NaBr is oxidized; Cl_2 is the oxidizing agent; NaBr is the reducing agent.**

Strategy and Explanation: Take the reactions written in Question 60 and identify what substance is oxidized and what substance is reduced. The reactant that is losing electrons is oxidized. The reactant that is gaining electrons is reduced. The reactant reduced is the oxidizing agent and the reactant oxidized is the reducing agent. Notice: (a), (e) and (f) produced no reaction.

(b) Br_2 is gaining electrons, so it is reduced and represents the oxidizing agent. NaI is losing electrons, so it is oxidized and represents the reducing agent.

(c) F_2 is gaining electrons, so it is reduced and represents the oxidizing agent. NaCl is losing electrons, so it is oxidized and represents the reducing agent.

(d) Cl_2 is gaining electrons, so it is reduced and represents the oxidizing agent. NaBr is losing electrons, so it is oxidized and represents the reducing agent.

✓ *Reasonable Answer Check:* The more active elemental halogen is always reduced.

Activity Series (Section 5.5)

64. *Answer/Explanation:* An example of a displacement reaction that is also a redox reaction:

$$Fe(s) + 2\ HCl(aq) \longrightarrow FeCl_2(aq) + H_2(g)$$

(a) The species that is oxidized is Fe, since iron loses electrons in the half reaction:

$$Fe(s) \longrightarrow Fe^{2+}(aq) + 2\ e^-$$

(b) The species that is reduced is HCl, since hydrogen gains electrons in the half reaction:

$$2\ H^+(aq) + 2\ e^- \longrightarrow H_2(g)$$

(c) The species that is oxidized functions as the reducing agent, see (a): Fe(s)

(d) The species that is reduced functions as the oxidizing agent, see (b): HCl(aq)

Notice, many other answers are possible.

66. *Answer:* **(a) "NR." (b) "NR." (c) "NR." (d) 3 Ag^+(aq) + Au(s)**

Strategy and Explanation:

(a) Na is more active than Zn, so: $Na^+(aq) + Zn(s) \longrightarrow$ "NR."

(b) H_2 is more active than Pt, so: $HCl(aq) + Pt(s) \longrightarrow$ "NR."

(c) Ag is more active than Au, so: $Ag^+(aq) + Au(s) \longrightarrow$ "NR."

(d) Au is less active than Ag, so: $Au^{3+}(aq) + 3\ Ag(s) \longrightarrow 3\ Ag^+(aq) + Au(s)$

Solution Concentrations (Section 5.6)

68. *Answer:* **Ba^{2+} and Cl^-; 0.12 M Ba^{2+} and 0.24 M Cl^-**

Strategy and Explanation: Formula stoichiometry: 1 mol of $BaCl_2$ contains 1 mol Ba^{2+} ions and 2 mol Cl^- ions. So, the 0.12 M $BaCl_2$ solutions contains 0.12 M Ba^{2+} ion and 2 × (0.12 M) Cl^- ion = 0.24 M Cl^- ion.

70. *Answer:* **(a) 0.254 M Na_2CO_3 (b) 0.508 M Na^+, 0.254 M CO_3^{2-}**

Strategy and Explanation: Given the mass of the solute and the volume of the solution, find the molarity of the solute, and the concentrations of the ions.

Use the molar mass to find moles of solute, convert the volume to liters from milliliters, and divide the moles of solute by the volume in liters to get molarity. Use formula stoichiometry to find the concentrations of the ions.

(a) Find moles of solute: $6.73\ g\ Na_2CO_3 \times \dfrac{1\ mol\ Na_2CO_3}{105.9885\ g\ Na_2CO_3} = 6.35 \times 10^{-2}\ mol\ Na_2CO_3$

$$250.\ mL\ solution \times \frac{1\ L}{1000\ mL} = 0.250\ L\ solution$$

$$Molarity = \frac{6.35 \times 10^{-2}\ mol\ Na_2CO_3}{0.250\ L\ solution} = 0.254\ M\ Na_2CO_3$$

Notice that the three sequential calculations shown above can be combined into one calculation:

$$\frac{6.73\ \text{g}\ Na_2CO_3}{250.\ \text{mL solution}} \times \frac{1\ \text{mol}\ Na_2CO_3}{105.9885\ \text{g}\ Na_2CO_3} \times \frac{1000\ \text{mL}}{1\ \text{L}} = 0.254\ \text{M}\ Na_2CO_3$$

This combined version prevents having to write unnecessary intermediate answers and helps cut down on round-off errors in significant figures. Both ways give the right answer, but it is helpful to consolidate your work, when you can.

(b) Na_2CO_3 has a formula stoichiometry that looks like this:

$$1\ \text{mol}\ Na_2CO_3 : 2\ \text{mol}\ Na^+\ \text{ions}: 1\ \text{mol}\ CO_3^{2-}\ \text{ions}.$$

So, the 0.254 M Na_2CO_3 contains 2 × (0.254 M) Na^+ ion = 0.508 M Na^+ and 0.254 M CO_3^{2-}.

✓ *Reasonable Answer Check:* The number 0.0635 is about one quarter of .250; this value looks right. The concentration of Na^+ is twice than the concentration of CO_3^{2-}.

72. *Answer:* **0.494 g $KMnO_4$**

Strategy and Explanation: Given the volume of the solution and its molarity, find the mass of solute in the solution.

Convert the volume of solution to liters from milliliters. Then use the molarity as a conversion factor to determine moles of solute. Then use the molar mass to find grams of solute.

$$250.\ \text{mL solution} \times \frac{1\ \text{L}}{1000\ \text{mL}} \times \frac{0.0125\ \text{mol}\ KMnO_4}{1\ \text{L solution}} \times \frac{158.0339\ \text{g}\ KMnO_4}{1\ \text{mol}\ KMnO_4} = 0.494\ \text{g}\ KMnO_4$$

✓ *Reasonable Answer Check:* The units cancel appropriately. The relative size of the mass seems appropriate.

74. *Answer:* **5.08×10^3 mL**

Strategy and Explanation: Given the mass of the solute and the solution's molarity, find the volume of the solution.

Use the molar mass to find determine moles of solute. Then use the molarity as a conversion factor to determine volume of the solute in liters. Then convert the volume of solution to milliliters from liters.

$$25.0\ \text{g NaOH} \times \frac{1\ \text{mol NaOH}}{39.9971\ \text{g NaOH}} \times \frac{1\ \text{L solution}}{0.123\ \text{mol NaOH}} \times \frac{1000\ \text{mL}}{1\ \text{L}} = 5.08 \times 10^3\ \text{mL solution}$$

✓ *Reasonable Answer Check:* The units cancel appropriately. The relatively large number of milliliters seems appropriate, since this relatively dilute solution contains a relatively large mass of solute.

76. *Answer:* **0.0150 M $CuSO_4$**

Strategy and Explanation: Given the volume of a concentrated solution, the concentrated solution's molarity, and the final volume of the dilute solution, find the concentration of the dilute solution.

Moles of solute do not change when water is added. So we can equate the moles of solute in the concentrated solution with the moles of solute in the dilute solution. Get the moles of solute in the concentrated solution by using the molarity of the concentrated solution as a conversion factor to convert the volume into moles. Divide the moles of solute by the new dilute solution's volume to get the dilute solution's concentration. The logical plan outlined here is built into the equation given in Section 5.6 on page 193 for doing dilution calculations:

$$\textit{Molarity}(\text{conc}) \times V(\text{conc}) = \textit{Molarity}(\text{dil}) \times V(\text{dil})$$

$$\frac{\textit{Molarity}(\text{conc}) \times V(\text{conc})}{V(\text{dil})} = \frac{(0.0250\ \text{M}\ CuSO_4\ \text{conc}) \times (6.00\ \text{mL conc})}{(10.0\ \text{mL dil})} = 0.0150\ \text{M}\ CuSO_4$$

✓ *Reasonable Answer Check:* The dilute solution has a smaller concentration than the concentrated solution. That is a sensible result, since water was added to make the new solution.

78. *Answer:* **Method (b)**

Strategy and Explanation: Given the desired volume and the molarity of a dilute solution, determine which of several dilution methods produces this solution.

Looking at each choice, it's easy to see that each solution results in a total volume of 1.00 L, which is the desired volume. So, we should focus our attention on which of the choices provides a solution that contains the proper number of moles. Convert milliliters to liters. Then calculate the moles in each choice by multiplying the volume in liters by the concentration of the concentrated solution.

We want 1.00 L of 0.125 M H_2SO_4

$$1.00 \text{ L dil} \times \frac{0.125 \text{ mol } H_2SO_4}{1 \text{ L dil}} = 0.125 \text{ moles of } H_2SO_4$$

So, determine moles H_2SO_4 in each choice and compare it to the 0.125 moles H_2SO_4 desired.

(a) $36.0 \text{ mL conc} \times \frac{1 \text{ L}}{1000 \text{ mL}} \times \frac{1.25 \text{ mol } H_2SO_4}{1 \text{ L conc}} = 0.00450 \text{ mol } H_2SO_4$ No

(b) $20.8 \text{ mL conc} \times \frac{1 \text{ L}}{1000 \text{ mL}} \times \frac{6.00 \text{ mol } H_2SO_4}{1 \text{ L conc}} = 0.125 \text{ mol } H_2SO_4$ Yes

(c) $50.0 \text{ mL conc} \times \frac{1 \text{ L}}{1000 \text{ mL}} \times \frac{3.00 \text{ mol } H_2SO_4}{1 \text{ L conc}} = 0.150 \text{ mol } H_2SO_4$ No

(d) $500. \text{ mL conc} \times \frac{1 \text{ L}}{1000 \text{ mL}} \times \frac{0.500 \text{ mol } H_2SO_4}{1 \text{ L conc}} = 0.250 \text{ mol } H_2SO_4$ No

Only choice (b) will make the desired solution.

Notice: You also could have used the *Molarity*(conc) x *V*(conc) = *Molarity*(dil) x *V*(dil) dilution equation to answer this question.

✓ *Reasonable Answer Check:* Compare moles of solute in the concentrated and dilute solutions:

$$Molarity(\text{conc}) \times V(\text{conc}) = (6.00 \text{ M } H_2SO_4 \text{ conc}) \times (20.8 \text{ mL conc}) \times \frac{1 \text{ L}}{1000 \text{ mL}} = 0.125 \text{ mol } H_2SO_4$$

$$Molarity(\text{dil}) \times V(\text{dil}) = (0.125 \text{ M } H_2SO_4 \text{ dil}) \times (1.00 \text{ L dil}) = 0.125 \text{ mol } H_2SO_4$$

Within three significant figures, the moles in the concentrated solution (0.125 mol) are the same as the moles in the dilute solution (0.125 mol), as expected.

80. *Answer:* **39.4 g $NiSO_4 \cdot 6H_2O$**

Strategy and Explanation: Given the volume of the solution and its molarity, find the mass of solute in the solution.

Use the molarity as a conversion factor to determine moles of solute. Then use formula stoichiometry to relate the solute to the solid hydrate, then use the molar mass of the hydrate to find grams of the hydrate.

The molar mass of $NiSO_4 \cdot 6\,H_2O$

$$= 58.6034 \text{ g Ni} + 32.065 \text{ g S} + 4 \times (15.9994 \text{ g}) + 6 \times [2 \times (1.0079 \text{ g}) + 15.9994 \text{ g}] = 262.757 \text{ g/mol}$$

$$0.500 \text{ L solution} \times \frac{0.300 \text{ mol } NiSO_4}{1 \text{ L solution}} \times \frac{1 \text{ mol } NiSO_4 \cdot 6H_2O}{1 \text{ mol } NiSO_4} \times \frac{262.847 \text{ g } NiSO_4 \cdot 6H_2O}{1 \text{ mol } NiSO_4 \cdot 6H_2O}$$

$$= 39.4 \text{ g } NiSO_4 \cdot 6H_2O$$

Calculations for Reactions in Solution (Sections 5.7 & 5.8)

83. *Answer:* **0.205 g Na_2CO_3**

Strategy and Explanation: Given the volume and molarity of a solution containing one reactant and the balanced chemical equation for a reaction, determine the mass of another reactant required for complete reaction.

The stoichiometry of a balanced chemical equation dictates how the moles of reactants combine, so we will commonly look for ways to calculate moles. Here, the volume and molarity can be used to find the moles of one reactant. Then we will use the equation stoichiometry to find out moles of the other reactant needed. Finally, we will use the molar mass to find the grams.

Notice: It is NOT appropriate to use the dilution equation when working with reactions!

We learn from the balanced equation that 2 mol of HNO_3 reacts with 1 mol Na_2CO_3.

$$25.0 \text{ mL } HNO_3 \text{ solution} \times \frac{1 \text{ L}}{1000 \text{ mL}} \times \frac{0.155 \text{ mol } HNO_3}{1 \text{ L } HNO_3 \text{ solution}} \times \frac{1 \text{ mol } Na_2CO_3}{2 \text{ mol } HNO_3} \times \frac{105.9885 \text{ g } Na_2CO_3}{1 \text{ mol } Na_2CO_3} = 0.205 \text{ g } Na_2CO_3$$

85. *Answer:* **121 mL HNO_3**

Strategy and Explanation: Given the mass of one reactant, the balanced chemical equation for a reaction, and the molarity of a solution containing the other reactant, determine the volume of the second solution for a complete reaction.

The stoichiometry of a balanced chemical equation dictates how the moles of reactants combine, so we will commonly look for ways to calculate moles. Here, the mass and molar mass can be used to find the moles of one reactant. Then we will use the equation stoichiometry to find out moles of the other reactant needed. Then we will use the moles and molarity to find volume in liters and convert liters into milliliters.

Notice: It is NOT appropriate to use the dilution equation when working with reactions!

We learn from the balanced equation that 1 mol of $Ba(OH)_2$ reacts with 2 mol HNO_3.

$$1.30 \text{ g } Ba(OH)_2 \times \frac{1 \text{ mol } Ba(OH)_2}{171.3416 \text{ g } Ba(OH)_2} \times \frac{2 \text{ mol } HNO_3}{1 \text{ mol } Ba(OH)_2} \times \frac{1 \text{ L } HNO_3 \text{ solution}}{0.125 \text{ mol } HNO_3} \times \frac{1000 \text{ mL}}{1 \text{ L}} = 121 \text{ mL } HNO_3 \text{ solution}$$

87. *Answer:* **22.9 mL NaOH solution**

Strategy and Explanation: Given the volume and molarity of a solution containing one reactant, and the molarity of a solution containing the other reactant, determine the volume of the second solution for a complete reaction.

First, determine what chemical reaction occurs between the reactants, write the appropriate products, and balance the equation. Then use the volume and molarity to find the moles of one reactant. Then use the equation stoichiometry to find out moles of the other reactant needed. Then use the moles and molarity to find volume in liters, and convert liters into milliliters.

Notice: It is NOT appropriate to use the dilution equation when working with reactions!

NaOH and H_2SO_4 experience an acid-base neutralization reaction.

$$2 \text{ NaOH(aq)} + H_2SO_4\text{(aq)} \longrightarrow 2 \text{ } H_2O(\ell) + Na_2SO_4\text{(aq)}$$

We learn from this balanced equation that 2 mol of NaOH reacts with 1 mol H_2SO_4.

$$25.0 \text{ mL } H_2SO_4 \text{ solution} \times \frac{1 \text{ L}}{1000 \text{ mL}} \times \frac{0.234 \text{ mol } H_2SO_4}{1 \text{ L } H_2SO_4 \text{ solution}} \times \frac{2 \text{ mol NaOH}}{1 \text{ mol } H_2SO_4} \times \frac{1 \text{ L NaOH solution}}{0.512 \text{ mol NaOH}} \times \frac{1000 \text{ mL}}{1 \text{ L}} = 22.9 \text{ mL NaOH solution}$$

Notice that, if we worked the solution in millimoles, the conversion between mL and L and then back to mL can be avoided.

$$25.0 \text{ mL } H_2SO_4 \text{ solution} \times \frac{0.234 \text{ mmol } H_2SO_4}{1 \text{ mL } H_2SO_4 \text{ solution}} \times \frac{2 \text{ mmol NaOH}}{1 \text{ mmol } H_2SO_4}$$

$$\times \frac{1 \text{ mL NaOH solution}}{0.512 \text{ mmol NaOH}} = 22.9 \text{ mL NaOH solution}$$

✓ *Reasonable Answer Check:* The volumes of the two reactants are similar in size, as expected.

89. *Answer:* **0.18 g AgCl; NaCl is the excess reactant; 0.0080 M NaCl**

Strategy and Explanation: Given the volumes and molarities of separate solutions containing each reactant of a precipitation reaction, determine the maximum mass of product produced and the identity and concentration of the excess reactant.

First, complete and balance the precipitation equation. Then use the volumes, molarities, and the equation stoichiometry to find the moles of product produced for each reactant. The reactant that produces the smallest number of moles is the limiting reactant. Use the moles of product produced by the limiting reactant to determine the maximum mass of product formed. Determine the number of moles of excess reactant in the solution, using stoichiometry. Determine the excess reactant's final concentration by dividing by the new solution's volume in liters.

Notice: It is NOT appropriate to use the dilution equation when working with reactions!

The precipitation equation looks like this:

$$AgNO_3(aq) + NaCl(aq) \longrightarrow AgCl(s) + NaNO_3(aq)$$

We learn from this balanced equation that 1 mol of NaCl produces 1 mol AgCl.

$$50.0 \text{ mL } AgNO_3 \text{ solution} \times \frac{1 \text{ L}}{1000 \text{ mL}} \times \frac{0.025 \text{ mol } AgNO_3}{1 \text{ L } AgNO_3 \text{ solution}} \times \frac{1 \text{ mol AgCl}}{1 \text{ mol } AgNO_3} = 0.0013 \text{ mol AgCl}$$

$$100.0 \text{ mL NaCl solution} \times \frac{1 \text{ L}}{1000 \text{ mL}} \times \frac{0.025 \text{ mol NaCl}}{1 \text{ L NaCl solution}} \times \frac{1 \text{ mol AgCl}}{1 \text{ mol NaCl}} = 0.0025 \text{ mol AgCl}$$

AgCl is the limiting reactant, and NaCl is the excess reactant. Use 0.0013 mol AgCl to determine the maximum grams of AgCl produced:

$$0.0013 \text{ mol AgCl} \times \frac{143.321 \text{ g AgCl}}{1 \text{ mol AgCl}} = 0.18 \text{ g AgCl}$$

If 0.0013 mol AgCl was formed and 0.0025 mol AgCl was expected from NaCl, the difference is how much AgCl was not formed:

$$0.0025 \text{ mol AgCl expected} - 0.0013 \text{ mol AgCl formed} = 0.0012 \text{ mol AgCl not formed}$$

$$0.0012 \text{ mol AgCl not formed} \times \frac{1 \text{ mol NaCl}}{1 \text{ mol AgCl}} = 0.0012 \text{ mol NaCl not reacted}$$

$$\text{Total Volume} = (50.0 \text{ mL} + 100.0 \text{ mL}) \times \frac{1 \text{ L}}{1000 \text{ mL}} = 0.1500 \text{ L}$$

Molarity of NaCl after the AgCl has completely precipitated:

$$\text{Molarity} = \frac{0.013 \text{ mol NaCl}}{0.1500 \text{ L solution}} = 0.0080 \text{ M NaCl}$$

✓ *Reasonable Answer Check:* The concentration of the excess reactant after the reaction is over is smaller than it was at the beginning. This is an expected result, because some of it reacted, and what was left over got diluted. The maximum mass of AgCl produced is a reasonable number.

91. *Answer:* **(a) Step (ii) is not correct. (Steps (iii) and (iv) show correct calculations but use the wrong numbers.) (b) 3.94×10^{-3} g citric acid**

Strategy and Explanation: Given the volume and molarity of a solution containing one reactant, the balanced chemical equation, and a series of steps describing the calculation, determine which of the steps is not correct, and correctly determine the mass of the other reactant in a specific volume of its solution.

Check each step to see if it is right or wrong. If it is wrong, correct it.

(a) **Step (i) is correct.** $6.42 \text{ mL} \times \frac{1 \text{ L}}{1000 \text{ mL}} \times \frac{9.580 \times 10^{-2} \text{ mol NaOH}}{1 \text{ L}} = 6.15 \times 10^{-4} \text{ mol NaOH}$

Step (ii) is not correct. The equation stoichiometry gives 1 mol citric acid reacting with 3 mol of NaOH, not 3 mol citric acid reacting with 1 mol of NaOH.

$$6.15 \times 10^{-4} \text{ mol NaOH} \times \frac{1 \text{ mol citric acid}}{3 \text{ mol NaOH}} = 2.05 \times 10^{-4} \text{ mol citric acid}$$

Step (ii) has a 3 in front of the moles of citric acid and a 1 in front of the NaOH, resulting in multiplication by 3 instead of division by 3, so the moles of citric acid calculated there are wrong.

Step (iii) is not correct, because it uses the erroneous answer from Step (ii); however, the calculation it shows is correct:

$$2.05 \times 10^{-4} \text{ mol citric acid} \times \frac{192.12 \text{ g citric acid}}{1 \text{ mol citric acid}} = 0.0394 \text{ g citric acid}$$

Step (iv) is not correct, because it uses a different answer than the correct one from Step (iii) and the significant figures on the volume are incorrect; however, the calculation it shows is correct:

$$\frac{0.0394 \text{ g citric acid}}{10.0 \text{ mL}} = 3.94 \times 10^{-3} \text{ g citric acid in 1 mL of soft drink}$$

(b) The right answer is 3.94×10^{-3} g citric acid in 1 mL of soft drink.

✓ *Reasonable Answer Check:* The moles of citric acid that react must be less than the moles of NaOH that react.

93. *Answer:* **1.192 M HCl**

Strategy and Explanation: Given the mass of one reactant, the balanced chemical equation for a reaction, and the volume of a solution containing the other reactant needed for a complete reaction, determine the molarity of the second solution.

The mass and molar mass can be used to find the moles of one reactant. Then we will use the equation stoichiometry to find out moles of the other reactant needed. Then we will use the moles and volume in liters to determine the molarity. *Note: It is NOT appropriate to use the dilution equation when working with reactions!*

We learn from the balanced equation that 1 mol of Na_2CO_3 reacts with 2 mol HCl.

$$2.050 \text{ g } Na_2CO_3 \times \frac{1 \text{ mol } Na_2CO_3}{105.9885 \text{ g } Na_2CO_3} \times \frac{2 \text{ mol HCl}}{1 \text{ mol } Na_2CO_3} = 3.868 \times 10^{-2} \text{ mol HCl}$$

$$32.45 \text{ mL HCl solution} \times \frac{1 \text{ L}}{1000 \text{ mL}} = 0.03245 \text{ L HCl solution}$$

$$\frac{3.868 \times 10^{-2} \text{ mol HCl}}{0.03245 \text{ L HCl solution}} = 1.192 \text{ M HCl solution}$$

The three separate calculations above can be consolidated into one calculation as follows:

$$\frac{2.050 \text{ g } Na_2CO_3}{32.45 \text{ mL HCl solution}} \times \frac{1 \text{ mol } Na_2CO_3}{105.9885 \text{ g } Na_2CO_3} \times \frac{2 \text{ mol HCl}}{1 \text{ mol } Na_2CO_3} \times \frac{1000 \text{ mL}}{1 \text{ L}} = 1.192 \text{ M HCl solution}$$

Both methods will give the right answer, however the consolidated calculation eliminates the need to write down unnecessary intermediate answers and helps eliminate round-off errors.

✓ *Reasonable Answer Check:* The units cancel appropriately, and the moles and liters are nearly the same value so it makes sense that the molarity is near 1.

95. *Answer:* **96.8% pure**

Strategy and Explanation: Given the volume and concentration of a solution containing one reactant, the balanced chemical equation for a reaction, and the mass of an impure sample of a salt of the second reactant, determine the percent purity of the impure sample.

The volume and molarity of the first reactant are used to calculate the moles. Equation and formula stoichiometry can be used to find the moles of the salt. Then we will use the molar mass of the salt to find the mass of the salt. Comparing this mass to the mass of the impure sample we can determine the percent purity. *Notice: It is NOT appropriate to use the dilution equation when working with reactions!*

We learn from the balanced equation that 2 mol of $S_2O_3^{2-}$ reacts with 1 mol I_2. Formula stoichiometry tells us that 1 mol of $S_2O_3^{2-}$ comes from 1 mol of $Na_2S_2O_3$.

$$40.21 \text{ mL } I_2 \text{ solution} \times \frac{1 \text{ L}}{1000 \text{ mL}} \times \frac{0.246 \text{ mol } I_2}{1 \text{ L } I_2 \text{ solution}} = 9.89 \times 10^{-3} \text{ mol } I_2$$

$$9.89 \times 10^{-3} \text{ mol } I_2 \times \frac{2 \text{ mol } S_2O_3^{2-}}{1 \text{ mol } I_2} \times \frac{1 \text{ mol } Na_2S_2O_3}{1 \text{ mol } S_2O_3^{2-}} \times \frac{158.1098 \text{ g } Na_2S_2O_3}{1 \text{ mol } Na_2S_2O_3} = 3.13 \text{ g } Na_2S_2O_3$$

The percent purity is calculated by dividing the mass of $Na_2S_2O_3$ by the mass of the impure sample and multiplying by 100%.

$$\frac{3.13 \text{ g } Na_2S_2O_3}{3.232 \text{ g impure sample}} \times 100\% = 96.8\% \text{ pure}$$

✓ *Reasonable Answer Check:* The units cancel appropriately, and the two masses are very similar, so it makes sense that the percentage is nearly 100%.

96. *Answer:* 16.1% $H_2C_2O_4$

Strategy and Explanation: Given the volume and concentration of a solution containing one reactant, the balanced chemical equation for a reaction, and the mass of a mixture containing two salts—one of which is the salt of the second reactant, determine the weight percent of the reactive salt in the mixture.

The volume and molarity of the first reactant are used to calculate the moles. Equation stoichiometry can be used to find the moles of the reactive compound. Then we will use the molar mass of the reactive compound to find the mass. Then we will compare that to the mass of the mixture to determine the weight percent. *Notice: It is NOT appropriate to use the dilution equation when working with reactions!*

We learn from the balanced equation that 1 mol of $H_2C_2O_4$ reacts with 2 mol NaOH.

$$29.58 \text{ mL NaOH solution} \times \frac{1 \text{ L}}{1000 \text{ mL}} \times \frac{0.550 \text{ mol NaOH}}{1 \text{ L NaOH solution}}$$
$$\times \frac{1 \text{ mol } H_2C_2O_4}{2 \text{ mol NaOH}} \times \frac{90.0348 \text{ g } H_2C_2O_4}{1 \text{ mol } H_2C_2O_4} = 0.732 \text{ g } H_2C_2O_4$$

Weight percent is calculated by dividing the mass of $H_2C_2O_4$ by the mixture mass and multiplying by 100%.

$$\frac{0.732 \text{ g } H_2C_2O_4}{4.554 \text{ g mixture}} \times 100\% = 16.1\% \ H_2C_2O_4$$

✓ *Reasonable Answer Check:* The mass of $H_2C_2O_4$ is significantly smaller than the mass of the sample, so it makes sense that the percentage is so low.

97. *Answer:* **Use NaOH; precipitate if Sr^{2+} present, no precipitate if Ca^{2+} present. Use H_2S; precipitate if Sr^{2+} present, no precipitate if Ca^{2+} present.**

Strategy and Explanation: Determine which of three reagents could be used to distinguish whether a solution contains calcium ions or strontium ions.

Determine the anion of the added reagent. Determine the products of its reaction with Ca^{2+} and Sr^{2+}. If the reactions are visibly different, then the reagent could be used to distinguish them.

First reagent: NaOH(aq) would provide the anion $OH^-(aq)$ for a precipitation reaction.

$Ca(OH)_2$ is slightly soluble, Table 5.1, Rule 10 $2\ OH^-(aq) + Ca^{2+}(aq) \longrightarrow$ NR.

$Sr(OH)_2$ is insoluble, Table 5.1, Rule 10 $2\ OH^-(aq) + Sr^{2+}(aq) \longrightarrow Sr(OH)_2(s)$

An insoluble precipitate of strontium hydroxide would form if we add a small amount of this reagent to an unknown contained Sr^{2+}, but no visual change (or only slight precipitation) would be seen if the unknown contained only Ca^{2+}. This reagent could be used to distinguish them.

Second reagent: $H_2SO_4(aq)$ would provide the $SO_4^{2-}(aq)$ anion to a precipitation reaction.

$CaSO_4$ is insoluble, Table 5.1, Rule 4 $SO_4^{2-}(aq) + Ca^{2+}(aq) \longrightarrow CaSO_4(s)$

$SrSO_4$ is insoluble, Table 5.1, Rule 4 $SO_4^{2-}(aq) + Sr^{2+}(aq) \longrightarrow SrSO_4(s)$

Both reactions produce insoluble sulfate precipitate, so this reagent could NOT be used to distinguish them.

Third reagent: $H_2S(aq)$ would provide the $S^{2-}(aq)$ anion to a precipitation reaction.

CaS is sparingly soluble, Table 5.1, Rule 12 $S^{2-}(aq) + Ca^{2+}(aq) \longrightarrow CaS(s)$

SrS is insoluble, Table 5.1, Rule 12 $S^{2-}(aq) + Sr^{2+}(aq) \longrightarrow SrS(s)$

An insoluble precipitate of strontium sulfide would form if we add a small amount of this reagent to an unknown contained Sr^{2+}, but no visual change (or only slight precipitation) would be seen if the unknown contained only Ca^{2+}. This reagent could be used to distinguish them.

✓ *Reasonable Answer Check:* The slight and sparingly soluble nature of the hydroxide and sulfide compounds of calcium ion can be used to distinguish it from strontium. This is not very satisfying, but it makes sense, because these two ions are both in the same group on the periodic table. Their chemical reactions will certainly be similar. So, we must be satisfied with these tests.

General Questions

99. *Answer:* **Cl^-; $CaCO_3(s) + 2\ H^+(aq) \longrightarrow CO_2(g) + Ca^{2+}(aq) + H_2O(\ell)$; gas-forming exchange reaction**

Strategy and Explanation: Given the complete ionic equation, locate the ions that are identical in the reactants and the products—these are spectator ions. Here, on the reactant side of the equation is "2 Cl^-." This makes Cl^- a spectator ion. Everything else is an active participant in the chemical reaction.

$$CaCO_3(s) + 2\ H^+(aq) \longrightarrow CO_2(g) + Ca^{2+}(aq) + H_2O(\ell) \quad \text{net ionic equation}$$

This is a gas-forming exchange reaction where an acid assists in the dissolving of an insoluble ionic solid.

101. *Answer/Explanation:*

(a) Using Table 5.1 page 163, we find that ammonium compounds and nitrate compounds are soluble. The HgS compound is insoluble.

$$(NH_4)_2S(aq) + Hg(NO_3)_2(aq) \longrightarrow HgS(s) + 2\ NH_4NO_3(aq) \quad \text{balanced equation}$$

(b) The reactants are **ammonium sulfide** and **mercury(II) nitrate**. The products are **mercury(II) sulfide** and **ammonium nitrate**.

(c) The aqueous compounds are ionized. The solid remains unionized. Spectator ions are NH_4^+ and NO_3^-.

$$S^{2-}(aq) + Hg^{2+}(aq) \longrightarrow HgS(s) \quad \text{balanced net ionic equation}$$

(d) This is a **precipitation** reaction.

103. *Answer/Explanation:*

(a) Combination reaction: $SO_3(g) + H_2O(\ell) \longrightarrow H_2SO_4(aq)$

(b) Combination reaction: $Sr(s) + H_2(g) \longrightarrow SrH_2(s)$

(c) Displacement reaction: $Mg(s) + H_2SO_4(aq) \longrightarrow MgSO_4(aq) + H_2(g)$

(d) Exchange (precipitation) reaction: $Na_3PO_4(aq) + 3\ AgNO_3(aq) \longrightarrow Ag_3PO_4(s) + 3\ NaNO_3(aq)$

(e) Decomposition and gas-forming reaction: $Ca(HCO_3)_2(s) \longrightarrow CaO(s) + H_2O(\ell) + 2\ CO_2(g)$

(f) Oxidation-reduction reaction: $2\ Fe^{3+}(aq) + Sn^{2+}(aq) \xrightarrow{\text{heat}} 2\ Fe^{2+}(aq) + Sn^{4+}(aq)$

105. *Answer:* **(a) H_2O and NH_3 molecules, plus a small amount of NH_4^+ and OH^- (b) H_2O and CH_3CO_2H molecules, plus a small amount of $CH_3CO_2^-$ and H^+ (c) H_2O, Na^+, and OH^- (d) H_2O, H^+, and Br^-**

Strategy and Explanation:

(a) $NH_3(aq)$ is a weak base (according to Table 5.2). That means the solution's primary components are H_2O and NH_3 molecules. A small amount of NH_4^+ and OH^- also exists in the solution.

(b) $CH_3CO_2H(aq)$ is a weak acid (according to Table 5.2). That means the solution's primary components are H_2O and CH_3CO_2H molecules. A small amount of $CH_3CO_2^-$ and H^+ also exists in the solution.

(c) $NaOH(aq)$ is a strong base (according to Table 5.2). That means the solution's primary components are H_2O, Na^+, and OH^-.

(d) $HBr(aq)$ is a strong acid (according to Table 5.2). That means the solution's primary components are H_2O, H^+, and Br^-.

107. *Answer:* **The only redox reaction is reaction (c); oxidizing agent is Ti; reducing agent is Mg.**

Strategy and Explanation: Given several reactions, determine if they are redox (oxidation-reduction) reactions and identify the oxidizing and reducing agents in those that are redox reactions.

To decide if a reaction is a redox reaction, we need to see if any of the elements change oxidation state. In redox reactions, the atoms change oxidation states during the reaction. If no change in oxidation state is observed, then the reaction is not a redox reaction. The oxidizing agent is the reactant that assists an oxidation by being reduced, so it will be the reactant whose atoms gain electrons, and end up with a lower (more negative or less positive) oxidation number. The reducing agent is the reactant that assists a reduction by being oxidized, so it will be the reactant whose atoms lose electrons, and end up with a higher (more positive or less negative) oxidation number.

(a) Look at the oxidation numbers for the reactants:

NaOH contains a monatomic cation, Na^+, and a diatomic anion, OH^-. Rule 2 gives us Ox. # Na = +1. Rule 3 gives us Ox. # O = –2 and Ox. # H = +1.

H_3PO_4 contains oxygen and hydrogen. Rule 3 gives us Ox. # O = –2 and Ox. # H = +1. We use the sum rule to find the Ox. # P. For a molecule, the sum of the oxidation numbers is equal to zero.

$$0 = 3 \times (+1) + 1 \times (\text{Ox. \# P}) + 4 \times (-2)$$

Therefore, Ox. # P = +5.

Look at the oxidation numbers for the products:

NaH_2PO_4 contains a monatomic cation, Na^+, and a polyatomic anion, $H_2PO_4^-$. Rule 2 gives us Ox. # Na = +1. $H_2PO_4^-$ contains oxygen and hydrogen. Rule 3 gives us Ox. # O = –2 and Ox. # H = +1. We use the sum rule to find the Ox. # P. For a polyatomic anion, the sum of the oxidation numbers is equal to the ion's charge of 1–.

$$-1 = 2 \times (+1) + 1 \times (\text{Ox. \# P}) + 4 \times (-2)$$

Therefore, Ox. # P = +5.

H_2O contains oxygen and hydrogen. Rule 3 gives us Ox. # O = –2 and Ox. # H = +1.

The oxidation numbers don't change, so this is NOT a redox reaction.

(b) Look at the oxidation numbers for the reactants:

NH_3 contains hydrogen. Rule 3 gives us Ox. # H = +1. We use the sum rule to find the Ox. # N. For a molecule, the sum of the oxidation numbers is equal to zero.

$$0 = 1 \times (\text{Ox. \# N}) + 3 \times (+1)$$

Therefore, Ox. # N = –3

CO_2 contains oxygen. Rule 3 gives us Ox. # O = –2. We use the sum rule to find the Ox. # C. For a molecule, the sum of the oxidation numbers is equal to zero.

$$0 = 1 \times (\text{Ox. \# C}) + 2 \times (-2)$$

Therefore, Ox. # C = +4.

H_2O contains oxygen and hydrogen. Rule 3 gives us Ox. # O = –2 and Ox. # H = +1.

Look at the oxidation numbers for the products:

NH_4HCO_3 contains a polyatomic cation, NH_4^+, and a polyatomic anion, HCO_3^-. NH_4^+ contains hydrogen. Rule 3 gives us Ox. # H = +1. We use the sum rule to find the Ox. # N. For a polyatomic cation, the sum of the oxidation numbers is equal to the ion's charge of 1+.

$$+1 = 1 \times (\text{Ox. \# N}) + 4 \times (+1)$$

Therefore, Ox. # N = –3.

HCO_3^- contains oxygen and hydrogen. Rule 3 gives us Ox. # O = –2 and Ox. # H = +1. We use the sum rule to find the Ox. # C. For a polyatomic anion, the sum of the oxidation numbers is equal to the ion's charge of 1–.

$$-1 = 1 \times (+1) + 1 \times (\text{Ox. \# C}) + 3 \times (-2)$$

Therefore, Ox. # C = +4.

The oxidation numbers don't change, so this is NOT a redox reaction.

(c) Look at the oxidation numbers for the reactants:

$TiCl_4$ contains a monatomic cation, Ti^{4+}, and a monatomic anion, Cl^-. Rule 2 gives us Ox. # Ti = +4 and Ox. # Cl = –1.

Rule 1 indicates the oxidation number of elements is zero, so Ox. # Mg = 0.

Look at the oxidation numbers for the products:

Rule 1 indicates the oxidation number of elements is zero, so Ox. # Ti = 0.

$MgCl_2$ contains a monatomic cation, Mg^{2+}, and a monatomic anion, Cl^-. Rule 2 gives us Ox. # Mg = +2 and Ox. # Cl = –1.

The oxidation numbers of Mg and Ti do change, so this IS a redox reaction.

The oxidizing agent is Ti since its Ox. # goes from +4 to zero.

The reducing agent is Mg since its Ox. # goes from zero to +2.

(d) Look at the oxidation numbers for the reactants:

NaCl contains a monatomic cation, Na^+, and a monatomic anion, Cl^-. Rule 2 gives us Ox. # Na = +1 and Ox. # Cl = –1.

$NaHSO_4$ contains a monatomic cation, Na^+, and a polyatomic anion, HSO_4^-. Rule 2 gives us Ox. # Na = +1. HSO_4^- contains oxygen and hydrogen. Rule 3 gives us Ox. # O = –2 and Ox. # H = +1. We use the sum rule to find the Ox. # S. For a polyatomic anion, the sum of the oxidation numbers is equal to the ion's charge of 1–.

$$-1 = 1 \times (+1) + 1 \times (\text{Ox. \# S}) + 4 \times (-2)$$

Therefore, Ox. # S = +6.

Look at the oxidation numbers for the products:

HCl contains hydrogen. Rule 3 gives us Ox. # H = +1. We use the sum rule to find the Ox. # Cl. For a molecule, the sum of the oxidation numbers is equal to zero.

$$0 = 1 \times (+1) + 1 \times (\text{Ox. \# Cl})$$

Therefore, Ox. # Cl = –1.

Na_2SO_4 contains a monatomic cation, Na^+, and a polyatomic anion, SO_4^{2-}. Rule 2 gives us Ox. # Na = +1. SO_4^{2-} contains oxygen. Rule 3 gives us Ox. # O = –2. We use the sum rule to find the Ox. # S. For a polyatomic anion, the sum of the oxidation numbers is equal to the ion's charge of 2–.

$$-2 = 1 \times (\text{Ox. \# S}) + 4 \times (-2)$$

Therefore, Ox. # S = +6.

The oxidation numbers don't change, so this is NOT a redox reaction.

✓ *Reasonable Answer Check:* The oxidation numbers are consistent with typical oxidation states of these elements. The reactants and products are common acids, ionic compounds and ions.

109. *Answer:* **(a) $CaF_2(s) + H_2SO_4(aq) \longrightarrow 2\ HF(g) + CaSO_4(s)$; calcium fluoride, sulfuric acid, hydrogen fluoride, calcium sulfate (b) precipitation (c) carbon tetrachloride, antimony(V) pentachloride, hydrogen chloride (d) CCl_3F**

Strategy and Explanation:

(a) $CaF_2(s) + H_2SO_4(aq) \longrightarrow 2\ HF(g) + CaSO_4(s)$

The reactants are called calcium fluoride and sulfuric acid.

The products are called hydrogen fluoride and calcium sulfate.

(b) Determine if a reaction is an acid base reaction, an oxidation-reduction reaction, or a precipitation reaction.

To decide if a reaction is an acid-base reaction, we need to see if an acid is reacting with a base. To decide if a reaction is an oxidation-reduction reaction, we need to see if any of the elements change oxidation state. To decide if a reaction is a precipitation reaction, look for insoluble ionic compound as a product.

First, check to see if it is an acid-base reaction: The strong reactant acid, H_2SO_4, reacts to form a weak product acid, HF. We don't see a hydroxide compound in the reaction, but in some sense of the word, the ionic fluoride compound is serving as a base. So, there is an acid-base reaction happening here. We'll learn more about these kinds of acid-base reactions in Chapter 16.

Second, check to see if it is an oxidation-reduction reaction: Look at the oxidation numbers for the reactants:

CaF_2 contains a monatomic cation, Ca^{2+}, and a monatomic anion, F^-. Rule 2 gives us Ox. # Ca = +2 and Ox. # F = –1.

H_2SO_4 contains oxygen and hydrogen. Rule 3 gives us Ox. # O = –2 and Ox. # H = +1. We use the sum rule to find the Ox. # S. For a molecule, the sum of the oxidation numbers is equal to zero.

$$0 = 2 \times (+1) + 1 \times (\text{Ox. \# S}) + 4 \times (-2)$$

Therefore, Ox. # S = +6.

Look at the oxidation numbers for the products:

HF contains hydrogen and fluorine. Rule 3 gives us Ox. # H = +1 and Ox. # F = –1.

$CaSO_4$ contains a monatomic cation, Ca^{2+}, and a polyatomic anion, SO_4^{2-}. Rule 2 gives Ox. # Ca = +2. SO_4^{2-} contains oxygen. Rule 3 gives Ox. # O = –2. We use the sum rule to find the Ox. # S. For a polyatomic anion, the sum of the oxidation numbers is equal to the ion's charge of 2–.

$$-2 = 1 \times (\text{Ox. \# S}) + 4 \times (-2)$$

Therefore, Ox. # S = +6.

The oxidation numbers don't change, so this is NOT an oxidation-reduction reaction.

Third, check to see if it is a precipitation reaction: An insoluble solid, $CaSO_4(s)$ (Table 5.1, Rule 4), is produced in a solution, so the reaction can also be considered a precipitation reaction.

In conclusion, most people studying Chapter 5 would probably decide this was a precipitation reaction.

✓ *Reasonable Answer Check:* The oxidation states determined are typical for these atoms in these compounds. With constant oxidation states, it is clearly not an oxidation-reduction reaction. It is logical to call this either a precipitation reaction or an acid-base reaction, but NOT an oxidation-reduction reaction. Because these kinds of acid-base reactions are described more thoroughly in a later chapter, most students will probably choose the precipitation reaction answer.

(c) CCl_4 is carbon tetrachloride, $SbCl_5$ is antimony pentachloride, HCl is hydrogen chloride.

(d) Given the identity and the percent mass of each element in a compound, find the empirical formula of the compound.

This Question might require a review of the "Empirical Formula" calculations introduced in Chapter 3, Section 3.10. Choose a convenient mass sample of the chlorofluorocarbon, such as 100.00 g. Using the given mass percents, determine the mass of C, Cl, and F in the sample. Using molar masses of the elements, determine the moles of each element in the sample. Find the whole number mole ratio of the elements C, Cl, and F, to determine the subscripts in the empirical formula.

A 100.00 gram sample will have 8.74 grams of C, 77.43 grams of Cl, and 13.83 grams of H.

Find moles of C, Cl, and F in the sample:

$$8.74\text{ g C} \times \frac{1\text{ mol C}}{12.0107\text{ g C}} = 0.728\text{ mol C} \qquad 77.43\text{ g Cl} \times \frac{1\text{ mol Cl}}{35.453\text{ g Cl}} = 2.184\text{ mol Cl}$$

$$13.83\text{ g F} \times \frac{1\text{ mol F}}{18.9984\text{ g F}} = 0.7279\text{ mol F}$$

Mole ratio: 0.728 mol C : 2.184 mol Cl : 0.7279 mol F

Divide each term by the smallest number of moles, 0.7279 mol, to get atom ratio:

Atom ratio: 1 C : 3 Cl : 1 F

The empirical formula is CCl_3F.

✓ *Reasonable Answer Check:* This empirical formula makes sense, because four halogens can be bonded to one carbon.

111. *Answer:* **6.28% impurity**

Strategy and Explanation: Given the mass of a tablet containing vitamin C, the balanced neutralization equation, and the volume and molarity of a base solution used for neutralization, determine what percentage of the tablet is impurity.

Use the volume, molarity, stoichiometry of the chemical equation and molar mass of the compound to calculate the mass of the vitamin C in the sample. Subtract the mass of vitamin C from the mass of the tablet, to determine that mass of the impurity. Divide the mass of the impurity into the mass of the soil sample and multiply by 100% to get percentage mass of the impurity.

$$21.30\text{ mL NaOH} \times \frac{1\text{ L NaOH}}{1000\text{ mL NaOH}} \times \frac{0.1250\text{ mol NaOH}}{1\text{ L NaOH}} \times \frac{1\text{ mol } HC_6H_7H_6}{1\text{ mol HCl}}$$

$$\times \frac{176.1238\text{ g } HC_6H_7O_6}{1\text{ mol } HC_6H_7O_6} \times \frac{1000\text{ mg } HC_6H_7O_6}{1\text{ g } HC_6H_7O_6} = 468.6\text{ mg } HC_6H_7O_6$$

$$500.0\text{ mg tablet} - 468.6\text{ mg } HC_6H_7O_6 = 468.6\text{ mg impurity}$$

$$\frac{31.4 \text{ g impurity}}{500.0 \text{ g tablet}} \times 100\% = 6.28\% \text{ impurity}$$

✓ *Reasonable Answer Check:* A vitamin C tablet should be mostly vitamin C, so it makes sense that the mass percent if the impurity is small.

Applying Concepts

112. For the first case: $LiCl(aq)$ and $AgNO_3(aq)$

(a) The two separate solutions of soluble salts would be clear and colorless like water. Once combined, insoluble white AgCl (Table 5.1, Rule 3) would precipitate. We would probably see it eventually sink to the bottom of the beaker, leaving a clear, probably colorless liquid containing aqueous $LiNO_3$ above it.

(b) Notice, for proper proportions, these diagrams really need many more water molecules.

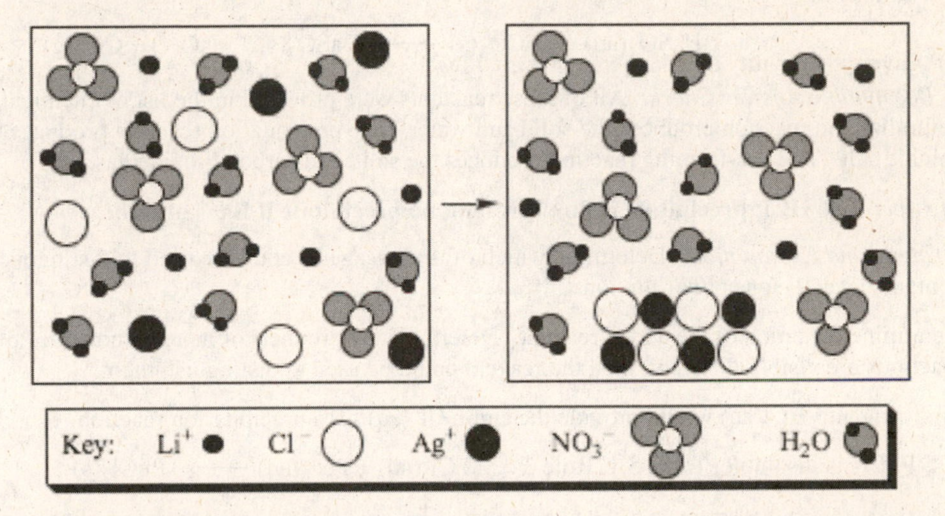

Key: Li^+ ● Cl^- ○ Ag^+ ● NO_3^- H_2O

(c) $Li^+(aq) + Cl^-(aq) + Ag^+(aq) + NO_3^-(aq) \longrightarrow AgCl(s) + Li^+(aq) + NO_3^-(aq)$

For the second case: $NaOH(aq)$ and $HCl(aq)$

(a) The separate acid and base solutions would be clear and colorless like water. Once combined, the solution would still be clear and colorless.

(b) Notice, for proper proportions, these diagrams really need many more water molecules.

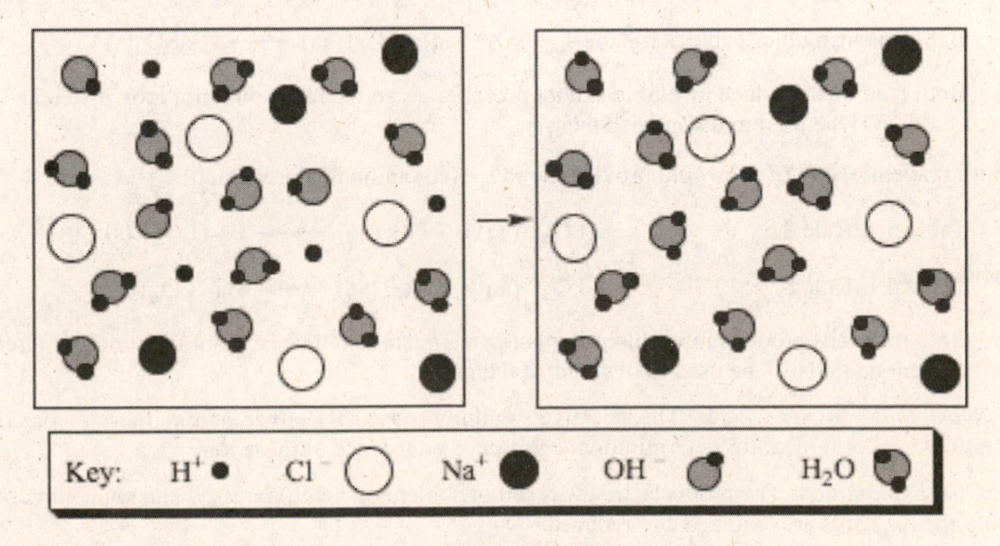

Key: H^+ ● Cl^- ○ Na^+ ● OH^- H_2O

(c) $Na^+(aq) + OH^-(aq) + H^+(aq) + Cl^-(aq) \longrightarrow H_2O(\ell) + Na^+(aq) + Cl^-(aq)$

114. *Answer:* **(a) combine H_2SO_4 and $Ba(OH)_2$ (b) combine Na_2SO_4 and $Ba(NO_3)_2$ (c) combine $H_2SO_4(aq)$ and $BaCO_3(s)$**

Strategy and Explanation: Prepare barium sulfate from a given list of chemicals by various means.

(a) To make $BaSO_4$ from an acid-base reaction, use a base with the cation and an acid with the anion:

$$H_2SO_4(aq) + Ba(OH)_2(aq) \longrightarrow BaSO_4(s) + 2\ H_2O(\ell)$$

(b) To make $BaSO_4$ from a precipitation reaction, use a soluble salt containing the anion and a soluble salt containing the cation:

$$Na_2SO_4(aq) + Ba(NO_3)_2(aq) \longrightarrow BaSO_4(s) + 2\ NaNO_3(aq)$$

(c) To make $BaSO_4$ from a gas-forming reaction, use an acid with the anion and the carbonate salt of the cation:

$$H_2SO_4(aq) + BaCO_3(s) \longrightarrow BaSO_4(s) + H_2O(\ell) + CO_2(g)$$

✓ *Reasonable Answer Check:* All of these reactants were provided in the list of chemicals. The neutralization reaction produces the solid and water. The precipitation reaction produces the solid and a soluble salt. The gas-forming reaction produces the solid and carbon dioxide gas.

116. *Answer:* **Use HCl; precipitate if Pb^{2+} present, no precipitate if Ba^{2+} present.**

Strategy and Explanation: Determine which of three reagents could be used to distinguish whether a solution contains lead(II) ions or barium ions.

Determine the anion of the added reagent. Determine the products of its reaction with Pb^{2+} and Ba^{2+}. If the reactions are visibly different, then the reagent could be used to distinguish them.

First reagent: HCl(aq) would provide the anion $Cl^-(aq)$ for a precipitation reaction.

$PbCl_2$ is insoluble, Table 5.1, Rule 3 $\quad 2\ Cl^-(aq) + Pb^{2+}(aq) \longrightarrow PbCl_2(s)$

$BaCl_2$ is soluble, Table 5.1, Rule 3 $\quad 2\ Cl^-(aq) + Ba^{2+}(aq) \longrightarrow$ N.R.

Insoluble lead(II) chloride precipitate would form if the unknown contained Pb^{2+}, but no precipitate would be seen if the unknown contained only Ba^{2+}. This reagent could be used to distinguish them.

Second reagent: $H_2SO_4(aq)$ would provide the $SO_4^{2-}(aq)$ anion to a precipitation reaction.

$PbSO_4$ is insoluble, Table 5.1, Rule 4 $\quad SO_4^{2-}(aq) + Pb^{2+}(aq) \longrightarrow PbSO_4(s)$

$BaSO_4$ is insoluble, Table 5.1, Rule 4 $\quad SO_4^{2-}(aq) + Ba^{2+}(aq) \longrightarrow BaSO_4(s)$

Both reactions produce insoluble sulfate precipitates, so the test would not look different. This reagent could NOT be used to distinguish them.

Third reagent: $H_3PO_4(aq)$ would provide the $PO_4^{3-}(aq)$ anion to a precipitation reaction.

Table 5.1, Rule 8 $\quad 2\ PO_4^{3-}(aq) + 3\ Pb^{2+}(aq) \longrightarrow Pb_3(PO_4)_2(s)$

Table 5.1, Rule 8 $\quad 2\ PO_4^{3-}(aq) + 3\ Ba^{2+}(aq) \longrightarrow Ba_3(PO_4)_2(s)$

Both reactions produce insoluble phosphate precipitates, so the test would not look different. This reagent could NOT be used to distinguish them.

✓ *Reasonable Answer Check:* The selective solubility of lead(II), silver, and mercury(I) ions in chloride solutions makes it ideal for determining the presence or absence of these ions.

118. *Answer/Explanation:* The products are the result of something being oxidized and something being reduced. Only the reactants are oxidized and reduced.

120. *Answer:* **(d)**

Strategy and Explanation: Too much water was added, making the solution too dilute. So, (d) the concentration of the solution is less than 1 M because you added more solvent than necessary.

122. *Answer:* **(a) and (d) are correct.**

Strategy and Explanation:

(a) Since the solution is acidic, there are more H^+ ions than OH^- ions in the mixture. So, this statement is TRUE.

(b) This statement is FALSE, for the same reason (a) was true.

(c) Only equal quantities of strong acid and strong base make a neutral solution. If either the acid or the base are weak or if the molarities are unequal, then the resulting solution will be basic or acidic, so this statement is FALSE.

(d) Since the resulting solution was acidic, and equal volumes were added, that means the acid's concentration must have been greater than the base's concentration. This statement is TRUE.

(e) While the concentration of H_2SO_4 might have been greater, it is not necessarily true that it MUST have been greater, since only half as many moles of the diprotic acid is needed to neutralize the NaOH base; therefore, it's concentration need only have been anything more than half as large. This statement is FALSE.

124. *Answer:* **(a) magnesium bromide, $MgBr_2(s)$; calcium bromide, $CaBr_2(s)$; strontium bromide, $SrBr_2(s)$ (b) $Mg(s) + Br_2(\ell) \longrightarrow MgBr_2(s)$; $Ca(s) + Br_2(\ell) \longrightarrow CaBr_2(s)$; $Sr(s) + Br_2(\ell) \longrightarrow SrBr_2(s)$ (c) oxidation-reduction (d) The point where increase stops gives: $\frac{\text{g metal}}{\text{g product}}$. The different metals have different molar masses, so the ratios will be different. Use grams to find moles, then set up a ratio.**

Strategy and Explanation:

(a) The reaction of magnesium with bromine will produce magnesium bromide, $MgBr_2(s)$. The reaction of calcium with bromine will produce calcium bromide, $CaBr_2(s)$. The reaction of strontium with bromine will produce strontium bromide, $SrBr_2(s)$.

(b)

$$Mg(s) + Br_2(\ell) \longrightarrow MgBr_2(s)$$

$$Ca(s) + Br_2(\ell) \longrightarrow CaBr_2(s)$$

$$Sr(s) + Br_2(\ell) \longrightarrow SrBr_2(s)$$

(c) The reactions here are oxidation-reduction reactions. The oxidation states of the reactants are all zero (Rule 1) and the ionic compounds produced have atoms with non-zero oxidation states. These should definitely not be called gas-forming reactions since no gas is formed. They should also not be called precipitation reactions, since the solid products would be soluble in water if water were present.

(d) Use the graph of mass of product vs. mass of metal to confirm the predicted formula of the metal halide produced in these reactions.

The "crossover" point, where the line changes from linearly rising to horizontal, describes the stoichiometric equivalence of the metal and the product. In other words, that's where we relate the grams of metal reacted to the grams of metal in a known mass of product. Determine the mass of bromine in a specific sample of the product by subtracting the mass of metal (extrapolated on the "Mass of metal" axis at the crossover point) from the sample mass (extrapolated on the "Mass of product" axis at the crossover point). Calculate moles of each and set up a mole ratio to determine empirical formula.

To find the crossover point, (1) extrapolate the linearly-rising part of the line upward and to the right, (2) extrapolate the horizontal part of the line to the "Mass of product" axis, (3) determine where those two straight lines cross each other, and (4) draw a vertical line down from that crossing point to the "Mass of metal" axis:

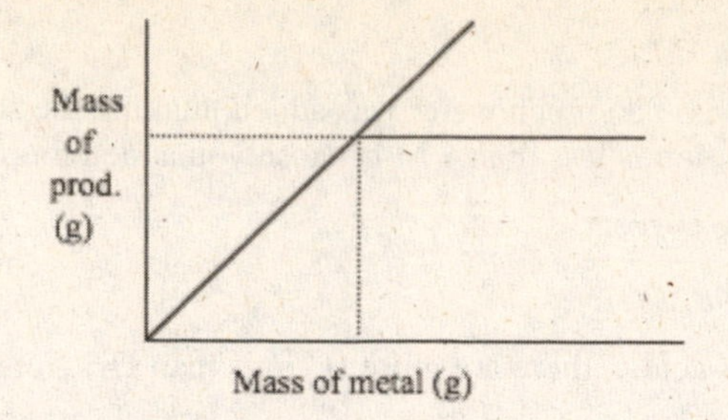

The crossover point on the magnesium curve is estimated at 1.6 g Mg and 11.4 g product. That means, in a product sample weighting 11.4 grams, the mass of Br is:

$$11.4 \text{ g sample} - 1.6 \text{ g Mg} = 9.8 \text{ g Br}$$

$$1.6 \text{ g Mg} \times \frac{1 \text{ mol Mg}}{24.3050 \text{ g Mg}} = 0.066 \text{ mol Mg} \qquad 9.8 \text{ g Br} \times \frac{1 \text{ mol Br}}{79.904 \text{ g Br}} = 0.12 \text{ mol Br}$$

$$\frac{0.12 \text{ mol Br}}{0.066 \text{ mol Mg}} = 1.9 \approx 2$$

So, the empirical formula of the product is $MgBr_2$, as predicted.

The crossover point on the calcium curve is estimated at 2.5 g Ca and 12.4 g product. That means, in a product sample weighing 12.4 grams, the mass of Br is:

$$12.4 \text{ g sample} - 2.5 \text{ g Ca} = 9.9 \text{ g Br}$$

$$2.5 \text{ g Ca} \times \frac{1 \text{ mol Ca}}{40.078 \text{ g Ca}} = 0.062 \text{ mol Ca} \qquad 9.9 \text{ g Br} \times \frac{1 \text{ mol Br}}{79.904 \text{ g Br}} = 0.12 \text{ mol Br}$$

$$\frac{0.12 \text{ mol Br}}{0.062 \text{ mol Ca}} = 2.0 \approx 2$$

So, the empirical formula of the product is $CaBr_2$, as predicted.

The crossover point on the strontium curve is estimated at 5.5 g Ca and 15.3 g product. That means, in a product sample weighing 15.3 grams, the mass of Br is:

$$15.3 \text{ g sample} - 5.5 \text{ g Sr} = 9.8 \text{ g Br}$$

$$5.5 \text{ g Sr} \times \frac{1 \text{ mol Sr}}{87.62 \text{ g Sr}} = 0.063 \text{ mol Sr} \qquad 9.8 \text{ g Br} \times \frac{1 \text{ mol Br}}{79.904 \text{ g Br}} = 0.12 \text{ mol Br}$$

$$\frac{0.12 \text{ mol Br}}{0.063 \text{ mol Sr}} = 1.9 \approx 2$$

So, the empirical formula of the product is $SrBr_2$, as predicted.

✓ *Reasonable Answer Check:* It makes sense that the mass of bromine from all three samples is approximately the same, since all three compounds were determined to contain the same number of bromine atoms. The only common ionic charge found in Group (II) ions is +2 and the only common ionic charge in halide ions is –1, so it makes sense that they would end up combining with one metal ion and two bromide ions. The magnesium and calcium data required obvious rounding to get an integer ratio. The approximations made from the very small graph could have contributed to this degree of imprecision. In a real experiment, the graph would be larger with better-scaled axes so that results extracted from it would be more reliable. The shape of the curve also lends to some approximations. However, even in a real experiment, the data themselves might provide such imprecise results.

126. *Answer:* **(a) Groups C&D: $Ag^+ + Cl^- \longrightarrow AgCl\ (s)$; Groups A&B: $Ag^+ + Br^- \longrightarrow AgBr\ (s)$ (b) silver halide product is the same for A&B and different from C or D (c) Curve has upward slope while the Ag^+ is the limiting reactant. Bromide is heavier than chloride, so the curve levels out at different masses of product.**

Strategy and Explanation:

(a) Na^+ and NO_3^- are spectator ions.

Net Ionic for groups C and D	Net Ionic for groups A and B
$Ag^+ + Cl^- \longrightarrow AgCl\ (s)$	$Ag^+ + Br^- \longrightarrow AgBr\ (s)$

(b) The silver halide solid (AgCl) produced by groups C and D is the same. The silver halide solid (AgBr) produced by groups A and B is the same.

(c) Below the mass of 0.75 g, the graph line increases with increasing masses of $AgNO_3$. In this region, Ag^+ is the limiting reactant. (See Question 124 for a more complete explanation.) These reactions both require the same mass of $AgNO_3$ to make their respective silver halide. However, since bromide ion is heavier than chloride ion, the mass of the product will be different and the mass of product where the graph levels out will be different because AgBr is heavier than AgCl. That means the products in groups A and B (AgCl) weigh less than the products in groups C and D (AgBr).

More Challenging Questions

130. *Answer:* **0.0154 M $CaSO_4$; 0.341 g $CaSO_4$ undissolved**

Strategy and Explanation: Given the maximum amount of an ionic solid that will dissolve in a fixed volume of a solution, the mass of the ionic solid added to a specific volume of a solution, determine the molarity of the solute in the solution and the mass of the solid that does not dissolve.

Use the molar mass to determine moles of solute dissolved in the solution, and determine the number of liters of solution. Divide these two numbers to get the molarity. Subtract the mass of solid that will dissolve from the mass of the solid added to the solution to determine how much does not dissolve.

The most $CaSO_4$ that can dissolve in 100.0 mL of water is 0.209 grams. Since more than that amount of solid was added, only 0.209 grams of it will dissolve.

$$\frac{0.209\text{ g CaSO}_4}{100.0\text{ mL}} \times \frac{1\text{ mol CaSO}_4}{136.141\text{ g CaSO}_4} \times \frac{1000\text{ mL}}{1\text{ L}} = 0.0154\text{ M CaSO}_4$$

0.550 g $CaSO_4$ added – 0.209 g $CaSO_4$ dissolved = 0.341 g $CaSO_4$ remain undissolved

✓ *Reasonable Answer Check:* There is more than the maximum amount of solid added, so it makes sense that there would be some left over.

132. *Answer:* **184 mL $K_2Cr_2O_7$**

Strategy and Explanation: Given the molarity of a reactant in a solution and the moles of product generated, determine the volume of the solution for a complete reaction with excess amounts of the other reactants available.

First, use the moles of product and the equation stoichiometry to find out moles of the reactant needed. Then use the moles and molarity to find volume in liters, and convert liters into milliliters. *Note: It is NOT appropriate to use the dilution equation when working with reactions!*

The balanced equation says 3 mol of CH_3COOH are produced from 2 mol $K_2Cr_2O_7$.

$$0.166\text{ mol CH}_3\text{COOH} \times \frac{2\text{ mol K}_2\text{Cr}_2\text{O}_7}{3\text{ mol CH}_3\text{COOH}} \frac{1\text{ L K}_2\text{Cr}_2\text{O}_7\text{ solution}}{0.600\text{ mol K}_2\text{Cr}_2\text{O}_7} \times \frac{1000\text{ mL}}{1\text{ L}} = 184\text{ mL K}_2\text{Cr}_2\text{O}_7$$

135. *Answer:* **3**

Strategy and Explanation: Given the mass of a sample of an acid dissolved in a given amount of water and neutralized with a given volume of a known concentration base, determine the number of acidic hydrogen's in a molecule of the acid.

Use the molar mass of the acid to calculate the moles of acid. Use the volume and molarity of the base, and the stoichiometry of acid base reactions (1 OH^- reacts with 1 H^+) base to determine how many acidic hydrogens were neutralized. Calculate the mole ratio to determine the stoichiometric relationship of acidic hydrogens in the acid.

$$0.400 \text{ g } C_6H_8O_7 \times \frac{1 \text{ mol } C_6H_8O_7}{192.123 \text{ g } C_6H_8O_7} = 0.00208 \text{ mol } C_6H_8O_7$$

$$31.2 \text{ mL NaOH} \times \frac{1 \text{ L NaOH}}{1000 \text{ mL NaOH}} \times \frac{0.200 \text{ mol NaOH}}{1 \text{ L NaOH}} \times \frac{1 \text{ mol } OH^-}{1 \text{ mol NaOH}} \times \frac{1 \text{ mol } H^+}{1 \text{ mol } OH^-} = 0.00624 \text{ mo } H^+$$

$$\frac{0.00624 \text{ mol } H^+}{0.00208 \text{ mol } C_6H_8O_7} = 3$$

✓ *Reasonable Answer Check:* The ratio is a whole number.

Chapter 6: Energy and Chemical Reactions

Solutions for Blue-Numbered Questions for Review and Thought

Topical Questions

The Nature of Energy (Section 6.1)

11. *Answer:* **(a) 399 Cal (b) 5.0×10^6 J/day**

Strategy and Explanation: Convert a quantity of kilojoules (provided by a piece of cake) into food Calories, and convert a quantity of food Calories into joules.

Use metric and energy conversion factors to achieve the conversions.

(a) $$1670\ \text{kJ} \times \frac{1000\ \text{J}}{1\ \text{kJ}} \times \frac{1\ \text{cal}}{4.184\ \text{J}} \times \frac{1\ \text{kcal}}{1000\ \text{cal}} \times \frac{1\ \text{Cal}}{1\ \text{kcal}} = 399\ \text{Cal}$$

(b) $$\frac{1200\ \text{Cal}}{1\ \text{day}} \times \frac{1\ \text{kcal}}{1\ \text{Cal}} \times \frac{1000\ \text{cal}}{1\ \text{kcal}} \times \frac{4.184\ \text{J}}{1\ \text{cal}} = \frac{5.0 \times 10^6\ \text{J}}{1\ \text{day}}$$

✓ *Reasonable Answer Check:* The food Calorie is about four times bigger than a kilojoule. So, the energy quantity in Calories should be about four times smaller than in kilojoules.

13. *Answer:* **1.12×10^4 J**

Strategy and Explanation: Given the calories required to melt one gram of lead, determine the number of joules needed to melt a larger mass of lead.

Start with the sample, the mass of the lead. Use metric and energy conversion factors to achieve the conversions.

$$454\ \text{g} \times \frac{5.91\ \text{cal}}{1\ \text{g}} \times \frac{4.184\ \text{J}}{1\ \text{cal}} = 1.12 \times 10^4\ \text{J}$$

✓ *Reasonable Answer Check:* A larger mass of lead will require more energy, and the smaller unit of joules will also make the answer larger in size, so it makes sense that the result ends up large.

15. *Answer:* **3.60×10^6 J, \$0.03**

Strategy and Explanation: Determine the number of joules in a kilowatt-hour and the cost of a megajoule of electricity given the cost per kilowatt-hour.

Use metric and energy conversion factors to achieve the conversions.

$$1\ \text{kW-hr} \times \frac{1000\ \text{W}}{1\ \text{kW}} \times \frac{1\ \frac{\text{J}}{\text{s}}}{1\ \text{W}} \times \frac{3600\ \text{s}}{1\ \text{hr}} = 3.60 \times 10^6\ \text{J}$$

$$1\ \text{MJ} \times \frac{10^6\ \text{J}}{1\ \text{MJ}} \times \frac{1\ \text{kW-hr}}{3.60 \times 10^6\ \text{J}} \times \frac{\$0.09}{1\ \text{kW-hr}} = \$0.03$$

✓ *Reasonable Answer Check:* A kilowatt-hour is much bigger than a joule. So, the energy quantity in kilojoules should be much smaller than in kW-hr. A megajoule is similar in size but smaller than a kilowatt-hour, so the cost would be slightly lower.

17. *Answer:* **the first product**

Strategy and Explanation: Given the energy content of one food product and the kJ per serving of another, determine which food provides the greater energy per serving.

Use energy conversion factors to get comparable results.

$$170\ \text{kcal} \times \frac{4.184\ \text{kJ}}{1\ \text{kcal}} = 710\ \text{kJ in one serving of the first product}$$

Compared to 280 kJ per serving of the second product, the first product provides the greater energy per serving.

✓ *Reasonable Answer Check:* A kcal is larger than a kJ, so it makes sense that the number of kJ is greater than the number of kcals.

Conservation of Energy (Section 6.2)

19. (a) In the process of lighting the match, the kinetic energy of moving the match across the striking surface is converted into thermal energy due to friction. This thermal energy causes the match to "light." This "lighting" is the result of a combustion process, where the chemical energy stored in the reactants is converted into heat energy and light energy. The heat energy of the match is used to light the fuse, and the chemical energy of the fuse is converted into heat energy and light energy. The heat from the fuse ignites the chemical propellants in the rocket. The chemical energy here is converted into heat, light, and the kinetic energy and potential energy of the rocket as the rocket's speed and altitude increase. When the rocket explodes, more chemical energy is converted into light, heat, and kinetic energy.

(b) As the fuel is pumped from the underground storage tank, its potential energy is increased by the mechanical energy of the pump. By using the fuel to drive 25 miles, some of the chemical potential energy stored in the fuel is converted into kinetic energy that moves the car and into heat energy energy that warms the engine and passenger compartment, when necessary.

21. *Answer/Explanation:* Describe and explain your choice of the system and the surroundings, and describe transfer of energy and materials into and out of the system.

The system is identified as precisely what we are studying. The surroundings are everything else. The important aspects of the surroundings are usually those things in contact with the system or in close proximity.

(a) The System: The plant (stem, leaves, roots, etc.)

The Surroundings: Anything not the plant (air, soil, water, sun, etc.)

(b) To study the plant growing, we must isolate it and see how it interacts with its surroundings.

(c) Light energy and carbon dioxide are absorbed by the leaves and are converted to other molecules storing the energy as chemical energy and using it to increase the size of the plant. Nutrients are absorbed through the soil (minerals and water) to assist in the chemical processes. The plant expels oxygen and other waste materials into the surroundings.

✓ *Reasonable Answer Check:* This definition of the system differentiates between the live organism and the materials and energy required for it to stay alive.

23. *Strategy and Explanation:* Describe and explain your choice of the system and the surroundings, describe transfer of energy and materials into and out of the system, and determine if the process is exothermic or endothermic.

The system is identified as precisely what we are studying. The surroundings are everything else. The important aspects of the surroundings are usually those things in contact with the system or in close proximity. The process is exothermic if the system loses energy. The process is endothermic if the system gains energy.

(a) The System: NH_4Cl

This choice was made following the discussion in Section 6.2 (page 218) when describing a reaction, "the system is usually defined as all the atoms that make up the reactants." Here the equation for the reaction being studied is: $NH_4Cl(s) \longrightarrow NH_4^+(aq) + Cl^-(aq)$

The Surroundings: Anything not NH_4Cl, including the water.

The choice to consider water as part of the surroundings instead of part of the system is based upon the fact that it is neither a reactant nor a product in given the dissolving equation. Although water undeniably has a strong interaction with the products of the reaction, there are still no H_2O molecules in this equation. Defining the system to include the water can also be done. Data collected from dissolving experiments will not actually be able to functionally separate the water from the material dissolving.

(b) To study the release of energy during the phase change of this ionic compound, we must isolate it and see how it interacts with the surroundings.

(c) The system's interaction with the surroundings causes heat energy to be transferred into the surroundings and out of the system. There is no material transfer in this process, but there is a change in the specific interaction between the water and system.

(d) Since changes in the system cause energy to be gained by the system, the process is endothermic.

✓ *Reasonable Answer Check:* This definition of the system is restrictive enough to allow us to learn more about the relationship between the energy required to break the ionic bonds in the solid and the energy involved with the products' increased interaction with the surrounding solvent molecules.

25. *Answer:* **$\Delta E = +32$ J**

Strategy and Explanation: Given descriptions and numerical values for work and heat energy changes, determine the ΔE for the system.

The change in energy of the system, ΔE, is calculated using Equation 6.1 on page 219: $\Delta E = q + w$, where q and w (work energy and heat energy) cause energy to flow across the boundary between the system and the surroundings. In chemisty, both q and w are chosen to be positive when the direction of energy flow is from the surroundings into the system. Alternatively, w and q will carry a negative sign, if energy flows out of the system into the surroundings.

The system does work, so energy flows out of the system and w is negative: $w = -75.4$ J.

Heat energy is transferred into the system, so energy flows into the syste and q is positive: $q = +25.7$ cal

Notice: Make sure that the units of energy are the same before adding the energy values.

$$\Delta E = (+25.7\text{ cal}) \times \frac{4.184\text{ J}}{1\text{ cal}} + (-75.4\text{ J}) = 32\text{ J}$$

✓ *Reasonable Answer Check:* The heat energy input is more than the work energy output, so it is reasonable that ΔE is positive.

27. *Answer:* **(see diagram below), $\Delta E_{system} = 715.6$ kJ**

Strategy and Explanation: Make heat flow diagram for given heat energy and work energy changes in a system and use that to help determine the ΔE_{system} using the method described in the solution to Question 25.

Work is done on the surroundings, so energy flows out of the system and w is negative: $w = -127.6$ kJ

Heat energy is transferred into the system, so energy flows into the system and q is positive: $q = 843.2$ kJ

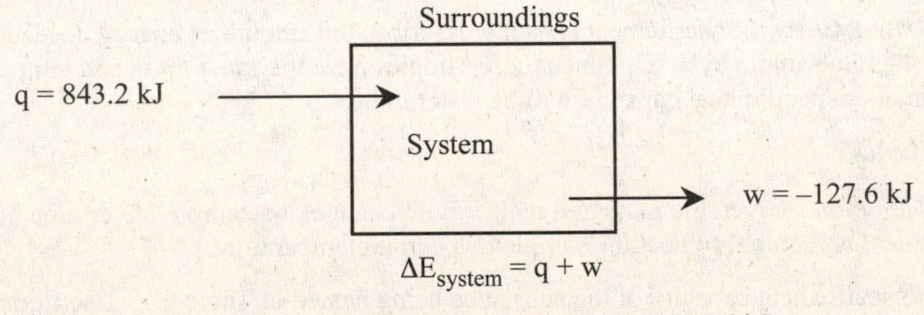

$$\Delta E_{system} = q + w$$

$$\Delta E_{system} = 843.2\text{ kJ} + (-127.6\text{ kJ}) = 715.6\text{ kJ}$$

✓ *Reasonable Answer Check:* More heat energy is entering the system than work energy leaving the system, so it makes sense that the ΔE is positive.

Heat Capacity (Section 6.3)

29. *Answer:* **Process (a) requires more energy than (b)**

Strategy and Explanation: Given the mass and temperature change for two samples, determine which process requires a greater transfer of energy.

Get the specific heat capacity of water from Table 6.1 on page 222, then use Equation 6.2' on page 221 to calculate the heat energy required in each scenario.

Table 6.1 tells us that the specific heat capacity of liquid water is 4.184 J $g^{-1}°C^{-1}$. Use Equation 6.2':

$$q = c \times m \times \Delta T.$$

(a) $\Delta T = T_f - T_i = 50\text{ °C} - 20\text{°C} = 30\text{ °C}$ *(must be rounded to tens place)*

$$q_{H_2O} = (4.184\text{ J g}^{-1}\text{°C}^{-1}) \times (10.0\text{ g}) \times (30\text{ °C}) = 1 \times 10^3\text{ J}$$

(b) $\Delta T = T_f - T_i = 37\text{ °C} - 25\text{°C} = 12\text{ °C}$

$$q_{Cu} = (0.385\text{ J g}^{-1}\text{°C}^{-1}) \times (20.0\text{ g}) \times (12\text{ °C}) = 92\text{ J}$$

The cooling described in (a) requires a greater transfer of energy than that described in (b).

✓ *Reasonable Answer Check:* The specific heat capacity of water is more that ten times the specific heat capacity of copper and the mass and the temperature change for the copper sample was also smaller. The water sample should thus require much greater transfer of energy.

31. *Answer:* **It takes less time to raise the Cu sample to body temperature.**

Strategy and Explanation: Given the mass and initial temperature of two samples made out of two different metals, also given the assumption that heat energy is transferred at the same rate when held in your hand, determine which metal will warm up to body temperature faster.

First, look up the specific heat capacities of the substances being heated in Table 6.1. The rate of absorption of heat energy can be related to heat capacity, mass, temperature increase, and elapsed time using Equation 6.2. The elapsed time can then be related to the specific heat capacity, the mass, the change in temperature, and the rate of absorption of heat energy. The fastest elapsed time will reach body temperature first.

The specific heat capacity (c) of Al is 0.902 J $g^{-1}°C^{-1}$ and of Cu is 0.385 J $g^{-1}°C^{-1}$ from Table 6.1. Use Equation 6.2, then divide by Δt to write an equation relating the rate of heat energy absorption to the specific heat capacity, mass, temperature increase, and elapsed time:

$$\text{rate of heat energy absorption} = \frac{q}{\Delta t} = c \times m \times \frac{\Delta T}{\Delta t}$$

Solve this equation for the elapsed time, Δt:

$$\Delta t = c \times \frac{m \times \Delta T}{\text{rate}}$$

Comparing the two samples, their mass, change in temperature, and the rate of heat energy absorption are all the same, simplifying the equation above to the following: Δt = c × constant. This equation demonstrates that the elapsed heating time is proportional to the specific heat capacity. That means the copper sample will warm to body temperature before the aluminum sample at a constant rate over a common temperature range.

✓ *Reasonable Answer Check:* Specific heat capacity describes the amount of energy needed to increase the temperature of a 1-gram sample by 1 °C. Since these samples have the same mass and temperature change, the object with the smaller specific heat capacity will be faster to heat.

33. *Answer:* **1.0×10^2 kJ**

Strategy and Explanation: Given the mass and temperature change of a sample, determine how much thermal energy (in kilojoules) is required to heat the sample to a certain temperature.

First, look up the specific heat capacity of the substance being heated in Table 6.1. Use Equation 6.2' to calculate the heat energy required.

Table 6.1 tells us that the specific heat capacity (c) of Al is 0.902 J $g^{-1}°C^{-1}$.

$$q = c \times m \times \Delta T.$$

$$\Delta T = T_f - T_i = 250\ °C - 25\ °C = 225\ °C \cong 230\ °C$$

Notice: because of the limited precision in the given oven temperature, the result must be rounded to tens place.

$$q_{Al} = (0.902\ J\ g^{-1}°C^{-1}) \times (500.\ g) \times (230\ °C) = 1.0 \times 10^5\ J$$

$$1.0 \times 10^5\ J \times \frac{1\ kJ}{1000\ J} = 1.0 \times 10^2\ kJ$$

✓ *Reasonable Answer Check:* This is a relatively large amount of aluminum, and the temperature increase is also fairly large, so it makes sense that the amount of energy is conveniently expressed in units of kilojoules.

35. *Answer:* **136 J mol^{-1} K^{-1}**

Strategy and Explanation: Given the specific heat capacity of a known substance, determine the molar heat capacity.

Calculate the molar mass of C_6H_6, and use that to calculate the molar heat capacity

The molar mass of C_6H_6 = 6 × (12.0107 g) + 6 × (1.0079 g) = 78.1116 g/mol

$$1.74\ \frac{J}{g\ K} \times \frac{78.1116\ g}{1\ mol} = 136\frac{J}{mol\ K}$$

✓ *Reasonable Answer Check:* The molar mass is larger than 1 gram, so it makes sense that the molar heat capacity is larger than the specific heat capacity.

37. *Answer:* **0.270 J g^{-1} °C^{-1}**

Strategy and Explanation: Given the masses and initial temperatures of a piece of metal and a quantity of water, and given the final temperature of the water after the metal is added, determine the specific heat capacity of the metal.

Thermal equilibrium is reached when the water and the metal reach the same temperature, so we know the final temperature of the metal is the same as the final temperature of the water. Use Equation 6.2' ($q = c \times m \times \Delta T$) to calculate the heat gained by the water, which also represents the heat lost by the metal. Use Equation 6.2 to calculate the specific heat capacity of the metal.

Table 6.1 tells us that the specific heat capacity (c) of water is 4.184 J g^{-1}°C^{-1}.

$$\Delta T = T_f - T_i = 15.3\ °C - 10.0\ °C = 5.3\ °C$$

$$q_{water} = (4.184\ J\ g^{-1}°C^{-1}) \times (244\ g) \times (5.3\ °C) = 5400\ J\ \ \text{(two sig figs)}$$

$$q_{water} = -\,q_{Mo}$$

$$c_{Mo} = \frac{q_{Mo}}{m_{Mo} \times \Delta T} = \frac{-5400\ J}{237\ g \times (15.3\ °C - 100.0\ °C)} = 0.270\ \frac{J}{g\ °C}$$

✓ *Reasonable Answer Check:* Because water as a much larger specific heat capacity than Mo, we expect a smaller temperature increase in the water than the decrease experienced by the molybdenum. Most metals have a heat capacity in this range.

39. *Answer:* **More energy (1.48 × 10^6 J) is absorbed by the water sample than by the ethylene glycol sample (9.56 × 10^5 J).**

Strategy and Explanation: Given the volume of a cooling system, the densities of water and ethylene glycol, and the temperature change for two different samples, compare the thermal energy increase in the two samples.

First convert the volumes into cubic centimeters, then use the densities to determine the masses of each sample. Then use Equation 6.2' ($q = c \times m \times \Delta T$) to calculate the heat energy required in each process and compare them.

Volume of the cooling system in cubic centimeters:

$$5.00\ \text{quarts} \times \frac{0.946\ L}{1\ \text{quart}} \times \frac{1000\ mL}{1\ L} \times \frac{1\ cm^3}{1\ mL} = 4730\ cm^3$$

Mass of the two samples:

$$4730\ cm^3 \times \frac{1.113\ g}{1\ cm^3} = 5260\ g\ \text{ethylene glycol}$$

$$4730\ cm^3 \times \frac{1.00\ g}{1\ cm^3} = 4730\ g\ \text{water}$$

In both samples, the change in temperature is the same:

$$\Delta T = T_f - T_i = 100.0\ °C - 25.0\ °C = 75.0\ °C$$

$$q_{ethylene\ glycol} = (2.42\ J\ g^{-1}°C^{-1}) \times (5260\ g) \times (75.0\ °C) = 9.56 \times 10^5\ J$$

$$q_{water} = (4.184\ J\ g^{-1}°C^{-1}) \times (4730\ g) \times (75.0\ °C) = 1.48 \times 10^6\ J$$

More energy is absorbed (1.48 × 10^6 J) by the water sample than is absorbed (9.56 × 10^5 J) by the ethylene glycol sample.

✓ *Reasonable Answer Check:* Water has a much larger specific heat capacity than ethylene glycol, so a somewhat smaller mass of water will still absorb more thermal energy than a larger mass of ethylene glycol.

41. *Answer:* **330. °C**

Strategy and Explanation: Given the mass of a piece of hot metal, the mass and temperature of a sample of cool water, and the final temperature after the two are combined and heat energy is transferred, determine the initial temperature of the metal.

First, look up the specific heat capacities of the metal and the water. Assuming that the temperature stopped dropping once the water and the metal reached the same temperature, set the final temperature of the metal to be the same as the final temperature of the water. Use Equation 6.2' to calculate the heat energy gained by the water from the metal. The heat energy gained by the water is the heat energy lost by the metal. Finally, rearrange Equation 6.2 to calculate the initial temperature of the metal.

The specific heat capacity (c) of Fe is 0.451 J $g^{-1}°C^{-1}$ and of water is 4.184 J $g^{-1}°C^{-1}$ according to Table 6.1.

$$\Delta T_{water} = T_f - T_i = 32.8\ °C - 20.0\ °C = 12.8\ °C$$

$$q_{water} = (4.184\ J\ g^{-1}°C^{-1}) \times (1.00\ kg) \times \left(\frac{1000\ g}{1\ kg}\right) \times (12.8\ °C) = 5.36 \times 10^4\ J$$

Heat energy is gained by the water, so q_{water} is positive. Heat energy is lost by the iron, so q_{iron} is negative. The quantity of heat energy lost by the hot iron is absorbed by the cold water: $q_{iron} = -\,q_{water} = -5.36 \times 10^4$ J

$$\Delta T_{iron} = \frac{q_{iron}}{c \times m} = \frac{-5.36 \times 10^4\ J}{0.451\ J\ g^{-1}°C^{-1} \times 400.\ g} = -297\ °C$$

$$T_i = T_f - \Delta T_{iron} = 32.8\ °C - (-297\ °C) = 330.\ °C$$

✓ *Reasonable Answer Check:* Because water has a much larger specific heat capacity than iron, a smaller temperature rise is experienced by the water than temperature drop experienced by the iron.

43. *Answer:* **(a) 0.45 J $g^{-1}°C^{-1}$ (b) 25 J $mol^{-1}°C^{-1}$**

Strategy and Explanation: Given the mass, initial and final temperatures, and energy needed to heat a piece of metal, determine the specific heat capacity and the molar heat capacity.

Plug known values into Equation 6.2. Convert grams to moles with molar mass to get the molar heat capacity

(a)
$$c = \frac{q}{m \times \Delta T} = \frac{41.0\ J}{12.3\ g \times (24.7\ °C - 17.3\ °C)} = \frac{41.0\ J}{12.3\ g \times 7.4\ °C} = 0.45\ \frac{J}{g\ °C}$$

(b) The molar mass of iron is 55.845 g/mol. Use that to calculate the molar heat capacity:

$$0.45 \frac{J}{g\ °C} \times \frac{55.845\ g}{1\ mol} = 25\ \frac{J}{mol\ °C}$$

✓ *Reasonable Answer Check:* The specific heat capacity of iron is 0.451 J $g^{-1}°C^{-1}$ according to Table 6.1, so the specific heat capacity calculated here makes sense. The molar mass is larger than 1 gram, so it makes sense that the molar heat capacity is larger than the specific heat capacity.

45. *Answer:* **Gold**

Strategy and Explanation: Given the mass, initial and final temperatures, and energy needed to heat an unknown element, determine its most probable identity using Table 6.1.

Adapt the method described in the solution to Question 43.

$$c = \frac{q}{m \times \Delta T} = \frac{34.7\ J}{23.4\ g \times (28.9\ °C - 17.3\ °C)} = 0.128\ \frac{J}{g\ °C}$$

Looking at Table 6.1, the element whose specific heat capacity is closest to this is Au.

✓ *Reasonable Answer Check:* The specific heat capacity of Au ($0.128\ J\ g^{-1}{}^{\circ}C^{-1}$) matches the calculated value to three significant figures ($0.128\ J\ g^{-1}{}^{\circ}C^{-1}$). The similarity in the two values gives us confidence in the answer.

47. *Answer:* **160 °C**

Strategy and Explanation: Given the mass of a piece of hot metal, the volume, density, and temperature of a sample of cool water, and the final temperature after the two are combined and heat energy is transferred, determine the initial temperature of the metal.

First, look up the specific heat capacities of the metal and the water. Use Equation 6.2' to calculate the heat energy gained by the water from the metal. The heat energy gained by the water is the heat energy lost by the metal. Finally, rearrange Equation 6.2 to calculate the initial temperature of the metal.

The specific heat capacity (c) of Al is $0.902\ J\ g^{-1}{}^{\circ}C^{-1}$ and of water is $4.184\ J\ g^{-1}{}^{\circ}C^{-1}$ according to Table 6.1.

$$\Delta T_{water} = T_f - T_i = 33.6\ ^{\circ}C - 22.0\ ^{\circ}C = 11.6\ ^{\circ}C$$

$$q_{water} = (4.184\ J\ g^{-1}{}^{\circ}C^{-1}) \times (500.\ mL) \times \left(\frac{0.98\ g}{1\ ml}\right) \times (11.6\ ^{\circ}C) = 2.4 \times 10^4\ J$$

Heat energy is gained by the water, so q_{water} is positive. Heat energy is lost by the aluminum, so $q_{aluminum}$ is negative. The quantity of heat energy lost by the hot aluminum is absorbed by the cold water. So,

$$q_{aluminum} = -\ q_{water} = -2.4 \times 10^4\ J$$

$$\Delta T_{aluminum} = \frac{q_{aluminum}}{c \times m} = \frac{-2.4 \times 10^4\ J}{0.902\ J\ g^{-1}{}^{\circ}C^{-1} \times 200.\ g} = -130\ ^{\circ}C$$

$$T_i = T_f - \Delta T_{aluminum} - T_f = 33.6\ ^{\circ}C - (-130\ ^{\circ}C) = 160\ ^{\circ}C$$

✓ *Reasonable Answer Check:* Because water has a much larger specific heat capacity than aluminum, a smaller temperature increase is experienced by the water than temperature decrease experienced by the aluminum. Quantitatively, we can use $q = c \times m \times \Delta T$ to confirm that the aluminum is losing 2.3×10^4 J at the same time as the water is gaining 2.4×10^4 J. They are the same within the uncertainty limit of $\pm\ 0.1 \times 10^4$ J.

49. *Answer:* **$\Delta T_{surroundings}$ = positive, ΔE_{system} = negative**

Strategy and Explanation: Given the direction of transfer of thermal energy between a system and the surroundings with no work done, determine the algebraic sign of $\Delta T_{surroundings}$ and ΔE_{system}.

If thermal energy enters the surroundings as heat energy, then the temperature in the surroundings, $T_{surroundings}$, rises. If thermal energy enters the system as heat energy, then the temperature in the surroundings, $T_{surroundings}$, drops. If the system gets energy from the surroundings, then the internal energy of the system, E_{system}, rises. If the surroundings gets energy from the system, then the internal energy of the system, E_{system}, lowers.

The thermal energy enters the surroundings as heat energy, so the temperature in the surroundings, $T_{surroundings}$, rises:

$$T_{f,surroundings} > T_{i,surroundings}$$

$$\Delta T_{surroundings} = T_{f,surroundings} - T_{i,surroundings} = \text{positive}$$

Here, the surroundings gets energy from the system, so the internal energy of the system, E_{system}, lowers:

$$E_{f,system} < E_{i,system}$$

$$\Delta E_{system} = E_{f,system} - E_{i,system} = \text{negative}$$

✓ *Reasonable Answer Check:* Energy leaves the system, so a negative ΔE_{system} makes sense. The energy must show up in the surroundings, so the fact that these two algebraic signs are opposite signs also makes sense.

Energy and Enthalpy (Section 6.4)

51. *Answer:* **330. J**

Strategy and Explanation: Given a tray of ice, the number of cubes in the tray, the mass of each cube, and the energy required to melt a sample of ice at the melting point, determine the energy required to melt the tray of ice cubes to form liquid water at the same temperature.

Use the given information as unit factors to determine the quantity of energy.

$$1\text{ tray} \times \frac{16\text{ cubes}}{1\text{ tray}} \times \frac{62.0\text{ g ice}}{1\text{ cube}} \times \frac{333\text{ J}}{1\text{ g ice}} \times \frac{1\text{ kJ}}{1000\text{ J}} = 330.\text{ kJ}$$

✓ *Reasonable Answer Check:* The tray has over 1000 grams of ice, so this large number of joules makes sense. Notice, the tray and the number of cubes are countable objects; hence, those numbers are exact with infinite significant figures, so their values do not limit the significant figures of the answer.

53. *Answer:* **10.3 kJ**

Strategy and Explanation: Given the moles of a sample of mercury and the heat of fusion of mercury at the freezing point, determine the quantity of energy transferred when freezing the sample at the freezing point.

Calculate the grams then use the heat of fusion to find energy.

$$4.37\text{ mol Hg} \times \frac{200.59\text{ g Hg}}{1\text{ mol Hg}} \times \frac{2.82\text{ cal}}{1\text{ mol}} = 2.47 \times 10^3\text{ cal}$$

$$2.47 \times 10^3\text{ cal} \times \frac{4.184\text{ J}}{1\text{ cal}} \times \frac{1\text{ kJ}}{1000\text{ J}} = 10.3\text{ kJ}$$

✓ *Reasonable Answer Check:* This quantity of heat seems reasonable.

55. *Answer:* **273 J New 55 replaces 49**

Strategy and Explanation: Given the volume and initial temperature of a liquid substance, the freezing point of the liquid, the final temperature of the solid substance, the density of the liquid, the specific heat capacity, and the enthalpy of fusion, determine the thermal energy (in joules) that must be released to the surroundings to complete the transition.

Use Equation 6.2', determine the heat energy that must be released when lowering the temperature to the freezing point. Then use the enthalpy of fusion (ΔH_{fus}) to determine the thermal energy that must be released from the liquid to form the solid at the freezing point.

Define the system as the liquid mercury.

$$m_{mercury} = (1.00\text{ mL}) \times \frac{1\text{ cm}^3}{1\text{ mL}} \times \frac{13.6\text{ g Hg}}{1\text{ cm}^3} = 13.6\text{ g Hg}$$

To change the temperature of the system, use $q = c \times m \times \Delta T$

$$q_{T\text{-drop}} = c_{mercury} \times m_{mercury} \times \Delta T_{mercury} = (0.140\text{ J g}^{-1}{}^\circ\text{C}^{-1}) \times (13.6\text{ g Hg}) \times (-38.8\text{ °C} - 23.0\text{ °C}) = -117.7\text{ J}$$

To change a phase from liquid to solid in the system, use $q_{freeze} = -\,q_{fusion} = -\,m \times \Delta H_{fus}$.

$$q_{freeze} = -\,m_{mercury} \times \Delta H_{fus,mercury} = -\,(13.6\text{ g Hg}) \times (11.4\text{ J/g}) = -155\text{ J}$$

The sum gives the total heat energy required.

$$q_{total} = q_{T\text{-drop}} + q_{freeze} = (-117.7\text{ J}) + (-155\text{ J}) = -273\text{ J}$$

The amount of thermal energy that must be transferred to the surroundings is 273 J.

✓ *Reasonable Answer Check:* To make the final product, the liquid's temperature needed to be lowered, and then the liquid needed to be frozen. Both of these changes require energy to be removed from the system.

57. A cooling curve shows how the temperature drops as the heat energy is removed from the system. In that respect, the lower part of this graph will look like the reverse of Figure 6.11.

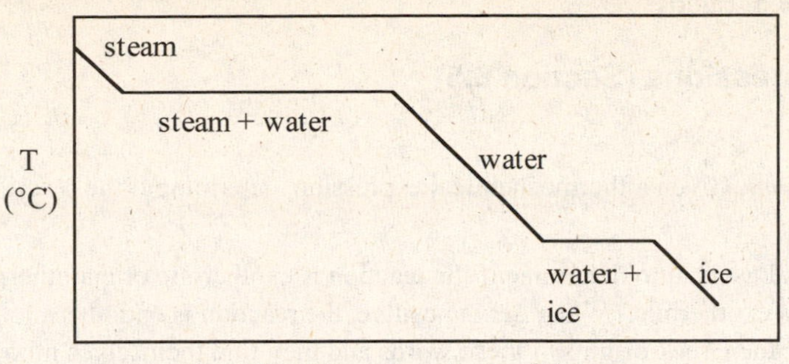

Quantity of heat transferred out of the system

59. *Answer:* **(a) X(s) (b) the enthalpy of fusion (c) positive**

Strategy and Explanation:

(a) The line on the graph with the steepest vertical slope has the largest temperature change when a fixed amount of heat energy is added. The substantial temperature increase means the specific heat capacity of the substance in that phase is smaller than in the other phases. Since the slope of the line representing the heating of the solid is steeper than the slopes of the lines representing the heating of the liquid or the gas, X(s) has the largest specific heat capacity.

(b) The horizontal lines on the graph represent phase transitions. The longest horizontal line on the graph indicates that more energy is required for that phase transition than the other does. Since the line representing the melting of the solid to liquid is shorter than the line representing the vaporization of the liquid to gas, the enthalpy of fusion is smaller than the enthalpy of vaporization. This is not surprising, since the vaporization of a liquid requires enough energy to overcome the all the attractions between the molecules in the liquid state; whereas the melting process only requires some of the attractive forces be overcome. We always find that the enthalpy of fusion is smaller than the enthalpy of vaporization.

(c) The algebraic sign of the enthalpy of vaporization is always positive, since heat energy is required to overcome all of the attractions between the molecules in the liquid state.

61. *Answer:* **(a) negative (b) positive**

Strategy and Explanation: Work is described in Section 6.4 on page 226 of the textbook. At constant pressure, $w = -P\Delta V$, hence it is sometimes referred to as PV work When chemical reactions involving gases occur, there may be noticeable changes in volume since the density of gases are typically much smaller than those of condensed states. Therefore, we will focus on the consumption or production of gases in the reaction.

(a) This reaction produces gaseous H_2S from condensed-phase reactants, so the volume of the system is greater than that of the reactants at fixed temperature and pressure. Work is done on the surroundings as the system's volume expands, so the sign of w is **negative**.

(b) This reaction consumes five moles of O_2 when it produces three moles of CO_2. All other reactants and products are condensed phases. This means the volume of the products is less than that of the reactants at fixed temperature and pressure. Work is done by the surroundings as a result of the system's volume decrease, so the sign of w is **positive**.

63. *Answer/Explanation:* Use equation 6.1, $\Delta E = q + w = 150.\ J + 0\ J =$ **150 J**

65. *Answer:* **271 J; endothermic**

Strategy and Explanation: Given the number of calories transferred to the system, calculate the number of Joules and identify whether the reaction is exothermic or endothermic.

Use energy conversion factors.

$$64.7\ \text{cal} \times \frac{4.184\ \text{J}}{1\ \text{cal}} = 271\ \text{J}$$

Energy is transferred to the system, so the reaction is endothermic.

✓ *Reasonable Answer Check:* A calorie is larger than a Joule, so it makes sense that the number of joules is greater than the number of calories.

Thermochemical Expressions (Section 6.5)

67. *Answer:* **endothermic**

Strategy and Explanation: Given a thermochemical expression, determine if the reaction is exothermic or endothermic.

Find the sign of ΔH, and use that to determine if the reaction is exothermic or endothermic. When ΔH is negative, the reaction is exothermic. When ΔH is positive, the reaction is endothermic. Notice: If students can't always remember the Greek origins of these words and they find themselves mixing which is positive and which is negative, ask them to think of the words "exhale" and "inhale." When you **exhale**, air goes **out** of your lungs (exo = negative); when you **in**hale, air goes **into** your lungs (endo = positive).

ΔH = +38 kJ, so this reaction is **endothermic**.

✓ *Reasonable Answer Check:* Endothermic reactions have positive ΔH.

69. *Answer:* **$CaO(s) + H_2O(\ell) \longrightarrow Ca^{2+}(aq) + 2OH^-(aq)$; exothermic**

Strategy and Explanation: Given the reactants of chemical reaction and an observation of temperature change, write a balanced chemical equation for the reaction and indicate if it is exothermic or endothermic.

The reactant CaO contains a base that reacts with water molecules to make hydroxide ions.

$$CaO(s) + H_2O(\ell) \longrightarrow Ca^{2+}(aq) + 2OH^-(aq)$$

The reaction mixture gets hot, meaning that heat energy is being transferred to the surroundings. That means the reaction is **exothermic**.

✓ *Reasonable Answer Check:* Acid-base reactions often give off heat. The exothermic themochemical expression for this reaction is given in Question 96.

71. *Answer:* **49.3 kJ**

Strategy and Explanation: Given the mass of a sample of a compound, the compounds formula, and the heat of vaporization of the compound at the boiling point, determine the energy required to vaporize the sample at the same temperature.

Calculate the molar mass of C_6H_6. $6 \times (12.0107\text{ g}) + 6 \times (1.0079\text{ g}) = 78.1116\text{ g/mol}$

$$125\text{ g} \times \frac{1\text{ mol}}{78.1116\text{ g}} \times \frac{30.8\text{ kJ}}{1\text{ mol}} = 49.3\text{ kJ}$$

✓ *Reasonable Answer Check:* This quantity of heat seems reasonable.

73. The quantity of 6.0 kJ of energy is used to convert one mole of solid water (ice) into liquid water. This energy is measured as heat energy absorbed at constant standard pressure, which is why we are using the symbol ΔH°.

75. *Answer:* **(a) -2.1×10^2 kJ (b) –33 kJ**

Strategy and Explanation: Given a thermochemical expression for a phase change, determine the quantity of energy transferred to the surroundings when two different samples undergo that phase change.

When necessary, convert the sample quantity into moles. Then use the thermochemical expression to create a unit conversion factor relating energy to moles of reactant.

Reversing the given balanced thermochemical expression to show the freezing reaction, causes the ΔH° to change sign:

$$H_2O(\ell) \longrightarrow H_2O(s) \qquad \Delta H° = -6.0\text{ kJ}$$

The thermochemical expression tells us when 1 mol $H_2O(\ell)$ is frozen, that 6.0 kJ are transferred to the surroundings.

(a) $$34.2\text{ mol } H_2O(s) \times \frac{-6.0\text{ kJ}}{1\text{ mol } H_2O(s)} = -2.1 \times 10^2\text{ kJ}$$

(b) $$100.0 \text{ g } H_2O(s) \times \frac{1 \text{ mol } H_2O(s)}{18.0152 \text{ g } H_2O(s)} \times \frac{-6.0 \text{ kJ}}{1 \text{ mol } H_2O(s)} = -33 \text{ kJ}$$

✓ *Reasonable Answer Check:* The freezing reaction is exothermic, since heat energy must be removed from the reactants to make the products. Also, 34 mol of water weigh more than 100 grams, so it makes sense that the answer in (a) is larger than the answer in (b).

Enthalpy Changes for Chemical Reactions (Section 6.6)

77. *Answer:* **(a)** $\mathbf{0.500\ C_8H_{18}(\ell) + 6.25\ O_2(g) \longrightarrow 4.00\ CO_2(g) + 4.50\ H_2O(\ell);\ \Delta H° = -2.75 \times 10^3\ kJ}$

(b) $\mathbf{100.\ C_8H_{18}(\ell) + 1250\ O_2(g) \longrightarrow 800.\ CO_2(g) + 900.\ H_2O(\ell);\ \Delta H° = -5.50 \times 10^5\ kJ}$

(c) $\mathbf{1.00\ C_8H_{18}(\ell) + 12.5\ O_2(g) \longrightarrow 8.00\ CO_2(g) + 9.00\ H_2O(\ell);\ \Delta H° = -5.50 \times 10^3\ kJ}$

Strategy and Explanation: Given a thermochemical expression for a reaction, write a thermochemical expression for multiples of the reaction.

Determine what number the expression must be multiplied by to get the desired number of reactants or products given. To determine the multiplier, divide the stoichiometric coefficient we want to have in the new expression by the stoichiometric coefficient given in the original expression. Multiply every coefficient in the expression and the ΔH° by that multiplier to set up the new thermochemical expression.

In each of the samples, a number of moles is given. Assume these are exact numbers.

(a) The original expression makes 16 mol CO_2. We want to make 4.00 mol CO_2. So,

$$\text{Multiplier} = \frac{4.00 \text{ mol}}{16 \text{ mol}} = 0.250$$

$0.250 \times [2\ C_8H_{18}(\ell) + 25\ O_2(g) \longrightarrow 16\ CO_2(g) + 18\ H_2O(\ell)]$ $\qquad \Delta H° = 0.250 \times (-10{,}992 \text{ kJ})$

$0.500\ C_8H_{18}(\ell) + 6.25\ O_2(g) \longrightarrow 4.00\ CO_2(g) + 4.50\ H_2O(\ell)$ $\qquad \Delta H° = -2.75 \times 10^3 \text{ kJ}$

(b) The original expression burns 2 mol isooctane. We want to burn 100. mol isooctane. So,

$$\text{Multiplier} = \frac{100. \text{ mol}}{2 \text{ mol}} = 50.0$$

$50.0 \times [2\ C_8H_{18}(\ell) + 25\ O_2(g) \longrightarrow 16\ CO_2(g) + 18\ H_2O(\ell)]$ $\qquad \Delta H° = 50.0 \times (-10{,}992 \text{ kJ})$

$100.\ C_8H_{18}(\ell) + 1250\ O_2(g) \longrightarrow 800.\ CO_2(g) + 900.\ H_2O(\ell)$ $\qquad \Delta H° = -5.50 \times 10^5 \text{ kJ}$

(c) The original expression burns 2 mol isooctane. We want to burn 1.00 mol isooctane. So,

$$\text{Multiplier} = \frac{1.00 \text{ mol}}{2 \text{ mol}} = 0.500$$

$0.500 \times [2\ C_8H_{18}(\ell) + 25\ O_2(g) \longrightarrow 16\ CO_2(g) + 18\ H_2O(\ell)]$ $\qquad \Delta H° = 0.500 \times (-10{,}992 \text{ kJ})$

$1.00\ C_8H_{18}(\ell) + 12.5\ O_2(g) \longrightarrow 8.00\ CO_2(g) + 9.00\ H_2O(\ell)$ $\qquad \Delta H° = -5.50 \times 10^3 \text{ kJ}$

✓ *Reasonable Answer Check:* It makes sense that when more moles are involved, the ΔH° is larger, and when fewer moles are involved, the ΔH° is smaller.

79. *Answer:* **see thermostoichiometric factors below**

Strategy and Explanation: Given a thermochemical expression for a reaction, write all the thermostoichiometric factors that can be derived.

The expression's stoichiometric coefficients are interpreted in units of moles. The thermostoichiometric factors are ratios between the enthalpy of reaction and the moles of all the different reactants and products

The balanced expression says that 1 mol CaO reacts with 3 mol C to form 1 mol CaC_2 and 1 mol CO with the consumption of 464.8 kJ of thermal energy.

$$\frac{464.8\ \text{kJ}}{1\ \text{mol CaO}} \quad \frac{464.8\ \text{kJ}}{3\ \text{mol C}} \quad \frac{464.8\ \text{kJ}}{1\ \text{mol CaC}_2} \quad \frac{464.8\ \text{kJ}}{1\ \text{mol CO}}$$

The reciprocal of the four factors above are also appropriate factors.

$$\frac{1\ \text{mol CaO}}{464.8\ \text{kJ}} \quad \frac{3\ \text{mol C}}{464.8\ \text{kJ}} \quad \frac{1\ \text{mol CaC}_2}{464.8\ \text{kJ}} \quad \frac{1\ \text{mol CO}}{464.8\ \text{kJ}}$$

✓ *Reasonable Answer Check:* In each factor, the enthalpy change is related to the moles of one of the reactants or products. There are four different reactants and products and two ways to set up the ratio, so it makes sense that there are eight factors total.

81. *Answer:* **–3.3 × 10⁴ kJ**

Strategy and Explanation: Given a thermochemical expression for a reaction and a specific volume of a liquid reactant and its density, determine the enthalpy change for the reaction.

Convert the volume to mass, then the mass into moles. Then use the thermochemical expression to create a unit conversion factor relating energy to moles of reactants or products.

The balanced thermochemical expression tells us that the burning of exactly 2 mol isooctane results in the evolution of 10,922 kJ of energy at constant pressure.

$$1.00\ \text{L C}_8\text{H}_{18} \times \frac{1000\ \text{mL}}{1\ \text{L}} \times \frac{0.69\ \text{g}}{1\ \text{mL C}_8\text{H}_{18}} \times \frac{1\ \text{mol C}_8\text{H}_{18}}{114.2278\ \text{g C}_8\text{H}_{18}} \times \frac{10922\ \text{kJ}}{2\ \text{mol CH}_3\text{OH}} = 3.3 \times 10^4\ \text{kJ evolved}$$

The reaction is exothermic, so the enthalpy change is -3.3×10^4 kJ.

✓ *Reasonable Answer Check:* The combustion of a fuel should produce a large amount of energy.

83. *Answer:* **-1.45×10^3 kJ/mol**

Strategy and Explanation: Given a chemical equation for the combustion of a fuel, the mass of fuel burned, and the thermal energy evolved at constant pressure for the reaction, determine the molar enthalpy of combustion of the fuel.

Molar enthalpy change ($\Delta H°$) is identical to the thermal energy released at constant pressure per mol of substance. So, convert the mass into moles, then divide the thermal energy by the moles to get the molar enthalpy of combustion.

The balanced thermochemical expression tells us that exactly 1 mol C_2H_5OH burns to produce heat energy.

$$q = -3.62\ \text{kJ (It is negative since heat energy is evolved, rather than absorbed.)}$$

$$n = 0.115\ \text{g C}_2\text{H}_5\text{OH} \times \frac{1\ \text{mol C}_2\text{H}_5\text{OH}}{46.0682\ \text{g C}_2\text{H}_5\text{OH}} = 0.00250\ \text{mol C}_2\text{H}_5\text{OH}$$

$$\Delta H° = \frac{q}{n} = \frac{-3.62\ \text{kJ}}{0.00250\ \text{mol}} = -1.45 \times 10^3\ \text{kJ/mol}$$

✓ *Reasonable Answer Check:* Ethanol is used as a fuel, so it makes sense that a large amount of heat energy is evolved.

85. *Answer:* **6.74×10^3 kJ**

Strategy and Explanation: Given a thermochemical expression for a reaction and a specific volume of a liquid product and its density, determine the how much energy is transferred out of the system during its production.

Convert the volume to mass, then the mass into moles. Then use the thermochemical expression to create a unit conversion factor relating energy to moles of reactants or products.

The balanced thermochemical expression tells us that the production of one mol acetic acid results in the evolution of 355.9 kJ of energy at constant pressure.

Calculate the molar mass of CH_3COOH

$$2 \times (12.0107 \text{ g}) + 4 \times (1.0079 \text{ g}) + 2 \times (15.9994 \text{ g}) = 60.0518 \text{ g/mol}$$

$$1.00 \text{ L } CH_3CO_2H \times \frac{1000 \text{ mL}}{1 \text{ L}} \times \frac{1.044 \text{ g } CH_3CO_2H}{1 \text{ mL } CH_3CO_2H} \times \frac{1 \text{ mol } CH_3CO_2H}{60.0518 \text{ g } CH_3CO_2H} \times \frac{355.9 \text{ kJ}}{1 \text{ mol } CH_3CO_2H} = 6.74 \times 10^3 \text{ kJ}$$

6.74×10^3 kJ are transferred out of the system.

✓ *Reasonable Answer Check:* The production of more than 15 moles of acetic acid at about 400 kJ per mole is going to take more than 6000 kJ.

87. *Answer:* **6×10^4 kJ released**

Strategy and Explanation: Given a thermochemical expression for a reaction and a specific quantity of a reactant, determine the quantity of energy released.

Convert the mass into moles. Then use the thermochemical expression to create a unit conversion factor relating energy to moles of reactant.

The balanced chemical equation tells us that when exactly 1 mol CH_2O react, 425 kJ of thermal energy are evolved by the reaction.

$$10 \text{ lb } CH_2O \times \frac{454 \text{ g}}{1 \text{ lb}} \times \frac{1 \text{ mol } CH_2O}{30.0259 \text{ g } CH_2O} \times \frac{425 \text{ kJ}}{1 \text{ mol } CH_2O} = 6 \times 10^4 \text{ kJ evolved}$$

✓ *Reasonable Answer Check:* About 150 moles should produce about 150 times the molar enthalpy change.

Where Does the Energy Come From? (Section 6.7)

89. *Answer:* **HF**

Strategy and Explanation: Refer to the chart given above Question 89. The bond with the largest bond enthalpy is the strongest, so H–F (566 kJ/mol) is the strongest of the four hydrogen halide bonds. The others are weaker with smaller bond enthalpies: H–Cl (431 kJ/mol), H–Br (366 kJ/mol), and H–I (299 kJ/mol).

91. *Answer:* **For reaction with fluorine: (a) 594 kJ (b) –1132 kJ (c) –538 kJ; For reaction with chlorine: (a) 678 kJ (b) –862 kJ (c) –184 kJ (d) Reaction of fluorine with hydrogen is more exothermic.**

Strategy and Explanation: Given a table of bond enthalpies (above Question 89) and the description of two chemical reactions, determine for each reaction (a) the enthalpy change for breaking all the bonds in the reactants, (b) the enthalpy change for forming all the bonds in the products, (c) the enthalpy change for the reaction, and (d) determine which reaction is most exothermic.

Balance the equations and determine how many moles of bonds are broken and formed. Use each bond's bond enthalpy (ΔH_{bond}) as a conversion factor to determine energy per bond type, then add up all the energies.

The balanced chemical equations look like this:

$$H_2 + F_2 \longrightarrow 2\text{ HF} \qquad\qquad H_2 + Cl_2 \longrightarrow 2\text{ HCl}$$

(a) In both reactions, one H–H bond and one halogen–halogen bond is broken.

$\Delta H_{reactants} = 1 \text{ mol} \times (\Delta H_{H\text{–}H\ bond}) + 1 \text{ mol} \times (\Delta H_{halogen\text{–}halogen\ bond})$

For fluorine, $\Delta H_{reactants} = 1 \text{ mol} \times (436 \text{ kJ/mol}) + 1 \text{ mol} \times (158 \text{ kJ/mol}) = 594 \text{ kJ}$

For chlorine, $\Delta H_{reactants} = 1 \text{ mol} \times (436 \text{ kJ/mol}) + 1 \text{ mol} \times (242 \text{ kJ/mol}) = 678 \text{ kJ}$

(b) In both reactions, two H–halogen bonds form. The enthalpy of forming a bond is the opposite sign of the enthalpy for breaking a bond, so we add a minus sign in the equation.

$$\Delta H_{products} = -2 \text{ mol} \times (\Delta H_{H\text{–}halogen\ bond})$$

For fluorine, $\Delta H_{products} = -2 \text{ mol} \times (566 \text{ kJ/mol}) = -1132 \text{ kJ}$

For chlorine, $\Delta H_{products} = -2 \text{ mol} \times (431 \text{ kJ/mol}) = -862 \text{ kJ}$

(c) To get the enthalpy change for the reaction, add the enthalpy change of the reactants to the enthalpy change of the products

$$\Delta H_{total} = \Delta H_{reactants} + \Delta H_{products}$$

For fluorine, $\Delta H_{total} = (594\text{ kJ}) + (-1132\text{ kJ}) = -538\text{ kJ}$

For chlorine, $\Delta H_{total} = (678\text{ kJ}) + (-862\text{ kJ}) = -184\text{ kJ}$

(d) The reaction of fluorine with hydrogen is more exothermic (–538 kJ is more negative) than the reaction of chlorine with hydrogen (–184 kJ is less negative).

✓ *Reasonable Answer Check:* We expect the fluorine to be more reactive than chlorine, so more energy would be released.

93. *Answer/Explanation:* The breaking of a C-C bond is **endothermic**. Separating two atoms that are bonded together requires a transfer of energy into the system, because work must be done against the force holding the pair of atoms together, as described in Section 6.7.

Measuring Enthalpy Changes: Calorimetry (Section 6.8)

94. *Answer:* **0.52 J g^{-1} °C^{-1}**

Strategy and Explanation: Given the masses and initial temperatures of a piece of metal and a quantity of water, and given the final temperature of the water after the metal is added, determine the specific heat capacity of the metal.

Thermal equilibrium is reached when the water and the metal reach the same temperature, so we know the final temperature of the metal is the same as the final temperature of the water. Use Equation 6.2' ($q = c \times m \times \Delta T$) to calculate the heat gained by the water, which also represents the heat lost by the metal. Use Equation 6.2 to calculate the specific heat capacity of the metal.

Table 6.1 tells us that the specific heat capacity (c) of water is 4.184 J g^{-1}°C^{-1}.

$$\Delta T = T_f - T_i = 24.3\text{ °C} - 21.7\text{ °C} = 2.6\text{ °C}$$

$$q_{water} = (4.184\text{ J g}^{-1}\text{°C}^{-1}) \times (75.0\text{ g}) \times (2.6\text{ °C}) = 816\text{ J} \quad \text{(two sig figs)}$$

$$q_{water} = -q_{Ti}$$

$$c_{Ti} = \frac{q_{Ti}}{m_{Ti} \times \Delta T} = \frac{-816\text{ J}}{20.8\text{ g} \times \left(24.3\text{ °C} - 99.5\text{ °C}\right)} = 0.522\ \frac{\text{J}}{\text{g °C}} = 0.52\text{ J g}^{-1}\text{ °C}^{-1} \quad \text{(two sig figs)}$$

✓ *Reasonable Answer Check:* Because water has a much larger specific heat capacity than Ti, we expect a smaller temperature increase in the water than the decrease experienced by the molybdenum. Most metals have a heat capacity in this range.

96. *Answer:* **23.9 °C**

Strategy and Explanation: Given an exothermic chemical reaction, the mass of a reactant, and the mass and temperature of a sample of water, determine the final temperature after the two are combined causing the reaction to occur.

Use the mass and molar mass of the reactant to calculate the moles of reactant. Heat energy from exothermic the reaction raises the temperature of the water to the final temperature. Use the balanced thermochemical expression to set up a relationship between the moles of reactant to the quantity of heat transferred to water. Then, calculate the amount of heat energy transferred to the water during the reaction. Look up the specific heat capacity of water and use Equation 6.2, to calculate the new temperature.

$$q_{water} = 0.100\text{ g CaO} \times \frac{1\text{ mol CaO}}{56.077\text{ g CaO}} \times \frac{81.9\text{ kJ transfered to water}}{1\text{ mol CaO}} \times \frac{1000\text{ J}}{1\text{ kJ}} = 146\text{ J}$$

The specific heat capacity (c) of water is 4.184 J g^{-1}°C^{-1} according to Table 6.1.

For a temperature change, rearrange Equation 6.2 to solve for ΔT_{water}

$$\Delta T = \frac{q_{water}}{m_{water} \times c_{water}} = \frac{-146\ J}{125\ g \times \left(4.184\ \frac{J}{g\ ^\circ C}\right)} = 0.279\ ^\circ C$$

$$\Delta T_{water} = T_f - T_i$$

$$T_f = 23.6\ ^\circ C + 0.279\ ^\circ C = 23.9\ ^\circ C$$

✓ *Reasonable Answer Check:* The final temperature is greater than initial temperature because the reaction was exothermic.

98. *Answer:* **(a) 1.4 × 10^4 J transferred (other answers are possible depending on the choice of system) (b) – 42 kJ**

Strategy and Explanation: Given the mass of a soluble ionic solid and the volume and temperature of a sample of water, determine the heat energy transfer from the system to the surroundings and the enthalpy change ($\Delta H°$) by calculating the energy change per mol of ionic solid.

We can define the system as the NaOH, as described in Question 21 and in Section 6.2 (page 218), with $\Delta H_{dissolving} = -q_{water}$ at constant pressure. Use the temperature of the surroundings (the water solution) and the specific heat capacity of water solution to calculate q_{water}. First, look up the specific heat capacity of water. Use Equation 6.2', find heat energy gained by the water. Heat energy from the reaction is used to raise the temperature of the water, so relate the heat energy gained by the water to that lost in the reaction, $\Delta H_{dissolving}$. Finally, divide the $\Delta H_{dissolving}$ by the mass of the ionic compound and convert grams to moles.

(a) The specific heat capacity (c) of water is 4.184 J $g^{-1}°C^{-1}$ according to Table 6.1.

For a temperature change: $q = c \times m \times \Delta T$. The water represents the part of the surroundings affected by a change in the system.

$$q_{water} = c_{water} \times m_{water} \times \Delta T_{water}$$

$$\Delta T_{water} = T_{f,water} - T_{i,water} = 30.7\ °C - 22.6\ °C = 8.1\ °C$$

$$q_{water} = (4.184\ J\ g^{-1}°C^{-1}) \times (400.0\ mL) \times \frac{1.00\ g\ H_2O}{1\ mL\ H_2O} \times (8.1\ °C)$$

$= 1.4 \times 10^4$ J transferred from the system to the surroundings

(b) Heat energy is lost by the reaction, so $q_{dissolving}$ is negative. Heat energy is gained by the water, so $q_{dissolving}$ is positive. The quantity of heat energy gained by the water is produced by the reaction.

$$q_{dissolving} = -\ q_{water} = -1.4 \times 10^4\ J$$

The reaction occurs at constant pressure and there is no work done, so $\Delta H = q_{dissolving}$ per mol of NaOH.

$$\Delta H° = \frac{-1.4 \times 10^4\ J}{13.0\ g\ NaOH} \times \frac{39.9971\ g\ NaOH}{1\ mol\ NaOH} \times \frac{1\ kJ}{1000\ J} = -42\ \frac{kJ}{mol\ NaOH}$$

✓ *Reasonable Answer Check:* The ionization of an ionic compound involves separating the cations and anions, then hydrating them. The enthalpy of this change could conceivably be positive or negative, but it is expect that it will be small compared to reactions where more significant bond rearrangement is happening. –42 kJ/mol is smaller (closer to zero) than the rest of the $\Delta H°$ values for other kinds of reactions recently studied. Note that the cooling of the solute in the process has been neglected. This may or may not be appropriate. It would be better if we had the specific heat capacity of the solution, then we could use the entire known mass of the solution, rather than just the water.

Notice: A very common, more-empirical approach to solving this question is to identify the water and the salt as the system. With that definition for the system: $\Delta E_{system} = q_{dissolving} + q_{water}$. If that is the case, the answer to the question in (a) will be different. (a) If the system is insolated, $\Delta E = 0$, so there is no transfer of energy from the system to the surroundings. (b) $q_{dissolving} = -q_{water}$. Measuring the system's temperature change indicates a gain of thermal energy by the water and q_{water} is positive. Since the reaction occurs at the same time and no heat energy escapes the insolated system, it proves that the dissolving reaction produces energy, $q_{reaction}$ is negative, and the reaction is exothermic. The numerical result is the same as provided above.

100. *Answer:* **ΔE per mole is –2.20 × 10³ kJ**

Strategy and Explanation: Given the mass of a reactant from a known reaction in a bomb calorimeter, the mass of water in the calorimeter, the initial and final temperatures, and the heat capacity of the bomb, determine the ΔE per mol of reactant.

$$\Delta E = q + w = q + 0 \text{ at constant volume}$$

Use the specific heat capacity of water, the mass of water and the temperature changes, use Equation 6.2', to find heat energy gained by the water. Calculate the heat energy gained by the bomb using the heat capacity (C_{bomb}). Equate the heat energy gained, by the water and the bomb, to that lost in the reaction. Convert grams to moles of the reactant. Divide the heat by the moles of reactant to get ΔE per mole.

The specific heat capacity (c) of water is 4.184 J $g^{-1}°C^{-1}$ according to Table 6.1.

For a temperature change in water: $q = c \times m \times \Delta T$. For a change in temperature in the bomb with the heat capacity: $q = C_{bomb} \times \Delta T$.

$$q_{total} = c_{water} \times m_{water} \times \Delta T + C_{bomb} \times \Delta T$$

$$\Delta T = T_f - T_i = 25.22\ °C - 21.70\ °C = 3.52\ °C$$

$$q_{total} = (4.184\ J\ g^{-1}°C^{-1}) \times (815\ g) \times (3.52\ °C) + (923\ J°C^{-1}) \times (3.52\ °C)$$

$$q_{total} = 8.47 \times 10^3\ J = -q_{reaction}$$

$$\Delta E = -8.47 \times 10^3\ J \times \frac{1\ kJ}{1000\ J} = -8.47\ kJ$$

$$0.692\ g\ C_6H_{12}O_6 \times \frac{180.1554\ g\ C_6H_{12}O_6}{1\ mol\ C_6H_{12}O_6} = 3.84 \times 10^{-3}\ mol\ C_6H_{12}O_6$$

$$\Delta E \text{ per mol} = \frac{-8.47\ kJ}{3.84 \times 10^{-3}\ mol\ C_6H_{12}O_6} = -2.20 \times 10^3 \frac{kJ}{mol\ C_6H_{12}O_6}$$

✓ *Reasonable Answer Check:* The enthalpy is negative, since the reaction's energy loss causes the temperature in the water and the bomb to increase. This reaction is the formation reaction for SO_2. Table 6.2 says that it is ΔH° is –296.830 kJ/mol.

102. *Answer:* **–394 kJ/mol**

Strategy and Explanation: Given the mass of a reactant of a given reaction in a bomb calorimeter, the mass of water in the calorimeter, the initial and final temperatures, and the heat capacity of the bomb, determine the heat energy evolved per mol of reactant.

Adapt the method described in the solution to Question 100.

The specific heat capacity (c) of water is 4.184 J $g^{-1}°C^{-1}$ according to Table 6.1.

For a temperature change: $q = c \times m \times \Delta T$. For a change in temperature in the bomb with the heat capacity: $q = C_{bomb} \times \Delta T$.

$$q_{total} = c_{water} \times m_{water} \times \Delta T + C_{bomb} \times \Delta T$$

$$\Delta T = T_f - T_i = 27.38\ °C - 25.00\ °C = 2.38\ °C$$

$$q_{total} = (4.184\ J\ g^{-1}°C^{-1}) \times (775\ g) \times (2.38\ °C) + (893\ J°C^{-1}) \times (2.38\ °C)$$

$$q_{total} = 7.72 \times 10^3\ J + 2.13 \times 10^3\ J$$

$$q_{total} = 9.84 \times 10^3 \text{ J}$$

$$\text{heat energy evolved by the reaction} = 9.84 \times 10^3 \text{ J} \times \frac{1 \text{ kJ}}{1000 \text{ J}} = 9.84 \text{ kJ}$$

$$\text{heat energy per mol} = \frac{-9.84 \text{ kJ}}{0.300 \text{ g C}} \times \frac{12.0107 \text{ g C}}{1 \text{ mol C}} = -394 \frac{\text{kJ}}{\text{mol C}}$$

✓ *Reasonable Answer Check:* This is the formation reaction for CO_2. Table 6.2 gives $\Delta H° = -393.5$ kJ/mol.

Hess's Law (Section 6.9)

104. *Answer:* **$\Delta H° = -43.6$ kJ**

Strategy and Explanation: Given two thermochemical expressions, determine the enthalpy change of a reaction.

Identify unique occurrences of the reactants and/or products in the given equations to help you determine which and how many of a given equation to use. If an equation's reactant shows in the products of the desired equation (or vice versa) then reverse the equation and change the sign of $\Delta H°$. If the substance has a different stoichiometric coefficient than the desired coefficient, determine an appropriate multiplier for the equation and multiply the $\Delta H°$ by the same multiplier (as was done in Question 77).

Look at the first reactant: $C_2H_4(g)$. The only given equation that has $C_2H_4(g)$ is the first one, where $C_2H_4(g)$ is also a reactant, so we must include that equation, as written, with its given $\Delta H°$. Look at the second reactant: $H_2O(\ell)$. This chemical is found in both of the given reactions, so let's skip it. Look at the product: $C_2H_5OH(\ell)$. The only given equation that has $C_2H_5OH(\ell)$ is the second one, however, $C_2H_5OH(\ell)$ is a reactant, and we need it to be a product, so we must reverse that equation and change the sign of the given $\Delta H°$. We have now a plan to use both equations, so let's do it:

Add the equations, and eliminate reactants that also show up as products: all $O_2(g)$, all $CO_2(g)$, and two of the $H_2O(g)$ molecules. The remaining reactants and products should form the net equation for the reaction, and the sum of the individual $\Delta H°$ values will give us the $\Delta H°$ for that reaction:

$$C_2H_4(g) + \cancel{3\,O_2(g)} \longrightarrow \cancel{2\,CO_2(g)} + \cancel{2\,H_2O(\ell)} \qquad \Delta H° = -1411.1 \text{ kJ}$$

$$+ \quad \cancel{2\,CO_2(g)} + \cancel{3}\,H_2O(\ell) \longrightarrow C_2H_5OH(\ell) + \cancel{3\,O_2(g)} \qquad \Delta H° = -(-1367.5 \text{ kJ})$$

$$C_2H_4(g) + H_2O(\ell) \longrightarrow C_2H_5OH(\ell) \qquad \Delta H° = -43.6 \text{ kJ}$$

✓ *Reasonable Answer Check:* The net equation adds up the equation we are looking for.

106. *Answer:* **– 320. kJ**

Strategy and Explanation: Given a chemical reaction with unknown $\Delta H°$ and two thermochemical expressions, determine the enthalpy change of a named reaction for a given quantity of product.

Write the equation for the named reaction. Identify unique occurrences of the reactants and/or products in the given equations to help you determine which and how many of a given equation to use. If an equation's reactant shows in the products of the desired equation (or vice versa) then reverse the equation and change the sign of $\Delta H°$. If the substance has a different stoichiometric coefficient than the desired coefficient, determine an appropriate multiplier for the equation and multiply the $\Delta H°$ by the same multiplier (as was done in Question 104).

Look at the first reactant: $P_4(s)$. The only given equation that has $P_4(s)$ is the first one which uses one mole of $P_4(s)$ as a reactant, so you must include that equation, as written, with its given $\Delta H°$. Look at the second reactant: Cl_2. This chemical is in both of the given reactions, so let's skip it. Look at the product: $PCl_3(\ell)$. The only given equation that has $PCl_3(\ell)$ is the second one, but one mole of $PCl_3(\ell)$ is a product in the equation and we need four of them. Therefore, we write the reverse of the equation with the sign of $\Delta H°$ changed, and multiply the equation and the $\Delta H°$ by four. We have now planned how we will use both the given reactions, so let's do it:

Add all the reactions as planned, eliminate reactants that also show up as products: four of the $Cl_2(g)$ and all the $PCl_5(s)$. Add all the remaining reactants and products to form the net equation for the desired reaction, and add all the individual ΔH° values to get the ΔH° for the reaction:

$$P_4(s) + \cancel{10}\,Cl_2(g) \longrightarrow 4\,\cancel{PCl_5}(s) \qquad \Delta H° = -1774.0 \text{ kJ}$$

$$+ \quad 4\,\cancel{PCl_5}(s) \longrightarrow 4\,PbCl_3(\ell) + \cancel{4\,Cl_2}(g) \qquad + \ \Delta H° = -4\,(-123.8 \text{ kJ})$$

$$P_4(s) + 6\,Cl_2(g) \longrightarrow 4\,PCl_3(\ell) \qquad \Delta H° = -1278.8 \text{ kJ}$$

Formation reaction is the production of one mole of product from standard state elements.

$$\frac{1}{4}P_4(s) + \frac{3}{2}Cl_2(g) \longrightarrow PCl_3(\ell)$$

The given equation is four times the formation reaction. To get the enthalpy of formation, we'll divide the calculated ΔH by four.

$$\Delta H_f^\circ = \frac{-1278.8 \text{ kJ}}{4 \text{ mol } PCl_3} = -319.7 \text{ kJ}$$

$$1.00 \text{ mol } PCl_3 \times \frac{-319.7 \text{ kJ}}{1 \text{ mol } PCl_3} = -320. \text{ kJ}$$

✓ *Reasonable Answer Check:* The net equation is the desired reaction. The enthalpy of formation of $PCl_3(\ell)$ is given in Appendix J as –319.7 kJ.

108. *Answer:* **–217.3 kJ/mol, 2.6 × 10^2 kJ evolved**

Strategy and Explanation: Given two thermochemical expressions and the mass of a formation reactant sample, determine the enthalpy change of the formation reaction and the heat energy evolved or absorbed when that sample reacts.

Write the equation for the desired net reaction. Follow the procedure described in Question 104. After that, calculate the heat energy evolved or absorbed by finding the moles and using the thermochemical expression to provide the molar energy change.

The formation equation for PbO is written with the standard state elements as reactants and one mole of compound as the product:

$$Pb(s) + \frac{1}{2}O_2(g) \longrightarrow PbO(s)$$

Look at the first reactant: Pb(s). The only given equation that has Pb(s) is the first one, but Pb(s) is a product, so the equation must be reversed. Therefore, we must write that equation reversed with the sign of its ΔH° changed. Look at the second reactant: $O_2(g)$. The only given equation that has $O_2(g)$ is the second one. It is a reactant, but the stoichiometric coefficient is not what we want it to be.

$$\text{Multiplier} = \frac{\text{coefficient we want}}{\text{coefficient we have}} = \frac{\frac{1}{2}}{1} = \frac{1}{2}$$

Therefore, we must include that reaction multiplied by $\frac{1}{2}$ going forward with $\frac{1}{2} \times \Delta H°$. We have now planned how we will use all the given reactions, so let's do it:

Add all the reactions as planned, eliminate reactants that also show up as products: the CO and the C(graphite). Add all the remaining reactants and products to form the net reaction, and add all the individual ΔH° values to get the ΔH° for the net reaction:

$$Pb(s) + CO(g) \longrightarrow PbO(s) + C(\text{graphite}) \qquad \Delta H° = -(106.8 \text{ kJ})$$

$$\frac{1}{2} \times [2\,C(\text{graphite}) + O_2(g) \longrightarrow 2\,CO(s)] \qquad \frac{1}{2} \times [\Delta H° = -221.0 \text{ kJ}]$$

Multiplying the second equation by $\frac{1}{2}$, then simplifying and adding the two equations gives:

$$\begin{array}{lll} & Pb(s) + \cancel{CO(g)} \longrightarrow PbO(s) + \cancel{C(graphite)} & \Delta H° = -106.8 \text{ kJ} \\ + & \cancel{C(graphite)} + \frac{1}{2} O_2(g) \longrightarrow \cancel{CO(s)} & + \ \Delta H° = -110.5 \text{ kJ} \\ \hline & Pb(s) + \frac{1}{2} O_2(g) \longrightarrow PbO(s) & \Delta H° = -217.3 \text{ kJ} \end{array}$$

The enthalpy change is negative, so heat energy is evolved during the reaction:

$$250 \text{ g Pb} \times \frac{1 \text{ mol Pb}}{207.2 \text{ g Pb}} \times \frac{217.3 \text{ kJ evolved}}{1 \text{ mol Pb}} = 2.6 \times 10^2 \text{ kJ evolved}$$

✓ *Reasonable Answer Check:* The net equation is the desired reaction. The enthalpy of formation of this lead(II) oxide is given in Appendix J with a ΔH° = –217.32 kJ/mol.

Standard Molar Enthalpies of Formation (Section 6.10)

110. *Answer/Explanation:* Decomposition of one mole of a compound into elements at standard state is the reverse of the formation reactions; hence the enthalpy of decomposition has the opposite sign of the enthalpy of formation.

Write the reverse of the formation equation, look up ΔH_f° from Appendix J and change its sign.

(a) For $C_2H_5OH(\ell)$, $\Delta H_f^\circ = -277.69$ kJ

$$C_2H_5OH(\ell) \longrightarrow 2\ C(s) + 2\ H_2(g) + \frac{1}{2} O_2(g) \quad \Delta H° = +277.69 \text{ kJ}$$

(b) For NaF(s), $\Delta H_f^\circ = -573.647$ kJ

$$NaF(s) \longrightarrow 2\ Na(s) + \frac{1}{2} F_2(g) \quad \Delta H° = +573.647 \text{ kJ}$$

(c) For $MgSO_4(s)$, $\Delta H_f^\circ = -1284.9$ kJ

$$MgSO_4(s) \longrightarrow Mg(s) + S(s, \text{rhombic}) + 2\ O_2(g) \quad \Delta H° = +1284.9 \text{ kJ}$$

(d) For $NH_4NO_3(s)$, $\Delta H_f^\circ = -365.56$ kJ

$$NH_4NO_3(s) \longrightarrow N_2(g) + 2\ H_2(g) + \frac{3}{2} O_2(g) \quad \Delta H° = +365.56 \text{ kJ}$$

112. *Answer:* **(a)** $\mathbf{2\ Al\ (s) + \frac{3}{2}\ O_2\ (g) \longrightarrow Al_2O_3\ (s)\ \Delta H° = -1675.7\ kJ}$ **(b)** $\mathbf{Ti(s) + 2\ Cl_2\ (g) \longrightarrow TiCl_4(\ell)\ \Delta H° = -804.2\ kJ}$ **(c)** $\mathbf{N_2(g) + 2\ H_2\ (g) + \frac{3}{2}\ O_2\ (g) \longrightarrow NH_4NO_3(s)\ \Delta H° = -365.56\ kJ}$

Strategy and Explanation:

(a) The formation of $Al_2O_3(s)$ is written with the reactants as standard state elements, Al(s) and $O_2(g)$, and the product as one mole of standard state compound:

$$2\ Al\ (s) + \frac{3}{2} O_2\ (g) \longrightarrow Al_2O_3\ (s) \qquad \Delta H° = -1675.7 \text{ kJ}$$

(b) The formation of $TiCl_4(\ell)$ is written with the reactants as standard state elements, Ti(s) and $Cl_2(g)$, and the product as one mole of standard state compound:

$$Ti(s) + 2\ Cl_2\ (g) \longrightarrow TiCl_4(\ell) \qquad \Delta H° = -804.2 \text{ kJ}$$

(c) The formation of $NH_4NO_3(s)$ is written with the reactants as standard state elements, $N_2(g)$, $H_2(g)$, and $O_2(g)$, and the product as one mole of standard state compound:

$$N_2(g) + 2\ H_2\ (g) + \frac{3}{2} O_2\ (g) \longrightarrow NH_4NO_3(s) \qquad \Delta H° = -365.56 \text{ kJ}$$

114. *Answer:* **(a) –1675.7 kJ (b) –205.4 kJ**

Strategy and Explanation:

(a) The given equation is for the decomposition of two moles of $Al_2O_3(s)$. The formation reaction is the reverse of the decomposition. When the equation is reversed, the sign of ΔH° changes. The molar formation reaction produces one mole of $Al_2O_3(s)$, so divide the ΔH° by two.

$$2\ Al(s) + \frac{3}{2}\ O_2(g) \longrightarrow Al_2O_3(s) \qquad \Delta H° = \frac{1}{2}\ (-3351.4\ kJ) = -1675.7\ kJ$$

(b) Calculate the moles of $Al_2O_3(s)$ formed, then use the enthalpy of formation to determine the ΔH° for this sample.

$$12.50\ g\ Al_2O_3 \times \frac{1\ mol\ Al_2O_3}{101.9612\ g\ Al_2O_3} \times \frac{3351.4\ kJ}{2\ mol\ Al_2O_3} = -205.4\ kJ$$

✓ *Reasonable Answer Check:* The calculated molar enthalpy of formation in (a) matches that given in Appendix J. The sample in (b) is smaller than a mole, so it makes sense that that enthalpy change is also smaller than the molar enthalpy change.

116. *Answer:* **–584 kJ**

Strategy and Explanation: Given the mass of a reactant in a described reaction and the heat given off during the process, calculate the molar enthalpy of formation of the product.

Balance the chemical equation. Use the molar mass of lithium to determine the moles of lithium in the sample, then use the stoichiometry of the balanced equation to determine the moles of Li_2O formed. Divide the heat by the moles of Li_2O to get the molar enthalpy of the reaction.

$$2\ Li(s) + \frac{1}{2}\ O_2(g) \longrightarrow Li_2O(s)$$

$$3.47\ g\ Li \times \frac{1\ mol\ Li}{6.941\ g\ Li} \times \frac{1\ mol\ Li_2O}{2\ mol\ Li} = 0.250\ mol\ Li_2O$$

$$\Delta H°_f = \frac{-146\ kJ}{0.250\ mol\ Li_2O} = -584\ kJ/mol$$

✓ *Reasonable Answer Check:* The calculated molar enthalpy of formation is not given in Appendix J, but a quick search on the internet gives the value –20.01 kJ/g, which translates to -597.9 kJ/mol. This answer is in the same range, but it is not within the stated uncertainty.

118. *Answer:* **ΔH° = –98.89 kJ**

Strategy and Explanation: Given a balanced chemical equation for a reaction and a table of molar enthalpies of formation, determine the enthalpy change of the reaction.

Using molar enthalpies of formation is a common variation of Hess's Law. We will form all the products (using their molar enthalpies, as written) and we will destroy—the opposite of formation—the reactants (using their molar enthalpies, with opposite sign). That logic is the basis of Equation 6.11.

$$\Delta H° = \sum\Big[(\text{moles of product}) \times \Delta H°_f(\text{product})\Big] - \sum\Big[(\text{moles of reactant}) \times \Delta H°_f(\text{reactant})\Big]$$

Use the stoichiometric coefficient of the balanced equation to describe the moles of each of the reactants and products. Set up a specific version of Equation 6.11. Look up the $\Delta H°_f$ for each; remember to check the physical phase and remember that $\Delta H°_f$ for standard state elements is exactly zero. Plug them into the equation and solve for ΔH°.

The product is SO_3. The reactants are SO_2 and O_2. O_2 is the elemental form for oxygen.

$$\Delta H° = (1\ mol) \times \Delta H°_f(SO_3) - [(1\ mol) \times \Delta H°_f(SO_2) + (1\ mol) \times \Delta H°_f(O_2)]$$

Look up the $\Delta H°_f$ values in Table 6.2.

$\Delta H° = (1 \text{ mol}) \times (-395.72 \text{ kJ/mol}) - [(1 \text{ mol}) \times (-296.830 \text{ kJ/mol}) + (1 \text{ mol}) \times (0 \text{ kJ/mol})] = -98.89 \text{ kJ}$

✓ *Reasonable Answer Check:* These two sulfur oxide formation energies are not very different, so making one of them from the other doesn't produce very much energy.

120. *Answer:* **(a) 178.32 kJ (b) –595.2 kJ (c) 44 kJ**

Strategy and Explanation: Given a balanced chemical equation for a reaction and a table of molar enthalpies of formation, determine the enthalpy change of the reaction.

Use the stoichiometric coefficient of the balanced equation to describe the moles of each of the reactants and products. Set up a specific version of Equation 6.11. Look up the ΔH_f° for each. Plug them into the equation and solve for $\Delta H°$.

(a) The products are CaO(s) and CO_2(g). The reactant is $CaCO_3$(s).

$\Delta H° = (1 \text{ mol}) \times \Delta H_f^\circ(CaO(s)) + (1 \text{ mol}) \times \Delta H_f^\circ(CO_2(g)) - (1 \text{ mol}) \times \Delta H_f^\circ(CaCO_3(s))$

Look up the ΔH_f° values in Table 6.2 and Appendix J.

$$\Delta H° = (1 \text{ mol}) \times (-635.09 \text{ kJ/mol}) + (1 \text{ mol}) \times (-393.509 \text{ kJ/mol}) - (1 \text{ mol}) \times (-1206.92 \text{ kJ/mol}) = 178.32 \text{ kJ}$$

(b) The products are HF(g) and I_2(s). The reactants are HI(g) and F_2(g).

$\Delta H° = (2 \text{ mol}) \times \Delta H_f^\circ(HF(g)) + (1 \text{ mol}) \times \Delta H_f^\circ(I_2(s)) - [(2 \text{ mol}) \times (HI(g)) + (1 \text{ mol}) \times (F_2(g))]$

Look up the ΔH_f° values in Table 6.2 and Appendix J.

$$\Delta H° = (2 \text{ mol}) \times \Delta H_f^\circ(-271.1 \text{ kJ/mol}) + (1 \text{ mol}) \times \Delta H_f^\circ(0 \text{ kJ/mol}) - [(2 \text{ mol}) \times (26.48 \text{ kJ/mol}) + (1 \text{ mol}) \times (0 \text{ kJ/mol})] = -595.2 \text{ kJ}$$

(c) The products are HF(g) and SO_3(g). The reactants are SF_6(g) and $H_2O(\ell)$.

$\Delta H° = (6 \text{ mol}) \times \Delta H_f^\circ(HF(g)) + (1 \text{ mol}) \times \Delta H_f^\circ(SO_3(g)) - [(1 \text{ mol}) \times (SF_6(g)) + (3 \text{ mol}) \times (H_2O(\ell))]$

Look up the ΔH_f° values in Table 6.2 and Appendix J.

$$\Delta H° = (6 \text{ mol}) \times \Delta H_f^\circ(-271.1 \text{ kJ/mol}) + (1 \text{ mol}) \times \Delta H_f^\circ(-395.72 \text{ kJ/mol}) - [(1 \text{ mol}) \times (-1209 \text{ kJ/mol}) + (3 \text{ mol}) \times (-285.83 \text{ kJ/mol})] = 44 \text{ kJ}$$

122. *Answer:* **(a) –0.50 kJ (b) +18 kJ (c) –5.60 kJ (d) –4.13 kJ**

Strategy and Explanation: Determine the equation for the reaction occurring and use Appendix J values to calculate the appropriate $\Delta H°$. Use the given quantity and the thermochemical stoichiometry to calculate the enthalpy change.

(a) $S(s) + O_2(g) \longrightarrow SO_2(g)$

This equation represents the formation of SO_2(g), with $\Delta H_f^\circ = -1206.92$ kJ/mol.

$$0.054 \text{ g S} \times \frac{1 \text{ mol S}}{32.065 \text{ g S}} \times \frac{-296.830 \text{ kJ}}{1 \text{ mol S}} = -0.50 \text{ kJ}$$

(b) $HgO(s) \longrightarrow Hg(\ell) + O_2(g)$

This equation represents the decomposition of HgO(s), which is the reverse of the formation reaction. Since $\Delta H_f^\circ = -90.83$ kJ/mol, the reaction has $\Delta H° = +90.83$ kJ/mol

$$0.20 \text{ mol HgO} \times \frac{+90.83 \text{ kJ}}{1 \text{ mol HgO}} = +18 \text{ kJ}$$

(c) $\frac{1}{2} N_2(g) + \frac{3}{2} H_2(g) \longrightarrow NH_3(g)$

This equation represents the formation of NH_3(g), with $\Delta H_f^\circ = -46.11$ kJ/mol.

$$2.40 \text{ g } NH_3 \times \frac{1 \text{ mol } NH_3}{17.0304 \text{ g } NH_3} \times \frac{-46.11 \text{ kJ}}{1 \text{ mol } NH_3} = -6.50 \text{ kJ}$$

(d) $C(s) + O_2(g) \longrightarrow CO_2(g)$

This equation represents the formation of $CO_2(g)$, with $\Delta H^\circ_f = -393.509$ kJ/mol.

$$1.05 \times 10^{-2} \text{ mol C} \times \frac{-393.509 \text{ kJ}}{1 \text{ mol C}} = -4.13 \text{ kJ}$$

124. *Answer:* **$\Delta H^\circ_f = 103.6$ kJ**

Strategy and Explanation: Given a balanced thermochemical expression for a reaction and a table of molar enthalpies of formation, determine the enthalpy of formation of one of the reactants.

Use the stoichiometric coefficient of the balanced equation to describe the moles of each of the reactants and products. Set up a specific version of Equation 6.11. Look up the ΔH°_f for each. Plug the numbers into the equation and solve for ΔH°_f.

The products are $CO_2(g)$ and $H_2O(\ell)$. The reactants are $C_8H_8(\ell)$ and $O_2(g)$.

$\Delta H^\circ = (8 \text{ mol}) \times \Delta H^\circ_f(CO_2(g)) + (4 \text{ mol}) \times \Delta H^\circ_f(H_2O(\ell)) - [(1 \text{ mol}) \times \Delta H^\circ_f(C_8H_8) + (10 \text{ mol}) \times \Delta H^\circ_f(O_2)]$

Look up the ΔH°_f values in Table 6.2.

$-4395.0 \text{ kJ} = (8 \text{ mol}) \times (-393.509 \text{ kJ/mol}) + (4 \text{ mol}) \times (-285.83 \text{ kJ/mol})$

$- [(1 \text{ mol}) \times \Delta H^\circ_f(C_8H_8) + (10 \text{ mol}) \times (0 \text{ kJ/mol})]$

Solve for $\Delta H^\circ_f(C_8H_8)$

$\Delta H^\circ_f(C_8H_8) = [+4395.0 \text{ kJ} + (8 \text{ mol}) \times (-393.509 \text{ kJ/mol}) + (4 \text{ mol}) \times (-285.83 \text{ kJ/mol})] / (1 \text{ mol})$

$$\Delta H^\circ_f(C_8H_8) = 103.6 \text{ kJ/mol}$$

✓ *Reasonable Answer Check:* The enthalpy of formation of styrene is not in Appendix J, and an internet search may not be useful. However, we find using the index of the textbook that styrene is a benzene ring, C_6H_6, attached to an ethene molecule, C_2H_2 (See Table 12.6 on page 564). Looking at Appendix J, $\Delta H^\circ_f(C_6H_6) = 49.03$ kJ/mol and $\Delta H^\circ_f(C_2H_2) = 226.73$ kJ/mol. It is satisfying to find the enthalpy of formation of styrene between these two values.

126. *Answer:* **41.2 kJ evolved**

Strategy and Explanation: Given the mass of a reactant, the description of a reaction, and molar enthalpies of formation, determine the thermal energy transferred out of the system at constant pressure.

Balance the equation and use it to describe the moles of each of the reactants and products. Set up a specific version of Equation 6.11. Look up the ΔH°_f for each. Plug them into the equation and solve for the molar enthalpy (ΔH°) – this represents the molar heat energy evolved at constant pressure. Convert mass to moles, and use the molar enthalpy as a conversion factor to get heat energy evolved at constant pressure for the reactant sample.

The product is Fe_2O_3. The reactants are both elements: Fe and O_2.

$$2 \text{ Fe(s)} + \frac{3}{2} O_2(g) \longrightarrow Fe_2O_3(s)$$

$$\Delta H^\circ = (1 \text{ mol}) \times \Delta H^\circ_f(Fe_2O_3) - [(2 \text{ mol}) \times \Delta H^\circ_f(Fe) + (\frac{3}{2} \text{ mol}) \times \Delta H^\circ_f(O_2)]$$

Look up the ΔH°_f values in Table 6.2.

$$\Delta H^\circ = (1 \text{ mol}) \times (-824.2 \text{ kJ/mol}) - (2 \text{ mol}) \times (0 \text{ kJ/mol}) - (\frac{3}{2} \text{ mol}) \times (0 \text{ kJ/mol}) = -824.2 \text{ kJ}$$

That means the thermochemical expression looks like this:

$$2\ Fe(s) + \frac{3}{2}\ O_2(g) \longrightarrow Fe_2O_3(s) \qquad \Delta H° = -824.2\ kJ$$

For 2 mol of Fe, 824.2 kJ needs to be transferred out. Now, we'll start the calculation for how much thermal energy is transferred out using the sample given:

$$5.58\ g\ Fe \times \frac{1\ mol\ Fe}{55.845\ g\ Fe} \times \frac{824.2\ kJ}{2\ mol\ Fe} = 41.2\ kJ \text{ is transferred out}$$

✓ *Reasonable Answer Check:* The sample is a tenth of a mole. The equation shows two mol of Fe are needed. So the resulting heat energy is half of a tenth of the original molar enthalpy.

Chemical Fuels (Section 6.11)

128. *Answer:* **1.1×10^2 kg CH_4**

Strategy and Explanation: Determine the mass of fuel required to raise the temperature of the air in a house to a specified value, given the dimensions of rooms in the house, the molar heat capacity of air, the average molar mass of air and the density of the air.

Use the dimensions of the house, the molar mass, and the density of the air to determine the mass of air being heated. Then use Equation 6.2' to calculate the heat energy needed to raise the temperature of the air. Get the balanced equation describing the combustion of the fuel. Use Equation 6.11, a table of molar enthalpies of formation, and the stoichiometry of the balanced equations to determine the molar enthalpy change for each combustion reaction. Then use that enthalpy and the molar mass as conversion factors to determine the mass of fuel needed.

$$\text{Volume of air} = (275\ m^2) \times 2.50\ m \times \left(\frac{100\,cm}{1\ m}\right)^3 \times \frac{1\ L}{1000\ cm^3} = 6.88 \times 10^5\ L$$

$$\text{Mol of air} = 6.88 \times 10^5\ L \times \frac{1.22\ g\ air}{1\ L air} \times \frac{1\ mol\ air}{28.9\ g\ air} = 8.45 \times 10^5\ mol\ air$$

$$q_{air} = c_{air} \times n_{air} \times \Delta T_{air}$$

ΔT in Kelvin is the same as ΔT in °C.

$$q_{air} = (29.1\ J\ mol^{-1}K^{-1}) \times (8.45 \times 10^5\ mol) \times (22.0\ °C - 15.0\ °C) \times \frac{1\ K\ change}{1\ °C\ change} = 5.9 \times 10^6\ J$$

The equation for the reaction describing the combustion of methane is given in Section 6.11:

$$CH_4(g) + 4\ O_2(g) \longrightarrow CO_2(g) + 2\ H_2O(\ell)$$

$$\Delta H° = (1\ mol) \times \Delta H_f^\circ(CO_2) + (2\ mol) \times \Delta H_f^\circ(H_2O(g)) - (1\ mol) \times \Delta H_f^\circ(CH_4) - (4\ mol) \times \Delta H_f^\circ(O_2)$$

Look up the ΔH_f° values in Table 6.2.

$$\Delta H° = (1\ mol) \times (-393.509\ kJ/mol) + (2\ mol) \times (-285.83\ kJ/mol) - (1\ mol) \times (-74.81\ kJ/mol) - (4\ mol) \times (0\ kJ/mol) = -\ 890.36\ kJ$$

Now, we'll figure out the mass needed for warming the air sample:

$$5.9 \times 10^6\ J \times \frac{1\ kJ}{1000\ J} \times \frac{1\ mol\ CH_4}{890.36\ kJ} \times \frac{16.0423\ g\ CH_4}{1\ mol\ CH_4} \times \frac{1\,kg}{1000\,g} = 1.1 \times 10^2\ kg\ CH_4$$

✓ *Reasonable Answer Check:* The quantity of air in the house is relatively large, so the amount of joules needed to raise its temperature by seven degrees is also relatively large. The enthalpy of the combustion reaction is exothermic and large, which makes sense because methane is a good fuel.

130. *Answer:* **44.422 kJ/g octane > 19.927 kJ/g methanol**

Strategy and Explanation: Compare the quantity of thermal energy evolved per gram when burning two different organic fuels.

Balance the combustion equations. Then use Equation 6.11, a table of molar enthalpies of formation, and the stoichiometry of the balanced equations to determine the molar enthalpy change for each reaction. Then use the molar mass to determine enthalpy per gram of each fuel and compare them.

The equations for the reactions describing the combustion of methanol and octane are similar to those described in Section 6.11. We'll assume that both fuels are initially in the liquid state.

$$CH_3OH(\ell) + \frac{3}{2} O_2(g) \longrightarrow CO_2(g) + 2 H_2O(g)$$

$$C_8H_{18}(\ell) + \frac{25}{2} O_2(g) \longrightarrow 8 CO_2(g) + 9 H_2O(g)$$

$$\Delta H^\circ_{methanol} = (1 \text{ mol}) \times \Delta H^\circ_f(CO_2) + (2 \text{ mol}) \times \Delta H^\circ_f(H_2O(g)) - (1 \text{ mol}) \times \Delta H^\circ_f(CH_3OH) - (\tfrac{3}{2} \text{ mol}) \times \Delta H^\circ_f(O_2)$$

$$\Delta H^\circ_{octane} = (8 \text{ mol}) \times \Delta H^\circ_f(CO_2) + (9 \text{ mol}) \times \Delta H^\circ_f(H_2O(g)) - (1 \text{ mol}) \times \Delta H^\circ_f(C_8H_{18}) - (\tfrac{25}{2} \text{ mol}) \times \Delta H^\circ_f(O_2)$$

Look up the ΔH°_f values in Table 6.2 and Appendix J.

$$\Delta H^\circ_{methanol} = (1 \text{ mol}) \times (-393.509 \text{ kJ/mol}) + (2 \text{ mol}) \times (-241.818 \text{ kJ/mol}) - (1 \text{ mol}) \times (-238.66 \text{ kJ/mol}) - (\tfrac{3}{2} \text{ mol}) \times (0 \text{ kJ/mol}) = -638.49 \text{ kJ}$$

$$\Delta H^\circ_{octane} = (8 \text{ mol}) \times (-393.509 \text{ kJ/mol}) + (9 \text{ mol}) \times (-241.818 \text{ kJ/mol}) - (1 \text{ mol}) \times (-250.1 \text{ kJ/mol}) - (\tfrac{25}{2} \text{ mol}) \times (0 \text{ kJ/mol}) = -5074.3 \text{ kJ}$$

Now, we'll determine the enthalpy evolved per gram:

$$\frac{638.49 \text{ kJ produced}}{1 \text{ mol } CH_3OH} \times \frac{1 \text{ mol } CH_3OH}{32.0417 \text{ g } CH_3OH} = 19.927 \frac{\text{kJ produced}}{\text{g } CH_3OH}$$

$$\frac{5074.3 \text{ kJ produced}}{1 \text{ mol } C_8H_{18}} \times \frac{1 \text{ mol } C_8H_{18}}{114.2278 \text{ g } C_8H_{18}} = 44.422 \frac{\text{kJ produced}}{\text{g } C_8H_{18}}$$

The enthalpy evolved per gram of octane (44.422 kJ/g) is larger than that evolved per gram of methanol (19.927 kJ/g).

✓ *Reasonable Answer Check:* Since the combustion process is an oxidation process, and the methanol is somewhat more oxidized than the hydrocarbon, it makes sense that its combustion will produce a smaller amount of energy.

Food and Energy (Section 6.12)

132. *Answer:* **720 kJ**

Strategy and Explanation: Take the percentages (by mass) of carbohydrate, fat, and protein in some candy and determine the quantity of energy transfer that would occur a given mass of the candy was burned in a bomb calorimeter.

Start with the sample mass. Using the percentage as a ratio of masses, and the kJ/g for each type of food component given in Section 6.12, calculate the energy related to each of the three components. Then add those three numbers for the total energy.

Carbohydrate: $$34.5 \text{ g M \& M} \times \frac{70. \text{ g carbohydrates}}{100 \text{ g M\&M}} \times \frac{17 \text{ kJ}}{1 \text{ g carbohydrates}} = 410 \text{ kJ}$$

Fat: $$34.5 \text{ g M \& M} \times \frac{21 \text{ g fats}}{100 \text{ g M\&M}} \times \frac{38 \text{ kJ}}{1 \text{ g fats}} = 280 \text{ kJ}$$

Protein: $$34.5 \text{ g M \& M} \times \frac{4.6 \text{ g proteins}}{100 \text{ g M\&M}} \times \frac{17 \text{ kJ}}{1 \text{ g proteins}} = 27 \text{ kJ}$$

Total: $$410 \text{ kJ} + 280 \text{ kJ} + 27 \text{ kJ} = 720 \text{ kJ}$$

✓ *Reasonable Answer Check:* We can convert this energy value to food Calories:

$$720 \text{ kJ} \times \frac{1 \text{ kcal}}{4.184 \text{ kJ}} \times \frac{1 \text{ Cal}}{1 \text{ kcal}} = 170 \text{ Cal}$$

Some vending-machine-size packets of M&M's list the contents as having 400 Calories. So, 170 Cal represents a little less than half the packet of M&Ms. This seems a reasonable sample size for this experiment, and therefore a reasonable amount of energy calculated.

134. *Answer:* **2.2 hours walking**

Strategy and Explanation: Determine how long you must walk to burn off the Calories of a quarter-pound hamburger.

Use the caloric value of hamburger (from Table 6.3) determine the Calories in the hamburger. Use the basal metabolic rate (BMR) for a 70 kg person and the multiplier for walking given in Section 6.12, determine the energy burned for that activity per hour. Then use that result to determine how many hours the person would need to walk.

$$0.25 \text{ lb hamburger} \times \frac{454 \text{ g}}{1 \text{ lb}} \times \frac{3.60 \text{ Cal}}{1 \text{ g hamburger}} = 410 \text{ Cal}$$

Section 6.12 gives the following information: A 70-kg person has a BMR of 1750 Cal/day. Walking gives him a multiplier of 2.5 × BMR.

$$2.5 \times \frac{1750 \text{ Cal}}{1 \text{ day}} \times \frac{1 \text{ day}}{24 \text{ hr}} = 182 \frac{\text{Cal}}{\text{hr}}$$

$$410 \text{ Cal} \times \frac{1 \text{ hr}}{182 \text{ Cal}} = 2.2 \text{ hr}$$

✓ *Reasonable Answer Check:* Every dieter knows that hamburgers (and most high Calorie junk food) requires significant exercise to work off.

General Questions

136. *Answer:* **Gold reaches 100 °C first.**

Strategy and Explanation: The metal with the smaller specific heat capacity will increase in temperature faster than a metal with a larger specific heat capacity. From Table 6.1, $c_{Cu} = 0.385 \text{ J g}^{-1}{}^{\circ}\text{C}^{-1}$ and $c_{Au} = 0.128 \text{ J g}^{-1}{}^{\circ}\text{C}^{-1}$, so the Au will reach 100 °C first.

138. *Answer:* 4.82×10^4 J

Strategy and Explanation: Given the mass and initial temperature of a sample of tin, the melting point of tin, the specific heat capacity, and the heat of fusion, determine the quantity of energy (in Joules) required to complete the transition.

Use Equation 6.2' to determine the heat energy required to increase the temperature of the tin to the melting point. Then use the enthalpy of fusion (ΔH_{fus}) to determine the thermal energy required to melt the solid at the melting point.

To change the temperature of the solid, use $q = c \times m \times \Delta T$

ΔT in Kelvin is the same as ΔT in °C.

$$q_{T\text{-rise}} = c_{Sn} \times m_{Sn} \times \Delta T_{Sn} = (0.227 \text{ J g}^{-1}\text{K}^{-1}) \times \frac{1 \text{ K change}}{1 \text{ °C change}} \times (454 \text{ g}) \times (231.9 \text{ °C} - 25.0 \text{ °C}) = 2.13 \times 10^4 \text{ J}$$

To change solid to liquid in the system, use $q = m \times \Delta H_{fus}$.

$$q_{melt} = m_{Sn} \times \Delta H_{fus,Sn} = (454 \text{ g}) \times (59.2 \text{ J/g}) = 2.69 \times 10^4 \text{ J}$$

The sum gives the total heat energy required.

$$q_{total} = q_{T\text{-rise}} + q_{melt} = (2.13 \times 10^4 \text{ J}) + (2.69 \times 10^4 \text{ J}) = 4.82 \times 10^4 \text{ J}$$

✓ *Reasonable Answer Check:* To reach the final state, the tin's temperature needed to be raised and then the solid needed to be melted. Both of these changes require energy to be added to the system.

140. *Answer:* **75.4 g ice melted**

Strategy and Explanation: Given the mass and initial temperature of a sample of ice, the mass and initial temperature of a sample of water, determine how much ice melted after the mixture reaches a common final temperature.

Using Equation 6.2' and Table 6.1, determine the heat energy required to lower the temperature of the water. Since the heat energy lost by the water comes from melting ice, relate those energies. Use the enthalpy of fusion (ΔH_{fus}) from page 228 to determine amount of ice that melted. Subtract the calculated mass from the initial mass of ice.

To change the temperature of the water, use $q = c \times m \times \Delta T$:

$$q_{water} = c_{water} \times m_{water} \times \Delta T_{water}$$

Table 6.1 gives the specific heat capacity of water to be 4.184 J $g^{-1}°C^{-1}$.

$$q_{water} = (100.0 \text{ g}) \times (4.184 \text{ J g}^{-1}°\text{C}^{-1}) \times (0.00\ °\text{C} - 60.0\ °\text{C}) = -2.51 \times 10^4 \text{ J}$$

$$-q_{water} = q_{ice} = 2.51 \times 10^4 \text{ J}$$

To melt the ice, 333 J/g of heat energy are needed to melt ice:

$$2.51 \times 10^4 \text{ J} \times \frac{1 \text{ g ice melted}}{333 \text{ J}} = 75.4 \text{ g melted}$$

✓ *Reasonable Answer Check:* The ice did stop melting before it was all gone, because the thermal energy needed to reach the freezing point was reached in the solution and thermal equilibrium was established.

142. *Answer:* $\mathbf{\Delta H_f^\circ(B_2H_6) = 36}$ **kJ/mol**

Strategy and Explanation: Given a balanced thermochemical expression for a reaction and some molar enthalpies of formation, determine an unknown molar enthalpy of formation.

Use the stoichiometric coefficient of the balanced equation to describe the moles of each of the reactants and products. Set up a specific version of Equation 6.11, and solve it for the unknown ΔH_f°. Look up the ΔH_f° for the rest of the reactants and products. Plug them into the equation and solve for ΔH_f°.

The products are B_2O_3 and $H_2O(\ell)$. The reactants are B_2H_6 and O_2. O_2 is the elemental form for oxygen.

$$\Delta H° = [(1 \text{ mol}) \times \Delta H_f^\circ(B_2O_3) + (3 \text{ mol}) \times \Delta H_f^\circ(H_2O(\ell))] - (1 \text{ mol}) \times \Delta H_f^\circ(B_2H_6)$$

$$\Delta H_f^\circ(B_2H_6) = -\frac{\Delta H°}{(1 \text{ mol})} + \Delta H_f^\circ(B_2O_3) + 3 \times \Delta H_f^\circ(H_2O(\ell))$$

Look up the ΔH_f° value for $H_2O(\ell)$ in Table 6.2.

$$\Delta H_f^\circ(B_2H_6) = -\frac{(-2166 \text{ kJ})}{(1 \text{ mol})} + (-1273 \text{ kJ/mol}) + 3 \times (-285.830) = 36 \text{ kJ/mol}$$

✓ *Reasonable Answer Check:* The high exothermicity of the reaction can be almost completely accounted for by the formation of the products. So, it makes sense that the molar enthalpy of formation of the reactant B_2H_6 is small.

144. *Answer:* $\mathbf{\Delta H° = -96.4}$ **kJ**

Strategy and Explanation: Given three thermochemical expressions, determine the molar enthalpy of another reaction.

Write the equation for the desired net reaction. Follow the procedure in Question 104.

Look at the first reactant: Ca^{2+}(aq). The only given equation that has Ca^{2+}(aq) is the third one, but we need Ca^{2+} to be a reactant, so the equation must be reversed. Therefore, we must write the reverse equation with the sign of $\Delta H°$ changed. Look at the second reactant: OH^-(aq). This reactant is also in equation 3, so let's skip it. Look at third reactant: CO_2(g). The only given equation that has CO_2(g) is the first one, but CO_2(g) is a product, so the equation must be reversed. Therefore, we must also write the reverse equation with the sign of ΔH° changed. Look at first product: $CaCO_3(s)$. The only given equation that has $CaCO_3(s)$ is the first one, and we already have that one so skip this product. Look at second product: $H_2O(\ell)$. The only given equation that has $H_2O(\ell)$ is the second one, but $H_2O(\ell)$ is a reactant, so the equation must be reversed. Therefore, we must also write the reverse equation with the sign of ΔH° changed. We have now planned how to use all the equations, so let's do it:

Add all the equations as planned, eliminate reactants that also show up as products: $Ca(OH)_2(s)$ and $CaO(s)$. Add all the remaining reactants and products to form the net equation for the reaction, and add all the individual ΔH° values to get the ΔH° for the reaction:

$$Ca^{2+}(aq) + 2\ OH^-(aq) \longrightarrow \cancel{Ca(OH)_2(s)} \qquad \Delta H° = -(-16.7\ kJ)$$

$$\cancel{CaO(s)} + CO_2(g) \longrightarrow CaCO_3(s) \qquad \Delta H° = -(+178.3\ kJ)$$

$$+\ \cancel{Ca(OH)_2(s)} \longrightarrow \cancel{CaO(s)} + H_2O(\ell) \qquad +\ \Delta H° = -(-41.2\ kJ)$$

$$Ca^{2+}(aq) + 2\ OH^-(aq) + CO_2(g) \longrightarrow CaCO_3(s) + H_2O(\ell) \qquad \Delta H° = -96.4\ kJ$$

✓ *Reasonable Answer Check:* The appropriate sum of the given reactions does indeed recreate the desired reaction, and Hess's law allows us to calculate the ΔH° from that sum.

146. *Answer:* **(a) –69.14 kJ (b) 933 kJ**

Strategy and Explanation: Determine the equation for the reaction occurring and use Appendix J values to calculate the appropriate ΔH°. Use the given quantity and the thermochemical stoichiometry to calculate the enthalpy change.

(a) The products are $CaCO_3(s)$ and $H_2O(g)$. The reactants are $Ca(OH)_2(s)$ and $CO_2(g)$.

$$\Delta H° = [(1\ mol) \times \Delta H_f^\circ(CaCO_3(s)) + (1\ mol) \times \Delta H_f^\circ(H_2O(g))] - [(1\ mol) \times \Delta H_f^\circ(Ca(OH)_2(s)) + (1\ mol) \times \Delta H_f^\circ(CO_2(g))]$$

Look up the ΔH_f° values.

$$\Delta H° = [(1\ mol) \times (-1206.92\ kJ/mol) + (1\ mol) \times (-241.818\ kJ/mol)] - [(1\ mol) \times (-986.09\ kJ/mol) + (1\ mol) \times (-393.509\ kJ/mol)] = -69.14\ kJ$$

(b) Heat is evolved when this reaction occurs as indicated by the sign of ΔH°. Calculate the energy evolved for the reaction of the sample.

$$1.00\ kg\ Ca(OH)_2 \times \frac{1000\ g}{1\ kg}\ \frac{1\ mol\ Ca(OH)_2}{74.093\ g\ Ca(OH)_2} \times \frac{69.14\ kJ\ \ evolved}{1\ mol\ \ Ca(OH)_2} = 933\ kJ\ \ evolved$$

✓ *Reasonable Answer Check:* The products collectively have more negative enthalpies of formation than the reactants, so it makes sense that the reaction is exothermic. Because 74 g $Ca(OH)_2$ produces 70 kJ, that means each gram of $Ca(OH)_2$ produces close to one kJ, so it makes sense that the 1000 g sample would produce approximately 1000 kJ.

148. *Answer:* $\mathbf{\Delta H_f^\circ(CH_3OH) = -200.660\ kJ/mol}$

Strategy and Explanation: Given a formula of a compound undergoing complete combustion to form carbon dioxide and water vapor and some molar enthalpies, determine an unknown molar enthalpy of formation.

Balance the complete combustion equation. Use the stoichiometric coefficient of the balanced equation to describe the moles of each of the reactants and products. Set up a specific version of Equation 6.11, and solve it for the unknown ΔH_f°. Look up the ΔH_f° for the rest of the reactants and products. Plug them into the equation and solve for ΔH_f°.

$$?\ CH_3OH(\ell) + ?\ O_2(g) \longrightarrow ?\ CO_2(g) + ?\ H_2O(g) \qquad \text{Balance C, H, O}$$

$$CH_3OH(\ell) + \frac{3}{2}\ O_2(g) \longrightarrow CO_2(g) + 2\ H_2O(g)$$

The products areCO_2 and $H_2O(\ell)$. The reactants are CH_3OH and O_2. O_2 is the elemental form for oxygen.

$$\Delta H^\circ = [(1 \text{ mol}) \times \Delta H^\circ_f(CO_2) + (2 \text{ mol}) \times \Delta H^\circ_f(H_2O(g))] - (1 \text{ mol}) \times \Delta H^\circ_f(CH_3OH)$$

$$\Delta H^\circ_f(CH_3OH) = -\frac{\Delta H^\circ}{(1 \text{ mol})} + \Delta H^\circ_f(CO_2) + 2 \times \Delta H^\circ_f(H_2O(g))$$

Look up the ΔH°_f values for $CO_2(g)$ and $H_2O(g)$ in Table 6.2.

$$\Delta H^\circ_f(CH_3OH) = -\frac{(-676.485 \text{ kJ})}{(1 \text{ mol})} + (-393.509 \text{ kJ/mol}) + 2 \times (-241.818) = -200.660 \text{ kJ/mol}$$

✓ *Reasonable Answer Check:* 1 C, 4 H, and 4 O means equation balances. Methanol can be used as a fuel, so it makes sense that this reaction is exothermic. The size of this enthalpy of formation is similar to that seen for other organic compounds of this size.

150. *Answer:* **0.745 m/s**

Strategy and Explanation: Given the mass and kinetic energy of an object, determine its speed in m/s.

Note: The density of the object was given in the question, but it is not needed.

Kinetic energy is discussed in Section 6.1, page 212-13. The equation relating the kinetic energy, mass and speed of an object is.

$$E_k = \frac{1}{2}mv^2$$

Solve this equation for v, using the unit relationship described on page 213, $1 \text{ J} = 1 \text{ kg m}^2/\text{s}^2$

$$v = \sqrt{\frac{2E}{m}} = \sqrt{\frac{2 \times 15.7 \text{ J} \times \left(\frac{\text{kg m}^2/\text{s}^2}{1 \text{ J}}\right)}{56.6 \text{ g} \times \left(\frac{1 \text{ kg}}{1000 \text{ g}}\right)}} = 0.745 \text{ m/s}$$

✓ *Reasonable Answer Check:* A macroscopic, massive object will not need to be moving very fast to have this energy.

151. *Answer:* **(a) 36.03 kJ evolved (b) 1.18 × 10⁴ kJ evolved**

Strategy and Explanation: Given a balanced chemical equation for two reactions and a table of molar enthalpies of formation, determine the enthalpy change of the first reaction and the thermal energy evolved by the second.

Use the stoichiometric coefficient of the balanced equation to describe the moles of each of the reactants and products. Set up a specific version of Equation 6.11. Look up the ΔH°_f for each. Plug them into the equation and solve for ΔH°.

(a) The products of the first reaction are N_2O and $H_2O(g)$. The reactant is $NH_4NO_3(s)$.

$$\Delta H^\circ = (1 \text{ mol}) \times \Delta H^\circ_f(N_2O) + (2 \text{ mol}) \times \Delta H^\circ_f(H_2O(g)) - (1 \text{ mol}) \times \Delta H^\circ_f(NH_4NO_3(s))$$

Look up the ΔH°_f values in Table 6.2 and Appendix J.

$$\Delta H^\circ = (1 \text{ mol}) \times (82.05 \text{ kJ/mol}) + (2 \text{ mol}) \times (-241.818 \text{ kJ/mol}) - (1 \text{ mol}) \times (-365.56 \text{ kJ/mol}) = -36.03 \text{ kJ}$$

(b) The products of the second reaction are N_2, $H_2O(g)$, and O_2. The reactant is $NH_4NO_3(s)$. N_2 and O_2 are elemental forms, with ΔH°_f exactly zero.

$$\Delta H^\circ = (4 \text{ mol}) \times \Delta H^\circ_f(H_2O(g)) - (2 \text{ mol}) \times \Delta H^\circ_f(NH_4NO_3(s))$$

Look up the ΔH°_f values in Table 6.2 and Appendix J.

$$\Delta H^\circ = (4 \text{ mol}) \times (-241.818 \text{ kJ/mol}) - (2 \text{ mol}) \times (-365.56 \text{ kJ/mol}) = -236.15 \text{ kJ}$$

The balanced chemical equation says that 2 mol NH_4NO_3 react with the evolution of 236.15 kJ of thermal energy.

$$8.00 \text{ kg } NH_4NO_3 \times \frac{1000 \text{ g}}{1 \text{ kg}} \times \frac{1 \text{ mol } NH_4NO_3}{80.0432 \text{ g } NH_4NO_3} \times \frac{236.15 \text{ kJ}}{2 \text{ mol } NH_4NO_3} = 1.18 \times 10^4 \text{ kJ}$$

✓ *Reasonable Answer Check:* These explosive reactions produce large amounts of heat energy. The sign and size of these answers make sense.

153. *Answer:* **Step 1: –137.23 kJ, Step 2: 275.341 kJ, Step 3: 103.71 kJ, $H_2O(g) \longrightarrow H_2(g) + \frac{1}{2}O_2(g)$; $\Delta H° = 241.82$ kJ; endothermic**

Strategy and Explanation: Given a balanced chemical equation for two reactions and a table of molar enthalpies of formation, determine the enthalpy change of the first reaction and the thermal energy evolved by the second.

Use the stoichiometric coefficient of the balanced equations to describe the moles of each of the reactants and products. Set up specific versions of Equation 6.11. Look up the ΔH_f° for each species. Plug them into the equations and solve for the three ΔH°. Then add the three chemical equations together to make the overall equation for the reaction and determine the overall ΔH°.

Step 1: The products of the reaction are $H_2SO_4(\ell)$ and HBr(g). The reactants are SO_2, $H_2O(g)$, and $Br_2(g)$. $Br_2(g)$ is NOT elemental bromine.

$$\Delta H° = (1 \text{ mol}) \times \Delta H_f^\circ(H_2SO_4(\ell)) + (2 \text{ mol}) \times \Delta H_f^\circ(HBr(g)) - [(1 \text{ mol}) \times \Delta H_f^\circ(SO_2) + (2 \text{ mol}) \times \Delta H_f^\circ(H_2O(g)) + (1 \text{ mol}) \times \Delta H_f^\circ(Br_2(g))]$$

Look up the ΔH_f° values in Table 6.2 and Appendix J.

$$\Delta H° = (1 \text{ mol}) \times (-813.989 \text{ kJ/mol}) + (2 \text{ mol}) \times (-36.40 \text{ kJ/mol}) - (1 \text{ mol}) \times (-296.830\text{kJ/mol}) - (2 \text{ mol}) \times (-241.818\text{kJ/mol}) - (1 \text{ mol}) \times (30.907\text{kJ/mol}] = -137.23 \text{ kJ}$$

Step 2: The products of the reaction are $H_2O(g)$, SO_2, and O_2. The reactant is $H_2SO_4(\ell)$. $O_2(g)$ is in elemental form.

$$\Delta H° = (1 \text{ mol}) \times \Delta H_f^\circ(H_2O(g)) + (1 \text{ mol}) \times \Delta H_f^\circ(SO_2) - (1 \text{ mol}) \times \Delta H_f^\circ(H_2SO_4(\ell))$$

Look up the ΔH_f° values in Table 6.2 and Appendix J.

$$\Delta H° = (1 \text{ mol}) \times (-241.818 \text{ kJ/mol}) + (1 \text{ mol}) \times (-296.830 \text{ kJ/mol}) - (1 \text{ mol}) \times (-813.989 \text{ kJ/mol}) = 275.341 \text{ kJ}$$

Step 3: The products of the reaction are $H_2(g)$ and $Br_2(g)$. The reactant is HBr(g). $H_2(g)$ is in elemental form, but $Br_2(g)$ is not.

$$\Delta H° = (1 \text{ mol}) \times \Delta H_f^\circ(Br_2(g)) - (2 \text{ mol}) \times \Delta H_f^\circ(HBr(g))$$

Look up the ΔH_f° values in Table 6.2 and Appendix J.

$$\Delta H° = (1 \text{ mol}) \times (30.907 \text{ kJ/mol}) - (2 \text{ mol}) \times (-36.40 \text{ kJ/mol}) = 103.71 \text{ kJ}$$

Add the three thermochemical expressions to get the net equation:

$$\cancel{SO_2(g)} + \cancel{2}\,H_2O(g) + \cancel{Br_2(g)} \longrightarrow \cancel{H_2SO_4(\ell)} + \cancel{2\,HBr(g)} \qquad \Delta H° = -137.23 \text{ kJ}$$

$$\cancel{H_2SO_4(\ell)} \longrightarrow \cancel{H_2O(g)} + \cancel{SO_2(g)} + \tfrac{1}{2}O_2(g) \qquad \Delta H° = 275.341 \text{ kJ}$$

$$+ \quad \cancel{2\,HBr(g)} \longrightarrow H_2(g) + \cancel{Br_2(g)} \qquad + \ \Delta H° = 103.71 \text{ kJ}$$

$$H_2O(g) \longrightarrow H_2(g) + \tfrac{1}{2}O_2(g) \qquad \Delta H° = 241.82 \text{ kJ}$$

The net equation is endothermic.

✓ *Reasonable Answer Check:* The net reaction is the reverse of the reaction for $\Delta H^\circ_f(H_2O(g))$, which Appendix J shows with a value of –241.818 kJ/mol. The positive value makes sense because the reaction is reversed, and the size of these two numbers is the same within given significant figures.

Applying Concepts

155. *Answer:* **endothermic, exothermic**

Strategy and Explanation:

Ice in a system melts in warm conditions and the molecules are moving around more quickly in the liquid state than in the solid state, so energy must be absorbed by the surroundings. Hence, melting ice is an endothermic process.

Freezing water in a system happens only in cold conditions and the molecules are moving more slowly in the solid state than in the liquid, so energy must be lost by the water into the surroundings to freeze it. Hence, freezing water is an exothermic process.

157. *Answer:* **Substance A**

Strategy and Explanation: Use Equation 6.2': $q = m \times c \times \Delta T$. The equation says that energy transferred is proportional to ΔT for constant mass samples, using the specific heat capacity, c. So, the slower the temperature rises with the transfer of energy, the larger the specific heat capacity. On the graph, the shallowest line (the line with the least steep slope) has the highest specific heat capacity; here, that is Substance A.

159. Thermal energy content is greater in Beaker 1 than in Beaker 2. A larger mass of water will contain larger quantity of thermal energy at a given temperature.

161. Enthalpy change is an extensive property. The given equation produces 2 mol SO_3. Formation enthalpy from Table 6.2 is for the production of 1 mol SO_3, so the enthalpy values must be different by a factor of two.

163. *Answer:* **$\Delta E = 310$ J; w = 0 J**

Strategy and Explanation: If the balloon is fully inflated, the volume will not change and PV work cannot be done; hence, w = 0 J. The heating of the contents of the balloon increases its thermal energy, so q = 310 J. Because $\Delta E = q + w$, $\Delta E = 310\text{ J} + 0\text{ J} = 310\text{ J}$.

More Challenging Questions

165. *Answer:* **$\Delta H^\circ_f(OF_2) = 18$ kJ/mol**

Strategy and Explanation: Given a balanced thermochemical expression for a reaction and some molar enthalpies of formation, determine an unknown molar enthalpy of formation.

Use the stoichiometric coefficient of the balanced equation to describe the moles of each of the reactants and products. Set up a specific version of Equation 6.11, and solve it for the unknown ΔH°_f. Look up the ΔH°_f for the rest of the reactants and products. Plug them into the equation and solve for ΔH°_f.

The products are HF and O_2. The reactants are OF_2 and $H_2O(g)$. O_2 is the elemental form for oxygen.

$$\Delta H^\circ = [(2\text{ mol}) \times \Delta H^\circ_f(HF) + (1\text{ mol}) \times \Delta H^\circ_f(O_2)] - [(1\text{ mol}) \times \Delta H^\circ_f(OF_2) + (1\text{ mol}) \times \Delta H^\circ_f(H_2O(g))]$$

$$\Delta H^\circ = (2\text{ mol}) \times \Delta H^\circ_f(HF) + (1\text{ mol}) \times \Delta H^\circ_f(O_2) - (1\text{ mol}) \times \Delta H^\circ_f(OF_2) - (1\text{ mol}) \times \Delta H^\circ_f(H_2O(g))$$

$$\Delta H^\circ_f(OF_2) = -\frac{\Delta H^\circ}{(1\text{ mol})} + 2 \times \Delta H^\circ_f(HF) + \Delta H^\circ_f(O_2) - \Delta H^\circ_f(H_2O(g))$$

Look up the ΔH°_f values for $H_2O(g)$ and HF in Table 6.2.

$\Delta H^\circ_f(OF_2) = -(-318\text{ kJ/mol}) + 2 \times (-271.1\text{ kJ/mol}) + (0\text{ kJ/mol}) - (-241.818\text{ kJ/mol}) = 18\text{ kJ/mol}$

✓ *Reasonable Answer Check:* The high exothermicity of the reaction can be almost completely accounted for with the destruction of the other reactant and formation of the products. So, it makes sense that the molar enthalpy of formation of OF_2, the reactant, is small. This enthalpy of formation is given in Question 125 to be 18 kJ/mol.

167. *Answer:* **50.014 kJ/g methane, 47.484 kJ/g ethane, 46.354 kJ/g propane, 45.7140 kJ/g butane; methane > ethane > propane > butane**

Strategy and Explanation: Compare the quantity of thermal energy evolved per gram (also called the fuel value) when burning four different organic fuels.

Write the formulas for the four smallest hydrocarbons. Balance their combustion equations. Then use Equation 6.11, a table of molar enthalpies of formation, and the stoichiometry of the balanced equations to determine the molar enthalpy change for each combustion reaction. Then use the molar mass to determine the enthalpy per gram of each fuel and rank them.

The reaction for the combustion of methane is described in Section 6.11, and those describing the combustion of ethane, propane, and butane are similar to that one. We'll assume that all fuels are initially in the gas state.

Look up the ΔH°_f values in Appendix J.

Methane: $$CH_4 + 2\,O_2 \longrightarrow CO_2 + 2\,H_2O$$

$$\Delta H^\circ_{methane} = (1\text{ mol}) \times \Delta H^\circ_f(CO_2) + (2\text{ mol}) \times \Delta H^\circ_f(H_2O(g)) - (1\text{ mol}) \times \Delta H^\circ_f(CH_4) - (2\text{ mol}) \times \Delta H^\circ_f(O_2)$$
$$= (1\text{ mol}) \times (-393.509\text{ kJ/mol}) + (2\text{ mol}) \times (-241.818\text{ kJ/mol}) - (1\text{ mol}) \times (-74.81\text{ kJ/mol}) - (2\text{ mol}) \times (0\text{ kJ/mol}) = -802.34\text{ kJ}$$

$$\frac{802.34\text{ kJ produced}}{1\text{ mol }CH_4} \times \frac{1\text{ mol }CH_4}{16.0423\text{ g }CH_4} = 50.014\ \frac{\text{kJ produced}}{\text{g }CH_4}$$

Ethane: $$C_2H_6 + \frac{7}{2}\,O_2 \longrightarrow 2\,CO_2 + 3\,H_2O$$

$$\Delta H^\circ_{ethane} = (2\text{ mol}) \times \Delta H^\circ_f(CO_2) + (3\text{ mol}) \times \Delta H^\circ_f(H_2O(g)) - (1\text{ mol}) \times \Delta H^\circ_f(C_2H_6) - (\tfrac{7}{2}\text{ mol}) \times \Delta H^\circ_f(O_2)$$
$$= (2\text{ mol}) \times (-393.509\text{ kJ/mol}) + (3\text{ mol}) \times (-241.818\text{ kJ/mol}) - (1\text{ mol}) \times (-84.68\text{ kJ/mol}) - (\tfrac{7}{2}\text{ mol}) \times (0\text{ kJ/mol}) = -1427.79\text{ kJ}$$

$$\frac{1427.79\text{ kJ produced}}{1\text{ mol }C_2H_6} \times \frac{1\text{ mol }C_2H_6}{30.0688\text{ g }C_2H_6} = 47.484\ \frac{\text{kJ produced}}{\text{g }C_2H_6}$$

Propane: $$C_3H_8 + 5\,O_2 \longrightarrow 3\,CO_2 + 4\,H_2O$$

$$\Delta H^\circ = (3\text{ mol}) \times \Delta H^\circ_f(CO_2) + (4\text{ mol}) \times \Delta H^\circ_f(H_2O(g)) - (1\text{ mol}) \times \Delta H^\circ_f(C_3H_8) - (5\text{ mol}) \times \Delta H^\circ_f(O_2)$$
$$\Delta H^\circ = (3\text{ mol}) \times (-393.509\text{ kJ/mol}) + (4\text{ mol}) \times (-241.818\text{ kJ/mol}) - (1\text{ mol}) \times (-103.8\text{ kJ/mol}) - (5\text{ mol}) \times (0\text{ kJ/mol}) = -2044.0\text{ kJ}$$

$$\frac{2044.0\text{ kJ produced}}{1\text{ mol }C_3H_8} \times \frac{1\text{ mol }C_3H_8}{44.0953\text{ g }C_3H_8} = 46.354\ \frac{\text{kJ produced}}{\text{g }C_3H_8}$$

Butane: $$C_4H_{10} + \frac{13}{2}\,O_2 \longrightarrow 4\,CO_2 + 5\,H_2O$$

$$\Delta H^\circ_{butane} = (4\text{ mol}) \times \Delta H^\circ_f(CO_2) + (5\text{ mol}) \times \Delta H^\circ_f(H_2O(g)) - (1\text{ mol}) \times \Delta H^\circ_f(C_4H_{10}) - (\tfrac{13}{2}\text{ mol}) \times \Delta H^\circ_f(O_2)$$
$$= (4\text{ mol}) \times (-393.509\text{ kJ/mol}) + (5\text{ mol}) \times (-241.818\text{ kJ/mol}) - (1\text{ mol}) \times (-126.148\text{ kJ/mol}) - (\tfrac{13}{2}\text{ mol}) \times (0\text{ kJ/mol}) = -2656.978\text{ kJ}$$

$$\frac{2656.978\text{ kJ produced}}{1\text{ mol }C_4H_{10}} \times \frac{1\text{ mol }C_4H_{10}}{58.1218\text{ g }C_4H_{10}} = 45.7140\ \frac{\text{kJ produced}}{\text{g }C_4H_{10}}$$

Methane has the highest fuel value, followed by ethane, then propane, and then butane.

✓ *Reasonable Answer Check:* While we might not have predicted which of these would be the highest, seeing a consistent trend between these numbers is satisfying.

169. *Answer/Explanation:* Given a formation reaction that cannot be done directly, outline a method to combine other reactions and use Hess' Law and Appendix J to confirm the given enthalpy of formation.

Calorimetrically, determine the heat the reaction of the basic oxide, CaO, and the acidic oxide, SO_3 and the heats of formation of the oxides by reacting known quantities of the elements to form the calcium and sulfur oxides, CaO and SO_3. Each experiment is done with known quantities of reactants. Temperature changes are carefully measured at constant pressure so that heat evolved or absorbed is the enthalpy change.

$$\cancel{CaO(s)} + \cancel{SO_3(g)} \longrightarrow CaSO_4(s) \qquad \Delta H° = -403.3\text{ kJ}$$

$$Ca(s) + \tfrac{1}{2}O_2(g) \longrightarrow \cancel{CaO(s)} \qquad \Delta H° = -635.09\text{ kJ}$$

$$+ \quad S(s) + \tfrac{3}{2}O_2(g) \longrightarrow \cancel{SO_3(g)} \qquad + \ \Delta H° = -395.72\text{ kJ}$$

$$Ca(s) + S(s) + 2\,O_2(g) \longrightarrow CaSO_4(s) \qquad \Delta H° = -1434.1\text{ kJ}$$

✓ *Reasonable Answer Check:* The appropriate sum of the given reactions does indeed recreate the desired

170. *Answer:* **0.15 kJ; 0.060 g apples**

Strategy and Explanation: Given the mass and speed of an object, determine its kinetic energy. Given the caloric value of a specific food, determine the minimum amount of energy needed to produce that kinetic energy.

Kinetic energy is discussed in Section 6.1, page 212-13. The equation relating the kinetic energy, mass and speed of an object is.

$$E_k = \frac{1}{2}mv^2$$

Use the unit relationship described on page 213, $1\text{ J} = 1\text{ kg m}^2/\text{s}^2$

$$v = \frac{180\text{ mi}}{1\text{ h}} \times \frac{1\text{ h}}{3600\text{ s}} \times \frac{1\text{ km}}{0.62137\text{ mi}} \times \frac{1000\text{ m}}{1\text{ km}} = 80.5\text{ m/s}$$

$$E_k = \frac{1}{2} \times (46\text{ g}) \times \frac{1\text{ kg}}{1000\text{ g}} \times (80.5\text{ m/s})^2 \times \frac{1\text{ J}}{1\text{ kg m}^2/\text{s}^2} \times \frac{1\text{ kJ}}{1000\text{ J}} = 0.15\text{ kJ}$$

Table 6.3 in Section 6.12 give the caloric value of apples to be 2.47 kJ/g

$$0.15\text{ kJ} \times \frac{1\text{ g}}{2.47\text{ kJ}} = 0.060\text{ g apples}$$

✓ *Reasonable Answer Check:* Hitting one golf ball should not take very much energy, so it make sense that a very small amount of food is required to provide that caloric content.

172. *Answer:* **(a) 26.6 °C, temperature leveled off from Exp. 3 on (b) ascorbic acid is limiting in Exp. 1, 2, and 3; sodium hydroxide is limiting in Exp. 3,4, and 5 (c) One; equal quantities of each reactant are present in Exp. 3 at the stoichiometric equivalence point**

Strategy and Explanation:

(a) We predict the temperature of experiment to be 26.6 °C, because, above $C_6H_8O_6$ masses of 8.81g, the other reactant NaOH appears to be the limiting reactant.

(b) $C_6H_8O_6$ limits in Experiments 1, 2, and 3, which is why the final temperature increases proportionally to the mass of $C_6H_8O_6$ from experiment to experiment in those three. NaOH limits in Experiments 3,4, and 5, as seen by the constant temperature increase in the available data for those three experiments, even though larger masses of $C_6H_8O_6$ are used.

(c) Experiment 3 seems to be the stoichiometric equivalence point, where both reactants run out at the same time. Before that experiment $C_6H_8O_6$ limits, and after that experiment, NaOH limits. Use the mass of $C_6H_8O_6$ and the volume and molarity of the NaOH to determine the moles present in that experiment.

$$100.\text{ mL NaOH} \times \frac{1\text{ L}}{1000\text{ mL}} \times \frac{0.500\text{ mol NaOH}}{1\text{ L NaOH}} = 0.0500\text{ mol NaOH}$$

$$8.81 \text{ g } C_6H_8O_6 \times \frac{1 \text{ mol } C_6H_8O_6}{176.1238 \text{ g } C_6H_8O_6} = 0.0500 \text{ mol } C_6H_8O_6$$

Because equal quantities of each reactant are present at the stoichiometric equivalence point, ascorbic acid, $C_6H_8O_6$, must have just one hydrogen ion.

174. *Answer:* **(a) Approx. 1.0×10^{10} kJ (b) Approx. 2.1 metric kilotons TNT (c) 15 kilotons for Hiroshima bomb, 20 kilotons for Nagasaki; max ever = 50,000 kilotons; approx. 14% Hiroshima bomb or 10% Nagasaki bomb (d) Approx. 20 hours of continuous hurricane damage**

Strategy and Explanation: Given five railroad tank cars full of liquid propane undergoing complete combustion at room temperature. (a) Estimate the necessary details (such as tank car volume and density of liquid propane near the boiling point), to determine the energy transferred if all the propane burned at once. (b) Given the enthalpy of decomposition of TNT, $C_7H_5N_3O_6$, determine how many metric toms of TNT would provide the same energy as in (a). (c) Find the energy transfer (in kilotons of TNT) resulting from the nuclear fission bombs dropped on Hiroshima and Nagasaki. Find the largest nuclear weapon thought to have been detonated to date. Compare these explosions to the answer in (b). Determine if it was wise to evacuate the town. (d) Find the energy of a hurricane and compare that energy to the answer in (a).

(a) Determine the mass of propane in the tank cars using the volume and the density. Use Hess' Law to determine the enthalpy of combustion for propane, and use molar mass of propane the themochemical stoichiometric relationship to determine the total energy released when the propane reacts. (b) Use the answer in (a) and the energy of the TNT per mol to determine the moles of TNT, then use the molar mass and metric mass relationships to determine the metric kilotons (m.k.t.) of TNT responsible for generating such energy. (c) Look up the energy output of the two bombs released above Japan and find how the energy reported in (b) compares by calculating percentages. (d) Look up the energy associated with a hurricane and determine how many hours a hurricane must sit at one location to generate the same energy calculated in (a).

Many different answers are possible for this question, depending on the various assumptions made. One of the biggest assumptions here is the volume of the tanker cars. A survey of the internet web sites (available on-line in early 2004) describing railroad tanker cars resulted in volumes from 11,000 gallons to 34,500 gallons.

(a) Here, a typical railroad tank car volume was assumed: 20,000 gallons. Please note that there is nothing substantially more accurate about this volume than any other volume in the range of 11,000 to 34,500 gallons. *(Note: Students wishing a more accurate estimate for volume should plan to do research on tank cars used for explosive and/or toxic chemicals, which may be smaller volume than a typical tank car.)* Some Web references located early in 2004 used to justify this estimate: http://www.ntsb.gov/publictn/2001/HZM0101.pdf and http://www.propanesafety.com/scene10.htm

The density of liquid propane at its boiling point is cited as 582 kg/m³ on the following web site: http://www.airliquide.com/en/business/products/gases/gasdata/index.asp?GasID=53

$$5 \text{ cars} \times \frac{20{,}000 \text{ gal } C_3H_8}{1 \text{ car}} \times \frac{4 \text{ quart}}{1 \text{ gal}} \times \frac{1.00 \times 10^{-3} \text{ m}^3}{1.056710 \text{ quart}} \times \frac{582 \text{ kg } C_3H_8}{1 \text{ m}^3 \text{ } C_3H_8} = 2.2 \times 10^5 \text{ kg } C_3H_8$$

Balance the chemical equation for complete combustion of propane:

$$C_3H_8(g) + 5\ O_2(g) \longrightarrow 3\ CO_2(g) + 4\ H_2O(g)$$

$\Delta H° = (3 \text{ mol}) \times \Delta H_f°(CO_2) + (4 \text{ mol}) \times \Delta H_f°(H_2O(g)) - (1 \text{ mol}) \times \Delta H_f°(C_3H_8) - (5 \text{ mol}) \times \Delta H_f°(O_2)$

Look up the $\Delta H_f°$ values in Table 6.2.

$\Delta H° = (3 \text{ mol}) \times (-393.509 \text{ kJ/mol}) + (4 \text{ mol}) \times (-241.818 \text{ kJ/mol})$

$- (1 \text{ mol}) \times (-103.8 \text{ kJ/mol}) - (5 \text{ mol}) \times (0 \text{ kJ/mol}) = -2044.0 \text{ kJ}$

Now, get the energy evolved by the mass of propane in the 5 tank cars:

$$2.2 \times 10^5 \text{ kg } C_3H_8 \times \frac{1000 \text{ g}}{1 \text{ kg}} \times \frac{1 \text{ mol } C_3H_8}{44.0953 \text{ g } C_3H_8} \times \frac{2044.0 \text{ kJ}}{1 \text{ mol } C_3H_8} = 1.0 \times 10^{10} \text{ kJ}$$

(b) $1.0\times10^{10}\text{ kJ}\times\frac{1\text{ mol TNT}}{1066.1\text{ kJ}}\times\frac{227.1309\text{ g TNT}}{1\text{ mol TNT}}\times\frac{1\text{ Mg}}{1\times10^{6}\text{ g}}\times\frac{1\text{ metric ton}}{1\text{ Mg}}\times\frac{1\text{ m.k.t.}}{1000\text{ metric ton}}=$

2.1 metric kilotons.

(c) Reported on the following web sites: http://www.grolier.com/wwii/wwii_atom.html and http://hypertextbook.com/facts/2000/MuhammadKaleem.shtml, the bomb that exploded above Hiroshima was about 15 kilotons and the one that exploded above Nagasaki was about 20 kilotons.

The biggest hydrogen bomb ever tested released an energy over 50 million tons of TNT according to the following web site: http://muller.lbl.gov/teaching/Physics10/chapters/6-NuclearWeapons.html.

Use percentage to compare the energy from the propane in the railroad tank cars with these bombs:

$$\frac{2.1\text{ m.k.t.}}{15\text{ m.k.t.}}\times100\%=14\%\text{ the energy of the Hiroshima bomb}$$

$$\frac{2.1\text{ m.k.t.}}{20\text{ m.k.t.}}\times100\%=10\%\text{ the energy of the Nagasaki bomb}$$

(d) Reported on the following web site: http://octopus.gma.org/surfing/weather/andypwer.pdf, the energy evolved by a hurricane is about 5×10^{11} J/h

$$1.0\times10^{10}\text{ kJ}\times\frac{1000\text{ J}}{1\text{ kJ}}\times\frac{1\text{ hour of hurricane damage}}{5\times10^{11}\text{ J}}=20\text{ hours of hurricane damage}$$

✓ *Reasonable Answer Check:* There are many different answers for this question depending on the assumptions made. The assumptions may be as varied and controversial as the results. Checking an answer against the one presented here may be fruitless.

Chapter 7: Electron Configurations and the Periodic Table

Solutions for Blue-Numbered Questions for Review and Thought

Topical Questions

Electromagnetic Radiation (Sections 7.1, 7.2)

10. *Answer/Explanation:* Electromagnetic radiation that is high in energy consists of waves that have **short** wavelengths and **high** frequencies. For example, the high-energy end of the spectrum includes γ rays and x-rays as seen in Figure 7.1.

12. *Answer:* **(a) Radio waves (b) Microwaves**

Strategy and Explanation: Use Figure 7.1

(a) **Radio waves** are lower frequency, thus have less energy, than infrared light.

(b) **Microwaves** are higher frequency than radio waves.

14. *Answer:* **3.00×10^{-3} m, 6.63×10^{-23} J/photon, 39.9 J/mol**

Strategy and Explanation: Given the frequency of electromagnetic radiation, determine the wavelength in meters, the energy of one photon, and the energy of one mole of photons. Use equations described in Section 7.1 and 7.2.

$$\nu = 1.00 \times 10^{11}\ \text{s}^{-1}$$

(a)
$$\lambda = \frac{c}{\nu} = \frac{2.998 \times 10^{8}\ \text{m/s}}{1.00 \times 10^{11}\ \text{s}^{-1}} = 3.00 \times 10^{-3}\ \text{m}$$

(b) $E = h\nu = (6.626 \times 10^{-34}\ \text{J·s}) \times (1.00 \times 10^{11}\ \text{s}^{-1}) = 6.63 \times 10^{-23}$ J for one photon

(c)
$$\frac{6.63 \times 10^{-23}\ \text{J}}{1\ \text{photon}} \times \frac{6.022 \times 10^{23}\ \text{photons}}{1\ \text{mol photons}} = 39.9\ \text{J/mol}$$

✓*Reasonable Answer Check*: High frequency has short wavelength. The energy for one photon is a tiny number, but a mole of photons has a sizable energy.

16. *Answer:* **(d) < (c) < (a) < (b)**

Strategy and Explanation: Use Figure 7.1. The energy of a photon is directly proportional to the frequency (ν):

Lowest energy (d) radio waves < (c) microwaves < (a) green light < (b) X-rays Highest energy

18. *Answer:* **6.06×10^{14} Hz**

Strategy and Explanation: Given the wavelength of light, determine the frequency. Use appropriate length conversions and the equation in Section 7.1. 1 Hz is defined as 1 s^{-1}. The wavelength is $\lambda = 495$ nm.

$$\nu = \frac{c}{\lambda} = \frac{2.998 \times 10^{8}\ \text{m/s}}{495\ \text{nm} \times \dfrac{1 \times 10^{-9}\ \text{m}}{1\ \text{nm}}} \times \frac{1\ \text{Hz}}{1\ \text{s}^{-1}} = 6.06 \times 10^{14}\ \text{Hz}$$

✓*Reasonable Answer Check*: Nanometer wavelengths are fairly small, so it makes sense that the frequency is high. Figure 7.1 also shows visible light in the 10^{14} - 10^{16} Hz frequency range.

20. *Answer:* **4.4×10^{-19} J/photon**

Strategy and Explanation: Given the wavelength of light, determine the energy of one photon. Use appropriate length conversions and the equation in Section 7.2. $\lambda = 450$ nm

$$E_{photon} = \frac{hc}{\lambda} = \frac{6.626\times10^{-34}\ \text{J}\cdot\text{s}\times2.998\times10^{8}\,\text{m/s}}{450\ \text{nm}\times\dfrac{1\times10^{-9}\ \text{m}}{1\ \text{nm}}} = 4.4\times10^{-19}\ \text{J}$$

✓*Reasonable Answer Check*: The energy of this blue photon is similar to that calculated in Section 7.2 for another blue photon (with a similar wavelength).

22. *Answer/Explanation:* This question requires some interpretation and some assumptions. Certain visible light interacts with the cells of our eyes. Ultraviolet light causes sunburns, which is probably in an interaction with skin cells. Infrared light feels like heat to us that might also be a cellular response of the skin cells. Microwave energy is used to cook food, which is made from cells, too; therefore it appears to interact with cells also. X-rays interact with matter, that's how we get X-ray images. Radio waves interact with matter, that's how we get MRI images. Gamma rays can interact with matter, that's one of the ways people die from radiation sickness. In conclusion, **many types of electromagnetic radiation** interact with molecules at the cellular level. There is a discussion of effects of electromagnetic radiation on matter at the end of Section 7.2 on page 278.

24. *Answer:* **1.1×10^{15} Hz, 7.4×10^{-19} J/photon**

Strategy and Explanation: Given the wavelength of electromagnetic radiation, determine the frequency and energy of this radiation. Use equations in Sections 7.1 and 7.2, and appropriate metric conversion factors. The energy calculated is the energy of one photon of this light.

$$\lambda = 270\ \text{nm} \qquad \nu = \frac{c}{\lambda} = \frac{2.998\times10^{8}\,\text{m/s}}{270\ \text{nm}\times\dfrac{1\times10^{-9}\ \text{m}}{1\ \text{nm}}} = 1.1\times10^{15}\ \text{s}^{-1} = 1.1\times10^{15}\ \text{Hz}$$

$$E_{photon} = h\nu = \left(6.626\times10^{-34}\ \text{J}\cdot\text{s}\right)\times\left(1.1\times10^{15}\ \text{s}^{-1}\right) = 7.4\times10^{-19}\ \text{J}$$

✓*Reasonable Answer Check*: The wavelength calculated coincides with that of ultraviolet light, according to Figure 7.1. This ultraviolet photon's energy is also more than those of visible light calculated in Section 7.2.

26. *Answer:* **8.42×10^{-19} J, it is larger**

Strategy and Explanation: Given the wavelength of electromagnetic radiation, determine the energy of one photon of this radiation and compare it to the given energy of another kind of photon. We use appropriate length conversions and the equation in Section 7.2. The wavelength is $\lambda = 2.36$ nm.

$$E_{photon} = \frac{hc}{\lambda} = \frac{6.626\times10^{-34}\ \text{J}\cdot\text{s}\times2.998\times10^{8}\,\text{m/s}}{2.36\ \text{nm}\times\dfrac{1\times10^{-9}\ \text{m}}{1\ \text{nm}}} = 8.42\times10^{-17}\ \text{J}$$

$$8.42\times10^{-17}\ \text{J} > 3.18\times10^{-19}\ \text{J}$$

✓*Reasonable Answer Check*: The energy of an x-ray photon is more than 250 times more than that of a visible photon. That make sense, since the wavelength of the x-ray is about 250 times shorter than the wavelengths of visible light (400 nm - 700 nm).

28. *Answer:* **1.20×10^{8} J/mol**

Strategy and Explanation: Given the wavelength of electromagnetic radiation, determine the energy possessed by one mole of photons of this radiation. Use appropriate length conversions and the equation in Section 7.2 to find the energy of the photon. The wavelength is $\lambda = 1.00\times10^{9}$ m. According to Figure 7.1, this is a typical wavelength for an x-ray.

$$E_{photon} = \frac{hc}{\lambda} = \frac{6.626\times10^{-34}\ \text{J}\cdot\text{s}\times2.998\times10^{8}\,\text{m/s}}{1.00\times10^{-9}\ \text{m}} = 1.99\times10^{-16}\ \text{J}$$

$$\frac{1.99\times10^{-16}\ \text{J}\times}{1\ \text{photon}}\times\frac{6.022\times10^{23}\ \text{photons}}{1\ \text{mol photons}} = 1.20\times10^{8}\ \frac{\text{J}}{\text{mol}}$$

✓*Reasonable Answer Check*: X-rays have high energy, so this large amount of energy per mole makes sense.

Photoelectric Effect (Section 7.2)

30. *Answer/Explanation:* Photons of light with long wavelength are low in energy. It is clear from the description that the energy of these photons is insufficient to cause electrons to be ejected. Increasing the intensity only increases the number of photons, not their energy.

32. *Answer:* **no**

Strategy and Explanation: Given the wavelength of light and the minimum energy for ejecting electrons from a metal surface, determine if the light has sufficient energy to cause the photoelectric effect (i.e., eject electrons from that metal's surface). We use appropriate length conversions and the equation in Section 7.2 to find the energy of the photon. Compare the photon energy to the minimum energy. If the photon energy is larger, then the photon can eject electrons; if the energy is smaller, it cannot. The wavelength is $\lambda = 600.$ nm.

$$E_{photon} = \frac{hc}{\lambda} = \frac{6.626\times10^{-34}\ \text{J}\cdot\text{s}\times 2.998\times10^{8}\ \text{m/s}}{600.\ \text{nm}\times\dfrac{1\times10^{-9}\ \text{m}}{1\ \text{nm}}} = 3.31\times10^{-19}\ \text{J}$$

$$E_{minimum} = 3.69\times10^{-19}\ \text{J} > E_{photon}$$

Therefore it has insufficient energy to eject electrons.

✓*Reasonable Answer Check*: The minimum energy is similar to ones calculated for visible light photons in Section 7.2. These two energies are close, so we probably would not have been able to predict this result before doing the calculation.

Atomic Spectra and the Bohr Atom (Section 7.3)

34. *Answer/Explanation:* Line emission spectra are mostly dark, with discrete bands of light. Sunlight is a continuous rainbow of color. This is described in detail in Section 7.3 starting on page 280.

36. *Answer/Explanation:* Energy is emitted from an atom when an electron moves from the **higher-energy** (excited) state to a **lower-energy** state (maybe the ground state, maybe another less-excited state). The energy of the emitted radiation corresponds to the **difference** between the two energy states.

38. *Answer:* **(a) absorbed (b) emitted (c) absorbed (d) emitted**

Strategy and Explanation: Given the values of n for the initial and final states of an electron in hydrogen, determine whether energy is absorbed or emitted. The size of n indicates the relative energy of the state. If the final state has a larger n value than the initial state, then energy must be absorbed. If the final state has a smaller n value than the initial state, then energy must be emitted.

(a) $n = 1$ to $n = 3$. Energy is absorbed.

(b) $n = 5$ to $n = 1$. Energy is emitted.

(c) $n = 2$ to $n = 4$. Energy is absorbed.

(d) $n = 5$ to $n = 4$. Energy is emitted.

✓*Reasonable Answer Check*: Energy needs to be used up to get the electron to a higher n value state. Energy is lost when moving the electron to a lower n value state.

40. *Answer:* **(a), (b), and (d)**

Strategy and Explanation: The difference between the energies of the two levels is the energy emitted. The smaller the energy, the longer the wavelength. When we look at Figure 7.8, we need to find which of the given levels is closer together than the energy levels represented by $n = 1$ and $n = 4$. The energies levels represented by (a) $n = 2$ and $n = 4$, (b) $n = 1$ and $n = 3$, and (d) $n = 3$ and $n = 5$, are closer together than $n = 1$ and $n = 4$. So, **(a), (b), and (d)** will require radiation with longer wavelength.

42. *Answer:* **-6.198×10^{-19} J, 320.5 nm, ultraviolet**

Strategy and Explanation: Given the values of n for the initial and final states of an electron in a helium cation, He^{+}, and an equation describing the relationship between the electron energy level, the atomic number, Z, and the quantum number, n, determine the transition energy of the electron and what region of the electromagnetic spectrum it is in. Adapt the method shown in Section 7.3 on page 283. Subtract the initial energy ($n = 5$) of the electron from the final energy (n = 3) of the electron in helium. Z = 2 for helium. Use Figure 7.1 to determine what part of the spectrum it is in.

$$\Delta E = E_3 - E_5 = \left[-\frac{2^2}{3^2}\left(2.179\times10^{-18}\text{ J}\right)\right] - \left[-\frac{2^2}{5^2}\left(2.179\times10^{-18}\text{ J}\right)\right] = -6.198\times10^{-19}\text{ J}$$

$$\nu = \frac{E_{photon}}{h} = \frac{-\Delta E}{h} = \frac{-6.198\times10^{-19}\text{ J}}{6.626\times10^{-34}\text{ J}\cdot\text{s}} = 9.354\times10^{14}\text{ s}^{-1}$$

$$\lambda = \frac{c}{\nu} = \frac{2.998\times10^{8}\text{ m/s}}{9.354\times10^{14}\text{ s}^{-1}} \times \frac{1\text{ nm}}{10^{-9}\text{ m}} = 320.5\text{ nm}$$

This wavelength indicates that the emission produces **ultraviolet** light.

✓*Reasonable Answer Check*: The electron levels in helium are closer together than those in hydrogen because the nucleus has a more-positive charge.

44. *Answer:* **-1.839×10^{-17} J, 10.80 nm, ultraviolet**

Strategy and Explanation: Given the values of n for the initial and final states of an electron in a lithium cation, Li^{2+}, and an equation describing the relationship between the electron energy level, the atomic number, Z, and the quantum number, n, determine the transition energy of the electron and determine what region of the electromagnetic spectrum it is in. Adapt the method shown in Section 7.3 on page 284. Subtract the initial energy (n = 4) of the electron from the final energy (n = 1) of the electron in lithium. Z = 3 for lithium. Use Figure 7.1 to determine what part of the spectrum it is in.

$$\Delta E = E_1 - E_4 = \left[-\frac{3^2}{1^2}\left(2.179\times10^{-18}\text{ J}\right)\right] - \left[-\frac{3^2}{4^2}\left(2.179\times10^{-18}\text{ J}\right)\right] = -1.839\times10^{-17}\text{ J}$$

$$\nu = \frac{E_{photon}}{h} = \frac{-\Delta E}{h} = \frac{-1.839\times10^{-17}}{6.626\times10^{-34}\text{ J}\cdot\text{s}} = 2.775\times10^{16}\text{ s}^{-1}$$

$$\lambda = \frac{c}{\nu} = \frac{2.998\times10^{8}\text{ m/s}}{2.775\times10^{16}\text{ s}^{-1}} \times \frac{1\text{ nm}}{10^{-9}\text{ m}} = 10.80\text{ nm}$$

This wavelength indicates that the emission produces **ultraviolet** light.

✓*Reasonable Answer Check*: The electron levels in lithium ion are closer together than those in hydrogen because the nucleus has a more-positive charge.

46. *Answer:* **4.576×10^{-19} J absorbed and 434.0 nm**

Strategy and Explanation: Given the values of n for the initial and final states of an electron in hydrogen, determine the energy and wavelength of the emission.

Using equations given in Section 7.3, calculate the energy of the photon emitted during the transition. Use the equation given in Section 7.1 and appropriate metric conversions to calculate the wavelength in nanometers.

Here, $n_i = 2$ and $n_f = 5$:

$$\Delta E_{e^-} = \left(2.179\times10^{-18}\text{ J}\right)\left(\frac{1}{n_i^2} - \frac{1}{n_f^2}\right) = \left(2.179\times10^{-18}\text{ J}\right)\left(\frac{1}{2^2} - \frac{1}{5^2}\right) = 4.576\times10^{-19}\text{ J}$$

The energy of the photon absorbed is equal to the energy gained by the electron as it moves to a higher-energy state. $E_{photon} = \Delta E_{e^-}$. The wavelength of the photon absorbed relates to the energy:

$$\lambda = \frac{hc}{E_{photon}} = \frac{\left(6.626\times10^{-34}\text{ J}\cdot\text{s}\right)\left(2.998\times10^{8}\text{ m/s}\right)}{4.576\times10^{-19}\text{ J}} \times \frac{1\text{ nm}}{10^{-9}\text{ m}} = 434.0\text{ nm}$$

✓ *Reasonable Answer Check*: The wavelength of radiation produced from the reverse of this transition is reported in Figure 7.8.

de Broglie Wavelength (Section 7.4)

48. *Answer:* **0.05 nm**

Strategy and Explanation: Given the mass and speed of a particle, determine the de Broglie wavelength. Use the equation in Section 7.4, the method described in Problem-Solving Example 7.5, and appropriate unit conversions.

Problem-Solving Example 7.5 shows how Planck's constant can be represented using a mass of kilograms: $h = 6.626 \times 10^{-34}$ kg m^{2}s^{-1}

Find speed in meters per second: $v = (0.05) \times (2.998 \times 10^8 \text{ m/s}) = 1 \times 10^7$ m/s

Now, find the de Broglie wavelength: $$\lambda_{DeBroglie} = \frac{h}{mv} = \frac{6.626 \times 10^{-34}\ \text{kg} \cdot \text{m}^2\text{s}^{-1}}{\left(9.11 \times 10^{-31}\ \text{kg}\right)\left(1 \times 10^7\ \frac{\text{m}}{\text{s}}\right)} = 5 \times 10^{-11}\ \text{m}$$

This wavelength is 0.05 nm.

✓ *Reasonable Answer Check*: The electron is a sub-nanoscale particle, so it makes sense that its de Broglie wavelength is similar (50 pm) to its size dimensions.

Quantum Numbers (Section 7.5)

50. *Answer:* **(a) $1,0,0,\frac{1}{2}$; $1,0,0,-\frac{1}{2}$; $2,0,0,\frac{1}{2}$; $2,0,0,-\frac{1}{2}$; $2,1,1,\frac{1}{2}$ (other answers with different m_ℓ and m_s are also possible); (b) $3,0,0,\frac{1}{2}$; $3,0,0,-\frac{1}{2}$; (c) $3,2,2,\frac{1}{2}$ (other answers with different m_ℓ and m_s are also possible)**

Strategy and Explanation: Give the four quantum numbers for various electrons. The set of three quantum numbers representing each orbital in an atom must be different (Pauli Exclusion Principle). ℓ values must be less than n, but not negative. m_ℓ values must be between $-\ell$ and $+\ell$. m_s values must be $+\frac{1}{2}$ or $-\frac{1}{2}$. Electrons with the same n, ℓ, and m_ℓ must have opposite signs for m_s. Electrons with the same n and ℓ, but different m_ℓ must have the same m_s value (Hund's rule) until all values of m_ℓ have been used once. Be systematic.

(a) There are five electrons in boron:

First electron	n = 1	$\ell = 0$	$m_\ell = 0$	$m_s = +\frac{1}{2}$
Second electron	n = 1	$\ell = 0$	$m_\ell = 0$	$m_s = -\frac{1}{2}$
Third electron	n = 2	$\ell = 0$	$m_\ell = 0$	$m_s = +\frac{1}{2}$
Fourth electron	n = 2	$\ell = 0$	$m_\ell = 0$	$m_s = -\frac{1}{2}$
Fifth electron	n = 2	$\ell = 1$	$m_\ell = 1$	$m_s = +\frac{1}{2}$

(The fifth electron could have different m_ℓ and m_s values also, as long as they are within the appropriate ranges.)

(b) Valence electrons are described in Sections 7.6 and 7.7 as the outermost electrons. In elements without partially filled d-shells, they will be the electrons with the highest n-value. Magnesium has two valence electrons:

First valence electron of Mg	n = 3	$\ell = 0$	$m_\ell = 0$	$m_s = +\frac{1}{2}$
Second valence electron of Mg	n = 3	$\ell = 0$	$m_\ell = 0$	$m_s = -\frac{1}{2}$

(c) A 3d electron in iron atom:

A 3d electron in Fe	$n = 3$	$\ell = 2$	$m_\ell = 2$	$m_s = +\frac{1}{2}$

(The values of m_ℓ and m_s can be different here, as long as they are within the appropriate ranges.)

52. *Answer:* **(a) cannot occur, m_ℓ must be between $-\ell$ and $+\ell$ (b) can occur (c) cannot occur, m_s cannot be one, here (d) cannot occur, ℓ must be less than n (e) can occur**

Strategy and Explanation:

(a) $n = 2, \ell = 1, m_\ell = 2, m_s = +\frac{1}{2}$ could not occur because m_ℓ must be between $-\ell$ and $+\ell$.

(b) $n = 3, \ell = 2, m_\ell = 0, m_s = -\frac{1}{2}$ can occur.

(c) $n = 1, \ell = 0, m_\ell = 0, m_s = 1$ could not occur because m_s must be $+\frac{1}{2}$ or $-\frac{1}{2}$.

(d) $n = 3, \ell = 3, m_\ell = 2, m_s = -\frac{1}{2}$ could not occur because ℓ must be less than n.

(e) $n = 2, \ell = 0, m_\ell = 0, m_s = +\frac{1}{2}$ can occur.

54. *Answer/Explanation:*

(a) The up-spin electron in the 4s orbital has the following four quantum numbers:

$n = 4$	$\ell = 0$	$m_\ell = 0$	$m_s = +\frac{1}{2}$

(b) The down-spin electron in the first of three 3p orbitals has the following four quantum numbers:

$n = 3$	$\ell = 1$	$m_\ell = 1$	$m_s = -\frac{1}{2}$

Notice: Two other values of m_ℓ (0 and –1) would also be valid.

(c) The up-spin electron in the third of three 3d orbitals has the following four quantum numbers:

$n = 3$	$\ell = 2$	$m_\ell = 0$	$m_s = +\frac{1}{2}$

Notice: Four other values of m_ℓ (+2, +1, –1, and –2) and one other value of m_s ($-\frac{1}{2}$) would also be valid.

Quantum Mechanics (Section 7.6)

56. *Answer/Explanation:* There are n subshells in the nth energy level; they have $\ell = n - 1, \ldots, 0$. That means **four** different subshells (designated $\ell = 3, 2, 1$, and 0) are found in the 4th energy level.

58. *Answer/Explanation:* The wave mechanical model of the atom tells us that electrons do not follow simple paths as do planets. Rather, they occupy regions of space having certain shapes and varying distances around the nucleus. Hence there are subshells and orbitals that were not part of the Bohr model.

60. *Answer/Explanation:* Orbits have predetermined paths – position and momentum are both exactly known at all times. Heisenberg's uncertainty principle says that we cannot know both position and momentum simultaneously.

62. *Answer/Explanation:* The n = 3 shell has orbitals with $\ell = 2, 1$, and 0. That means it has **d, p, and s orbitals**. There are n^2 orbitals in the nth shell. That means there are **nine** orbitals in the n = 3 shell.

Electron Configurations (Section 7.7)

63. *Answer/Explanation:* Find the atomic number and make sure you use that number of electrons:

$_{12}Mg$ electron configuration: $\mathbf{1s^22s^22p^63s^2}$

$_{17}Cl$ electron configuration: $\mathbf{1s^22s^22p^63s^23p^5}$

64. *Answer/Explanation:* Find the atomic number and make sure you use that number of electrons:

$_{13}Al$ electron configuration: $\mathbf{1s^22s^22p^63s^23p^1}$

$_{16}S$ electron configuration: $\mathbf{1s^22s^22p^63s^23p^4}$

66. *Answer/Explanation:* $_{32}Ge$ electron configuration: $\mathbf{1s^22s^22p^63s^23p^63d^{10}4s^24p^2}$

68. *Answer/Explanation:* $_8O$, **oxygen**, is an element in Group 6A. (There are several other possible elements that can answer this question.) The valance electron configuration of this element is $2s^22p^4$. All elements in this group have six valence electrons, so their electron configurations are: $\mathbf{ns^2np^4}$

69. *Answer:* **(a) Cs^+, Se^{2-}; (b) Cs^+, Se^{2-}**

Srategy and Explanation:

(a) Cations likely to form are ones that have lost only valence electrons. Any ion that has a more positive charge than its number of valence electrons is not likely to form. Anions likely to form are ones that add enough electrons to the atom to increase the number of valence electrons to eight. Any ion that has a more negative charge than the difference between eight and the number of valence electrons is not likely to form.

K has one valence electron; K^{2+} is **not** likely to form because a core electron must be removed.

Cs has one valence electron; Cs^+ **is** likely to form.

Al has three valence electrons; Al^{4+} is **not** likely to form because a core electron must be removed.

F has seven valence electrons. It needs 8 - 7 = 1 electron. F^{2-} is **not** likely to form because a new valence shell must be started to add the second electron.

Se has six valence electrons. It needs 8 – 6 = 2 electrons. Se^{2-} **is** likely to form.

(b) The only element above that loses its valence electrons and no core electrons is Cs. Cs^+ has a [Xe] electron configuration. The only element above that gains just enough electrons to fill its valence shell without adding any more is Se. Se^{2-} has a [Kr] electron configuration.

70. *Answer/Explanation:*

(a) The diagram in the book shows total of 28 electrons and six 3d sublevels, but Fe atom should have only 26 electrons and the 3d sublevel should have only five orbitals. Also, in the ground state of Fe atom, the lower-energy **4s orbital must be full.**

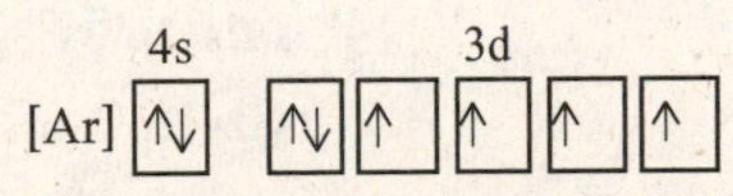

(b) The **orbital labels must be 3, not 2**. The electrons in 3p subshell of P should be in separate orbitals with parallel spin (Hund's Rule).

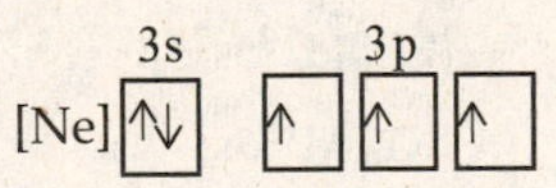

(c) The diagram in the book shows total of 50 electrons and six 3d sublevels, but the cation, Sn^{2+} should have only 48 electrons and the 3d sublevel should have only five orbitals. Also, electrons should be removed from the higher-energy valence 5p orbitals of Sn to make the cation, Sn^{2+}, not from the lower-energy 4d orbitals. The **4d orbital must be completely filled before the 5p orbitals start filling**.

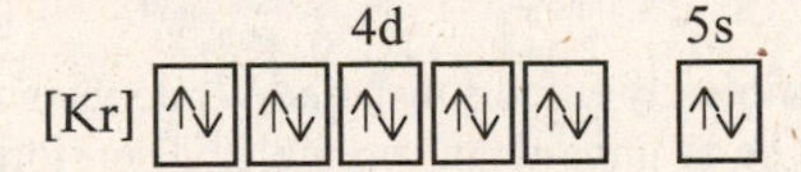

72. *Answer/Explanation:* Find the number of electrons in the subshell, and add arrows according to Hund's rule.

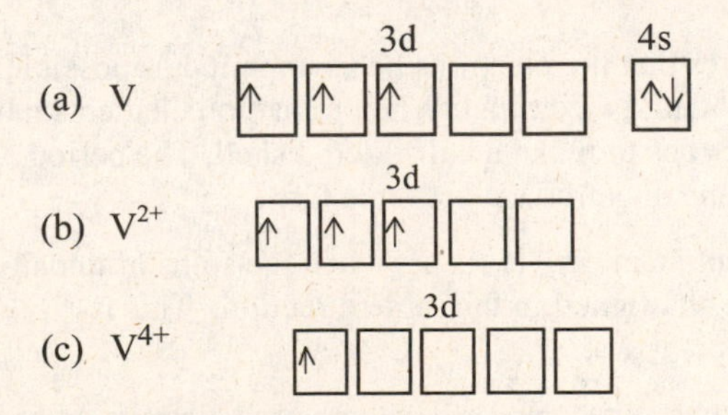

74. *Answer/Explanation:* **18** elements are in the fourth period of the periodic table. It is not possible for there to be another element in this period because **all possible orbital electron combinations are already used**.

76. *Answer/Explanation:* When transition metals form cations, the electrons lost first are those from their valence (highest n) s orbitals.

$_{25}$Mn electron configuration: $\mathbf{1s^22s^22p^63s^23p^63d^54s^2}$. Mn has **5 unpaired electrons**:

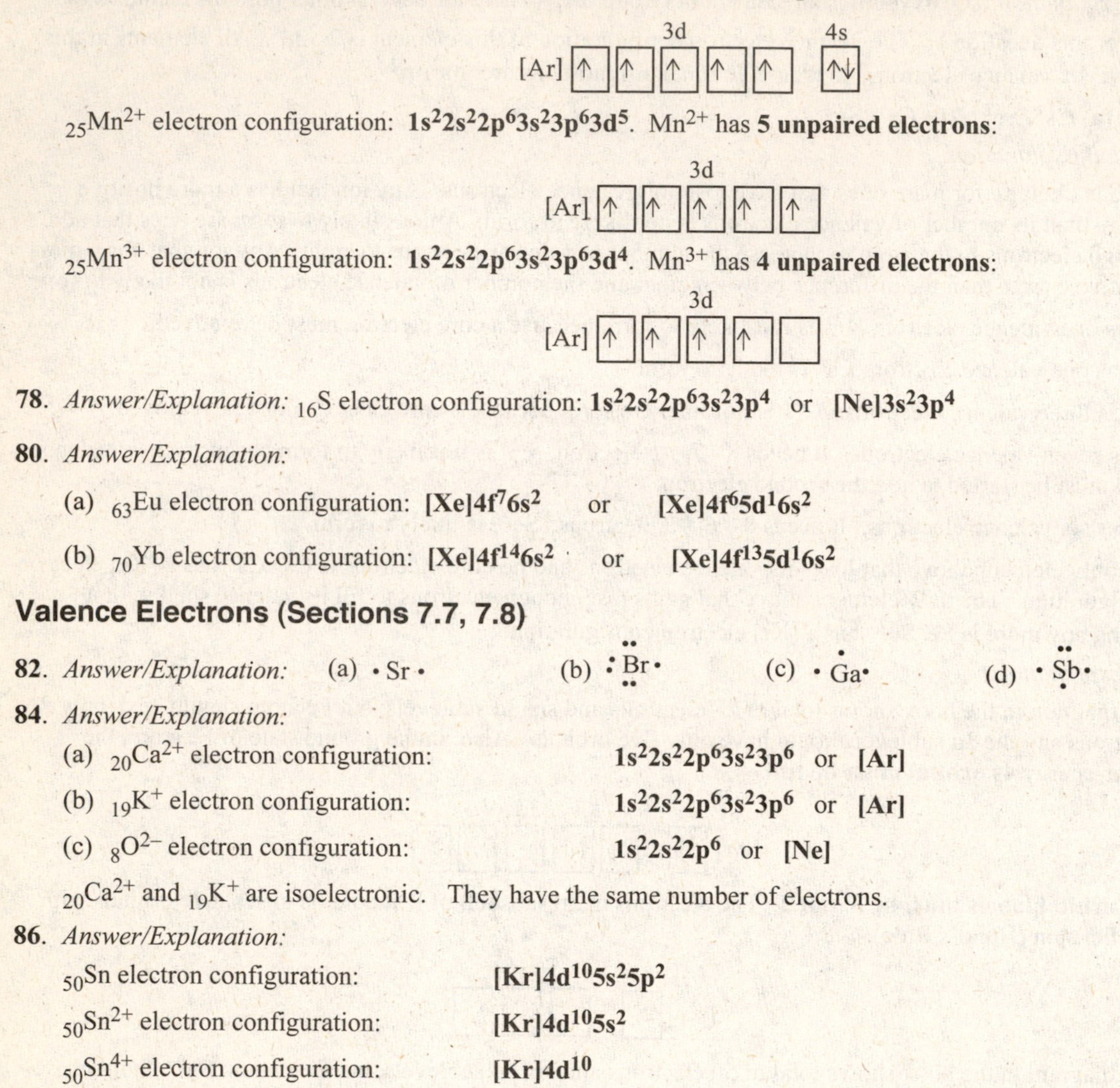

[Ar] ↑ ↑ ↑ ↑ ↑ (3d) ↑↓ (4s)

$_{25}Mn^{2+}$ electron configuration: $\mathbf{1s^22s^22p^63s^23p^63d^5}$. Mn^{2+} has **5 unpaired electrons**:

[Ar] ↑ ↑ ↑ ↑ ↑ (3d)

$_{25}Mn^{3+}$ electron configuration: $\mathbf{1s^22s^22p^63s^23p^63d^4}$. Mn^{3+} has **4 unpaired electrons**:

[Ar] ↑ ↑ ↑ ↑ ☐ (3d)

78. *Answer/Explanation:* $_{16}$S electron configuration: $\mathbf{1s^22s^22p^63s^23p^4}$ or $\mathbf{[Ne]3s^23p^4}$

80. *Answer/Explanation:*

(a) $_{63}$Eu electron configuration: $\mathbf{[Xe]4f^76s^2}$ or $\mathbf{[Xe]4f^65d^16s^2}$

(b) $_{70}$Yb electron configuration: $\mathbf{[Xe]4f^{14}6s^2}$ or $\mathbf{[Xe]4f^{13}5d^16s^2}$

Valence Electrons (Sections 7.7, 7.8)

82. *Answer/Explanation:* (a) ·Sr· (b) :Br: with one unpaired dot (7 dots) (c) ·Ġa· (d) ·Sb· (5 dots)

84. *Answer/Explanation:*

(a) $_{20}Ca^{2+}$ electron configuration: $\mathbf{1s^22s^22p^63s^23p^6}$ or **[Ar]**

(b) $_{19}K^{+}$ electron configuration: $\mathbf{1s^22s^22p^63s^23p^6}$ or **[Ar]**

(c) $_{8}O^{2-}$ electron configuration: $\mathbf{1s^22s^22p^6}$ or **[Ne]**

$_{20}Ca^{2+}$ and $_{19}K^{+}$ are isoelectronic. They have the same number of electrons.

86. *Answer/Explanation:*

$_{50}$Sn electron configuration: $\mathbf{[Kr]4d^{10}5s^25p^2}$

$_{50}Sn^{2+}$ electron configuration: $\mathbf{[Kr]4d^{10}5s^2}$

$_{50}Sn^{4+}$ electron configuration: $\mathbf{[Kr]4d^{10}}$

Paramagnetism and Unpaired Electrons (Section 7.8)

87. *Answer/Explanation:*

(a) Diamagnetic elements have completely filled subshells. While the subshell is incompletely full, Hund's Rule requires that the electrons be as unpaired as possible. Therefore, only the element with completely filled subshells will be diamagnetic. The only period four transition element with completely filled s and d subshell is **Zn** ($[Ar]3d^{10}4s^2$).

(b) While the subshell is incompletely full, Hund's Rule requires that the electrons be as unpaired as possible. That means a transition element will have at most five unpaired d electrons. When transition elements have six outer electrons, they have one s electron and five d electrons to make a half filled d shell. The period four transition element with five unpaired d electrons and one unpaired s electron is **Cr**.

88. *Answer/Explanation:* Ferromagnetism is a property of permanent magnets. It occurs when the spins of unpaired electrons in a cluster of atoms (called a domain) in the solid are all aligned in the same direction. This is described at the end of Section 7.8.

90. *Answer/Explanation:* In both paramagnetic and ferromagnetic substances, atoms have unpaired elctrons with spin and so are attracted to magnets. Ferromagnetic substances retain their aligned spins after an external magnetic field has been removed, so they can function as magnets. Paramagnetic substances lose their aligned spins after a time and therefore cannot be used as permanent magnets.

92. *Answer:* **(a) TiO_2; it has no unpaired electrons (diamagnetic) (b) TiO; it has unpaired electrons (paramagnetic)**

Strategy and Explanation: The oxide compounds contain different titanium ions: Ti^{2+} and Ti^{4+} Ions with no unpaired electrons are diamagnetic and those that have unpaired electrons are attracted to magnets. The one with more unpaired electrons will have the greater attraction to a magnetic field.

$_{22}Ti^{2+}$ electron configuration: $\mathbf{1s^22s^22p^63s^23p^63d^2}$. Fe^{2+} has **2 unpaired electrons.**

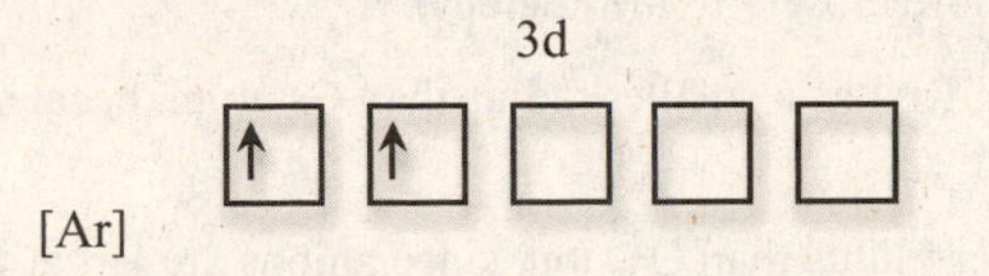

$_{22}Ti^{4+}$ electron configuration: $\mathbf{1s^22s^22p^63s^23p^6}$. Fe^{3+} has **no unpaired electrons.**

(a) TiO_2 is **diamagnetic**, because the Ti^{4+} ion has **no unpaired electrons**.

(b) **TiO** has more attraction to a magnetic field, because the Ti^{2+} ion has **unpaired electrons,** which makes it **paramagnetic**.

Periodic Trends (Sections 7.9 - 7.11)

94. *Answer:* **P < Ge < Ca < Sr < Rb**

Strategy and Explanation: Without using the table of numerical sizes, list elements in order of increasing size. Without using the table that gives numerical sizes, we are asked to list elements in order of increasing size. Periodic Trends in atom size are related to the radius of the atoms:

- Comparing atoms across the period (row) of the periodic table, atom sizes increase from right to left.
- Comparing atoms down a group (column) of the periodic table, atom sizes increase from top to bottom.

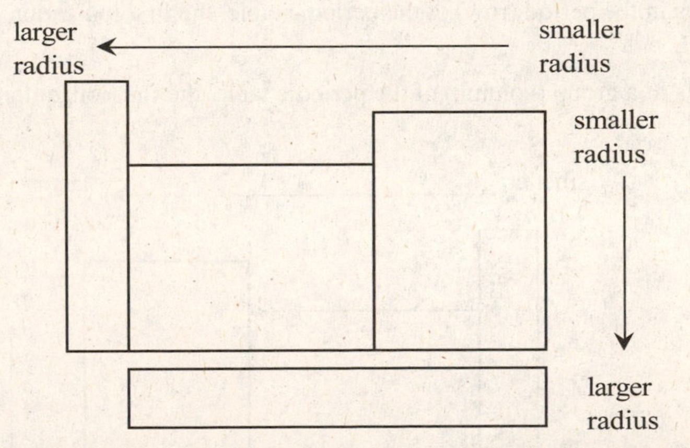

Looking at a periodic table, the smallest of the ions given is P, since its the closest to the top right corner. Next is Ge, diagonally down and to the left of P. Then comes Ca, in the same period as Ge, but further to the left. Then comes Sr, in Group 2A, below Ca. Then comes Rb, in the same period as Sr, but further to the left. So, the order is: P < Ge < Ca < Sr < Rb.

✓*Reasonable Answer Check:* Using Figures 7.22 and 7.23, we can confirm these predictions.

96. *Answer:* **(a) Rb (b) O (c) Br (d) Ba^{2+} (e) Ca^{2+}**

Strategy and Explanation: Without using the table that gives numerical sizes, determine which element of a pair of atoms and ions has larger radius. Use the periodic table for atom size trends (as described in the solution to Question 94) and the following periodic trends.

Periodic trends comparing atoms to ions:

- The cation of an element will always be smaller than the atom of the same element (since fewer electrons are attracted more closely to the nucleus).
- The anion of an element will always be larger than the atom of the same element (since more electrons will be less attracted to the nucleus).

Periodic trend comparing ions to ions:

- When comparing isoelectronic ions, the element with the larger atomic number is smaller (since it has more protons to attract the electrons).
- When comparing ions of the same element, the ion with the larger positive charge is smaller (since fewer electrons are attracted more closely to the nucleus).

(a) Using the periodic table, Rb has a smaller radius than Cs atom, because they are both in the same group and Cs is below Rb.

(b) The O atom has a smaller radius than O^{2-} ion, since anions are larger than the neutral atom. (8 protons attract 8 electrons better than they can attract 10 electrons.)

(c) Using the periodic table, Br has a smaller radius than As atom, because they are both in the same period and Br is further to the right.

(d) The Ba^{2+} ion has a smaller radius than Ba atom, since cations are smaller than the neutral atom. (56 protons attract 54 electrons better than they can attract 56 electrons.)

(e) The Ca^{2+} ion has a smaller radius than Cl^{-} ion, since they are isoelectronic and the atomic number of Ca (20) is larger than the atomic number of Cl (17). (20 protons can attract 18 electrons better than 17 protons can.)

✓*Reasonable Answer Check:* Using Figures 7.22 - 7.24, we can confirm these predictions.

98. *Answer:* **Al < Mg < P < F**

Strategy and Explanation: List elements in order of increasing first ionization energy. Sometimes we will not have a chart with numbers like Figure 7.24, so let's get an idea of what periodic trends are exhibited in first ionization energies:

- Comparing atoms in the period (row) of the periodic table, the first ionization energy increases from left to right.
- Comparing atoms in a group (column) of the periodic table, the first ionization energy increases from bottom to top.

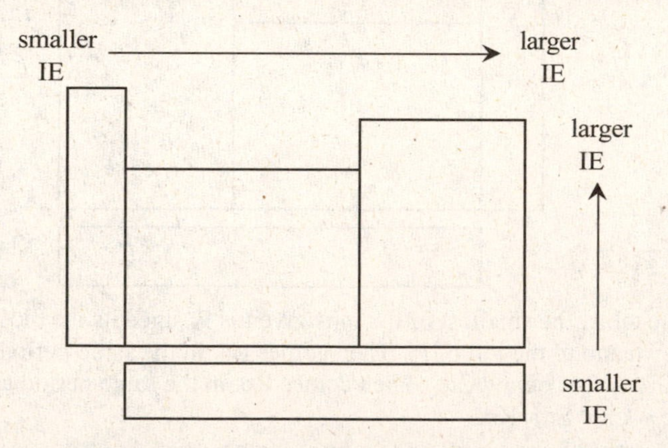

- Elements in group 2A, 2B, and 5A seem to be exceptions to this general trend, having higher IE than elements to their immediate right.

Looking at a periodic table, the element the furthest to the left in period 3 is Mg; however, it is in one of the exception groups (Group 2A). So, we'll predict that Al has the lowest first ionization energy, then Mg in the same period. Then comes P, to the right of Al in the same period. Then comes F, to the right and in the period above the others. So the order is: Al < Mg < P < F

✓*Reasonable Answer Check*: Using Figure 7.24, we can confirm these predictions.

100. *Answer:* **Choice (c)**

Strategy and Explanation: List elements in order of increasing first ionization energy. Sometimes we will not have a chart with numbers like Figure 7.24, so follow the periodic trends described in the solution to Question 98.

Looking at a periodic table, the element the furthest to the left is Li, so, we'll predict that it has the lowest first ionization energy. Then comes Si, to the right of the metal. Then comes C, above Si in the same group. Then comes Ne, to the right of C in the same period. So the correct order is: Li < Si < C < Ne. That is choice (c).

✓ *Reasonable Answer Check:* Using Figure 7.24, we can confirm these predictions.

102. Looking at Table 7.8, it is clear that the second ionization energies are all higher than the first ionization energies and once core electrons are being removed, the ionization energy rises dramatically. To have a large gap between the first and second ionization energies, the element **must have one valence electron**. The only element given that has one valence electron is **Na**, from Group 1A.

104. *Answer:* **(a) Al (b) Al (c) Al < B < C**

Strategy and Explanation: Given a list of elements, determine which is most metallic, determine which has largest radius, and place the list in order of increasing of ionization energy. Metallic character follows these general trends:

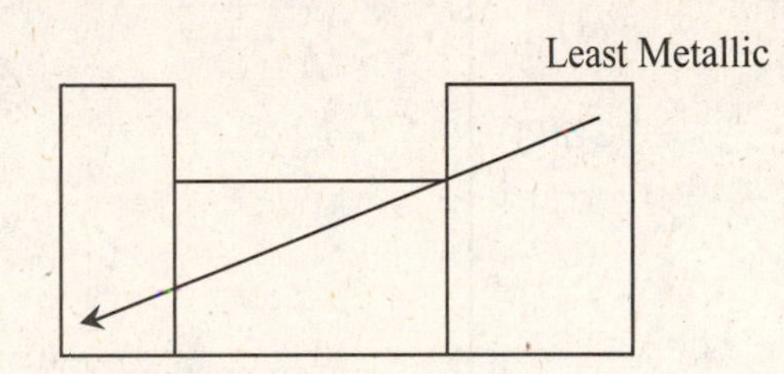

Use the periodic table for atom size trends and ionization energy trends (as described in the solutions to Questions 94 and 98).

Locate the elements: B, Al, C, and Si on a periodic table.

(a) We find Al closest to the bottom left corner. That means Al is most metallic.

(b) We find Al closest to the bottom left corner. That means Al is the largest of these four atoms.

(c) We find Al closest to the bottom left corner. That means Al has the smallest first ionization energy. B is next, because it is above Al in the same group. Last comes C, to the right of B in the same period. So the order is: Al < B < C

✓ *Reasonable Answer Check:* Using Figures 7.23, 7.24, and 7.25, we can confirm these predictions.

106. *Answer:* **(a) H^- (b) N^{3-} (c) F^-**

Strategy and Explanation: Given pairs of ions and/or atoms, determine which member of the pair has the larger radius. Use the periodic table and atom and ion size trends (as described in the solutions to Questions 93 - 98).

(a) H^- is larger than H^+. The anion has two electrons and the cation has none!

(b) N^{3-} is larger than N, since anions are larger than the neutral atom. (7 protons attract 7 electrons better than they can attract 10 electrons.)

(c) The F^- ion has a larger radius than an F atom, since anions are larger than the neutral atom. (9 protons attract 9 electrons better than they can attract 10 electrons.) Using the periodic table, an F atom has a larger radius than an Ne atom, because they are both in the same period and Ne is further to the right. Hence, an F^- ion also has a larger radius than an Ne atom.

✓ *Reasonable Answer Check:* Using Figures 7.23 and 7.24, we can confirm these predictions.

108. *Answer/Explanation:* Adding a free electron to the oxygen atom is exothermic and produces an anion. When the second electron is add, its negative charge is repelled by the anion's negative charge, according to Coulomb's Law. Additional energy is required to overcome the coulombic charge repulsion, making the second electron affinity endothermic.

Born-Haber Cycle (Section 7.13)

110. *Answer:* **–862 kJ**

Strategy and Explanation: Given sublimation, ionization energy, bond energy, electron affinity, and formation enthalpies, determine the lattice energy. Adapt the method in Section 7.13, using Hess's Law.

Construct a diagram similar to Figure 7.29, using the enthalpy values given.

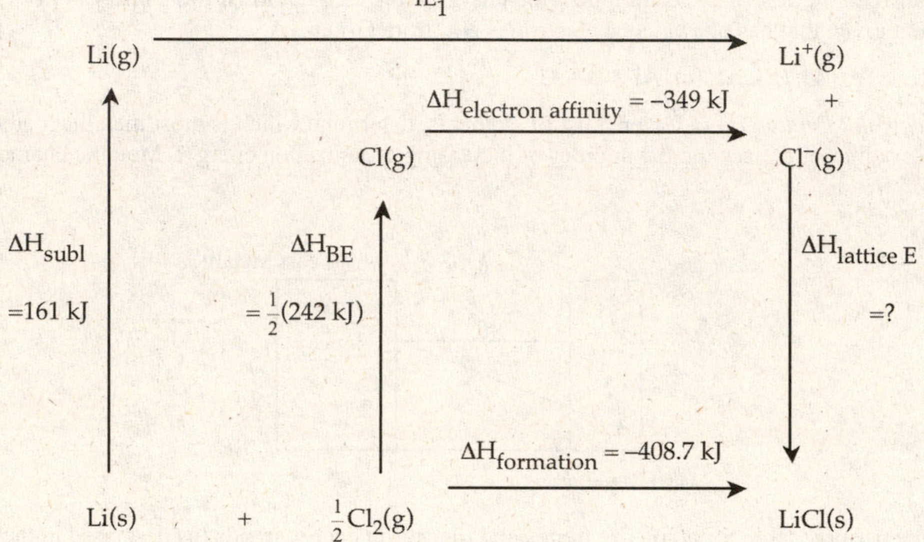

Use Hess's Law to combine the given chemical equations given to make the desired chemical equation:

	$Li(s) + \frac{1}{2}Cl_2(g) \longrightarrow LiCl(s)$	$\Delta H_f = -408.7\ kJ$
	$Li(g) \longrightarrow Li(s)$	$-\Delta H_{subl} = -(161\ kJ)$
	$Cl(g) \longrightarrow \frac{1}{2}Cl_2(g)$	$\Delta H_{BE} = -\frac{1}{2}(242\ kJ)$
	$e^- + Li^+(g) \longrightarrow Li(g)$	$-\Delta H_{IE_1} = -(520.\ kJ)$
+	$Cl^-(g) \longrightarrow e^- + Cl(g)$	$-\Delta H_{electron\ affinity} = -(-349\ kJ)$
	$Li^+(g) + Cl^-(g) \longrightarrow LiCl(s)$	$\Delta H_{Lattice\ Energy} = -862\ kJ$

✓ *Reasonable Answer Check*: Lithium ion is smaller than sodium ion and both chloride compounds are composed by combining a +1-cation and a –1-ion, so, as expected, this lattice energy is similar in size but slightly more negative than that seen for sodium chloride (which is given in the text (–786 kJ).

General Questions

112. *Answer/Explanation:* Using Figure 7.1, and the fact that the photon energy is proportional to frequency:

Lowest-energy photons: **red < yellow < green < violet** :Highest-energy photons

114. *Answer/Explanation:* Selenium has the following electron configuration: $1s^22s^22p^63s^23p^63d^{10}4s^24p^4$. That means the first two p sublevels are full, and the third one has 4 electrons. Look at the orbital diagrams of the p sublevels to find out the number of pairs:

2p: [↑↓][↑↓][↑↓] ... 3p: [↑↓][↑↓][↑↓] ... 4p: [↑↓][↑][↑]

From this, it is very easy to tell that there are **seven pairs** of electrons in p orbitals.

116. *Answer:* **(a) He (b) Sc (c) Na**

Strategy and Explanation:

(a) For a ground state element to be in Group 8A but have no p electrons, it must have fewer than five electrons, because the fifth electron goes into a 2p orbital (i.e., $1s^22s^22p^1$). This Group 8 element is **He** ($1s^2$).

(b) For a ground state element to have a single electron in the 3d subshell, it must have this electron configuration: $1s^22s^22p^63s^23p^63d^1$. This element is **Sc**.

(c) For a ground state element's cation to have a +1 charge and an electron configuration of $1s^22s^22p^6$, the element's electron configuration would have one more electron: $1s^22s^22p^63s^1$. This element is **Na**.

118. *Answer:* **(a) F < O < S (b) S**

Strategy and Explanation: Given lists of elements, place one list in order of increasing atomic radius and determine which member of another list has the largest ionization energy. Use the periodic table, atom size trends, and ionization energy trends (as described in the solutions to Questions 94 and 98).

(a) Locate the elements: O, S, and F on a periodic table. We find F closest to the top right corner. That means F is the smallest of these three atoms. Then comes O, to the left of the F in the same period. Then comes S, below O in the same group. So the order is: F < O < S

(b) Locate the elements: P, Si, S, and Se on a periodic table. We find S closest to the top right corner. That means S has the largest first ionization energy.

✓*Reasonable Answer Check:* Using Figures 7.23, 7.24, and 7.25, we can confirm these predictions.

120. *Answer:* **(a) S (b) Ra (c) N (d) Ru (e) Cu**

Strategy and Explanation:

(a) The ground state element whose atoms have an electron configuration of $1s^22s^22p^63s^23p^4$ has 16 electrons and 16 protons. This is sulfur, **S**, with the atomic number of 16.

(b) The largest *known* element in the alkaline earth group (Group 2A) is the one at the bottom of the group on the periodic table. This is radium, **Ra**.

(c) The largest first ionization energy of elements in the group is the group member at the top of the group on the periodic table. In Group 5A, this is nitrogen, **N**.

(d) For a ground state element's cation to have a +2 charge and an electron configuration of $[Kr]4d^6$, the element's electron configuration would have two more electrons: $[Kr]4d^65s^2$. This element is ruthenium, **Ru**.

(e) The ground state element whose neutral atoms have an electron configuration of $[Ar]3d^{10}4s^1$ has 29 electrons and 29 protons. This is copper, **Cu**, with the atomic number of 29.

122. *Answer:* **(a) Z (b) Z**

Strategy and Explanation: Considering $X = [Ar]3d^84s^2 = Ni$ and $Z = [Ar]3d^{10}4s^24p^5 = Br$

(a) Using the periodic trends for sizes of first ionization energies (as described in the solution to Question 98), we find that Br has a larger first ionization energy than Ni, because they are both in period 4, and Br is found further to the right. That means **Z** is the answer.

(b) Using the periodic trends for atom sizes (as described in the solution to Question 93), we find that Br has a smaller radius than the Ni, because they are both in period 4, and Br is found further to the right. That means **Z** is the answer.

124. *Answer:* **In^{4+}, Fe^{6+}, Sn^{5+}, very high successive ionization energies**

Strategy and Explanation: Cations likely to form are ones that have lost only valence electrons. Any ion that has a more positive charge than its number of valence electrons is not likely to form. Anions likely to form are ones that add enough electrons to the atom to increase the number of valence electrons to eight. Any ion that has a more negative charge than the difference between eight and the number of valence electrons is not likely to form. Free ions with charges larger than ±3 are rare, because making the ion requires us to remove electrons from ions that are already very positive or add electrons to ions that are already very negative (i.e., Notice how large the high successive ionization energies get in Table 7.7).

Cs has one valence electrons; Cs^{+} **is** likely to form.

In has three valence electrons; In^{4+} is **unlikely** to form because a core electron must be removed.

Fe has two valence electrons and six d electrons, so core electrons are being removed; however, Fe^{6+} is **unlikely** to form, because the positive charge is too large.

Te has six valence electrons. It needs 8 – 6 = 2 electrons. Te^{2-} **is** likely to form.

Sn has four valence electrons; Sn^{5+} is **unlikely** to form because a core electron must be removed and the positive charge is too large.

I has seven valence electrons. It needs 8 – 7 = 1 electron. I^{-} **is** likely to form.

126. *Answer:* **(a) false, wavelength is shorter (b) true (c) true (d) false, they are inversely related**

Strategy and Explanation: Use Figure 7.1 and the relationships between wavelength, frequency, and photon energies described in Sections 7.1 and 7.2.

(a) "The wavelength of green light is longer than that of red light." This statement is **false**. We can fix it in several ways, such as with this related true statement: "The wavelength of green light is **shorter** than that of red light." Red light has the longest wavelength of all visible light.

(b) "Photons of green light have greater energy than those of red light." This statement is **true**. Red photons are the lowest energy visible photons.

(c) "The frequency of green light is greater than that of red light." This statement is **true**. Red light has the lowest frequency of all visible light.

(d) "In the electromagnetic spectrum, frequency and wavelength are directly related." This statement is **false**. We can fix it in several ways, such as with this related true statement: "In the electromagnetic spectrum, frequency and wavelength are **inversely** related." The word "directly" implies directly proportional, and these two quantities are inversely proportional.

128. *Answer:* **(a) Boxes B, D, and I (b) Box E (c) Box D (d) Boxes A and G (e) Boxes H and C**

Strategy and Explanation: Infrared light has wavelengths greater than 700 nm (Box C). Light with wavelengths between 400 nm and 700 nm are visible (e.g., Box G). Ultraviolet light has a wavelengths less than 300 nm (Box I).

Figure 7.8 in Section 7.3 on page 281 shows several transitions of electrons in the hydrogen atom from higher energy states to lower energy states resulting in emission of different wavelengths of light. From it we learn the following:

Box A ($n_4 \rightarrow n_2$) represents a visible emission.

Box B ($n_3 \rightarrow n_1$) represents an ultraviolet emission.

Box D ($n_4 \rightarrow n_1$) represents an ultraviolet emission.

Box H ($n_6 \rightarrow n_3$) represents an infrared emission.

When an electron undergoes a transition from a lower energy state to a higher energy state, photons of various frequencies must be absorbed.

Box E ($n_2 \rightarrow n_7$) represents a visible absorption.

Box F ($n_3 \rightarrow n_5$) represents an infrared absorption.

(a) The ultraviolet emissions are Boxes B and D. Emissions less than 400 nm could also be ultraviolet (Box I), provided they are low enough energy to not be X-ray or gama-rays emissions.

(b) There are two absorptions identified (Boxes E and F), the highest energy of the two is Box E, because visible light is higher energy than infrared light.

(c) There are four specific emissions identified (Boxes A, B, D, and H) and two of them (Boxes B and D) have untraviolet wavelengths. The highest energy of that pair is Box D, because the energy level represented by n_4 is higher energy that that with n_3 and they are both making a transition to the energy level represented by n_1.

(d) The visible emission is Box A. Any emissions between 400 nm and 700 nm will also be visible (e.g., Box G), so if the wavelength described here is the result of an emission, it is visible.

(e) The infrared emission is Box H. Any emissions above 700 nm could also be infrared (e.g., Box C), provided they are high enough energy to not be called mircrowave or radiowaves.

130. *Answer:* **(a) replace "inversely" with "directly" (b) it is inversely proportional to the square of n (c) replace "as soon as" with "before" (d) replace "frequency" with "wavelength"**

Strategy and Explanation:

(a) This statement: "The energy of a photon is inversely related to its frequency." is **false**. Actually, the energy of a photon is **directly** related to its frequency according to the equation: E = hν.

(b) This statement: "The energy of the hydrogen electron is inversely proportional to its principle quantum number n." is **false**. Actually, the energy of the hydrogen electron is **inversely proportional to the square of** its principle quantum number, n.

$$E = -\frac{2.179 \times 10^{-18}\ \text{J}}{n^2}$$

(c) This statement: "Electrons start to fill the fourth energy level as soon as the third level is full." is **false**. Several different slight changes can make it into a true statement, for example: "Electrons start to fill the fourth energy level **before** the third level is full" This can be seen in the electron configuration for potassium and calcium: $1s^22s^22p^63s^23p^63d^44s^1$ and $1s^22s^22p^63s^23p^63d^44s^2$ where the 4s sublevel fills before the 3d sublevel starts to fill.

(d) This statement: "Light emitted by an n = 4 to n = 2 transition will have a longer frequency than that from an n = 5 to n = 2 transition." is **false**. Several different slight changes can make it into a true statement, for example: "Light emitted by an n = 4 to n = 2 transition will have a longer **wavelength** than that from an n = 5 to n = 2 transition." The energy difference is smaller in the first transition than the second, so the frequency of the emitted photon is smaller, and its wavelength is longer.

132. *Answer:* **ultraviolet; 91.18 nm or shorter is needed to ionize hydrogen.**

Strategy and Explanation: Given a list of types of radiation, determine which is needed to ionize hydrogen. Use the ionization energy of hydrogen, from Figure 7.25 in kJ/mol, and Avogadro's number to determine the energy required to ionize one hydrogen atom, then determine the wavelength of light with that energy using the equation from Section 7.2.

$$\frac{1312\ \text{kJ}}{1\ \text{mol}} \times \frac{1000\ \text{J}}{1\ \text{kJ}} \times \frac{1\ \text{mol}}{6.022 \times 10^{23}\ \text{atoms}} \times \frac{1\ \text{atom}}{1\ \text{photon}} = 2.179 \times 10^{-18}\ \text{J / photon}$$

$$\lambda = \frac{hc}{E_{photon}} = \frac{6.626 \times 10^{-34}\ \text{J} \cdot \text{s} \times 2.998 \times 10^{8}\ \text{m/s}}{2.179 \times 10^{-18}\ \text{J}} \times \frac{1\ \text{nm}}{1 \times 10^{-9}\ \text{m}} = 91.18\ \text{nm}$$

To ionize hydrogen, we need light with a wavelength of 91.18 nm or shorter. This is the ultraviolet region of the electromagnetic spectrum.

✓*Reasonable Answer Check*: It is satisfying to get 2.179 × 101^{-18} J/photon, since that is the number used in the Bohr model calculations for electrons changing quantum states in hydrogen:

$$\Delta E_{e-} = \left(2.179 \times 10^{-18}\ \text{J}\right)\left(\frac{1}{n_i^2} - \frac{1}{n_f^2}\right)$$

To ionize the electron, we must move it from n = 1 (the ground state) to above n = ∞ (the last state in the atom). This should be the minimum photon energy needed for ionization:

$$\Delta E_{e-} = \left(2.179 \times 10^{-18}\ \text{J}\right)\left(\frac{1}{1^2} - \frac{1}{\infty^2}\right) = 2.179 \times 10^{-18}\ \text{J} = E_{\text{ionization photon}}$$

134. *Answer:* **(a) [Rn]$5f^{14}6d^{10}7s^27p^68s^2$ (b) magnesium (or any other Group 2A elements) (c) EtO, $EtCl_2$**

Strategy and Explanation: This question asks us to use the predictive power of the periodic table. Suppose an element, Et, has atomic number 120. That means atoms of the neutral element have 120 protons and 120 electrons.

(a) The ground state electron configuration has 120 electrons: **[Rn]$5f^{14}6d^{10}7s^27p^68s^2$**. The Rn noble gas core has 86 electrons, and we add 14 + 10 + 2 + 6 + 2 outer electrons, to make a total of 120 electrons.

(b) The valence electrons are the n = 8 electrons ($8s^2$) of which there are two. That means we will predict this element to be a member of Group 2A. There are several elements in this group, including **magnesium**.

(c) Group 2A metals will combine with nonmetals forming ionic compounds. The metallic elements in that group can be predicted to form cations with a 2+ charge. They will combine with the anions of oxygen and chlorine (oxide, O^{2-}, and chloride, Cl^-) to form **EtO** and **$EtCl_2$**, respectively.

Applying Concepts

136. *Answer/Explanation:* The smallest halogen is the element at the top of Group 7A, fluorine. The neutral atom loses an electron during the process of ionization:

$$F\ (1s^22s^22p^5) \longrightarrow F^+\ (1s^22s^22p^4) + e^- \qquad \text{First ionization}$$

So the electron configuration is: **$1s^22s^22p^4$**

138. *Answer/Explanation:* The element, X, with electron configuration: $1s^22s^22p^63s^1$ has one valence electron ($3s^1$), placing it in Group 1A. Group 1A metals will combine with nonmetals forming ionic compounds. The Group 1A elements form cations with a 1+ charge. They will combine with the anion of chlorine (chloride, Cl^-) to form **XCl**.

140. *Answer:* **(a) ground state (b) either ground state or excited state (c) excited state (d) impossible (e) excited state (f) excited state**

Strategy and Explanation:

(a) $1s^22s^1$ is a ground state electron configuration. The 1s sublevel has the lowest energy it is full. The 2s sublevel has the second lowest energy and the remaining electron is there. This is the lowest energy electron configuration, so it would be called the **ground state**.

(b) $1s^22s^22p^3$ could be **ground state or excited state**. If each of the p electrons is in a separate orbital with the same spin, that is the ground state. However, if any of the spins are reversed or if any of the electrons are paired, this is a different and higher energy state, hence an excited state. We would need to see an orbital energy diagram of this state to determine an unambiguous answer:

Ground state 2p orbital energy diagram: 2p [↑][↑][↑]

Some excited state 2p orbital energy diagrams: 2p [↓][↑][↑]   2p [↑↓][↑][ ]

(c) $[Ne]3s^23p^34s^2$ is an **excited state**. The lower-energy 3p sublevel needs to be full ($3p^6$) before the 4s sublevel gets any electrons; therefore, this possible electron configuration is not the lowest energy state, and would be called an excited state. The ground state electron configuration would be $[Ne]3s^23p^5$.

(d) $[Ne]3s^23p^64s^33d^2$ is **impossible**. The 4s sublevel has only one orbital, so the maximum number of electrons is 2. This electron configuration has three electrons in that orbital.

(e) $[Ne]3s^23p^64f^4$ is an **excited state**. Several lower-energy sublevels need to be full before the 4f sublevel gets any electrons; therefore, this possible electron configuration is not the lowest energy state, and would be called an excited state. The ground state electron configuration would be $[Ne]3s^23p^63d^24s^2$.

(f) $1s^22s^22p^43s^2$ is an **excited state**. The lower-energy 2p sublevel needs to be full ($2p^6$) before the 3s sublevel gets any electrons; therefore, this possible electron configuration is not the lowest energy state, and would be called an excited state. The ground state electron configuration would be $1s^22s^22p^6$.

142. *Answer:* **$[Rn]5f^{14}5g^{18}6d^{10}6f^{14}7s^27p^67d^{10}8s^28p^2$; bottom of Group 4A**

Strategy and Explanation: This question asks us to use the predictive power of the periodic table. Appendix D can be of some assistance, also. Since the atomic number of our undiscovered element is 164, the number of protons and electrons in a neutral atom is also 164. We need to extend Figure 7.12 using Table 7.4, to predict what sublevels the new electrons fill. After 7p, the 8s orbitals fill. Then we fill the first g orbitals, 5g. Then we fill 6f, 7d, 8p, 9s, and so on. Table 7.4 tells us the maximum capacity of the g sublevel is 18. Constructing the electron configuration in the order in which the electrons fill looks like this:

$$[Rn]7s^25f^{14}6d^{10}7p^68s^25g^{18}6f^{14}7d^{10}8p^2$$

(86 + 2 + 14 + 10 + 6 + 2 + 18 + 14 + 10 + 2 =) 164 electrons. We have added enough electrons now. To conform to those shown in Appendix D, we then rearrange the sublevels into shell order:

$$[Rn]5f^{14}5g^{18}6d^{10}6f^{14}7s^27p^67d^{10}8s^28p^2$$

There are four valence electrons in this element ($8s^28p^2$), so we will predict it is at the bottom of Group 4A.

143. *Answer:* **(a) increase, decrease (b) helium (c) 5 and 13 (d) no electrons left (e) It is a core electron. (f) $Mg^{2+}(g) \longrightarrow Mg^{3+}(g) + e^-$**

Strategy and Explanation:

(a) Based in the graphical data, ionization energies **increase** left to right and **decrease** top to bottom on the periodic table.

(b) The element with the largest first ionization energy is **helium** (atomic number = 2).

(c) The peaks of the fourth ionization energies will occur at the atomic number of boron (**5**) and at the atomic number of aluminum (**13**).

(d) He has only two electrons, so after two ionizations, there are **no electrons left**.

(e) First electron in lithium is a valence electron, but the **second electron is a core electron**.

(f) The atomic number of 12 belongs to magnesium, $_{12}Mg$.

$$Mg^{2+}(g) \longrightarrow Mg^{3+}(g) + e^-$$

More Challenging Questions

146. *Answer:* **(a) 4.34×10^{-19} J (b) 6.54×10^{14} Hz**

Strategy and Explanation: Given wavelengths of light absorbed by an electron ejected from a metal and the kinetic energy of the ejected electron, determine the binding energy and the minimum frequency required for the photoelectric effect.

Use the equation in Section 7.2 to find the energy of each photon. Subtract the kinetic energy of the electron from the photon energy to find the binding energy. Use $E = h\nu$ to find the lowest frequency photon.

(a) $\lambda = 4.00\times10^{-7}$ m $\quad E_{photon} = \frac{hc}{\lambda} = \frac{6.626\times10^{-34}\ J\cdot s\times 2.998\times10^{8}\ m/s}{4.00\times10^{-7}\ m} = 4.97\times10^{-19}\ J$

Binding Energy = Photon Energy – Kinetic Energy $= 4.97\times10^{-19}\ J - 0.63\times10^{-19}\ J = 4.34\times10^{-19}\ J$

(b) $\quad \nu_{photon} = \frac{E}{h} = \frac{4.34\times10^{-19}\ J}{6.626\times10^{-34}\ J\cdot s} = 6.55\times10^{14}\ s^{-1} = 6.55\times10^{14}\ Hz$

✓*Reasonable Answer Check*: The binding energy is lower than the photon energy, since the photon required to eject a zero kinetic energy electron needs to have less energy than one with 6.3×10^{-20} J.

148. *Answer:* **2.18×10^{-18} J**

Strategy and Explanation: Given an equation describing the relationship between the electron energy level, the atomic number, Z, and the quantum number, n, determine the ionization energy of the electron. Subtract the ionization energy state (n = ∞) of the electron from ground state (n = 1) of the electron in hydrogen.

Execute the plan: Z = 1 for hydrogen

$$E_{ionization} = E_{\infty} - E_1 = -\frac{1^2}{(\infty)^2}\left(2.18\times10^{-18}\ J\right) - \left[-\frac{1^2}{(1)^2}\left(2.18\times10^{-18}\ J\right)\right] = 2.18\times10^{-18}\ J$$

✓*Reasonable Answer Check*: The ionization of hydrogen is given in Figure 7.8 as the energy required to go from n = 1 to n= = ∞.

150. *Answer/Explanation:* To remove an electron from an N atom requires disruption of a half filled sub-shell (p^3), which is relatively stable, so the ionization of oxygen will be less energy than the ionization of nitrogen.

152. *Answer:* **(a) Ir (b) [Rn]$7s^2 5f^{14} 6d^7$**

Strategy and Explanation:

(a) Another transition element with the parallel outer electron configuration as Mt, is iridum, **Ir**. It is found directly above Mt on the periodic table. These two elements have the same number of electrons in the same types of orbitals. (Note, of course, that the core is different, and the values of n for electrons in Ir are one less than the n values in Mt.) The ground state electron configuration of Ir using the xenon core is $[_{54}Xe]6s^2 4f^{14} 5d^7$.

(b) The ground state electron configuration of Mt using the radon core is **$[_{86}Rn]7s^2 5f^{14} 6d^7$**

154. *Answer/Explanation:*

(a) Element 112 will reside in Group 2B, as one of the **transition metals**, directly below mercury.

(b) The ground state electron configuration of element 112 using the radon core is **$[_{86}Rn]7s^2 5f^{14} 6d^{10}$**.

Chapter 8: Covalent Bonding

Solutions for Blue-Numbered Questions for Review and Thought

Topical Questions

Lewis Structures (Sections 8.2, 8.4)

14. *Answer:* **(a)** :Cl—F: **(b)** H—Se—H **(c)** [:F—B(—F:)(—F:)—F:]$^{-}$ **(d)** [:O—P(—O:)(—O:)—O:]$^{3-}$

Strategy and Explanation: Write Lewis structures for a list of ions and molecules.

Following a systematic plan will give you reliable results every time. Trial and error often works for small molecules, but as molecules get more complex, it's better to follow a procedure than to try to guess where electrons will end up. There are a number of methods. The one described here is the same as that described in the text and it works all the time.

[A] Count the total number of valence electrons. If there is a nonzero charge, adjust the electron count appropriately. Add electrons for negative charges and subtract electrons for positive charges.

[B] Determine the skeleton structure, which often means figure out which atom is the central atom. As a general rule, the central atom is usually the first element in the formula, and it will often have a smaller electronegativity (see Section 8.7) than the rest of the atoms in the formula. There will be exceptions, for example: H will **never** be the central atom. The tendency of certain atoms to make certain numbers of bonds will also help inform you of how to arrange the skeleton structure (e.g., we often see elements in Group 4A make four bonds, Group 5A make 3 bonds, Group 6A make two bonds, and Group 7A and H make one bond.) Connect each atom in the skeleton structure with a single bond.

[C] Each single bond (–) is formed using two electrons. Subtract the electrons used for single bonds in the skeleton structure from the total number of electrons.

[D] Complete the octets of all of the terminal "outer" atoms by adding pairs of dots (called lone pairs) so that each of them ends up with a total of eight electrons, including the two shared with the central atom. There is an exception: H only need two electrons, so **never** put dots on H.

[E] Subtract the electrons used for lone pairs from the total.

[F] If you still have unused electrons, distribute all the remaining electrons on the central atom (or the inner atoms), preferentially with a goal to complete all their octets. If you have extra electrons after completing all the octets, put the extra electrons as dots on the central atom so that only one atom gets more than eight electrons.

[G] Check the octet of the inner, central atom(s).

(i) If each has eight or more electrons, then the structure is complete. The only reason an atom would have more than eight electrons is if there were too many electrons for single bonds and the octet rule to use up all the electrons.

(ii) If any atom has less than eight electrons, move a lone pair of electrons from one of the outer atoms to make a new shared pair, forming a multiple bond to the central/inner atom that needs electrons. Repeat this procedure until all atoms have an octet. Never make more than the minimum number of multiple bonds unless you have a really good reason you can explain.

[H] If the structure is an ion, put it in brackets and designate the net charge outside the bracket at the upper right corner.

✓ *Reasonable Answer Check*: Check the octets of every atom in the structure (and make sure H atoms have only two electrons) then count the electrons and make sure the total is right.

(a) [A] The atoms Cl and F are both in Group 7A: $7 + 7 = 14\ e^-$

[B] Choose the Cl atom is the central atom with the F atom bonded to it, since it has the lower electronegativity.

Cl—F

[C] Connecting the two atoms uses two electrons, so subtract 2 electrons from the current electron count: $14\ e^- - 2\ e^- = 12\ e^-$

[D] F has two electrons already (in the single bond, so add six dots to the F atom: Cl—$\ddot{\underset{..}{F}}$:

Notice that F now has a complete octet:

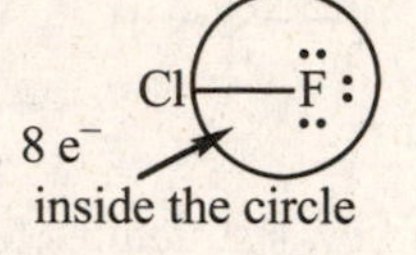

[E] Completing the octet of the F atom uses 6 electrons, so subtract 6 electrons from the current electron count: $12\ e^- - 6\ e^- = 6\ e^-$

[F] Put the last six electrons on the Cl. :$\ddot{\underset{..}{Cl}}$—$\ddot{\underset{..}{F}}$:

[G] Cl now has a complete octet: :$\ddot{\underset{..}{Cl}}$—$\ddot{\underset{..}{F}}$:

8 e⁻ inside the circle

So, the structure is complete. :$\ddot{\underset{..}{Cl}}$—$\ddot{\underset{..}{F}}$:

✓ *Reasonable Answer Check*: The Cl has eight electrons (one pair in a single bond and six dots). The F has eight electrons (one pair in a single bond and six dots). Looking at the structure from left to right, the total number of electrons can be counted: 6 dots + 2 in a bond + 6 dots = 14 total.

(b) [A] The H atoms are in Group 1A and the Se atom is in Group 6A:

$2 \times (1) + 6 = 8\ e^-$

[B] The Se atom is the central atom with the H atoms around it.

H—Se—H

[C] Connecting the H and Se atoms uses four electrons: $8\ e^- - 4\ e^- = 4\ e^-$

[D] H atoms need no more electrons.

H—Se—H

[E] No electrons used in this step. So we still have 4 e⁻.

[F] Put the last four electrons on the Se.

H—$\ddot{\underset{..}{Se}}$—H

[G] Se has eight electrons, so the structure is complete.

(c) [A] The B atom is in Group 3A and the F atoms are in Group 7A. The 1– charge **adds** one extra electron: $3 + 4 \times (7) + 1 = 32\ e^-$

[B] The B atom is the central atom with the F atoms around it.

[C] Connecting four F atoms uses eight electrons: $32\ e^- - 8\ e^- = 24\ e^-$

[D] Complete the octets of all the F atoms.

[E] Completing the octets of the four F atoms uses 24 electrons: $24\ e^- - 24\ e^- = 0\ e^-$

[F] No more electrons are available.

[G] B has eight electrons, so the structure is complete.

[H] The structure is an ion, so put it in brackets and add the – charge.

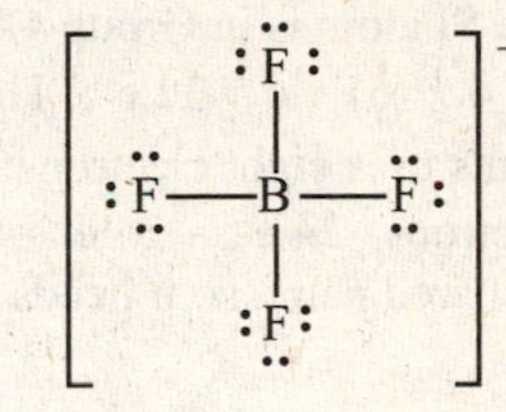

(d) [A] The P atom is in Group 5A and the O atoms are in Group 6A. The 3– charge **adds** three extra electrons: $5 + 4 \times (6) + 3 = 32\ e^-$

[B] The P atom is the central atom with the O atoms around it.

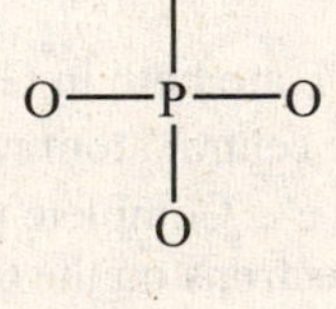

[C] Connecting four O atoms uses eight electrons: $32\ e^- - 8\ e^- = 24\ e^-$

[D] Complete the octets of all the O atoms.

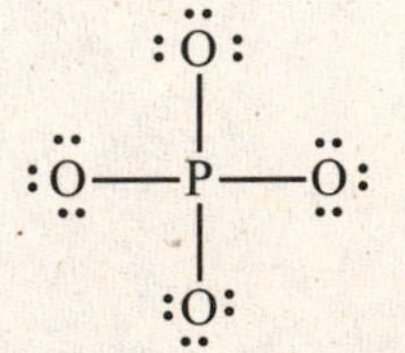

[E] Completing the octets of the four O atoms uses 24 electrons: $24\ e^- - 24\ e^- = 0\ e^-$

[F] No more electrons are available.

[G] P has eight electrons, so the structure is complete.

[H] The structure is an ion, so put it in brackets and add the 3– charge.

✓ *Reasonable Answer Check:* All the Period 2 and 3 elements have an octet of electrons. The H atoms have two electrons. All valance electrons are accounted for in the structures, either as members of shared pairs or of lone pairs.

16. *Answer:* **(a)** **(b)** **(c)** **(d)**

Strategy and Explanation: Write Lewis structures for a list of ions and molecules.

Follow the systematic plan given in the solution to Question 14. Use what you learned about organic molecules in Chapter 3 to determine how atoms are bonded in the organic structures.

(a) The C atom is in Group 4A, the H atom is in Group 1A, and the Cl atom is in Group 7A: $4 + 3 \times (1) + 7 = 14\ e^-$. The C atom is the central atom with the other atoms around it. Connecting four atoms uses eight electrons: $14\ e^- - 8\ e^- = 6\ e^-$. H needs no more electrons. Complete the octet of the Cl atom using six electrons: $6\ e^- - 6\ e^- = 0\ e^-$. No more electrons are available. The central atom, C, has eight electrons in shared pairs, so it needs no more. We get the following structure:

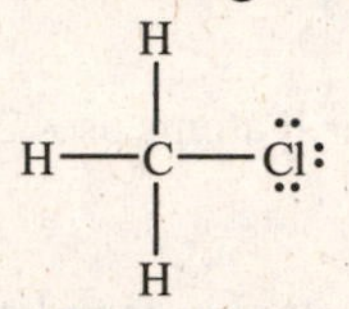

(b) The Si atom is in Group 4A and the O atoms are in Group 6A. The 4– charge **adds** four extra electrons: $4 + 4 \times (6) + 4 = 32\ e^-$. The Si atom is the central atom with the O atoms around it. Connecting four O atoms uses eight electrons: $32\ e^- - 8\ e^- = 24\ e^-$. Complete the octets of all the O atoms using 24 electrons. $24\ e^- - 24\ e^- = 0\ e^-$. No more electrons are available. The central atom, Si, has eight electrons in shared pairs, so it needs no more. The structure is an ion, so put it in brackets and add the 4– charge:

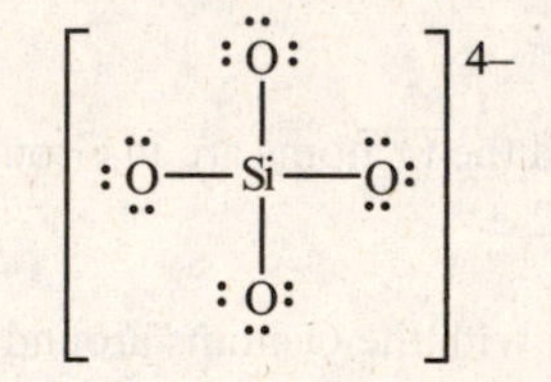

(c) The atoms Cl and F are both in Group 7A. The + charge **removes** one electron: $7 + 4 \times (7) - 1 = 34\ e^-$. The Cl atom is the central atom with the F atoms around it. Connecting four F atoms uses eight electrons: $34\ e^- - 8\ e^- = 26\ e^-$. Complete the octets of all the F atoms, using 24 electrons: $26\ e^- - 24\ e^- = 2\ e^-$. Put the last two electrons on the Cl. The central atom, Cl, has ten electrons already, so it needs no more. The structure is an ion, so put it in brackets and add the + charge.

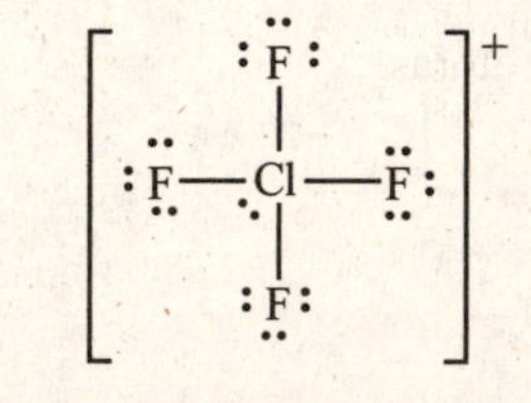

(d) The C atom is in Group 4A and the H atoms are in Group 1A: $2 \times (4) + 6 \times (1) = 14\ e^-$. As described in Chapter 3 Section 3.3, the ethane molecule has three H atoms bonded to each C, and the two carbons bonded to each other. Connecting eight atoms uses 14 electrons: $14\ e^- - 14\ e^- = 0\ e^-$ No more electrons are available. Each C has eight electrons in shared pairs, so they need no more.

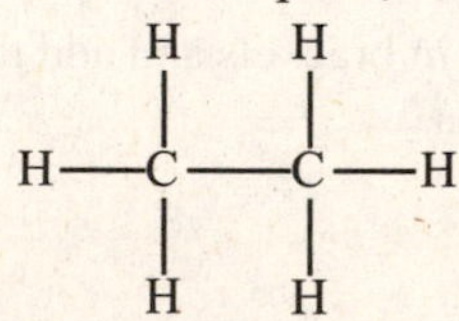

✓ *Reasonable Answer Check:* All the Period 2 elements have an octet of electrons. The H atoms have two electrons. All the Period 3 elements have eight or more electrons. All valance electrons are accounted for in the structures, either as members of shared pairs or of lone pairs.

18. *Answer:* **(a)** **(b)**

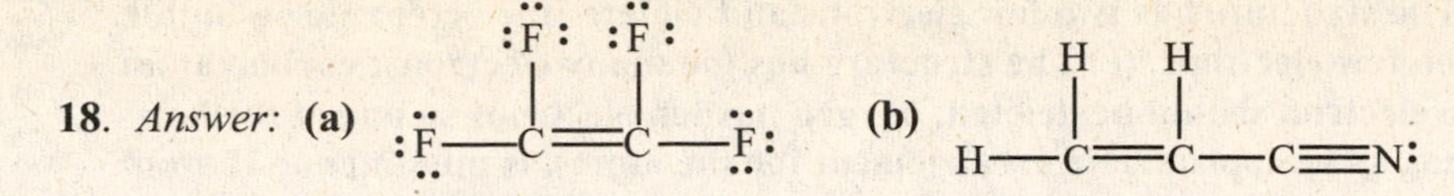

Strategy and Explanation: Write Lewis structures for a list of ions and molecules.

Follow the systematic plan given in the solution to Question 14. Use what you learned about organic molecules in Chapter 3 to determine how atoms are bonded in the organic structures.

(a) $2 \times (4) + 4 \times (7) = 36\ e^-$. This organic molecule looks like ethene with all the H atoms changed to F atoms. The two carbon atoms are bonded to each other, and each has two F atoms bonded to it. Any other arrangement would make F–F bonds. F has the highest electronegativity (see Section 8.7) and won't bond with another F atom if any other element is present. Connecting six atoms with single bonds uses 10 electrons: $36\ e^- - 10\ e^- = 26\ e^-$. For the purposes of this method, let's call the first C atom the central atom. Complete the octets of the F atoms and the second C atom, using 26 electrons: $26\ e^- - 26\ e^- = 0\ e^-$. At this point, the structure looks like this:

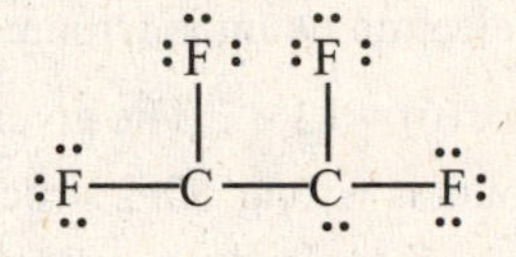

The first C atom has only six electrons in shared pairs, so it needs two more, but there are no more electrons available for lone pairs. Therefore, we must move one lone pair of electrons from the C atom to make a new shared pair, forming a double bond to the C atom:

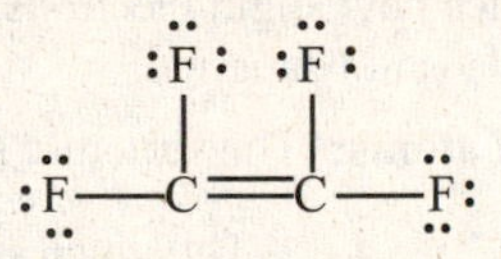

The first C atom now has eight electrons in shared pairs and this is the proper Lewis structure.

(b) $3 \times (4) + 3 \times (1) + 5 = 20\ e^-$. As described in Chapter 3 Section 3.1, the structural formula given here tells us that this organic compound has two H atoms and a C atom bonded to the first C atom, an H atom and a C atom bonded to the second C atom, and an N atom bonded to the third C atom. Connecting seven atoms uses 12 electrons: $20\ e^- - 12\ e^- = 8\ e^-$. For the purposes of this method, let's call the second and third C atoms "central atoms" so we can start from the ends and work toward the middle. H needs no more electrons. Complete the octet of the N atom and the octet of the first carbon using eight electrons: $8\ e^- - 8\ e^- = 0\ e^-$. At this point, we have the following structure:

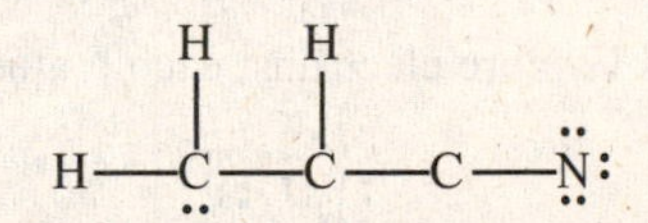

The second C atom has only six electrons in shared pairs and the third C atom has only four electrons in shared pairs, but there are no more electrons available. Therefore, we must move one lone pair of electrons from the first C atom to make a new shared pair between the first and second C atoms, making a double bond. We must also move two lone pairs of electrons from the N atom to the third C atom to make two new shared pairs, forming a triple bond between the N atom and C atom:

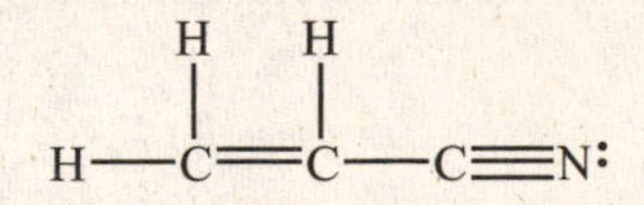

The second and third C atoms now both have eight electrons in shared pairs and this is the proper Lewis structure.

✓ *Reasonable Answer Check:* All the Period 2 elements have an octet of electrons. The H atoms have two electrons. All valance electrons are accounted for in the structures, either as members of shared pairs or of lone pairs.

22. *Answer:* **All are incorrect. (a) The structure has two few electrons and violates the octet rule on both F atoms. (b) The structure has too few electrons. (c) The structure has too many electrons; carbon atom has nine electrons, so the single electron should be deleted; oxygen has ten electrons so one of the lone pairs should be removed. (d) has an in appropriate arrangement for the atoms, is missing one H atom, and violates the rule of two for hydrogen. (e) NO_2^- has too few electrons and violates the octet rule.**

Strategy and Explanation: Determine if given Lewis structures are correct and explain what is wrong with the incorrect ones.

A Lewis structure is correct if it has the correct type and number of atoms, if all valance electrons are accounted for in the structures, either as members of shared pairs or of lone pairs; if all the Period 2 elements have an octet of electrons; if the H atoms have two electrons; and if all the Period 3 and higher elements have eight or more electrons. So, first count the atoms and compare with the formula. Then count the electrons in the structure and compare that with the total number of valence electrons. If the count is correct, then check the octets of all the elements and check H atoms for two electrons. Only elements in period 3 and higher are allowed to exceed eight electrons.

(a) OF_2 **Check electron count**: $6 + 2 \times (7) = 20\ e^-$. The given structure has 8 electrons (two lone pairs and two single bonds), so the total electron count is wrong. This structure is incorrect.

(b) O_2 **Check electron count**: $2 \times (6) = 12\ e^-$. The given structure has 10 electrons (two lone pairs and a triple bond), so the total electron count is wrong. This structure is incorrect.

(c) CCl_2O **Check electron count**: $4 + 2 \times (7) + 6 = 24\ e^-$. The given structure also has 27 electrons (nine lone pairs, one unpaired electron, two single bonds, and one double bond), so the total electron count is wrong.

Check octets, etc.: Both Cl atoms each have eight electrons, but the C atom has 9 electrons and the O atom has 10 electrons, so the octet rule is not satisfied.

(d) CH_3Cl **Check type and number of atoms:** The structure has only two H atoms, not three.

Check electron count: $4 + 3 \times (1) + 7 = 14\ e^-$. The given structure has 16 electrons (four lone pairs, and four single bonds), so the total electron count is wrong.

Check octets, etc.: The structure also has one hydrogen atom with too many electrons (4, not 2). This structure is incorrect.

(e) NO_2^- **Check electron count**: $5 + 2 \times (6) + 1 = 18\ e^-$. The given structure has 16 electrons (five lone pairs, one single bond, and one double bond), so the total electron count is wrong. This structure is incorrect. It is also missing the brackets and the net charge.

✓ *Reasonable Answer Check:* We used what we know about Lewis structures to determine the incorrect structures. Let's draw correct Lewis structures for those that were incorrect, (a), (b), (d) and (e).

The correct structure for (a) OF_2 needs 12 more electrons; each F atom needs three more lone pairs:

:F:O:F:

The correct structure for (b) O_2 needs 2 more electrons, two more lone pairs, and one less shared pair:

:O=O:

The correct structure for (d) has two fewer electrons and the carbon bonded to the chlorine without an H atom between them. Remember, H atom only shares two electrons:

H
H:C:Cl:
H

The correct structure for (e) has two more electrons on N (and brackets with a charge):

[:O—N=O]⁻

The corrected structures are different from the incorrect ones, so these answers look right.

Bonding in Hydrocarbons (Sections 8.3, 8.5)

24. *Answer:* **(see structures below)**

Strategy and Explanation: Write structural formulas for all the branched-chain compounds with a given formula.

Be systematic. Start with a long chain that has only one methyl branch. Move the methyl around (but don't put it on the end carbon and don't put it on a carbon past the first half of the chain). Then make two methyl branches, and move them around similarly. Continue this process until the main chain is too short for methyl branches. Then make an ethyl branch and move it around (but don't put it on the end carbon or the carbon next to the end carbon and don't put it on a carbon past the first half of the chain). Follow a similar pattern with ethyls as with methyls.

The straight chain isomer of C_6H_{14} has six carbons, so start with a five-carbon chain.

One methyl branch can go on the five-carbon chain in two different ways. The methyl branch can go on the second carbon or on the third carbon.

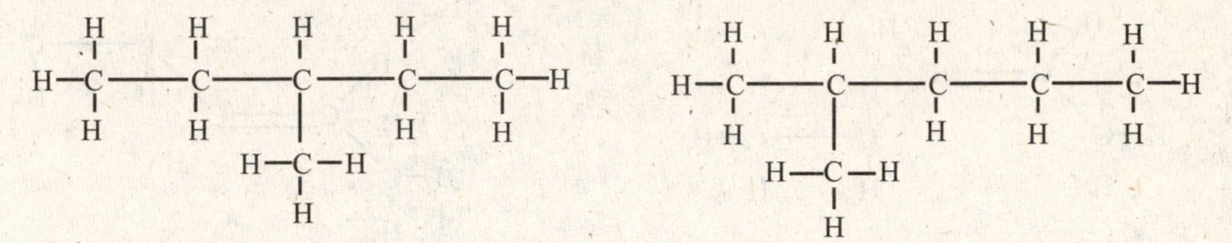

Two methyl branches can go on the four-carbon chain in two ways. They can both go on the second carbon or one can go on the second carbon and one can go on the third carbon.

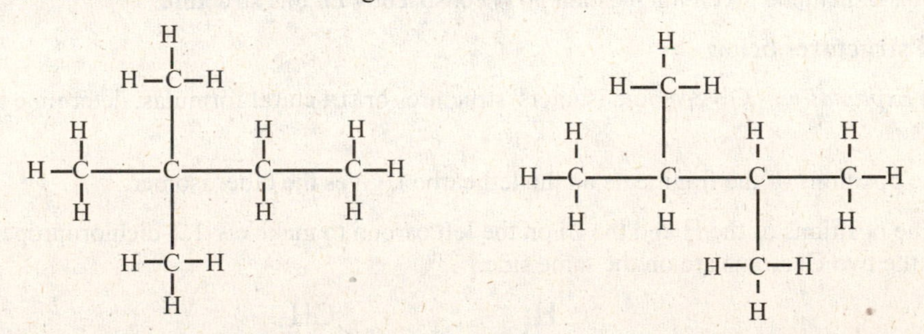

Three methyl branches cannot all three be attached to the one middle carbon in a three-carbon chain, so we're done with methyl branches.

We could try to make a four carbon chain with an ethyl branch. However, ethyl branches can't be attached to chain-carbons within two carbons of the end of the chain. (If you do that, the molecules "longest" chain will actually include your branch!) A four-carbon chain isn't long enough and therefore we can't use ethyl branches. So we have found all the branched isomers of the hydrocarbon with the formula C_6H_{14}.

✓ *Reasonable Answer Check:* The molecules have one or more branches off the three-carbon chain. They also all have six C atoms and 14 H atoms.

26. *Answer:* **(a) alkyne (b) alkane (c) alkene**

Strategy and Explanation: Given formulas of hydrocarbons determine if they are alkane, alkene, or alkyne.

If we assume that each of these straight chain molecules has at most one multiple bond, we can use the formula pattern to determine the class. The formulas of alkanes are C_nH_{2n+2}, for n = 1 and higher. Each multiple bond removes two electrons. The formulas of alkenes are C_nH_{2n}, for n = 2 and higher. The formulas of alkynes are C_nH_{2n-2}, for n = 2 and higher. Using the number of C atoms, set n. Then determine 2n+2, 2n, or 2n–2. Compare these numbers to the number of H atoms. Form a conclusion based on that comparison.

(a) C_5H_8 n = 5, with this n, 2n+2 = 12, 2n = 10, 2n–2 = 8. This hydrocarbon is an alkyne. Notice, that it could also be an alkene, if there are two double bonds in the molecule.

(b) $C_{24}H_{50}$ n = 24, with this n, 2n+2 = 50, 2n = 48, 2n–2 = 46. This hydrocarbon is an alkane.

(c) C_7H_{14} n = 7, with this n, 2n+2 = 16, 2n = 14, 2n–2 = 12. This hydrocarbon is an alkene.

✓ *Reasonable Answer Check:* One and only one of the three calculations using n gave a matching number.

28. *Answer:* **See structures below**

Strategy and Explanation: Given the name of a hydrocarbon, write *cis*- and *trans*-isomers for it.

Use the prefix to determine number of carbons, and the numeral provided in the name to determine where the double bond starts along the chain. Add H atoms to complete the octets. Look at the fragments attached at the double bond put the two larger ones on the same side for the *cis*-isomer, and on opposite sides for the *trans*-isomer.

2-pentene has a five-carbon chain with the double bond starting at the second carbon.

cis-2-pentene (methyl and ethyl on the same side)

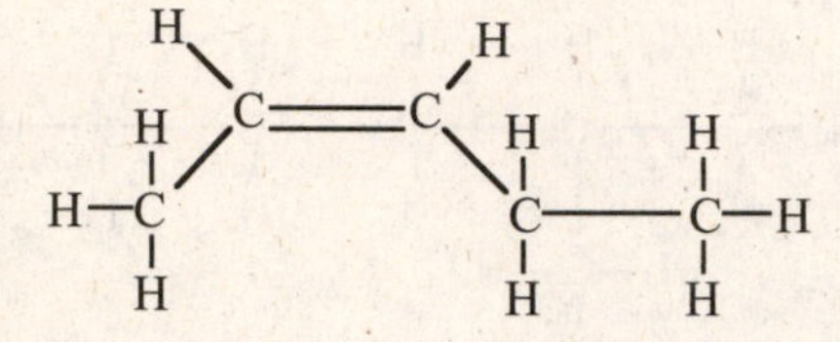

trans-2-pentene (methyl and ethyl on the opposite sides)

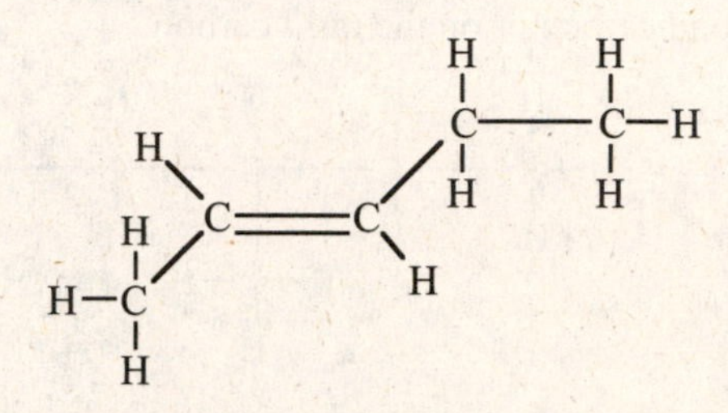

✓ *Reasonable Answer Check:* These are the same isomers discovered when answering Question 25. The structure for *cis*-2-pentene given in Question 30 is consistent with this structure.

30. *Answer:* **See structures below**

Strategy and Explanation: Given some isomers' structures or structural formulas, determine the other *cis*-or *trans*-isomers.

Switching the positions of the fragments on the left carbon, gives the other isomer.

(a) Switch the positions of the H and the Cl on the left carbon to make *cis*-1,2-dichloropropene. (It's *cis*- because the two Cl atoms are on the same side.)

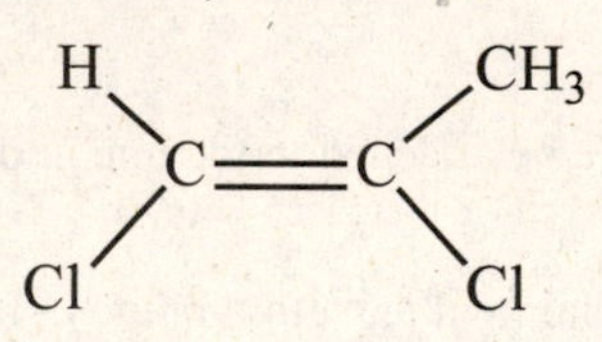

(b) Switch the positions of the H and the CH_3 on the left carbon to make *trans*-2-pentene. (It's *trans*- because the ethyl and methyl fragments are opposite sides.)

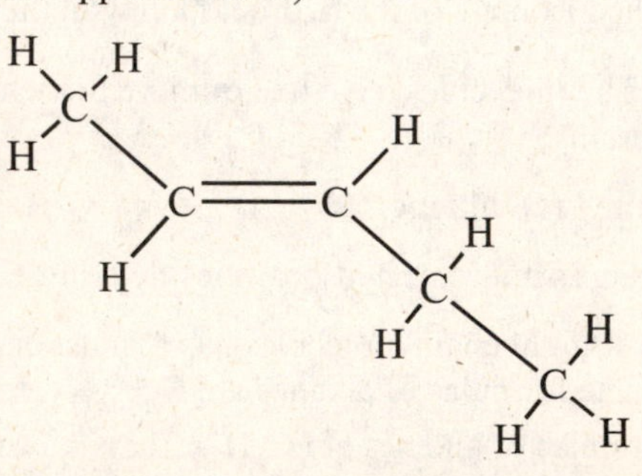

(c) Switch the positions of the H and the CH_3 on the left carbon to make *trans*-3-hexene. (It's *trans*- because the two ethyl fragments are on opposite sides.)

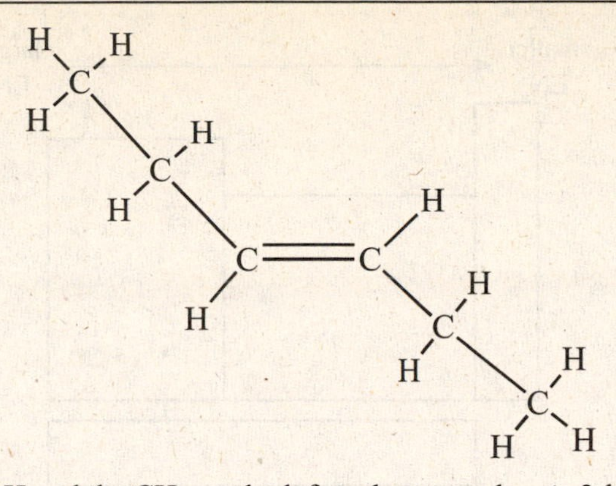

(d) Switch the positions of the H and the CH_3 on the left carbon to make *cis*-2-hexene. (It's *cis*- because the methyl and propyl fragments are on the same side.)

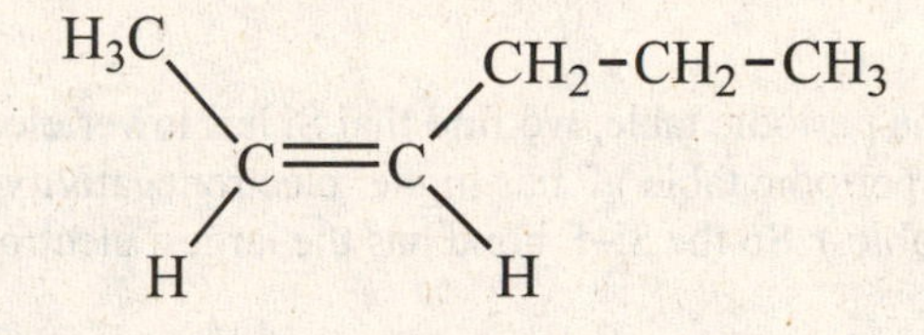

✓ *Reasonable Answer Check:* We've seen some of these isomers in previous examples. The particular isomers formed can be correlated to the name change.

32. *Answer/Explanation:* It is not possible for oxalic acid to have *cis-trans* isomerization since the double bonds are only to the oxygen atoms and the C–C bond is a single bond. That means free rotation about the C–C bond can occur.

Bond Properties (Sections 8.6, 8.7)

34. *Answer:* **(a) B–Cl (b) C–O (c) P–O (d) C=O**

Strategy and Explanation: Given a series of pairs of bonds, predict which of the bonds will be shorter.

Use periodic trends in atomic radii to identify the smaller atoms. The bond with the smaller atoms, will have a shorter bond. In cases where bonds between the same atoms are compared, triple bonds are shorter than double bonds, which are shorter than single bonds.

(a) Both bonds have Cl, so compare the sizes of B and Ga. B is smaller than Ga. (It is higher in the same group of the periodic table.) So, B–Cl is shorter than Ga–Cl.

(b) Both bonds have O, so compare the sizes of C and Sn. C is smaller than Sn. (It is higher in the same group of the periodic table.) So, C–O is shorter than Sn–O.

(c) Both bonds have P, so compare the sizes of O and S. O is smaller than S. (It is higher in the same group of the periodic table.) So, P–O is shorter than P–S.

(d) Both bonds have C, so compare the sizes of C and O. O is smaller than C. (It is further to the right in the same period of the periodic table.) So, C=O is shorter than C=C.

✓ *Reasonable Answer Check:* Several of these predictions are confirmed in Table 8.1.

36. *Answer:* **(a)**

Strategy and Explanation: Given some bonds and only a periodic table, predict which of the bonds will be strongest.

Ionic bonds are stronger than polar covalent bonds. Polar covalent bonds are stronger than purely covalent bonds. Use the periodic trend for electronegativity (EN):

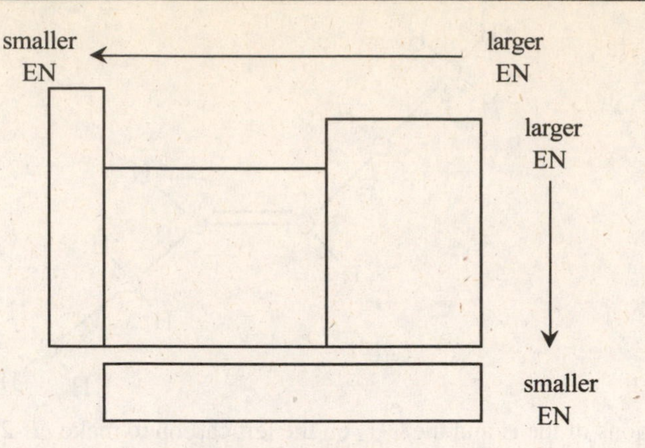

The larger the difference in electronegativity, the stronger the bond will be.

(a) Si–F, (b) P–S, (c) P–O

Looking up these atoms on the periodic table, we find that Si has lower electronegativity than P. (It is further to the left in the same period of the periodic table.) F has higher electronegativity than O. (It is further to the left in the same period of the periodic table.) So the Si–F bond has the largest electronegativity difference; hence, it is the strongest.

✓ *Reasonable Answer Check:* These predictions are confirmed in Figure 8.6.

38. *Answer:* **CO**

Strategy and Explanation: Given two chemical formulas, predict which has the shorter carbon-oxygen bond.

First, write Lewis structures for the two molecules. In cases where bonds between the same two atoms are compared, triple bonds are shorter and stronger than double bonds, which are shorter and stronger than single bonds. There is only one plausible Lewis structure for formaldehyde, H_2CO, with 12 valence electrons.

H
|
H—C=O:

In formaldehyde, the carbon-oxygen bond is a double bond.

There is only one plausible Lewis structure for carbon monoxide, CO, with 10 valence electrons.

:C≡O:

In carbon monoxide, the carbon-oxygen bond is a triple bond.

That means the shorter bond is the bond in CO.

✓ *Reasonable Answer Check:* The Lewis structures obey the octet rule and have the right number of valance electrons. These predictions are upheld in general with values given in Table 8.1.

40. *Answer:* **CO_3^{2-}; it has more resonance structures with single C–O bonds.**

Strategy and Explanation: Given two chemical formulas, predict which has the shorter carbon-oxygen bond.

First, write resonance structures for the two molecules. In cases where bonds between the same two atoms are compared, triple bonds are shorter and stronger than double bonds, which are shorter and stronger than single bonds. If more than one resonance structure is possible, average their contribution.

Consider formate ion first. Two equivalent plausible Lewis structures exist for HCO_2^- with 18 valence electrons:

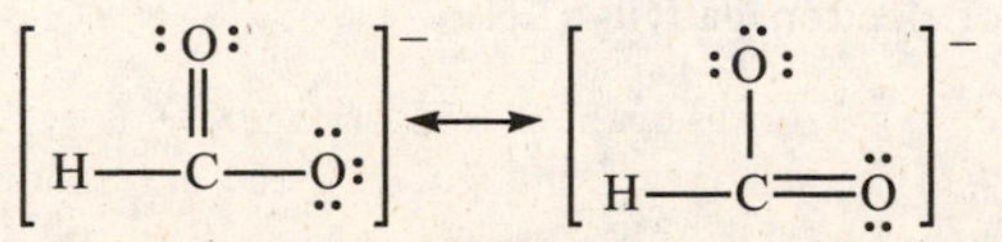

In the formate ion, one resonance structure has a single carbon-oxygen bond and one has a double carbon-oxygen bond. We predict that the carbon-oxygen bond will be halfway between a single and double bond.

There are three plausible equivalent Lewis structures for carbonate ion, CO_3^{2-}, with 24 valence electrons:

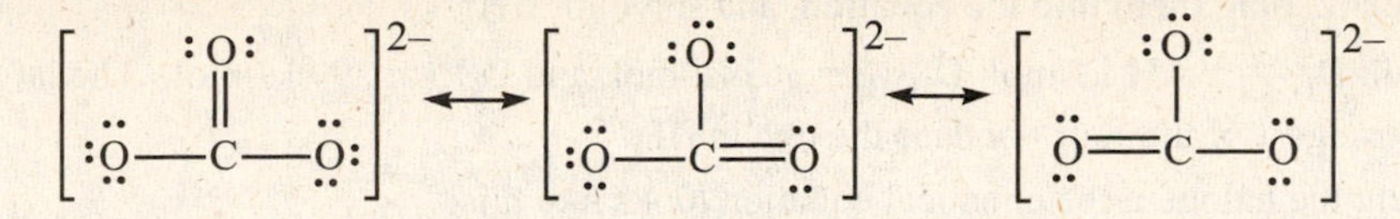

In the carbonate, one resonance structure has a double carbon-oxygen bond and two have single carbon-oxygen bond. We predict that the carbon-oxygen bond will be closer to a single bond than a double bond.

That means the longer bond is the bond in CO_3^{2-}.

✓ *Reasonable Answer Check:* The Lewis structures obey the octet rule and have the right number of valance electrons. These predictions are upheld in general with values given in Table 8.1.

Bond Energies and Enthalpy Changes (Section 8.6)

42. *Answer:* **–92 kJ; the reaction is exothermic.**

Strategy and Explanation: Given a description of a chemical equation for a reaction and a table of bond enthalpies, estimate the standard enthalpy change of the reaction and determine whether the reaction is exothermic or endothermic.

First, balance the chemical equation. Then use another variation of Hess's Law to estimate ΔH°. We will break all the bonds in the reactants (by putting their bond enthalpies into the system) and then we will form all the bonds—the opposite of breaking—in the products (by removing their bond enthalpy from the system). That is the logic behind Equation 8.1.

$$\Delta H° = \sum\left[(\text{moles of bond}) \times D(\text{bond broken})\right] - \sum\left[(\text{moles of bond}) \times D(\text{bond formed})\right]$$

Get the balanced equation and use it to describe the moles of each of the reactants and products. Count the moles of bonds of each type that break and form. Set up a specific version of the above equation. Look up the D for each bond in Table 8.2, plug them into the equation and solve for ΔH°.

The reactants are nitrogen, N_2, and hydrogen, H_2. The product is ammonia, NH_3. The balanced equation is

$$N_2 \quad + \quad 3\,H_2 \quad \longrightarrow \quad 2\,NH_3$$

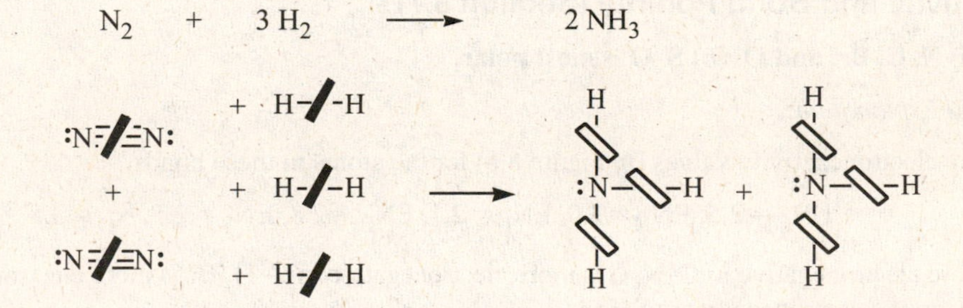

Hence, we break one mol of N≡N bond and three mol of H–H bonds then form six mol of N–H bonds:

$$\Delta H° = (1 \text{ mol of N≡N}) \times D_{N≡N} + (3 \text{ mol H–H}) \times D_{H\text{–}H} - (6 \text{ mol N–H}) \times D_{N\text{–}H}$$

Look up the D values in Table 8.2.

$$\Delta H° = (1 \text{ mol}) \times (946 \text{ kJ/mol}) + (3 \text{ mol}) \times (436 \text{ kJ/mol}) - (6 \text{ mol}) \times (391 \text{ kJ/mol}) = -92 \text{ kJ}$$

This reaction is exothermic.

✓ *Reasonable Answer Check:* The reaction is twice the formation reaction for NH_3. Appendix J tells us that NH_3 has $\Delta H_f^\circ = -46.11$ kJ/mol. Twice this value produces a ΔH° = –92.22 kJ/mol. This is very close to the estimate calculated here.

44. *Answer:* **H–F bond; –538 kJ; –184 kJ; –103 kJ; –11 kJ; The reaction of H_2 with F_2 is most exothermic.**

Strategy and Explanation: Given a list of molecules, a description of their chemical reactions and a table of bond enthalpies, identify which has the strongest bond and estimate the standard enthalpy change of the reaction and which is the most exothermic.

The strongest chemical bond has the largest bond enthalpy. Balance the chemical equations for the reactions. Use the equations to describe the moles of each of the reactants and products. Count the moles of bonds of each type that break and form. Set up an equation as described in the solution to Question 42. Look up the D for each bond in Table 8.2, plug them into the equation, and solve for $\Delta H°$.

D_{H-F} = 566 kJ/mol, D_{H-Cl} = 431 kJ/mol, D_{H-Br} = 363 kJ/mol, and D_{H-I} = 299 kJ/mol. The largest D is for the H–F bond so the strongest of the four bonds is the one in HF.

Using X to represent the halogen, the chemical equation looks like this:

$$H_2 + X_2 \longrightarrow 2\ HX$$

So, we break one mol of H–H bonds and one mol of X–X bonds. We form two mol H–X bonds:

$\Delta H° = (1\text{ mol of H–H}) \times D_{H-H} + (1\text{ mol X–X}) \times D_{X-X} - (2\text{ mol H–X}) \times D_{H-X}$

Look up the D values in Table 8.2.

When X = F

$$\Delta H° = (1\text{ mol}) \times (436\text{ kJ/mol}) + (1\text{ mol}) \times (158\text{ kJ/mol}) - (2\text{ mol}) \times (566\text{ kJ/mol}) = -538\text{ kJ}$$

When X = Cl

$$\Delta H° = (1\text{ mol}) \times (436\text{ kJ/mol}) + (1\text{ mol}) \times (242\text{ kJ/mol}) - (2\text{ mol}) \times (431\text{ kJ/mol}) = -184\text{ kJ}$$

When X = Br

$$\Delta H° = (1\text{ mol}) \times (436\text{ kJ/mol}) + (1\text{ mol}) \times (193\text{ kJ/mol}) - (2\text{ mol}) \times (366\text{ kJ/mol}) = -103\text{ kJ}$$

When X =I

$$\Delta H° = (1\text{ mol}) \times (436\text{ kJ/mol}) + (1\text{ mol}) \times (151\text{ kJ/mol}) - (2\text{ mol}) \times (299\text{ kJ/mol}) = -11\text{ kJ}$$

The $H_2 + F_2$ reaction is the most exothermic.

✓ *Reasonable Answer Check:* The reaction is twice the formation reaction for HX. Appendix J tells us that HF has ΔH_f° = –271.1 kJ/mol, HCl has ΔH_f° = –92.307 kJ/mol, and HBr has ΔH_f° = –36.40 kJ/mol. Twice these values produce $\Delta H°$ = –542.2 kJ/mol, $\Delta H°$ = –184.614 kJ/mol, and $\Delta H°$ = –72.80 kJ/mol. These are reasonably close to the estimates calculated here. The bromine value is farther away because Appendix J uses standard state liquid bromine, not gas-phase as assumed in the bond enthalpy numbers. HI would be far off since the standard state of I_2 is solid.

Electronegativity and Bond Polarity (Section 8.7)

46. *Answer:* **(a) N, C, Br, and O (b) S–O is most polar.**

Strategy and Explanation:

(a) Look up electronegativity values (in Figure 8.6) for the atoms in these bonds.

$$EN_C = 2.5,\ EN_N = 3.0,\ EN_H = 2.1,\ EN_{Br} = 2.8,\ EN_S = 2.5,\ N_O = 3.5$$

N is more electronegative in C–N, C is more electronegative in C–H, Br is more electronegative in C–Br, and O is more electronegative in S–O.

(b) Bonds are more polar when the electronegativity difference is larger. Get ΔEN to find most polar.

$\Delta EN_{C-N} = EN_N - EN_C = 3.0 - 2.5 = 0.5$

$\Delta EN_{C-H} = EN_C - EN_H = 2.4 - 2.1 = 0.4$

$\Delta EN_{C-Br} = EN_{Br} - EN_C = 2.8 - 2.5 = 0.3$

$\Delta EN_{S-O} = EN_O - EN_S = 3.5 - 2.5 = 1.0$ most polar

48. *Answer:* **(a) all bonds are somewhat polar (b) C=O bond is the most polar; O atom is partial negative**

Strategy and Explanation:

(a) Bonds are polar if the atoms' electronegativities are different.

$$EN_H = 2.1,\ EN_N = 3.0,\ EN_C = 2.5,\ EN_O = 3.5$$

All electronegativities are different, the bonds are all somewhat polar. None of the bonds are nonpolar.

(b) Bonds are more polar when the electronegativity difference is larger.

$$\Delta EN_{N\text{–}H} = EN_N - EN_H = 3.0 - 2.1 = 0.9$$

$$\Delta EN_{C\text{–}N} = EN_N - EN_C = 3.0 - 2.5 = 0.5$$

$$\Delta EN_{C\text{–}O} = EN_O - EN_C = 3.5 - 2.5 = 1.0$$

The largest electronegativity difference is for the C=O bond, so it is the most polar.

C — O
δ^+ δ^-

The O atom is the partial negative end of this bond, because it has the larger electronegativity.

Formal Charge (Section 8.8)

50. *Answer:* **Zero; net ionic charge**

Strategy and Explanation: The total of the formal charges in a molecule must be zero, because molecules have no net charge. The total of the formal charges in an ion must be its ionic charge.

52. *Answer:* **(a)** O=S(–O)–O with formal charges −1 (top O), 0 (=O), 2+ (S), −1 (–O) **(b)** :N≡C—C≡N: with formal charges 0 0 0 0 **(c)** [O=N—O:]⁻ with formal charges 0 0 −1

Strategy and Explanation: Given the formulas of molecules or ions, write the correct Lewis structure and assign formal charges to each atom.

Write the Lewis structures. Then determine the number of lone pair electrons and bonding electrons around each atom. Use the method described in Section 8.8 on page 350 to determine the formal charges on each atom.

Formal charge = (number of valence electrons in an atom)
– [(number of lone pair electrons) + ($\frac{1}{2}$ number of bonding electrons)]

To get the *(number of lone pair electrons)* just count all the dots. To get *(number of bonding electrons)* just count two times the number lines representing covalent bonds.

(a) Lewis structure for SO_3 molecule: 24 electrons total

:O:
|
O=S—O:

The formal charges are calculated for each atom using the number of valence electrons, the number of lone pair electrons and the number of bonding electrons. Let's set up a chart for these values:

	S	=O	–O
Valence electrons	6	6	6
Lone pair electrons	0	4	6
Bonding electrons	8	4	2
Formal charge	6 – (0 + 4) = +2	6 – (4 + 2) = 0	6 – (6 + 1) = –1

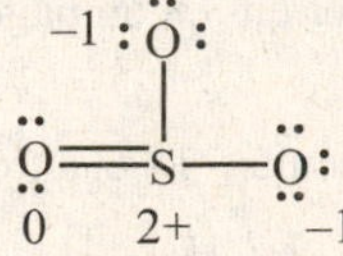

There are other resonance structures that could be written for SO_3 with the double bond moved to each of the other two O atoms and with more than one double bond to the sulfur atom. (A more involved description of writing and judging the feasibility of resonance structures is found in the solutions to Questions 60-66.)

(b) Lewis structure for NCCN molecule: 18 electrons total.

$$:N\equiv C-C\equiv N:$$

The formal charges are calculated for each atom using the number of valence electrons, the number of lone pair electrons and the number of bonding electrons. Let's set up a chart for these values:

	C	N
Valence electrons	4	5
Lone pair electrons	0	2
Bonding electrons	8	6
Formal charge	4 – (0 + 4) = 0	5 – (2 + 3) = 0

$$\underset{0}{:N}\equiv\underset{0}{C}-\underset{0}{C}\equiv\underset{0}{N:}$$

(c) Lewis structure for NO_2^- ion: 18 electrons total

$$\left[\ddot{\underset{..}{O}}=\ddot{N}-\ddot{\underset{..}{O}}:\right]^-$$

The formal charges are calculated for each atom using the number of valence electrons, the number of lone pair electrons and the number of bonding electrons. Let's set up a chart for these values:

	N	=O	–O
Valence electrons	5	6	6
Lone pair electrons	2	4	6
Bonding electrons	6	4	2
Formal charge	5 – (2 + 3) = 0	6 – (4 + 2) = 0	6 – (6 + 1) = –1

$$\left[\underset{0}{\ddot{\underset{..}{O}}}=\underset{0}{\ddot{N}}-\underset{-1}{\ddot{\underset{..}{O}}:}\right]^-$$

One other resonance structure could be written for NO_2^- with the double bond moved to the other O atom. (A more involved description of writing resonance structures is found in the solutions to Questions 60-66.)

✓ *Reasonable Answer Check:* The sum of the formal charges is zero for the neutral molecules and the ionic charge for the charged ion. The atoms with more bonds have more positive formal charges than those with fewer bonds and more lone pairs.

54. *Answer:* **(a)** H–C(H)(H)–C(H)=Ö with formal charges all 0 (H 0, C 0, H 0, H 0, C 0, H 0, O 0) **(b)** $[\ddot{\underset{..}{N}}=N=\ddot{\underset{..}{N}}]^-$ with formal charges –1, +1, –1 **(c)** H–C(H)(H)–C≡N: with formal charges all 0

Strategy and Explanation: Given the formulas of molecules or ions, write the correct Lewis structure and assign formal charges to each atom.

Write the Lewis structures. Then determine the number of lone pair electrons and bonding electrons around each atom. Use the method described in Section 8.8 on page 350 and the solution to Question 52 to determine the formal charges on each atom.

(a) The Lewis structure for CH_3CHO molecule: 18 electrons total.

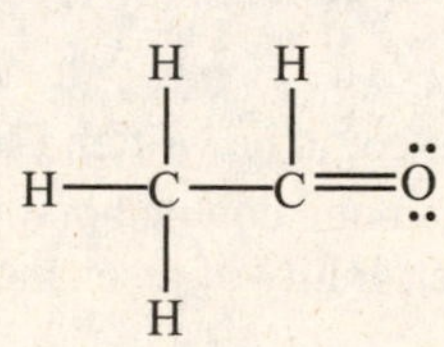

Set up the chart:

	C–	C=	H	O
Valence electrons	4	4	1	6
Lone pair electrons	0	0	0	4
Bonding electrons	8	8	2	4
Formal charge	4 – (0 + 4) = 0	4 – (0 + 4) = 0	1 – (0 + 1) = 0	6 – (4 + 2) = 0

0 H 0 H
H—C—C=Ö
0 0 0 0
H 0

(b) There are three possible Lewis structures for N_3^- ion: 16 electrons total.

First structure:

$[\ddot{N}{=}N{=}\ddot{N}]^-$

Set up the chart:

	N=	=N=
Valence electrons	5	5
Lone pair electrons	4	0
Bonding electrons	4	8
Formal charge	5 – (4 + 2) = –1	5 – (0 + 4) = +1

$[\ddot{N}{=}N{=}\ddot{N}]^-$
–1 +1 –1

Second structure:

$[:N{\equiv}N{-}\ddot{N}:]^-$

Set up the chart:

	Left-most N	Middle N	Right-most N
Valence electrons	5	5	5
Lone pair electrons	2	0	6
Bonding electrons	6	8	2
Formal charge	5 – (2 + 3) = 0	5 – (0 + 4) = +1	5 – (6 + 1) = –2

$[:N{\equiv}N{-}\ddot{N}:]^-$
0 +1 –2

Third structure:

$[:\ddot{N}{-}N{\equiv}N:]^-$

Set up the chart:

	Left-most N	Middle N	Right-most N
Valence electrons	5	5	5
Lone pair electrons	6	0	2
Bonding electrons	2	8	6
Formal charge	5 – (6 + 1) = –2	5 – (0 + 4) = +1	5 – (2 + 3) = 0

$$\left[:\ddot{\underset{\cdot\cdot}{N}}—N\equiv N:\right]^-$$

–2 +1 0

The first of these three structures is the best, since the formal charges on each atom are closer to zero.

(c) Lewis structure for CH_3CN molecule: 16 electrons total

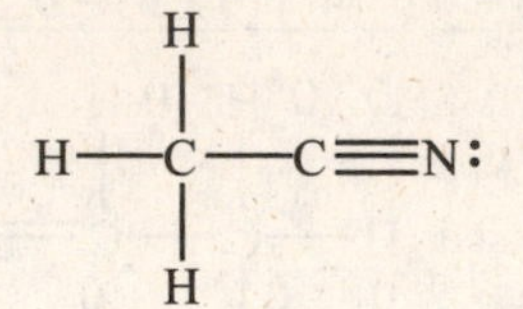

Set up the chart:

	C	H	N
Valence electrons	4	1	5
Lone pair electrons	0	0	2
Bonding electrons	8	2	6
Formal charge	4 – (0 + 4) = 0	1 – (0 + 1) = 0	5 – (2 + 3) = 0

0 H

H—C—C≡N:

0 0 0 0

H 0

✓ *Reasonable Answer Check:* The sum of the formal charges is zero for the neutral molecules and the ionic charge for the charged ion. The atoms with more bonds have more positive formal charges than those with fewer bonds and more lone pairs.

56. *Answer:* -1 +1 -1 $[:C\equiv N—\ddot{\underset{\cdot\cdot}{O}}:]^-$

Strategy and Explanation: Given the formula of an ion, write the correct Lewis structure and assign formal charges to each atom. Use the method described in the solution to Question 52 to get the formal charges.

With 16 electrons total, the central N atom needs two multiple bonds:

$[:C\equiv N—\ddot{\underset{\cdot\cdot}{O}}:]^-$ (I) $[:\underset{\cdot\cdot}{C}=N=\underset{\cdot\cdot}{O}:]^-$ (II) $[:\ddot{\underset{\cdot\cdot}{C}}—N\equiv O:]^-$ (III)

Set up a chart:

	C			N			O		
	I	II	III	I	II	III	I	II	III
Valence electrons	4	4	4	5	5	5	6	6	6
Lone pair electrons	2	4	6	0	0	0	6	4	2
Bonding electrons	6	4	2	8	8	8	2	4	6
Formal charge	–1	–2	–3	+1	+1	+1	–1	0	+1

I: -1 +1 -1 $[:C\equiv N—\ddot{\underset{\cdot\cdot}{O}}:]^-$

II: -2 +1 0 $[:\underset{\cdot\cdot}{C}=N=\underset{\cdot\cdot}{O}:]^-$

III: -3 +1 +1 $[:\ddot{\underset{\cdot\cdot}{C}}—N\equiv O:]^-$

The lowest number formal charge structure is the best Lewis structure, so the right answer is structure (I).

58. *Answer:* (formal charges −1, 0, +1) :Ö—N̈=C̈l and (formal charges 0, 0, 0) Ö=N̈—C̈l:

Strategy and Explanation: Given the atoms in a molecule, write two correct Lewis structures and assign formal charges to each atom. Use the method described in the solution to Question 52 to get the formal charges.

With 18 electrons total, the central N atom needs one multiple bond:

:Ö—N̈=C̈l (I) Ö=N̈—C̈l: (II)

Set up a chart:

	O I	O II	N I	N II	Cl I	Cl II
Valence electrons	6	6	5	5	7	7
Lone pair electrons	6	4	2	2	4	6
Bonding electrons	2	4	6	6	4	2
Formal charge	−1	0	0	0	0	+1

(formal charges −1, 0, +1) :Ö—N̈=C̈l (I) (formal charges 0, 0, 0) Ö=N̈—C̈l: (II)

The lowest number formal charge structure is the best Lewis structure, so the right answer is structure (II).

Resonance (Section 8.9)

60. *Answer:* **See structures below**

Strategy and Explanation: Given the formulas of molecules or ions, write all the resonance structures.

Write the Lewis structure. Each resonance structure differs only by where the electrons for a multiple bond come from. When two or more atoms with lone pairs are bonded to an atom that needs more electrons, any one of them can supply the electron pair for a multiple bond. To write all resonance structures, systematically and sequentially supply the central atom with needed electrons from each of the possible sources. Separate these different structures with a double-headed arrow to show that they are resonance structures.

(a) There are three plausible Lewis structures for nitric acid, HNO_3, with 24 valence electrons. They are formed using one lone pair from a different one of the outer O atoms to make the second bond in the double bond to complete the octet of the N atom.

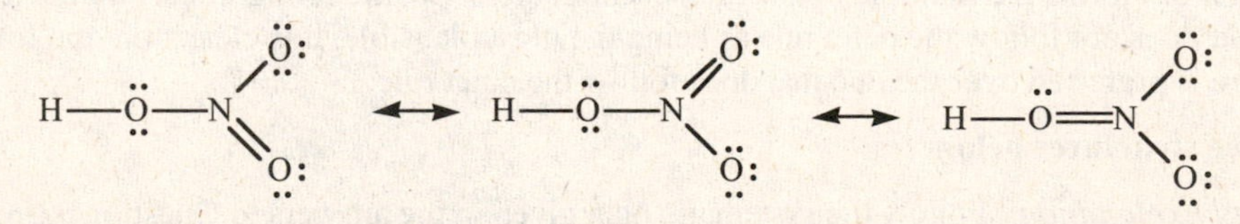

These structures are not equally plausible.

(b) There are three plausible Lewis structures for nitrate ion, NO_3^-, with 24 valence electrons. They are formed using one lone pair from a different one of the outer O atoms to make the second bond in the double bond to complete the octet of the N atom.

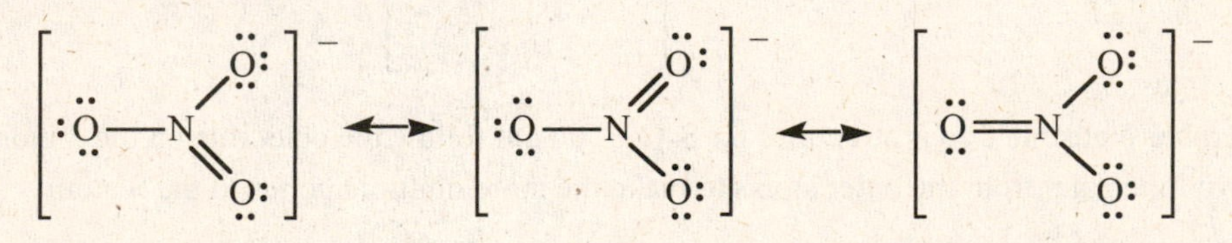

✓ *Reasonable Answer Check:* The structures drawn all follow the octet rule and have the right number of valence electrons. They differ by which outer atom is double bonded to the N atom.

62. *Answer:* **see structures below; the fourth one is the most plausible.**

Strategy and Explanation: Write resonance structures using all single bonds, one, two and three double bonds, and use formal charges to predict the most plausible one.

Write the Lewis structure. To write all resonance structures, systematically and sequentially supply the central atom with needed electrons from each of the possible sources. Keep in mind that Period 2 elements must have an octet, but Period 3 elements can have 8 or more electrons. Separate these different structures with a double-headed arrow to show that they are resonance structures. To determine the relative plausibility of the structures, determine the formal charges (as described in the answers to Question 52) then use the rules described in Section 8.8:

- Smaller formal charges are more favorable than larger ones.
- Negative formal charges should reside on the more electronegative atoms. Conversely, positive formal charges should reside on the least electronegative atoms.
- Like charges should not be on adjacent atoms.

BrO_4^- has 32 electrons. Formal charges for single bonded O atoms are always –1. Formal charges for double bonded O are always 0. Each time electrons are moved from lone pairs into bonding pairs the positive formal charge on Br goes down: The Lewis structure that follows the octet rule for all the atoms is the first one:

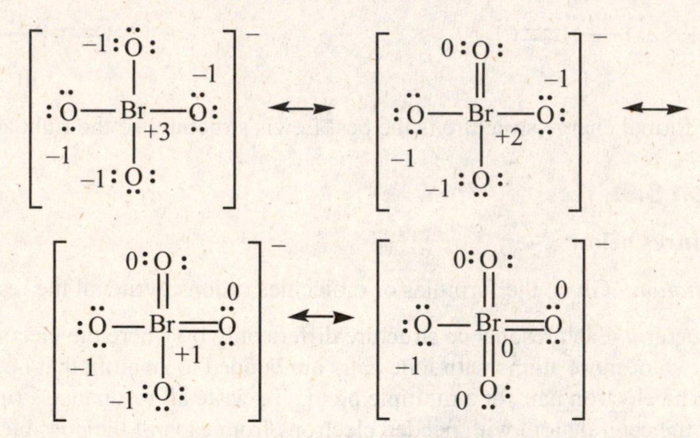

Since smaller formal charges are more favorable than larger ones, we predict that this **fourth resonance structure**, with three double bonds and the most zero formal charges, is the most plausible.

✓ *Reasonable Answer Check:* The Period 2 O atoms in all these structures follow the octet rule. The Period 3 Br atom has eight or more electrons. All structures have the right number of valence electrons. They differ by which outer atom forms the multiple bonds to the central atom. While it might feel strange to select a resonance structure that does not follow the octet rule as being the most plausible, it is clear from the formal charges that this structure is preferred over the one that does follow the octet rule.

65. *Answer:* **See structures below**

Strategy and Explanation: Follow the systematic plan given in the answers to Question 62-63. One Lewis structures for $S_2O_3^{2-}$ follows the octet rule. The molecule has 32 valence electrons.

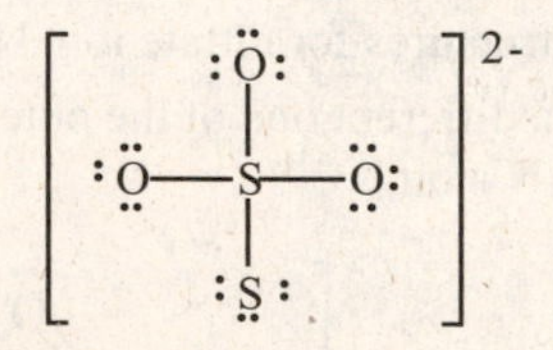

Several other plausible Lewis structures for $S_2O_3^{2-}$ do not follow the octet rule on the S atom from Period 3, using two lone pairs from the outer atoms to make the more multiple bonds to the S atom:

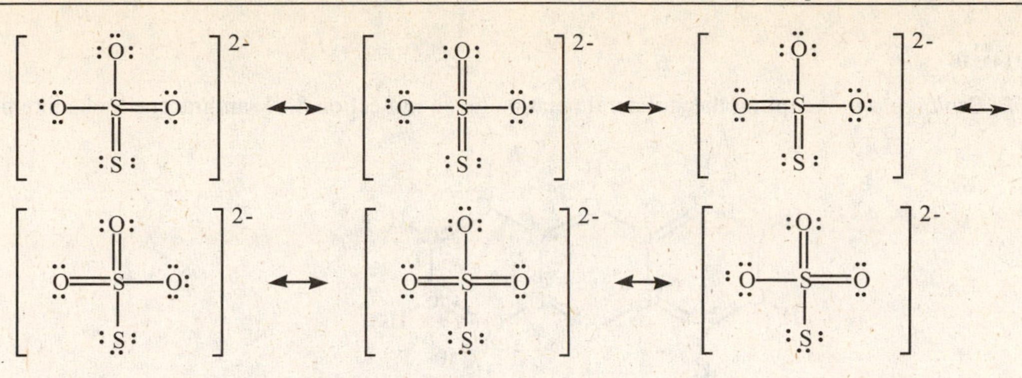

Exceptions to the Octet Rule (Section 8.10)

67. *Answer:* **See structures below**

Strategy and Explanation: Follow the systematic plan given in the solutions to Question 13.

(a) BrF_5 has 42 valance electrons. Use ten of the electrons to connect the six F atoms to the central Br atom. Use 30 more of them to fill the octets of the F atoms. Put the last two on Br.

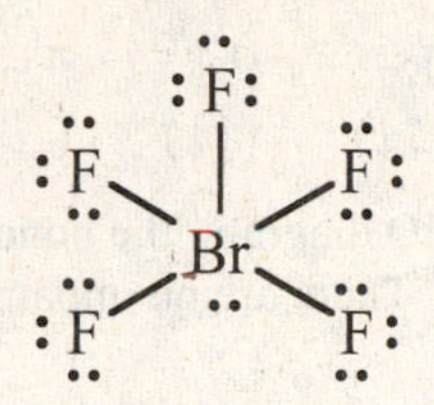

(b) IF_5 has 42 valance electrons. Use ten of the electrons to connect the five F atoms to the central I atom. Use 30 more of them to fill the octets of the F atoms. Put the last two on I.

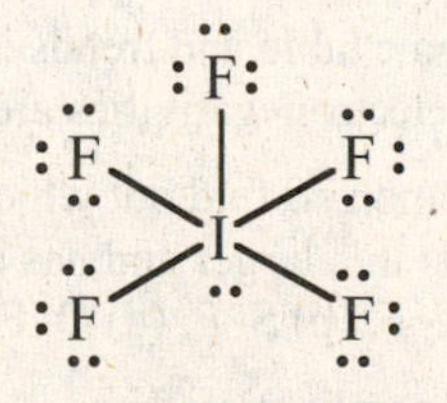

(c) IBr_2^- has 22 valance electrons. Use four of the electrons to connect the Br atoms to the central I atom. Use 12 more of them to fill the octets of the Br atoms. Put the last six on I.

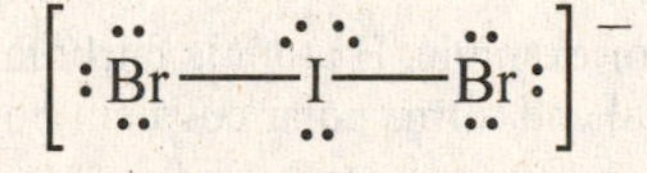

✓ *Reasonable Answer Check:* All the Period 2 elements have an octet of electrons. All the Period 3 elements have eight or more electrons. All valance electrons are accounted for in the structures, either as members of shared pairs or of lone pairs.

69. *Answer:* **(b) P, (e) Cl, (g) Se, and (h) Sn**

Strategy and Explanation: Elements in Periods 3 or higher can form compounds with five or six pairs of valence electrons surrounding their atoms. The Period 2 elements cannot. Using a periodic table, we find that (b) P, (e) Cl, (g) Se, and (h) Sn can, and (a) C, (c) O, (d) F and (f) B cannot.

Aromatic Compounds (Section 8.11)

71. *Answer/Explanation:* The statement "All carbon-to-carbon bond lengths are the same." argues against the presence of C=C bonds in benzene, since the C=C bonds would be shorter than the C–C bonds.

73. *Answer:* **$C_{14}H_{10}$**

Strategy and Explanation: Adapting the structural notation given in Section 8.11, anthracene looks like this:

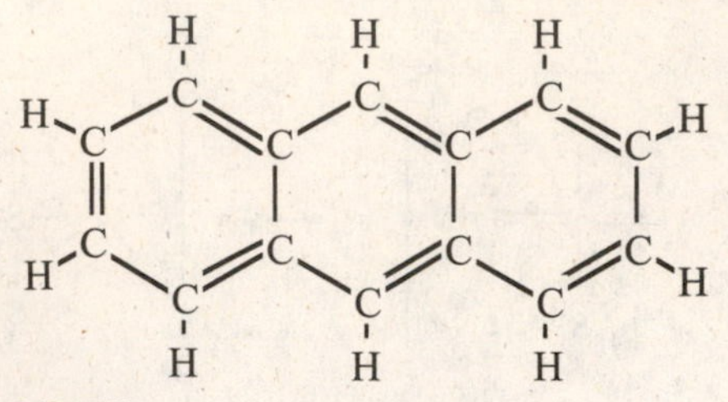

That means it has a formula of $C_{14}H_{10}$.

Molecular Orbital Theory (Section 8.12)

75. *Answer:* **see MO diagram below; three bonds, no unpaired electrons**

Strategy and Explanation: The nitrosyl ion, NO^+, has 10 electrons. Using the method described in Problem-Solving Example 8.12.

	σ_{2s}	σ_{2s}^*	π_{2p}	π_{2p}	σ_{2p}	π_{2p}^*	π_{2p}^*	σ_{2p}^*
NO^+	(↑↓)	(↑↓)	(↑↓)	(↑↓)	(↑↓)	( )	( )	( )

As a result of how the electrons fill the MO diagram, the bond order is (8-4)/2 = 3, meaning that the bond between the N and the O is a triple bond. There are no unpaired electrons.

General Questions

78. *Answer:* **The bond in (c) Si–F, Si is farthest from F on the periodic table, so it is larger and has a lower electronegativity.**

Strategy and Explanation: Use the periodic table and trends in electronegativities to determine which pair is farthest apart. That will mean that their electronegativities are most different and the bond will be most polar.

Here, all the choices have F atom in common so find out which of the other elements is the farthest from F. Si is farthest from F on the periodic table, so it is larger and has a lower electronegativity. Therefore, (c) Si–F is more polar than these other choices (a) C–F, (b) S–F, (d) O–F

80. *Answer/Explanation:* **Yes**, it is a good generalization, because elements close together in the periodic table have similar electronegativities and the bonds would be formed by shared electrons (polar covalent). If they are far apart on the periodic table, their electronegativities will likely be very different and the bond more likely to be ionic.

A few exceptions exist, of course. For example, H atom is fairly far away from most of the nonmetals on the periodic table, yet it is also a nonmetal and forms polar covalent bonds.

82. *Answer:* **(a) C=C (b) C=C (c) C≡N; N is the partial negative end.**

Strategy and Explanation: The resonance structure given in the question is the only resonance structure with zero formal changes, so it is the dominant form for the molecule.

(a) There are two carbon-carbon bonds, a C=C and a C–C. Double bonds are shorter than single bonds, so C=C is the shortest.

(b) Double bonds are stronger than single bonds, so C=C is the strongest.

(c) The most polar bond is the bond with atoms that have the largest electronegativity difference. The difference between the electronegativities of C (EN = 2.5) and H (EN = 2.1) is 0.4. The difference between the electronegativities of C and N (EN = 3.0) is 0.5. So, the carbon-nitrogen bond is slightly more polar. Nitrogen has the higher electronegativity, so it is the partial negative end of the bond.

$$\underset{\delta^+}{C}\equiv\underset{\delta^-}{N}$$

84. *Answer:* **(a)** :C̈l—S̈—C̈l: **(b)** $[$:C̈l—C̈l—C̈l:$]^+$ **(c)** :C̈l—Ö—Cl(=O)(=O)=Ö: **(d)** Ö=S(—C̈l:)—C̈l:

Strategy and Explanation: Use the Lewis structure method outlined in the answers to Question 13.

(a) SCl_2 has 20 valence electrons. Use four of them to attach the two Cl atoms to the central S atom. Use the 12 of them for the lone pairs on the Cl atoms. Use the remaining four of them for the lone pairs on the S atom.

:C̈l—S̈—C̈l:

(b) The Cl_3^+ ion has 20 valence electrons, just as in (a). One of the Cl atoms is the central atom, while the other two Cl atoms are single bonded to the central Cl atom:

$[$:C̈l—C̈l—C̈l:$]^+$

(c) $ClOClO_3$ has 38 valence electrons. We are told there is a Cl–O–Cl bond, so we will put the last three O atoms on the second Cl. Use ten electrons to attach the atoms together using single bonds. Use the remaining 28 of them to complete the octets.

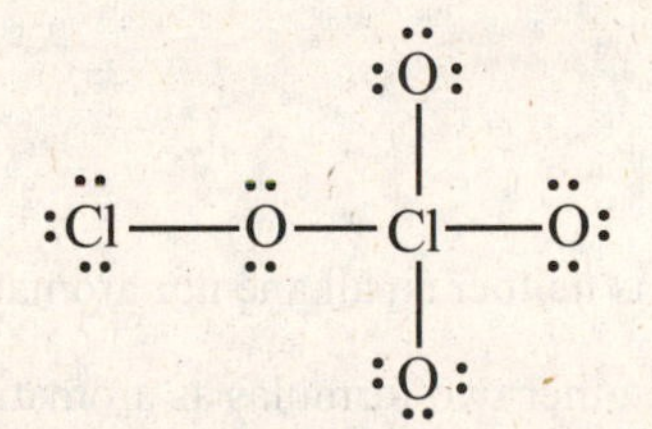

The second Cl atom has a +3 formal charge in this Lewis structure, so we can make a lower formal charge resonance structure for this molecule by expanding the octet of the Period 3 Cl atom, as we did for Br atom in Question 67.

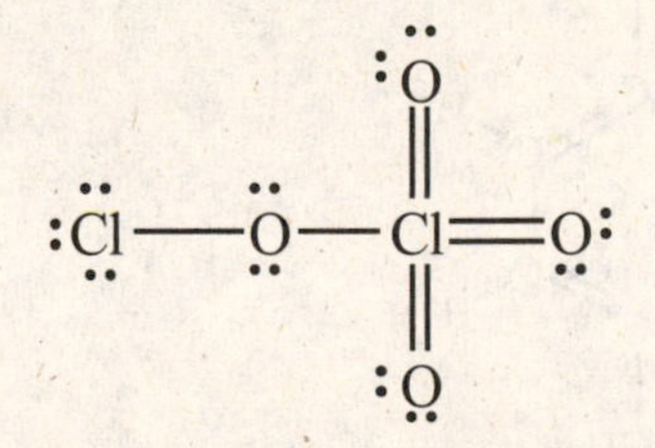

Now, all the formal charges are zero.

(d) $SOCl_2$ has 26 valence electrons. Use six of them to attach the O atoms to the central Cl atom. Use the remaining 18 of them complete the octets of the O atoms. Use the remaining 2 to complete the octet on S.

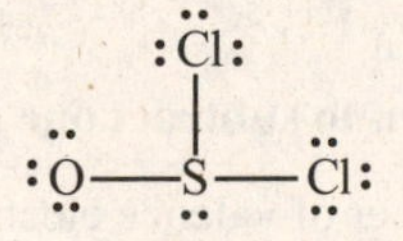

The S atom has a +1 formal charge in this Lewis structure, so we can make a lower formal charge resonance structure for this molecule by expanding the octet of the Period 3 S atom and making one double bond to the S atom, similarly to what we did in (c) for Cl.

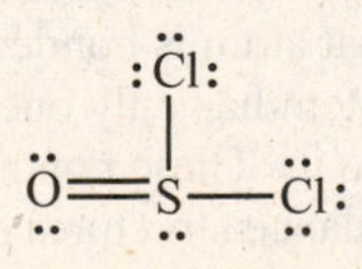

86. *Answer:* **O–O < Cl–O < O–H < O=O < O=C**

Strategy and Explanation: The strongest chemical bond has the largest bond enthalpy. Look up the bond enthalpies for the bonds in Table 8.2: D_{O-H} = 467 kJ/mol, $D_{O=O}$ = 498 kJ/mol, $D_{O=C}$ = 695 to 803 kJ/mol, D_{O-O} = 146 kJ/mol, and D_{O-Cl} = 205 kJ/mol. The order of increasing strength is:

O–O < Cl–O < O–H < O=O < O=C

88. *Answer:* **(a) aromatic (b) aromatic (c) neither category (d) alkane (e) alkane (f) neither category**

Strategy and Explanation: Given formulas of hydrocarbons, determine if they are alkane, aromatic, or neither. We can try what was done in the solution to Question 26.

	n	2n+2 (alkane)	2n (alkene)	2n–2 (alkyne)
$C_6H_?$	6	? = 14	? = 12	? = 10
$C_8H_?$	8	? = 18	? = 16	? = 14
$C_{10}H_?$	10	? = 22	? = 20	? = 18

It's not clear how (a) and (b) fit, yet, so save them for later. (See Section 8.5) We find the answers to (c) - (f) in this chart:

(c) C_6H_{12} has the formula of an alkene, so this is neither an alkane nor aromatic (though it could be a cyclic alkane).

(d) C_6H_{14} is an alkane.

(e) C_6H_{18} is an alkane.

(f) C_6H_{10} is an alkyne, so this is neither an alkane nor aromatic.

Let's draw structures that fit the other two formulas as aromatic compounds with help from Section 8.11:

(a) Xylene is an aromatic compound given in Section 8.11 on page 359, and its formula is C_8H_{10} (*para*-xylene shown here).

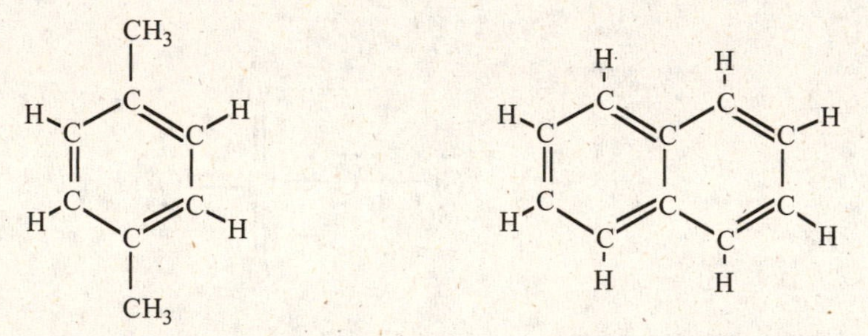

(b) Naphthalene is an aromatic compound given in Section 8.11 on page 359, and its formula is $C_{10}H_8$.

Applying Concepts

90. *Answer:* **student may have forgotten to subtract one electron for the positive charge**

Strategy and Explanation: The number of valence electrons in SF_5^+ is 6 + 5 × (7) – 1 = 40 e^-. It looks like the student forgot to subtract one electron for the positive charge, since the given structure has 6 + 5 × (7) = 41 e^-.

94. *Answer:* **their bonding arrangements differ**

Strategy and Explanation: Resonance structures must only be different by where the electrons are. The bonding arrangement of the atoms (what atom is bonded to what atom) must be the same from structure to structure. In the first structure, the C atom has only one S atom bonded to it. In the second structure the C atom has an N atom and an S atom bonded to it. These bonding arrangements differ; therefore these structures represent different molecules, not resonance structures of one molecule.

96. *Answer:* **(a)**

```
H : O : H
    ...
    H
```

(b)

```
H : F :
    ...
    H
```

Strategy and Explanation: In this fictional universe, a nonet is nine electrons. We will assume that the number of electrons in the atom called "O" in the other universe is still six, the number of electrons in the atom called "H" is still one, and the number of electrons in the atom called "F" is still seven.

(a) The 6 electrons in the O atom would need three more electrons to make a total of nine. That means it would combine with three H atoms to make this molecule:

```
H : O : H
    ...
     H
```

(b) The 7 electrons in the F atom would need two more electrons to make a total of nine. That means it would combine with two H atoms to make this molecule:

```
H : F :
    ...
     H
```

98. *Answer:* **Cl: 3.0; S: 2.5; Br: 2.5; Se: 2.4; As: 2.1**

Strategy and Explanation: Using the periodic trend for electronegativity (EN)

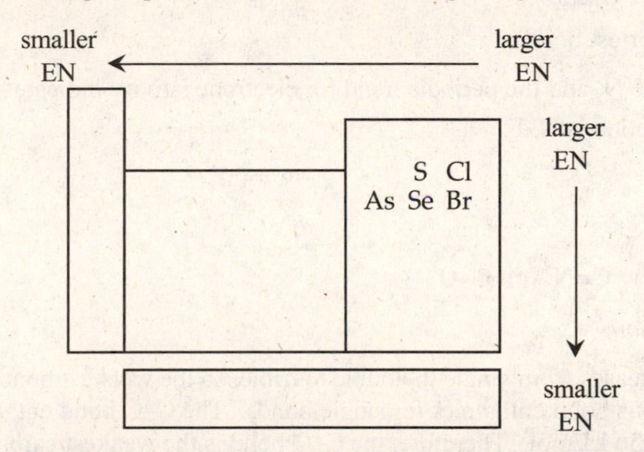

We certainly know that Cl's EN is the largest, so we'll assign it the value of 3.0. We certainly know that As's EN is the lowest, so we'll assign it value of 2.1 The trend shows that EN's of S and Br are both larger than that of Se, so we'll assign them both the value of 2.5. That leaves the EN value of 2.4 for Se. With the values of 2.5 and 2.4 so close together, the assignment of those values are uncertain.

100. *Answer:* **(a) Box 7 (b) Box 1 (c) Boxes 4 and 9 (d) Box 10 (e) Box 12 (f) Box 2**

Strategy and Explanation:

(a) Three boxes have Si-containing bonds, Box 2, 7 and 8. H is a much smaller atom than F or Si, so from periodic trends, we predict that Si-H has the shortest bond. Table 8.1 can assist in confirming that Si-H (145 pm) is shorter than Si-F (181 pm) or Si-Si (234 pm). Si-H is in **Box 7**.

(b) The two choices for this answer are Box 1 and Box 5. The atoms get slightly smaller as you look across the period, so the prediction is that the Cl-Br bond will be the smallest and the Si-Br bond will be the longest. Table 8.1 can assist in confirming that Cl-Br (213 pm) is the shortest and Si-Br (231 pm) is the longest. The correct order is given in **Box 1**.

(c) Two boxes have structures of $FClO_2$: Boxes 4 and 9. The two structures both have 26 electrons. The structure in Box 4 follows the octet rule and the structure in Box 9 shows the expanded octet on the central Cl atom. Expanded octet is legitimate way to lower formal charges for elements in Periods 3 or higher, so both of these structures are resonance hybrids for $FClO_2$. **Boxes 4 and 9** are correct.

(d) The two choices for this answer are Box 3 and Box 10. Polarity is related to the difference in electronegativities. The electronegativity from Figure 8.6 are: EN(Cl) = 2.7, EN(S) = 2.3, EN(Se) = 2.4, and EN(O) = 3.4. So, the most polar will be Cl-O (ΔEN = 0.7), then S-Cl(ΔEN = 0.4), then Se-Cl(ΔEN = 0.3). The correct order is given in **Box 10**.

(e) The two choices for this answer are Box 11 and Box 12. Bond enthalpy of multiple bonds tends to be quite high, so the best prediction would be Box 12 where the double bond is listed as the highest. Table 8.2 can assist in confirming that C=N (616 kJ/mol) is stronger than C-F (486 kJ/mol) or C-P (264 kJ/mol). This order is listed in **Box 12**.

(f) Three boxes have Si-containing bonds, Box 2, 7 and 8. The bond enthalpies are given in Table 8.2. It shows that Si-F (582 kJ/mol) is shorter than Si-H (323 kJ/mol) or Si-Si (226 kJ/mol). Si-F is in **Box 2**.

102. *Answer:* **NF_5; N cannot expand its octet**

Strategy and Explanation: All of these compounds require the central atom to have more than eight electrons around it in order to have bonds to five atoms. Three of the four of the central atoms are in Periods 3 or higher, making it possible to expand their octets. The Period 2 N atom cannot have more than eight electrons, so NF_5 is least likely to exist.

104. *Answer:* :F̤̈—N̈=Ö: **(a) N–F (b) N=O (c) N–F**

Strategy and Explanation: FNO has 18 electrons. The best Lewis structure for this molecule has a double bond on the oxygen, since the formal charges are all zero for this resonance structure.

:F̤̈—N̈=Ö:

(a) Single bonds are longer than double bonds so, N–F is the longest bond.

(b) N=O is the the shortest bond.

(c) Both bonds include N, and the periodic trend for electronegativity indicates that EN_F is greater than EN_O, so the most polar bond is N–F.

N——F

δ^+ δ^-

106. *Answer:* **(a) C–O (b) C≡N (c) C–O**

Strategy and Explanation:

(a) Bond strength increases from single to double to triple, so the weakest bond will be a single bond. Table 8.2 gives the various bond enthalpies for single bonds. The C–C bond enthalpy is 356 kJ/mol. The C–O bond enthalpy is 336 kJ/mol. Therefore, the C–O bond is the weakest carbon-containing bond.

(b) The triple C≡N bond is the strongest carbon-containing bond.

(c) The electronegativity of O atom is greater than that of any other element in this structure, so the C–O bond is the most polar.

More Challenging Questions

108. *Answer:* **(a) see structure below (b)** H—Ö̤—N̈=N̈—Ö̤—H

Strategy and Explanation:

(a) B_4Cl_4 cannot follow the octet rule; there are too few electrons. The structure has 40 valence electrons. We must use 8 electrons to attach the Cl atoms to the B atoms and we must use 24 electrons to fill the octets of the Cl atoms. If the B atoms cluster together and use a total of four electrons to bond with each other, then all the electrons could be used. Dashed lines (- - - - -) are used between B and B in the structure below because they do not represent typical two electron bonds.

(Notice: At the time of production of this textbook, the three-dimensional ball-and-stick model of the B_4H_4 molecule is provided at http://www.chem.ox.ac.uk/inorganicchemistry3/index.html, as are several of the others in this section of questions.)

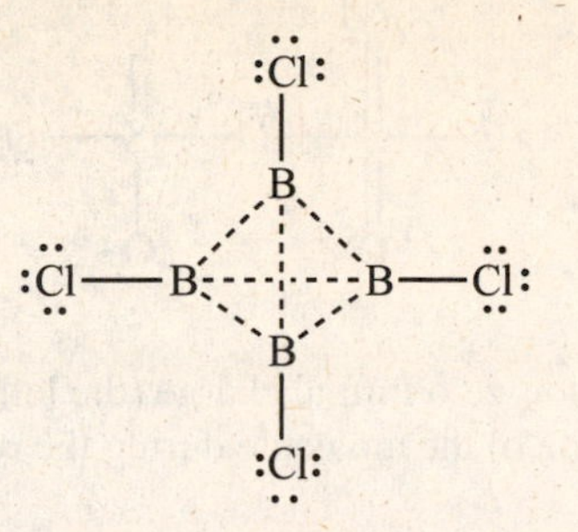

(b) This molecule is an acid so we will attach the H atoms to the O atom as we have seen with other acids. The molecule has 24 electrons. The best Lewis structure would look like this:

H—O—N=N—O—H

110. *Answer:* **(a) see structure below (b) see structure below**

Strategy and Explanation: Many different and creative guesses could result from attempting to write Lewis structures of this molecule. There is limited information provided, so several structures will be equally good at answering the question. The N atoms must follow the octet rule. It is also not likely that the N atoms will bond to each other or to Cl atoms, since their electronegativity is relatively high.

(Notice: At the time of production of this textbook, the three-dimensional ball-and-stick model of the $(Cl_2PN)_3$ and $(Cl_2PN)_4$ molecules are provided at http://www.chem.ox.ac.uk/inorganicchemistry3/index.html.)

(a) There are 72 valence electrons in $(Cl_2PN)_3$.

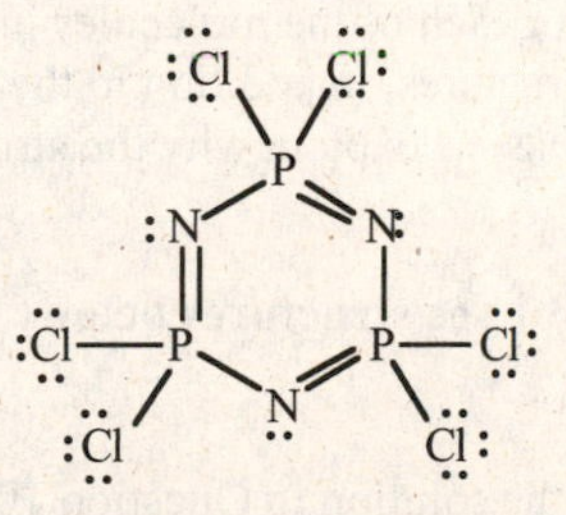

(b) There are 96 valence electrons in $(Cl_2PN)_4$.

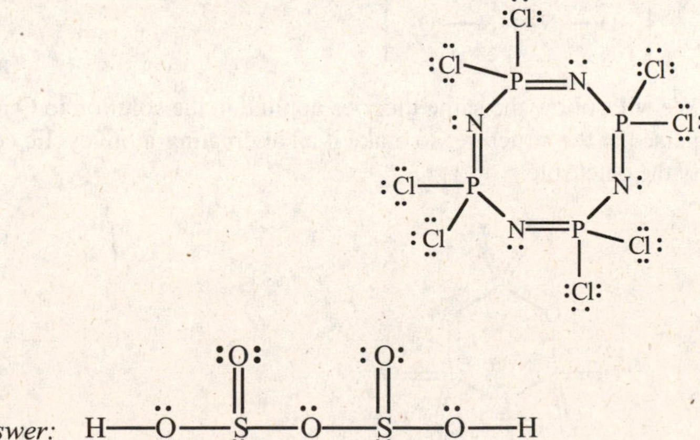

112. *Answer:* H—O—S(=O)(=O)—O—S(=O)(=O)—O—H

Strategy and Explanation: The molecule has 56 electrons. If we imagine an SO_3 molecule bonding with a sulfuric acid molecule, by having one of the O atoms on H_2SO_4 attached to the S atom of SO_3, then the molecule would look like this.

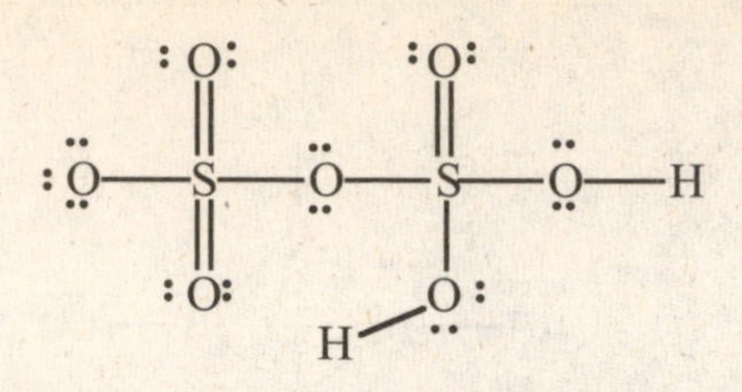

This structure has two atoms with non-zero formal charge (the left-most O and the right S), so it is likely that the H atom will shift to the other side of the molecule during the reaction to form the more stable (all zero-formal-charge) molecule.

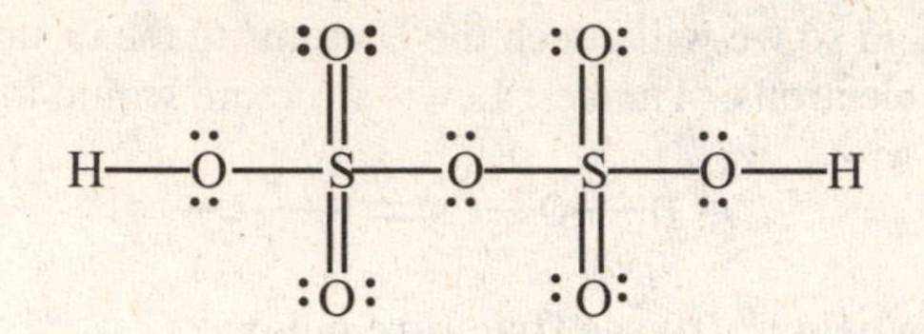

114. *Answer:* **The NON structures have higher formal charges than the NNO structures.**

Strategy and Explanation:

Draw the best Lewis structures for the NNO and NON structures of N_2O, and calculate the formal charges:

One other resonance structure exists for each of the molecules in question. The NON structures all have higher formal charge than the NNO structures, in addition to the positive formal charge being on the more electronegative atom. Both of these reasons explain why the structure of N_2O is not NON, with the N atoms each bonded to the O atom.

117. *Answer:* **(a) see structures below (b) see structures below**

Strategy and Explanation:

(a) Having done hyponitrous acid in the solution to Question 108(b) makes hyponitrite anion fairly easy. Remove the H atoms from the acid and add the charge for this 24 balance electron structure:

$$\left[:\ddot{O}-\ddot{N}=\ddot{N}-\ddot{O}: \right]^{2-}$$

(b) This molecule has 72 electrons. We will follow the same theories applied in the solution to Question 108. The S atoms must be interspersed in the structure, so make it an alternating-atom cyclic compound. Make sure that the O atoms satisfy the octet rule.

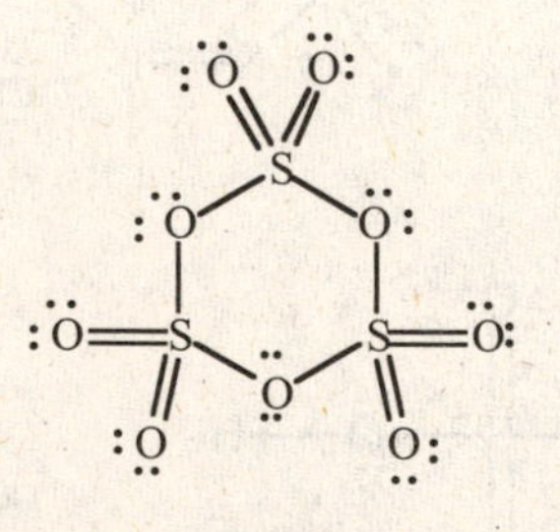

119. *Answer:* **see structure below**

Strategy and Explanation: The C and N atoms in this molecule must all follow the octet rule. Make a straight chain, and add as many multiple bonds to fulfill the octet rule for both period two atoms.

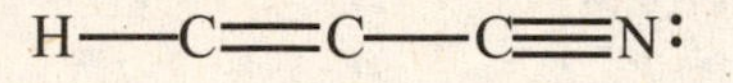

121. *Answer:* **see structure below**

Strategy and Explanation: The aldehyde functional group is present, along with the thio meaning replace an O atom with an S atom. The valence electron count is 30. All atoms must satisfy the octet rule.

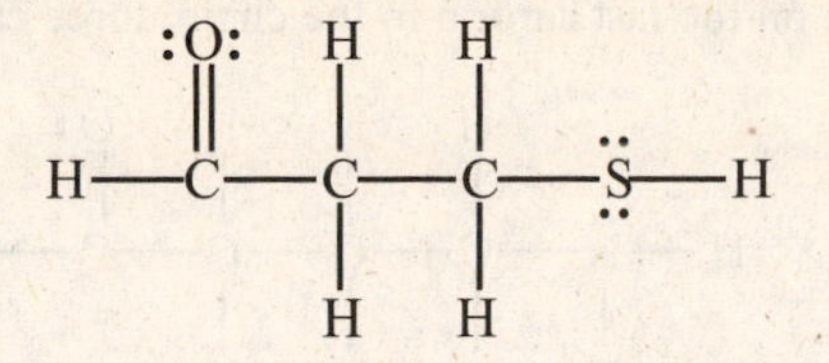

123. *Answer:* **see structure below**

Strategy and Explanation: Tetraphosphrous trisulfide is P_4S_3, with 38 valance electrons.

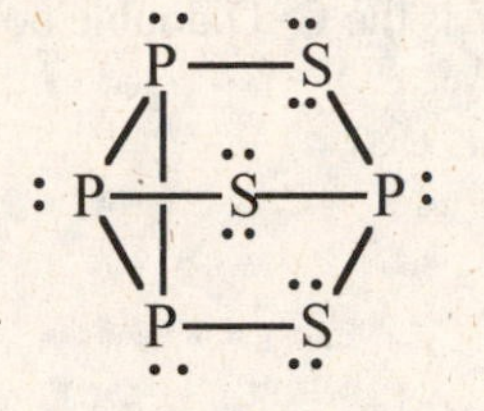

125. To write correct Lewis structures for $S_2O_4^{2-}$, $S_2O_6^{2-}$ and $S_4O_6^{2-}$ bond the S atoms in a line and add O atoms around the terminal S atoms.

The $S_2O_4^{2-}$ ion has 38 electrons. It must have a bond between the two sulfur atoms and two O atoms on each S atom. The first bonds use 10 electrons. 24 electrons fill the octets of the O atoms and each S atom needs two more electrons to use the last two electrons. Forming double bonds between O and S reduces the formal charges:

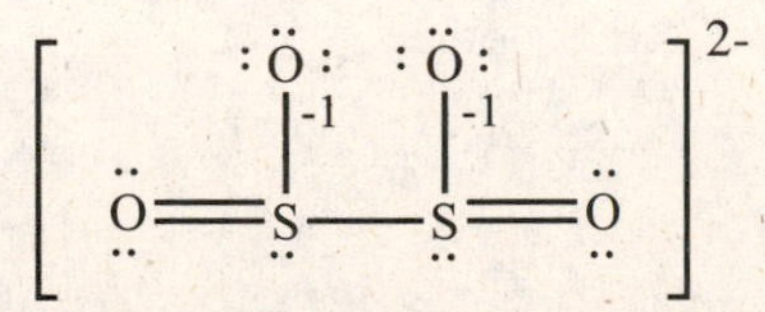

The $S_2O_6^{2-}$ ion has 50 electrons. It must have a bond between the two sulfur atoms and three O atoms on each S atom. The first bonds use 14 electrons. 36 electrons fill the octets of the O atoms. Forming double bonds between O and S reduces the formal charges:

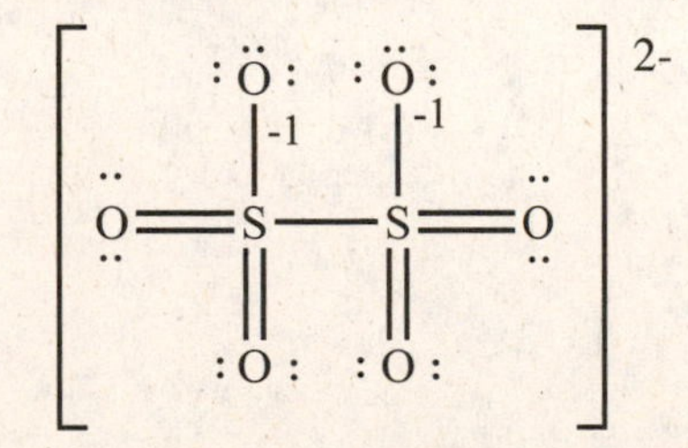

The $S_4O_6^{2-}$ ion has 62 electrons. It must have a chain of four sulfur atoms and three O atoms on each end S atom. The first bonds use 18 electrons. 36 electrons fill the octets of the O atoms. Each of the two middle S atoms needs four more electrons to use the last eight electrons. Forming double bonds between O and S reduces the formal charges:

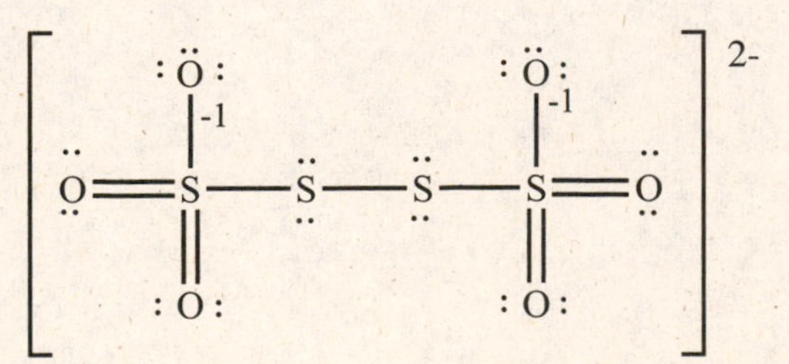

128. *Answer:* **(a) see structure below (b) C–C (c) C=O (d) C=O**

Strategy and Explanation:

(a) This four-carbon chain as has an amine group and an organic acid. Gamma is the third letter of the Greek alphabet, so put the amine on the last carbon in the chain, three carbons away from the carboxylic acid.

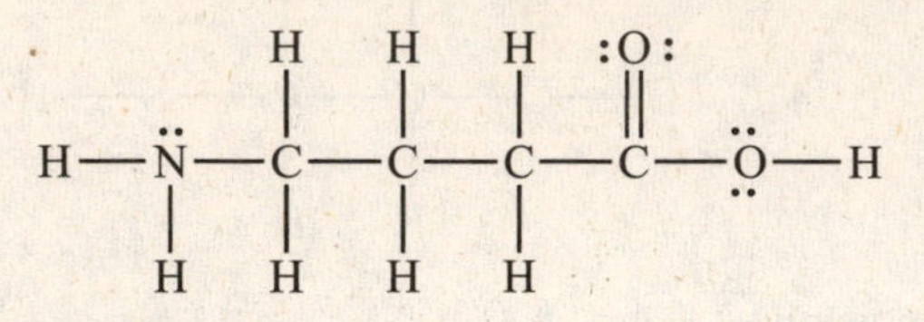

(b) The longest bond C–C single bond.

(c) The most polar bond in this molecule is the C=O double bond.

(d) The strongest bond in this molecule is the C=O double bond.

Chapter 9: Molecular Structures
Solutions for Blue-Numbered Questions for Review and Thought

Topical Questions

Molecular Shape (Section 9.2)

13. *Answer:* **See structures below; (a) linear (b) triangular planar (c) octahedral (d) triangular bipyramidal**

Strategy and Explanation: Write Lewis structures for a list of formulas and identify their shape.

Follow the systematic plan for Lewis structures given in the solution to Question 8.13, then determine the number of bonded atoms and lone pairs on the central atom, determine the designated type (AX_nE_m) and use Table 9.1. *Notice: It is not important to expand the octet of a central atom solely for the purposes of lowering its formal charge, because the shape of the molecule will not change. Hence, we will write Lewis structures that follow the octet rule unless the atom needs more than eight electrons.*

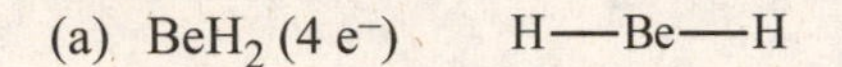

(a) BeH_2 (4 e$^-$) H—Be—H The type is AX_2E_0, so it is linear.

(b) CH_2Cl_2 (20 e$^-$) The type is AX_4E_0, so it is tetrahedral.

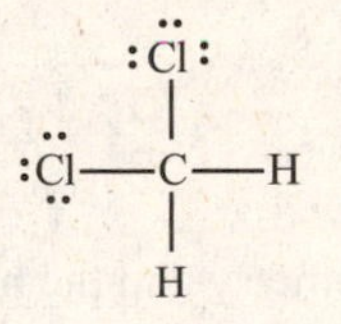

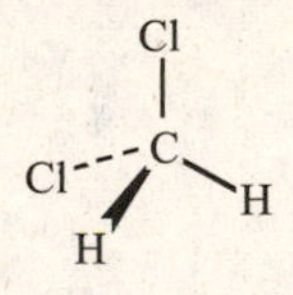

(c) BH_3 (6 e$^-$) The type is AX_3E_0, so it is triangular planar.

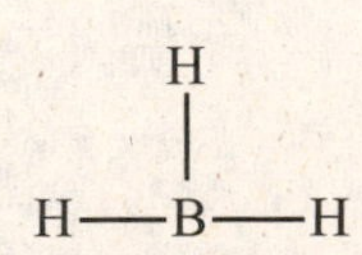

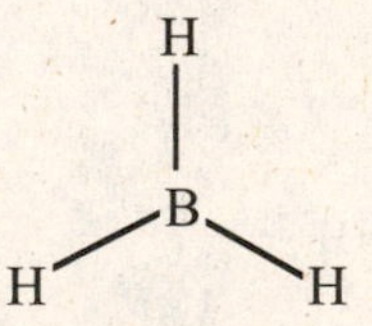

(d) SCl_6 (48 e$^-$) The type is AX_6E_0, so it is octahedral.

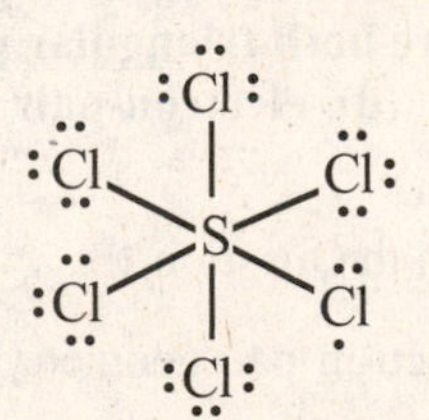

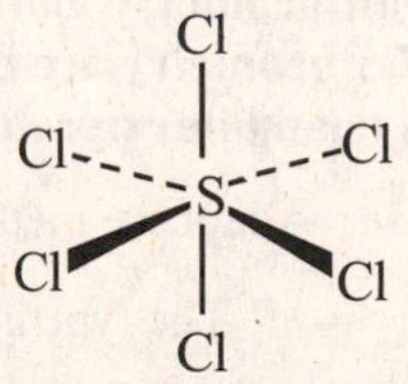

(e) PF_5 (40 e$^-$)

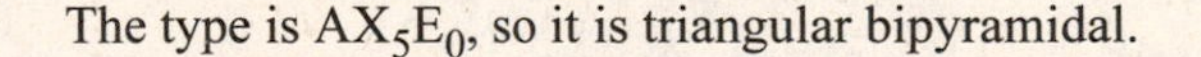

The type is AX_5E_0, so it is triangular bipyramidal.

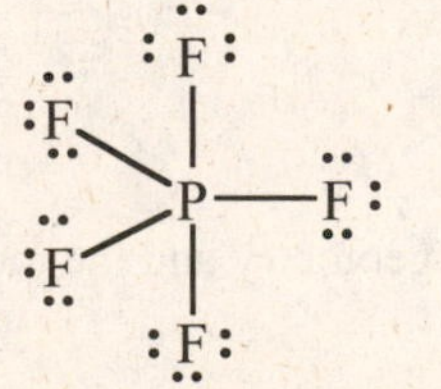

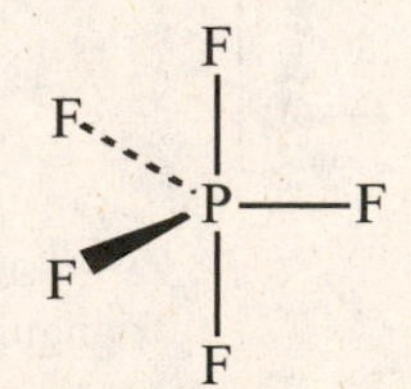

15. *Answer:* **(a) electron-pair geometry and molecular geometry are tetrahedral (b) electron-pair geometry is tetrahedral; molecular geometry is angular (109.5°) (c) electron-pair geometry and molecular geometry are both triangular planar (d) electron-pair geometry and molecular geometry are both triangular planar**

Strategy and Explanation: Adapt the method given in the solution to Question 13.

(a) PH_4^+ (8 e^-) The type is AX_4E_0, so electron-pair geometry and the molecular geometry are both tetrahedral.

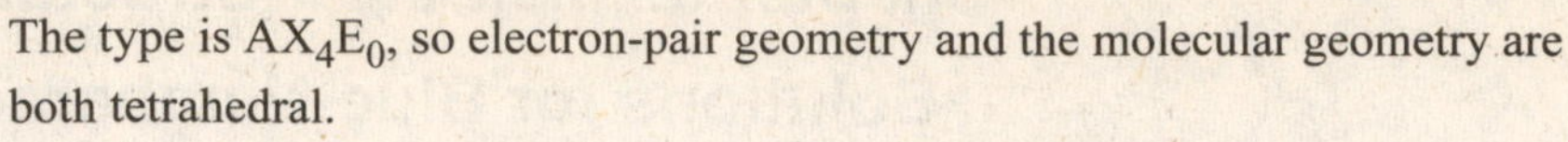

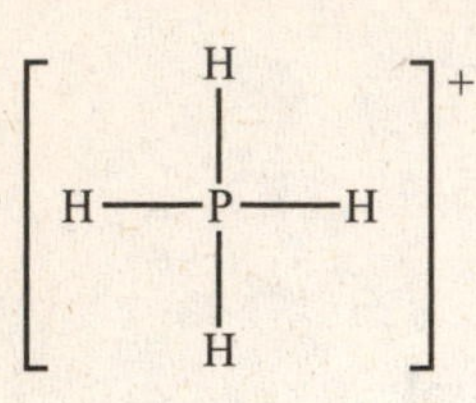

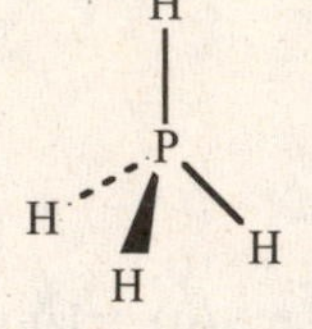

(b) OCl_2 (20 e^-) The type is AX_2E_2, so the electron-pair geometry is tetrahedral and the molecular geometry is angular (109.5°).

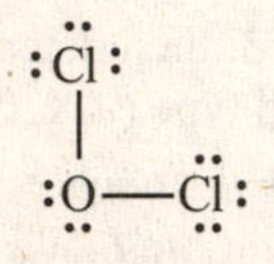

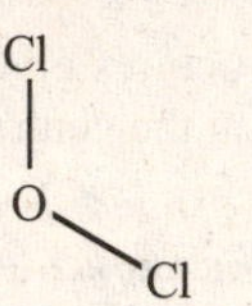

(c) SO_3 (24 e^-) The type is AX_3E_0, so electron-pair geometry and the molecular geometry are both triangular planar.

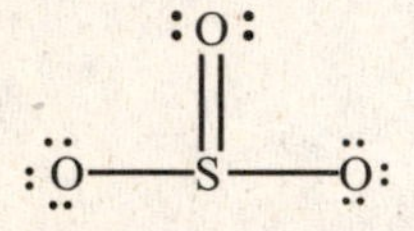

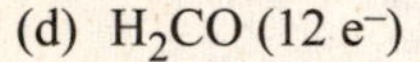

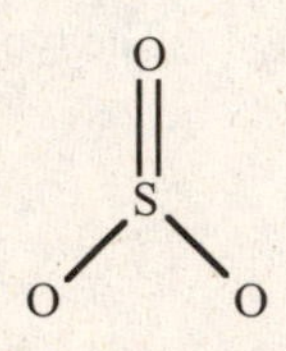

(d) H_2CO (12 e^-) The type is AX_3E_0, so electron-pair geometry and the molecular geometry are both triangular planar.

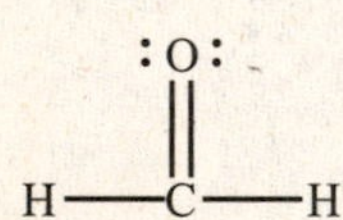

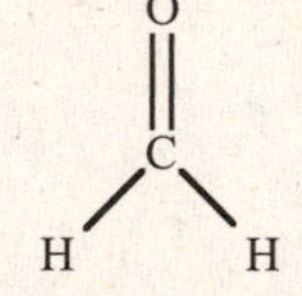

17. *Answer:* **See structures below; (a) electron-pair geometry and molecular geometry are both triangular planar (b) electron-pair geometry and molecular geometry are both triangular planar (c) electron-pair geometry and molecular geometry are both triangular planar (d) electron-pair geometry is tetrahedral; molecular geometry is triangular pyramidal**

Strategy and Explanation: Adapt the method given in the solution to Question 13.

(a) BO_3^{3-} (24 e^-) The type is AX_3E_0, so both the electron-pair geometry and the molecular geometry are triangular planar.

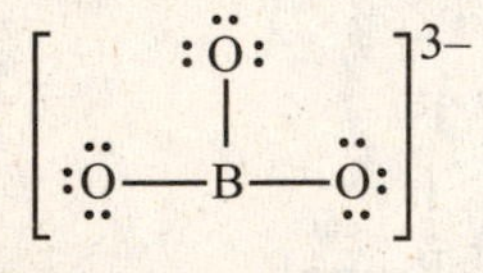

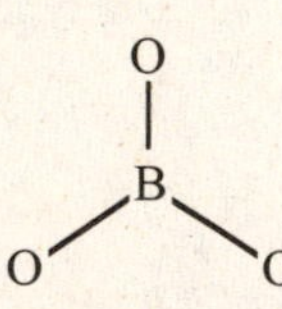

(b) CO_3^{2-} (24 e^-) AX_3E_0, so both the electron-pair geometry and the molecular geometry are triangular planar.

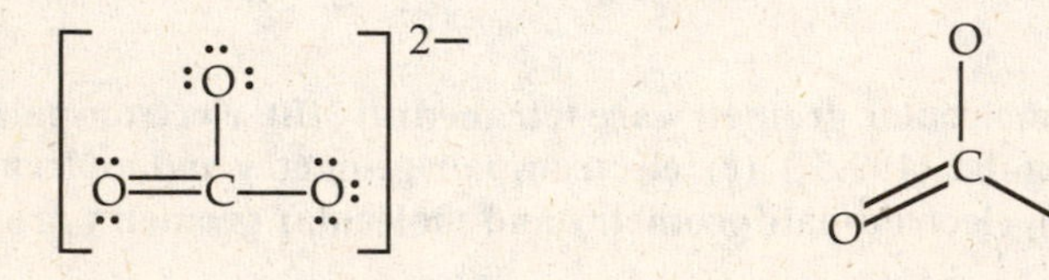

(c) SO_3^{2-} (26 e^-) AX_3E_1, so the electron-pair geometry is tetrahedral and the molecular geometry is triangular pyramidal.

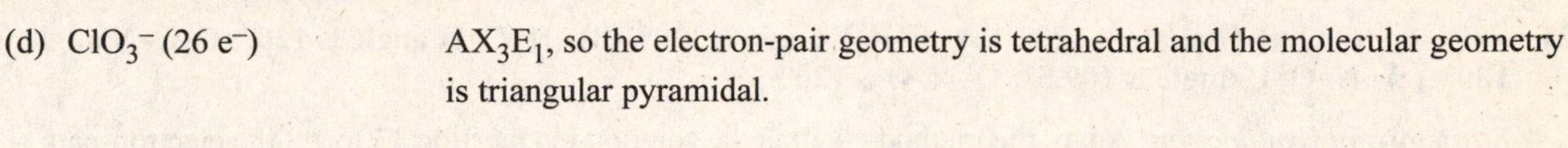

(d) ClO_3^- (26 e^-) AX_3E_1, so the electron-pair geometry is tetrahedral and the molecular geometry is triangular pyramidal.

All of these ions and molecules have one central atom with three O atoms bonded to it. The number and type of bonded atoms is constant. The geometries vary depending on how many lone pairs are on the central atom. The structures with the same number of valance electrons all have the same geometry.

19. *Answer:* **See structures below; (a) electron-pair geometry and the molecular geometry are both octahedral (b) electron-pair geometry is triangular bipyramid; molecular geometry is seesaw (c) electron-pair geometry is triangular bipyramidal; molecular geometry is triangular bipyramidal (d) electron-pair geometry is octahedral; molecular geometry is square pyramidal**

Strategy and Explanation: Adapt the method given in the solution to Question 13.

(a) SiF_6^{2-} (48 e^-) The type is AX_6E_0, so both the electron-pair geometry and the molecular geometry are octahedral.

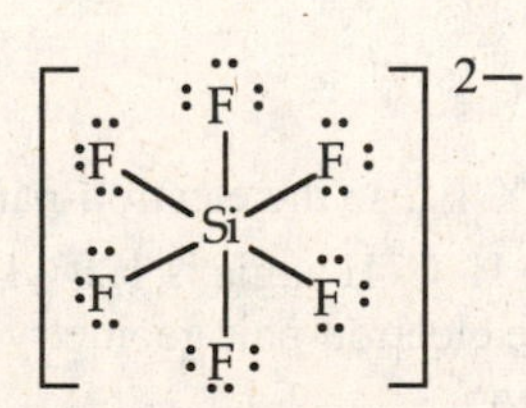

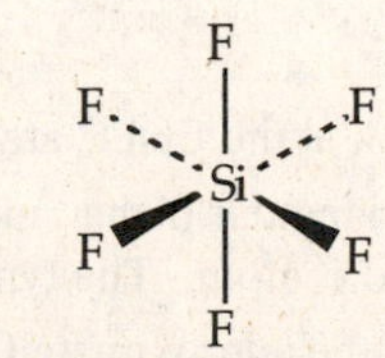

(b) SF_4 (34 e^-) The type is AX_4E_1, so the electron-pair geometry is triangular bipyramid and the molecular geometry is seesaw.

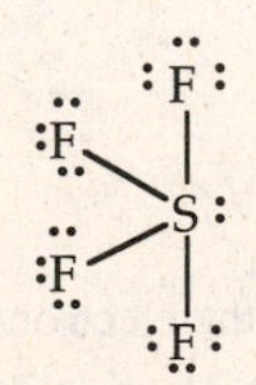

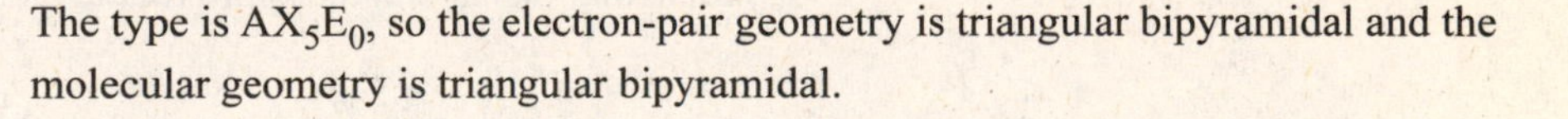

(c) PF_5 (40 e^-) The type is AX_5E_0, so the electron-pair geometry is triangular bipyramidal and the molecular geometry is triangular bipyramidal.

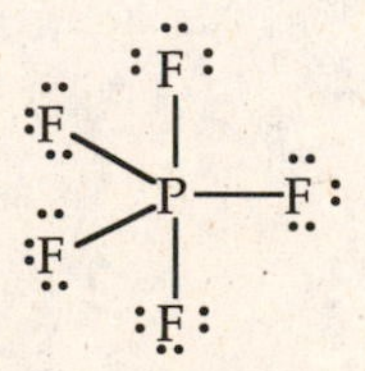

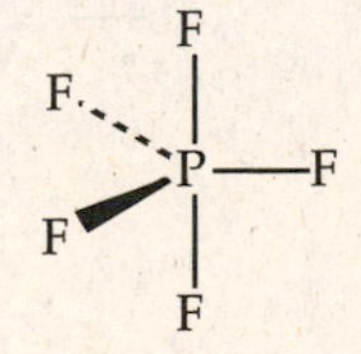

(d) XeF_4 (36 e^-) The type is AX_4E_2, so the electron-pair geometry is octahedral and the molecular geometry is square pyramidal.

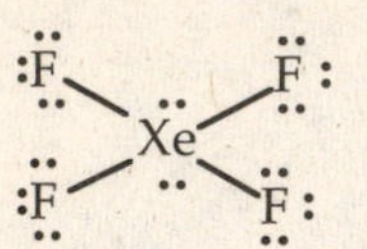

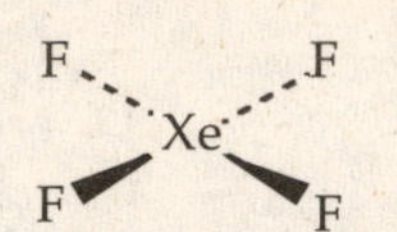

21. *Answer:* **(a) O–S–O angle is 120° (b) F–B–F angle is 120° (d) H–C–H angle is 120°; C–C–N angle is 180° (c) N–O–H angle is 109.5°; O–N–O is 120°**

Strategy and Explanation: Adapt the method given in the solution to Question 13 to get the electron-pair geometry of the second atom in the bond. Use this to predict the approximate bond angle.

(a) SO_2 (18 e^-) The type is AX_2E_1, so the electron-pair geometry is triangular planar and the approximate O–S–O angle is 120°.

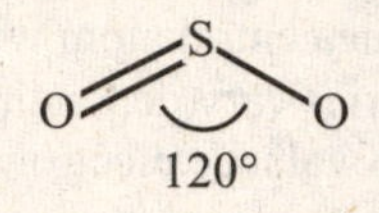

(b) BF_3 (24 e^-) The type is AX_3E_0, so the electron-pair geometry is triangular planar and the approximate F–B–F angle is 120°.

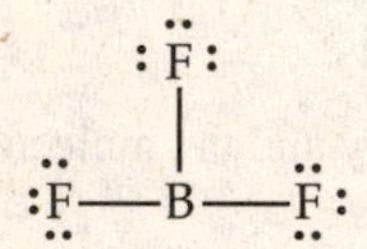

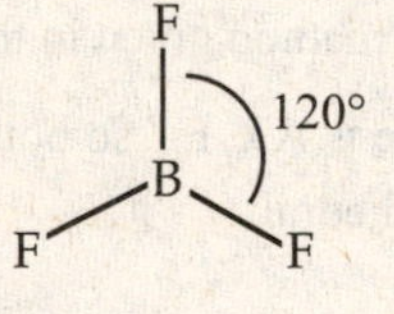

(c) CH_2CHCN (20 e^-) Look at the first C atom: The type is AX_3E_0, so the electron-pair geometry is triangular planar and the approximate H–C–H angle is 120°. Look at the third C atom: The type is AX_2E_0, so the electron-pair geometry is linear and the approximate C–C–N angle is 180°.

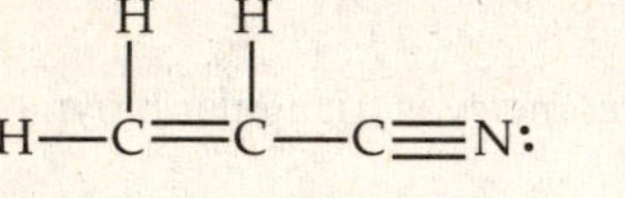

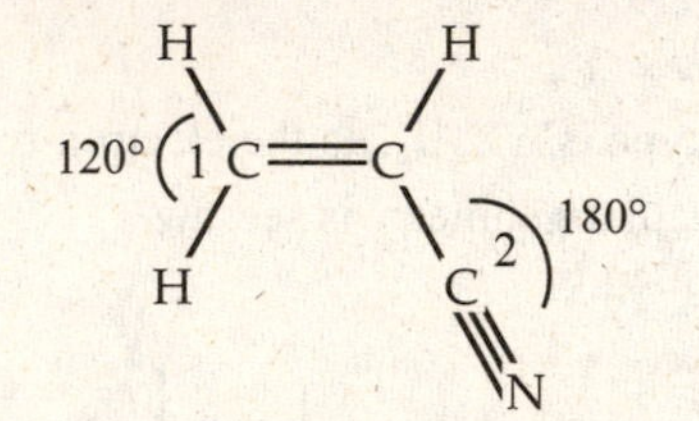

(d) HNO_3 (24 e^-) Look at the first O atom: The type is AX_2E_2, so the electron-pair geometry is tetrahedral and the approximate N–O–H angle is 109.5°. Look at the N atom: The type is AX_3E_0, so the electron-pair geometry is triangular planar and the approximate O–N–O angle is 120°.

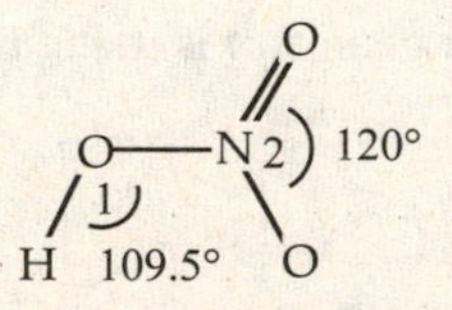

23. *Answer:* **(a) four at 90°, one at 120°, and one at 180° (b) 90° and 120° (c) eight at 90° and two at 180°**

Strategy and Explanation: Adapt the method given in the solution to Question 21.

(a) SeF_4 (34 e^-)

The type is AX_4E_1, so the electron-pair geometry is triangular bipyramid. The $F_{equitorial}$–Se–$F_{equitorial}$ angle is 120°, the $F_{equitorial}$–Se–F_{axial} angles are 90° and F_{axial}–Se–F_{axial} angle is 180°.

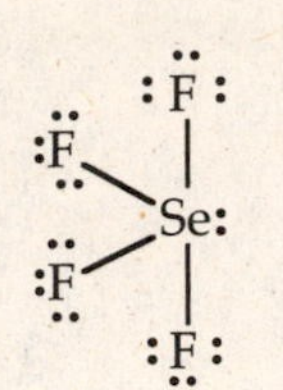

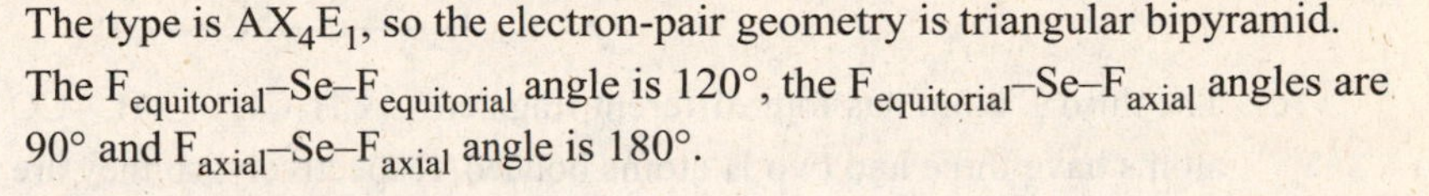

(b) SOF_4 (40 e^-)

The type is AX_5E_0, so the electron-pair geometry is triangular bipyramidal, equatorial-F–S–O angles are 120° and the axial-F–S–O angles are 90°.

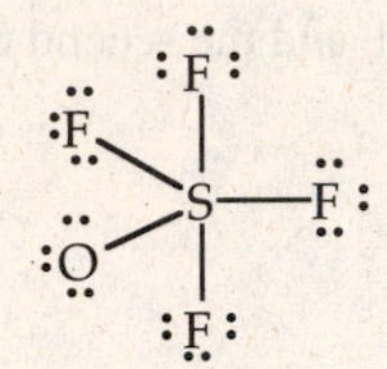

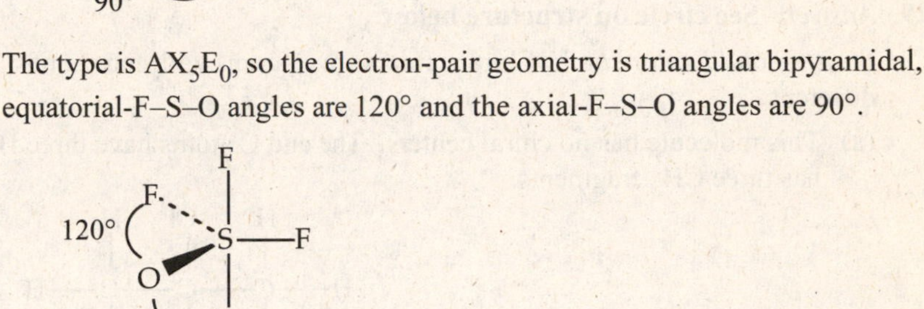

(c) BrF_5 (42 e^-)

The type is AX_5E_1, so the electron-pair geometry is octahedral and all the F–Br–F angles are 90°. The angles across the bottom of the structure will be 180°.

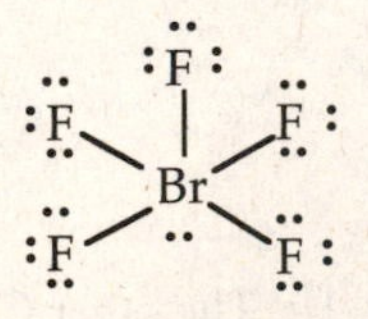

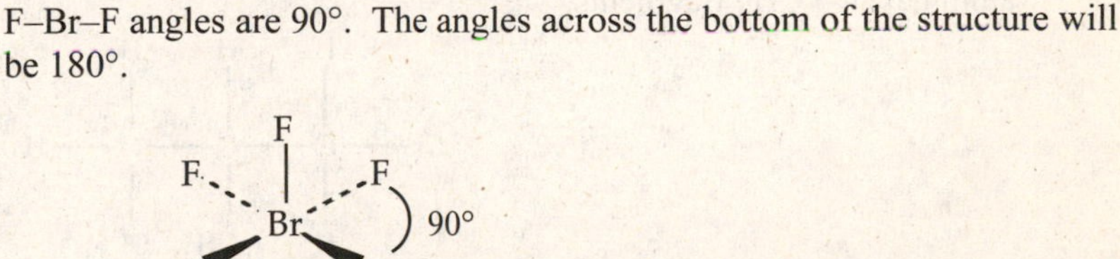

25. *Answer:* $\mathbf{NO_2^+}$

Strategy and Explanation: NO_2 molecule and NO_2^+ ion differ by only one electron. NO_2 has 17 electrons and NO_2^+ has 16 electrons. Their Lewis structures are quite similar:

O=N–O [O=N=O]$^+$

According to VESPR, NO_2 is triangular planar (AX_2E_1) and NO_2^+ is linear (AX_2). The predict O–N–O angle in NO_2 is approximately 120°. The predict O–N–O angle in NO_2^+ ion is 180°. So, $\mathbf{NO_2^+}$ has a greater O–N–O angle.

Chirality in Organic Compounds (Section 9.2)

27. *Answer:* **See circles on structures below**

Strategy and Explanation: Chiral centers are C atoms with four bonds, where each fragment bonded is different.

(a) The second and third C atoms have four different fragments: HOOC–, HO–, –H, and –CH(OH)COOH.

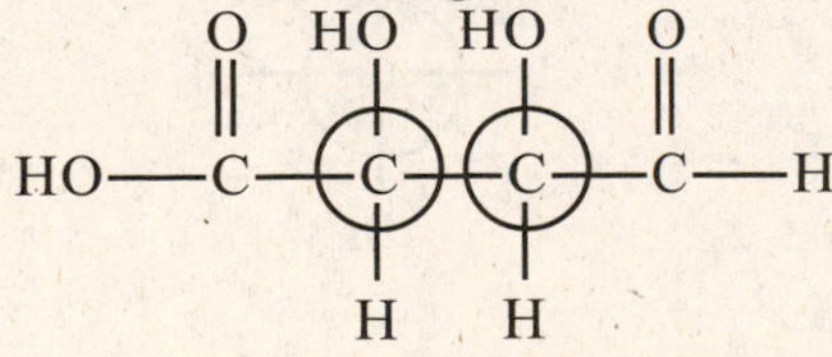

(b) This molecule has no chiral centers. The two inner C atoms do not have four bonds. The first C atom has three H atoms bonded.

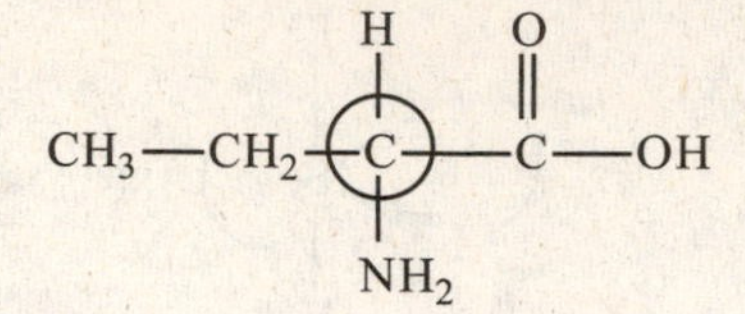

(c) The third C atom has four different fragments: CH_3CH_2–, –H, –COOH, and –NH_2. The first and second C atoms have three and two H atoms bonded, respectively, so they are not chiral centers.

$$CH_3—CH_2—\overset{H}{\underset{NH_2}{C}}—\overset{O}{\overset{||}{C}}—OH$$

29. *Answer:* **See circle on structure below**

Strategy and Explanation: Chiral centers are C atoms with four bonds, where each fragment bonded is different.

(a) This molecule has no chiral centers. The end C atoms have three H atoms bonded, and the second C atom has three CH_3 fragments.

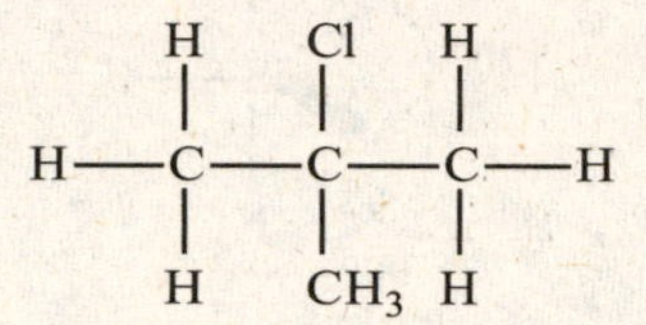

(b) This molecule has no chiral centers. The end C atoms have two or three H atoms bonded, and the second C atom has two CH_3 fragments.

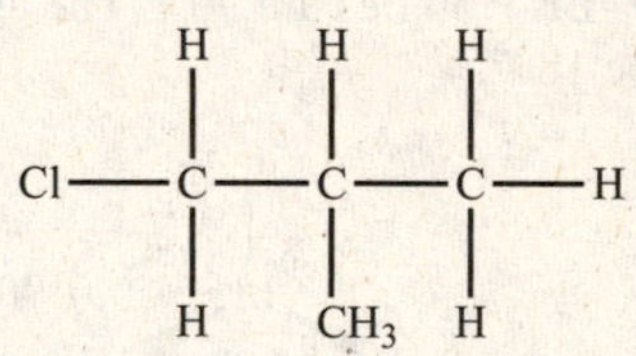

(c) The second C atom has four different fragments: CH_3–, Br–, –$CH_2CH_2CH_3$, and –H. The other C atoms have more than one H atom bonded, so they are not chiral centers.

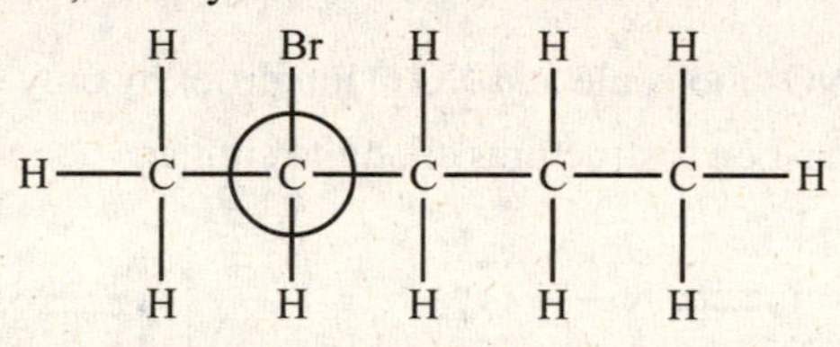

(d) This molecule has no chiral centers. Most of the C atoms have two or three H atoms, and the second C atom has two –CH_2CH_3 fragments.

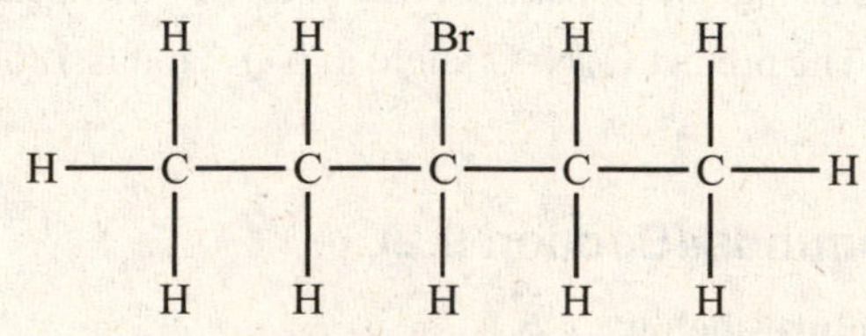

31. *Answer/Explanation:* A compound can exist as a pair of enantiomers if it has one or more chiral centers, so look at the structural formula and find an atom with four different fragments bonded to it. For example:

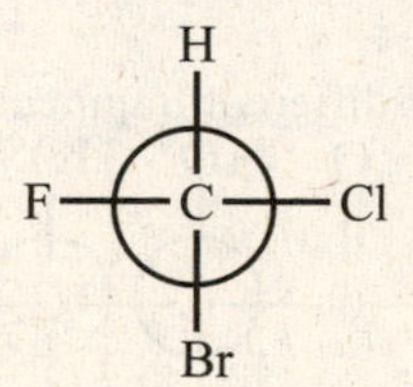

The C in this formula (circled) has four different fragments: F–, H–, –Cl, and –Br. This chiral center allows for the creation of two nonsuperimposeable mirror images of the same molecule, making them enantiomers of each other:

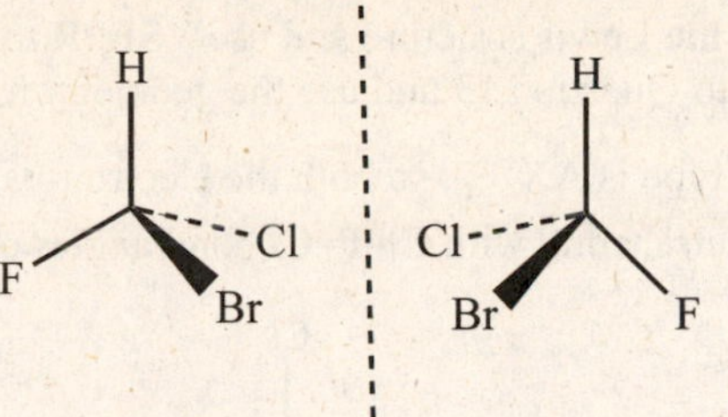

Hybridization (Section 9.3)

33. *Answer:* **(a)** ***sp³*** **(a)** ***sp³d²*** **(a)** ***sp²***

Strategy and Explanation:

(a) One s and three p orbitals combine to make sp^3-hybrid orbitals.

(b) One s, three p, and two d orbitals combine to make sp^3d^2-hybrid orbitals.

(c) One s and two p orbitals combine to make sp^2-hybrid orbitals.

35. *Answer:* **tetrahedral, *sp³*-hybridization**

Strategy and Explanation: Write a Lewis structure for $HCCl_3$ and use VSEPR to determine the molecular geometry as described in the solution to Question 13. Use Table 9.2 to determine the hybridization of the central atom using the electron-pair geometry. The molecule is AX_4E_0 type, so its electron-pair geometry and its molecular geometry are tetrahedral.

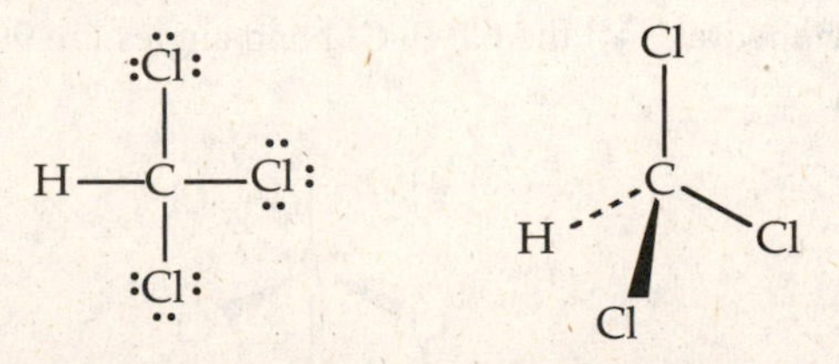

To make four equal bonds with an electron-pair geometry of tetrahedral, the C atom must be sp^3-hybridized, according to Table 9.2. The H atom and Cl atoms are not hybridized.

37. *Answer:* **The central S atom in SCl_2 is *sp³*-hybridized; the central C atom in OCS is *sp*-hybridized. The bonding is polar covalent in both molecules.**

Strategy and Explanation: Follow the procedure described in the solution to Question 35.

SCl_2 (20 e⁻) The type is AX_2E_2, so the electron-pair geometry is tetrahedral and the molecular geometry is bent (109.5°).

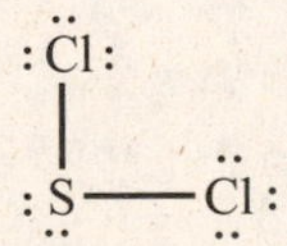

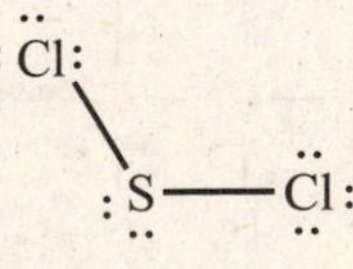

The S atom must be sp^3-hybridized, according to Table 9.2.

OCS (16 e⁻) The C atom must be the central atom. The type is AX_2E_0, so the electron-pair geometry is linear and the O–C–S angle is 180°.

:Ö=C=S̈:

The C atom must be *sp*-hybridized, according to Table 9.2.

The bonding in these molecules is polar covalent.

38. *Answer:* **PCl_4^+ is tetrahedral with 109.5° bond angles. PCl_5 is triangular bipyramidal with 120° and 90° bond angles. PCl_6^- is octahedral with 90° bond angles.**

Strategy and Explanation: Draw the Lewis structure and use VSEPR to determine the geometry of the central atom as described in the solution to Question 13 and use the geometry to predict the bond angle.

(a) PCl_4^+ (32 e^-) The type is AX_4E_0, so both the electron-pair geometry and the molecular geometry are tetrahedral with Cl–P–Cl bond angles of 109.5°.

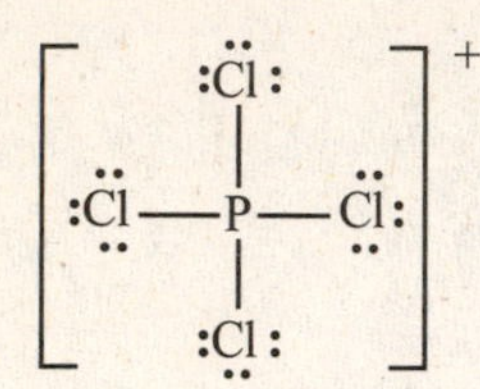

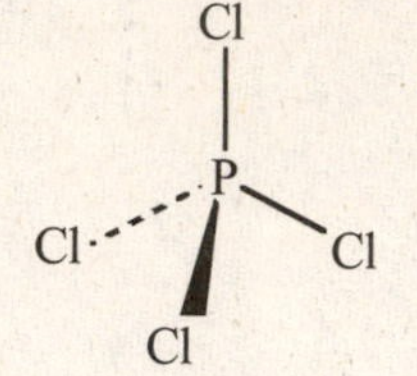

(b) PCl_5 (40 e^-) The type is AX_5E_0, so the electron-pair geometry and the molecular geometry are triangular bipyramidal. The equatorial-equatorial Cl–P–Cl bond angles are 120° and the axial-equatorial Cl–P–Cl bond angles are 90°.

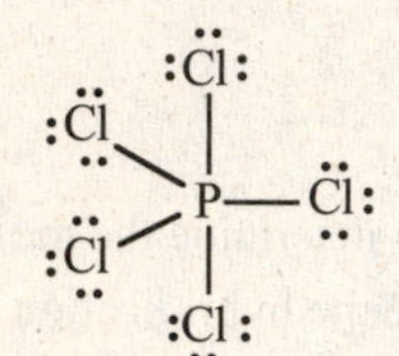

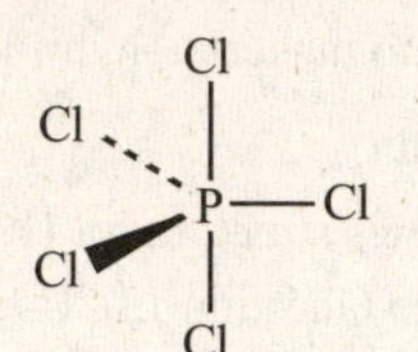

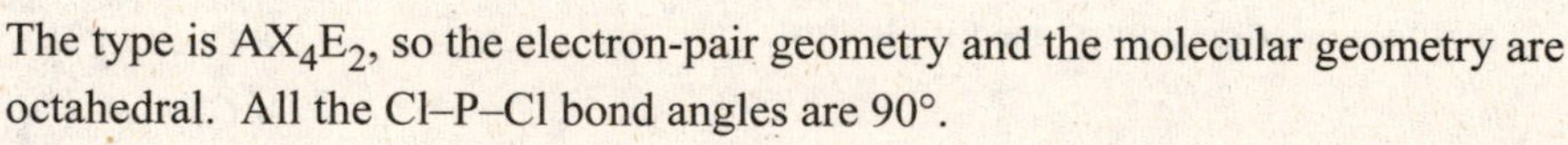

(c) PCl_6^- (48 e^-) The type is AX_4E_2, so the electron-pair geometry and the molecular geometry are octahedral. All the Cl–P–Cl bond angles are 90°.

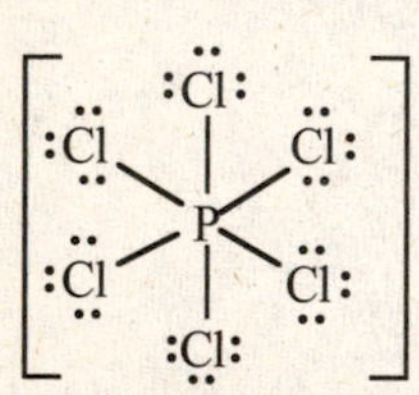

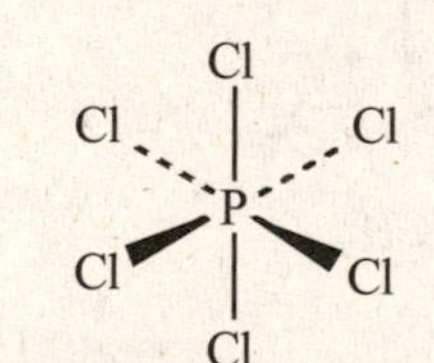

42. *Answer:* **N: *sp*^{3}, 109.5°; first two C's (from the left): *sp*^{3}, 109.5°; third C: *sp*^{2}, 120°; top O: *sp*^{2}, 120°; right O: *sp*^{3}, 109.5°**

Strategy and Explanation: Use the Lewis structure of alanine and the VSEPR model to determine the electron-pair geometry of each of the atoms. Use Table 9.2 to determine the hybridization of the central atom using the electron-pair geometry. The bond angles are associated with the electron-pair geometry also, so use Figure 9.4. For reference, the atoms present in multiple quantities are numbered here left to right: C_1, C_2, C_3, O_1, and O_2.

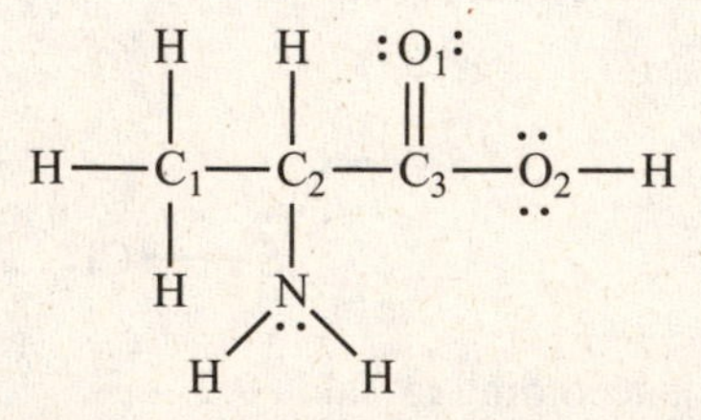

- Look at C_1: The type is AX_4E_0, so the electron-pair geometry is tetrahedral and this C atom's hybridization is *sp*^{3}. The *sp*^{3}-hybridized C atom has tetrahedral bond angles of approximately 109.5°.
- Look at C_2: The type is AX_4E_0, so the electron-pair geometry is tetrahedral and this C atom's hybridization is *sp*^{3}. The *sp*^{3}-hybridized C atom has tetrahedral bond angles of approximately 109.5°.
- Look at the N atom: The type is AX_3E_1, so the electron-pair geometry is tetrahedral and its hybridization is

- Look at C_3: The type is AX_3E_0, so the electron-pair geometry is triangular planar and this C atom's hybridization is sp^2. The sp^2-hybridized C atom has triangular planar bond angles of approximately 120°.
- Look at the N atom: The type is AX_3E_1, so the electron-pair geometry is tetrahedral and its hybridization is sp^3. The sp^3-hybridized N atom has tetrahedral bond angles of approximately 109.5°.
- Look at O_1: The type is AX_1E_2, so the electron-pair geometry is triangular planar and this C atom's hybridization is sp^2. The sp^2-hybridized C atom has triangular planar bond angles of approximately 120°.
- Look at O_2: The type is AX_2E_2, so the electron-pair geometry is tetrahedral and this O atom's hybridization is sp^3. The sp^3-hybridized O atom has tetrahedral bond angles of approximately 109.5°.

44. *Answer:* **(a) first two C's (from the left): *sp*^{3}, 109.5°; third and fourth C's (with triple bond): *sp*, 180° (b) shortest: C≡C (c) strongest: C≡C**

Strategy and Explanation:

(a) Write the Lewis structure. Use the VSEPR model to determine the electron-pair geometry of each of the atoms. Use Table 9.2 to determine the hybridization of the central atom using the electron-pair geometry. The bond angles are associated with the electron-pair geometry also, so use Figure 9.4.

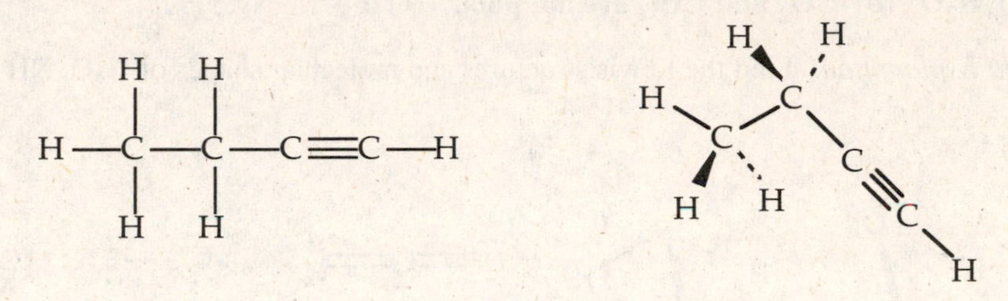

Look at the first two C atoms (the two farthest to the left): They are both of type AX_4E_0, so the electron-pair geometry is tetrahedral and these C atoms have sp^3-hybridization. The sp^3-hybridized C atoms have tetrahedral bond angles of approximately 109.5°.

Look at the second two C atoms (the two farthest to the right): They are both of type AX_2E_0, so the electron-pair geometry is linear and these C atoms have *sp*-hybridization. The *sp*-hybridized C atoms have linear bond angles of approximately 180°.

(b) The shortest carbon-carbon bond is the triple bond. That can be confirmed by looking up the bond lengths in Table 8.1.

(c) The strongest carbon-carbon bond is the triple bond. That can be confirmed by looking up the bond enthalpies in Table 8.2.

46. *Answer:* **see structures with σ and π bonds designated below**

Strategy and Explanation: The first bond between two atoms must always be a σ bond. When more than one pair of electrons are shared between atoms, they are always part of π bonds. So the second bond in a double bond and the second and third bonds in a triple bond are always π bonds.

(a) (b) (c)

:O=C=S: (π and σ in each double bond)

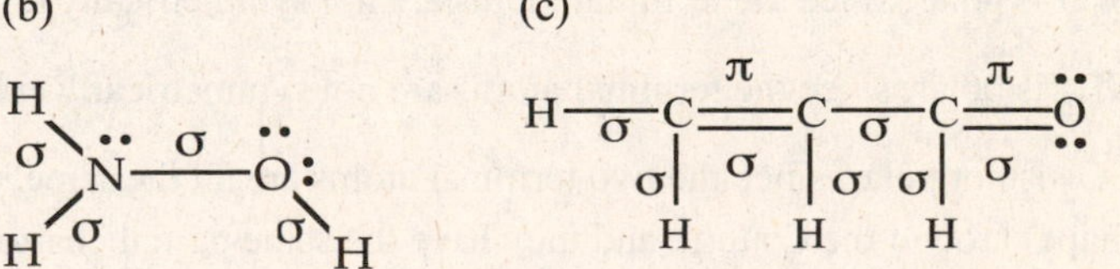

48. *Answer:* **see structure below for location of sigma and pi bonds; (a) 6 sigma bonds (b) 3 pi bonds (c) *sp*-hybridization (d) *sp*-hybridization (e) both have *sp*^{2}-hybridization**

Strategy and Explanation: Use the method described in the solution to Question 46.

(a) As seen above, the structure has six sigma bonds.

(b) As seen above, the structure has three pi bonds.

(c) Look at the C atom bonded to the N atom: It is of type AX_2E_0, so the electron-pair geometry is linear and these C atoms have *sp*-hybridization.

(d) Look at the N atom: It is of type AX_1E_1, so the electron-pair geometry is linear and it has *sp*-hybridization.

(e) Both H-bearing C atoms are type AX_3E_0, so the electron-pair geometry is triangular planar and both of them have a hybridization of sp^2.

Molecular Polarity (Section 9.5)

50. *Answer:* **(a) H_2O (b) CO_2 and CCl_4 are not polar (c) F**

Strategy and Explanation: Find the Lewis structures and molecular shapes of H_2O, NH_3, CS_2, ClF, and CCl_4.

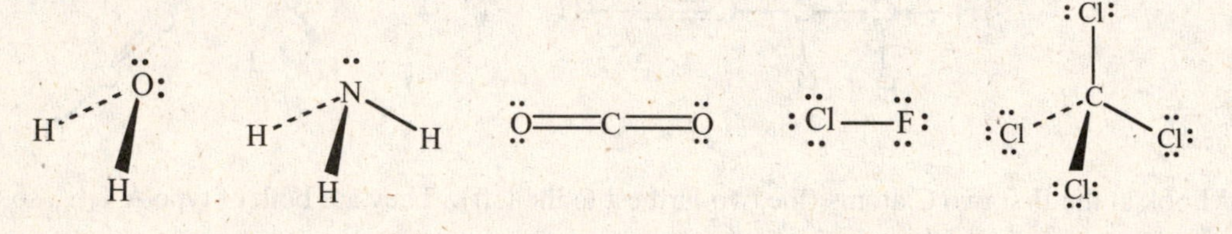

(a) The bond polarity is related to the difference in electronegativity. Use Figure 8.6 to get electronegativity (EN) values:

$$EN_O = 3.5,\ EN_H = 2.1,\ EN_N = 3.0,\ EN_C = 2.5,\ EN_{Cl} = 3.0$$

$$\Delta EN_{O\text{–}H} = EN_O - EN_H = 3.5 - 2.1 = 1.4 \qquad \text{O–H is most polar.}$$

$$\Delta EN_{N\text{–}H} = EN_N - EN_H = 3.0 - 2.1 = 0.9$$

$$\Delta EN_{C\text{–}O} = EN_O - EN_C = 3.5 - 2.5 = 1.0$$

$$\Delta EN_{Cl\text{–}F} = EN_F - EN_{Cl} = 3.5 - 3.0 = 0.5$$

$$\Delta EN_{C\text{–}Cl} = EN_{Cl} - EN_C = 3.0 - 2.5 = 0.5$$

The most polar bonds are in H_2O.

(b) Use the description given in Section 9.5 to determine if the molecule is polar:

H_2O is polar, since the terminal atoms are not symmetrically arranged around the O atom.

NH_3 is polar, since the terminal atoms are not symmetrically arranged around the N atom.

CO_2 is not polar, since the two terminal atoms are all the same, they are symmetrically arranged (in a l shape) around the C atom, and they have the same partial charge.

C —— O
δ^+ δ^-

CCl_4 is not polar, since the four terminal atoms are all the same, they are symmetrically arranged (in a tetrahedral shape) around the C atom, and they all have the same partial charge.

C —— Cl
δ^+ δ^-

So, CO_2 and CCl_4 are the molecules in the list that are not polar.

(c) The more negatively charged atom of a pair of bonded atoms is the atom with the largest electronegativity. In ClF, the F atom is more negatively charged.

52. *Answer:* **Molecules (b) and (c) are polar; HBF_2 has the F atoms on partial negative end and the H atom on the partial positive end; CH_3Cl has the Cl atom on partial negative end and the H atoms on the partial positive end. (see dipole moment diagrams below)**

Strategy and Explanation: We need the Lewis structures and molecular shapes of the molecules:

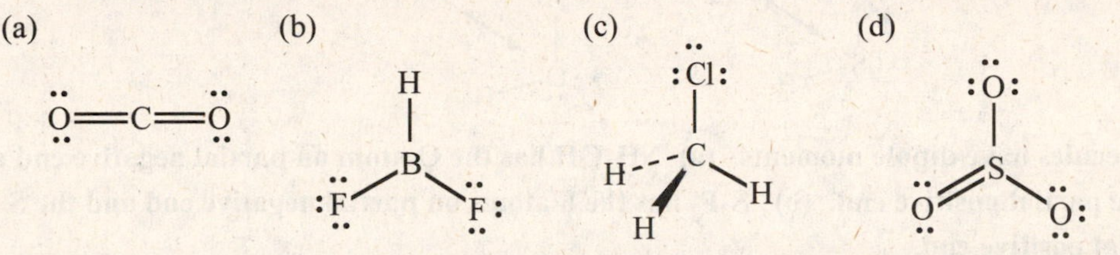

The molecules that are polar have asymmetrical atom arrangements; here, they are (b) HBF_2 and (c) CH_3Cl. The others, (a) CO_2 and (d) SO_3, have symmetrical arrangements of identical atoms with the same partial charge. (Notice: Three equivalent resonance structures can be written for SO_3, making each S–O bond the same length and strength.)

The bond polarities related to the difference in electronegativity. Use Figure 8.6 to get electronegativity (EN) values: $EN_B = 2.0$, $EN_H = 2.1$, $EN_F = 4.0$, $EN_C = 2.5$, $EN_{Cl} = 3.0$

$$\Delta EN_{B\text{–}H} = EN_H - EN_B = 2.1 - 2.0 = 0.1$$

$$\Delta EN_{B\text{–}F} = EN_F - EN_B = 4.0 - 2.0 = 2.0$$

$$\Delta EN_{C\text{–}H} = EN_C - EN_H = 2.5 - 2.1 = 0.4$$

$$\Delta EN_{C\text{–}Cl} = EN_{Cl} - EN_C = 3.0 - 2.5 = 0.5$$

The bond pole arrows' lengths are related to their ΔEN, and points toward the atom with the more negative EN. So, the B–F arrows are much longer than the HB arrow, and a net dipole points toward the F atoms' side of the molecule, making the F atoms' side of the molecule the partial negative end and the H atom's side of the molecule the partial positive end.

The C–Cl arrow points toward Cl, and the C–H points toward the C, so all the arrows point toward the C. (Left right, and back forward cancel due to the symmetry of the triangular orientation of the H atoms). That means a net dipole points toward the Cl atom's side of the molecule, making the Cl atom's side of the molecule the partial negative end and the H atoms' side of the molecule the partial positive end.

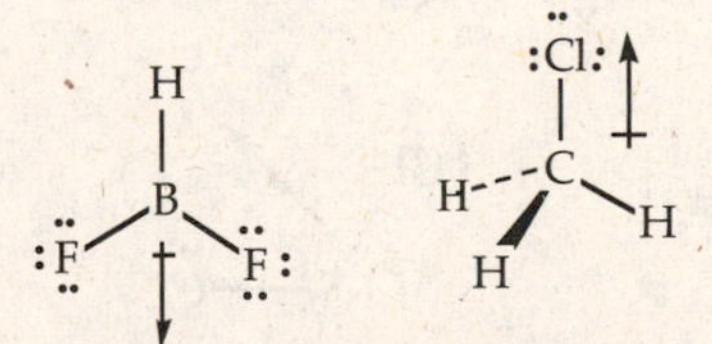

54. *Answer:* **(a) The Br–F bond has a larger electronegativity difference. (b) The H–O bond has a larger electronegativity difference.**

Strategy and Explanation: The dipole moment relates both to the strength of the bond poles and their directionality.

The bond polarities are related to the difference in electronegativity. Use Figure 8.6 to get electronegativity (EN) values: $EN_{Br} = 2.8$, $EN_F = 4.0$, $EN_{Cl} = 3.0$

$$\Delta EN_{Br\text{–}F} = EN_F - EN_{Br} = 4.0 - 2.8 = 1.2$$

$$\Delta EN_{Cl\text{–}F} = EN_F - EN_{Cl} = 4.0 - 3.5 = 0.5$$

The bond polarity of the bond in BrF is much greater than that in ClF, which explains the significant dipole moment difference. $EN_H = 2.1$, $EN_O = 3.5$, $EN_S = 2.5$

$$\Delta EN_{O\text{–}H} = EN_O - EN_H = 3.5 - 2.1 = 1.4$$

$$\Delta EN_{S\text{–}H} = EN_S - EN_H = 2.5 - 2.1 = 0.4$$

The bond polarity of the O–H bonds is greater than that of the S–H bonds, which explains the dipole moment difference.

56. *Answer:* **Both molecules have dipole moments. (a) NH_2OH has the O atom on partial negative end and the H atoms on the partial positive end. (b) S_2F_2 has the F atoms on partial negative end and the S atoms on the partial positive end.**

Strategy and Explanation: Follow the method described in the solution to Question 51 - 53.

(a) 14 e^-. The Lewis structure looks like this:

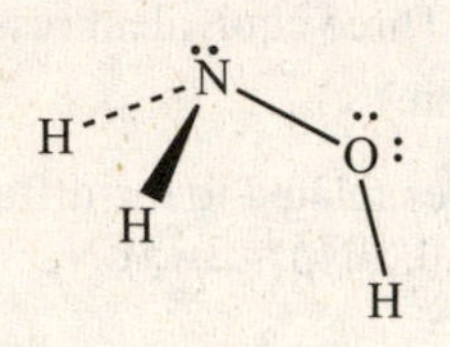

$$\Delta EN_{O\text{–}H} = EN_O - EN_H = 3.5 - 2.1 = 1.4$$

$$\Delta EN_{N\text{–}H} = EN_N - EN_H = 3.0 - 2.1 = 0.9$$

$$\Delta EN_{N\text{–}O} = EN_O - EN_N = 3.5 - 3.0 = 0.5$$

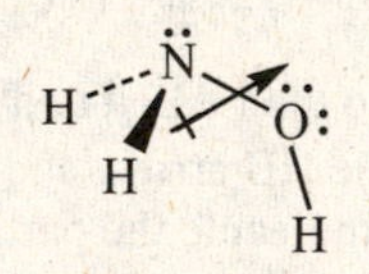

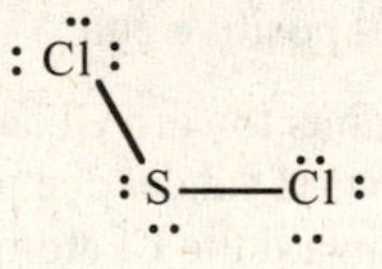

(b) 20 e^-. The Lewis structure looks like this:

$$\Delta EN_{Cl\text{–}S} = EN_{Cl} - EN_S = 3.5 - 2.5 = 1.0$$

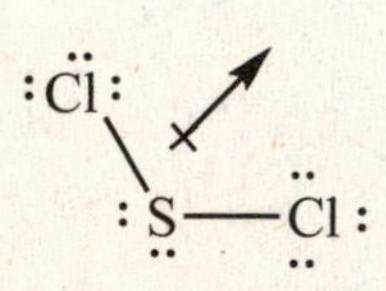

Noncovalent Interactions (Section 9.6)

58. *Answer/Explanation:*

Interaction	Distance	Example
ion-ion	longest range	Na^+ interaction with Cl^-
ion-dipole	long range	Na^+ ions in H_2O
dipole-dipole	medium range	H_2O interaction with H_2O
dipole-induced dipole	short range	H_2O interaction with Br_2
induced dipole-induced dipole	shortest range	Br_2 interaction with Br_2

60. *Answer:* **Wax is non-polar. Water droplets will form so that the polar water molecules can avoid interaction with the wax. Dirty, unwaxed cars have surface soils and salts that interact well with water.**

Strategy and Explanation: Wax is made up of nonpolar molecules that interact almost exclusively by London forces. The water molecules are highly polar and would have to give up hydrogen bonds between other water molecules to interact with wax using only much weaker London forces. These two substances would not interact very well. The water "beads up" in an attempt to have the smallest possible necessary surface interaction with the wax. On a dirty, unwaxed car, the soils and salts that are often found on cars contain ions and polar compounds. They interact much more readily with water, so the water would not "bead up" on the dirty surface.

62. *Answer:* **Molecules (c), (d) and (e) will form hydrogen bonds.**

Strategy and Explanation: Hydrogen bonds form between very electronegative atoms in one molecule to H atoms bonded to a very electronegative atom (EN $\geq$ 3.0) in another molecule. If H atoms are present in a molecule but they are bonded to lower electronegativity atoms, such as C atoms, the molecule cannot use those H atoms for hydrogen bonding.

(a) The H atoms are bonded to C atoms and the highest electronegativity atom in the molecule is Br (EN = 2.8), so this molecule cannot form hydrogen bonds.

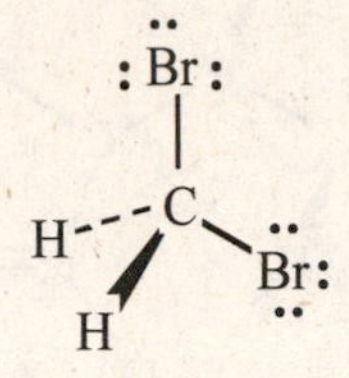

(b) The H atoms are bonded to C atoms, so this molecule cannot form hydrogen bonds with other molecules of the same compound. It does have a high electronegativity O atom (EN = 3.5), so this molecule could interact with other molecules, such as H_2O, which can provide the H atoms for hydrogen bonding to the O atom on this molecule.

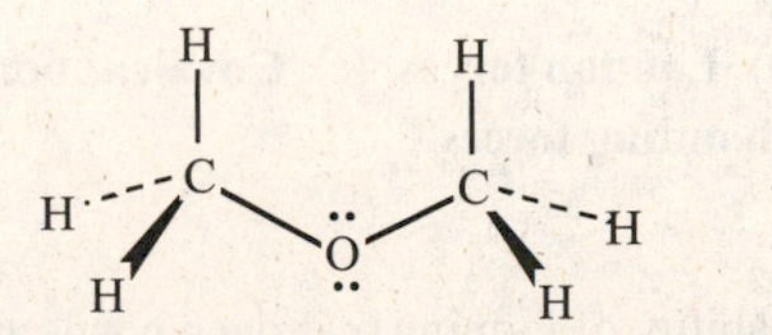

(c) Three H atoms are bonded to high electronegativity atoms (EN_O = 3.5 and EN_N = 3.0). Those H atoms (circled in the structure below) can form hydrogen bonds with the N and O atoms in neighboring molecules.

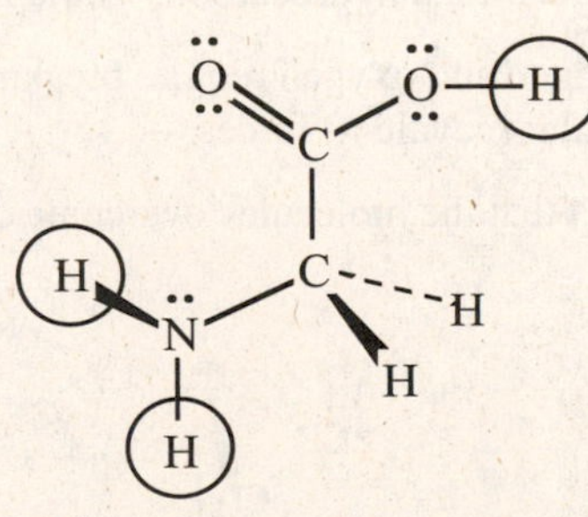

(d) Two H atoms are bonded to high electronegativity O atoms (EN = 3.5). Those H atoms (circled in the structure below) can form hydrogen bonds with the O atoms in neighboring molecules.

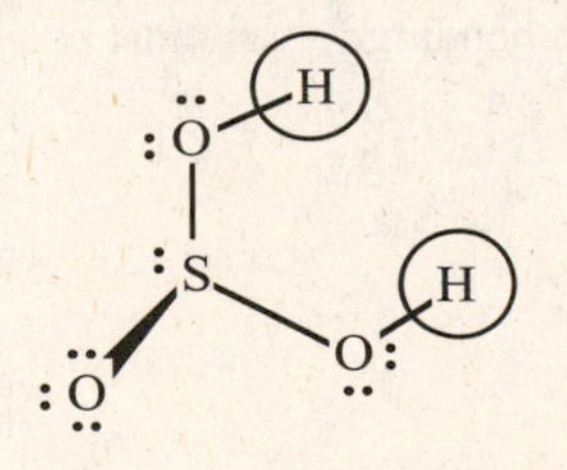

(e) One H atom is bonded to a high electronegativity O atom (EN = 3.5). That H atom (circled in the structure below) can form hydrogen bonds with the O atoms in neighboring molecules.

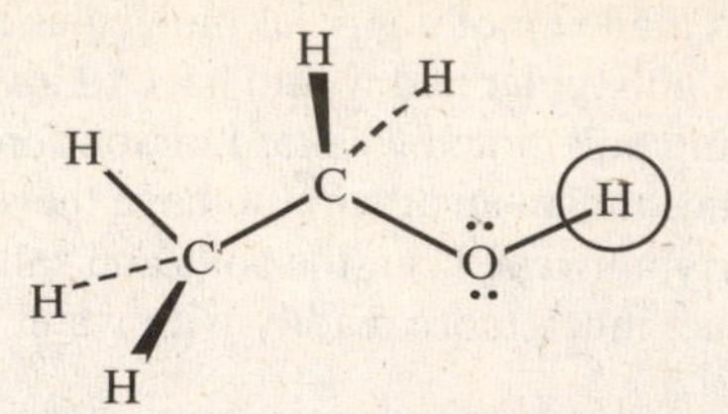

64. *Answer:* **Vitamin C is capable of forming hydrogen bonds with water, thus vitamin is miscible in H_2O.**

Strategy and Explanation: Four H atoms are bonded to high electronegativity O atoms (EN = 3.5). Those H atoms (circled in the structure below) can form hydrogen bonds with H_2O molecules.

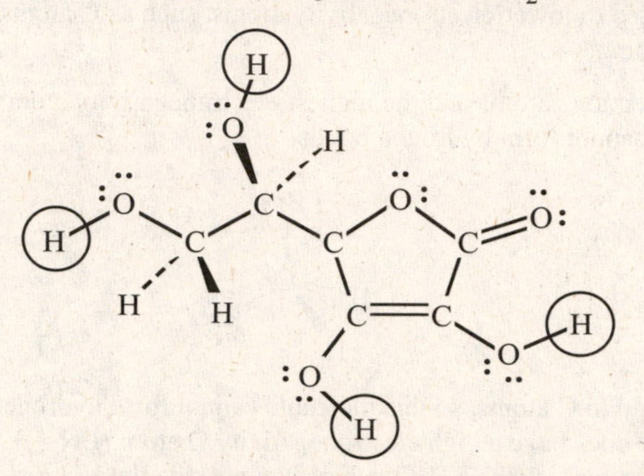

This molecule is quite polar and would not interact well with fats, because fats interact primarily using weaker London forces.

66. *Answer:* **(a) London forces (b) London forces (c) Covalent bonds (d) Dipole-dipole forces (e) Intermolecular hydrogen bonding forces**

Strategy and Explanation:

(a) Hydrocarbons have no capability of forming hydrogen bonds and the bonds are close to non-polar. The molecules interact primarily with London forces. The London forces between the molecules must be overcome to sublime $C_{10}H_8$.

(b) As described in (a), propane molecules, a hydrocarbon, would need to overcome London forces to melt.

(c) Decomposing molecules of nitrogen and oxygen require breaking covalent bonds, so the forces that must be overcome are the intramolecular (covalent) forces.

(d) To evaporate polar PCl_3 requires that the molecules overcome dipole-dipole forces.

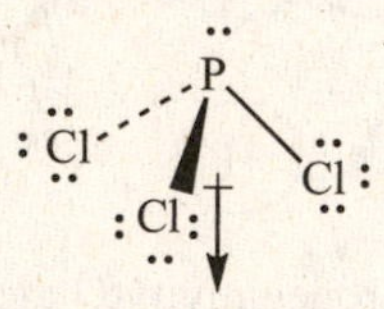

(e) The double strands of DNA are shown in Figure 9.27. The interaction between the two strands are hydrogen bonds. These hydrogen bonding forces must be overcome to "unzip" the DNA double helix.

Molecular Structure Determination by Spectroscopy (Sections 9.2, 9.5)

69. *Answer:* **(a) C–C stretch (b) O–H stretch**

Strategy and Explanation: The "Tools of Chemistry" box in Section 9.2 on pages 386-7 describes infrared spectroscopy. Wavelength, λ, is the variable on the horizontal axis of the spectrum shown. Wavelength is inversely proportional to energy. So, longer wavelength light has lower energy and shorter wavelength light has higher energy.

(a) The lowest energy motion described in the spectrum is C–C stretch.

(b) The highest energy motion described in the spectrum is O–H stretch.

70. *Answer:* **Compared to UV: (a) IR has lower energy, (b) IR has longer wavelength, (c) IR has lower frequency.**

Strategy and Explanation:

(a) IR radiation is lower in energy than UV radiation. (IR is below visible spectrum and UV is above).

(b) IR radiation have longer wavelength than UV radiation. (Lower energy waves have longer wavelength than higher energy waves.).

(c) IR radiation has lower frequency than UV radiation. (Lower energy waves have smaller frequency than higher energy waves).

71. *Answer:* **427 nm and 483 nm; 427 nm is the more energetic transition**

Strategy and Explanation: Two "shoulders" on either side of the main peak can be estimated to be found at 427 nm and 483 nm. Because shorter wavelength indicates higher energy light, the 427 nm peak is the more energetic transition.

Biomolecules (Section 9.7)

73. *Answer/Explanation:* **The structure of a section of DNA with T-C-G**

(SEE NEXT PAGE)

General Questions

75. *Answer:* **(a) Angle 1: 180° Angle 2: 109.5° Angle 3: 109.5° (b) C=O (c) C=O**

Strategy and Explanation:

(a) Follow the method described in the solution to Questions 41 - 43.

Angle 1: Look at the C atom bonded to the N atom: It is of type AX_2E_0, so the electron-pair geometry is linear and this C atom has *sp*-hybridization. The *sp*-hybridized C atoms have linear bond angles of approximately 180°. So, the CCN angle is 180°.

Angles 2 and 3: Look at the C atom bonded to only one O atom: The type is AX_4E_0, so the electron-pair geometry is tetrahedral this C atom has sp^3-hybridization. The sp^3-hybridized C atom has tetrahedral bond angles of approximately 109.5°. So, the HCH angle (Angle 2) and the OCH angle (Angle 3) are both 109.5°.

(b) Follow the method described in the solution to Question 49. The most polar bond is the C=O bond.

(c) Multiple bonds are shorter than single bonds, so the C=O bond is the shortest CO bond:

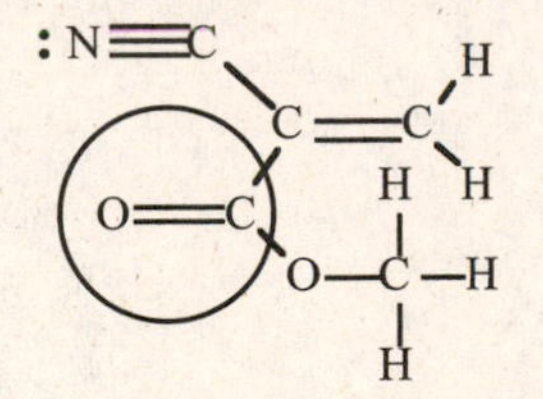

The structure of a section of DNA with T-C-G
(The structure for the solution to Question **73**, page 201)

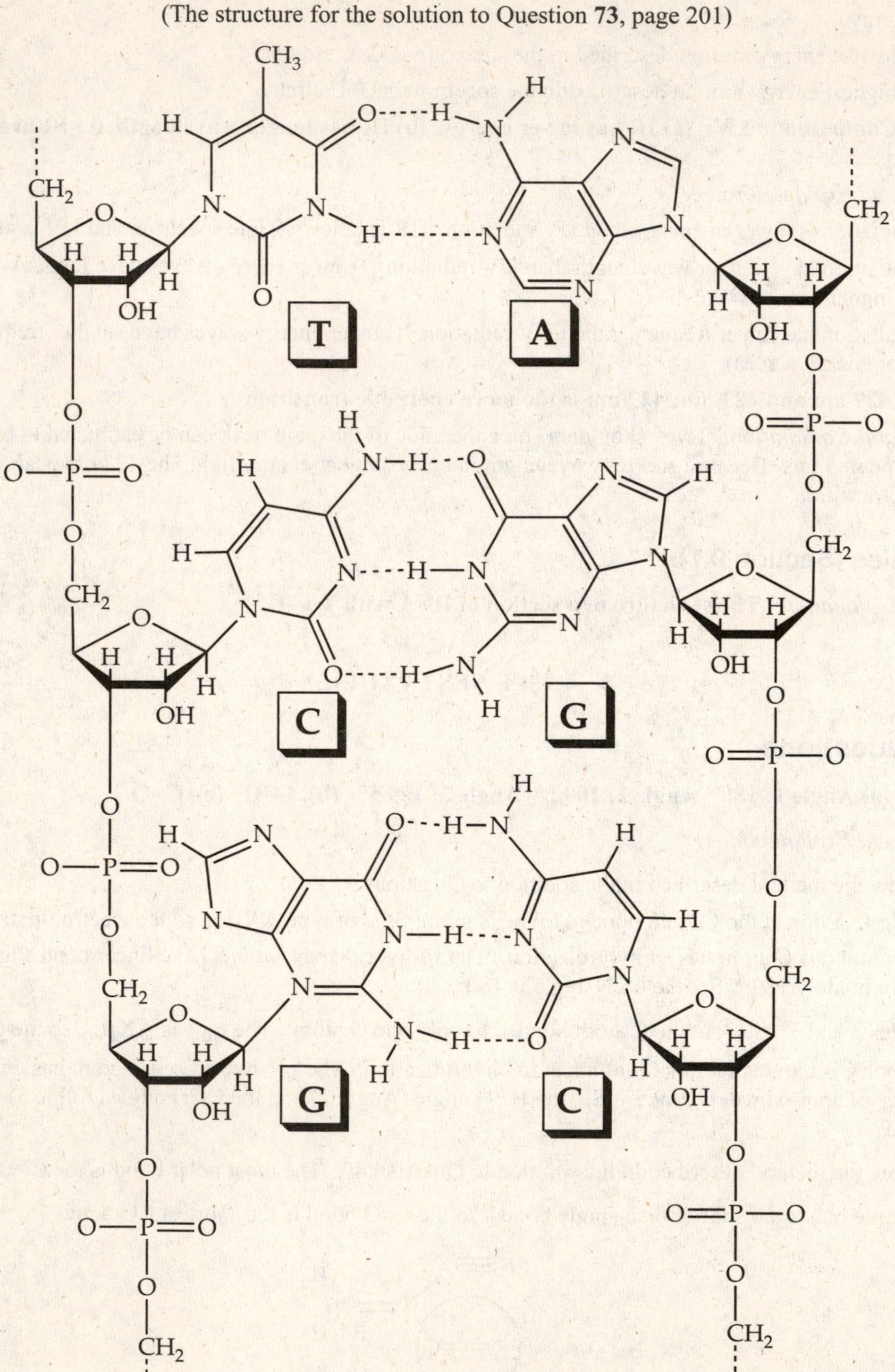

77. *Answer:* **(a) see structure below (b) Angle 1: 180° (c) Angle 2: 180°**

Strategy and Explanation:

(a) :Ö═C═C═C═Ö:

Follow the method described in the solution to Questions 41 - 43.

(b) Angle 1: Look at the C atom bonded to the O atom: It is of type AX_2E_0, so the electron-pair geometry is linear and this C atom has *sp*-hybridization. The *sp*-hybridized C atoms have linear bond angles of approximately 180°. So, the CCO angle is 180°.

(c) Angle 2: Look at the C atom bonded only to the C atoms: It is of type AX_2E_0, so the electron-pair geometry is linear and this C atom has *sp*-hybridization. The *sp*-hybridized C atoms have linear bond angles of approximately 180°. So, the CCO angle is 180°.

79. *Answer:* **It will absorb visible light because it has an extended pi-system.**

Strategy and Explanation: The "Tools of Chemistry" box in Section 9.5 (page 401) describes ultraviolet-visible spectroscopy. Compounds with an extended pi system formed by conjugation between a series of alternating double bonds absorb visible light. This molecule also has an extended pi system, so we would predict it absorbs visible light.

81. *Answer:* **(a) see structure below (b) each C has sp^2-hybridization.**

Strategy and Explanation:

(a) $C_4O_4^{2-}$ (42 e^-) has four equivalent resonance structures. One of these resonance structures looks like this:

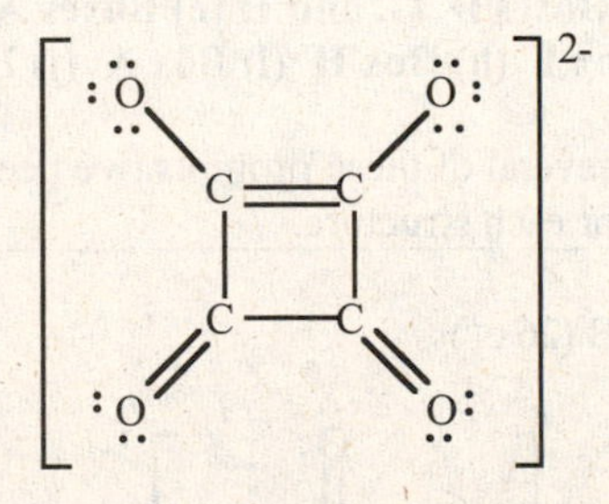

(b) Look at each C atom: The type for each C atom is AX_3E_0, so the electron-pair geometry is triangular planar and each of these C atom has sp^2-hybridization.

83. *Answer:* **(a) For HCl: 2.69×10^{-20} C; For HF: 6.95×10^{-20} C (b) partial negative charge on F is larger than that on Cl, so F has a higher EN.**

Strategy and Explanation: Dipole moment is determined by multiplying the partial charge by the bond length.

$$\mu = (\delta^+) \times (\text{bond length})$$

For HCl: $$\delta^+ = \frac{\mu}{\text{bond length}} = \frac{3.43 \times 10^{-30}\ \text{C}\cdot\text{m}}{127.4\ \text{pm}} \times \frac{1\ \text{pm}}{10^{-12}\ \text{m}} = 2.69 \times 10^{-20}\ \text{C}$$

For HF: $$\delta^+ = \frac{\mu}{\text{bond length}} = \frac{6.37 \times 10^{-30}\ \text{C}\cdot\text{m}}{91.68\ \text{pm}} \times \frac{1\ \text{pm}}{10^{-12}\ \text{m}} = 6.95 \times 10^{-20}\ \text{C}$$

The partial charges in the HF bond are larger than those of the HCl bond, indicating that the fluorine is more electronegative than chlorine.

85. *Answer:* **1.32×10^{-19} C; this partial charge is 81.6% of a full negative charge of an electron, so KF is not really completely ionic.**

Strategy and Explanation: The definition of dipole moment was given in Question 83:

$$\mu = (\delta^+) \times (\text{bond length})$$

For KF: $\delta^+ = \dfrac{\mu}{\text{bond length}} = \dfrac{28.7\times10^{-30}\ \text{C}\cdot\text{m}}{217.2\ \text{pm}} \times \dfrac{1\ \text{pm}}{10^{-12}\ \text{m}} = 1.32\times10^{-19}\ \text{C}$

$$\delta^- = -1.32\times10^{-19}\ \text{C}$$

If the bond were completely ionic, then the partial negative charge on fluorine would be the same as the charge on the electron ($-1.62\text{x}10^{-19}$ C). So, KF is not completely ionic.

Percent of full negative charge $= \dfrac{-1.32\times10^{-19}\ \text{C}}{-1.62\times10^{-19}\ \text{C}} \times 100\,\% = 81.6\,\% = 81.6\%$ of a full negative charge.

87. *Answer:* **Bonds are polar, if the atom's EN's are different. If equal bond poles point in opposite or symmetrically balanced directions such that they cancel each other to give a zero dipole moment for the molecule, the molecule will be nonpolar.**

Strategy and Explanation: A molecule with polar bonds can have a dipole moment of zero if the bond poles are equal and point in opposite directions. CO_2 is a good example of a nonpolar molecule with polar bonds.

$$\ddot{\text{O}}{=}\text{C}{=}\ddot{\text{O}}$$

89. *Answer:* **(a) Box A, B, D, and H (b) Boxes D. G, and H (c) Boxes A and G (d) Boxes E and I (e) Boxes B, C, D, E,F, H, and I (f) Box A (g) Box F (h) Box H (i) Box A (j) Box F and G**

Strategy and Explanation: To answer several of these prompts, we need to determine the Lewis Structures and the electron and molecular geometry for each structure.

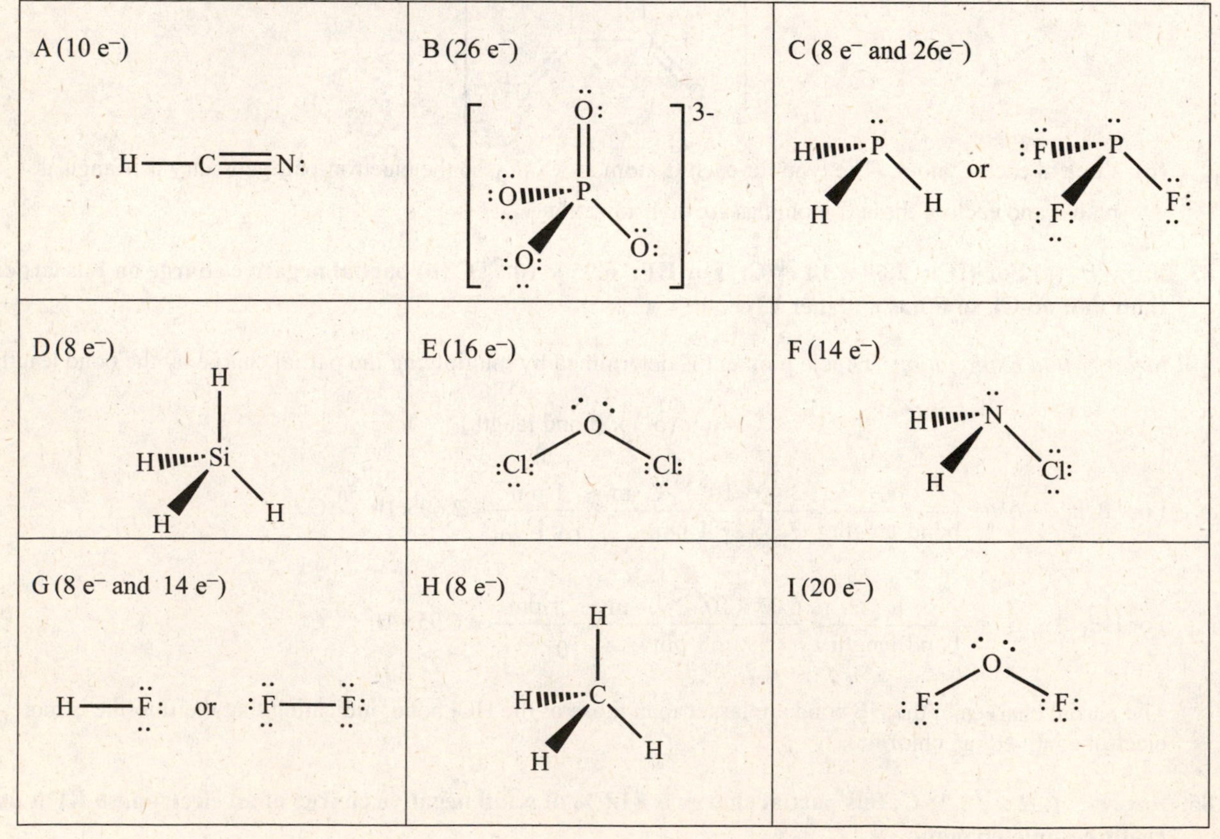

(a) The structures that have electron-pair geometry the same as the molecular geometry are the structures with no lone pairs on the central atom. These are seen in **Box A** with a linear geometry (AX_2E_0) and **Boxes B, D and H** with a tetrahedral geometry (AX_4E_0). In Box G, both of the two structures qualify (with a linear geometry), so it is appropriate to include **Box G** in the answer, as well.

(b) Non-polar molecular structures have the same atoms bonded to the central atom and no lone pairs on the central atom, so that all the bond poles balance and result in a zero net dipole moment. These are represented in **Boxes D and H** with the bonding type of AX_4E_0. One of the two structures in Box G also qualifies (F_2). Since that box indicates that <u>one or the other</u> structure could answer the prompt, it is logical to include **Box G** in the answer, as well.

(c) The structures that have linear geometry have bonding type of AX_2E_0 or are diatomic molecules. These are seen in **Box A** with bonding type of AX_2E_0 and **Box G** with two diatomic molecules.

(d) The structures that have angular (bent) geometry have bonding type of AX_2E_2 or AX_2E_1. These are seen in **Box E and I**, both with bonding type of AX_2E_2.

(e) The central atom is sp^3-hybridized in structures with the bonding type of AX_4E_0, AX_3E_1, or AX_2E_2. **Boxes B, D and H** have the central atom with the AX_4E_0 bonding type. **Boxes C and F** have the central atom with the AX_3E_1 bonding type. **Boxes E and I** have the central atom with the AX_2E_2 bonding type.

(f) The central atom has *sp*-hybridization in structures with the bonding type of AX_2E_0. **Box A** has the central C atom with an AX_2E_0 bonding type.

(g) Take each pair of molecules given in the grid and compare them. The liquid with the lowest boiling point will be composed of molecules with the strongest intermolecular forces. Box F has a molecule that would interact using both dipole-dipole intermolecular forces and hydrogen bonding intermolecular forces. The only other molecule in the grid that can undergo hydrogen bonding forces is HF (one of the choices in Box G). Since HF is smaller and has fewer atom, NH_2Cl would compose the liquid with the lower boiling point when compared to any other molecule in the chart. The answer is **Box F**.

(h) Take each pair of molecules given in the grid and compare them. The liquid with the highest vapor pressure will be composed of molecules with the weakest intermolecular forces. Boxes D and H have molecules that would interact using only London intermolecular forces. The only other nonpolar molecule in the box that is non-polar is F_2 (one of the choices in Box G). Since CH_4 has fewer electrons than SiH_4 or F_2 it probably has weaker London forces, and so would compose the liquid with the higher vapor pressure when compared to any other molecule in the chart. The answer is **Box H**.

(i) Take each pair of molecules given in the grid and compare them. To determine relative dipole moments, start with the change in electronegativity ΔEN to estimate relative bond polarity. Boxes A and F contain the widest range of electronegativities. $EN_F = 4.0$, $EN_N = 3.0$, $EN_{Cl} = 2.7$, $EN_C = 2.4$, $EN_H = 2.1$. While the N-H bonds are more polar (in Box F), the pair together partially cancel each other out, pulling in different directions.

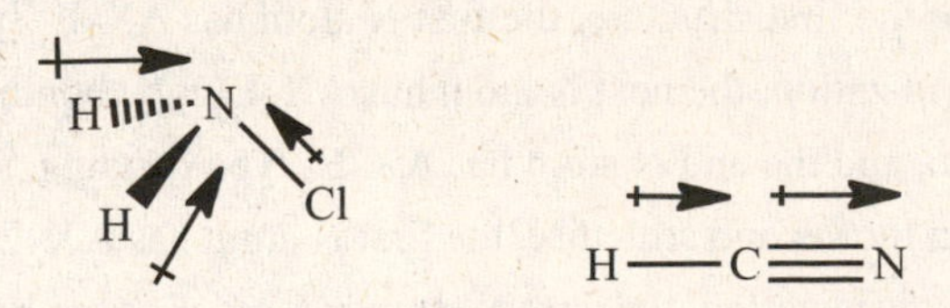

There is one other highly polar atom in Box G: H-F has the largest ΔEN, and would therefore be predicted to have the higher dipole moment.

H—F

So, technically the most polar molecule in the grid is found in Box G. However, there is a non-polar molecule also in Box G that has a zero dipole moment, F_2, and the molecules in both Box A and F have greater dipole moments than a molecule in Box G. Therefore, aside from HF, the most polar molecule in the grid is HCN in **Box A**.

(j) Several of the structures are polar, which is the prerequisite for dipole-dipole intermolecular forces; however only two molecules, HF (one of the two molecules in Box G) and NH_2Cl (in Box F) have an H atom bonded to a high-electronegativity atom; therefore, both **Boxes F and G** qualify to answer this prompt.

Applying Concepts

91. *Answer:* **(a) see structure below, 120° (b) 180° (c) sp^2, sp^3**

Strategy and Explanation:

(a) Determine Lewis structure knowing that C-N-N-N is part of the structure (based on angles requested in this and later parts). The structure has 40 e^-.

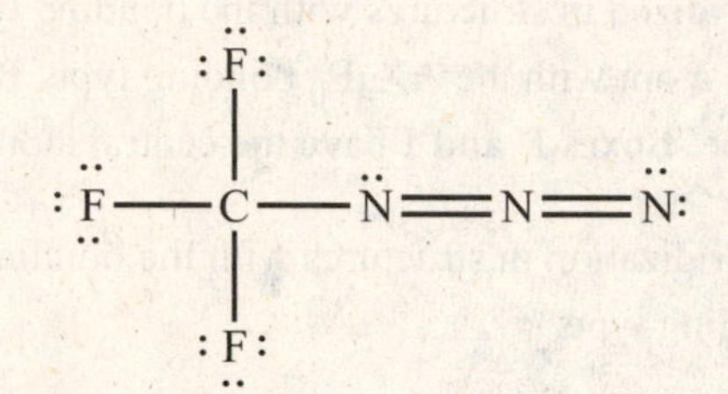

The central N in the N-N-C angle has AX_2E_1 type bonding, so the bond angle is 120°.

(b) The central N in the N-N-N angle has AX_2E_0 type bonding, so the bond angle is 180°.

(c) The N atoms next to the C atom has AX_2E_1 type bonding, so that means sp^2-hybridization is needed. The C atom has AX_4E_0 type bonding, so it needs sp^3-hybridization.

(d) The new H atoms will react with the electrons of a π-bond. The product resonance structures with smaller formal charges is shown here:

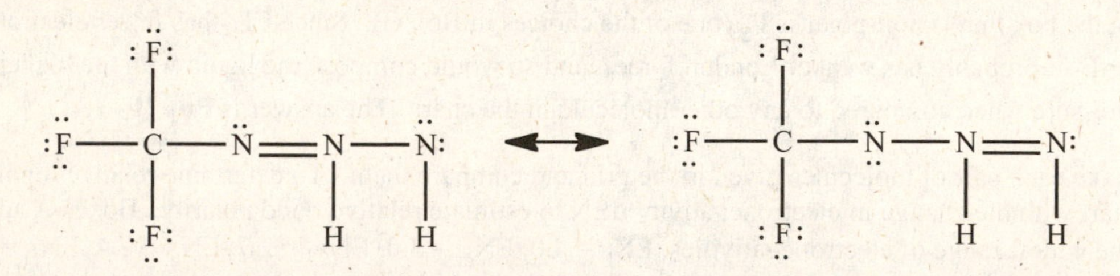

There are other answers.

From left to right: In the first structure, the first N atom has AX_2E_1 type bonding, so that means the N atom has sp^2-hybridization, the next N atom has AX_3E_0 type bonding, so that means the N atom has sp^2-hybridization, and the end N atom has AX_2E_2 type bonding, so that means the N atom has sp^3-hybridization. In the second structure, the first N atom has AX_2E_2 type bonding, so that means the N atom has sp^3-hybridization, the next N atom has AX_3E_0 type bonding, so that means the N atom has sp^2-hybridization, and the end N atom has AX_2E_1 type bonding, so that means the N atom has sp^2-hybridization.

95. *Answer:* **Many correct answers exist for this question. These are examples:**
(a) nitrogen (b) boron (c) phosphorus (d) iodine

Strategy and Explanation: The answers given here are not the only correct answers. They are only examples.

(a) XH_3 would have a central atom with one lone pair of electrons, if X = N, nitrogen.

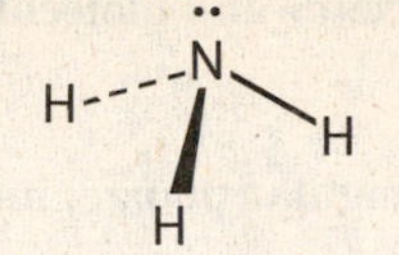

(b) XCl_3 would have no lone pairs on the central atom, if X = B, boron.

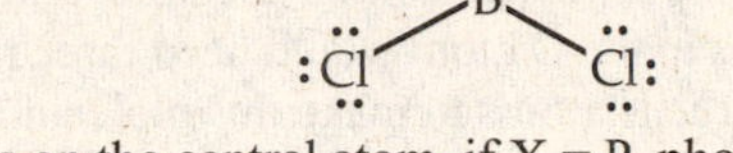

(c) XF_5 would have no lone pairs on the central atom, if X = P, phosphorus.

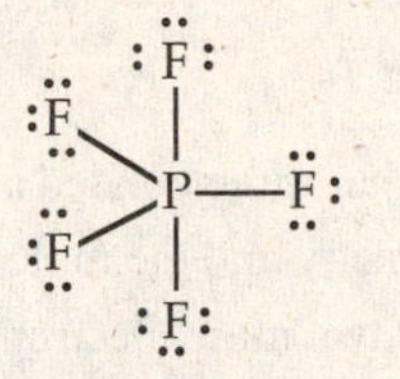

(d) XCl_3 would have two lone pairs on the central atom, if X = I, iodine.

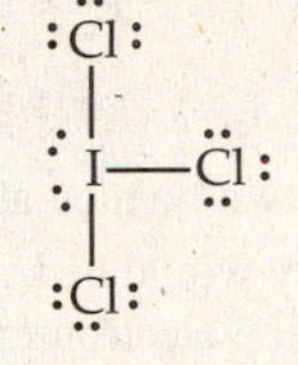

97. *Answer:* **Five; see diagram below with dashed lines (– – – – – –) showing intermolecular interactions.**

Strategy and Explanation: Five water molecules could hydrogen bond to an acetic acid molecule: one to each of the four lone pairs on O atoms in acetic acid and one to the H atom bonded to an O atom in acetic acid.

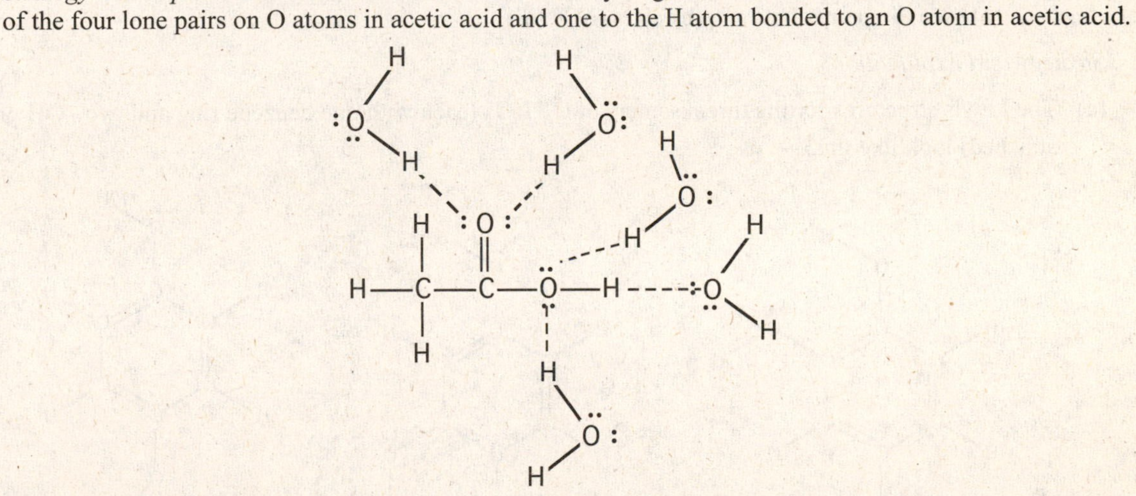

99. *Answer:* **Diagram (d)**

Strategy and Explanation:

(a) This is incorrect, because the H atoms are not bonded to highly electronegative atoms.

(b) This is incorrect. The H atoms are both bonded to O atom (EN = 3.5), but two partial-positive H atoms will not form a hydrogen bond to each other!

(c) This is incorrect. The H atom is bonded F. This is a covalent bond, not an example of the noncovalent interactive force called hydrogen bonding.

(d) This is correct. The H atom is bonded to an O atom (EN = 3.5) and is hydrogen bonding to another O atom.

The only answer that is correct is (d).

101. *Answer:*

H : O : H
•••
H

; electron-trio geometry and molecular geometry both are triangular planar;

H : F :
•••
H

; electron-trio geometry is triangular planar; molecular geometry is angular (120°).

Strategy and Explanation: In this fictional universe, a nonet is nine electrons. We will assume that the number of electrons in the atom called "O" in the other universe is still six, the number of electrons in the atom called "H" is still one, and the number of electrons in the atom called "F" is still seven. The equivalent to our VSEPR model is the VSTPR model (Valence Shell Electron Trio Repulsion Theory), where electron trios repel each other. The 6 electrons in the O atom would need three more electrons to make a total of nine. That means it would combine with three H atoms to make this molecule:

H : O : H
•••
H

In this H_3O molecule, the central atom has three bonded atoms and no lone trios. That would make it AX_3E_0 type. As a result, the electron-trio geometry and the molecular geometry would both be triangular planar.

The 7 electrons in the F atom would need two more electrons to make a total of nine. That means it would combine with two H atoms to make this molecule:

H : F :
•••
H

In this H_3F molecule, the central atom has two bonded atoms and one lone trio. That would make it AX_2E_1 type. As a result, the electron-trio geometry would be triangular planar and the molecular geometry would be angular at somewhat less than 120° due to the electron trio repelling the bonding trios somewhat better.

More Challenging Questions

103. *Answer:* **(a) See diagrams below (b) (i) < (ii) < (iii)**

Strategy and Explanation:

(a) The Lewis structures for the three isomers of $C_6H_6O_2$ (each with one benzene ring and two –OH groups attached) look like this:

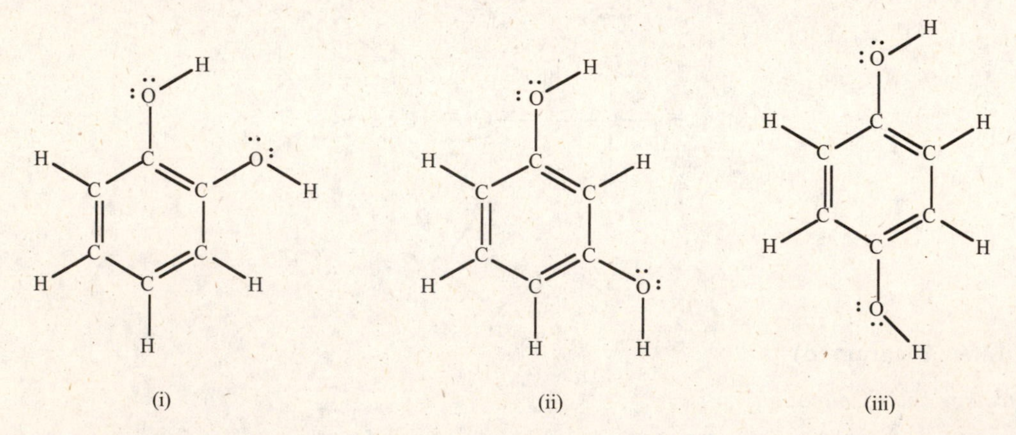

(i) (ii) (iii)

(b) All of these molecules have the same kind and strength of hydrogen bonding; however, because of the proximity of the two –OH groups in structure (i), these molecules are able to experience intramolecular hydrogen bonding (that is hydrogen bonding between H atoms and O atoms within the same molecule). Less hydrogen bonding disruption will be experienced when the molecules undergo a transition to the liquid state; hence, (i) will melt at a lower temperature than the other two. The asymmetry of structure (ii) might suggest that the intermolecular forces could be a little bit weaker in the solid, compared to those experienced among the higher-symmetry molecules in structure (iii). Therefore, the predicted order of for melting points of these three solids would be (i) < (ii) < (iii).

105. *Answer:* **(a) see structure below (b) First C: triangular planar, H–C–C and H–C–H bond angles are**

120°; Second C: linear, C–C–O bond angle is 180° (c) First C: sp^2-hybridized; Second C: sp-hybridized (d) polar, because the polar C=O bond contributes to a nonzero dipole moment.

Strategy and Explanation:

(a) Ketene, C_2H_2O (16 e^-) has no –OH bond, so it must have the following Lewis structure.

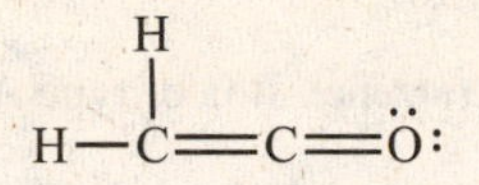

(b) 1st C atom: It is type AX_3E_0, so the electron-pair geometry is triangular planar with H–C–C and H–C–H bond angles of 120°.

2nd C atom: It is type AX_2E_0, so the electron-pair geometry is linear with a C–C–O bond angle of 180°.

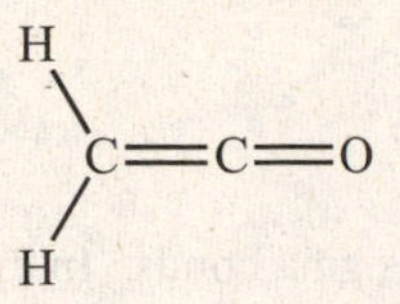

(c) Look at the 1st C atom: It is type AX_3E_0, so the C atom must be sp^2-hybridized.

Look at the 2nd C atom: It is of type AX_2E_0, so the C atom must be sp-hybridized.

Look at the O atom: It is of type AX_1E_2, so the O atom must be sp^2-hybridized.

(d) The C=O bond in this asymmetric molecule is a significantly polar bond which makes the molecule polar.

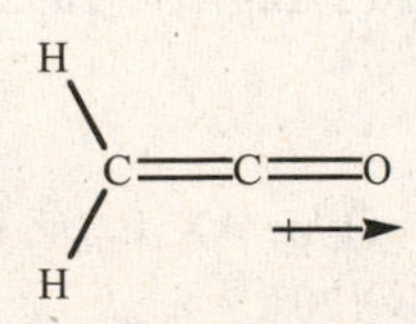

107. *Answer:* **(a) see diagrams below (b) sp; sp; sp^2 (c) sp^2, sp^2, sp^3 (d) hydrogen azide: 3 σ bonds, cyclotriazene: 4 σ bonds (e) hydrogen azide: 2 π bonds; cyclotriazene: 1 π bond (f) hydrogen azide: 180°; cyclotriazene: 60°.**

Strategy and Explanation:

(a) The two isomers of N_3H (16 e^-) have the following Lewis structures.

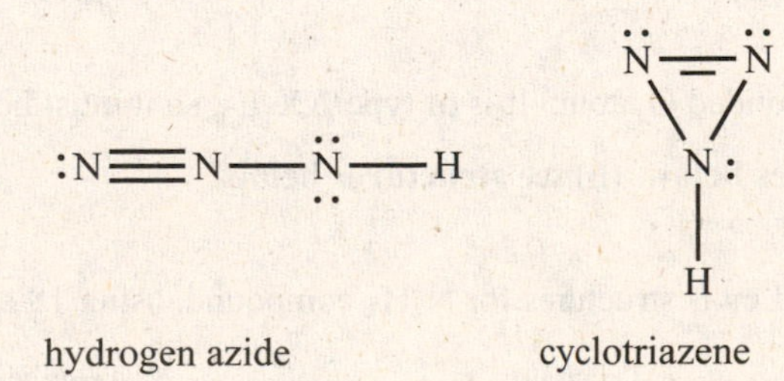

(b) Look at the 1st N atom (the left-most one) in hydrogen azide: It is of type AX_1E_1, so this N must be sp-hybridized.

Look at the 2nd N atom in hydrogen azide: It is of type AX_2E_0, so this N is sp-hybridized.

Look at the 3rd N atom in hydrogen azide: It is of type AX_2E_2, so we'd think that the N is sp^3-hybridized. However, to accommodate the other resonance structure (shown in (c) with two double bonds between the N atoms), one p-orbital must remain unhybridized, so the N is sp^2-hybridized

(c) Look at the first two N atom (the two on top of the triangle) in cyclotriazene: Each N is of type AX_2E_1, so each N atom must be sp^2-hybridized.

Look at the 3rd N atom in cyclotriazene: It is of type AX_3E_1, so this N is sp^3-hybridized.

For answering questions (d) and (e), recall that every single bond is a sigma-bond; every double bond is one sigma bond and one pi bond. So the sigma and pi bonds in this molecule are shown here:

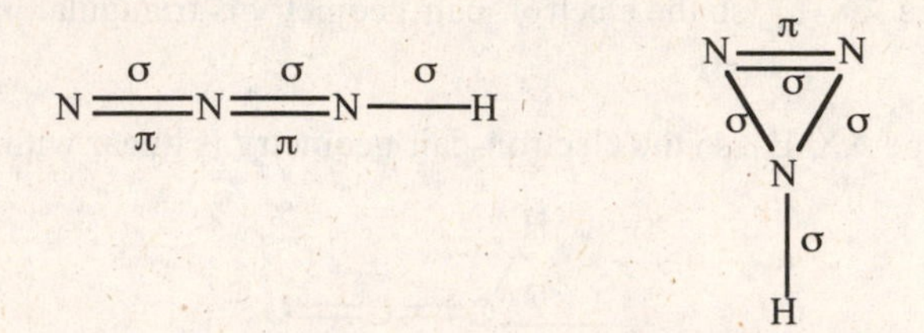

(d) In hydrogen azide, there are three sigma bonds. In cyclotriazene, there are four sigma bonds.

(e) In hydrogen azide, there are two pi bonds. In cyclotriazene, there is one pi bond.

(f) In hydrogen azide, the second N dictates the N–N–N bond angle. In (b), we ascertained that this N atom is sp-hybridized, so the bond angle must be 180°.

In cyclotriazene, the methods of this chapter indicate that the bond angles should be 120° (for the two angles represented by [left N-right N-bottom N], or [right N-left N- bottom N]) and 109.5° (for the angle represented by [left N-bottom N-right N]); however, since the three N atoms form a triangle, geometric constraints dictate that the bond angles will be approximately 60°.

109. *Answer:* **(a) Angle 1: 120°, Angle 2: 120°, Angle 3: 109.5° (b) sp^3 (c) single-bonded O: sp^3, double-bonded O: sp^2**

Strategy and Explanation:

(a) Angle 1: Look at the C atom: It is of type AX_3E_0, so it must have triangular planar electron geometry, and the bond angle is approximately 120°.

Angle 2: Look at the second of the three carbon atoms: It is of type AX_3E_0, so it must have triangular planar electron geometry, and the bond angle is approximately 120°.

Angle 3: Look at the C atom: It is of type AX_4E_0, so it must have tetrahedral electron geometry, and the bond angle is approximately 109.5°.

(b) The N atom is of type AX_3E_1, so it must be sp^3-hybridized.

(c) Look at the O atoms with two single bonds: Each of them are of type AX_2E_2, so they must each be sp^3-hybridized.

Look at the double bonded O atom: It is of type AX_1E_2, so it must be sp^2-hybridized.

111. *Answer:* **(a) see structures below (b) see structures below**

Strategy and Explanation:

(a) Write two Lewis structures for N_3H_3 compound, using 18 electrons.

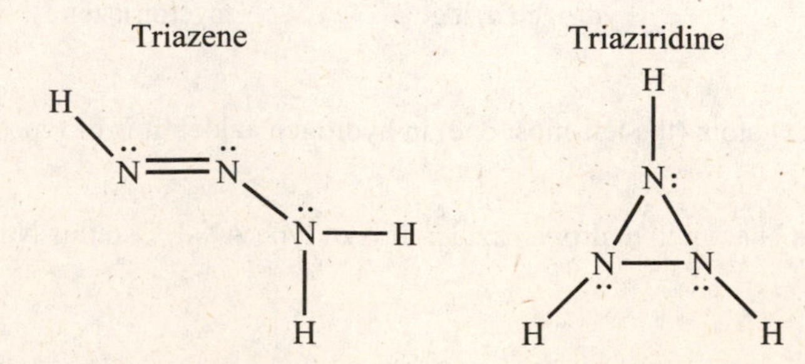

(b) The central N atom in the N-N-N bond of triazene is of type AX_2E_1, so its bond angle is approximately 120°. The N atoms in triaziridine form a three-membered ring, so they are geometrically constrained to has a 60° angle, and not the VSEPR predicted bond angles.

113. *Answer:* **(a) *sp*^{3} (b) *sp*^{2} (c) *sp*^{2} (d) Eight σ-bonds and two π-bonds (e) See structure below (f) See structure below.**

Strategy and Explanation:

(a) Looking at the structure, the single-bonded N atoms are of type AX_2E_2, so the N atoms must be *sp*^{3}-hybridized.

(b) Looking at the structure, the double-bonded S atoms are of type AX_2E_1, so the S atoms must be *sp*^{2}-hybridized.

(c) Looking at the structure, the double-bonded N atoms are of type AX_2E_1, so the N atoms must be *sp*^{2}-hybridized.

(d) Each single bond is one σ-bond. Each double bond is one π-bond and one σ-bond.

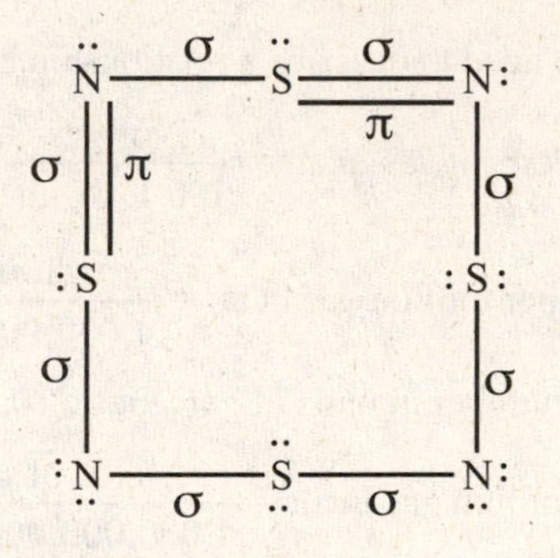

The molecule has eight σ-bonds and two π-bonds.

(e) See structure drawn in (d).

(f) Make a new double bond with one electron pair on a single-bonded N atom with a double bonded S atom and move one pair of the original double bond electrons to a lone pair on the next N atom in the chain, as shown here on the left side of the structure.

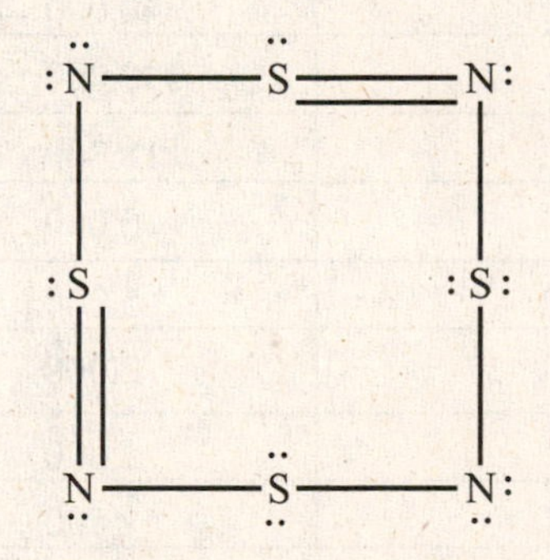

Chapter 10: Gases and the Atmosphere

Solutions for Blue-Numbered Questions for Review and Thought

Topical Questions

The Atmosphere (Section 10.1)

11. *Answer/Explanation:* Nitrogen serves to moderate the reactiveness of oxygen by diluting it. Oxygen sustains animal life as a reactant in the conversion of food to energy. Oxygen is produced by plants in the process of photosynthesis.

13. *Answer:* **See table below**

Strategy and Explanation: Convert all the numbers in Table 10.1 to parts per million (ppm) and parts per billion (ppb)

To accomplish these conversions, we need to describe a relationship between percent and ppm and ppb:

$$\text{Percent gas in air} = \frac{\text{L of gas}}{\text{100 L of air}}$$

$$\text{Parts per million gas in air} = \frac{\text{L of gas}}{\text{1,000,000 L of air}}$$

So, 1% = 10,000 ppm. Multiply the number in units of percent by 10,000 to get ppm.

$$\text{Parts per billion gas in air} = \frac{\text{L of gas}}{\text{1,000,000,000 L of air}}$$

So, 1ppm = 1,000 ppb. Multiply the number in units of ppm by 1,000 to get ppm.

Molecule	ppm	ppb	
N_2	780,840	780,840,000	↑
O_2	209,480	209,480,000	
Ar	9,340	9,340,000	
CO_2	330	330,000	
Ne	18.2	18,200	> 1 ppm
H_2	10.	10,000	
He	5.2	5,200	
CH_4	2	2,000	↓
Kr	1	1,000	↑
CO	0.1	100	between
Xe	0.08	80	1ppm
O_3	0.02	20	and
NH_3	0.01	10	1ppb
NO_2	0.001	1	↓
SO_2	0.0002	0.2	< 1ppb

✓ *Reasonable Answer Check:* The numbers are different only by the appropriate factor of 10.

15. *Answer:* **1.5×10^8 metric tons added, 2×10^6 metric tons total**

Strategy and Explanation: Given the mass of a sample of coal, the percentage of sulfur in the coal, the weight fraction of SO_2 in the atmosphere, and (given in Question 14) the mass of the atmosphere, determine the mass of SO_2 added to the atmosphere, and the total amount of SO_2 in the atmosphere.

To answer the first question, use metric conversions and the mass fraction as a conversion factor to determine the mass of sulfur. Use mole and molar mass conversion factors to determine the mass of SO_2, assuming that all the sulfur in the coal is converted to SO_2 and released into the atmosphere. Convert the mass back to metric tons. To answer the second question, use the mass of the atmosphere (given in Question 14) and the mass fraction of SO_2 in terms of metric tons as a conversion factor to determine the total mass of SO_2.

First find the mass of S:

$$3.1 \times 10^9 \text{ metric tons coal} \times \frac{1000 \text{ kg coal}}{1 \text{ metric ton coal}} \times \frac{2.5 \text{ kg S}}{100 \text{ kg coal}} \times \frac{1000 \text{ g S}}{1 \text{ kg S}} = 7.8 \times 10^{13} \text{ g S}$$

Then, find the mass of SO_2:

$$7.8 \times 10^{13} \text{ g S} \times \frac{1 \text{ mol S}}{32.066 \text{ g S}} \times \frac{1 \text{ mol } SO_2}{1 \text{ mol S}} \times \frac{64.0638 \text{ g } SO_2}{1 \text{ mol } SO_2} = 1.5 \times 10^{14} \text{ g } SO_2$$

Then, convert the mass of SO_2 back to metric tons:

$$1.5 \times 10^{14} \text{ g } SO_2 \times \frac{1 \text{ kg } SO_2}{1000 \text{ g } SO_2} \times \frac{1 \text{ metric ton } SO_2}{1000 \text{ kg } SO_2}$$

$$= 1.5 \times 10^8 \text{ metric tons } SO_2 \text{ was added to the atmosphere in 1980}$$

Get the total mass of SO_2 currently in the atmosphere

$$5.3 \times 10^{15} \text{ metric tons air} \times \frac{0.4 \text{ metric tons } SO_2}{1{,}000{,}000{,}000 \text{ metric tons air}} = 2 \times 10^6 \text{ metric tons } SO_2$$

✓ *Reasonable Answer Check:* It is clear that some of the SO_2 presumably released into the atmosphere in 1980 is no longer there, since the total mass of SO_2 is less than what was introduced that year. SO_2 is a reactive gas, getting oxidized to SO_3 in the presence of air and then producing sulfuric acid when reacting with rainwater. This removes the sulfur from the air. See Section 10.10 for more details on acid rain and SO_2 as a primary pollutant.

Properties of Gases (Section 10.2)

16. *Answer:* **(a) 0.947 atm (b) 950. mm Hg (c) 542 torr (d) 98.7 kPa (e) 6.91 atm**

Strategy and Explanation: Convert a series of pressure quantities into other pressure units.

Use Table 10.2 to design conversion factors to achieve the conversions.

(a) $$720. \text{ mmHg} \times \frac{1 \text{ atm}}{760. \text{ mmHg}} = 0.947 \text{ atm}$$

(b) $$1.25 \text{ atm} \times \frac{760 \text{ mmHg}}{1 \text{ atm}} = 950. \text{ mmHg}$$

(c) $$542 \text{ mmHg} \times \frac{760 \text{ torr}}{760 \text{ mmHg}} = 542 \text{ torr}$$

(d) $$740. \text{ mmHg} \times \frac{1 \text{ atm}}{760. \text{ mmHg}} \times \frac{101.325 \text{ kPa}}{1 \text{ atm}} = 98.7 \text{ kPa}$$

(e) $$700. \text{ kPa} \times \frac{1 \text{ atm}}{101.325 \text{ kPa}} = 6.91 \text{ atm}$$

✓ *Reasonable Answer Check:* The unit of atm represents a lot more pressure than the units of kPa, torr, and mm Hg, so it makes sense that the numbers of atmospheres are always much smaller than the pressure expressed in these other units. The unit of kPa represents more pressure than the units of torr and mm Hg, so it makes sense that the numbers of kPa are always smaller than the pressure expressed in the other units of torr and mm Hg. Torr and mm Hg are the same size, so their quantities should be identical.

18. *Answer:* **14 m**

Strategy and Explanation: Given the density of mercury, the density of an oil used to construct a barometer, and the atmospheric pressure, determine the height in meters of the oil column in the oil barometer.

Use Table 10.2 to convert the pressure into mm Hg. Pressure on the liquid pushes the liquid up the barometer until its mass exerts the same force per unit area as the air pressure: P_{liquid} = Force/Area = mg^2/Area. Since g is a constant, the force per unit area of the liquid counteracting the air pressure is proportional to just the mass per unit area. Relate the mass per unit area of mercury to the mass per unit area of the oil for 1 atm pressure. Relate the mass of each liquid to its respective density and volume. Relate the volume of each liquid to the dimensions of its respective barometers, including the height. Relate the height of mercury in the mercury barometer to the height of the oil in the oil barometer.

According to Table 10.2, at 1.0 atm the height of a column of mercury in a mercury barometer is 760 mm. Let m = mass of the liquid, A = cylindrical area of the barometer's column, d = density of the liquid, V = volume of the liquid, and h = the height of the liquid in the barometer. Now, equate the mass per unit area of each barometer and derives an equation relating their heights:

h

A

$$m_{Hg}/A_{Hg} = m_{oil}/A_{oil}$$

$$d_{Hg}V_{Hg}/A_{Hg} = d_{oil}V_{oil}/A_{oil}$$

$$V = (\text{Area}) \times (\text{height}) = Ah$$

$$d_{Hg}A_{Hg}h_{Hg}/A_{Hg} = d_{oil}A_{oil}h_{oil}/A_{oil}$$

$$d_{Hg}h_{Hg} = d_{oil}h_{oil}$$

$$h_{oil} = \frac{d_{Hg}h_{Hg}}{d_{oil}} = \frac{(13.596\ g/cm^3)(760\ mm)}{(0.75\ g/cm^3)} \times \frac{1\ m}{1000\ mm} = 14\ m$$

✓ *Reasonable Answer Check:* A larger mass of oil will be needed to counterbalance the atmospheric pressure because it is less dense than mercury. A higher column of oil makes sense.

20. *Answer/Explanation:* With a perfect vacuum at the top of the well, this system would resemble a water barometer. Using the equation derived in the solution to Question 18:

$$h_{water} = \frac{d_{Hg}h_{Hg}}{d_{water}} = \frac{(13.596\ g/cm^3)(760\ mm)}{(1.00\ g/cm^3)} \times \frac{1\ m}{1000\ mm} \times \frac{3.281\ feet}{1\ m} = 34\ feet$$

That means atmospheric pressure can only push water up to about 34 feet. The well cannot be deeper than that, not even using a high quality vacuum pump. So, it would not help much at all to have such a vacuum pump.

Kinetic-Molecular Theory (Section 10.3)

22. *Answer:* **See I-V below; assumption III and IV become false at high P or low T; assumption I becomes false at high P; assumption II is most nearly correct.**

Strategy and Explanation: The five basic concepts of kinetic-molecular theory are given in Section 10.3:

I. A gas is composed of molecules whose size is much smaller than the distances between them.

II. Gas molecules move randomly at various speeds and in every possible direction.

III. Except when molecules collide, forces of attraction and repulsion between them are negligible.

IV. When collisions occur, they are elastic.

V. The average kinetic energy of gas molecules is proportional to the absolute temperature.

A discussion of nonideal behavior is provided in Section 10.8. Of these five basic assumptions, the ones that become false at very high pressures or very low temperatures are assumptions III and IV. Slow molecules crowded together are much more likely to interact even when they are not colliding. Collisions may not be elastic under these circumstances, because colliding molecules might stick together due to large enough interactive forces between them. Assumption I may become false at very high pressures, because the molecules are crowded together reducing the distance between them.

The assumption that seems most likely to always be most nearly correct is assumption II. As long as a substance is a gas, it retains the capacity to disperse to uniformly fill any container it is introduced into, though the rate of dispersal may vary.

24. *Answer:* $\mathbf{CH_2Cl_2 < Kr < N_2 < CH_4}$

Strategy and Explanation: Given the formulas of various atoms and molecules and their common temperature, put their gases in order of increasing average molecular speed.

Kinetic energy is proportional to temperature. With all of the samples at the same temperature, their average kinetic energies are the same. Kinetic energy is related to mass and velocity. $E_{kin} = \frac{1}{2}mv^2$. Velocity is related to kinetic energy and mass: $v^2 = \frac{2E}{m}$. Therefore, molecules with smaller mass have the faster molecular speed. To rank the molecules with increasing speed, rank them from the largest molar mass to the smallest.

Estimate the molar masses: Kr molar mass 83.8 g/mol, CH_4 molar mass = 16.0 g/mol, N_2 molar mass is 28.0 g/mol, CH_2Cl_2 molar mass = 84.9 g/mol.

$$\text{slowest speed: } CH_2Cl_2 < Kr < N_2 < CH_4 \text{ fastest speed}$$

✓ *Reasonable Answer Check:* It makes sense that lightweight things go faster.

26. *Answer:* **Ne**

Strategy and Explanation: Given the formulas of various gases introduced into one end of a tube, determine which gas will reach the end of the tube first.

Velocity is related to kinetic energy and mass: $v^2 = \frac{2E}{m}$. Therefore, molecules with smaller mass have the faster molecular speed. The molecule with the fastest speed will travel a fixed distance the quickest. To determine which molecule arrives first, rank them from the largest molar mass to the smallest.

Estimate the molar masses: Ar molar mass = 40 g/mol, Ne molar mass = 20 g/mol, Kr molar mass = 84 g/mol, Xe molar mass = 131 g/mol. Ne will arrive first.

✓ *Reasonable Answer Check:* It makes sense, because the lightest weight things go faster, they will arrive sooner.

Gas Behavior and the Ideal Gas Law (Section 10.4)

29. *Answer:* $\mathbf{4.2 \times 10^{-5}}$ **mol**

Strategy and Explanation: Given the volume of a sample of air at STP and the volume fraction of CO in ppm, determine the moles of CO.

Use the volume fraction in terms of liters as a conversion factor to determine the liters of CO. Use the molar volume of a gas at STP as a conversion factor to determine the moles of CO.

$$1.0 \text{ L air} \times \frac{950 \text{ L CO}}{1{,}000{,}000 \text{ L air}} \times \frac{1 \text{ mol CO}}{22.414 \text{ L CO}} = 4.2 \times 10^{-5} \text{ mol CO}$$

✓ *Reasonable Answer Check:* Air has a very small proportion of CO. This small sample of air has a very small amount of CO.

30. *Answer:* **25.5 L**

Strategy and Explanation: Given the mass, volume, temperature and pressure of a sample of gas, determine the volume of a different mass of the gas at the same temperature and pressure).

Use Avogadro's law to relate volume to the number of moles.

$$\frac{V_1}{n_1} = \frac{V_2}{n_2} \qquad \text{(P and T constant)}$$

For any given substance, the number of moles is directly proportional to the mass. (because n = m/M, where m is the mass and M is the molar mass). Therefore, there is a similar relationship between volume and mass:

$$\frac{V_1}{m_1} = \frac{V_2}{m_2} \quad \text{(if the samples are made of the same substance.)}$$

Solve for V_2
$$V_2 = \frac{V_1 \times m_2}{m_1} = (10.0\ \text{L}) \times \frac{(129.3\ \text{g})}{(50.75\ \text{g})} = 25.5\ \text{L}$$

✓ *Reasonable Answer Check:* With all the information given, it is possible to calculate the molar mass of the gaseous substance from the first set of data, and then use that information to determine the new volume for the second sample; however, this calculation takes a lot longer to complete:

$$\frac{50.75\ \text{g}}{10.0\ \text{L @ STP}} \times \frac{22.414\ \text{L @ STP}}{1\ \text{mol}} = 114\ \text{g/mol}$$

$$129.3\ \text{g} \times \frac{1\ \text{mol}}{114\ \text{g}} \times \frac{22.414\ \text{L @ STP}}{1\ \text{mol}} = 25.5\ \text{L @ STP}$$

It makes sense that the sample with the larger mass of gas present has a larger volume.

32. *Answer:* **62.5 mm Hg**

Strategy and Explanation: Given the volume and pressure of a sample of gas in one flask and the volume of a flask it is transferred to at the same temperature, determine the new pressure.

Use Boyle's law to relate volume to pressure. $P_1V_1 = P_2V_2$ (unchanging T and n)

$$P_2 = \frac{P_1V_1}{V_2} = \frac{(100.\ \text{mm Hg}) \times (125\ \text{mL})}{(200.\ \text{mL})} = 62.5\ \text{mm Hg}$$

✓ *Reasonable Answer Check:* The larger volume should have a smaller pressure.

35. *Answer:* **172 mm Hg**

Strategy and Explanation: Use Boyle's law: $P_1V_1 = P_2V_2$ (unchanging T and n).

$$P_2 = \frac{P_1V_1}{V_2} = \frac{(735\ \text{mm Hg}) \times (3.50\ \text{L})}{(15.0\ \text{L})} = 172\ \text{mm Hg}$$

37. *Answer:* **26.5 mL**

Strategy and Explanation: Given the original volume and temperature of a sample of gas in a syringe (presumably at atmospheric pressure) and the new temperature of the sample, determine the new volume (presumably still at atmospheric pressure).

Convert the temperatures to Kelvin. Use Charles' law to relate volume to absolute temperature.

$$\frac{V_1}{T_1} = \frac{V_2}{T_2} \quad \text{(P and n constant)}$$

$$T_1 = 20.\ ^\circ\text{C} + 273.15 = 293\ \text{K} \qquad T_2 = 37\ ^\circ\text{C} + 273.15 = 310.\ \text{K}$$

$$V_2 = V_1 \times \frac{T_2}{T_1} = (25.0\ \text{mL}) \times \frac{(310.\ \text{K})}{(293\ \text{K})} = 26.5\ \text{mL}$$

✓ *Reasonable Answer Check:* Gas at higher temperature should have a larger volume.

39. *Answer:* **– 96 °C**

Strategy and Explanation: Use Charles' law to relate volume to absolute temperature, as described in the solution to Question 37.
$$T_1 = 80.\ ^\circ\text{C} + 273.15 = 353\ \text{K}$$

$$T_2 = T_1 \times \frac{V_2}{V_1} = (353\ \text{K}) \times \frac{(1.25\ \text{L})}{(2.50\ \text{L})} = 177\ \text{K}$$

$$177\ \text{K} - 273.15 = -96\ ^\circ\text{C}$$

✓ *Reasonable Answer Check:* Gas at smaller volume should have a lower temperature.

41. *Answer:* **4.00 atm**

Strategy and Explanation: Given the original pressure and temperature of gas in a tire, the assumption that volume is unchanged, and the new temperature, determine the new pressure exerted by the gas in the tire.

Convert the temperature to Kelvin. Use the combined gas law to relate pressure to absolute temperature.

$$\frac{P_1V_1}{T_1} = \frac{P_2V_2}{T_2} \qquad \text{(n constant)}$$

At constant volume $$\frac{P_1}{T_1} = \frac{P_2}{T_2} \qquad \text{(V and n constant)}$$

$$T_1 = 15\ °C + 273.15 = 288\ K \qquad T_2 = 35\ °C + 273.15 = 308\ K$$

$$P_2 = P_1 \times \frac{T_2}{T_1} = (3.74\ \text{atm}) \times \frac{(308\ K)}{(288\ K)} = 4.00\ \text{atm}$$

✓ *Reasonable Answer Check:* A gas with a higher temperature should exert a higher pressure.

43. *Answer:* **501 mL**

Strategy and Explanation: Adapt the combined gas law, described in the solution to Question 41.

$$T_1 = 22\ °C + 273.15 = 295\ K \qquad T_2 = 42\ °C + 273.15 = 315\ K$$

$$V_2 = V_1 \times \frac{T_2}{T_1} \times \frac{P_1}{P_2} = (754\ \text{mL}) \times \frac{(315\ K)}{(295\ K)} \times \frac{(165\ \text{mmHg})}{(265\ \text{mmHg})} = 501\ \text{mL}$$

✓ *Reasonable Answer Check:* The temperature fraction: $\frac{(315\ K)}{(295\ K)}$ is larger than one, consistent with increasing the volume due to the increased temperature. The pressure fraction: $\frac{(165\ \text{mmHg})}{(265\ \text{mmHg})}$ is smaller than one, consistent with decreasing the volume due to an increased pressure. Clearly these two effects counteract each other, but this pressure change affects the volume more than the temperature change does.

45. *Answer:* **0.507 atm**

Strategy and Explanation: Given the mass, identity, temperature, and volume of a gas sample, determine the pressure of the sample.

Use the ideal gas law: $PV = nRT \qquad R = 0.08206\ \frac{L \cdot atm}{mol \cdot K}$

The units of R remind us to determine the moles of gas (using mass and molar mass), to convert the temperature to Kelvin, and to convert the volume to liters.

$$1.55\ \text{g Xe} \times \frac{1\ \text{mol Xe}}{131.29\ \text{g Xe}} = 0.0118\ \text{mol Xe}$$

$$T = 20.\ °C + 273.15 = 293\ K$$

$$V = 560.\ \text{mL} \times \frac{1\ L}{1000\ \text{mL}} = 0.560\ L$$

$$P = \frac{nRT}{V} = \frac{(0.0118\ \text{mol}) \times \left(0.08206 \frac{L \cdot atm}{mol \cdot K}\right) \times (293\ K)}{(0.560\ L)} = 0.507\ \text{atm}$$

✓ *Reasonable Answer Check:* The small fraction of a mole makes sense with the small mass. The units in the pressure calculation cancel properly to give atm.

47. *Answer:* **Largest number in sample (d); smallest number in sample (c)**

Strategy and Explanation: Given a set of gas samples, determine which has the largest number of molecules and which has the smallest number of molecules.

Some of the samples are at STP, so use the molar volume of a gas at STP to determine the number of molecules (in units of moles).

$$\text{1 mol of **any** gas occupies 22.414 L.} \quad 1.0\ L \times \frac{1\ \text{mol gas}}{22.414\ L} = 0.045\ \text{mol gas}$$

In the other cases, use the ideal gas law as described in the solution to Question 45. Once all the moles are calculated, identify which sample has the most moles and which sample has the least moles.

(a) 0.045 mol H_2 at STP

(b) 0.045 mol N_2 at STP

(c) $T = 27\ °C + 273.15 = 300.\ K$ $\qquad P = 760.\ mmHg \times \frac{1\ atm}{760.\ mmHg} = 1.00\ atm$

$$n = \frac{PV}{RT} = \frac{(1.0\ atm)\times(1.00\ L)}{\left(0.08206\frac{L\cdot atm}{mol\cdot K}\right)\times(300.\ K)} = 0.0406\ mol$$

(d) $T = 0.\ °C + 273.15 = 273\ K$ $\qquad P = 800.\ mmHg \times \frac{1\ atm}{760.\ mmHg} = 1.05\ atm$

$$n = \frac{PV}{RT} = \frac{(1.05\ atm)\times(1.0\ L)}{\left(0.08206\frac{L\cdot atm}{mol\cdot K}\right)\times(273\ K)} = 0.047\ mol$$

Of these samples, sample (d) has the most molecules and sample (c) has the smallest number of molecules.

✓ *Reasonable Answer Check:* To keep a 1.0-L gas sample at standard temperature and still have larger than standard pressure suggests that there must be more molecules hitting the walls than a sample at STP. To have a 1.0-L gas sample at higher than standard temperature and still stay at standard pressure suggests that there must be fewer molecules hitting the walls harder, than a sample at STP.

Quantities of Gases in Chemical Reactions (Section 10.5)

49. *Answer:* **6.0 L H_2**

Strategy and Explanation: Given the balanced chemical equation for a chemical reaction, the mass of one reactant, and excess other reactant, determine the volume of the product produced at a specified pressure and temperature.

Convert from grams to moles. Use the stoichiometric relationship from the balanced equation to determine moles of product. Use the ideal gas law to determine the pressure of the product.

$$6.5\ g\ Al \times \frac{1\ mol\ Al}{26.9815\ g\ Al} \times \frac{2\ mol\ H_2}{2\ mol\ Al} = 0.24\ mol\ H_2$$

$T = 22\ °C + 273.15 = 295\ K$ $\qquad P = 742\ mmHg \times \frac{1\ atm}{760\ mmHg} = 0.976\ atm$

$$P = \frac{n_{H_2}RT}{V} = \frac{(0.24\ mol\ H_2)\times\left(0.08206\frac{L\cdot atm}{mol\cdot K}\right)\times(295\ K)}{(0.976\ atm)} = 6.0\ L\ H_2$$

✓ *Reasonable Answer Check:* All units cancel properly in the calculation of atmosphere. A quarter mole of gas occupies a volume is a little larger than one quarter of the molar volume of a gas at STP, which make sense since the sample's temperature is larger than the STP temperature.

51. *Answer:* **0.63 L CO_2; approximately 15% of the typical volume of two loaves of bread**

Strategy and Explanation: Given the mass of sucrose, the formula and the product of a reaction, determine the maximum volume of CO_2 produced at STP. Compare that volume with the typical volume of two loaves of bread. Let's use French bread.

Balance the equation. Determine the moles of sucrose from the molar mass, then use the given stoichiometric relationships to determine the moles of CO_2 produced. Last, use the molar volume of a gas at STP to determine the volume of CO_2. Estimate the total volume of two loaves of French bread assuming they are cylinders. Compare the two volumes.

$$2.4\ g\ C_{12}H_{22}O_{11} \times \frac{1\ mol\ C_{12}H_{22}O_{11}}{342.2956\ g\ C_{12}H_{22}O_{11}} \times \frac{4\ mol\ CO_2}{1\ mol\ C_{12}H_{22}O_{11}} \times \frac{22.414\ L\ CO_2\ at\ STP}{1\ mol\ CO_2} = 0.63\ L\ CO_2$$

Assume one loaf of French bread is a cylinder, 3.0 inches in diameter and 18 inches long.

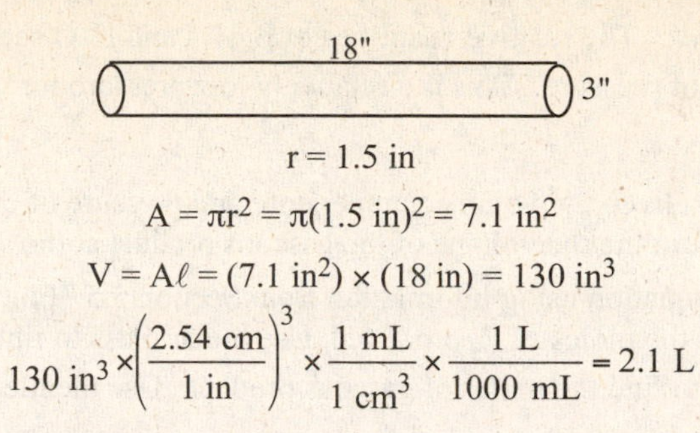

$$r = 1.5 \text{ in}$$

$$A = \pi r^2 = \pi(1.5 \text{ in})^2 = 7.1 \text{ in}^2$$

$$V = A\ell = (7.1 \text{ in}^2) \times (18 \text{ in}) = 130 \text{ in}^3$$

$$130 \text{ in}^3 \times \left(\frac{2.54 \text{ cm}}{1 \text{ in}}\right)^3 \times \frac{1 \text{ mL}}{1 \text{ cm}^3} \times \frac{1 \text{ L}}{1000 \text{ mL}} = 2.1 \text{ L}$$

Two loaves would have twice this volume: $2 \times (2.1 \text{ L}) = 4.2 \text{ L}$. The CO_2 bubbles produced in the bread are about 15% its volume.

✓ *Reasonable Answer Check:* Slicing open French bread we see that it has a vast "honeycomb" of bubble-shaped spaces in it. It makes sense that approximately half the loaf's volume can be associated with the CO_2 bubbles formed by the yeast when the bread was rising.

53. *Answer:* **10.4 L O_2; 10.4 L H_2O**

Strategy and Explanation: Given the balanced chemical equation for a chemical reaction and the volume of one reactant at a specified pressure and temperature, determine the volume of the other reactant at a specified pressure and temperature that will cause complete reaction, and the volume of one of the products at a specified pressure and temperature that will be produced.

Avogadro's law allows us to interpret a balanced equation with gas reactants and products in terms of gas volumes, as long as their temperatures and pressures are the same. Use the stoichiometry to relate liters of $SiH_4(g)$ that react with liters of $O_2(g)$ and liters of $H_2O(g)$.

The balanced equation tells us that one volume of $SiH_4(g)$ reacts with two volumes of $O_2(g)$ to make two volumes of $H_2O(g)$, since all the volumes are measured at the same temperature and pressure.

$$5.2 \text{ L } H_2(g) \times \frac{2 \text{ L} O_2(g)}{1 \text{ L} SiH_4(g)} = 10.4 \text{ L} O_2(g)$$

$$5.2 \text{ L } H_2(g) \times \frac{2 \text{ L} H_2O(g)}{1 \text{ L} SiH_4(g)} = 10.4 \text{ L} H_2O(g)$$

Notice: Multiplying a number by an exact whole number, n, is like adding that number to itself n times. 2 × (5.2) = 5.2 L + 5.2 L = 10.4 L, that is why we use the addition rule for assessing the significant figures, here.

✓ *Reasonable Answer Check:* Twice as many O_2 molecules are needed compared to the number of SiH_4 molecules, forming twice as many H_2O molecules. So, it makes sense that both the volume of O_2 and the volume of H_2O are twice the volume of SiH_4.

55. *Answer:* **21 mm Hg**

Strategy and Explanation: Given the balanced chemical equation for a chemical reaction, the mass of one reactant, and excess other reactant, determine the pressure of the product produced at a specified volume and temperature.

Convert from grams to moles. Use the stoichiometric relationship from the balanced equation to determine moles of product. Use the ideal gas law to determine the pressure of the product.

$$0.050 \text{ g } B_4H_{10} \times \frac{1 \text{ mol } B_4H_{10}}{53.32 \text{ g } B_4H_{10}} \times \frac{10 \text{ mol } H_2O}{2 \text{ mol } B_4H_{10}} = 0.0047 \text{ mol } H_2O$$

$$T = 30.\ ^\circ\text{C} + 273.15 = 303 \text{ K}$$

$$P = \frac{n_{H_2O}RT}{V} = \frac{(0.0047 \text{ mol } H_2O) \times \left(0.08206 \frac{\text{L} \cdot \text{atm}}{\text{mol} \cdot \text{K}}\right) \times (303 \text{ K})}{(4.25 \text{ L})} = 0.027 \text{ atm}$$

$$0.027 \text{ atm} \times \frac{760 \text{ mmHg}}{1 \text{ atm}} = 21 \text{ mmHg}$$

✓ *Reasonable Answer Check:* The relative quantities of B_4H_{10} and H_2O seem sensible. All units cancel properly in the calculation of pressure. This is a relatively low pressure for water but the sample is also small.

57. *Answer:* **10.0 L Br_2**

Strategy and Explanation: Given the volume, temperature and pressure of the gaseous reactant and the mass of the solid ionic reactant, determine the volume of the gaseous product at the same temperature and pressure.

Balance the displacement equation using information from Section 5.5. Find the limiting reaction. The ideal gas law can be used to find the moles of F_2 produced. Use molar mass to find the moles of the solid. The stoichiometry is then used to find the moles of gaseous product. Use the ideal gas law to find the volume of gaseous product.

Balance the displacement equation:

$$F_2(g) + CaBr_2(s) \longrightarrow Br_2\,(g) + CaF_2(s)$$

Calculate the moles of Br_2 formed, if all the F_2 reacts:

$$T = 100.\ °C + 273.15 = 373\ K$$

$$n_{F_2} = \frac{PV}{RT} = \frac{(1.00\ atm) \times (10.0\ L)}{\left(0.08206 \frac{L \cdot atm}{mol \cdot K}\right) \times (373\ K)} = 0.327\ mol\ F_2$$

$$0.327\ mol\ F_2 \times \frac{1\ mol\ Br_2}{1\ mol\ F_2} = 0.327\ mol\ Br_2$$

Calculate the moles of Br_2 formed, if all the $CaBr_2$ reacts:

$$99.9\ g\ CaBr_2 \times \frac{1\ mol\ CaBr_2}{199.986\ g\ CaBr_2} \times \frac{1\ mol\ Br_2}{1\ mol\ CaBr_2} = 0.500\ mol\ Br_2$$

The limiting reactant is F_2 and 0.327 mol Br_2 formed.

$$V = \frac{n_{Br_2}RT}{P} = \frac{(0.327\ mol\ Br_2) \times \left(0.08206 \frac{L \cdot atm}{mol \cdot K}\right) \times (373\ K)}{(1.00\ atm)} = 10.0\ L\ Br_2$$

✓ *Reasonable Answer Check:* Because volume of the Br_2 product sample is measured at the same temperature and pressure as the initial F_2sample, it makes sense that the volume is the same as the initial volume of F_2.

59. *Answer:* **1.44 g $Ni(CO)_4$**

Strategy and Explanation: Given the description of a chemical reaction and the volume of one reactant at a specified pressure and temperature, determine the mass of the product that can be produced.

Balance the equation. Use the ideal gas law to determine the moles of the reactant. Use the stoichiometric relationship from the balanced equation to determine moles of product. Then, convert from moles to grams.

$$Ni(s) + 4\ CO(g) \longrightarrow Ni(CO)_4$$

$$T = 25.0\ °C + 273.15 = 298.2\ K \qquad 418\ mmHg \times \frac{1\ atm}{760\ mmHg} = 0.550\ atm$$

$$n_{CO} = \frac{PV}{RT} = \frac{(0.550\ atm) \times (1.50\ L)}{\left(0.08206 \frac{L \cdot atm}{mol \cdot K}\right) \times (298.2\ K)} = 0.0337\ mol\ CO$$

$$0.0337\ mol\ CO \times \frac{1\ mol\ Ni(CO)_4}{4\ mol\ CO} \times \frac{170.7\ g\ Ni(CO)_4}{1\ mol\ Ni(CO)_4} = 1.44\ g\ Ni(CO)_4$$

✓ *Reasonable Answer Check:* The units all cancel properly in the calculation of moles of $Ni(CO)_4$. A reasonable mass of product is formed considering the molar quantities and the molar mass.

61. *Answer:* **(a) $2\ CH_3OH(\ell) + 3\ O_2(g) \longrightarrow 2\ CO_2(g) + 4\ H_2O(g)$ (b) 1.1×10^3 L CO_2**

Strategy and Explanation:

(a) The balanced equation: $2\ CH_3OH(\ell) + 3\ O_2(g) \longrightarrow 2\ CO_2(g) + 4\ H_2O(g)$

(b) Given the length in miles of a trip, the fuel efficiency of a car, the density of the liquid fuel, and the temperature and pressure, determine the volume of a gas-phase product produced during the trip.

Use the miles, the fuel efficiency, volume conversions, the density, and the stoichiometry to find the moles of the product. Use the ideal gas law to determine the volume of the product.

$$10.\ \text{miles} \times \frac{1\ \text{gal fuel}}{20.\ \text{miles}} \times \frac{3.785\ \text{L fuel}}{1\ \text{gal fuel}} \times \frac{1000\ \text{mL}}{1\ \text{L}} \times \frac{1\ \text{cm}^3}{1\ \text{mL}} = 1.9 \times 10^3\ \text{cm}^3\ \text{fuel}$$

$$1.9 \times 10^3\ \text{cm}^3\ \text{fuel} \times \frac{0.791\ \text{g}\ CH_3OH}{1\ \text{cm}^3\ \text{fuel}} \times \frac{1\ \text{mol}\ CH_3OH}{32.0417\ \text{g}\ CH_3OH} \times \frac{2\ \text{mol}\ CO_2}{2\ \text{mol}\ CH_3OH} = 47\ \text{mol}\ CO_2$$

$$T = 25\ °C + 273.15 = 298\ K$$

$$V = \frac{n_{CO_2}RT}{P} = \frac{(47\ \text{mol}\ CO_2) \times \left(0.08206 \frac{\text{L} \cdot \text{atm}}{\text{mol} \cdot \text{K}}\right) \times (298\ \text{K})}{(1.0\ \text{atm})} = 1.1 \times 10^3\ \text{L}$$

✓ *Reasonable Answer Check:* This is the same question as Chapter 12 Question 89 and produces the same result. This is a large volume of CO_2! However, it is not an unreasonable quantity considering how many gallons of methanol are used and how much CO_2 is generated from each methanol molecule. Comparing to the answer in Question 60, less CO_2 is generated by methanol as a fuel than octane.

Gas Density and Molar Mass (Section 10.6)

63. *Answer:* **130. g/mol**

Strategy and Explanation: Given the density of a gas at STP, determine the molar mass of the gas.

Density is related to mass and volume: $d = \frac{m}{V}$. Because we know the molar volume of a gas at STP (22.414 liters per mol), we can use this equation to get the molar mass (M = grams per mole).

$$d \times V = m$$

$$d \times (\text{molar volume}) = (\text{molar mass})$$

$$5.79 \frac{\text{g}}{\text{L}} \times 22.414 \frac{\text{L}}{\text{mol}} = 130. \frac{\text{g}}{\text{mol}}$$

65. *Answer:* **2.7 × 10³ mL**

Strategy and Explanation: Given the mass of a gas, the identity of the gas, its temperature, and its pressure, determine the volume occupied by the gas.

Convert from grams to moles. Use the ideal gas law to determine the volume of the gas.

$$4.4\ \text{g}\ CO_2 \times \frac{1\ \text{mol}\ CO_2}{44.0095\ \text{g}\ CO_2} = 0.10\ \text{mol}\ CO_2$$

$$T = 27\ °C + 273.15 = 300.\ K \qquad 730.\ \text{mmHg} \times \frac{1\ \text{atm}}{760.\ \text{mmHg}} = 0.961\ \text{atm}$$

$$V = \frac{nRT}{P} = \frac{(0.10\ \text{mol}) \times \left(0.08206 \frac{\text{L} \cdot \text{atm}}{\text{mol} \cdot \text{K}}\right) \times (300.\ \text{K})}{(0.961\ \text{atm})} = 2.7\ \text{L}$$

$$2.7\ \text{L} \times \frac{1000\ \text{mL}}{1\ \text{L}} = 2.7 \times 10^3\ \text{mL}$$

✓ *Reasonable Answer Check:* The molar volume of a gas under these conditions is going to be similar to that at STP (22.4 L/mol), because T is slightly higher than standard T and P is slightly lower than standard P. When using one tenth of a mole of CO_2, one would expect approximately one tenth of the molar volume, roughly 2.2 liters. This answer seems right, though it is unclear why it needed to be given in milliliters.

67. *Answer:* **P_{He} is 7.000 times greater than P_{N_2}**

Strategy and Explanation: Given the formulas of the compounds composing two gas samples with equal density, volume and temperature, determine the relationship between the pressures of the two gas samples.

We have insufficient information to take a straightforward approach to this task, so find a way to relate pressure to molar mass when the density, volume and temperature are equal: Density is mass per unit volume. Density can be calculated using the molar mass (M, the grams per mole, or m/n) and the molar volume (the volume per mole, or V/n).

$$d = \frac{m}{V} = \frac{m/n}{V/n} = \frac{M}{V/n} \qquad \text{so, } \frac{V}{n} = \frac{M}{d}$$

Rearranging the ideal gas law: $P\frac{V}{n} = RT$

Substituting the first equation into the second, gives: $P \times \frac{M}{d} = RT$ or $PM = dRT$

Therefore, at constant d and T: $P_2M_2 = M_1P_1$ or $\frac{P_2}{P_1} = \frac{M_1}{M_2}$

$$\frac{P_{He}}{P_{N_2}} = \frac{M_{N_2}}{M_{He}} = \frac{(28.0134 \text{ g/mol})}{(4.0026 \text{ g/mol})} = 7.000$$

The pressure of helium is seven times greater than the pressure of nitrogen.

✓ *Reasonable Answer Check:* A much larger number of atoms would need to be in the helium sample for it to have the same density as the nitrogen sample. Because pressure is proportional to number of particles, it makes sense that the pressure in the helium sample is much larger than in the nitrogen sample.

69. *Answer:* **3.7×10^{-4} g/L**

Strategy and Explanation: Given the molar mass of a gas and its pressure and temperature, determine the density of the gas.

We have insufficient information to take a straightforward approach to this task, so let's look at what we know: Density is mass per unit volume. If we can calculate the molar volume (the volume per mole) and divide that into the molar mass (M, the grams per mole), we can get the density. Use the ideal gas law to determine the molar volume of the gas.

$$\text{molar volume} = \frac{V}{n} = \frac{RT}{P} \qquad d = \frac{m}{V} = \frac{M}{\left(\frac{RT}{P}\right)} = \frac{MP}{RT}$$

$T = -23\ °C + 273.15 = 250.\ K$

$$0.20 \text{ mmHg} \times \frac{1 \text{ atm}}{760 \text{ mmHg}} = 2.6 \times 10^{-4} \text{ atm}$$

$$d = \frac{MP}{RT} = \frac{\left(29.0\frac{g}{mol}\right) \times (2.6 \times 10^{-4} \text{ atm})}{\left(0.08206\frac{L \cdot atm}{mol \cdot K}\right) \times (250.\ K)} = 3.7 \times 10^{-4}\ \frac{g}{L}$$

✓ *Reasonable Answer Check:* The low density makes sense at the very low pressure.

Partial Pressures of Gases (Section 10.7)

71. *Answer:* **4.51 atm total**

Strategy and Explanation: Given the masses of gases in a mixture at a specified volume and temperature, determine the total pressure.

Convert from grams to moles. Use the ideal gas law to determine the partial pressure of each gas. Use Dalton's law of partial pressures to determine the total pressure. Dalton's law and its applications are described in Section 10.7, page 446-7. Dalton's law of partial pressures states that the total pressure (P_{tot}) exerted by a mixture of gases is the sum of their partial pressures (P_1, P_2, P_3, etc.), if the volume (V) and temperature (T) are constant.

$$P_{tot} = P_1 + P_2 + P_3 + \ldots \qquad \text{(V and T constant)}$$

$$1.50 \text{ g } H_2 \times \frac{1 \text{ mol } H_2}{2.0158 \text{ g } H_2} = 0.744 \text{ mol } H_2$$

$$T = 25\ °C + 273.15 = 298 \text{ K}$$

$$P_{H_2} = \frac{n_{H_2}RT}{V} = \frac{(0.744 \text{ mol } H_2) \times \left(0.08206 \frac{L \cdot atm}{mol \cdot K}\right) \times (298 \text{ K})}{(5.00 \text{ L})} = 3.64 \text{ atm } H_2$$

$$5.00 \text{ g } N_2 \times \frac{1 \text{ mol } N_2}{28.0134 \text{ g } N_2} = 0.178 \text{ mol } N_2$$

$$P_{N_2} = \frac{n_{N_2}RT}{V} = \frac{(0.178 \text{ mol } H_2) \times \left(0.08206 \frac{L \cdot atm}{mol \cdot K}\right) \times (298 \text{ K})}{(5.00 \text{ L})} = 0.873 \text{ atm } N_2$$

Using Dalton's law: $P_{tot} = P_{H_2} + P_{N_2} = 3.64 \text{ atm } H_2 + 0.873 \text{ atm } N_2 = 4.51 \text{ atm total}$

✓ *Reasonable Answer Check:* The relative pressure of H_2 and N_2 makes sense from the relative number of moles. All un-needed units cancel properly in the calculation of atmosphere.

73. *Answer:* **(a) 154 mm Hg (b) X_{N_2}=0.777, X_{O_2} = 0.208, X_{Ar} = 0.0093, X_{CO_2} = 0.0003, X_{H_2O} = 0.0053 (c) 77.7% N_2, 20.8% O_2, 0.93% Ar, 0.03% CO_2, 0.54% H_2O; slight difference, since this sample is wet**

Strategy and Explanation: Given the partial pressure of several gases in a sample of the atmosphere, and the total pressure of the sample, determine the partial pressure of O_2, the mole fraction of each gas, and the percent by volume. Compare the percentages to Table 10.1.

Dalton's law and its applications are described in Section 10.7, page 456-7. Dalton's law of partial pressures states that the total pressure (P) exerted by a mixture of gases is the sum of their partial pressures (P_1, P_2, P_3, etc.), if the volume (V) and temperature (T) are constant.

$$P_{tot} = P_1 + P_2 + P_3 + \ldots \qquad \text{(V and T constant)}$$

Because $P_i = X_i P_{tot}$, we can use the total pressure and the partial pressure of a component to determine its mole fraction:

$$X_i = \frac{P_i}{P_{tot}}$$

According to Avogadro's law, moles and gas volumes are proportional, so the mole fraction is equal to the volume fraction. To get percent, multiply the volume fraction by 100%.

(a)

$$P_{tot} = P_{N_2} + P_{O_2} + P_{Ar} + P_{CO_2} + P_{H_2O}$$

$$P_{O_2} = P_{tot} - P_{N_2} - P_{Ar} - P_{CO_2} - P_{H_2O}$$

$$= (740.\text{ mm Hg}) - (575 \text{ mm Hg}) - (6.9 \text{ mm Hg}) - (0.2 \text{ mm Hg}) - (4.0 \text{ mm Hg}) = 154 \text{ mm Hg}$$

(b)

$$X_{N_2} = \frac{P_{N_2}}{P_{tot}} = \frac{575 \text{ mmHg}}{740.\text{ mmHg}} = 0.777 \qquad X_{O_2} = \frac{P_{O_2}}{P_{tot}} = \frac{154 \text{ mmHg}}{740.\text{ mmHg}} = 0.208$$

$$X_{Ar} = \frac{P_{Ar}}{P_{tot}} = \frac{6.9 \text{ mmHg}}{740.\text{ mmHg}} = 0.0093 \qquad X_{CO_2} = \frac{P_{CO_2}}{P_{tot}} = \frac{0.2 \text{ mmHg}}{740.\text{ mmHg}} = 0.0003$$

$$X_{H_2O} = \frac{P_{H_2O}}{P_{tot}} = \frac{4.0 \text{ mmHg}}{740.\text{ mmHg}} = 0.0054$$

(c)

$$\%\ N_2 = X_{N_2} \times 100\% = 0.777 \times 100\% = 77.7\%\ N_2$$

$$\% \, O_2 = X_{O_2} \times 100\% = 0.208 \times 100\% = 20.8\% \, O_2$$

$$\% \, Ar = X_{Ar} \times 100\% = 0.0093 \times 100\% = 0.93\% \, Ar$$

$$\% \, CO_2 = X_{CO_2} \times 100\% = 0.0003 \times 100\% = 0.03\% \, CO_2$$

$$\% \, H_2O = X_{H_2O} \times 100\% = 0.0054 \times 100\% = 0.54\% \, H_2O$$

The Table 10.1 figures are slightly different. This sample is wet, whereas the proportions given in Table 10.1 are for dry air.

✓ *Reasonable Answer Check:* The percentages are very close to those provided in the table, and the variations are explainable. The sum of the mole fractions is 1, and the sum of the percentages is 100%.

75. *Answer:* **(a) 1.98 atm (b) P_{O_2} = 0.438 atm, P_{N_2} = 0.182 atm, P_{Ar} = 1.36 atm**

Strategy and Explanation: Given three containers of gases, with known volume and pressure, determine the total pressure and partial pressures of the gases when the three containers are opened to each other.

Use Boyle's law to determine how the pressure of each gas changes with the increase in total volume and then use Dalton's law to determine the total pressures after mixing. Use the definition of partial pressure to show that the pressures calculated are the partial pressure.

The gas in one chamber is allowed to diffuse into all three chambers, so its final volume increases to the total of the three volumes: V_{tot} = 3.00 L + 2.00 + 5.00 L = 10.00 L

Boyle's law: $P_{f,gas} = \dfrac{P_{i,gas} V_{i,gas}}{V_{f,gas}}$ For O_2, $P_{f,O_2} = \dfrac{P_i V_i}{V_f} = \dfrac{(1.46 \text{ atm}) \times (3.00 \text{ L})}{(10.00 \text{ L})} = 0.438 \text{ atm}$

For N_2, $P_{f,N_2} = \dfrac{(0.908 \text{ atm}) \times (2.00 \text{ L})}{(10.00 \text{ L})} = 0.182 \text{ atm}$ For Ar, $P_{f,Ar} = \dfrac{(2.71 \text{ atm}) \times (5.00 \text{ L})}{(10.00 \text{ L})} = 1.36 \text{ atm}$

(a) $P_{f,tot} = P_{f,O_2} + P_{f,N_2} + P_{f,Ar} = 0.438 \text{ atm} + 0.182 \text{ atm} + 1.36 \text{ atm} = 1.98 \text{ atm}$

(b) Partial pressure is the pressure each gas would cause if it were alone in the container.
$P_{f,O_2} = P_{O_2} = 0.438 \text{ atm}$, $P_{f,N_2} = P_{N_2} = 0.182 \text{ atm}$, $P_{f,Ar} = P_{Ar} = 1.36 \text{ atm}$

✓ *Reasonable Answer Check:* All the individual final pressures are less than the initial pressures, which makes sense because the volume is larger. The total pressure is a weighted average of the three pressures, and is influenced most by the gas present in largest quantity (Ar).

77. *Answer:* **Membrane irritation: 1 × 10^{-4} atm; Fatal narcosis: 0.02 atm**

Strategy and Explanation: Given two concentrations of a gas as pressure ratios, determine the partial pressure of the gas at a given temperature and pressure.

Use the ppm pressure ratio values to get the partial pressure:

$$P_i = (\text{pressure ratio}) \times P_{tot}$$

$$P_{tot} = 1 \text{ atm at STP}$$

Membrane irritation: $P_{benzene} = \dfrac{100 \text{ atm}}{1{,}000{,}000 \text{ atm}} \times 1 \text{ atm} = 1 \times 10^{-4} \text{ atm}$

Fatal narcosis: $P_{benzene} = \dfrac{20{,}000 \text{ atm}}{1{,}000{,}000 \text{ atm}} \times 1 \text{ atm} = 0.02 \text{ atm}$

✓ *Reasonable Answer Check:* The larger pressure ratio has a larger partial pressure.

79. *Answer:* **X_{H_2O} = 0.0041; P_{H_2O} = 3.1 mm Hg; the mean partial pressure includes humid and dry air, summer and winter, worldwide**

Strategy and Explanation: Given the mean fraction by weight of water in the atmosphere and the molar mass of "air," determine the mean mole fraction of water in the atmosphere.

The mean fraction by weight describes the average number of grams of water per gram of air. Convert grams to moles in both cases, to get mole fraction. Use $P_i = X_i P_{tot}$, to get its partial pressure.

The molar mass of air is given at 29.2 g/mol. The molar mass of water is 18.0152 g/mol.

$$X_{H_2O} = \frac{0.0025 \text{ g } H_2O}{1 \text{ g air}} \times \frac{29.2 \text{ g air}}{1 \text{ mol air}} \times \frac{1 \text{ mol } H_2O}{18.0152 \text{ g } H_2O} = 0.0041 \frac{\text{mol } H_2O}{\text{mol air}} = 0.0041$$

Assume that the P_{tot} = 760 mm Hg, standard pressure.

$$P_{H_2O} = X_{H_2O}P_{tot} = (0.0041) \times (760 \text{ mm Hg}) = 3.1 \text{ mm Hg}$$

This number represents the mean partial pressure, both humid and dry air, summer and winter, worldwide is included in this average.

✓ *Reasonable Answer Check:* The mole fraction is small, meaning the partial pressure is small compared to the total pressure.

The Behavior of Real Gases (Section 10.8)

82. *Answer:* **18 mL $H_2O(\ell)$; 22.4 L $H_2O(g)$; No we cannot achieve a pressure of 1 atm at this temperature, because the vapor pressure of water at 0 °C to be 4.6 mm Hg, at pressures higher than this, the water would liquefy.**

Strategy and Explanation: Use standard conversion factors:

$$1 \text{ mol } H_2O \times \frac{18.02 \text{ g } H_2O}{1 \text{ mol } H_2O} \times \frac{1 \text{ mL } H_2O}{1.0 \text{ g } H_2O} = 18 \text{ mL } H_2O \text{ liquid}$$

1 mol H_2O at STP occupies 22.4 L.

Table 10.4 gives the vapor pressure of water at 0 °C to be 4.6 mm Hg. At pressures higher than this, water would liquefy. We cannot achieve 1 atm pressure of water vapor at this low temperature, so we cannot achieve the standard state condition for water vapor.

84. *Answer/Explanation:* The behavior of real gases is discussed in Section 10.8. At low temperatures, the molecules are moving relatively slowly; however, when the pressure is very low, they are still quite far apart. As external pressures increases, the gas volume decreases, the slow molecules are squeezed closer together, and the attractions among the molecules get stronger. Figure 10.17 shows that a gas molecule strikes the walls of the container with less force due to the attractive forces between it and its neighbors. This makes the mathematical product PV smaller than the mathematical product nRT.

86. *Answer:* **N_2**

Strategy and Explanation: Table 10.5 in Section 10.8 gives the values the van der Waals constants, a and b. The a constant is related to the pressure correction. The smaller the value of a, the closer to ideal the gas is.

$a_{N_2} = 1.39$ $a_{CO_2} = 3.59$

N_2 is more like an ideal gas at high pressures.

Ozone and Ozone Depletion (Section 10.9)

88. *Answer:* **(a) $\cdot CF_3 + \cdot Cl$ (b) $ClO\cdot$ (c) $\cdot Cl + O_2$**

Strategy and Explanation: The equations for these reactions can be found in Section 10.9.

(a) $CF_3Cl \xrightarrow{h\nu} \cdot CF_3 + \cdot Cl$

(b) $\cdot Cl + \cdot O\cdot \longrightarrow ClO\cdot$

(c) $ClO\cdot + \cdot O\cdot \longrightarrow \cdot Cl + O_2$

90. *Answer/Explanation:* CH_3F has no C–Cl bonds, which in CH_3Cl are readily broken when exposed to UV light. Looking at the bond enthalpies, C–Cl (327 kJ/mol) is much weaker than C–F (486 kJ/mol). In fact, the bond enthalpy of C–F is very close to the bond enthalpy of O=O (498 kJ/mol)!

92. *Answer/Explanation:* CFCs are not toxic. Refrigerants used before CFCs were very dangerous. One example is NH_3, a strong-smelling, reactive chemical. In any web browser, type the keywords "CFCs" or "refrigerants" and you will get a plethora of hits.

Chemistry and Pollution in the Troposphere

95. *Answer:* **Primary pollutants (e.g., particle pollutants, including aerosols and particulates; sulfur dioxide; nitrogen oxides; hydrocarbons) secondary pollutants (e.g., ozone). (see Section 10.10)**

Strategy and Explanation: Primary pollutants are substances that are introduced into the air directly from their source.

- Particle pollutants: pollutants made out of particles:
  - aerosols: particles incorporated into water droplets.
  - particulates: larger solid particles
- Sulfur dioxide: pollutant produced when sulfur or sulfur compounds are burned in air.
- Nitrogen oxides: pollutant produced when nitrogen and oxygen react at high temperatures.
- Hydrocarbons: pollutants produced from many organic sources; their identity is small hydrocarbons from CH_4 to ones with six or seven carbons.

Secondary pollutants are substances produced from reactions of a primary pollutants.

- Ozone: ozone in the troposphere is produced from a reaction of O_2 with O_2 in the presence of intense energy (e.g., spark, lightening, etc.)
- Sulfur trioxide: SO_3 is produced from the reaction of SO_2.
- PAN (peroxyacetylnitrate): produced by the reaction of various free radicals in urban air.

Read Section 10.10 for the specific ways these pollutants are harmful.

97. *Answer/Explanation:* Adsorption is the process of firmly attaching to a surface. Absorption is the process of drawing a substance into the bulk of a solid or liquid.

99. *Answer:* **1.6×10^9 metric tons; 2.3×10^6 hr**

Strategy and Explanation: Given the above information and the percentage of sulfur in the coal, determine the mass (in metric tons) of coal burned and determine the number of hours this quantity of coal will burn.

Determine the moles of sulfur in the product, then determine the mass of coal that contains this amount of sulfur. Use the metric tons of coal per hour to determine the hours one plant would need to burn this quantity.

$$65\times10^6 \text{ metric tons } SO_2 \times \frac{1000 \text{ kg } SO_2}{1 \text{ metric tons } SO_2} \times \frac{1000 \text{ g } SO_2}{1 \text{ kg } SO_2} = 6.5\times10^{13} \text{ g } SO_2$$

$$6.5\times10^{13} \text{ g } SO_2 \times \frac{1 \text{ mol } SO_2}{64.07 \text{ g } SO_2} \times \frac{1 \text{ mol S}}{1 \text{ mol } SO_2} \times \frac{32.065 \text{ g S}}{1 \text{ mol S}} = 3.3\times10^{13} \text{ g S}$$

$$3.3\times10^{13} \text{ g S} \times \frac{100 \text{ g coal}}{2.0 \text{ g S}} \times \frac{1 \text{ kg coal}}{1000 \text{ g coal}} \times \frac{1 \text{ metric tons coal}}{1000 \text{ kg coal}} = 1.6\times10^9 \text{ metric tons coal}$$

$$1.6\times10^9 \text{ metric tons coal} \times \frac{1 \text{ hr}}{700. \text{ metric tons coal}} = 2.3\times10^6 \text{ hr}$$

✓ *Reasonable Answer Check:* That's about 30 decades; that seems like a long time. It is clear that there are many power plants adding SO_2 to the atmosphere for this amount to be produced in one year.

Urban Air Pollution (Section 10.10)

101. *Answer/Explanation:* Photochemical reactions require the absorption of a photon of light. An example of a photochemical reaction is given in this equation:

$$O_2 \xrightarrow{h\nu} \cdot O \cdot + \cdot O \cdot$$

Not all photons of light have sufficient energy to cause photochemical reactions. Looking at the example above, according to the discussion of Section 10.10, we find that only photons with wavelengths of less than

242 nm are able to break this bond. Therefore, this reaction does not occur if the light is visible light, or even low-energy ultraviolet (with wavelength 242 - 300 nm).

103. *Answer:* **SO_2, coal and oil, $2\ SO_2 + O_2 \longrightarrow 2\ SO_3$**

Strategy and Explanation: The reducing nature of industrial (London) smog is due to a sulfur oxide, SO_2. The burning of coal and oil as fuels produces this oxide. Further oxidation happens in air according to this equation:

$$2\ SO_2 + O_2 \longrightarrow 2\ SO_3$$

105. *Answer/Explanation:* The atmospheric reaction that favors the formation of nitrogen monoxide, NO, is given in this chemical equation:

$$N_2 + O_2 \xrightarrow{\text{heat}} 2\ NO$$

The formation of NO in a combustion chamber is similar to the formation of NH_3 in a reactor designed to manufacture ammonia, because in both cases a reaction takes elemental nitrogen and makes a compound of nitrogen.

107. *Answer:* **(a) 1.33×10^{-4} mm Hg (b) $X_{SO_2} = 1.75 \times 10^{-7}$ (c) 500. μg SO_2**

Strategy and Explanation: Given the concentration of an air pollutant, determine the partial pressure, the mole fraction, and the mass in micrograms contained in a given volume of air at STP. (We are given the time of exposure, too, but that information is not needed to answer the questions asked.)

Do the calculation for (b) first by converting the concentration of SO_2 from ppm to mole fraction, using Avogadro's law to relate volume and moles. Then do the calculation for (a) by using $P_i = X_iP_{tot}$ and the standard air pressure to find the partial pressure of SO_2. (c) Use the molar volume of a gas to determine moles of SO_2 in the given volume at STP, then convert to micrograms.

Do (b) first, then (a), then (c).

(b)
$$\frac{0.175\ \text{L SO}_2}{1{,}000{,}000\ \text{L air}} = 1.75\times10^{-7}\ \frac{\text{L SO}_2}{\text{L air}}$$

A gas-volume ratio is equal to a mole ratio, according to Avogadro's law, so

$$X_{SO_2} = 1.75 \times 10^{-7}$$

(a) Assume that the P_{tot} = 760 mm Hg, standard pressure.

$$P_i = X_iP_{tot} = (1.75 \times 10^{-7}) \times (760\ \text{mm Hg}) = 1.33 \times 10^{-4}\ \text{mm Hg}$$

(c)
$$1\ \text{m}^3\ \text{air} \times \left(\frac{100\,\text{cm}}{1\ \text{m}}\right)^3 \times \frac{1\ \text{mL}}{1\ \text{cm}^3} \times \frac{1\ \text{L}}{1000\ \text{mL}} \times \frac{0.175\ \text{L SO}_2}{1{,}000{,}000\ \text{L air}} = 1.75\times10^{-4}\ \text{L SO}_2$$

$$1.75\times10^{-4}\ \text{L SO}_2 \times \frac{1\ \text{mol SO}_2}{22.414\ \text{L SO}_2} \times \frac{64.0638\ \text{g SO}_2}{1\ \text{mol SO}_2} \times \frac{1\ \mu\text{g SO}_2}{10^{-6}\ \text{g SO}_2} = 500.\ \mu\text{g SO}_2$$

✓ *Reasonable Answer Check:* The concentration is expressed in small units (ppm), so small partial pressure, the small mole fraction, and the small mass all make sense.

Greenhouse Gases and Global Warming (Section 10.11)

108. *Answer/Explanation:* Greenhouse effect is the trapping of heat radiation by atmospheric gases. Global warming is the increase of the average global temperature. Global warming is caused by an increase in the amount of greenhouse gases in the atmosphere.

110. *Answer/Explanation:* CO_2 gets into the atmosphere by animal respiration, by burning fossil fuels and other plant materials, and by the decomposition of organic matter. CO_2 gets removed from the atmosphere by plants during photosynthesis, when it is dissolved in rainwater, and when it is incorporated into carbonate and bicarbonate compounds in the oceans. Currently, atmospheric CO_2 production exceeds CO_2 removal.

General Questions

112. *Answer:* **(a) Before: p_{H_2} = 3.7 atm; P_{Cl_2} = 4.9 atm (b) Before: P_{tot} = 8.6 atm (c) After: P_{tot} = 8.6 atm (d) Cl_2; 0.5 mol remain (e) P_{HCl} = 7.4 atm; p_{Cl_2} = 1.2 atm (f) P = 8.9 atm**

Strategy and Explanation: Given the description of a reaction, the masses of two gaseous reactants, and the volume and temperature of the mixture, determine the partial pressure of the reactants, the total pressure due to the gases in the flask before and after the reaction, the excess reactant, the number of moles of it left over, the partial pressures of the gases in the flask, and the total pressure after the temperature is increased to a given value.

Balance the equation. Find moles of each reactant and use the ideal gas law to determine the pressures. Use Dalton's law for total pressures. Find limiting reactant and excess reactant as described in Chapter 4. Use the combined gas law to determine the pressure at a different temperature.

$$H_2(g) + Cl_2(g) \longrightarrow 2HCl(g)$$

(a) $$3.0\text{ g }H_2 \times \frac{1\text{ mol }H_2}{2.0158\text{ g }H_2} = 1.5\text{ mol }H_2 \qquad 140.\text{ g }Cl_2 \times \frac{1\text{ mol }Cl_2}{70.906\text{ g }Cl_2} = 1.97\text{ mol }Cl_2$$

$$T = 28\text{ °C} + 273.15 = 301\text{ K}$$

$$P_{H_2} = \frac{nRT}{V} = \frac{(1.5\text{ mol }H_2)\times\left(0.08206\frac{\text{L}\cdot\text{atm}}{\text{K}\cdot\text{mol}}\right)\times(301\text{ K})}{10.\text{ L}} = 3.7\text{ atm}$$

$$P_{Cl_2} = \frac{nRT}{V} = \frac{(1.97\text{ mol }Cl_2)\times\left(0.08206\frac{\text{L}\cdot\text{atm}}{\text{K}\cdot\text{mol}}\right)\times(301\text{ K})}{10.\text{ L}} = 4.9\text{ atm}$$

(b) $$P_{tot,\ before} = P_{H_2} + P_{Cl_2} = (3.7\text{ atm }H_2) + (4.9\text{ atm }Cl_2) = 8.6\text{ atm total}$$

(c) The number of moles of gas reactants is equal to the number of moles of gaseous products, so the total pressure will not change. $P_{tot,\ after}$ = 8.6 atm total

(d) The balanced chemical equation shows equal molar quantities of each reactant reacting, so the H_2 is the limiting reactant and Cl_2 is the excess reactant, and 1.5 mol of each reactant react. Subtracting the moles of Cl_2 that react from the moles of Cl_2 describes how many moles of Cl_2 remain.

$$2.0\text{ mol }Cl_2\text{ initial} - 1.5\text{ mol }Cl_2\text{ react} = 0.5\text{ mol }Cl_2\text{ remain}$$

(e) When 1.5 mol of each reactant react, 3.0 mol of HCl are formed.

$$P_{HCl} = \frac{nRT}{V} = \frac{(3.0\text{ mol HCl})\times\left(0.08206\frac{\text{L}\cdot\text{atm}}{\text{K}\cdot\text{mol}}\right)\times(301\text{ K})}{10.\text{ L}} = 7.4\text{ atm}$$

$$P_{Cl_2} = PP_{tot,\ after} - P_{HCl} = (8.6\text{ atm total}) - (7.4\text{ atm HCl}) = 1.2\text{ atm }Cl_2$$

(f) $T_1 = 301$ K, $T_2 = 40.$ °C + 273.15 = 313 K

$$P_2 = P_1 \times \frac{V_1}{V_2} \times \frac{T_2}{T_1} = P_1 \times \frac{T_2}{T_1}\text{ at constant volume}$$

$$P_2 = P_1 \times \frac{T_2}{T_1} = (8.6\text{ atm}) \times \frac{(313\text{K})}{(301\text{ K})} = 8.9\text{ atm}$$

✓ *Reasonable Answer Check:* The partial pressure of Cl_2 calculated in (e) can also be found using the ideal gas law:

$$P_{Cl_2} = \frac{nRT}{V} = \frac{(0.5\text{ mol }Cl_2)\times\left(0.08206\frac{\text{L}\cdot\text{atm}}{\text{K}\cdot\text{mol}}\right)\times(301\text{ K})}{10.\text{ L}} = 1.\text{ atm}$$

We could construct mole fractions of the products and multiply them by the total pressure to determine the partial pressures after the reaction was completed:

$$X_{HCl} = \frac{3.0 \text{ mol HCl}}{3.0 \text{ mol HCl} + 0.50 \text{ mol} Cl_2} = 0.86$$

$$P_{HCl} = X_{HCl}P_{tot} = (0.86) \times (8.6 \text{ atm}) = 7.4 \text{ atm}$$

$$X_{Cl_2} = \frac{0.50 \text{ mol } Cl_2}{3.0 \text{ mol HCl} + 0.5 \text{ mol} Cl_2} = 0.14$$

$$P_{Cl_2} = X_{Cl_2}P_{tot} = (0.14) \times (8.6 \text{ atm}) = 1.2 \text{ atm}$$

$$P_{tot,\ after} = P_{HCl} + P_{Cl_2} = 7.4 \text{ atm} + 1.2 \text{ atm} = 8.6 \text{ atm total}$$

These answers are the same as calculated above. The results are self-consistent and reasonable size (i.e., smaller number of moles produce smaller partial pressures and smaller mole fractions).

Applying Concepts

114. *Answer:* **Statements (a), (b), (c), and (d) are true.**

Strategy and Explanation: We are considering a real gas, N_2(g), whose conditions are such that it obeys the ideal gas law exactly.

(a) N_2 is a larger molecule than Ne is an atom and therefore N_2 is not as much like an ideal gas. If these conditions allow N_2 to behave as an ideal gas, they would certainly be sufficient for Ne to also behave like an ideal gas. Hence, this statement: "A sample of Ne(g) under the same conditions must obey the ideal gas law exactly." is true.

(b) All collisions are elastic, however, energy can be transferred during an elastic collision from one molecule to another. A faster molecule hitting a slower molecule might make the slow molecule go faster if the first one ends up slower. In addition, each time the molecule hits the wall, there is a split second when its speed is zero as it bounces off the wall and goes flying away in the opposite direction. Hence, the statement: "The speed at which one particular N_2 molecule is moving changes from time to time." is true.

(c) The average speed of the N_2 molecules will be faster than the average speed of the O_2 molecules, but ideal gas particles move at varying speeds, so some molecules in each sample will be moving very slowly and others very quickly. Hence, the statement: "Some of the N_2 molecules are moving more slowly than some of the molecules in a sample of O_2(g) under the same conditions." is true.

(d) The average speed of the N_2 molecules will be slower than the average speed of the Ne molecules; hence, the statement: "Some of the N_2 molecules are moving more slowly than some of the molecules in a sample of Ne(g) under the same conditions." is true.

(e) All collisions are elastic, so collisions must conserve energy. There is no way that both molecules could be going faster, since that implies that energy has increased. Hence, the statement: "When two N_2 molecules collide, it is possible that both may be moving faster after the collision than they were before." is false.

116. *Answer:* **See graph below**

Strategy and Explanation: The molar masses of C_2H_6 and F_2 are 30 g/mol and 38 g/mol. That means the average speed of F_2 is somewhat less than that of C_2H_6. The total pressure of the gases is 720. mm Hg, and the partial pressure of F_2 is 540. mm Hg. Dalton's law tells us that the sum of the partial pressures is the total pressure, and so the partial pressure of C_2H_6 is 180. mm Hg, or one third that of the F_2 molecules. The partial pressure is directly proportional to the mole fraction of the molecules, in the container, so there should be one third as many C_2H_6 molecules as F_2 molecules. So, the graph of number of molecules verses molecular speed will have the F_2 curve peaking at a slightly smaller speed value and the C_2H_6 curve will have one third of the vertical rise.

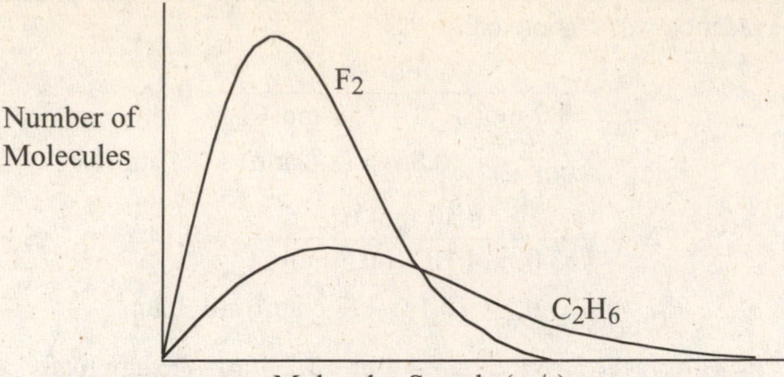

119. *Answer:* **(a) Pressure: Box (a) = Box (c) > Box (b) = Box (d) (b) Density: Box (c) > Box (a) = Box (d) > Box (b) (c) Average kinetic energy Box (a) = Box (b) = Box (c) = Box (d) (d) Average molecular speed Box (a) = Box (b) = Box (c) = Box (d)**

Strategy and Explanation:

(a) The pressure of a gas is related to how often the particles hit the walls of the container. Boxes (a) and (c) have equal volume and equal number of particles, so the pressure in those boxes are equal. . Boxes (b) and (d) have equal volume and equal number of particles, so the pressure in those boxes are equal. Boxes (a) and (c) are half the volume of Boxes (b) and (d) with the same number of atoms, so their pressures will be greater. Therefore, the samples rank in this order: Box (a) = Box (c) > Box (b) = Box (d)

(b) The density of a gas is the relationship between the mass of the gas particles and the volume of the container. The masses of the neon atoms are (20g/4g) five times larger than the masses of the helium atoms. The small boxes are half the volume of the large boxes.

Box (c) has the heavy-weight atoms in the smaller volume, so the gas density in that box would be the largest. Box (a) is half the size of Box (d) with a mass of atoms fives times smaller, so the density in Box (d) is greater than the density in Box (a). The same number of light-weight atoms in the larger box, Box (b), will result in the lowest density. Therefore, the samples rank in this order: Box (c) > Box (d) > Box (a) > Box (b)

(c) $E = \frac{1}{2} mv^2$ and all the samples have the same temperature, so the average kinetic energy in each sample is equal, so the samples rank in this order: Box (a) = Box (b) = Box (c) = Box (d)

(d) The speed of the atoms is related to the mass and the temperature: $T \propto mv^2$. All the samples have the same temperature and the average speed of the heavier atoms is slower than the average speed of the lighter-weight atoms. And the samples rank in this order: Box (a) = Box (b) > Box (c) = Box (d)

121. *Answer:* **See drawings below**

Strategy and Explanation: The speed of the molecules is related to the temperature. In particular, $T \propto mv^2$. So, when the temperature changes by a certain factor, the speed of the molecules (v) changes by the square root of that same factor. We'll indicate that fact by making the tails of the arrows proportionately shorter. The pressure describes how close together the molecules are. When the pressure is changed by a certain factor, the molecules will be represented as that much closer together. The volume will change the space occupied by the gas, and that will be indicated in the movement of the syringe plunger.

For reference, the initial state looks like this:

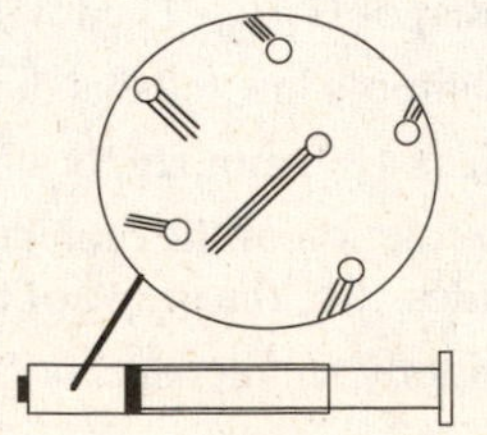

(a) When the temperature is decreased to one half of its original value, Charles' law tells us that the volume decreases by half. The molecules are just as far apart as they were (since the pressure is the same), but the plunger level goes down by half, and the tails on the molecules decrease in length by about 0.7 times.

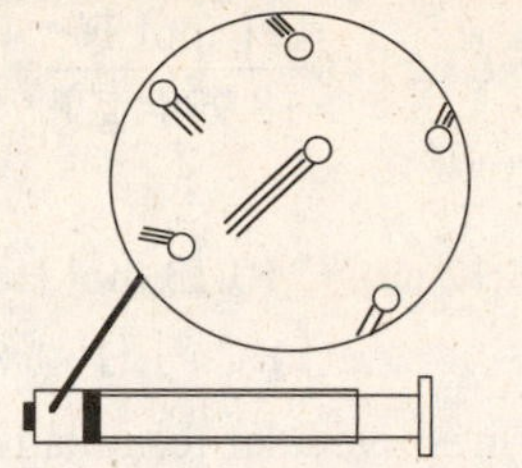

(b) When the pressure decreased to one half of its original value, Boyle's law tells us that the volume increases by half. The tails on the molecules are the same length, but the molecules are twice as far apart as they were (so we put half as many in this view), and the plunger level moves up by half.

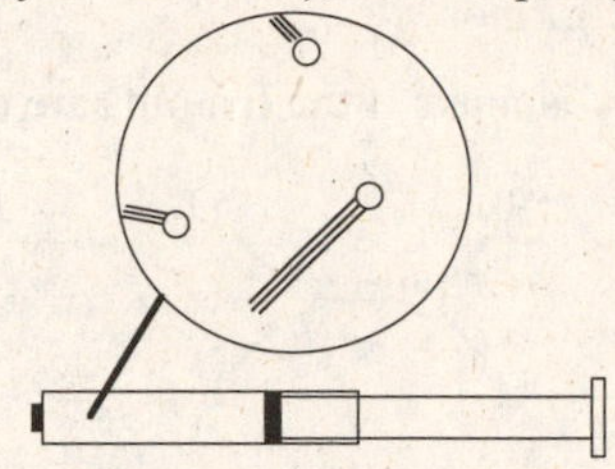

(c) When the temperature is tripled and the pressure is doubled, the combined gas law tells us that the volume will have a net increase by 3/2 (three times larger due to the temperature change and half as large due to the pressure change). The molecules are two times closer together than they were (so we'll add twice as many), the plunger level goes up by three halves, and the tails on the molecules increase in length by about 1.7 times.

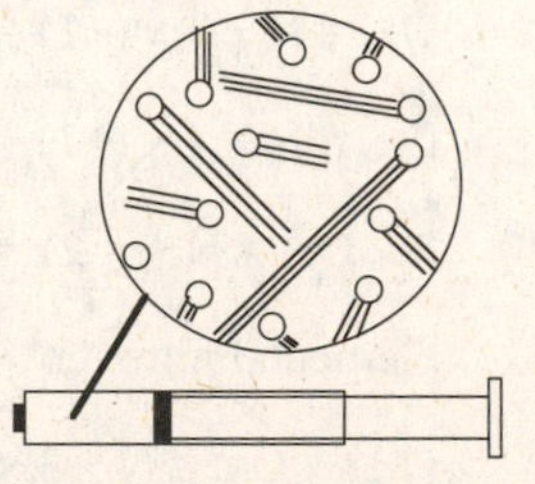

123. *Answer:* **Box (b)**

Strategy and Explanation: The initial volume is 1.8 L and the final volume is 0.9 L. This 2:1 ratio in the gas volumes means for every two molecules of gas reactants there must be one molecule of gas products. The reactant count is six, so the box that has three product molecules fits these observations. The correct box is Box (b):

$$6\ AB_2(g) \longrightarrow 3\ A_2B_4(g)$$

125. *Answer:* **(a) 64.1 g/mol (b) empirical: CHF, molecular: $C_2H_2F_2$ (c) see structures below**

Strategy and Explanation: Use the methods described in the solution to Question 68.

(a)

$$T = 50.0\ °C + 273.15 = 323.2\ K$$

$$750.\ mmHg \times \frac{1\ atm}{760.\ mmHg} = 0.987\ atm \qquad 125\ mL \times \frac{1\ L}{1000\ mL} = 0.125\ L$$

$$n_{C_xH_yF_z} = \frac{PV}{RT} = \frac{(0.987\ atm)\times(0.125\ L)}{\left(0.08206\frac{L\cdot atm}{mol\cdot K}\right)\times(323.2\ K)} = 4.65\times10^{-3}\ mol\ C_xH_yF_z$$

$$\text{Molar Mass} = \frac{0.298\ g\ C_xH_yF_z}{4.65\times10^{-3}\ mol\,C_xH_yF_z} = 64.1\ g/mol\ \ C_xH_yF_z$$

(b) $37.5 \text{ g C} \times \frac{1 \text{ mol C}}{12.0107 \text{ g C}} = 3.12 \text{ mol C}$ $3.15 \text{ g H} \times \frac{1 \text{ mol H}}{1.0079 \text{ g H}} = 3.13 \text{ mol H}$

$$59.3 \text{ g F} \times \frac{1 \text{ mol F}}{18.9984 \text{ g F}} = 3.12 \text{ mol F}$$

Set up a mole ratio and simplify:

$$3.12 \text{ mol C} : 3.13 \text{ mol H} : 3.12 \text{ mol F}$$

$$1 \text{ C} : 1 \text{ H} : 1 \text{ F}$$

The empirical formula is CHF and the molecular formula is $(CHF)_n$. So, n = x = y = z

The molar mass of the empirical formula = 32.02 g/mol

$n = \frac{63 \text{ g/mol}}{32.02 \text{ g/mol}} = 2.0$ The molecular formula is $C_2H_2F_2$.

(c) $C_2H_2F_2$ can have *cis-* and *trans-*isomers if the F atoms are on different C atoms. (described in Chapter 8)

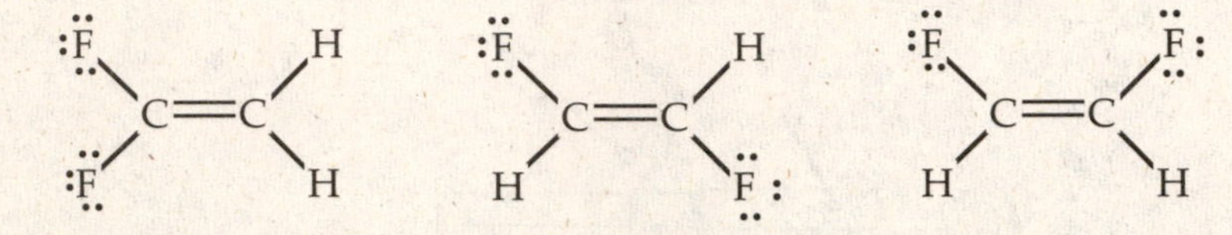

127. *Answer:* **(a) NH_3; Ox. # N = –3 (b) NH_4^+; Ox. # N = –3 (c) N_2O; Ox. # N = + 1 (d) N_2; Ox. # N = 0 (e) HNO_3; Ox. # N = + 5 (f) HNO_2; Ox. # N = + 3 (g) NO_2; Ox. # N = + 4**

Strategy and Explanation: Names related to formulas were described in Chapter 3. Oxidation numbers were described in Section 5.4.

(a) Ammonia is: NH_3 Ox. # N + 3 × (+1) = 0 Ox. # N = –3

(b) Ammonium is: NH_4^+ Ox. # N + 4 × (+1) = +1 Ox. # N = –3

(c) Dinitrogen monoxide is: N_2O 2 × Ox. # N + (–2) = 0 Ox. # N = + 1

(d) Gaseous nitrogen is: N_2 elemental nitrogen Ox. # N = 0

(e) Nitric acid is: HNO_3 (+1) + Ox. # N + 3 × (–2) = 0 Ox. # N = + 5

(f) Nitrous acid is; HNO_2 (+1) + Ox. # N + 2 × (–2) = 0 Ox. # N = + 3

(g) Nitrogen dioxide is: NO_2 Ox. # N + 2 × (–2) = 0 Ox. # N = + 4

129. *Answer:* **(a) •OH + •OH (b) HNO_3 (c) •R + H_2O**

Strategy and Explanation: The equations for these reactions can be found in Section 10.10.

(a) $\cdot O \cdot + H_2O \longrightarrow \cdot OH + \cdot OH$

(b) $\cdot NO_2 + \cdot OH \longrightarrow HNO_3$

(c) $RH + \cdot OH \longrightarrow \cdot R + H_2O$

More Challenging Questions

130. *Answer:* **0.88 atm**

Strategy and Explanation: Given the names and masses of three gases and the temperature and volume of the vessel they occupy, determine the total pressure in the vessel.

Calculate the moles of each substance, using the molar masses. Add the moles of each gas to determine the total moles of gas in the vessel. Use the volume and temperature in the ideal gas law to determine the total pressure.

Carbon dioxide is CO_2 and methane is CH_4. $V_{tot} = 8.7$ L

Find moles of CO_2, CH_4, and Ar in the sample:

$$3.44 \text{ g } CO_2 \times \frac{1 \text{ mol } CO_2}{44.009 \text{ g } CO_2} = 0.0782 \text{ mol } CO_2$$

$$1.88 \text{ g } CH_4 \times \frac{1 \text{ mol } CH_4}{16.0423 \text{ g } CH_4} = 0.117 \text{ mol } CH_4$$

$$4.28 \text{ g Ar} \times \frac{1 \text{ mol Ar}}{39.948 \text{ g Ar}} = 0.107 \text{ mol Ar}$$

$$n_{tot} = 0.0781 \text{ mol } CO_2 + 0.117 \text{ mol } CH_4 + 0.107 \text{ mol Ar} = 0.302 \text{ mol tot}$$

$$T = 37\ °C + 273.15 = 310.\ K$$

$$P = \frac{n_{tot}RT}{V_{tot}} = \frac{(0.302 \text{ mol}) \times \left(0.08206 \frac{L \cdot atm}{mol \cdot K}\right) \times (310.\ K)}{(8.7\ L)} = 0.88 \text{ atm}$$

✓ *Reasonable Answer Check:* The percentages are very close to those provided in the table, and the variations are explainable. The sum of the mole fractions is 1, and the sum of the percentages is 100%.

132. *Answer:* $\mathbf{\frac{m_{Ne}}{m_{Ar}} = 1.0}$

Strategy and Explanation: Given two known gaseous substances in separate balloons at the same temperature and pressure with one volume designated as double the other, determine the mass ratio of the two substances.

Use the ideal gas law and the relationship between mass and moles to determine the mass ratio in terms of volume and molar mass: M=m/n

$$PV = nRT = \left(\frac{m}{M}\right)RT$$

Rearranging to solve for m gives: $m = \frac{MPV}{RT}$

At equal T and P: $\frac{m_2}{m_1} = \frac{M_2V_2}{M_1V_1}$

$$\frac{m_{Ne}}{m_{Ar}} = \frac{M_{Ne}V_{Ne}}{M_{Ar}V_{Ar}}$$

The volume of the neon balloon is twice that of the argon balloon. $V_{Ne} = 2\ V_{Ar}$ so $\frac{V_{Ne}}{V_{Ar}} = 2$

$$\frac{m_{Ne}}{m_{Ar}} = \frac{20.1797 \text{ g/mol}}{39.948 \text{ g/mol}}(2) = 1.0 = \text{mass ratio}$$

The mass of the neon in the orange balloon is approximately equal the mass of the argon in the blue balloon.

✓ *Reasonable Answer Check:* Even though the neon balloon is twice the volume, the heavier argon atoms result in a more massive argon sample.

134. *Answer:* **(a) 29.1 mol CO_2; 14.6 mol N_2; 2.43 mol O_2 (b) 1.1×10^3 L (c) P_{CO_2}= 0.631 atm; P_{N_2}= 0.317 atm; P_{O_2} = 0.0527 atm**

Strategy and Explanation: Given the mass of a reactant and the balanced chemical equation for the explosive decomposition of the reactant, determine the total moles of gases produced, the volume occupied by the product gases at a given temperature and pressure, and the partial pressures of each gas.

Use molar mass to determine the moles of reactant, and use equation stoichiometry to calculate the moles of each *gas phase* reactant. (Notice: reactants that are not gas phase will not contribute this sum). Add the moles of each gas to determine the total moles of gas in the vessel. Use the pressure and temperature in the ideal gas law to determine the total volume. Calculate mole fraction with the following equation: $X_i = \frac{n_i}{n_{tot}}$ then partial pressure is calculated with $P_i = X_iP_{tot}$.

Calculate the moles of reactant, using the molar masses.

$$1.00 \text{ kg } C_3H_5(NO_3)_3 \times \frac{1000 \text{ g}}{1 \text{ kg}} \times \frac{1 \text{ mol } C_3H_5(NO_3)_3}{103.0765 \text{ g } C_3H_5(NO_3)_3} = 9.70 \text{ mol } C_3H_5(NO_3)_3$$

(a) Find moles of CO_2, N_2, and O_2 in the sample: (Water is a liquid.)

$$9.70 \text{ mol } C_3H_5(NO_3)_3 \times \frac{12 \text{ mol } CO_2}{4 \text{ mol } C_3H_5(NO_3)_3} = 29.1 \text{ mol } CO_2$$

$$9.70 \text{ mol } C_3H_5(NO_3)_3 \times \frac{6 \text{ mol } N_2}{4 \text{ mol } C_3H_5(NO_3)_3} = 14.6 \text{ mol } N_2$$

$$9.70 \text{ mol } C_3H_5(NO_3)_3 \times \frac{1 \text{ mol } O_2}{4 \text{ mol } C_3H_5(NO_3)_3} = 2.43 \text{ mol } O_2$$

(b)

$$n_{tot} = 29.1 \text{ mol } CO_2 + 14.6 \text{ mol } N_2 + 2.43 \text{ mol } O_2 = 46.1 \text{ mol tot}$$

$$T = 25\ °C + 273.15 = 298 \text{ K}$$

$$V_{tot} = \frac{n_{tot}RT}{P} = \frac{(46.1 \text{ mol}) \times \left(0.08206 \frac{L \cdot atm}{mol \cdot K}\right) \times (298 \text{ K})}{(1.0 \text{ atm})} = 1.1 \times 10^3 \text{ L}$$

(c) $X_{CO_2} = \frac{n_{CO_2}}{n_{tot}} = \frac{29.1 \text{ mole } CO_2}{46.1 \text{ mole tot}} = 0.631$ $\quad P_{CO_2} = (0.631)(1.00 \text{ atm}) = 0.631 \text{ atm}$

$X_{N_2} = \frac{n_{N_2}}{n_{tot}} = \frac{14.6 \text{ mole } N_2}{46.1 \text{ mole tot}} = 0.317$ $\quad P_{N_2} = (0.317)(1.00 \text{ atm}) = 0.317 \text{ atm}$

$X_{O_2} = \frac{n_{O_2}}{n_{tot}} = \frac{2.43 \text{ mole } O_2}{46.1 \text{ mole tot}} = 0.0527$ $\quad P_{O_2} = (0.0527)(1.00 \text{ atm}) = 0.0527 \text{ atm}$

✓ *Reasonable Answer Check:* The decomposition of TNT produces a large volume of gas product, which explains one reason why this decomposition is explosive. The sum of the mole fractions is 1, and the sum of the partial pressures is the total pressure.

136. *Answer:* **458 torr**

Strategy and Explanation: Given the initial pressure of a gas, the relative amount of it that undergoes reaction, and the stoichiometric relationship between the reactants and products, determine the new pressure in the container.

Determine the pressure of the gas reactant that reacts. Use the equation stoichiometry interpreted in units of pressure for the gases to determine pressure of the gas products. Then add this number to the pressure of the unreacted reactant gas to calculate the final pressure.

Half of the reactant reacts, so the pressure of the reactant that undergoes reaction is:

$$0.5 \times (550. \text{ torr}) = 275 \text{ torr.}$$

$$275 \text{ torr reactant} \times \frac{2 \text{ torr product}}{3 \text{ torr reactant}} = 183 \text{ torr product}$$

The unreacted reactant has a pressure of 550. torr – 275 torr = 275 torr

The total pressure after this reaction is complete = 275 torr + 183 torr = 458 torr

✓ *Reasonable Answer Check:* The reaction reduces the number of gasses, so it makes sense that the pressure after some of the reactant has reacted is less.

138. *Answer:* **4.5 mm^3**

Strategy and Explanation: Given the volume and pressure of a gas bubble at the bottom of a lake, determine the volume of the same bubble at the surface of the lake with given pressure.

Use Boyle's law to determine how the volume of gas changes with the decrease in pressure:

$$V_{f,bubble} = \frac{P_{i,bubble}V_{i,bubble}}{P_{f,bubble}} = \frac{\left(4.4\ \text{atm} \times \dfrac{760\ \text{torr}}{1\ \text{atm}}\right) \times \left(1\ \text{mm}^3\right)}{(740.\ \text{torr})} = 4.5\ \text{mm}^3$$

✓ *Reasonable Answer Check:* At the lower pressure, it makes sense that the bubble will increase in volume.

140. *Answer:* **(a) More significant, because of more collisions (b) More significant, because of more collisions (c) Less significant, because the molecules will move faster**

Strategy and Explanation:

(a) When the gas is compressed to a smaller volume, the molecules will be closer together and they will collide with each other more often. The temperature is fixed, so the molecules will hit each other at the same average speed; however, with more collisions the interactions between the molecules will be more significant.

(b) When more molecules of the same gas are added to the container, the molecules will be closer together and they will collide with each other more often. The temperature is fixed, so the molecules will hit each other at the same average speed; however, with more collisions the interactions between the molecules will be more signficant.

(c) When the temperature is increased, the average kinetic energy of the molecules is increased and the molecules will move faster. That means they will collide with each other more often. Because the speed has increased the time these molecules spend in proximity to each other will decrease, so, compared to the situation in (a) and (b), the effect of intermolecular interactions will less significant.

Chapter 11: Liquids, Solids, and Materials

Solutions for Blue-Numbered Questions for Review and Thought

Topical Questions

The Liquid State (Section 11.1)

13. *Answer and Explanation:* Surface tension is based on the ability of a liquid to interact with other particles in the liquid. At higher temperatures, the molecules move around more. The increased random motion disrupts the intermolecular interactions responsible for surface tension.

15. *Answer:* **Reduce the pressure**

Strategy and Explanation: A liquid can be converted to a vapor without changing the temperature by reducing the pressure above it. This can be accomplished by putting the liquid in a container whose pressure can be altered using a vacuum pump. Pumping the gases out of the container reduces the atmospheric pressure acting on the liquid allowing it to reach the boiling point at a much lower temperature. Incidentally, a liquid in an open container will eventually evaporate at temperatures much lower than its boiling point because it absorbs energy from the surroundings and its vapor pressure never reaches the equilibrium vapor pressure, therefore, evaporation continues.

17. *Answer/Explanation:* Evaporation is an endothermic process. The molecules of water in your sweat have a wide distribution of molecular speeds. The fastest of these molecules are more likely to escape the liquid state into the gas phase, removing thermal energy from your body, thus lowering the temperature.

19. *Answer:* **1.5×10^6 kJ**

Strategy and Explanation: Given the molar enthalpy of vaporization of a compound, determine the heat energy required to vaporize a given mass of a compound.

Convert the mass to moles, and use the molar enthalpy of vaporization to get total heat energy.

$$1.0 \text{ metric ton } NH_3 \times \frac{10^3 \text{ kg}}{1 \text{ metric ton}} \times \frac{1000 \text{ g}}{1 \text{ kg}} \times \frac{1 \text{ mol } NH_3}{17.0304 \text{ g } NH_3} \times \frac{25.1 \text{ kJ}}{1 \text{ mol } NH_3} = 1.5 \times 10^6 \text{ kJ}$$

✓ *Reasonable Answer Check:* Units cancel appropriately, and it makes sense that a large amount of ammonia needs a large amount of heat energy.

21. *Answer:* **233 kJ**

Strategy and Explanation: Given the molar enthalpy of vaporization of a compound and the density of the liquid, determine the heat energy required to vaporize a given volume of the liquid.

Convert the volume to moles, then use the molar enthalpy of vaporization to get total heat energy.

$$250. \text{ mL } CH_3OH \times \frac{0.787 \text{ g } CH_3OH}{1 \text{ mL } CH_3OH} \times \frac{1 \text{ mol } CH_3OH}{32.0417 \text{ g } CH_3OH} \times \frac{38.0 \text{ kJ}}{1 \text{ mol } CH_3OH} = 233 \text{ kJ}$$

✓ *Reasonable Answer Check:* Units cancel appropriately, and it makes sense that a quantity of about six moles needs about six times the molar enthalpy.

23. *Answer:* **2.00 kJ for Hg sample; 1.13 kJ for water sample; even though the heat of vaporization for water is greater than for Hg, the Hg sample takes more energy because of its greater mass.**

Strategy and Explanation: Given the enthalpy of vaporization of two compounds and the density of their liquids, determine the heat energy required to vaporize a given volume of each liquid and compare them.

Convert to grams and use the heat of vaporization for the mercury. Convert the volume of water to moles, then use the molar heat of vaporization to get total heat energy for the water.

$$0.500 \text{ mL Hg} \times \frac{13.6 \text{ g Hg}}{1 \text{ mL Hg}} \times \frac{294 \text{ J}}{1 \text{ g Hg}} \times \frac{1 \text{ kJ}}{1000 \text{ J}} = 2.00 \text{ kJ to vaporize Hg sample}$$

$$0.500 \text{ mL } H_2O \times \frac{1.00 \text{ g } H_2O}{1 \text{ mL } H_2O} \times \frac{1 \text{ mol } H_2O}{18.0152 \text{ g } H_2O} \times \frac{40.7 \text{ kJ}}{1 \text{ mol } H_2O} = 1.13 \text{ kJ to vaporize water sample}$$

It takes more energy to vaporize the Hg sample than the water sample. Even though the heat of vaporization for water is greater than for Hg, the Hg sample takes more energy because of its greater mass.

✓ *Reasonable Answer Check:* Units cancel appropriately. Equal volumes of water and Hg have very different masses, due to the large difference in their densities. So, while water has a larger heat of vaporization (2.26×10^3 J/g), the number of grams of water in the sample volume is quite a bit smaller.

25. *Answer/Explanation:* NH_3 has a relatively large boiling point because the molecules interact using relatively strong hydrogen bonding intermolecular forces. The increase in the boiling points of the series PH_3, AsH_3, and SbH_3 is related to the increasing London dispersion intermolecular forces experienced due to the larger central atom in the molecule and the greater polarizability of the valence electrons. (size: P < As < Sb)

Vapor Pressure (Section 11.2)

27. *Answer/Explanation:* Methanol molecules are capable of hydrogen bonding, whereas formaldehyde molecules use dipole-dipole forces to interact. Molecules experiencing stronger intermolecular forces (such as methanol here) will have higher boiling points and lower vapor pressures compared to molecules experiencing weaker intermolecular forces (such as formaldehyde here).

28. *Answer:* **Substance D; it has lower vapor pressure at any given temperature than A, B, or C.**

    Strategy and Explanation: Vapor pressure increases as more molecules are able to escape the liquid state, so the substance with the greatest intermolecular attractive forces at a given temperature is the substance with the lowest vapor pressure. Looking at the graph at 25°C, curve D has the lowest vapor pressure, so substance D has the greatest intermolecular attractive forces.

30. *Answer:* **0.21 atm, approximately 57 °C**

    Strategy and Explanation: Given the altitude on a mountain above sea level, a simple relationship between altitude and pressure changes, and Figure 11.5, determine the atmospheric pressure on the mountain and the boiling point of water at that altitude.

    Use the altitude/pressure relationship to get the atmospheric pressure on the mountain, then look up the boiling temperature on the graph.

$$22834 \text{ ft} \times \frac{3.5 \text{ mbar decrease}}{100 \text{ ft}} \times \frac{1 \text{ bar}}{1000 \text{ mbar}} \times \frac{10^5 \text{ Pa}}{1 \text{ bar}} \times \frac{1 \text{ kPa}}{1000 \text{ Pa}} \times \frac{1 \text{ atm}}{101.325 \text{ kPa}} = 0.79 \text{ atm decrease}$$

$$1.00 \text{ atm} - 0.79 \text{ atm} = 0.21 \text{ atm}$$

$$0.21 \text{ atm} \times \frac{760 \text{ mm Hg}}{1 \text{ atm}} = 160 \text{ mm Hg}$$

Looking at Figure 11.5, this corresponds to a boiling temperature between 40 °C and 60 °C, about 57 °C.

✓ *Reasonable Answer Check:* It makes sense that a lower pressure causes a liquid to have a lower boiling T.

32. *Answer:* **1600 mm Hg**

    Strategy and Explanation: Given the enthalpy of vaporization, and the vapor pressure at a given temperature, determine the vapor pressure at another given temperature.

    Use the two-set Clausius-Claypeyron equation: $\ln\left(\frac{P_2}{P_1}\right) = \frac{-\Delta H_{vap}}{R}\left(\frac{1}{T_2} - \frac{1}{T_1}\right)$

    R is given in Table 10.3. $T_1 = 90°C + 273.15 = 363.15 \text{ K} \cong 360 \text{ K}$ *(round to tens place)*

$$T_2 = 130°C + 273.15 = 403.15 \text{ K} \cong 4.0 \times 10^2 \text{ K}$$

$$\ln\left(\frac{P_2}{370 \text{ mmHg}}\right) = \frac{-(44.0 \text{ kJ/mol})}{\left(8.314 \frac{\text{J}}{\text{K}\cdot\text{mol}}\right)\left(\frac{1 \text{ kJ}}{1000 \text{ J}}\right)}\left(\frac{1}{400 \text{ K}} - \frac{1}{360 \text{ K}}\right)$$

$$\ln\left(\frac{P_2}{370 \text{ mmHg}}\right) = \frac{-(44.0)}{(0.008314)}(0.0025 - 0.0028) = \frac{-(44.0)}{(0.008314)}(-0.0003)$$

$$\ln\left(\frac{P_2}{370 \text{ mmHg}}\right) = 1.5$$

$$P_2 = (370 \text{ mmHg})\, e^{1.5} = 1600 \text{ mm Hg}$$

✓ *Reasonable Answer Check:* Since the temperature increases, the vapor pressure should be larger.

34. *Answer:* **70. kJ/mol**

Strategy and Explanation: Given two temperatures that represent the temperature changes required to double the vapor pressure. Determine the enthalpy of vaporization.

Use the two-set Clausius-Claypeyron equation (given in the solution to Question 32).

R is given in Table 10.3. The pressure doubles, so the pressure ratio is: $\frac{P_2}{P_1} = 2$

$$T_1 = 70.0°C + 273.15 = 343.2 \text{ K} \qquad T_2 = 80.0°C + 273.15 = 353.2 \text{ K}$$

$$\ln(2) = \frac{-\Delta H_{vap}}{\left(8.314 \frac{J}{K \cdot mol}\right)\left(\frac{1 \text{ kJ}}{1000 \text{ J}}\right)}\left(\frac{1}{353.2 \text{ K}} - \frac{1}{343.2 \text{ K}}\right)$$

$$\ln(2) = \frac{-\Delta H_{vap}}{\left(0.008314 \frac{kJ}{mol}\right)}(0.002914 - 0.002832) = \frac{-\Delta H_{vap}}{\left(0.008314 \frac{kJ}{mol}\right)}(-0.000083)$$

$$0.693 = -\Delta H_{vap}\left(-0.0099 \frac{mol}{kJ}\right)$$

$$70. \frac{kJ}{mol} = \Delta H_{vap}$$

✓ *Reasonable Answer Check:* The enthalpy of vaporization must be positive, since energy is required to take a liquid into the gas state. The size is similar to the enthalpy of vaporizations given in earlier questions.

Phase Changes: Solids, Liquids, and Gases (Section 11.3)

36. *Answer/Explanation:* A high melting point and a high heat of fusion tell us that a large amount of energy is required to melt a solid. That is the case when the interparticle interactions between the particles in the solid are very strong, such as in solids composed of ions.

38. *Answer/Explanation:* A higher heat of fusion occurs when the intermolecular forces are stronger. The intermolecular forces between the molecules of H_2O in the solid (hydrogen bonding) are stronger than the intermolecular forces between the molecules of H_2S (dipole-dipole).

40. *Answer:* **27 kJ**

Strategy and Explanation: Given the molar enthalpies of fusion and vaporization of a compound, determine the heat energy required to raise a given number of moles of solid to the melting point, melt it, raise the liquid to the boiling point and boil it.

Find the mass from the moles. Use the molar enthalpies to get total heat energy for both of the phase transitions. Use Equation 6.2' and Table 6.1 to get the heat energy needed to change the temperature of the solid and liquid water (as described in Chapter 6, Section 6.3). Add together the heat energy for each stage to get the total quantity of heat energy.

$$0.50 \text{ mol } H_2O \times \frac{18.0152 \text{ g } H_2O}{1 \text{ mol } H_2O} = 9.0 \text{ g } H_2O$$

Total heat energy is the sum of the heat energies to warm the ice to the melting point (0 °C), the heat energy required to melt the ice, the heat energy to warm the water to the boiling point (100. °C), and the heat energy required to vaporize the water.

$$q_{tot} = q_{heat\ ice} + q_{melt} + q_{heat\ water} + q_{boil}$$

$$q_{tot} = (c_{ice} \times m \times \Delta T) + (n\Delta H_{fus}) + (c_{liquid} \times m \times \Delta T) + (n\Delta H_{vap})$$

$$q_{tot} = \left(\frac{2.06\ J}{g\ °C}\right) \times \left(\frac{1\ kJ}{1000\ J}\right) \times (9.0\ g) \times [0\ °C - (-5\ °C)] + (0.50\ mol) \times \left(\frac{6.020\ kJ}{mol}\right)$$

$$+ \left(\frac{4.184\ J}{g\ °C}\right) \times \left(\frac{1\ kJ}{1000\ J}\right) \times (9.0\ g) \times [100.\ °C - (0\ °C)] + (0.50\ mol) \times \left(\frac{40.07\ kJ}{mol}\right)$$

$$q_{tot} = 0.09\ kJ + 3.0\ kJ + 3.8\ kJ + 20.\ kJ = 27\ kJ$$

✓ *Reasonable Answer Check:* The relative size of the four terms seems sensible, comparing the sizes of the heat capacities and the heats of the phase transitions.

42. *Answer:* **51.9 g CCl_2F_2**

Strategy and Explanation: Given the heat of vaporization of a compound, the heat of fusion for ice, and the specific heat capacity of water, determine what mass of the liquid compound must evaporate to lower the temperature of a sample of water to the freezing point and freeze it.

Find the mass from the moles of water. Use the heat for changing the temperature of the liquid water (as described in Chapter 6, Section 6.3). Determine the molar heat of crystallization from the molar heat of fusion and use it to get total heat energy for the phase transition. Add together the heat energy for each stage to get the total quantity of heat energy required. Use the heat of vaporization of the compound to determine the mass.

$$2.00\ mol\ H_2O \times \frac{18.0152\ g\ H_2O}{1\ mol\ H_2O} = 36.0\ g\ H_2O$$

Total heat energy for freezing the water is the sum of the heat energy to cool the ice to the freezing point (0 °C) and the heat energy required to freeze it.

$$q_{tot\ for\ water} = q_{cool\ ice} + q_{freeze}$$

$$q_{tot\ for\ water} = (c_{water} \times m \times \Delta T) + (n\Delta H_{cryst})$$

$$\Delta H_{cryst} = -\Delta H_{fus} = -6.02\ kJ/mol$$

$$q_{tot\ for\ water} = \left(\frac{4.184\ J}{g\ °C}\right) \times \left(\frac{1\ kJ}{1000\ J}\right) \times (36.0\ g) \times [0\ °C - (-20.\ °C)] + (2.00\ mol) \times \left(\frac{-6.02\ kJ}{mol}\right)$$

$$q_{tot\ for\ water} = -3.0\ kJ + (-12.0\ kJ) = -15.0\ kJ$$

15.0 kJ must be removed to freeze the water, so determine how many grams of CCl_2F_2 must evaporate to use up the 15.0 kJ of thermal energy.

$$15.0\ kJ \times \frac{1000\ J}{1\ kJ} \times \frac{1\ g\ CCl_2F_2}{289\ J} = 51.9\ g\ CCl_2F_2 \text{ must evaporate}$$

✓ *Reasonable Answer Check:* The relative size of the two terms seems sensible, comparing the sizes of the heat capacity and the heat energy of the phase transition. The calculated mass of CCl_2F_2 would definitely fit inside a typical freezer compressor.

44. *Answer/Explanation:* A higher melting point is a result of stronger interparticle forces. Coulomb's law describes the attraction between charged particles; the ion-ion coulombic interaction is stronger with smaller distance between the charges because the charges are more localized and closer together. According to the periodic trends described in Section 7.10, Li^+ and F^- are smaller than Cs^+ and I^-. Therefore, LiF experiences higher coulombic interactions than those in solid CsI, and LiF has a higher melting point.

46. *Answer:* **Highest melting point is (a) SiC. Lowest melting point is (d) $CH_3CH_2CH_2CH_3$.**

Strategy and Explanation: The highest melting point is a result of strongest interparticle forces. Covalent interactions are stronger forces in solids than intermolecular forces between separate molecules. SiC solid is described as a nonoxide ceramic in Section 11.11. It is held together by an extended covalent network. Rb is a metal from Group 1A, and alkali metals have relatively low melting points. The other two interact by London forces, so the highest melting point is (a) SiC.

The lowest melting point is a result of the weakest intermolecular forces. We need to compare the intermolecular forces in the two molecules, I_2 (molar mass 254 g/mol) and $CH_3CH_2CH_2CH_3$ (molar mass 54 g/mol). The larger I_2 molecules can experience more significant London forces than the smaller $CH_3CH_2CH_2CH_3$ molecules, so (d) $CH_3CH_2CH_2CH_3$ has the lowest melting point.

48. *Answer/Explanation:* The freezer compartment of a frost-free refrigerator keeps the air so cold and dry that ice inside the freezer compartment undergoes sublimation at normal pressures, the direct conversion of solid to gaseous form. The hailstones would eventually disappear.

49. *Answer:* **(a) A = solid, B = liquid, C = gas (b) Point 1: solid and gas; point 2: solid, liquid and gas; point 3: liquid and gas; point 5: solid and liquid**

Strategy and Explanation:

(a) The material has to be in the solid state at very low temperatures and very high pressures, so the area of the phase diagram identified by A represents the solid phase. The material has to be in the gas phase at very high temperatures and very low pressures, so the area of the phase diagram identified by C represents the gas phase. At other pressures and temperatures, the material can be in the liquid state, that leaves the area of the phase diagram identified by B represents the liquid phase.

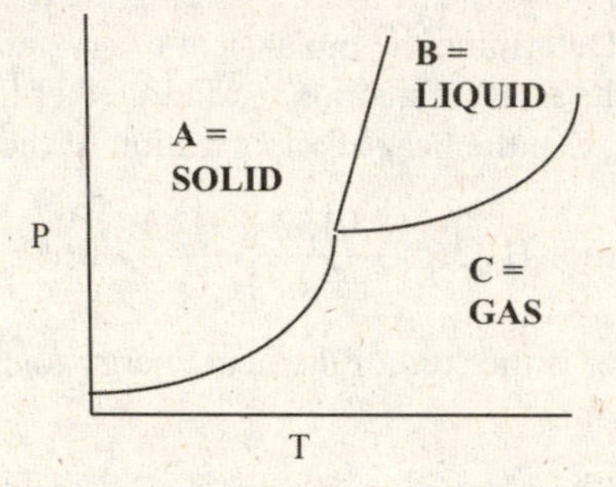

(b) The phases in equilibrium at points along the lines of the phase diagram are those that border the point. At point 1, the phases present will be solid and gas. At point 2 (the triple point), all three phases will be present, solid liquid and gas. At point 3, the phases present would be liquid and gas. At point 5, the phases present will be solid and liquid.

51. *Answer:* **(a) gas phase (b) liquid phase (c) solid phase**

Strategy and Explanation:

(a) The point at –70°C and 1 atm is just below and to the right of the solid/gas equilibrium line, so the phase present will be the gas phase.

(b) The point at –40°C and 15.5 atm is to the right of the solid/liquid equilibrium line and above the liquid/gas equilibrium line, so the phase present will be the liquid phase.

(c) The point at –80°C and 4.7 atm is below and to the left of the solid/liquid equilibrium line and above the solid/gas equilibrium line, so the phase present will be the solid phase.

Water (Section 11.4)

53. *Answer/Explanation:* Water molecules interact using relatively strong hydrogen-bonding intermolecular forces, so greater kinetic energy is needed for molecules to escape.

55. *Answer/Explanation:* The crystal structure in ice maximizes the hydrogen-bonding capacity and leaves considerable open spaces completed to the liquid. See discussion in Section 11.4 for more details.

Types of Solids (Section 11.5)

59. *Answer:* **(a) molecular solid (b) metallic solid (c) network solid (d) ionic solid**

Strategy and Explanation:

(a) P_4O_{10} is a molecule, so it forms a molecular solid.

(b) Brass is composed of various metals, so it is a metallic solid.

(c) Graphite is an extended array of covalently bonded carbon, so it is a network solid.

(d) $(NH_4)_3PO_4$ is composed of common ions, NH_4^+ and PO_4^{3-}, so it is an ionic solid.

61. *Answer:* **(a) amorphous solid (b) molecular solid (c) ionic solid (d) metallic solid; see explanations below.**

Strategy and Explanation: Pure substances have fixed melting temperatures, whereas mixtures and amorphous solids have ill-defined melting points. Network and ionic solids have much higher melting points, because of stronger interparticle bonds. Solids that conduct are made from metals. Liquids that conduct could be metals or ions.

(a) A soft, slippery solid that has no definite melting point is probably an amorphous solid.

(b) Violet crystals with moderate (not high) melting point that do not conduct in either phase probably represent a molecular solid.

(c) Hard colorless crystal with a high melting point and liquid that conducts is probably an ionic solid.

(d) A hard solid that melts at a high temperature and conducts in both solid and liquid states is a metallic solid.

63. *Answer:* **(a) molecular solid (b) ionic solid (c) metallic solid or network solid (d) amorphous solid**

Strategy and Explanation: Use the characterization conclusions as described in the solution to Question 56:

(a) A solid that melts below 100 °C and is insoluble in water is probably a nonpolar molecular solid.

(b) An ionic solid will conduct electricity only when melted.

(c) A solid that is insoluble in water and conducts electricity is probably a metallic solid, though it might be a network solid like graphite.

(d) A noncrystalline solid that has a wide melting point range is an amorphous solid.

Crystalline Solids (Section 11.6)

65. *Answer/Explanation:* See Figure 11.21 and its description.

67. *Answer:* **220 pm**

Strategy and Explanation: Given the length of the edge of a face-centered cubic unit cell of xenon, determine the radius of the xenon atoms.

Use the figure and method similar to those described in Problem-Solving Example 11.7. The face diagonal distance is the hypotenuse of an equilateral triangle. Diagonal distance = $\sqrt{2} \times (\text{edge})$. We can see from the figure that the diagonal distance represents four times the radius of the atom.

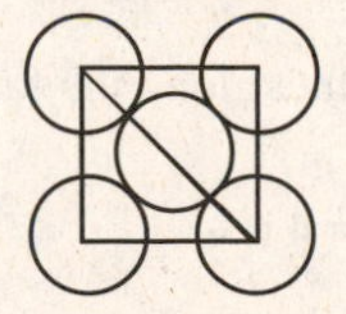

Use these two relationships to find the radius of the atom.

$$\text{Diagonal distance} = \sqrt{2} \times (\text{edge length}) = \sqrt{2} \times (620 \text{ pm}) = 880 \text{ pm}$$

$$\text{Radius} = \frac{\text{Diagonal distance}}{4} = \frac{880 \text{ pm}}{4} = 220 \text{ pm}$$

✓ *Reasonable Answer Check:* The geometric relationships are logical. We expect the radius to be less than the edge length. The van der Waals radius for Xenon is 216 pm.

70. *Answer:* **body diagonal length = 696. pm; side length = 401 pm.**

Strategy and Explanation: Given the radii of cesium and chloride ions, determine the length of the body diagonal and the length of the side of the unit cell.

Use Figure 11.22 showing the unit cell for CsI, which is identified as the same as CsCl. The body diagonal runs through one whole Cs^+ ion and from the center to the edge of two Cl^- ions. The body diagonal is the hypotenuse of a triangle with sides represented by the face diagonal and the edge, so:

$$(\text{Body diagonal length})^2 = (\text{edge length})^2 + (\text{face diagonal length})^2$$

The face diagonal is the hypotenuse of an equilateral triangle.

$$\text{Face diagonal length} = \sqrt{2} \times (\text{edge length})$$

Use these three relationships to find the length of the body diagonal and the length of the side of the unit cell.

$$\text{Diagonal distance} = 2 \times \text{Radius } Cs^+ + 2 \times \text{Radius } Cs^+$$

$$= 2 \times 167 \text{ pm} + 2 \times 181 \text{ pm} = 696. \text{ pm}$$

$$(\text{Body diagonal length})^2 = (\text{edge length})^2 + \left[\sqrt{2} \times (\text{edge length})\right]^2 = 3 \times (\text{edge length})^2$$

$$(\text{edge length})^2 = \frac{(\text{Body diagonal length})^2}{3} = \frac{(696\,\text{pm})^2}{3} = 1.61 \times 10^3 \text{ pm}^2$$

$$\text{edge length} = \sqrt{1.61 \times 10^3 \text{ pm}^3} = 401 \text{ pm}$$

72. *Answer:* **No; the ratio of the ions in the unit cell must reflect the empirical formula of the compound.**

Strategy and Explanation: The ratio of ions in the unit cell must reflect the empirical formula of the compound. Here, the compound $CaCl_2$ has a 1:2 ratio between Ca^{2+} and Cl^-. The compound NaCl has a 1:1 ratio between the Na^+ and Cl^- ions, so it is not possible to have the same structure.

74. *Answer:* **(a) 152 pm (b) r_{I^-} = 212 pm, r_{Li^+} = 88.0 pm (c) see discussion below**

Strategy and Explanation: Given the lengths of the sides of the unit cells for solid lithium metal and solid lithium iodide, the type of structure, and some assumptions of which atoms are touching, determine the radius of Li atom and the radii of Li^+ and I^-.

Use the geometrical relationships described in the solution to Question 70.

(a) Looking at Figure 11.21 at the body-centered cubic, and assuming that the atoms touch along the body diagonal, we get: Body diagonal length = $4r_{Li}$

Given the edge length of the Li unit cell, 351 pm, calculate the body diagonal length:

$$\text{Body diagonal length} = \sqrt{3 \times (\text{edge length})^2} = \sqrt{3} \times 351 \text{ pm} = 608 \text{ pm}$$

Use body diagonal length to get radius of Li atom:

$$r_{Li} = \frac{\text{Body diagonal length}}{4} = \frac{608 \text{ pm}}{4} = 152 \text{ pm}$$

(b) Assuming that the I^- ions touch each other along the face diagonal and the I^- ions touch the Na^+ ion along the edge:

$$\text{Face diagonal length} = 4\, r_{I^-} \qquad \text{Edge length} = 2\, r_{Li^+} + 2\, r_{I^-}$$

Given the edge length of the LiI unit cell, 600. pm, calculate the face diagonal length:

(For sig. figs, assume that this length is as precise as the length in (a), ± 1 pm.)

$$\text{Face diagonal length} = \sqrt{2} \times (\text{edge length}) = \sqrt{2} \times 600. \text{ pm} = 849 \text{ pm}$$

Use face diagonal length to get radius of I^- ion:

$$r_{I^-} = \frac{\text{Body diagonal length}}{4} = \frac{849 \text{ pm}}{4} = 212 \text{ pm}$$

Use the edge length of the LiI unit cell radius of I^- ion to get the radius of Li^+ ion:

$$2\, r_{Li^+} = \text{Edge length} - 2\, r_{I^-} = 600. \text{ pm} - 2 \times 212 \text{ pm} = 176 \text{ pm}$$

$$r_{Li^+} = 88.0 \text{ pm}$$

(c) It is reasonable that the Li atom is larger than the Li^+ cation. Figure 7.24 gives slightly larger value for the radius of the Li atom (157 pm), for Li^+ ion (90 pm) and for I^- (206 pm). The assumption that I^- anions touch each other is reasonable, since the very tiny Li^+ ions cannot span the entire gap between the I^- ions in the unit cell.

✓ *Reasonable Answer Check:* The geometric relationships are logical. The relative sizes of the atom and ion radii, the body diagonal length, and the edge length are all reasonable. The sum of the Figure 7.24 radii to back-calculate the edge length (2 × 90 pm + 2 × 206 pm) gives 592 pm, a little less than the actual measured length, suggesting that the ions do not touch. Similarly, the face diagonal length (4 × 206 pm) gives 824 pm, a little less than the actual measured length.

Network Solids (Section 11.7)

75. *Answer/Explanation:* Carbon atoms in diamond are sp^3 hybridized and are tetrahedrally bonded to four other carbon atoms. Carbon atoms in pure graphite are sp^2 hybridized and bonded with a trianglar planar shape to other carbon atoms. These bonds are partially double bonded so they are shorter than the single bonds in diamond. However, the planar sheets of sp^2 hybridized carbon atoms are only weakly attracted by intermolecular forces to adjacent layers, so these interplanar distances in graphite are much longer than the C–C single bonds in the diamond. The net result is that graphite is less dense than diamond.

76. *Answer/Explanation:* Some examples of network solids are:

    Graphite: A planar network solid. It is insoluble in water and nonpolar solvents.

    Diamond: A three-dimensional network solid. It is insoluble in all solvents.

    All of these solids are huge molecules held together by covalent bonds. Covalent bonds are much stronger than the intermolecular forces that can be formed with the solvent. The energy released on mixing is insufficient to overcome the energy of these intramolecular bonds. As a result, it is not surprising to find that network solids are insoluble.

77. *Answer/Explanation:* Diamond is an electrical insulator because all the electrons are in single bonds, which are shared between two specific atoms and cannot move around. However, graphite is a good conductor of electricity because its electrons are delocalized in conjugated double bonds, which allow the electrons to move easily through the graphite sheets.

Tools of Chemistry: X-Ray Crystallography (Section 11.8)

79. *Answer:* **$\nu = 5.30 \times 10^{17}\ s^{-1}$; (a) 3.51×10^{-16} J for one photon (b) 2.11×10^{8} J; x-ray**

    Strategy and Explanation: Using the length of the edge of a unit cell as the wavelength of light, determine the frequency of light, the energy per photon, and the energy per mole.

    Use methods described in Chapter 7 (Sections 7.1 and 7.2).

    In Problem-Solving Example 11.8, the length of a unit cell of NaCl is found to be 566 pm. Use that as the wavelength for the electromagnetic radiation, and calculate the frequency.

$$\nu = \frac{c}{\lambda} = \frac{2.998 \times 10^{8}\ \frac{m}{s}}{566\ pm \times \frac{10^{-12}\ m}{1\ pm}} = 5.30 \times 10^{17}\ s^{-1}$$

(a) $$E = h\nu = (6.626 \times 10^{-34}\ J{\cdot}s) \times (5.30 \times 10^{17}\ s^{-1}) = 3.51 \times 10^{-16}\ J \text{ for one photon}$$

(b) $$1\ mol\ photon \times \frac{3.51 \times 10^{-16}\ J}{1\ photon} \times \frac{6.022 \times 10^{23}\ photons}{1\ mol\ photon} = 2.11 \times 10^{8}\ J$$

This photon is in the x-ray region of the electromagnetic spectrum.

✓ *Reasonable Answer Check:* It makes sense that the electromagnetic radiation is an x-ray, since x-ray crystallography uses x-rays to examine crystal structures.

81. *Answer:* **361 pm**

    Strategy and Explanation: Given the second-order (n = 2) Bragg reflection angle and the wavelength of the X-ray beam, determine the spacing between the planes of the atoms in a metallic crystal.

    Use methods described in Section 11.8 in "Tools of Chemistry" about X-ray Crystallography on page 510-11:

$$n\lambda = 2d \sin\theta$$

The metal is copper. It is second order (n = 2). The angle of scattering (θ) is 27.35°. The wavelength (λ) is 166 pm.

$$d = \frac{n\lambda}{2\sin\theta} = \frac{(2)(166 \text{ pm})}{2\sin(27.35°)} = 361 \text{ pm}$$

✓ *Reasonable Answer Check:* The atomic radius of copper is 126 pm (see Chapter 7, Figure 7.23), and this distance is somewhat more than twice that number.

Metals, Semiconductors, and Insulators (Section 11.9)

84. *Answer/Explanation:* In a conductor, the valence band is only partially filled, whereas, in an insulator, the valence band is completely full, the conduction band is empty, and there is a wide energy gap between the two. In a semiconductor, the gap between the valence band and the conduction band is very small so that electrons are easily excited into the conduction band.

86. *Answer/Explanation:* Substance (c) Ag has the greatest electrical conductivity because it is a metal. Substance (d) P_4 has the smallest electrical conductivity because it is a nonmetal. (The other two are metalloids.)

88. *Answer/Explanation:* A superconductor is a substance that is able to conduct electricity with no resistance. Two examples are found near the end of Section 11.9: $YBa_2Cu_3O_7$ and $Hg_{0.8}Tl_{0.2}Ba_2Ca_2Cu_3O_{8.23}$. The transition temperatures are 90 K for the yttrium compound and 138 K for the mercury compound.

Silicon and the Chip (Section 11.10)

90. *Answer/Explanation:* Equations showing these reactions are given and described in Section 11.10:

$SiO_2(s) + 2\ C(s) \longrightarrow Si(s) + 2\ CO(g)$; Si is being reduced and C is being oxidized.

$SiCl_4(\ell) + 2\ Mg(s) \longrightarrow Si(s) + 2\ MgCl_2(s)$; Si is being reduced and Mg is being oxidized.

92. Doping is described in Section 11.10. It is the intentional addition of small amounts of specific impurities into very pure silicon. Group III elements are used because they have one less electron per atom than the group IV silicon. Group V elements are used because they have one more electron per atom.

Cements, Ceramics, and Glasses (Section 11.11)

95. *Answer/Explanation:* Amorphous solids are compared to crystalline solids in Section 11.5 and glasses are discussed in more detail in Section 11.11. The amorphous solids known as glasses are different from NaCl, because they lack symmetry or long-range order, whereas ionic solids such as NaCl are extremely symmetrical. NaCl must be heated to melting temperatures, then cooled very slowly, to make a glass.

97. *Answer:* Ceramics are described in Section 11.11. Two examples of oxide ceramics: Al_2O_3 and MgO. Two examples of nonoxide ceramics: Si_3N_2 and SiC. Other correct answers are possible.

General Questions

99. *Answer:* **780 kJ to heat the liquid, 1.6 × 10^4 kJ total to reach the final state and temperature.**

Strategy and Explanation: Given the normal boiling point, molar heat of vaporization, and the specific heat capacities of the gas and liquid states of a compound, determine the heat energy evolved when a given mass of the substance is cooled from an initial to a final temperature.

Use techniques similar to those described in the solution to Question 40.

$$10.\text{ kg NH}_3 \times \frac{1000\text{ g NH}_3}{1\text{ kg NH}_3} = 1.0 \times 10^4\text{ g NH}_3$$

$$10.\text{ kg NH}_3 \times \frac{1000\text{ g NH}_3}{1\text{ kg NH}_3} \times \frac{1\text{ mol NH}_3}{17.0304\text{ g NH}_3} = 5.9 \times 10^2\text{ mol NH}_3$$

A change in temperature in Kelvin degrees is the same as the change in temperature in Celsius degrees. So, use c_{gas} = 2.2 J/g °C and c_{liqiud} = 4.7 J/g °C

Find the heat energy absorbed when heating the liquid to the boiling point (– 33.4 °C)

$$q_{heating\ liquid} = c_{liquid} \times m \times \Delta T$$

$$q_{heating\ liquid} = \left(\frac{4.7\ J}{g\ °C}\right) \times (1.0 \times 10^4\ g\ NH_3) \times [-33.4\ °C - (-50.0\ °C)] \times \left(\frac{1\ kJ}{1000\ J}\right) = 780\ kJ$$

The total heat energy absorbed to get to 0.0 °C is the sum of the heat energy absorbed heating the liquid, the heat energy absorbed when vaporizing the liquid to a gas, and the heat energy absorbed when heating the gas to the final temperature (0.0 °C). $q_{tot} = q_{heating\ liquid} + (n \times \Delta H_{vap}) + (c_{gas} \times m \times \Delta T)$

$$q_{tot} = 780\ kJ + 5.9 \times 10^2\ mol \times \left(\frac{23.5\ kJ}{mol}\right) + \left(\frac{2.2\ J}{g\ °C}\right) \times (1.0 \times 10^4\ g\ NH_3) \times [0.0\ °C - (-33.4\ °C)] \times \left(\frac{1\ kJ}{1000\ J}\right)$$

$$q_{tot} = 780\ kJ + (14000\ kJ) + (730\ kJ) = 16000\ kJ = 1.6 \times 10^4\ kJ$$

✓ *Reasonable Answer Check:* The relative size of the three terms seems sensible, comparing the sizes of the heat capacities and the heat of vaporization.

101. *Answer:* **(a) dipole-dipole forces and London forces (b) $CH_4 < NH_3 < SO_2 < H_2O$**

Strategy and Explanation:

(a) The molecular geometry of the SO_2 molecule is determined in the solution to Question 9.16(c):

SO_2 (18 e^-) The type is AX_2E_1, so the electron-pair geometry is triangular planar and the molecular geometry is angular (120°).

Ö═S̈—Ö:

S
O O

The asymmetric shape of the SO_2 molecule means that it is polar. Therefore, the strongest intermolecular forces between molecules in solid and liquid SO_2 are dipole-dipole forces. All molecules experience London forces, so that force is part of the attractions in the liquid and solid states of SO_2, also.

(b) The normal boiling point is smaller when the intermolecular attractions are weaker. The normal boiling point is larger when the intermolecular attractions are stronger.

$$CH_4\ (-161.5\ °C) < NH_3\ (-33.4\ °C) < SO_2\ (-10\ °C) < H_2O\ (100\ °C)$$

✓ *Reasonable Answer Check:* Ordering the molecules by boiling points indicates the surprising result that the intermolecular attractions in SO_2 are stronger than the attractions in NH_3. This is probably a result of variable size. The molar mass of SO_2 is 64 g/mol, but the molar mass of NH_3 is only 17 g/mol. SO_2 has three atoms from period two and NH_3 has only one atom from period two. So it makes sense SO_2 molecules may have significant enough London forces to be more attractive to each other than are tiny, polar molecules of NH_3 undergoing hydrogen bonding.

Applying Concepts

102. *Answer:* **(a) 80 mm Hg (b) 18 °C (c) 640 mm Hg (d) diethyl ether and ethanol (e) diethyl ether would evaporate; ethanol and water would remain liquids (f) water**

Strategy and Explanation:

(a) To find the equilibrium vapor pressure for ethyl alcohol at room temperature, look at Figure 11.5 and find the pressure where the ethyl alcohol curve passes a temperature of 25 °C. This is about 80 mm Hg.

(b) To find the temperature when the equilibrium vapor pressure for diethyl ether is 400 mm Hg, look at Figure 11.5 and find the temperature where the diethyl ether curve passes a pressure of 400 mm Hg. This is about 18 °C.

(c) Find the equilibrium vapor pressure for water at 95 °. To do this, look at Figure 11.5 and find the pressure where the water curve passes a temperature of 95 °C. This is about 640 mm Hg.

(d) To find out which substances will be gases under specific condition of 200 mm Hg and 60 °C, look at Figure 11.5 and find the pressure where each curve passes a temperature of 60 °C. If their vapor pressure is above 200 mm Hg, they will be gases. Here, both diethyl ether and ethanol are gases.

(e) To find out which substance will immediately evaporate in your hand, look at Figure 11.5 and find the pressure where each curve passes body temperature of 37 °C. If their pressure is close to 760 mm Hg they will readily evaporate. Here, diethyl ether will evaporate readily. The ethanol and water would evaporate more slowly.

(f) Water has the strongest intermolecular attractions as empirically indicated by having a lower vapor pressure than the other two liquids at any common temperature.

104. *Answer/Explanation:* The butane in the lighter is under great enough pressure that the vapor pressure of butane at room temperature is less than the pressure inside the lighter. Hence, it exists as a liquid.

106. *Answer:* **Diagram 1 and region C; diagram 2 and line E, diagram 3 and region B, diagram 4 and line F, diagram 5 and line G, diagram 6 and point H, diagram 7 and region A**

Strategy and Explanation: Refer to Section 11.3, if you need to be reminded about the different parts of a phase diagram.

Nanoscale diagram 1 looks like the substance atoms are all in the gas phase. That is region C on the phase diagram.

Nanoscale diagram 2 looks like some of the substance atoms are in the solid phase (atoms piled up in a regular array) and some of the substance atoms are in the gas phase. That is a point on the line described by E on the phase diagram.

Nanoscale diagram 3 looks like the substance atoms are all in the liquid phase. That is region B on the phase diagram.

Nanoscale diagram 4 looks like some of the substance atoms are in the solid phase and some of the substance atoms are in the liquid phase. That is a point on the line described by F on the phase diagram.

Nanoscale diagram 5 looks like some of the substance atoms are in the liquid phase and some of the substance atoms are in the gas phase. That is a point on the line described by G on the phase diagram.

Nanoscale diagram 6 looks like some of the substance atoms are in the liquid phase, some of the substance atoms are in the liquid phase, and some of the substance atoms are in the gas phase. That is the point described by H on the phase diagram.

Nanoscale diagram 7 looks like the substance atoms are all in the solid phase. That is region A on the phase diagram.

108. *Answer/Explanation:* Each has the same fraction of filled space. The fraction of spaces filled by closest packed equal-sized spheres is the same, no matter what the size of the spheres.

More Challenging Questions

109. *Answer/Explanation:* Vapor-phase water condenses on contact with cool skin. The condensation of steam into water is exothermic. That energy is also absorbed by the skin causing more burning, along with the burn resulting from the heat energy given off as the temperature drops (common in each case).

112. *Answer:* **(a) condensation then freezing (b) triple point (c) melting point curve**

Strategy and Explanation: Phase diagrams are described in Section 11.3.

(a) The substance's initial state (F) is gaseous (at low-enough pressure and high-enough temperature). The temperature remains fixed as the transition toward point G, but the pressure is increased. At a sufficiently high pressure (point E), the gas undergoes **condensation** to a liquid. Just before arriving at point G, the liquid undergoes a change to solid by **freezing**.

(b) The point (A) where the vapor pressure curve meets the point curve and the sublimation point curve, is called the **triple point**.

(c) The AB line is a segment of the **melting point curve**, and indicates how the melting temperature change over a range of pressures, from the triple point, point A, to atmospheric pressure (1 atm), at point B.

114. *Answer:* **(a) 560 mm Hg (b) benzene (c) 73°C (d) methyl ethyl ether is 7°C; carbon disulfide is 47°C; benzene is 81°C (these are all approximate, due to the size of the graph provided)**

Strategy and Explanation: Answers to (a), (c) and (d) are obtained from the graph, by locating the given information on the appropriate axis and following that horizontal or vertical to the point on the appropriate curve.

(a) Finding 0°C on the horizontal axis and tracing a vertical line to the methyl ethyl ether curve shows that the vapor pressure of methyl ethyl ether at 0°C is approximately 560 mm Hg.

(b) A substance will remain in the liquid state at higher temperatures if its molecules experience higher intermolecular forces; hence, its vapor pressure will be lower. Therefore, of the three liquids compared here, molecules in benzene experience the greatest intermolecular forces.

(c) Finding 600 mm Hg on the vertical axis and tracing a horizontal line to the benzene curve gives a temperature of approximately 73°C.

(d) Normal boiling point is the boiling temperature at 1 atm (760 mm Hg). Finding 760 mm Hg on the vertical axis and tracing horizontal lines to each curve gives the following normal boiling points: for methyl ethyl ether, approximately 7°C; for carbon disulfide, approximately 47°C, for benzene, approximately 81°C.

116. *Answer:* **22.2°C**

Strategy and Explanation: Given the enthalpy of vaporization for water, determine the boiling point at a specific vapor pressure.

Use the two-set Clausius-Claypeyron equation $\ln\left(\frac{P_2}{P_1}\right) = \frac{-\Delta H_{vap}}{R}\left(\frac{1}{T_2} - \frac{1}{T_1}\right)$

and the normal boiling point of water (See Table 11.2). R is given in Table 10.3. $P_1 = 24$ mm Hg

$$T_2 = 100.00°C + 273.15 = 373.15 \text{ K}$$

$$P_2 = 1 \text{ atm} = 760.0 \text{ mm Hg}$$

$$\ln\left(\frac{760.0 \text{ mm Hg}}{24 \text{ mm Hg}}\right) = \frac{-(40.7 \text{ kJ/mol})}{\left(8.314 \frac{\text{J}}{\text{K}\cdot\text{mol}}\right)\left(\frac{1 \text{ kJ}}{1000 \text{ J}}\right)}\left(\frac{1}{373.15 \text{ K}} - \frac{1}{T_1}\right)$$

$$\ln(3\underline{1}.67) = 3.\underline{4}55 = \frac{-(40.7)}{(0.008314)}\left(0.0026799 - \frac{1}{T_1}\right)$$

$$-\frac{(3.4\underline{5}5)(0.008314)}{(40.7)} = 0.0026799 - \frac{1}{T_1}$$

$$-0.000706 = 0.0026799 - \frac{1}{T_1}$$

$$\frac{1}{T_1} = 0.0026799 + 0.000706 = 0.00339$$

$$T_1 = 295.3 \text{ K} \qquad T_1 = 295.3 \text{ K} - 273.15 = 22.2°C$$

✓ *Reasonable Answer Check:* Since the pressure is lower than the pressure for the normal boiling temperature, the boiling point must be lower than the normal boiling point.

118. *Answer:* **0.533 g/cm³**

Strategy and Explanation: Given the lengths of the edge of the unit cells for solid lithium metal, the type of structure, and the temperature, determine the density of the metal at this temperature.

Find the edge length in units of centimeters so that the density will come out in g/cm^3. Determine the volume of the unit cell, and the mass of the atoms in that cell. Then, divide mass by volume to get density.

$$351 \text{ pm} \times \frac{10^{-12} \text{m}}{1 \text{ pm}} \times \frac{1 \text{ cm}}{10^{-2}} = 3.51 \times 10^{-8} \text{ cm}$$

$$V = (\text{edge}) = (3.81 \times 10^{-8} \text{ cm})^3 = 4.32 \times 10^{-23} \text{ cm}^3$$

In Section 11.6 we learn that a body-centered cubic unit cell has 2 atoms in it.

$$2 \text{ atoms Li} \times \frac{1 \text{ mole Li}}{6.022 \times 10^{23} \text{ atoms Li}} \times \frac{6.941 \text{ g Li}}{1 \text{ mole Li}} = 2.305 \times 10^{-23} \text{ g Li}$$

$$\text{Density} = \frac{2.305 \times 10^{-23} \text{ g}}{4.32 \times 10^{-23} \text{ cm}^3} = 0.533 \text{ g/cm}^3$$

✓ *Reasonable Answer Check:* Lithium has a small molar mass. Table 1.1 shows various metals (not Li) and the metals with small molar masses have smaller densities.

Chapter 12: Fuels, Organic Chemicals, and Polymers

Solutions for Blue-Numbered Questions for Review and Thought

Topical Questions

Petroleum and Fuels (Sections 12.1, 12.2)

22. *Answer/Explanation:*

(a) The gasoline fraction, with the hydrocarbons that will provide fuel for most people's cars, has a temperature range of 20 - 200 °C.

(b) The octane rating of the straight-run fraction is 55.

(c) Do not use the straight-run fraction as a motor fuel. The octane rating is far lower than regular gasoline we buy at the pump (87 - 92), that means it would cause far more pre-ignition than we expect from the gasoline. It would need to be reformulated to make it an acceptable motor fuel.

24. *Answer/Explanation:* Gasolines contain molecules in the liquid phase that, at ambient temperatures, can easily overcome their intermolecular forces and escape into the vapor phase. That means all gasolines evaporate easily.

Alcohols (Section 12.4)

27. *Answer:* **see structures below (There are other correct answers; these are examples.)**

Strategy and Explanation:

(a) A primary alcohol is one that has the –OH group attached to a C atom that is only bonded to one other C atom. An example of a primary alcohol is 1-propanol:

$$CH_3-CH_2-CH_2-OH$$

(b) A secondary alcohol is one that has the –OH group attached to a C atom that is bonded to two other C atoms. An example of a secondary alcohol is 2-propanol:

$$\begin{array}{c} OH \\ | \\ CH_3-CH-CH_3 \end{array}$$

(c) A tertiary alcohol is one that has the –OH group attached to a C atom that is bonded to three other C atoms. An example of a tertiary alcohol is 2-methyl-2-propanol:

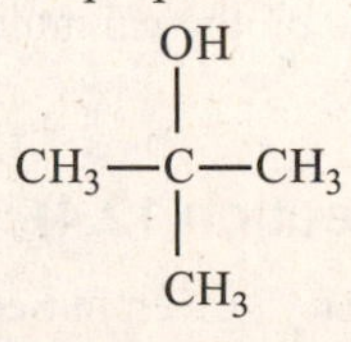

29. *Answer:* **(a) tertiary (b) primary (c) secondary (d) secondary (e) tertiary (f) secondary**

Strategy and Explanation:

(a) A tertiary alcohol is one that has the –OH group attached to a C atom that is bonded to three other C atoms; hence, the given structure is an example of a tertiary alcohol.

(b) A primary alcohol is one that has the –OH group attached to a C atom that is bonded to only one other C atom; hence, the given structure is an example of a primary alcohol.

(c) A secondary alcohol is one that has the –OH group attached to a C atom that is bonded to two other C atoms; hence, the given structure is an example of a secondary alcohol.

(d) As described in (c), the given structure is an example of a secondary alcohol.

(e) As described in (a), the given structure is an example of a tertiary alcohol.

(f) As described in (c), the given structure is an example of a secondary alcohol.

31. *Answer:* **see structures below**

Strategy and Explanation: Oxidation of a primary alcohol gives an aldehyde. Oxidation of an aldehyde gives a carboxylic acid.

(a)

$CH_3-\overset{\overset{O}{\|}}{C}-H$ first oxidation product

$CH_3-\overset{\overset{O}{\|}}{C}-OH$ second oxidation product

(b)

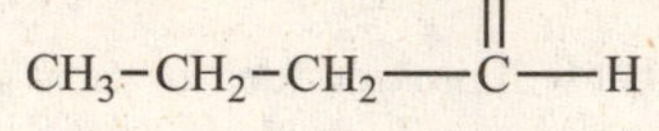

$CH_3-CH_2-CH_2-\overset{\overset{O}{\|}}{C}-H$ first oxidation product

$CH_3-CH_2-CH_2-\overset{\overset{O}{\|}}{C}-OH$ second oxidation product

33. *Answer:* **see structures below**

Strategy and Explanation: Oxidation of a secondary alcohol gives a ketone. Oxidation of a primary alcohol gives an aldehyde. Oxidation of an aldehyde gives a carboxylic acid.

(a) This aldehyde is produced from the oxidation of a primary alcohol:

$$CH_3-\underset{\underset{CH_3}{|}}{CH}-CH_2-CH_2-OH$$

(b) This ketone is produced from the oxidation of a secondary alcohol:

$$CH_3-CH_2-\overset{\overset{OH}{|}}{CH}-CH_2-CH_3$$

(c) This carboxylic acid is produced from the oxidation of a primary alcohol and subsequent oxidation of the resulting aldehyde. The original alcohol is this one:

$$CH_3-CH_2-\underset{\underset{CH_3}{|}}{CH}-CH_2-OH$$

35. *Answer/Explanation:* Wood alcohol (methanol) is made by heating hardwoods such as beech, hickory, maple, or birch. Grain alcohol (ethanol) is made from the fermentation of plant materials, such as grains.

37. *Answer/Explanation:* –OH groups are a common site of hydrogen bonding intermolecular forces. Their presence would increase the solubility of the biological molecule in water and create specific interactions with other biological molecules.

Tools of Chemistry: NMR and MRI (Section 12.4)

38. *Answer/Explanation:* The "Tools of Chemistry" given in Section 12.4 on pages 552-3 describes NMR as using **radiowave** frequencies.

Carboxylic Acids and Esters (Section 12.5)

41. *Answer:* **(a) $CH_3COOCH_2CH_3$ (b) $CH_3CH_2COOCH_2CH_2CH_3$ (c) $CH_3CH_2COOCH_3$**

Strategy and Explanation: The carboxylic acid group loses OH, and the alcohol group loses H (in the formation of water). Connect the remaining fragments to get the product of the condensation reaction produces the resulting ester.

(a) $CH_3CO\underline{OH} + \underline{H}OCH_2CH_3 \longrightarrow CH_3COOCH_2CH_3 + H_2O$

(b) $CH_3CH_2CO\underline{OH} + \underline{H}OCH_2CH_2CH_3 \longrightarrow CH_3CH_2COOCH_2CH_2CH_3 + H_2O$

(c) $CH_3CH_2CO\underline{OH} + \underline{H}OCH_3 \longrightarrow CH_3CH_2COOCH_3 + H_2O$

43. *Answer:* **(a) CH_3CH_2COOH, CH_3OH (b) HCOOH, CH_3CH_2OH (c) CH_3COOH, CH_3CH_2OH**

Strategy and Explanation: Break the ester product of the condensation reaction between the two O atoms. Add OH to the C=O to make the carboxylic acid, and add H to the other fragment to make the alcohol group.

(a) $CH_3CH_2CO\underline{OH} + \underline{H}OCH_3 \longrightarrow CH_3CH_2COOCH_3 + H_2O$

(b) $HCO\underline{OH} + \underline{H}OCH_2CH_3 \longrightarrow HCOOCH_2CH_3 + H_2O$

(c) $CH_3CO\underline{OH} + \underline{H}OCH_2CH_3 \longrightarrow CH_3COOCH_2CH_3 + H_2O$

Synthetic Organic Polymers (Section 12.6)

45. *Answer/Explanation:* Examples of thermoplastics are milk jugs (polyethylene), cheap sun glasses and toys (polystyrene) and CD audio discs (polycarbonates). Thermoplastics soften and flow when heated.

47. *Answer:* **see structures below**

Strategy and Explanation:

(a) 1-butene has a double bond between the first and second C atoms in the four-atom chain, CH_2=$CHCH_2CH_3$, where the addition polymerization occurs. The saturated two-carbon end of the butene will put an ethyl branch on every other carbon in the long polymer chain

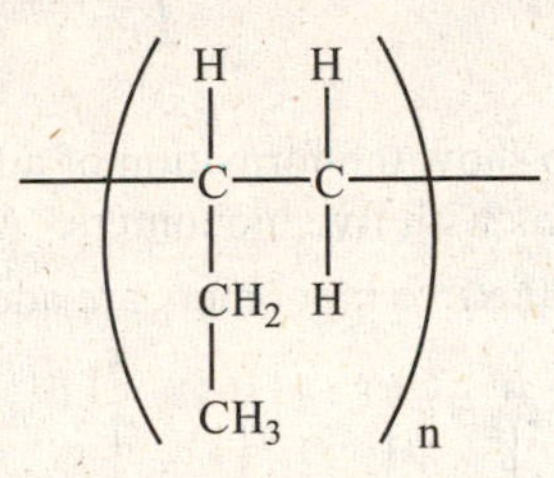

(b) 1,1-dichloroethylene has a double bond between the two C atoms where the addition polymerization occurs The two Cl atoms on the first C atom in the monomer mean that there will be two Cl atoms on every other carbon in the long polymer chain:

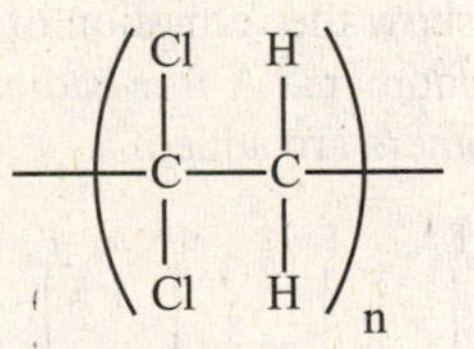

(c) Vinyl acetate, CH_3COOCH=CH_2, has a double bond between the two vinyl C atoms where the addition polymerization occurs. The acetate group in the monomer means that every other carbon in the long polymer chain will have an acetate group attached to it:

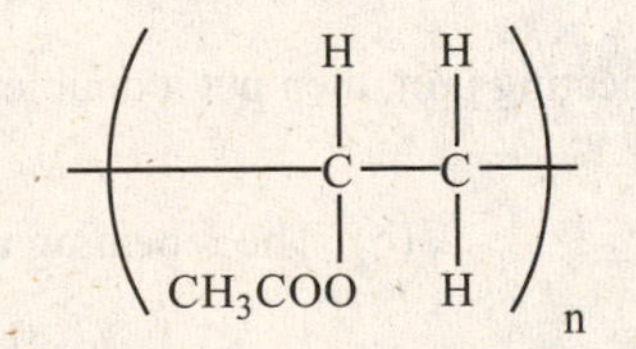

49. *Answer:* **see equation and structure below**

Strategy and Explanation: The monomer, methyl methacrylate, CH_2=$C(CH_3)COOCH_3$, has a double bond between the first two C atoms where the addition polymerization occurs. The methyl branch and the $-COOCH_3$ group on the first carbon in the monomer mean that every other carbon in the long polymer chain will have a methyl and a $-COOCH_3$ group attached to it:

(a) The equation for the reaction to form a four unit chain of the polymer looks like this.

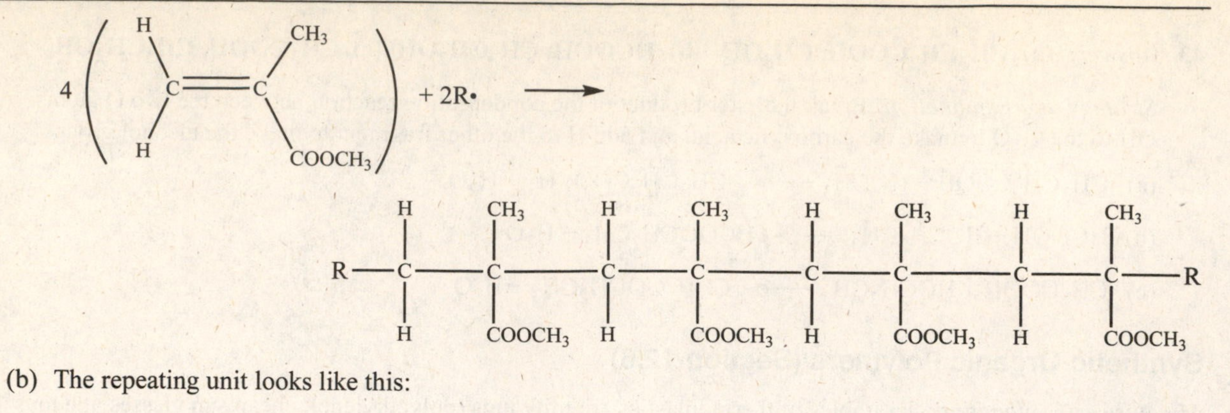

(b) The repeating unit looks like this:

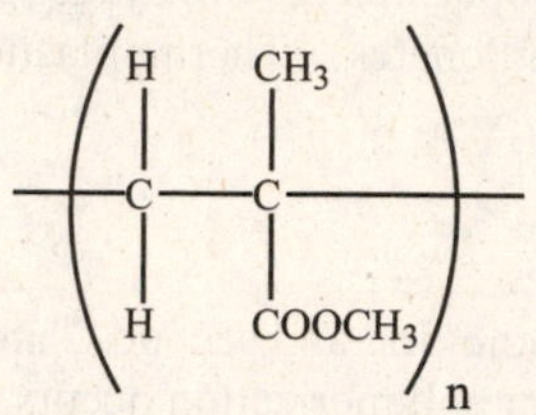

51. *Answer:* **see equations below**

Strategy and Explanation:

(a) The chemical equation that occurs to show the formation of a five-unit polymer chain with the monomer vinylidene chloride ($H_2C{=}CCl_2$) starts with five monomers. A free-radical initiator (•OR) is used to start the reaction and to terminate it after the five monomers are added.

$$5\ H_2C{=}CD + 2\ {\bullet}OR \longrightarrow RO{-}CH_2{-}CCl_2{-}CH_2{-}CCl_2{-}CH_2{-}CCl_2{-}CH_2{-}CCl_2{-}CH_2{-}CCl_2{-}OR$$

(b) The chemical equation that occurs to show the formation of a five-unit polymer chain with the monomer tetrafluoroethylene starts with five monomers. A free-radical initiator (•OR) is used to start the reaction and to terminate it after the five monomers are added.

$$5\ F_2C{=}CF_2 + 2\ {\bullet}OR \longrightarrow RO{-}CF_2{-}CF_2{-}CF_2{-}CF_2{-}CF_2{-}CF_2{-}CF_2{-}CF_2{-}CF_2{-}CF_2{-}OR$$

52. *Answer:* **see structures below**

Strategy and Explanation: Find the repeating unit, then put a double bond between the two C atoms at that point in the chain:

(a) The repeating unit looks like this.

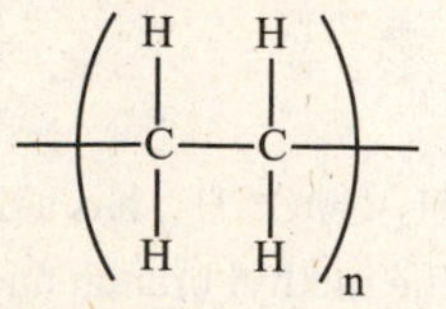

That means $CH_2{=}CH_2$ is the monomer.

(b) The repeating unit looks like this.

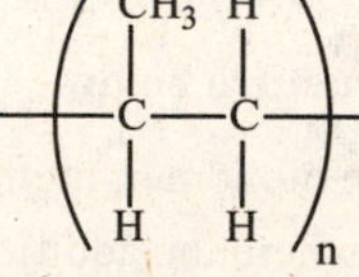

That means $CH_3CH{=}CH_2$ is the monomer.

(c) The repeating unit looks like this.

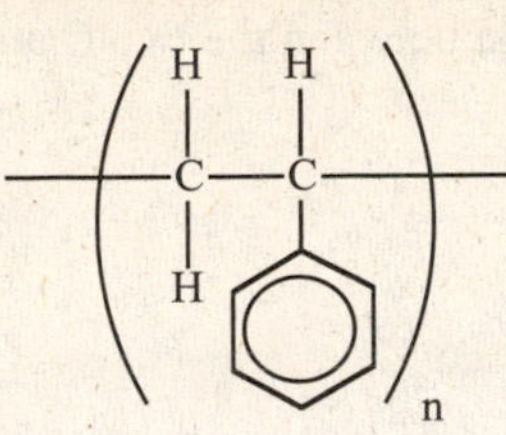

That means the monomer is: $CH_2{=}CH$

(d) The repeating unit looks like this.

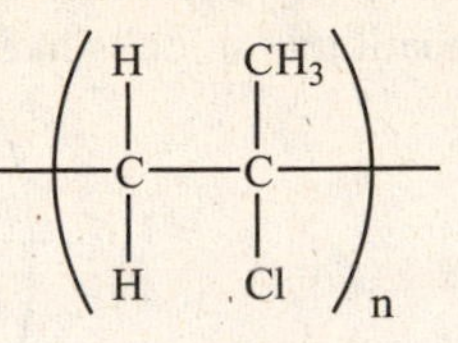

That means $CH_2{=}CClCH_3$ is the monomer.

(e) The repeating unit looks like this.

That means we need $CH_2{=}CHCOOC_2H_5$ for the monomer.

53. *Answer:* **isoprene, 2-methyl-1,3-butadiene; cis-isomer**

Strategy and Explanation: The monomer of natural rubber is isoprene, 2-methyl-1,3-butadiene. The isomer present in natural rubber is the *cis*-isomer.

55. *Answer:* **see structures below**

Strategy and Explanation: The polymer chain will form from making an ester bond between the OH group and the COOH group. The carboxylic acid group loses -OH, and the alcohol group loses -H (in the formation of water).

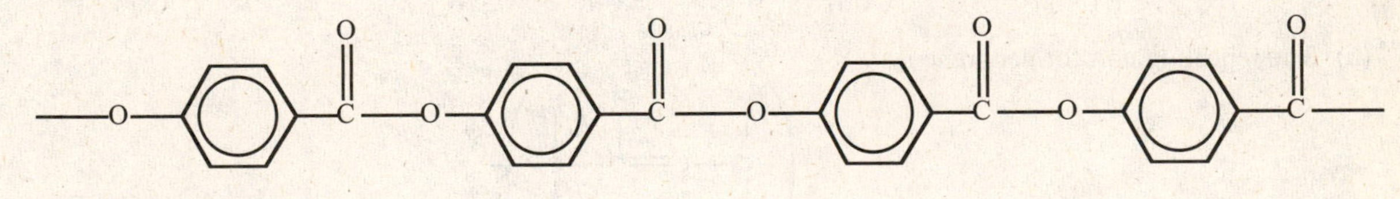

The repeating unit is:

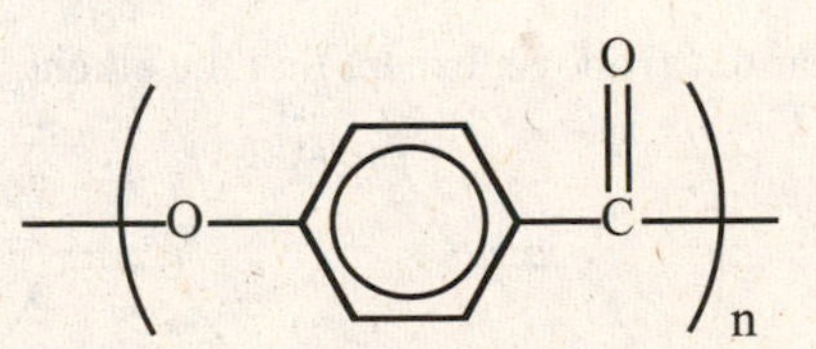

57. *Answer:* **carboxylic acid and alcohol**

Strategy and Explanation: The formation of polyesters involves the carboxylic acid and alcohol functional groups.

59. *Answer:* **carboxylic acid and amine; nylon**

Strategy and Explanation: The formation of polyamides involves the carboxylic acid and amine functional groups. The most common example of this class of polymer is nylon.

61. *Answer/Explanation:* One major difference between proteins and most other polyamides is that the protein polymer's monomers are not all alike. Different side chains on the amino acids change the properties of the protein.

63. *Answer:* **CH_2=CHCN**

Strategy and Explanation: Find the repeating unit, then put a double bond between the two C atoms at that point in the chain. The repeating unit looks like this.

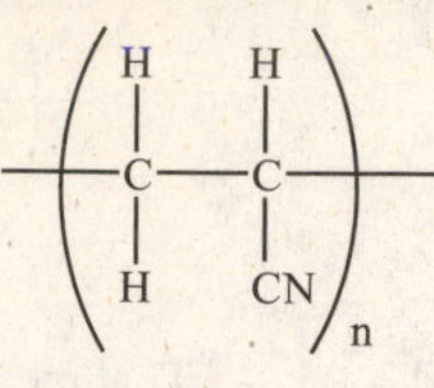

That means we need CH_2=CHCN for the monomer.

65. *Answer:* **16,000 monomers**

Strategy and Explanation: In Section 12.6, the degree of polymerization is identified as the number of repeating units in a polymer chain. Polyethylene is an addition polymer, meaning that the entire monomer molecule is incorporated into the polymer chain. First, calculate the molar mass of the monomer. Then divide the molar mass of the polymer (also known as "molecular weight" in units of grams per mole) by the molar mass of the monomer, we can estimate the number or monomers in the polymer chain.

$$\frac{450{,}000 \text{ g/mol polymer}}{28.0530 \text{ g/mol monomer}} = 16{,}000 \frac{\text{monomers}}{\text{polymer}}$$

67. *Answer:* **see structures below**

Strategy and Explanation: To get the repeating unit for the three polymers described we need to find the monomer used and then link the monomer appropriately.

(a) Natural rubber is described in Section 12.6 as poly-*cis*-2-methyl-1,3-butadiene. To show the cis-features of the polymer, two monomer units need to be identified in the repeating unit.

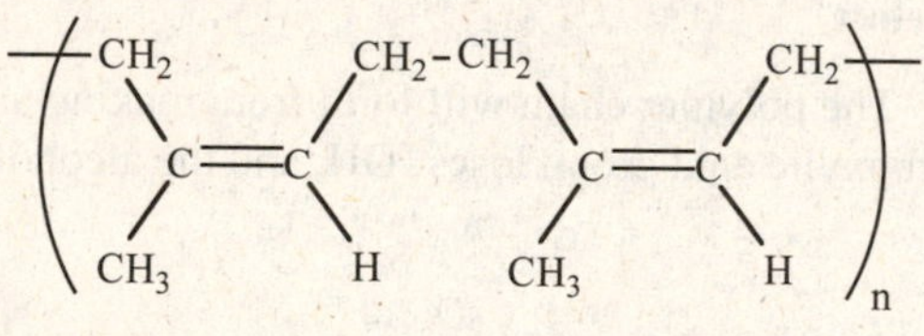

(b) The repeating unit for neoprene is:

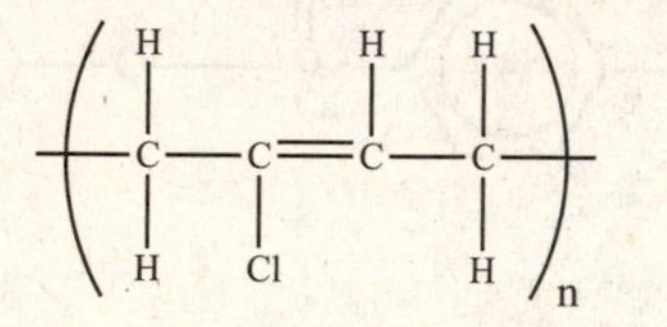

(c) Polybutadiene would be similar to (a) above, except that the alkene has no substitutions.

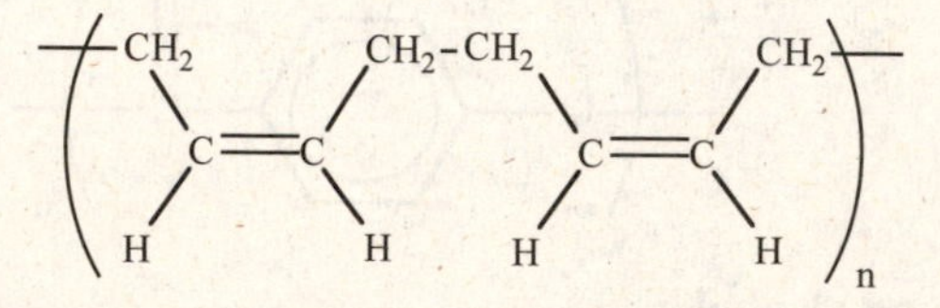

68. *Answer/Explanation:* Major end uses for recycled PET include fiberfill for ski jackets and sleeping bags, carpet fibers, and tennis balls. HDPE is converted into fiber used for sportswear, insulating wrap for new buildings, and very durable shipping containers.

Polysaccharides and Proteins (Section 12.7)

70. *Answer/Explanation:* The biological molecules that have monomer units that are not all alike are proteins, DNA, and RNA.

71. *Answer:* **amine, amide, alcohol, carboxylic acid**

Strategy and Explanation: Looking at the molecule from left to right we see several functional groups. NH_2 is an amine functional group. RCONHR' is an amide functional group. ROH is an alcohol functional group. RCOOH is a carboxylic acid. (A benzene ring with an –OH group on it is phenol and RSH is thiol, but these are not apt to be answers the students would give from studying this textbook.)

72. *Answer:* **see structure below**

Strategy and Explanation: The monomers alanine, glycine, and phenylalanine are given in Table 12.8. We will link them using peptide linkages.

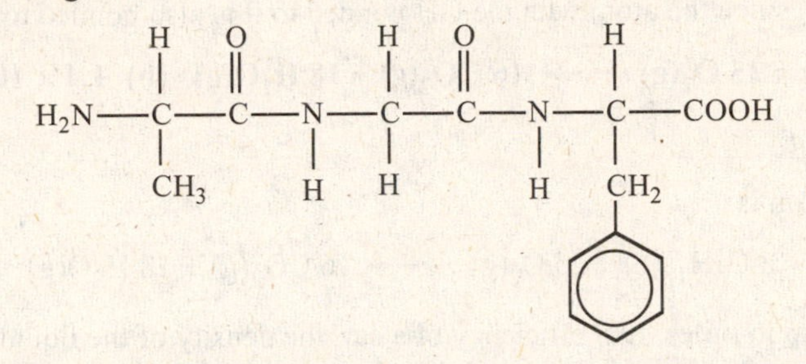

74. *Answer/Explanation:*

(a) A monosaccharide is a molecule composed of one simple sugar molecule, while disaccharides are molecules composed of two simple sugar molecules.

(b) Disaccharides have only two simple sugar molecules; whereas, polysaccharides have many.

76. *Answer/Explanation:* Starch and cellulose are polysaccarides that yield only D-glucose upon complete hydrolysis.

78. *Answer/Explanation:*

(a) Glycogen contains glucose linked together with the glycosidic linkages in "*cis*-positions" and cellulose contains glucose with the glycosidic linkages in "*trans*-positions." The difference is generally described as an α-glycosidic link or a β-glycosidic link. The β-glycosidic link found in amylose, amylopectin, and glycogen is produced when the β-D-glucose is linked together, producing what the textbook calls a "cis-looking" stereoisomer. The β-glycosidic link found in cellulose is produced when the β-D-glucose is linked together, producing what the textbook calls a "trans-looking" stereoisomer. Humans lack the enzyme to hydrolyze the β-glycosidic link.

(b) Cows (and other ruminants) contain microorganisms in their digestive tract that produce the appropriate enzyme, allowing digestion of cellulose to occur.

Tools of Chemistry: Gas Chromatography (Section 12.2)

80. *Answer/Explanation:* Polar molecules would not be attracted to the non-polar liquid stationary phase, so they will exit the chamber earlier. Hence, they will show up earlier on the chromatograph.

General Questions

82. *Answer/Explanation:* $CH_3CH_2CH_2CH_2CH_2CH_2CH_2CH_2CH_2CH_2OH$ is a larger molecule than CH_3CH_2OH. The polar end of the alcohol group will interact well with the water; however, the nonpolar end of the molecule will not. The longer nonpolar end of the decanol will not be miscible in water, lowering the solubility compared to smaller, more polar ethanol.

84. *Answer/Explanation:* Vulcanized rubber has short chains of sulfur atoms that bond together (crosslink) the polymer chains of natural rubber.

85. *Answer:* **see structure below; tertiary**

Strategy and Explanation: 3-ethyl-5-methyl-3-hexanol has a six-carbon main chain with an ethyl branch on the third C atom, a methyl branch attached to the fifth C atom, and an OH attached to the third C atom:

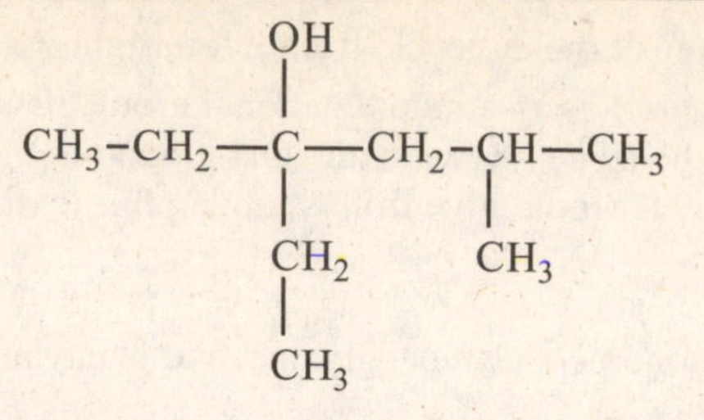

This is a tertiary alcohol, since the atom with the OH bonded to it is also bonded to three other C atoms.

88. *Answer:* **(a) $2\ C_8H_{18}(\ell) + 25\ O_2(g) \longrightarrow 16\ CO_2(g) + 18\ H_2O(g)$ (b) 1.4×10^3 L CO_2**

Strategy and Explanation:

(a) The balanced equation is:

$$2\ C_8H_{18}(\ell) + 25\ O_2(g) \longrightarrow 16\ CO_2(g) + 18\ H_2O(g)$$

(b) Given length of a trip in miles, fuel efficiency of a car, the density of the liquid fuel, temperature and pressure, determine the volume of a gas-phase product produced during the trip.

Use the miles, the fuel efficiency, volume conversions, the density, and the stoichiometry to find the moles of the product. Use the ideal gas law (from Chapter 10) to determine the volume of the product.

$$10.\text{ miles} \times \frac{1\text{ gal gasoline}}{32\text{ miles}} \times \frac{3.785\text{ L gasoline}}{1\text{ gal gasoline}} \times \frac{1000\text{ mL}}{1\text{ L}} \times \frac{1\text{ cm}^3}{1\text{ mL}} = 1.2 \times 10^3\text{ cm}^3$$

$$1.2 \times 10^3\text{ cm}^3\text{ gasoline} \times \frac{0.703\text{ g gasoline}}{1\text{ cm}^3\text{ gasoline}} \times \frac{1\text{ mol gasoline}}{114.22\text{ g gasoline}} \times \frac{16\text{ mol CO}_2}{2\text{ mol gasoline}} = 58\text{ mol CO}_2$$

$$58\text{ mol CO}_2 \times \left(\frac{24.5\text{ L}}{1\text{ mol}}\right) = 1.4 \times 10^3\text{ L}$$

✓ *Reasonable Answer Check:* This is the same question as Chapter 10 Question 60 and produces the same result. This is a large volume of CO_2. However, it is not an unreasonable quantity considering how many gallons of gasoline are used and how much CO_2 is generated from each octane molecule. Comparing the results to the answer in Question 89, less CO_2 is generated by methanol as a fuel than octane.

90. *Answer:* **see structures below**

Strategy and Explanation: Glycogen has the glycosidic linkages in "*cis*-positions," whereas cellulose has glycosidic linkages in "*trans*-positions." NOTICE: Shown below are simple skeletal diagrams; the many –OH functional groups on each of the glucose rings have been left off, so that we can focus on the linkages themselves.

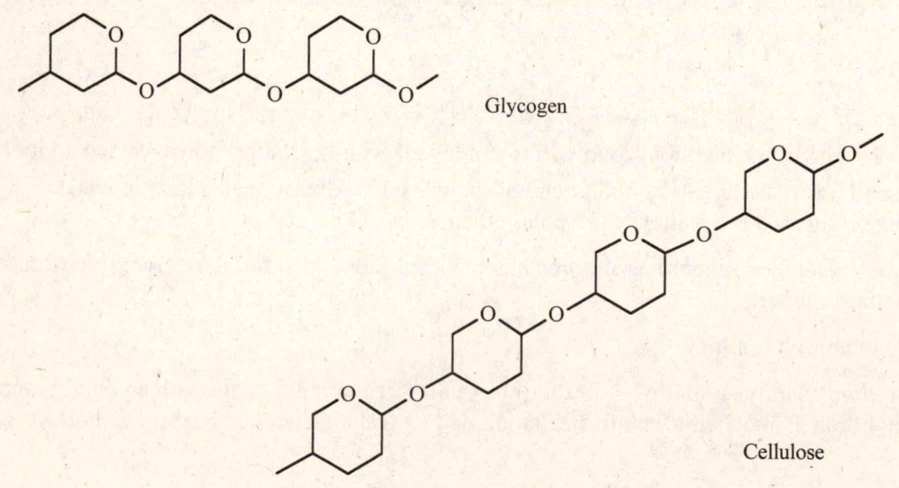

Because of the shape, the glycogen molecules can pack more densely and can curl into granules. The cellulose is more rigid and stretched out, so it forms sheets.

Applying Concepts

92. *Answer:* **see structures below; propanoic acid boils at higher T, due to more H-bonding.**

Strategy and Explanation: The hydrogen bonding between molecules occurs when a hydrogen-bonding H atom (one that is bonded to a highly electronegative element) is attracted to another highly electronegative element in a neighboring molecule.

In propanoic acid, there is one hydrogen-bonding H atom and two highly electronegative O atoms in the molecule. That means there will be four different kinds of hydrogen-bonding interactions between the given molecule (in the box below) and its neighbors: two ways for the H atom to be attracted to neighboring O atoms and two ways for O atoms to be attracted to neighboring H atoms:

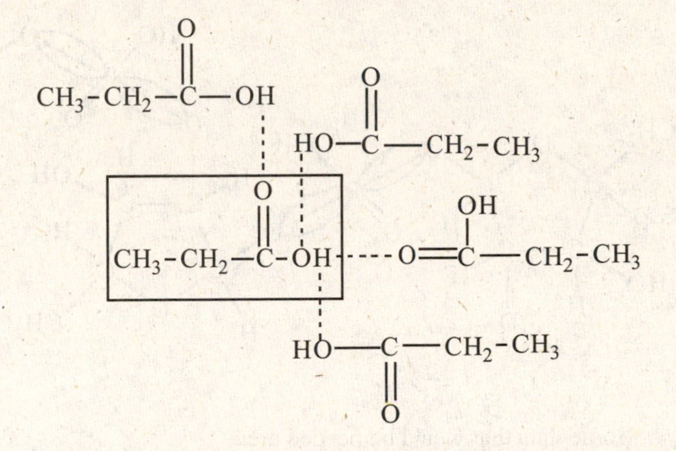

In 1-butanol, there is one hydrogen-bonding H atom and one highly electronegative O atom in the molecule. That means there will be two different kinds of hydrogen-bonding interactions between the given molecule (in the box below) and its neighbors; one way for the H atom to be attracted to neighboring O atoms and one way for O atoms to be attracted to neighboring H atoms:

$HO-CH_2-CH_2-CH_2-CH_3$

$CH_3-CH_2-CH_2-CH_2-OH$

$HO-CH_2-CH_2-CH_2-CH_3$

The more extensive hydrogen bonding in the propanoic acid suggests that it will have a higher boiling point, since the larger the intermolecular attractions, the higher the boiling point.

94. *Answer:* **CH_3–C≡C–H**

Strategy and Explanation: Find the repeating unit, then put an extra bond between the two C atoms at that point in the chain. The repeating unit looks like this.

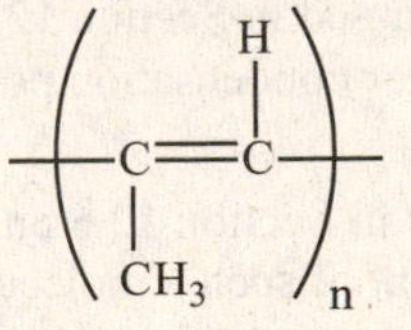

That means we need CH_3–C≡C–H for the monomer.

96. *Answer:* **see annotated structure, below**

Strategy and Explanation: The ester linkages are formed between the POH bonds on the phosphate (playing the role of the carboxylic acid) and the alcohol sites on the sugar. The phosphate ester linkages are the bonds formed where the condensation occurred.

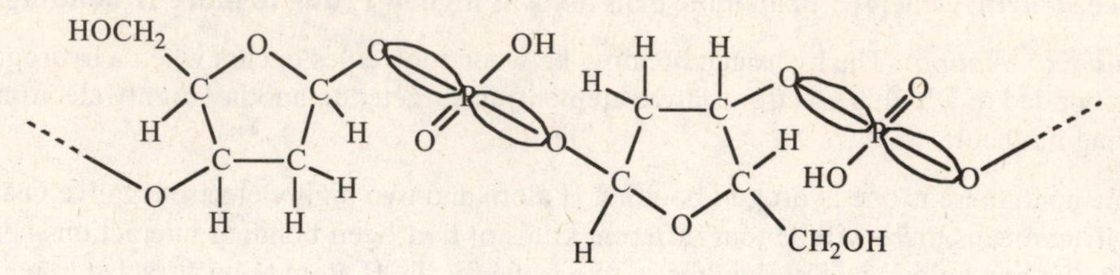

In the backbone of DNA, the phosphate ester linkages are made to the third and fifth –OH groups:

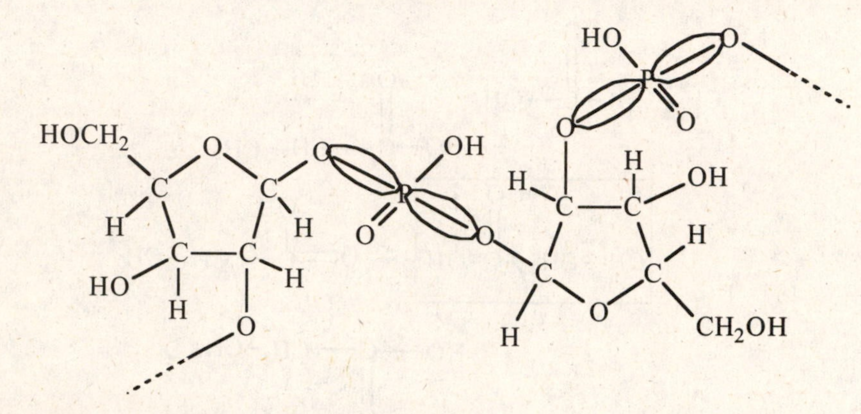

98. *Answer/Explanation:* Some data that would be needed are:

- Sources and amounts of CO_2 generated over time to determine additional CO_2.
- Photosynthesis rate of depletion per tree per year.
- Average number of trees per acre.
- The number of acres of land in Australia that could support trees.
- The allowable tree density.

Other information would also be necessary. Students should be asked to try to think of other things they would need to know to determine the validity of the assertion.

100. *Answer:* **(a) Box C (b) Box G (c) Box G (d) Box F (e) Box D (f) Box B (g) Boxes A and H (h) Box C (i) Box I (j) Box E**

Strategy and Explanation: Given several prompts and set of choices in a grid of boxes, identify which box(es) answer the prompt.

(a) Triglycerides are described in Section 12.5 as an ester formed between three fatty acids and glycerol (see page 557) **Box C** shows a triglyceride.

(b) Oxidation of alcohols is discussed in Section 12.4. The final oxidation of a primary alcohol is a carboxylic acid. **Box G** has a molecule that contains the functional group carboxylic acid, –COOH.

(c) Condensation polymerization is discussed in Section 12.6 on page 569. Molecules with two carboxylic acid groups can serve as monomers for condensation polymerization. The molecule in **Box G** fits this description.

(d) Addition polymerization is discussed in Section 12.6 on page 562. The addition polymer typically is a long C–C chain molecule. A fragment of such a molecule is shown in **Box F**.

(e) Oxidation of alcohols is discussed in Section 12.4. The initial oxidation of a primary alcohol is an aldehylde. **Box D** has a molecule that contains the aldehyde functional group, –CHO.

(f) Amide bonds are discussed in Section 12.6 on page 571. **Box B** contains a molecule with an amide bond, –CONH.

(g) Addition polymerization is discussed in Section 12.6 on page 562. The monomer is a molecule with a C=C bond. Two such molecules are shown in **Boxes A and H**.

(h) Ester linkage is the C–O bond formed in the condensation of a carboxylic acid with an alcohol, as shown in the formation of a triglyceride on page 557. Therefore **Box C** has three ester linkages.

(i) Oxidation of alcohols is discussed in Section 12.4. The final oxidation of a secondary alcohol is a ketone. **Box I** has a molecule that contains the ketone functional group, –CO–.

(j) Alcohols are classified according to the number of carbon atoms directly bonded to the –C–OH carbon. If three carbon atoms are bonded directly, the compound is a tertiary alcohol. (See page 549 in Section 12.4). **Box E** shows a tertiary alcohol.

More Challenging Questions

103. *Answer:* **(a) X is ethanol, Y is acetic acid, Z is acetaldehyde; see Lewis structures below (b) See equations below (c) See explanation below.**

Strategy and Explanation: Given three chemical names and some information about reactions characterizing them, determine which substances are reacting. Write Lewis structures and equations using structural formulas for the reactions observed. Explain information given regarding the oxidation of two of these substances.

(a) An ester is formed by the reaction of a carboxylic acid and an alcohol. So, X and Y must be acetic acid and ethanol. Because Y is also seen to product an acid aqueous solution, Y must be acetic acid. That leaves X to be ethanol and Z to be acetealdehyde.

X = Ethanol is CH_3CH_3OH. It's Lewis structure looks like this:

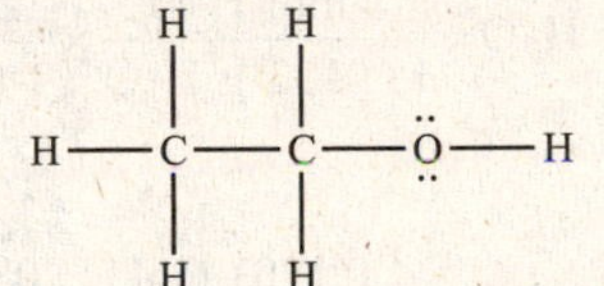

Y = Acetic acid is CH_3COOH. It's Lewis structure looks like this:

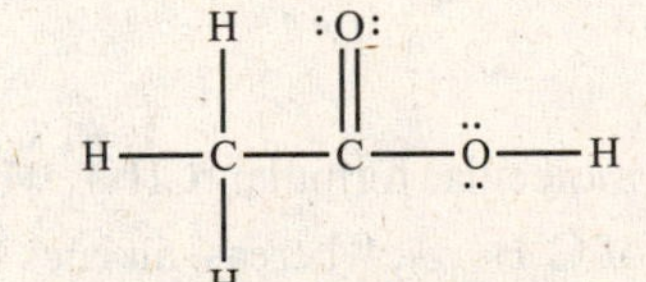

Z = Acetaldehyde is CH_3COH. It's Lewis structure looks like this:

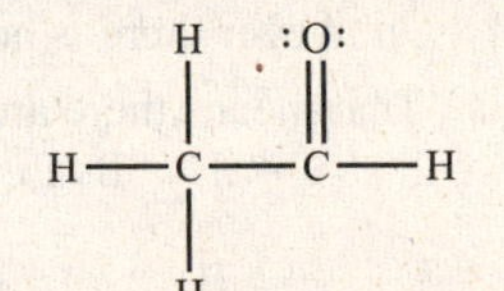

(b) The reaction of acetic acid and ethanol is a condensation reaction. This is the specific equation:

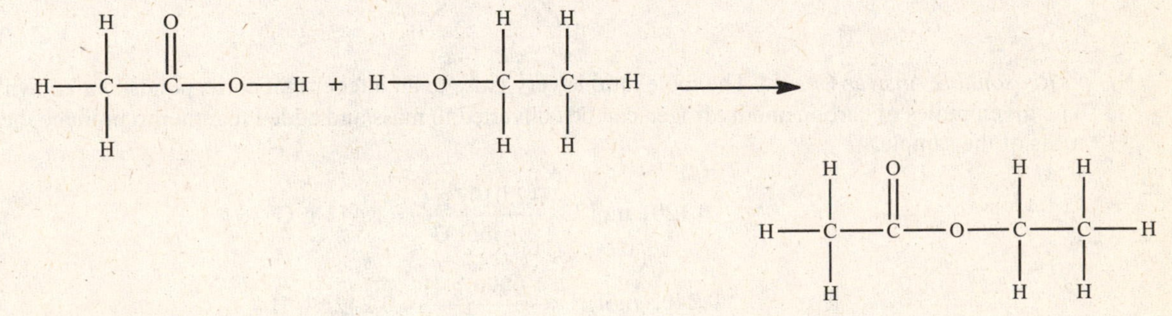

The ionization of acetic acid looks like this:

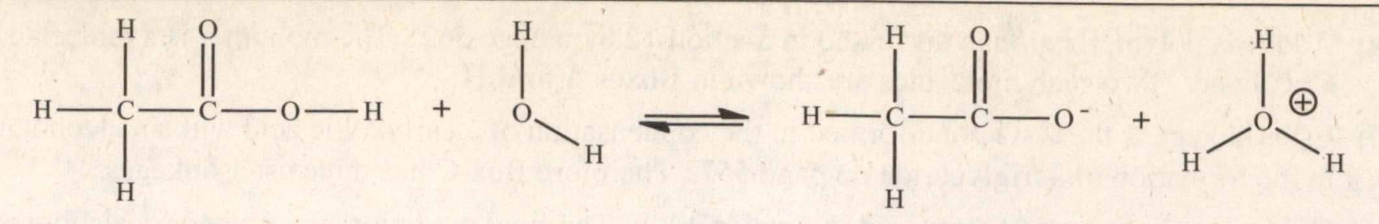

(c) Ethanol is a primary alcohol that is initially oxidized to acetaldehyde; further oxidation of acetaldehyde produces acetic acid. When acetaldehyde is the initial reactant, it is directly oxidized to acetic acid.

105. *Answer:* **(a) CH_2 (b) alkene (c) see structure below**

Strategy and Explanation: Given the mass of a hydrocarbon that undergoes complete combustion to form given masses of carbon dixode and water, determine the empirical formula, determine if the compound is an alkane or an alkene and write a plausible Lewis structure for it.

Calculate the moles of the two products, CO_2 and H_2O produced. Then use the stoichiometry of the product formulas to determine the moles of C and H, use the molar masses of C and O to find the moles of C and moles of O in the sample. Then set up a mole ratio and find the simplest whole number relationship between the atoms. Use the mass of one molecule to determine the molar mass of the compound. Use the molar mass and the empirical formula to determine the molecular formula.

(a) The combustion produces 5.287 g CO_2 and 2.164 g H_2O. Determine the moles of C and O.

$$5.287 \text{ g } CO_2 \times \frac{1 \text{ mol } CO_2}{44.0095 \text{ g } CO_2} \times \frac{1 \text{ mol C}}{1 \text{ mol } CO_2} = 0.1201 \text{ mol C}$$

$$2.164 \text{ g } H_2O \times \frac{1 \text{ mol } H_2O}{18.0152 \text{ g } H_2O} \times \frac{2 \text{ mol H}}{1 \text{ mol } H_2O} = 0.2402 \text{ mol H}$$

Set up mole ratio:

0.1201 mol C : 0.2402 mol O

Divide all the numbers in the ratio by the smallest number of moles, 0.1201 mol: 1 C : 2 H

Empirical formula: CH_2

(b) This compound will have a molecular formula: $(CH_2)_n$ which can also be written C_nH_{2n}. Alkanes have formulas fitting the pattern of C_nH_{2n+2}; whereas, alkenes have formulas fitting the pattern of C_nH_{2n}. (Refer to Chapter 3 and 8 in the textbook). So, the compound burned here is an alkene.

(c) The molecular formula is $(CH_2)_n$, however there is not enough information given in the question to determine what the molecule is. That means there are many answers to this part of the question. The smallest possible alkene has n = 2, H_2C=CH_2. Its Lewis Structure is:

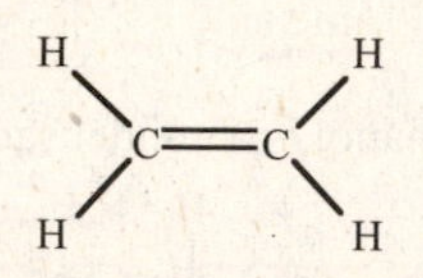

✓ *Reasonable Answer Check:* The mole ratio is very close to an integer that corresponds to a known alkene. The given moles of carbon and hydrogen can be converted to mass and added together to produce the initial mass of the sample,

$$0.1201 \text{ mol C} \times \frac{12.0107 \text{ g C}}{1 \text{ mol C}} = 1.442 \text{ g C}$$

$$0.2402 \text{ mol H} \times \frac{1.0079 \text{ g H}}{1 \text{ mol H}} = 0.2423 \text{ g H}$$

$$1.442 \text{ g C} + 0.2423 \text{ g H} = 1.6485 \text{ g compound}$$

So, we can be assured of 100% yield. With regard to the answer in (c), several other answers are just as plausible as the one that is published, including propylene, also given in Section 8.5 of the textbook. Student should be made aware that there is not one unique answer to this question.

107. *Answer:* **see graph below; for C_3H_8: 2200 kJ/mol; for C_9H_{20}: 6100 kJ/mol; for $C_{16}H_{34}$: 10700 kJ/mol**

Strategy and Explanation: Graph the enthalpy of combustion per mole of carbon atoms against the number of carbon atoms:

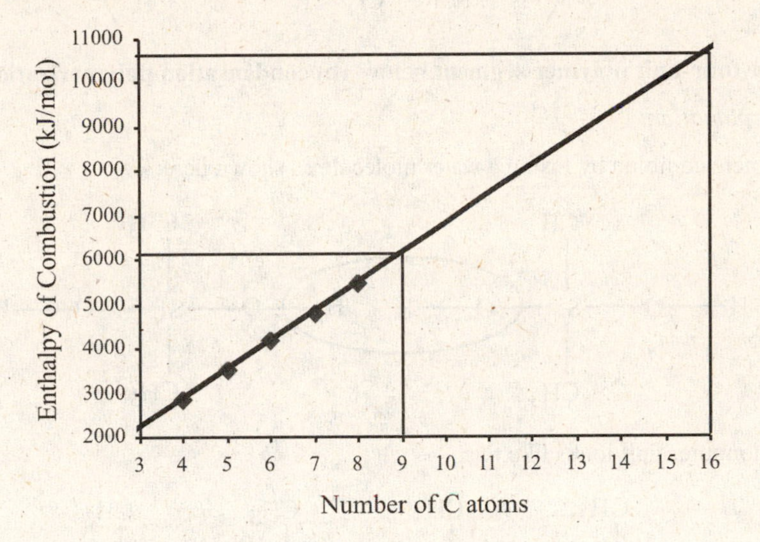

After extrapolating the straight line that best represents the linear relationship, we can determine an estimate for 3-carbons, 9-carbons and 16 carbons.

For three carbons (look at the edge of the graph on the left), an estimated enthalpy of vaporization per mole of carbon for three carbons is 2200 kJ/mol.

For nine carbons (see horizontal and vertical lines provided), an estimated enthalpy of vaporization per mole of carbon for three carbons is 6100 kJ/mol.

For 16 carbons (look at the edge of the graph on the right), an estimated enthalpy of vaporization per mole of carbon for three carbons is 10700 kJ/mol.

109. *Answer:* **see structures below**

Strategy and Explanation: The monomer for PVC, vinyl chloride, $H_2C{=}CHCl$ can combine in three different ways. The first forms a chain with chlorine atoms on alternating carbon atoms, formed when the monomers come together always leading with the same side of the molecule:

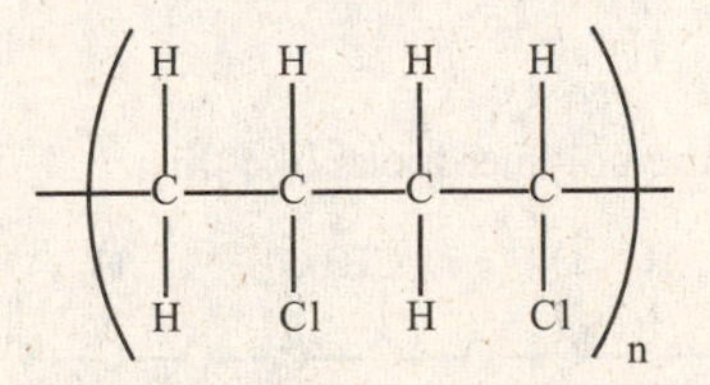

The second forms a chain with chlorine atoms on adjacent carbon atoms, formed when the monomers alternate which side of the monomer attaches to the polymer chain:

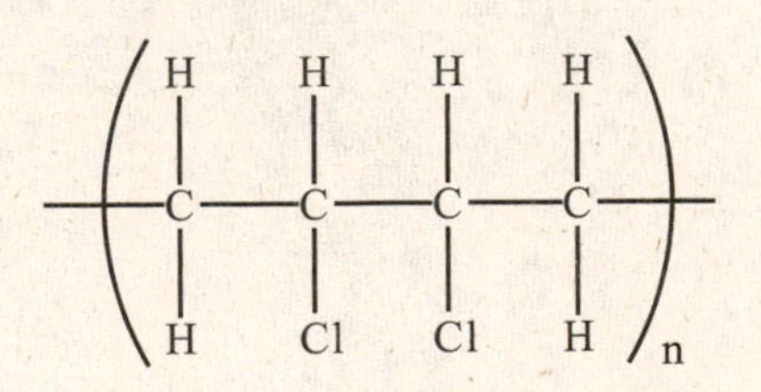

The third forms a chain with random orientation of the monomers and therefore variable orientation of the Cl atoms in the polymer chain:

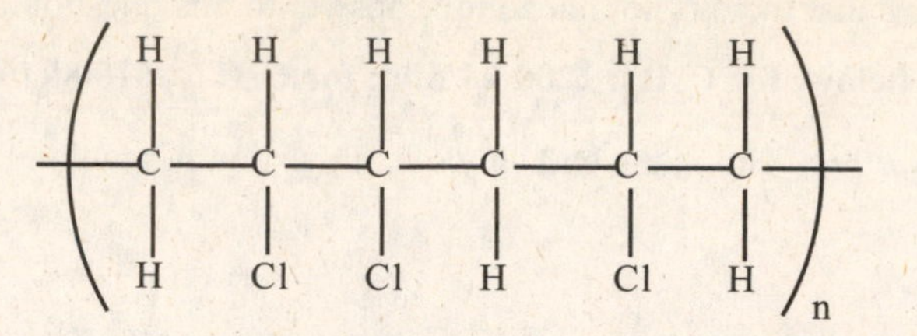

111. *Answer:* **(a) see four-unit polymer segment below (b) condensation polymerization**

Strategy and Explanation:

(a) The monomers combine by losing a water molecule as shown here:

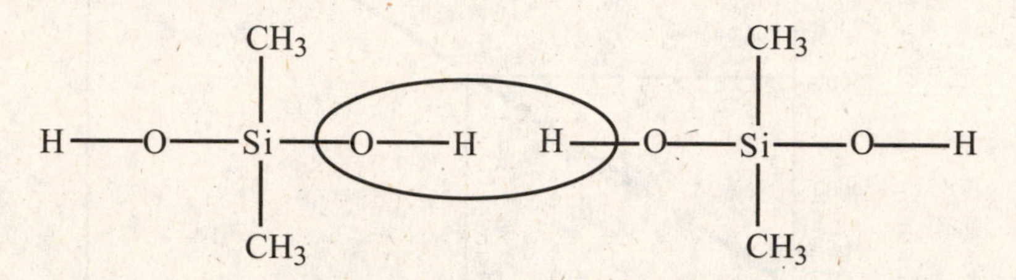

The four-monomer unit looks like this:

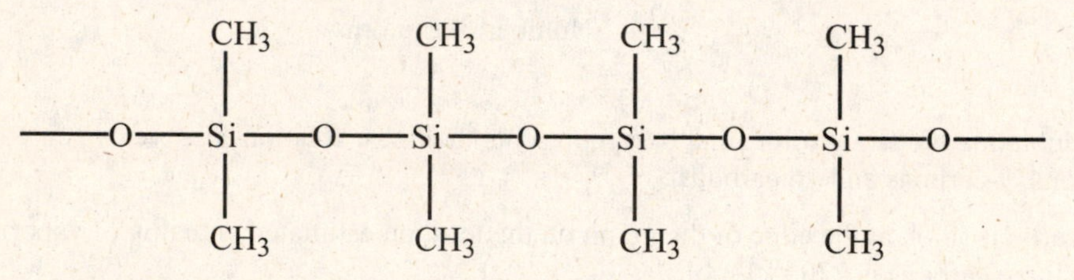

(b) This type of polymerization is called condensation polymerization.

113. *Answer:* **see four-unit polymer segment below**

Strategy and Explanation: Lactic acid is given in the textbook near the end of Section 9.2 on page 385, $CH_3CH(OH)COOH$. It forms a condensation copolymer with glycolic acid, $HOCH_2COOH$, by the elimination of water:

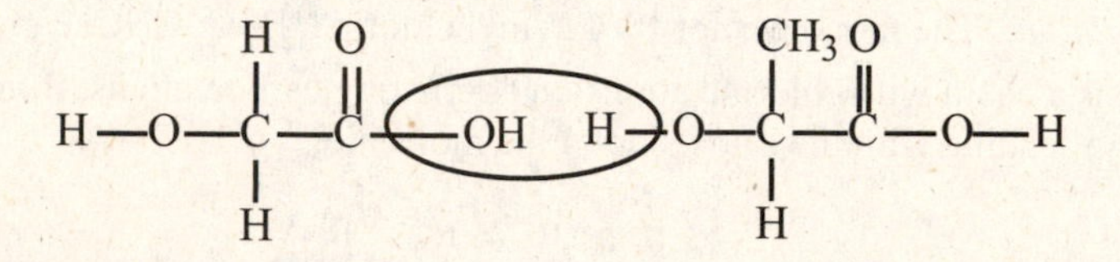

A four-monomer-unit chain of the polymer looks like this:

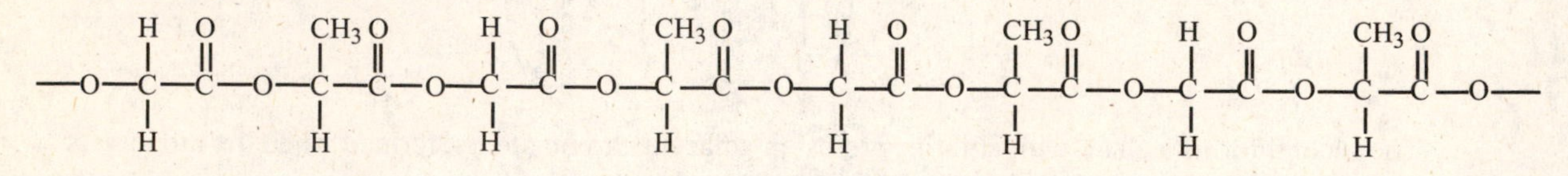

Chapter 13: Chemical Kinetics: Rates of Reactions

Solutions for Blue-Numbered Questions for Review and Thought

Topical Questions

Reaction Rate (Section 13.1)

9. The smaller the grains of sugar, the faster the sugar will dissolve, because the surface contact with the water is essential to the dissolving process. The most finely granulated sugar is (d) powdered sugar. The second most finely granulated sugar is (c) granulated sugar. Granulated sugar compressed into cubes, or (b) sugar cubes, would be somewhat slower than granulated sugar to dissolve. The slowest of the four to dissolve would be large sugar crystals in (a) rock candy sugar.

11. *Answer:* **(a) see graph below; 0.0167 M/s, 0.0119 M/s, 0.0089 M/s, 0.0070 M/s; [reactant] is decreasing (b) Rate of change of [B] is twice as fast as the rate of change of [A] (c) 0.0238 M/s**

Strategy and Explanation:

(a) A graph of concentration verses time look like this

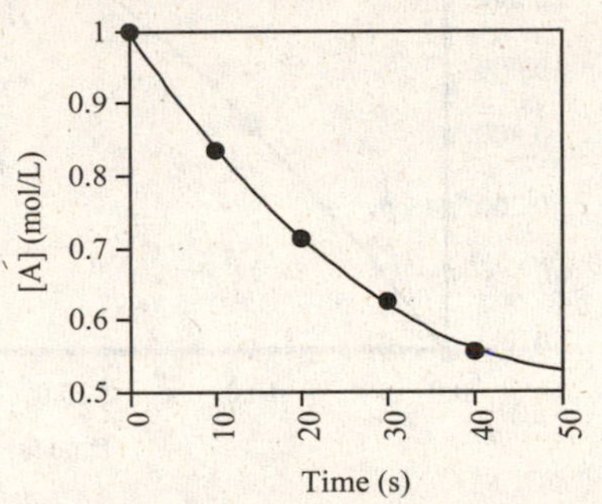

Given data of concentrations at various times during the course of a chemical reaction, find the rate of change for each time interval.

To calculate the average rate for each time interval, use the method described in Section 13.1, page 594.

$$\text{Rate} = \frac{-\Delta[\text{reactant}]}{\Delta t} = \frac{-\left([\text{reactant}]_{\text{final}} - [\text{reactant}]_{\text{initial}}\right)}{t_{\text{final}} - t_{\text{initial}}} = \frac{[\text{reactant}]_{\text{initial}} - [\text{reactant}]_{\text{final}}}{t_{\text{final}} - t_{\text{initial}}}$$

From 0.00 to 10.0 seconds:

$$\text{Rate}_{0\text{-}10} = \frac{(1.000\text{ mol/L}) - (0.833\text{ mol/L})}{10.0\text{ s} - 0.00\text{ s}} = \frac{0.167\text{ mol/L}}{10.0\text{ s}} = 0.0167\ \frac{\text{mol}}{\text{L}\cdot\text{s}}$$

From 10.0 to 20.0 seconds:

$$\text{Rate}_{10\text{-}20} = \frac{(0.833\text{ mol/L}) - (0.714\text{ mol/L})}{20.0\text{ s} - 10.0\text{ s}} = \frac{0.119\text{ mol/L}}{10.0\text{ s}} = 0.0119\ \frac{\text{mol}}{\text{L}\cdot\text{s}}$$

From 20.0 to 30.0 seconds:

$$\text{Rate}_{20\text{-}30} = \frac{(0.714\text{ mol/L}) - (0.625\text{ mol/L})}{30.0\text{ s} - 20.0\text{ s}} = \frac{0.089\text{ mol/L}}{10.0\text{ s}} = 0.0089\ \frac{\text{mol}}{\text{L}\cdot\text{s}}$$

From 30.0 to 40.0 seconds:

$$\text{Rate}_{30\text{-}40} = \frac{(0.625\text{ mol/L}) - (0.555\text{ mol/L})}{40.0\text{ s} - 30.0\text{ s}} = \frac{0.070\text{ mol/L}}{10.0\text{ s}} = 0.0070\ \frac{\text{mol}}{\text{L}\cdot\text{s}}$$

The rate of change decreases because the concentration of the reactant is decreasing.

(b) As described in Section 13.1, "Reaction Rates and Stoichiometry" on page 596, since there are two B products formed whenever one A reactant undergoes reaction, that means that B is forming twice as fast as A is being depleted. Therefore, the rate of change of [B] is twice as fast as the rate of change of [A].

(c) In (a) above, the rate of depletion of A calculated for the 10 - 20 second range is 0.0119 mol/L·s. According to (b), the rate of production of B is twice that number:

$$2 \times (0.0119 \text{ mol/L·s}) = 0.0238 \text{ M/s}$$

✓ *Reasonable Answer Check:* According to the discussion on the effect of concentration on the rate in Section 13.2, it makes sense that the rate decreases as the reactant concentration decreases.

13. *Answer:* **(a) see graph below; 0.0326 mol L^{-1} s^{-1}, 0.0246 mol L^{-1} s^{-1}, 0.0178 mol L^{-1} s^{-1}; 0.0140 mol L^{-1} s^{-1}; Yes, [A] is decreasing (b) Rate of change of [A] is half as fast as the rate of change of [B]; 0.0123 mol L^{-1} s^{-1} (c) 0.0160 mol L^{-1} s^{-1}**

Strategy and Explanation:

(a) A graph of concentration of [B] verses time look like this:

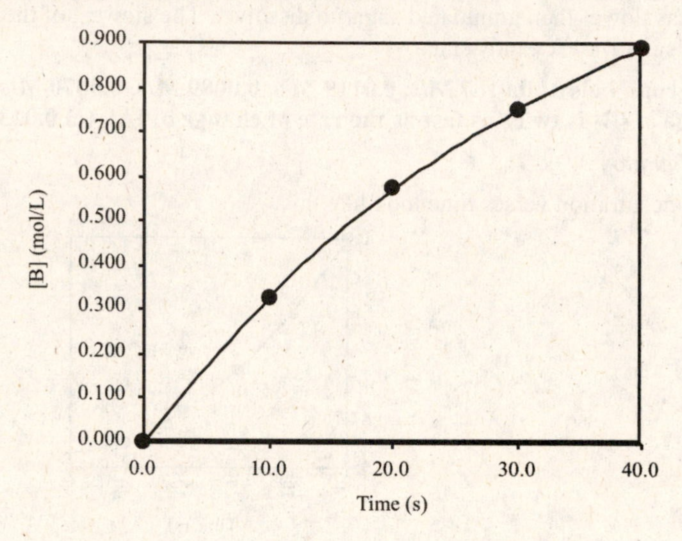

Given data of concentrations at various times during the course of a chemical reaction, find the rate of change for each time interval.

To calculate the average rate for each time interval, use the method described in Section 13.1, page 594.

$$\text{Rate} = \frac{\Delta[\text{product}]}{\Delta t} = \frac{[\text{product}]_{\text{final}} - [\text{product}]_{\text{initial}}}{t_{\text{final}} - t_{\text{initial}}}$$

From 0.00 to 10.0 seconds:

$$\text{Rate}_{0\text{-}10} = \frac{(0.326 \text{ mol/L}) - (0.000 \text{ mol/L})}{10.0 \text{ s} - 0.00 \text{ s}} = \frac{0.326 \text{ mol/L}}{10.0 \text{ s}} = 0.0326 \frac{\text{mol}}{\text{L·s}}$$

From 10.0 to 20.0 seconds:

$$\text{Rate}_{10\text{-}20} = \frac{(0.572 \text{ mol/L}) - (0.326 \text{ mol/L})}{20.0 \text{ s} - 10.0 \text{ s}} = \frac{0.246 \text{ mol/L}}{10.0 \text{ s}} = 0.0246 \frac{\text{mol}}{\text{L·s}}$$

From 20.0 to 30.0 seconds:

$$\text{Rate}_{20\text{-}30} = \frac{(0.750 \text{ mol/L}) - (0.572 \text{ mol/L})}{30.0 \text{ s} - 20.0 \text{ s}} = \frac{0.178 \text{ mol/L}}{10.0 \text{ s}} = 0.0178 \frac{\text{mol}}{\text{L·s}}$$

From 30.0 to 40.0 seconds:

$$\text{Rate}_{30\text{-}40} = \frac{(0.890 \text{ mol/L}) - (0.750 \text{ mol/L})}{40.0 \text{ s} - 30.0 \text{ s}} = \frac{0.140 \text{ mol/L}}{10.0 \text{ s}} = 0.0140 \frac{\text{mol}}{\text{L·s}}$$

Yes, the rate of change decreases because the concentration of the reactant is decreasing.

(b) As described in Section 13.1, "Reaction Rates and Stoichiometry" on page 596, since there are two B products formed whenever one A reactant undergoes reaction, that means that B is forming twice as fast as A is being depleted. Therefore, the rate of change of [A] is half as fast as the rate of change of [B].

In (a) above, the rate of depletion of B calculated for the 10.0 s to 20.0 s range is 0.0246 mol/L·s. So, the rate of depletion of A is half that:

$$\frac{1}{2} \times (0.0246 \text{ mol/L·s}) = 0.0123 \text{ mol/L·s}$$

(c) To get the instantaneous rate at a specific concentration of B, find the slope of the tangent of the curve at 0.750 mol/L:

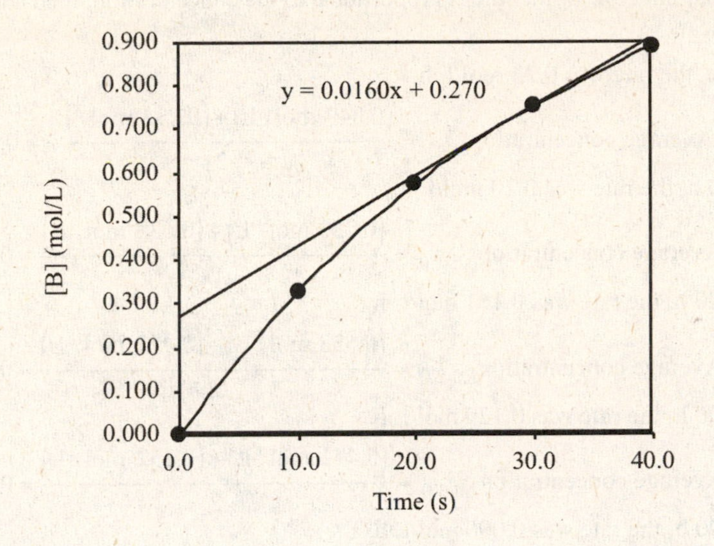

The slope of the tangent line = $0.0160 \text{ mol L}^{-1} \text{ s}^{-1}$ = instantaneous rate at 0.750 M B

✓ *Reasonable Answer Check:* According to the discussion on the effect of concentration on the rate in Section 13.2, it makes sense that the rate decreases as the reactant concentration decreases.

15. *Answer:* **(a) 0.23 mol/L·h (b) 0.20 mol/L·h (c) 0.161 mol/L·h (d) 0.12 mol/L·h (e) 0.090 mol/L·h (f) 0.066 mol/L·h**

Strategy and Explanation: Given data of concentrations at various times during the course of a chemical reaction, find the rate of change for each time interval.

To calculate the average rate for each time interval, use the method described in Section 13.1 and in the answers to Questions 11 and 12:

(a) From 0 to 0.50 h: $$\text{Rate}_{0\text{-}0.5} = \frac{(0.849 \text{ mol/L}) - (0.733 \text{ mol/L})}{0.50 \text{ h} - 0.00 \text{ h}} = \frac{0.116 \text{ mol/L}}{0.50 \text{ h}} = 0.23 \frac{\text{mol}}{\text{L} \cdot \text{h}}$$

(b) From 0.5 to 1.00 h: $$\text{Rate}_{0.5\text{-}1.0} = \frac{(0.733 \text{ mol/L}) - (0.633 \text{ mol/L})}{1.00 \text{ h} - 0.50 \text{ h}} = \frac{0.100 \text{ mol/L}}{0.50 \text{ h}} = 0.20 \frac{\text{mol}}{\text{L} \cdot \text{h}}$$

(c) From 1.00 to 2.00 h: $$\text{Rate}_{1.0\text{-}2.0} = \frac{(0.633 \text{ mol})/\text{L} - (0.472 \text{ mol/L})}{2.00 \text{ h} - 1.00 \text{ h}} = \frac{0.161 \text{ mol/L}}{1.00 \text{ h}} = 0.161 \frac{\text{mol}}{\text{L} \cdot \text{h}}$$

(d) From 2.00 to 3.00 h: $$\text{Rate}_{2.0\text{-}3.0} = \frac{(0.472 \text{ mol/L}) - (0.352 \text{ mol/L})}{3.00 \text{ h} - 2.00 \text{ h}} = \frac{0.120 \text{ mol/L}}{1.00 \text{ h}} = 0.120 \frac{\text{mol}}{\text{L} \cdot \text{h}}$$

(e) From 3.00 to 4.00 h: $$\text{Rate}_{3.0\text{-}4.0} = \frac{(0.352 \text{ mol/L}) - (0.262 \text{ mol/L})}{4.00 \text{ h} - 3.00 \text{ h}} = \frac{0.090 \text{ mol/L}}{1.00 \text{ h}} = 0.090 \frac{\text{mol}}{\text{L} \cdot \text{h}}$$

(f) From 4.00 to 5.00 h: $$\text{Rate}_{3.0\text{-}4.0} = \frac{(0.262 \text{ mol/L}) - (0.196 \text{ mol/L})}{5.00 \text{ h} - 4.00 \text{ h}} = \frac{0.066 \text{ mol/L}}{1.00 \text{ h}} = 0.066 \frac{\text{mol}}{\text{L} \cdot \text{h}}$$

✓ *Reasonable Answer Check:* According to the discussion of rate dependence on concentration in Section 13.2, it makes sense that the rate decreases as the reactant concentration decreases.

17. *Answer:* **(a) The average rate is directly proportional to average $[N_2O_5]$ for each pair. (b) $k = 0.29\ hr^{-1}$**

Strategy and Explanation: Given data of concentrations at various times during the course of a chemical reaction and the average rate calculated over different time intervals, show that the reaction obeys a specific given rate law, evaluate the rate constant by averaging.

The average rates over each time interval were calculated in the solution to Question 14. Find the average concentration over that interval. If the rate is proportional to the concentration, then a graph of concentration vs. rate should be linear.

(a) From 0 to 0.50 h, the rate was 0.23 mol/L·h.

$$\text{Average concentration}_{0\text{-}0.5} = \frac{(0.849\ \text{mol/L}) + (0.733\ \text{mol/L})}{2} = 0.791\ \frac{\text{mol}}{\text{L}}$$

From 0.5 to 1.00 h, the rate was 0.20 mol/L·h.

$$\text{Average concentration}_{0.5\text{-}1.0} = \frac{(0.733\ \text{mol/L}) + (0.633\ \text{mol/L})}{2} = 0.683\ \frac{\text{mol}}{\text{L}}$$

From 1.00 to 2.00 h, the rate was 0.161 mol/L·h

$$\text{Average concentration}_{1.0\text{-}2.0} = \frac{(0.633\ \text{mol/L}) + (0.472\ \text{mol/L})}{2} = 0.553\ \frac{\text{mol}}{\text{L}}$$

From 2.00 to 3.00 h, the rate was 0.120 mol/L·h

$$\text{Average concentration}_{2.0\text{-}3.0} = \frac{(0.472\ \text{mol/L}) + (0.352\ \text{mol/L})}{2} = 0.412\ \frac{\text{mol}}{\text{L}}$$

From 3.00 to 4.00 h, the rate was 0.090 mol/L·h.

$$\text{Average concentration}_{3.0\text{-}4.0} = \frac{(0.352\ \text{mol/L}) + (0.262\ \text{mol/L})}{2} = 0.307\ \frac{\text{mol}}{\text{L}}$$

From 4.00 to 5.00 h, the rate was 0.066 mol/L·h.

$$\text{Average concentration}_{4.0\text{-}5.0} = \frac{(0.262\ \text{mol/L}) + (0.196\ \text{mol/L})}{2} = 0.229\ \frac{\text{mol}}{\text{L}}$$

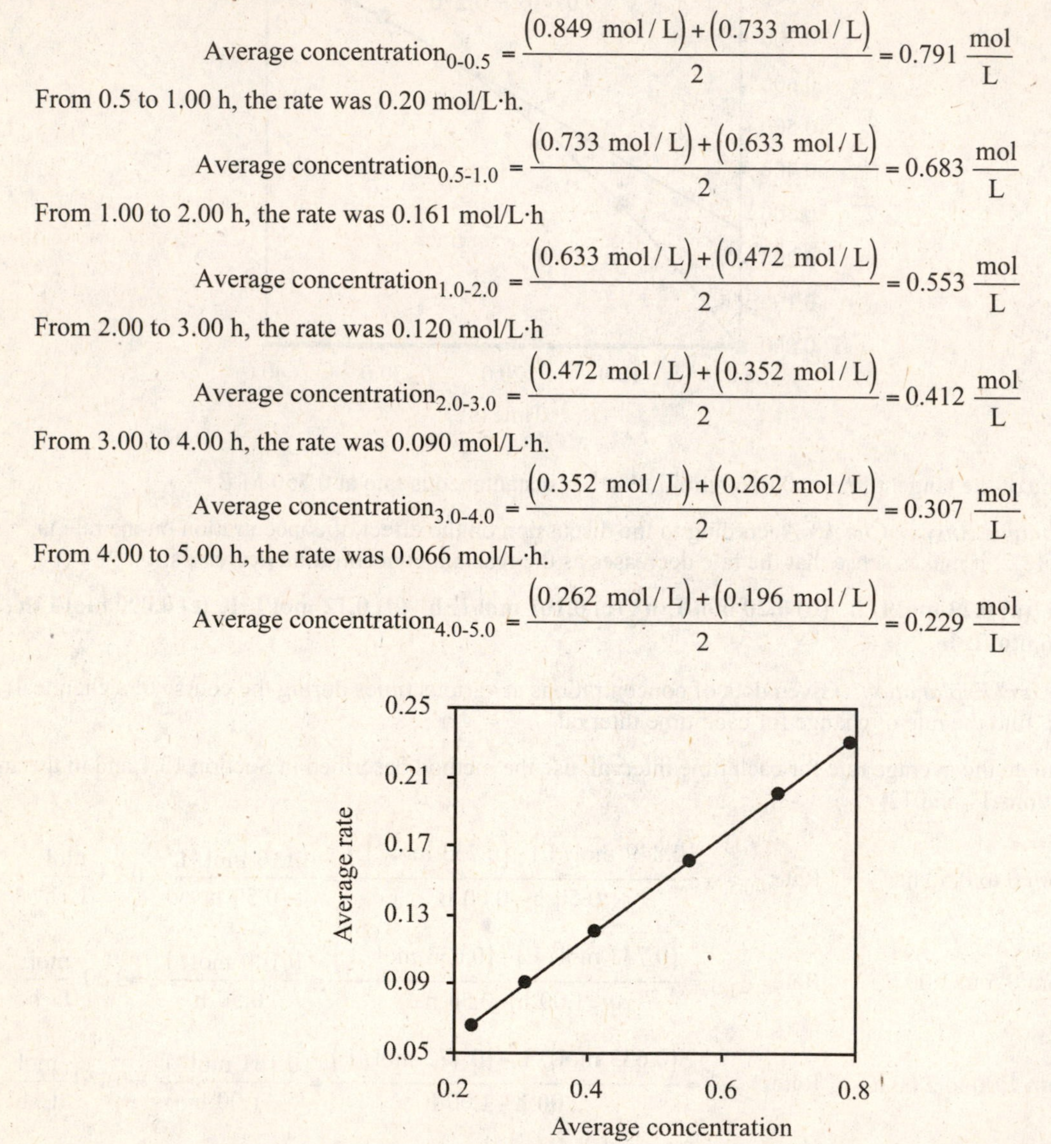

The linear relationship between rate and concentration satisfies the equation described in the question:

$$\text{Rate} = k\,[N_2O_5]$$

(b) From 0 to 0.50 h, the rate is 0.23 mol/L·h and the average concentration is 0.791 mol/L.

$$k = \frac{\text{Rate}}{[N_2O_5]} = \frac{0.23\dfrac{\text{mol}}{\text{L}\cdot\text{h}}}{0.791\dfrac{\text{mol}}{\text{L}}} = 0.29\text{h}^{-1}$$

Similarly, for each of the other time increments:

Average rate, mol/L·h	Average concentration, mol/L	k (h^{-1})
0.23	0.791	0.29
0.20	0.683	0.29
0.161	0.553	0.291
0.120	0.412	0.291
0.090	0.307	0.29
0.066	0.229	0.29

The average value of k is 0.29 h^{-1}.

✓ *Reasonable Answer Check:* According to the discussion of rate dependence on concentration in Section 13.2, it makes sense that the rate decreases as the reactant concentration decreases. It also makes sense that the value of k is the same each time it was calculated.

19. *Answer:* **See graph and explanations below**

Strategy and Explanation: Qualitatively, the concentration of the reactant will drop nonlinearly (first-order reaction) as the concentrations of the products increase. The rate of increase of NO concentration will mirror the rate of decrease of NO_2, and will be twice that of O_2, due to their stoichiometric ratios being 2 : 2 : 1. At late times, the concentrations of all species will level off to an unchanging value. (The dashed line on the graph is used to find the initial rate in (a).)

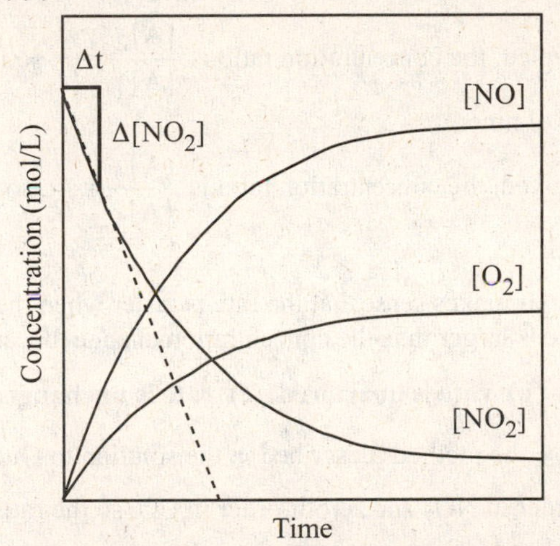

(a) As shown on the graph, the initial rate can be determined by taking a small time interval (Δt) near the initial time where the curve is still well approximated by a straight line (represented by the dashed line) and finding the slope of that line.

$$\text{Rate} = \frac{-\Delta[NO_2]}{\Delta t} = -\text{ slope of the straight line}$$

(b) As shown on the graph at very long times, the final rate will be zero because the concentration stops changing and the numerator of the rate expression is zero. That will happen for one of two reasons, either all the reactants will have been used up or the reaction will reach an equilibrium state.

21. *Answer:* **(a)** $\frac{-\Delta[N_2O_4]}{\Delta t}$ **(b)** $\frac{\Delta[N_2]}{\Delta t}$

Strategy and Explanation: Express the rate of the reaction in terms of changing concentrations of a reactant and a product. Rate must be a positive quantity. Δt is positive when time is increasing. Because the concentrations of reactants decrease with time, the rate of change of reactant concentration is negative; therefore, the numerator of the reactant rate expression will include a minus sign to make the rate a positive number. Because the concentrations of products increase with time, the rate of change of reactant concentration is positive as must be the rate.

(a) The rate of depletion of the reactant N_2O_4 is $\frac{-\Delta[N_2O_4]}{\Delta t}$

(a) The rate of formation of the product N_2 is $\frac{\Delta[N_2]}{\Delta t}$

Effect of Concentration on Reaction Rates (Section 13.2)

23. *Answer:* **(a) nine times faster (b) one quarter as fast**

Strategy and Explanation: Given the rate law of a reaction, determine what happens to the rate when the concentration is changed.

Knowing the rate law, the rate increase can be determined by creating a ratio to provide a simple relationship between the rate changes and the concentration changes.

The rate law is: rate = $k[A]^2$. Make a ratio of the rate law of the reaction under two different conditions: the initial condition (designated by a subscript "1", $rate_1 = k[A]_1^2$) compared to the final condition (designated by subscript "2", $rate_2 = k[A]_2^2$)

$$\text{rate change factor} = \frac{rate_2}{rate_1} = \frac{k[A]_2^2}{k[A]_1^2} = \left(\frac{[A]_2}{[A]_1}\right)^2$$

(a) When concentration is tripled, the concentration ratio is $\frac{[A]_2}{[A]_1} = 3$, so the rate change factor = $3^2 = 9$. The rate increases by a factor of nine.

(b) When concentration is halved, the concentration ratio is $\frac{[A]_2}{[A]_1} = \frac{1}{2}$, so the rate change factor = $\left(\frac{1}{2}\right)^2 = \frac{1}{4}$. The rate will be quartered.

✓ *Reasonable Answer Check:* It makes sense that the rate is faster when the concentration is larger and vice versa, also that the rate change is larger than the concentration change, because the reaction is second-order.

25. *Answer:* **(a) rate = $k[NO_2]^2$ (b) rate is quartered (c) rate is unchanged**

Strategy and Explanation: Use the methods described in the solution to Question 23.

(a) The reaction is second-order in NO_2 and zeroth-order in CO, so the rate law has powers of 2 and zero on the concentrations of NO_2 and CO, respectively.

$$\text{rate} = k[NO_2]^2[CO]^0$$

Because any number raised to the zero power results in 1, that says the rate is unaffected by the concentration of CO.

$$\text{rate} = k[NO_2]^2$$

(b) A ratio gives the rate change factor:

$$\text{rate change factor} = \frac{rate_2}{rate_1} = \frac{k[NO_2]_2^2}{k[NO_2]_1^2} = \left(\frac{[NO_2]_2}{[NO_2]_1}\right)^2$$

When the concentration of NO_2 is halved, the concentration ratio is $\frac{[NO_2]_2}{[NO_2]_1} = \frac{1}{2}$, so the rate change factor $= \left(\frac{1}{2}\right)^2 = \frac{1}{4}$. The rate will be quartered.

(c) Because the rate law does not include concentration dependence for CO, the rate will be unchanged when the CO concentration is doubled.

27. *Answer:* **(a) 9.0×10^{-4} M/h, 1.8×10^{-3} M/h, 3.6×10^{-3} M/h; (b)-(d) see explanations below**

Strategy and Explanation:

(a) Given the rate law of a reaction, the value of the rate constant, and several concentrations, determine the instantaneous rates at each of those concentrations.

Plug the concentration and the rate constant into the rate law: rate = $k[Pt(NH_3)_2Cl_2]$

(i) $rate_{0.010} = (0.090\ h^{-1})(0.010\ M) = 9.0 \times 10^{-4}$ M/h

(ii) $rate_{0.020} = (0.090\ h^{-1})(0.020\ M) = 1.8 \times 10^{-3}$ M/h

(iii) $rate_{0.040} = (0.090\ h^{-1})(0.040\ M) = 3.6 \times 10^{-3}$ M/h

(b) If initial concentration of $Pt(NH_3)_2Cl_2$ is high, the initial rate of disappearance of $Pt(NH_3)_2Cl_2$ will be high. If initial concentration of $Pt(NH_3)_2Cl_2$ is low, the initial rate of disappearance of $Pt(NH_3)_2Cl_2$ will be small. The rate of disappearance of $Pt(NH_3)_2Cl_2$ is directly proportional to the concentration of $Pt(NH_3)_2Cl_2$.

(c) The rate law shows direct proportionality between rate and the concentration of $Pt(NH_3)_2Cl_2$.

(d) When the initial concentration of $Pt(NH_3)_2Cl_2$ is high, the rate of appearance of Cl^- will be high. When initial concentration is low, rate of appearance Cl^- will be small. The rate of appearance of Cl^- is directly proportional to the concentration of $Pt(NH_3)_2Cl_2$.

✓ *Reasonable Answer Check:* The rate is faster when the concentration is higher in (a).

29. *Answer:* $\frac{\Delta[C]}{\Delta t}$ **= (3.0 × 10⁻³)[A]**

Strategy and Explanation: Given the initial concentrations and initial rates of a reaction for several different experimental conditions for the same chemical reaction, determine the rate law and the rate constant.

In Section 13.2, the method of finding the rate law from initial rates is described for getting the orders. Compare pairs of experiments where only one of the concentrations is different, and relate that to the changes in the rate. Once the orders are determined, plug the data into the rate law to determine the value of k.

Because A and B are reactants and are both varied in the different experiments, we will seek a rate law that looks like this: rate = $k[A]^x[B]^y$ where x and y are currently unknown orders, and k is the rate constant.

Looking at Experiments 1 and 3, the initial concentration of B triples, the initial concentration of A stays constant, and the initial rate doesn't change (3.0×10^{-4} M). That means the order with respect to B is zero.

Looking at Experiments 3 and 2, the initial concentration of B stays constant, the initial concentration of A triples, and the initial rate changes by a factor of $\frac{9.0\times10^{-4}\ M\ min^{-1}}{3.0\times10^{-4}\ M\ min^{-1}} = 3$. The rate change is the directly proportional to the concentration change, so the order with respect to A is one.

The rate law now looks like this: rate = k[A]

Solve the rate law for k: $k = \frac{rate}{[A]}$

Plug in each experiment's data where A is different. Here is an example of the first experiment's calculation:

$$k_1 = \frac{3.0\times10^{-4}\ M\ min^{-1}}{(0.10\ M)} = 0.0030\ min^{-1}$$

Experiment	[A] (mol/L)	Initial rate of formation of C ($M\ min^{-1}$)	Rate constant (min^{-1})
1	0.10	3.0×10^{-4}	3.0×10^{-3}
2	0.30	9.0×10^{-4}	3.0×10^{-3}
4	0.20	6.0×10^{-4}	3.0×10^{-3}

So, the rate law expression is: rate = $\frac{\Delta[C]}{\Delta t} = (3.0 \times 10^{-3})[A]$

30. *Answer:* **(a) rate = k [I][II] (b) k = 1.04 L/mol·s**

Strategy and Explanation: Given the initial concentrations and initial rates of a reaction for several different experimental conditions for the same chemical reaction, determine the rate law and the rate constant for the reaction.

In Section 13.2, the method of finding the rate law from initial rates is described for getting the orders. However, comparing pairs of experiments where only one of the concentrations is different and relating that to the changes in the rate does not give consistent results. So, here we will derive a linear equation relating the concentrations and the rates and graph the results. Once the orders are determined, plug the data into the rate law to determine the value of k.

Because I and II are reactants and are both varied in the different experiments, we will seek a rate law looks like this: rate = $k[I]^i[II]^j$ where k, i, and j are currently unknown.

(a) Comparing experiments with [I] constant, the rate law can be simplified to:

$$\text{rate} = k'[II]^j \qquad \text{where } k' = k[I]^i$$

If we take the log of both sides of the equation, we can derive a linear relationship:

$$\log(\text{rate}) = \log(k') + \log([II]^j)$$

$$\log(\text{rate}) = \log(k') + j\log[II]$$

$$\log(\text{rate}) = j\log[II] + \log(k')$$

Comparing to the equation of a line: y = m x + b, we see that if we plot

log(rate) against log[II], the slope of the line will be the reaction order, j.

Similarly comparing experiments with [II] constant, the rate law can be simplified to:

$$\text{rate} = k''[II]^I \qquad \text{where } k'' = k[II]^j$$

This gives a similar linear relationship: log(rate) = i log[I] + log(k")

Comparing to the equation of a line: y = m x + b, we see that if we plot log(rate) against log[I], the slope of the line will be the reaction order, i.

Four data sets have constant [I] and four data sets have constant [II].

log[II] with constant [I]	**log rate** with constant [I]	**log[I]** with constant [II]	**log rate** with constant [II]
-4.745	-8.509	-4.783	-8.824
-4.453	-8.201	-4.606	-8.569
-4.149	-7.951	-4.304	-8.345
-3.975	-7.752	-3.827	-7.752

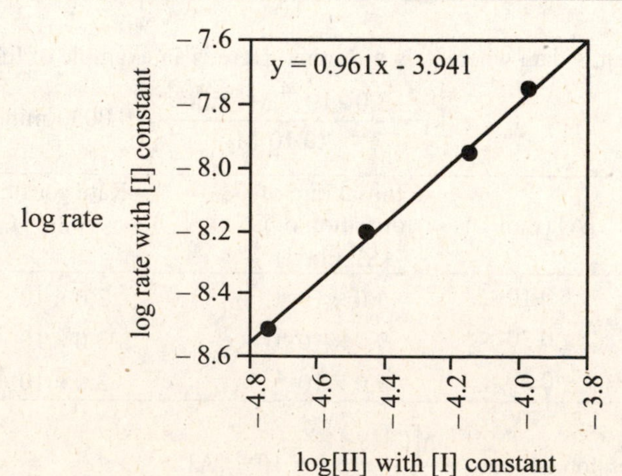

So, the reaction order of reactant II (j) is one.

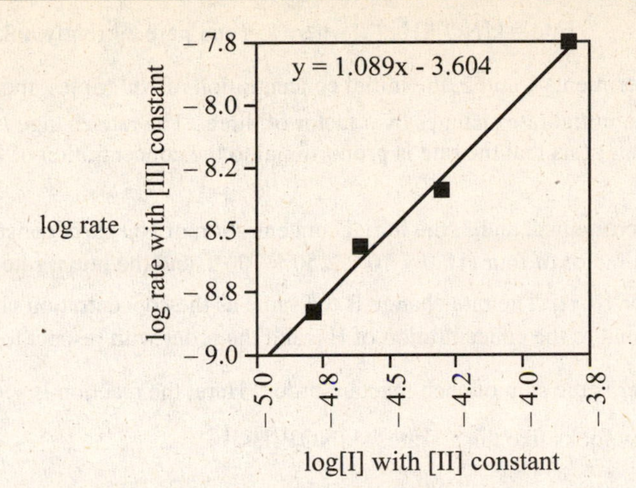

So, the reaction order of reactant I (i) is also one. So, rate = k[I][II]

(b) Solve the rate law for k: $k = \frac{\text{rate}}{[\text{I}][\text{II}]}$

Plug in each experiment's data. Here is an example of the first experiment's calculation:

$$k_1 = \frac{1.50 \times 10^{-9}\ \text{mol L}^{-1}\text{s}^{-1}}{(1.65 \times 10^{-5}\ \text{mol/L})(10.6 \times 10^{-5}\ \text{mol/L})} = 0.858\ \text{L mol}^{-1}\text{s}^{-1}$$

$[I] \times 10^5$ (mol/L)	$[I] \times 10^5$ (mol/L)	Initial rate $\times 10^9$ (mol $L^{-1}s^{-1}$)	Rate constant (L $mol^{-1}s^{-1}$)
1.65	10.6	1.50	0.858
14.9	10.6	17.7	1.12
14.9	7.10	11.2	1.06
14.9	3.52	6.30	1.20
14.9	1.76	3.10	1.18
4.97	10.6	4.52	0.858
2.48	10.6	2.70	1.03

The average of these seven rate constants is 1.04 L $mol^{-1}s^{-1}$

✓ *Reasonable Answer Check:* The variations in the values of k and the deviation of the slopes from the exact integer values expected for the orders make these results unsatisfying. In addition, we can try to get the value of k from the y-intercepts of the graphs:

First Graph: $b = \log(k') = \log(k[\text{I}]^i)$

$$k = \frac{10^b}{[\text{I}]^i} = \frac{10^{-3.941}}{\left(14.9 \times 10^{-5}\right)^1} = 0.769\ \text{L mol}^{-1}\text{s}^{-1}$$

Second Graph: $b = \log(k'') = \log(k[\text{II}]^j)$

$$k = \frac{10^b}{[\text{II}]^j} = \frac{10^{-3.604}}{\left(10.6 \times 10^{-5}\right)^1} = 2.35\ \text{L mol}^{-1}\text{s}^{-1}$$

These are not very close to each other either. These wide variations might suggest a systematic error in the collection of the data under various conditions.

32. *Answer:* **(a) The order with respect to NO is one and with respect to H_2 is one. (b) The reaction is second-order. (c) rate = k[NO][H_2] (d) k = 2.4×10^2 L $mol^{-1}s^{-1}$ (e) initial rate = 1.5×10^{-2} mol $L^{-1}s^{-1}$**

Strategy and Explanation: Follow the method described in the answers to Questions 23 and 25.

The reactants are NO and H_2. Data are available for the changes in each of these reactants' concentrations, so the rate law looks like this:

rate = $k[NO]^i[H_2]^j$ where k, i and j are currently unknown.

(a) Looking at Experiments 1 and 2, the initial concentration of NO triples, the initial concentration of H_2 stays constant, and the initial rate changes by a factor of three. The rate change is the same as the concentration change, which suggests that the rate is proportional to the concentration of NO, and the order with respect to NO is one.

Looking at Experiments 2 and 3, the initial concentration of NO stays constant, the initial concentration of H_2 goes up by a factor of four ($10.0 \times 10^{-3}/2.50 \times 10^{-3}$), and the initial rate changes by a factor of four ($3.6 \times 10^{-2}/9.0 \times 10^{-3}$). The rate change is the same as the concentration change, which suggests that the rate is proportional to the concentration of H_2, and the order with respect to H_2 is also one.

(b) The overall order is the sum of each reactant order. Here, the reaction is second-order.

(c) The rate law now looks like this: rate = $k[NO]^1[H_2]^1$

(d) Solve the rate law for k: $k = \dfrac{\text{rate}}{[NO][H_2]}$

Plug in each experiment's data. Here is an example of the calculation for Experiment 1:

$$k_1 = \frac{3.0 \times 10^{-3}\ \text{mol L}^{-1}\text{s}^{-1}}{(5.00 \times 10^{-3}\ \text{mol / L})(2.50 \times 10^{-3}\ \text{mol / L})} = 2.4 \times 10^2\ \text{L mol}^{-1}\text{s}^{-1}$$

[NO] (mol/L)	$[H_2]$ (mol/L)	Initial rate (mol $L^{-1}s^{-1}$)	Rate constant (L $mol^{-1}s^{-1}$)
5.00×10^{-3}	2.50×10^{-3}	3.0×10^{-3}	2.4×10^2
15.0×10^{-3}	2.50×10^{-3}	9.0×10^{-3}	2.4×10^2
15.0×10^{-3}	10.0×10^{-3}	3.6×10^{-2}	2.4×10^2

The average of these three rate constants is 2.4×10^2 L $mol^{-1}s^{-1}$.

(e) rate = $k[NO]^1[H_2]^1 = (2.4 \times 10^2\ \text{L mol}^{-1}\text{s}^{-1})(8.0 \times 10^{-3}\text{mol L}^{-1})^1(8.0 \times 10^{-3}\ \text{mol L}^{-1})^1$

$= 1.5 \times 10^{-2}$ mol $L^{-1}s^{-1}$

Rate Law and Order of Reaction (Section 13.3)

34. *Answer:* **(a) First-order in A, third-order in B, and fourth-order overall (b) First-order in A, first-order in B, and second-order overall (c) First-order in A, zero-order in B, and first-order overall (d) Third-order in A, first-order in B, and fourth-order overall**

Strategy and Explanation: Reaction order is the power to which the concentration of a component is raised. The overall order is the sum of all the individual orders.

(a) In the rate law: Rate = $k[A][B]^3$ the order with respect to A is one and the order with respect to B is three. The overall order is (1 + 3 =) four.

(b) In the rate law: Rate = k[A][B] the order with respect to A is one and the order with respect to B is one. The overall order is (1 + 1 =) two.

(c) In the rate law: Rate = k[A] the order with respect to A is one and the order with respect to B is zero. The overall order is (1 + 0 =) one.

(d) In the rate law: Rate = $k[A]^3[B]$ the order with respect to A is three and the order with respect to B is one. The overall order is (3+ 1 =) four.

36. *Answer:* **Equation (a) cannot be correct**

Strategy and Explanation: If a reaction is second-order in B, the rate law must include the concentration of B squared. That means "(a) Rate = k[A][B]" cannot be right, since it shows [B] raised to the first power.

38. *Answer:* **(a)** $\dfrac{\Delta[NH_3]}{\Delta t} = k$ **(b)** 2.5×10^{-4} mol $L^{-1}min^{-1}$ **(c) at every concentration above zero**

Strategy and Explanation: Given the order of a reaction, write the rate expression, calculate the rate at a given concentration, and determine at what concentration of the reactant would the rate be equal to the rate constant.

(a) The reactant is zero order, so $\frac{\Delta[NH_3]}{\Delta t} = k$

(b) Because the reaction is zero order, the rate is independent of the concentration.

$$\text{rate} = k = 2.5 \times 10^{-4} \text{ mol L}^{-1}\text{min}^{-1}$$

(c) Because the reaction is zero order, the rate is equal to the rate constant at any concentration above zero.

39. *Answer:* **(a) rate = k [phenyl acetate] (b) First-order in phenyl acetate (c) k = 1.259 s^{-1} (d) 0.13 mol/L·s**

Strategy and Explanation: Given concentration data as a function of time, find the rate law, the reaction order, the rate constant, and the rate at a specific concentration.

In Section 13.3, the method of finding the rate law from linear graphs is described. Construct the three graphs described in Table 13.2, representing zero, first and second-order functions. Determine which one of these three graphs is the linear graph. The linear graph describes the order of the reaction. Once the order is determined, the slope of the straight line is used to determine the value of k.

(a) The reactants are phenyl acetate and water, and the question tells us to assume that the concentration of water does not change. Data are available for the concentration changes with time for phenyl acetate, so the rate law looks like this:

$$\text{rate} = k[\text{phenyl acetate}]^i \quad \text{where k and i are currently unknown.}$$

Graph [phenyl acetate] vs. time, ln[phenyl acetate] vs. time, and 1/[phenyl acetate] vs. time, to see which one produces a straight line.

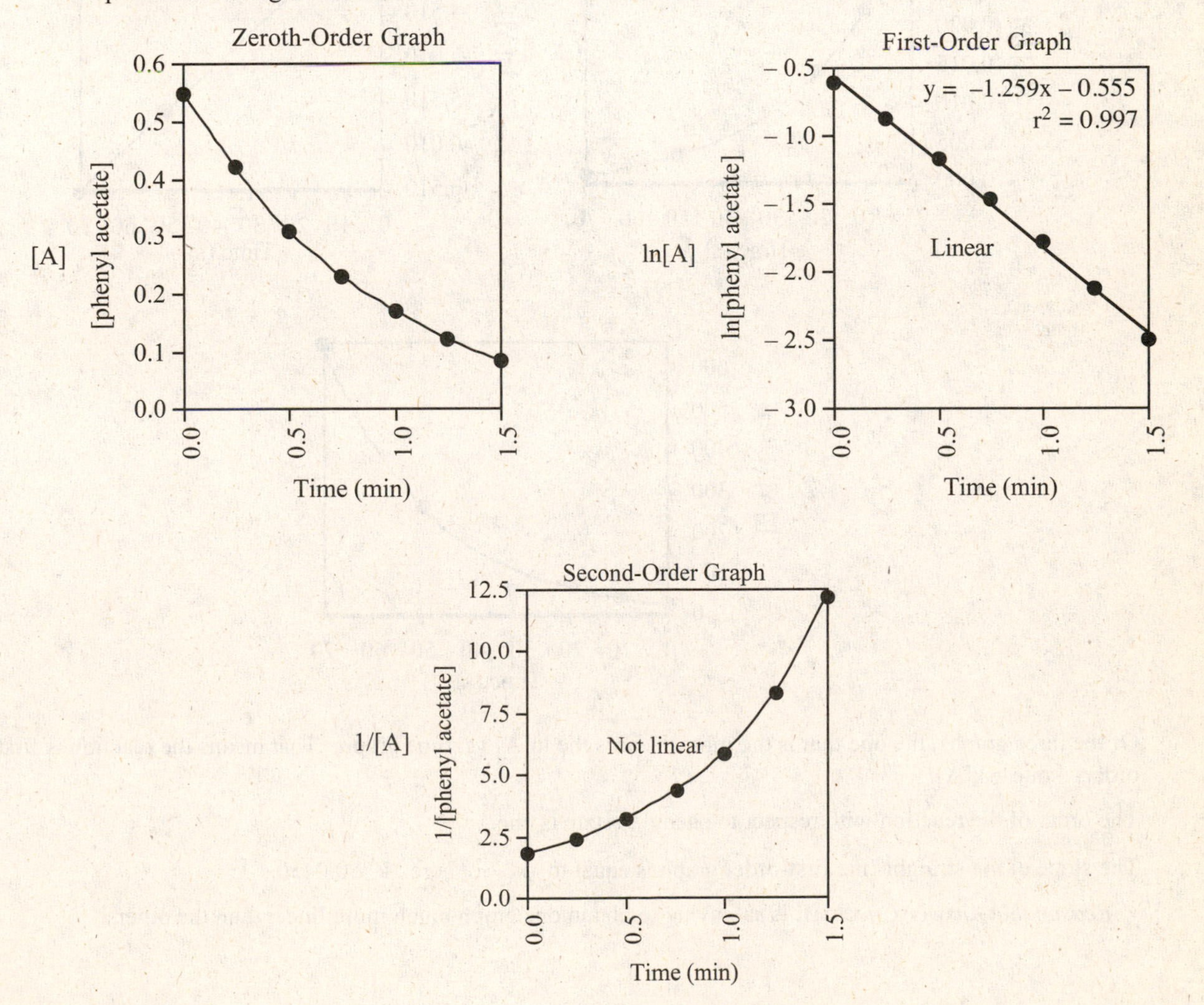

Of the three graphs, the one that is the most linear is the ln[phenyl acetate] vs. time graph. That means the reaction is first-order. rate = k[phenyl acetate]^1

(b) The order of the reaction with respect to phenyl acetate is one.

(c) The slope of the straight-line graph is equal to –k. So, here: k = 1.259 min^{-1}

(d) rate = k[phenyl acetate]^1 = (1.259 min^{-1})(0.10 mol/L)^1 = 0.13 mol L^{-1}min^{-1}

✓ *Reasonable Answer Check:* It is satisfying to obtain one graph much more linear than the others.

40. *Answer:* **k = 0.0450 s^{-1}; first-order**

Strategy and Explanation: Given concentration data as a function of time, find the rate constant and the order.

In Section 13.3, the method of finding the rate law from linear graphs is described. Construct the three graphs described in Table 13.2, representing zero, first and second-order functions. Determine which one of these three graphs is the linear graph. The linear graph describes the order of the reaction. Once the order is determined, find the slope of the straight line in that graph and use it to determine the value of k.

Graph [A] vs. time, ln[A] vs. time, and 1/[A] vs. time, to see which one produces a straight line. If the first of these graphs is linear, the reaction is zero order. If the second of these graphs is linear, the reaction is first order. If the third of these graphs is linear, then the reaction is second order. There is only one correct order, so two of the graphs must be non-linear.

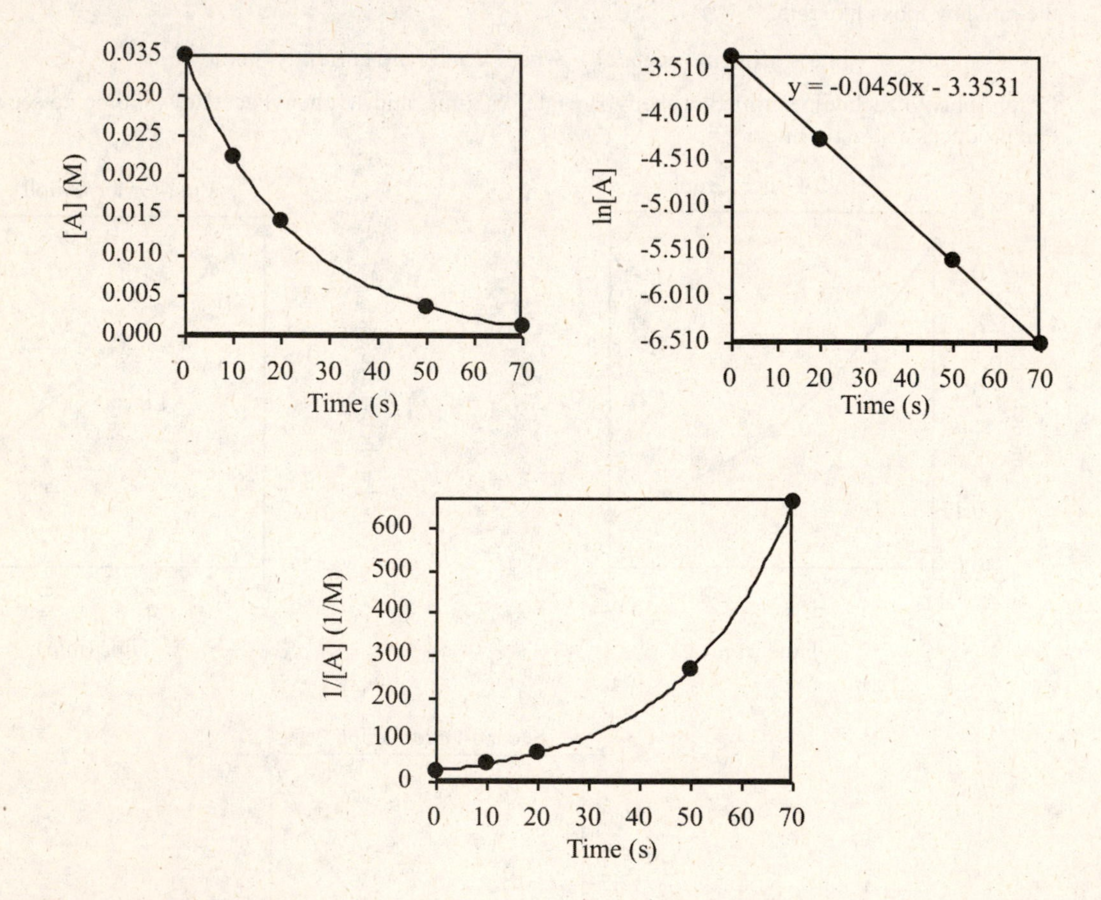

Of the three graphs, the one that is the most linear is the ln[A] vs. time graph. That means the reaction is first-order. rate = k[A]^1

The order of the reaction with respect to phenyl acetate is one.

The slope of the straight-line first-order graph is equal to –k. So, here: k = 0.0450 s^{-1}

✓ *Reasonable Answer Check:* It is satisfying to obtain one graph much more linear than the others.

42. *Answer:* **(a) rate = k[CO][NO_2] (b) first order in both NO_2 and CO (c) 4.2×10^8 L mol^{-1}h^{-1}**

Strategy and Explanation: Follow the method described in the solution to Question 29.

(a) The reactants are CO and NO_2. Data are available for the changes in each of these reactants' concentrations, so the rate law looks like this: rate = $k[CO]^i[NO_2]^j$ where k, i and j are currently unknown. The results of the calculations in (b) are needed to find the values of i and j.

(b) Looking at Experiments 1 and 4, the initial concentration of CO stays constant, the initial concentration of NO_2 doubles, and the initial rate changes by a factor of approximately two ($1.0 \times 10^{-3}/5.1 \times 10^{-4} = 2.0$). The rate change is the same as the concentration change, which suggests that the rate is proportional to the concentration of NO_2, and the order with respect to NO_2 is one.

Looking at Experiments 3 and 4, the initial concentration of CO doubles ($0.35 \times 10^{-3}/0.18 \times 10^{-4} = 1.9$) approximately, the initial concentration of NO_2 stays constant, and the initial rate changes by a factor of approximately two ($1.0 \times 10^{-3}/5.1 \times 10^{-4} = 2.0$). The rate change is the same as the concentration change, which suggests that rate is proportional to the concentration of CO, and the order with respect to CO is one.

The rate law now looks like this: $\text{rate} = k[CO]^1[NO_2]^1$

(c) Solve the rate law for k: $k = \dfrac{\text{rate}}{[CO][NO_2]}$

Plug in each experiment's data. Here is an example using Experiment 1:

$$k_1 = \frac{5.1 \times 10^{-4}\ \text{mol L}^{-1}\text{h}^{-1}}{(0.35 \times 10^{-4}\ \text{mol/L})(3.4 \times 10^{-8}\ \text{mol/L})} = 4.3 \times 10^8\ \text{L mol}^{-1}\text{h}^{-1}$$

[CO] (mol/L)	[NO_2] (mol/L)	Initial rate (mol L^{-1}h^{-1})	Rate constant (L mol^{-1}h^{-1})
0.35×10^{-4}	3.4×10^{-8}	5.1×10^{-4}	4.3×10^8
0.70×10^{-4}	1.7×10^{-8}	5.1×10^{-4}	4.3×10^8
0.18×10^{-4}	6.8×10^{-8}	5.1×10^{-4}	4.2×10^8
0.35×10^{-4}	6.8×10^{-8}	1.0×10^{-3}	4.2×10^8
0.35×10^{-4}	10.2×10^{-8}	1.5×10^{-3}	4.2×10^8

The average of these five rate constants is 4.2×10^8 L mol^{-1}h^{-1}.

44. *Answer:* **(a) rate = k[CH_3COCH_3][H_3O^+]; first order in both H_3O^+ and CH_3COCH_3, and zero order in Br_2 (b) 4×10^{-3} L mol^{-1}s^{-1} (c) 2×10^{-5} mol L^{-1}s^{-1}**

Strategy and Explanation: Follow the method described in the solution to Question 29.

(a) The reactants are CH_3COCH_3, Br_2 and water. The reaction is catalyzed by acid, so the rate is affected by the concentration of H_3O^+. Data are available for the changes in the first two reactants' concentrations and the catalyst concentration, so we will assume the quantity of water is relatively unchanged in this reaction. Hence, the rate law looks like this:

$$\text{rate} = k[CH_3COCH_3]^i[Br_2]^j[H_3O^+]^h \quad \text{where k, i, j and h are currently unknown.}$$

Looking at Experiments 1 and 2, the initial concentration of CH_3COCH_3 stays constant, the initial concentration of Br_2 doubles, the initial concentration of H_3O^+ stays constant, and the initial rate does not change. The rate is independent of the concentration of Br_2, and the order with respect to Br_2 is zero.

Looking at Experiments 1 and 3, the initial concentration of CH_3COCH_3 stays constant, the initial concentration of Br_2 stays constant, the initial concentration of H_3O^+ doubles, and the initial rate changes by a factor of approximately two ($12.0 \times 10^{-5}/5.7 \times 10^{-5} = 2.1$). The rate change is the same as the concentration change, which suggests that the rate is proportional to the concentration of H_3O^+, and the order with respect to H_3O^+ is one.

Looking at Experiments 1 and 5, the initial concentration of CH_3COCH_3 increases by a factor of 1.3 (0.40/0.30), the initial concentration of Br_2 stays constant, the initial concentration of H_3O^+ stays constant, and the initial rate changes by a factor of 1.3 ($7.6 \times 10^{-5}/5.7 \times 10^{-5}$). The rate change is the same as the concentration change, which suggests that the rate is proportional to the concentration of CH_3COCH_3, and the order with respect to CH_3COCH_3 is one.

The rate law now looks like this: rate = $k[CH_3COCH_3]^1[H_3O^+]^1$

(b) Solve the rate law for k: $$k = \frac{\text{rate}}{[CH_3COCH_3][H_3O^+]}$$

Plug in each experiment's data. Here is an example of the calculation for Experiment 1:

$$k_1 = \frac{5.7\times10^{-5}\ \text{mol L}^{-1}\text{s}^{-1}}{(0.30\ \text{mol/L})(0.05\ \text{mol/L})} = 4\times10^{-3}\ \text{L mol}^{-1}\text{s}^{-1}$$

$[CH_3COCH_3]$ (mol/L)	$[H_3O^+]$ (mol/L)	Initial rate (mol $L^{-1}s^{-1}$)	Rate constant (L $mol^{-1}s^{-1}$)
0.30	0.05	5.7×10^{-5}	4×10^{-3}
0.30	0.10	12.0×10^{-5}	4.0×10^{-3}
0.40	0.20	31.0×10^{-5}	3.9×10^{-3}
0.40	0.05	7.6×10^{-5}	4×10^{-3}

The first two experiments have the same rate (represented in the first row of the chart above). The average of these five rate constants is 4×10^{-3} L $mol^{-1}s^{-1}$.

(c) rate = $k[CH_3COCH_3]^1[H_3O^+]^1 = (4 \times 10^{-3}\ \text{L mol}^{-1}\text{s}^{-1})(0.10\ \text{mol/L})^1(0.050\ \text{mol/L})^1 = 2 \times 10^{-5}\ \text{mol L}^{-1}\text{s}^{-1}$

46. *Answer:* **see plots and comparison description below**

Strategy and Explanation: Using the integrated rate law equations from Table 13.2, and setting $[A]_0$ = 1.0 mol/L and k = 1.0 in appropriate units, the following values were determined:

Time(t) (s)	$[A]_t = -kt + [A]_0$ $[A]_t$ (Zeroth-order)	$\ln[A]_t = -kt + \ln[A]_0$ $[A]_t$ (First-order)	$\frac{1}{[A]_t} = kt + \frac{1}{[A]_0}$ $[A]_t$ (Second-order)
0.0	1.0	1.0	1.0
1.0	0.0	0.37	0.50
2.0	0.0	0.14	0.33
3.0	0.0	0.050	0.25
4.0	0.0	0.018	0.20
5.0	0.0	0.0067	0.17

These concentrations can now be plotted as a function time for each order:

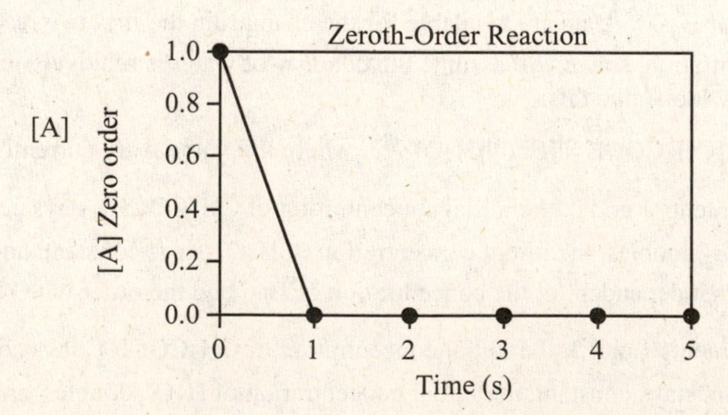

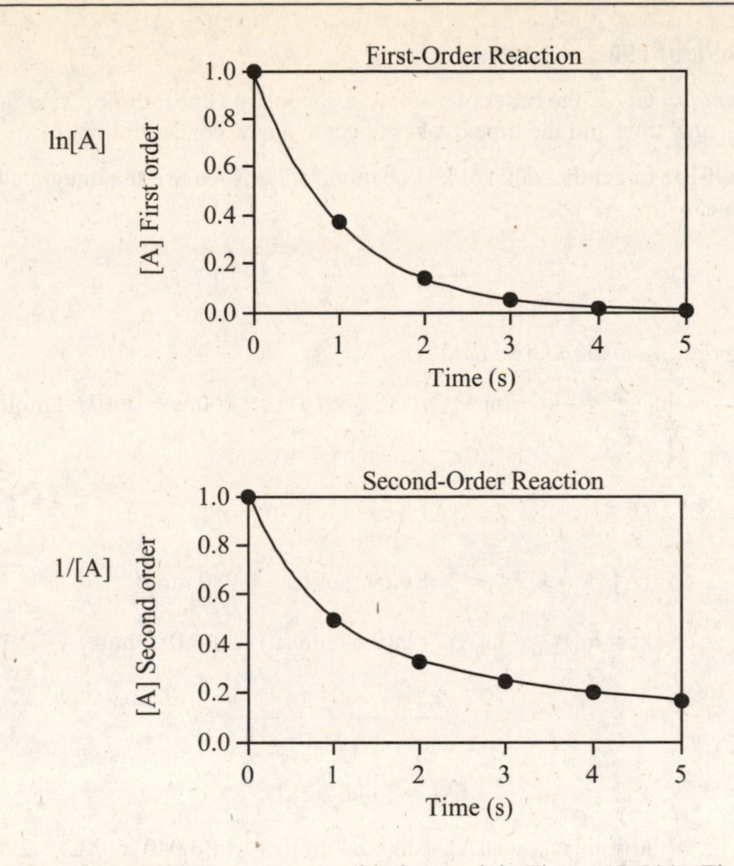

Only the second graph constructed here compares with the graph in Figure 13.5(a). They both show curved-down functional dependence of [A] verses time, characteristic of first-order reactions.

48. *Answer:* **4.3 mg**

Strategy and Explanation: Given the order of a reaction, the initial mass, and the half-life, determine the amount remaining at a new time.

Use the first-order half-life to get the value of k (Equation 13.7). Then use the integrated rate law to find concentration and time.

$$k = \frac{\ln 2}{t_{1/2}} = \frac{\ln 2}{2.7\ \text{d}} = 0.26\ \text{d}^{-1}$$

The first order integrated rate equation looks like this: $\ln[A]_t = -kt + \ln[A]_0$

The mass of Au is proportional to the concentration of Au by constant factors (molar mass and fixed volume). That means we can use the mass directly in the first-order integrated rate law:

$$\ln(m_t) = -kt + \ln(m_0) = -(0.26\ \text{d}^{-1})(1.0\ \text{d}) + \ln(5.6\ \text{mg}) = 1.5$$

$$m_t = e^{1.5} = 4.3\ \text{mg}$$

Notice: This kind of substitution will ONLY work with first-order reactions, because we only need the concentration ratio.

50. *Answer:* **11.4 d**

Strategy and Explanation: Given the order of a reaction and the rate constant, determine the half-life.

Use the first-order half-life to get the value of k (Equation 13.7).

$$t_{1/2} = \frac{\ln 2}{k} = \frac{\ln 2}{0.0606\,\text{d}^{-1}} = 11.4\ \text{d}$$

51. *Answer:* **(a) 0.16 mol/L (b) 90. s (c) 120 s**

Strategy and Explanation: Given the order of a reaction, the initial concentration, and the half-life, determine the concentration at a new time and the time it takes to get to a new concentration.

Use the first-order half-life to get the value of k (Equation 13.7). Then use the integrated rate law to find concentration and time.

$$k = \frac{\ln 2}{t_{1/2}} = \frac{\ln 2}{30.\ \text{s}} = 2.3 \times 10^{-2}\ \text{s}^{-1}$$

(a) *For enough sig figs, assume the time data is ± 1 s.*

$$\ln[A]_t = -kt + \ln[A]_0 = -(2.3 \times 10^{-2}\ \text{s}^{-1})(60.\ \text{s}) + \ln(0.64\ \text{mol/L})$$

$$\ln[A]_t = -1.8$$

$$[A]_t = e^{-1.8} = 0.16\ \text{mol/L}$$

(b)

$$[A]_t = \frac{1}{8} \times [A]_0 = \frac{1}{8} \times (0.64\ \text{mol/L}) = 0.080\ \text{mol/L}$$

$$kt = \ln[A]_0 - \ln[A]_t = \ln(0.64\ \text{mol/L}) - \ln(0.080\ \text{mol/L}) = 2.1$$

$$t = \frac{2.1}{k} = \frac{2.1}{2.3 \times 10^{-2}\ \text{s}^{-1}} = 90.\ \text{s}$$

(c)

$$[A]_t = 0.040\ \text{mol/L}$$

$$kt = \ln[A]_0 - \ln[A]_t = \ln(0.64\ \text{mol/L}) - \ln(0.040\ \text{mol/L}) = 2.8$$

$$t = \frac{2.8}{k} = \frac{2.8}{2.3 \times 10^{-2}\ \text{s}^{-1}} = 120\ \text{s}$$

✓ *Reasonable Answer Check:* The concentration after some time has passed is always smaller than the initial concentration. It also makes sense that the longer the time, the smaller the concentration.

53. *Answer:* **t = 5.49 × 10⁴ s**

Strategy and Explanation: Use the method described in the solution to Question 51.

$$k = \frac{\ln 2}{t_{1/2}} = \frac{\ln 2}{1.47 \times 10^{4}\ \text{s}} = 4.72 \times 10^{-5}\ \text{s}^{-1}$$

$$[A]_0 = \frac{1.6 \times 10^{-3}\ \text{mol}}{2.0\ \text{L}} = 8.0 \times 10^{-4}\ \text{mol/L}$$

$$[A]_t = \frac{1.2 \times 10^{-4}\ \text{mol}}{2.0\ \text{L}} = 6.0 \times 10^{-5}\ \text{mol/L}$$

$$kt = \ln[A]_0 - \ln[A]_t = \ln(8.0 \times 10^{-4}\ \text{mol/L}) - \ln(6.0 \times 10^{-5}\ \text{mol/L}) = 2.59$$

$$t = \frac{2.59}{k} = \frac{2.59}{4.72 \times 10^{-5}\ \text{s}^{-1}} = 5.49 \times 10^{4}\ \text{s}$$

55. *Answer:* **(a) 0.02 M (b) 14.0 y (c) 5.55 y**

Strategy and Explanation: Adapt the method described in the solution to Question 51.

(a)

$$\ln[A]_t = -kt + \ln[A]_0 = -(3.42 \times 10^{-4}\ \text{d}^{-1})(2\ \text{month}\ \frac{30\ \text{d}}{1\ \text{month}}) + \ln(0.0200\ \text{M})$$

$$\ln[A]_t = -3.9$$

$$[A]_t = e^{-3.9} = 0.02\ M$$

(b)

$$kt = \ln[A]_0 - \ln[A]_t = \ln(0.0200\ M) - \ln(0.00350\ M) = 1.74$$

$$t = \frac{1.74}{k} = \frac{1.74}{3.42 \times 10^{-4}\ d^{-1}} = 5.10 \times 10^3\ d$$

$$5.10 \times 10^3\ d \times \frac{1\ y}{365.25\ d} = 14.0\ y$$

(c)

$$t_{1/2} = \frac{\ln 2}{k} = \frac{\ln 2}{3.42 \times 10^{-4}\ d^{-1}} = 2.03 \times 10^3\ d$$

$$2.03 \times 10^3\ d \times \frac{1\ y}{365.25\ d} = 5.55\ y$$

A Nanoscale View: Elementary Reactions (Section 13.4)

56. *Answer:* **Reaction (d) is unimolecular and elementary; reaction (b) is bimolecular and elementary; reactions (a) and (c) are not elementary.**

Strategy and Explanation: A reaction with exactly one reactant is unimolecular and elementary. A reaction with exactly two reactants, either two of the same reactants or one of each of two different reactants, is bimolecular and elementary. A reaction that has more than two reactants, in any combination, is not elementary. We will assume that these reactions fit one of these three descriptions.

(a) The reaction has three reactants (one CH_4 and two O_2 molecules), so it is not elementary.

(b) The reaction has two reactants (one O_3 molecule and one O atom), so it is bimolecular and elementary.

(c) The reaction has three reactants (one Mg atom and two H_2O molecules), so it is not elementary.

(d) The reaction has one reactant (one O_3 molecule), so it is unimolecular and elementary.

58. *Answer:* **Reaction NO with O_3; NO is asymmetrical and Cl is symmetrical.**

Strategy and Explanation: The reaction of NO with O_3 will have a more important steric factor than the reaction of Cl with O_3, because NO is an asymmetrical molecule and Cl is a symmetrical atom. All collisions with a symmetrical atom could be effective, if they have enough energy. Collisions with the "wrong end" of the asymmetric molecule might be ineffective just because of which atoms came in contact during the collision.

Temperature and Reaction Rates (Section 13.5)

60. *Answer:* **$E_a = 19$ kJ/mol; $\frac{rate_1}{rate_2} = 1.8$**

Strategy and Explanation: Given activation energy of a reaction find the ratio of the rates at two different Ts.

A ratio of the rates when only the temperature changes must be equal to a ratio of the rate constants. Convert the temperatures to Kelvin, then use the equation derived in Problem-Solving Example 13.9 to calculate the rate constant ratio.

$E_a = 19$ kJ/mol (from Problem-Solving Example 13.8).

$T_1 = 50.\ °C + 273.15 = 323\ K$, $T_2 = 25\ °C + 273.15 = 298\ K$

$$\frac{rate_1}{rate_2} = \frac{k_1}{k_2} = e^{\frac{E_a}{R}\left(\frac{1}{T_2} - \frac{1}{T_1}\right)} = e^{\frac{(19\ kJ/mol)}{(0.008314\ kJ/mol\cdot K)}\left(\frac{1}{(298\ K)} - \frac{1}{(323\ K)}\right)} = e^{0.6} = 1.8$$

✓ *Reasonable Answer Check:* The rate increases at a higher temperature.

62. *Answer:* **10.7 times faster**

Strategy and Explanation: As directed in Question 61, use the equation derived there (and also in Problem-Solving Example 13.9) to answer this question. Use the method described in the solution to Question 60.

$E_a = 76$ kJ/mol, $T_1 = 50.\ °C + 273.15 = 323$ K, $T_2 = 25\ °C + 273.15 = 298$ K

$$\frac{rate_1}{rate_2} = \frac{k_1}{k_2} = e^{\frac{E_a}{R}\left(\frac{1}{T_2} - \frac{1}{T_1}\right)} = e^{\frac{(76\ kJ/mol)}{(0.008314\ kJ/mol\cdot K)}\left(\frac{1}{(298\ K)} - \frac{1}{(323\ K)}\right)}$$

$$\frac{rate_1}{rate_2} = e^{2.4} = 10.7$$ *With strict sig figs, this should be $1 \times 10^1 = 10$*

64. *Answer:* **3×10^2 kJ/mol**

Strategy and Explanation: As directed in Question 61, use the equation derived there (and also in Problem-Solving Example 13.9) to answer this question.

$T_1 = 600.$ K $T_2 = 610.$ K $k_2 = 3k_1$

$$\ln\left(\frac{k_1}{3k_1}\right) = \frac{E_a}{(0.008314\ kJ/mol \cdot K)}\left(\frac{1}{610.\ K} - \frac{1}{600.\ K}\right)$$

$$E_a = 3 \times 10^2\ kJ/mol$$

65. *Answer:* **(a) $E_a = 120.$ kJ/mol, $A = 1.22 \times 10^{14}\ s^{-1}$ (b) $k = 1.7 \times 10^{-3}\ s^{-1}$**

Strategy and Explanation: Given the values of the rate constant and several temperatures, determine the activation energy, the frequency factor and the rate constant at a new temperature.

As described in Section 13.5, we can make a linear graph by taking the logarithm of the rate constant and the reciprocal of the absolute temperature. The slope is related to the activation energy and the y-intercept is related to the frequency factor. The Arrhenius equation can then be used to find the rate constant at a new temperature.

(a) Don't forget to convert temperatures to Kelvin:

T (°C)	T (K)	$\frac{1}{T}$ (K^{-1})	ln(k)
25.0	298.2	0.003354	−16.348
30.0	303.2	0.003299	−15.255
56.2	329.4	0.003036	−11.474
78.2	351.4	0.002846	− 8.839

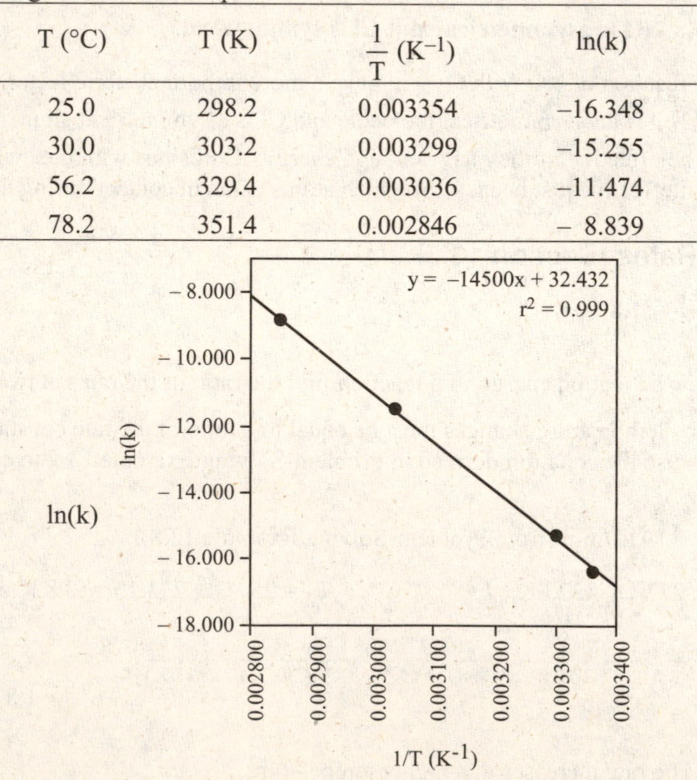

The slope, $m = -\frac{E_a}{R}$. Notice that E_a has units of kJ/mol, using R = 0.008314 kJ/mol·K

$$E_a = -mR = -(-14500)(0.008314) = 120.\ kJ/mol$$

The y-intercept, b = lnA. Notice that A has the same units as the rate constant, k.

$$A = e^{b} = e^{32.432} = 1.22 \times 10^{14}\ s^{-1}$$

(b) Use the Arrhenius equation to estimate the rate constant.

$$T = 100.0\ °C + 273.15\ K = 373.2\ K$$

$$k = A\,e^{-E_a/RT} = (1.22 \times 10^{14}\ s^{-1})\,e^{-(120.\ kJ/mol)/(0.008314\ kJ/mol\cdot K)(373.2\ K)} = 1.7 \times 10^{-3}\ s^{-1}$$

✓ *Reasonable Answer Check:* The linearity of the graph is satisfying. It also makes sense that the rate constant would be larger at a higher temperature.

67. *Answer:* **(a) 22.2 kJ/mol, 6.66 × 10^{7} L^{2}mol^{-2}s^{-1} (b) 8.39 × 10^{4} L^{2}mol^{-2}s^{-1}**

Strategy and Explanation: Given the values of the rate constant and several temperatures, determine the activation energy, the frequency factor and the rate constant at a new temperature.

As described in Section 13.5, we can make a linear graph by taking the logarithm of the rate constant and the reciprocal of the absolute temperature. The slope is related to the activation energy and the y-intercept is related to the frequency factor. The Arrhenius equation can then be used to find the rate constant at a new temperature.

(a) Notice, according to the table heading, that the k values all have a multiplier of 10^{-5}:

$\frac{1}{T}$ (K^{-1})	ln(k)
0.002393	11.626
0.002080	12.468
0.001923	12.889
0.001579	13.752
0.001500	13.955
0.001408	14.292
0.001355	14.433

Now make a graph of 1/T versus lnk:

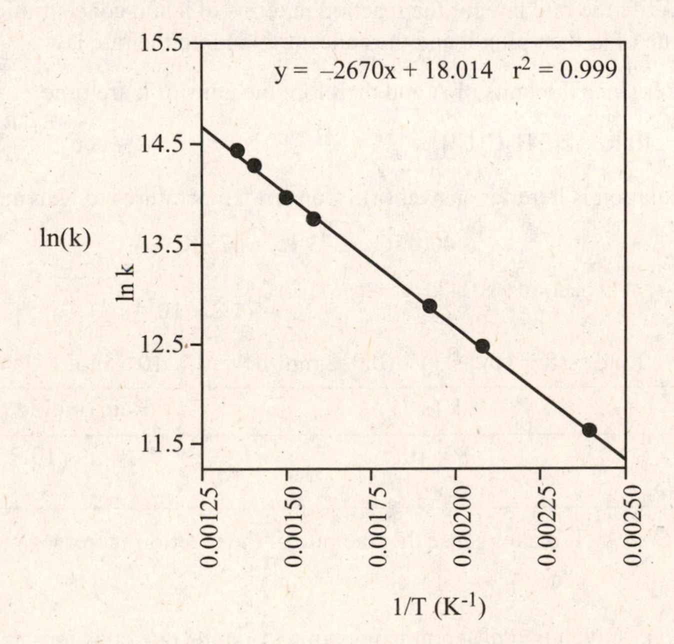

The slope, $m = -\frac{E_a}{R}$. $E_a = -mR = -(-2670)(0.008314) = 22.2$ kJ/mol

The y-intercept, $b = \ln A$. $A = e^b = e^{18.014} = 6.66 \times 10^7\ L^2mol^{-2}s^{-1}$

(b) Use the Arrhenius equation to estimate the rate constant. $k = A\,e^{-E_a/RT}$

$$k = (6.66 \times 10^7\ L^2mol^{-2}s^{-1})\,e^{-(22.2\,kJ/mol)/(0.008314\,kJ/mol\cdot K)(400.0\,K)} = 8.39 \times 10^4\ L^2mol^{-2}s^{-1}$$

69. *Answer:* **(a) 3×10^{-20} (b) 4×10^{-16} (c) 4×10^{-10} (d) 1.9×10^{-6}**

Strategy and Explanation: Given the activation energy and several temperatures, determine the fraction of molecules whose energies would be energetic enough to react.

In Equation 13.8 of Section 13.5, the exponential term, $e^{-E_a/RT}$, is described as the fraction of sufficiently energetic molecules.

Convert temperatures to Kelvin: °C + 273.15 = Kelvin. An example of the calculation is here for answer (a):

$$100.\ °C + 273.15 = 373\ K$$

$$\text{Fraction} = e^{-E_a/RT} = e^{-(139.7\,kJ/mol)/(0.008314\,kJ/mol\cdot K)(373\,K)} = e^{-45.0} = 3 \times 10^{-20}$$

	T (K)	Fraction of sufficiently energetic molecules
(a)	373	$e^{-45.0} = 3 \times 10^{-20}$
(b)	473	$e^{-35.5} = 4 \times 10^{-16}$
(c)	773	$e^{-21.7} = 4 \times 10^{-10}$
(d)	1273	$e^{-13.20} = 1.9 \times 10^{-6}$

✓ *Reasonable Answer Check:* The fraction of molecules with enough energy to react increases with increasing temperature.

71. *Answer:* **(a) 1×10^{-5} mol $L^{-1}s^{-1}$ (b) 25 mol $L^{-1}s^{-1}$**

Strategy and Explanation: Given the activation energy, the concentration of the reactant, the frequency factor, and two temperatures, determine the rates of the reaction at those two temperatures.

The units of the frequency factor are also the units of the rate constant. Use that information to determine the order of the reaction. Write the rate law for the reaction in terms of k and concentration. Use the Arrhenius equation to find the value of k, then plug it and the concentration into the rate law.

The reaction is first-order since the units of A and therefore the units of k are $time^{-1}$. (Refer to Table 13.2)

$$\text{Rate} = k[CH_3CH_2I]^1 \qquad k = A\,e^{-E_a/RT}$$

An example of the calculation is here for answer (a): Convert temperatures to Kelvin.

$$400.\ °C + 273.15 = 673\ K$$

$$k = (1.2 \times 10^{14}\,s^{-1}) \times e^{-(221\,kJ/mol)/(0.008314\,kJ/mol\cdot K)(673\,K)} = (1.2 \times 10^{14}\,s^{-1}) \times e^{-39.5} = 8 \times 10^{-4}\,s^{-1}$$

$$\text{Rate} = (8 \times 10^{-4}\,s^{-1}) \times (0.012\ mol/L)^1 = 1 \times 10^{-5}\ mol\,L^{-1}s^{-1}$$

	T (K)	k (s^{-1})	Rate (mol $L^{-1}s^{-1}$)
(a)	673	8×10^{-4}	1×10^{-5}
(b)	1073	3×10^{1}	25

✓ *Reasonable Answer Check:* It makes sense that the rate of the reaction increases with increasing temperature.

73. *Answer:* **3×10^2 kJ/mol**

Strategy and Explanation: Given two different temperatures and the rate constants at those temperatures, determine the activation energy of a reaction.

Convert the temperatures to Kelvin, then use the equation derived in Problem-Solving Example 13.9 to calculate the activation energy:

$$\ln\left(\frac{k_1}{k_2}\right) = \frac{E_a}{R}\left(\frac{1}{T_2} - \frac{1}{T_1}\right)$$

25. °C + 273.15 = 298 K 55. °C + 273.15 = 328 K

$$\ln\left(\frac{3.46\times10^{-5}\ s^{-1}}{1.5\times10^{-3}\ s^{-1}}\right) = \frac{E_a}{R}\left(\frac{1}{328\ K} - \frac{1}{298\ K}\right)$$

$$\ln\left(2.3\times10^{-2}\right) = \frac{E_a}{(0.008134 kJ / mol\cdot K)}\left(0.00305\ K^{-1} - 0.00336\ K^{-1}\right)$$

$$-3.77 = \frac{E_a}{(0.008134 kJ / mol\cdot K)}\left(0.00031\ K^{-1}\right)$$

$$-3.77 = E_a\,(-.037\ mol/kJ)$$

$$E_a = 3\times10^2\ kJ/mol$$

✓ *Reasonable Answer Check:* The calculated E_a has a reasonable size and sign, similar to those seen in previous questions. The result of the subtraction limits the number of significant figures we can report, suggesting that the temperature data should be kept more precisely.

Rate Laws for Elementary Reactions (Section 13.6)

75. *Answer* **(a) rate = k[Cl][ICl] (b) rate = k[O][O_3] (c) rate = k[NO_2]^2**

Strategy and Explanation: Section 13.6 shows that it is possible to write the rate laws for elementary reactions using the stoichiometry. Unimolecular equations are first-order. Bimolecular equations are second-order, when the reactants are the same, and first order in each reactant when the reactants are different.

(a) This is a bimolecular equation with reactants that are different: rate = k[Cl][ICl]

(b) This is a bimolecular equation with reactants that are different: rate = k[O][O_3]

(c) This is a bimolecular equation with reactants that are the same: rate = k[NO_2]^2

77. *Answer:* **See graphs below**

Strategy and Explanation: Draw these diagrams using the information described in the solution to Question 76, and in Section 13.4. ΔH and ΔE are identical for these diagrams.

(a) $E_{a,reverse} = E_{a,forward} - \Delta H = (75\ kJ\ mol^{-1}) - (-145\ kJ\ mol^{-1}) = 220.\ kJ\ mol^{-1}$

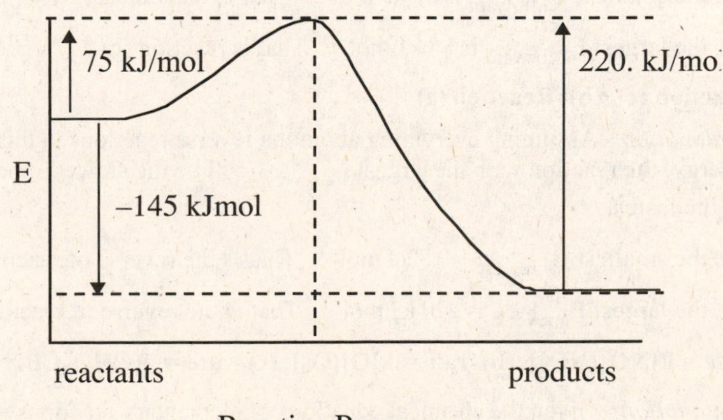

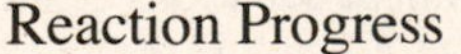

(b) $E_{a,reverse} = E_{a,forward} - \Delta H =$ (65 kJ mol^{-1}) – (–70. kJ mol^{-1}) = 135 kJ mol^{-1}

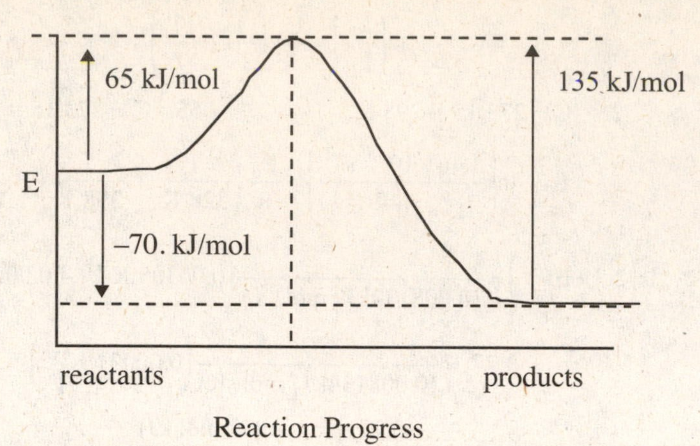

(c) $E_{a,reverse} = E_{a,forward} - \Delta H =$ (85 kJ mol^{-1}) – (+70. kJ mol^{-1}) = 15 kJ mol^{-1}

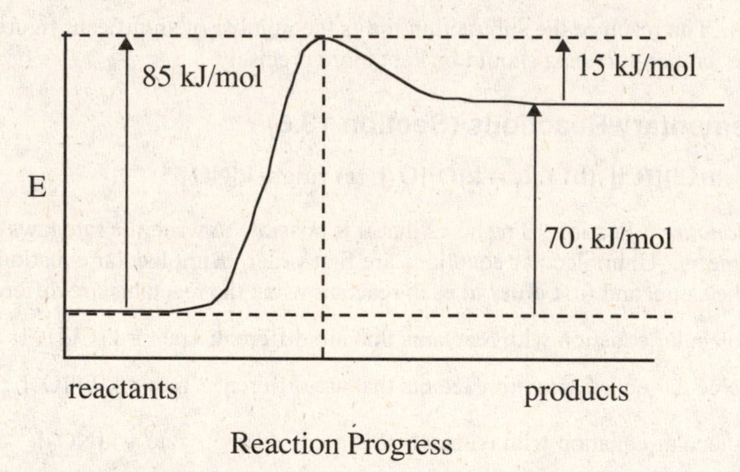

79. *Answer:* **(a) Reaction (b) (b) Reaction (c)**

Strategy and Explanation: Assuming everything about the forward reactions in this question is identical except the activation energy, the reaction with the largest $E_{a,forward}$ will be the slowest, and that with the smallest $E_{a,forward}$ will be the fastest.

(a) Of the three, the smallest $E_{a,forward}$ is 65 kJ mol^{-1}. That is reaction (b).

(b) Of the three, the largest $E_{a,forward}$ is 85 kJ mol^{-1}. That is reaction (c).

81. *Answer:* **(a) Reaction (c) (b) Reaction (a)**

Strategy and Explanation: Assuming everything about the reverse reactions in this question is identical except the activation energy, the reaction with the largest $E_{a,reverse}$ will be the slowest, and that with the smallest $E_{a,reverse}$ will be the fastest.

(a) Of the three, the smallest $E_{a,reverse}$ is 15 kJ mol^{-1}. That is the reverse of reaction (c).

(b) Of the three, the largest $E_{a,reverse}$ is 220 kJ mol^{-1}. That is the reverse of reaction (a).

83. *Answer:* **(a) rate = k[NO][NO_3] (b) rate = k[O][O_3] (c) rate = [$(CH_3)_3CBr$] (d) rate = k[HI]^2**

Strategy and Explanation: Given the chemical equations of elementary reactions write their rate laws.

As described in Section 13.6, the stoichiometric coefficient of a reactant in an elementary equation is the reaction order for that reactant.

(a) $NO + NO_3 \longrightarrow$ products rate = $k[NO]^1[NO_3]^1$

(b) $O + O_3 \longrightarrow$ products rate = $k[O]^1[O_3]^1$

(c) $(CH_3)_3CBr \longrightarrow$ products rate = $[(CH_3)_3CBr]^1$

(d) $2\ HI \longrightarrow$ products rate = $k[HI]^2$

Reaction Mechanisms (Section 13.7)

85. *Answer:* **(a) see equations below (b) the first step is rate-determining**

Strategy and Explanation:

(a) As we did in Chapter 6 with Hess's law questions, cancel anything that ends up on both sides of the equation and add up the rest.

$$\begin{array}{l} NO_2 + F_2 \longrightarrow FNO_2 + \cancel{F} \\ + NO_2 + \cancel{F} \longrightarrow FNO_2 \\ \hline 2\ NO_2 + F_2 \longrightarrow 2\ FNO_2 \end{array}$$

(b) The rate law looks like this: rate = $k[NO_2][F_2]$

That matches the stoichiometry of the first reaction, so it must be the rate-determining step.

✓ *Reasonable Answer Check:* It makes sense that the first reaction would be the rate limiting step, since a fluorine free radical is going to be extremely reactive.

87. *Answer:* **(a) see equations below (b) bimolecular (c) rate = $k[NO_2]^2$ (d) NO_3**

Strategy and Explanation:

(a) Use the method described in the solution to Question 85.

$$\begin{array}{l} NO_2 + \cancel{NO_2} \rightleftharpoons NO + \cancel{NO_3} \\ + \cancel{NO_3} + CO \longrightarrow \cancel{NO_2} + CO_2 \\ \hline NO_2 + CO \longrightarrow NO + CO_2 \end{array}$$

(b) Because both elementary steps have two reactants, they are bimolecular. The first step is second order with respect to NO_2. The second step is first order in NO_3 and first order in CO.

(c) The first step is slow and the second is fast, so the first step must be the rate-determining step and the experimental rate equation must match the rate equation of the first elementary step: rate = $k[NO_2]^2$

(d) An intermediate appears in a mechanism as a product in an early step and then again as a reactant in a later step. One intermediate is formed in this reaction: NO_3

✓ *Reasonable Answer Check:* The steps add up to the expected balanced equation. The slow reaction limits the rate of the experimental reaction. NO_3 is quite reactive.

89. *Answer:* **(a) rate = $k'[NO]^2[Cl_2]$ (b) see mechanism below (c) see mechanism below**

Strategy and Explanation:

(a) Given a mechanism, determine the rate expression in terms of concentrations of reactants (and possibly products).

This mechanism has a fast initial step. Following the description given in Section 13.7, write the rate law from the rate-determining step that includes an intermediate as a reactant. Set up an equation showing the steady state for the creation and destruction of the intermediate. Solve that equation for the intermediate concentration in terms of the concentrations of reactants (and possibly products). Use that equation to eliminate the intermediate concentration from the rate law.

The slow step is step 2, so the rate of the reaction is equal to the rate of this rate-determining step. Because this reaction is elementary, its rate law is related to the stoichiometry of its reactants:

$$\text{rate of reaction} = \text{rate}_{\text{reaction 2}} = k_2[NOCl_2][NO]$$

$NOCl_2$ is an intermediate, so we need to eliminate it from the proposed mechanism. Look for all the ways this intermediate is created and destroyed during the mechanism. It is only created in the forward reaction of step 1. It is destroyed in the reverse reaction of step 1 and also in the reaction of step 2. Because all three of these reactions are elementary, use their reactants' stoichiometric coefficients to write rate laws for each of these three reactions:

$$\text{rate}_{\text{forward reaction 1}} = k_1[NO][Cl_2]$$

$$\text{rate}_{\text{reverse reaction 1}} = k_{-1}[NOCl_2]^1$$

$$\text{rate}_{\text{reaction 2}} = k_2[NOCl_2][NO]$$

The rate of $NOCl_2$ creation is equal to the rate of its destruction once the steady state condition is reached:

$$\text{rate}_{\text{forward reaction 1}} = \text{rate}_{\text{reverse reaction 1}} + \text{rate}_{\text{reaction 2}}$$

$$k_1[NO][Cl_2] = k_{-1}[NOCl_2]^1 + k_2[NOCl_2][NO]$$

Because the rate of step 2 is presumed to be much smaller than the rate of step 1, the second term is presumed to be negligibly small compared to the first term:

$$\text{rate}_{\text{forward reaction 1}} \cong \text{rate}_{\text{reverse reaction 1}}$$

$$k_1[NO][Cl_2] = k_{-1}[NOCl_2]^1$$

$$[NOCl_2] = \frac{k_1}{k_{-1}}[NO][Cl_2]$$

Substitute the equality just derived into the rate law: $\text{rate} = k_2[NOCl_2][NO]$

$$\text{rate} = k_2\left(\frac{k_1}{k_{-1}}[NO][Cl_2]\right)[NO]$$

We will define the new group of rate constants by one variable, $k' = k_2\frac{k_1}{k_{-1}}$.

$$\text{rate} = k'[NO][Cl_2][NO]$$

$$\text{rate} = k'[NO]^2[Cl_2]$$

(b) Suggesting mechanisms involves some creativity. We need to make sure that the rate law is satisfied, but we often must use our imaginations to determine what may get formed during these reactions. Here is an example of a reaction that also satisfies the rate law: $\text{rate} = k'[NO]^2[Cl_2]$.

$$2\,NO \rightleftharpoons N_2O_2 \quad \text{Fast}$$

$$N_2O_2 + Cl_2 \longrightarrow 2\,NOCl_2 \quad \text{Slow}$$

To think this one up, we just switched the roles of one NO molecule and the Cl_2 molecule, putting both NO molecules in the fast reaction and making an intermediate, then having the Cl_2 react with that intermediate.

To confirm that it qualifies, let's rederive the rate law for this new mechanism as described in (a):

$$\text{rate of reaction} = \text{rate}_{\text{reaction 2}} = k_2[N_2O_2][Cl_2]$$

Setting up the steady state for the N_2O_2 intermediate gives this equation:

$$k_1[NO]^2 = k_{-1}[N_2O_2]^1 + k_2[N_2O_2][Cl_2]$$

Again the rate of step 2 is presumed to be much smaller than the rate of step 1:

$$k_1[NO]^2 = k_{-1}[N_2O_2]^1$$

$$[N_2O_2] = \frac{k_1}{k_{-1}}[NO]^2$$

Substitute the equality just derived into the rate law: $\text{rate} = k_2[N_2O_2][Cl_2]$

$$\text{rate} = k_2\left(\frac{k_1}{k_{-1}}[NO]^2\right)[Cl_2]$$

We again define the new group of rate constants by one variable, $k' = k_2\frac{k_1}{k_{-1}}$.

$$\text{rate} = k'[NO]^2[Cl_2]$$

(c) Many mechanisms will not satisfy this rate law: $\text{rate} = k'[NO]^2[Cl_2]$. Here is an example of a reaction that does not satisfy the rate law:

$$NO + Cl_2 \rightleftharpoons NOCl + Cl \quad \text{Slow}$$

$$NO + Cl \longrightarrow NOCl \quad \text{Fast}$$

This reaction has a rate law that looks like this: $\text{rate} = [NO][Cl_2]$

It is not the same as the observed rate law and cannot be the mechanism for this reaction.

✓ *Reasonable Answer Check:* (a) The first step involves the reaction of NO and Cl_2 molecules in the formation of the steady state of the intermediate, so it makes sense that the concentration of Cl_2 is involved in the reaction's rate along with the concentration of NO. (b) The mechanism does recreate the rate law. (c) The mechanism produces a rate law different from that observed.

91. *Answer:* **(a) $CH_3COOCH_3 + H_2O \rightleftharpoons CH_3COOH + CH_3OH$ (b) rate = $k'[CH_3COOCH_3][H_3O^+]$ (c) catalyst: H_3O^+ (d) intermediates: $H_3C(OH)OCH_3^+$, $H_3C(H_2O)(OH)OCH_3^+$, $H_3C(OH)_2OHCH_3^+$, and H_2O**

Strategy and Explanation:

(a) As we did in Chapter 6 with Hess's law questions, cancel anything that ends up on both sides of the equation and add up the rest.

$$CH_3COOCH_3 + \cancel{H_3O^+} \rightleftharpoons \cancel{CH_3C(OH)OCH_3^+} + \cancel{H_2O}$$

$$+ \cancel{CH_3C(OH)OCH_3^+} + \cancel{H_2O} \rightleftharpoons \cancel{CH_3C(H_2O)(OH)OCH_3^+}$$

$$+ \cancel{CH_3C(H_2O)(OH)OCH_3^+} \rightleftharpoons \cancel{CH_3C(OH)_2OHCH_3^+}$$

$$+ \cancel{CH_3C(OH)_2OHCH_3^+} + H_2O \rightleftharpoons CH_3COOH + CH_3OH + \cancel{H_3O^+}$$

$$CH_3COOCH_3 + H_2O \rightleftharpoons CH_3COOH + CH_3OH$$

(b) Follow the same method as described in the solution to Question 89.

The slow step is step 2, so the rate of the reaction is equal to the rate of this rate-determining step. Because this reaction is elementary, its rate law is related to the stoichiometry of its reactants:

$$\text{rate of reaction} = \text{rate}_{\text{reaction 2}} = k_2[CH_3C(OH)OCH_3^+][H_2O]$$

$CH_3C(OH)OCH_3^+$ is an intermediate, so we need to eliminate it from the proposed mechanism. Look for all the ways this intermediate is created and destroyed during the mechanism. It is only created in the forward reaction of step 1. It is destroyed in the reverse reaction of step 1 and also in the forward reaction of step 2. Because all three of these reactions are elementary, use their reactants' stoichiometric coefficients to write rate laws for each of these three reactions:

$$\text{rate}_{\text{forward reaction 1}} = k_1[CH_3COOCH_3][H_3O^+]$$

$$\text{rate}_{\text{reverse reaction 1}} = k_{-1}[CH_3C(OH)OCH_3^+][H_2O]$$

$$rate_{reaction\ 2} = k_2[CH_3C(OH)OCH_3^+][H_2O]$$

The rate of $CH_3C(OH)OCH_3^+$ creation is equal to the rate of its destruction once the steady state condition is reached:

$$rate_{forward\ reaction\ 1} = rate_{reverse\ reaction\ 1} + rate_{reaction\ 2}$$

$$k_1[CH_3COOCH_3][H_3O^+] = k_{-1}[CH_3C(OH)OCH_3^+][H_2O] + k_2[CH_3C(OH)OCH_3^+][H_2O]$$

Because the rate of step 2 is presumed to be much smaller than the rate of step 1, the second term is presumed to be negligibly small compared to the first term:

$$k_1[CH_3COOCH_3][H_3O^+] \cong k_{-1}[CH_3COHOCH_3^+][H_2O]$$

$$[CH_3C(OH)OCH_3^+] = \frac{k_1}{k_{-1}} \frac{[CH_3C(OH)OCH_3^+][H_3O^+]}{[H_2O]}$$

Substitute the equality just derived into the rate law:

$$rate = k_2[CH_3C(OH)OCH_3^+][H_2O]$$

$$rate = k_2\left(\frac{k_1}{k_{-1}} \frac{[CH_3C(OH)OCH_3^+][H_3O^+]}{[H_2O]}\right)[H_2O]$$

We will define the new group of rate constants by one variable, $k' = k_2\frac{k_1}{k_{-1}}$.

$$rate = k'[CH_3COOCH_3][H_3O^+]$$

(c) A catalyst shows up in a mechanism as a reactant in an early step and then again as a product in a later step. There is a catalyst in this reaction. It is H_3O^+, introduced in reaction 1 and reproduced in step 4.

(d) An intermediate appears in a mechanism as a product in an early step and then again as a reactant in a later step. Three intermediates are formed in this reaction: $H_3C(OH)OCH_3^+$, $H_3C(H_2O)(OH)OCH_3^+$, and $H_3C(OH)_2OHCH_3^+$. A molecule of H_2O is also created then destroyed in this reaction, but another H_2O molecule is consumed, so water may be technically considered to be an intermediate, as well as a reactant.

93. *Answer:* **Only mechanism (a) is compatible with the observed rate law.**

Strategy and Explanation: Follow the method described in Question 89. The observed rate law is given:

$$rate = k[(CH_3)_3CBr]$$

(a) The rate law for this mechanism comes from the rate of the slow first step:

rate = $k[(CH_3)_3CBr]$, so it is compatible with the observed rate law.

(b) The single step mechanism suggests an elementary bimolecular reaction with a rate law of that looks like this: rate = $k[(CH_3)_3CBr][OH^-]$ It is incompatible with the observed rate law.

(c) The rate law for this mechanism comes from the rate of the slow second step:

$$rate = k_2[(CH_3)_2(CH_2)CBr^-]$$

The steady state conditions for $(CH_3)_2(CH_2)CBr^-$ must be derived to replace the $[(CH_3)_2(CH_2)CBr^-]$ in this expression with reactant concentrations. The steady state equation looks like this:

$$k_1[(CH_3)_3CBr][OH^-] = k_{-1}[(CH_3)_2(CH_2)CBr^-][H_2O] + k_2[(CH_3)_2(CH_2)CBr^-]$$

The second term on the right side is negligible, since the rate of the second step is considered to be small compared to the first step.

$$k_1[(CH_3)_3CBr][OH^-] = k_{-1}[(CH_3)_2(CH_2)CBr^-][H_2O]$$

Solve for $[(CH_3)_2(CH_2)CBr^-]$ in terms of in terms of reactant concentrations,

$$[(CH_3)_2(CH_2)CBr^-] = \frac{k_1}{k_{-1}} \frac{[(CH_3)_3CBr][OH^-]}{[H_2O]}$$

Substituting this equality back into the rate law gives: $\quad$ rate $= k_2[(CH_3)_2(CH_2)CBr^-]$

$$\text{rate} = k_2 \frac{k_1}{k_{-1}} \frac{[(CH_3)_3CBr][OH^-]}{[H_2O]} = k' \frac{[(CH_3)_3CBr][OH^-]}{[H_2O]}$$

This is incompatible with the observed rate law.

So, only mechanism (a) is compatible.

95. *Answer:* **(a) $CH_3OH + H^+ + Br^- \longrightarrow CH_3Br + H_2O$ (b) See diagram below (c) rate = $k'[CH_3OH][H^+][Br^-]$**

Strategy and Explanation:

(a) Use the method described in the solution to Question 85.

$$CH_3OH + H^+ \rightleftharpoons CH_3OH_2^+$$

$$+ \quad CH_3OH_2^+ + Br^- \longrightarrow CH_3Br + H_2O$$

$$CH_3OH + H^+ + Br^- \longrightarrow CH_3Br + H_2O$$

(b) The reaction energy diagram will resemble the enzyme-catalyzed reaction energy diagram shown in Figure 13.17 (the green curve). The diagram here has two humps representing each of the two steps in the mechanism. The first of these is endothermic, so the energy of the intermediate will be higher than the energy of the reactants. Because the second step is rate-determining, the second hump is higher energy than the first hump. Because the overall reaction is exothermic, the products are lower in energy than the reactants.

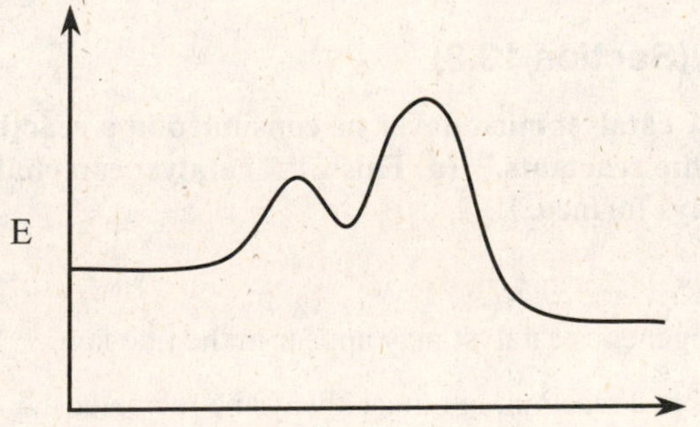
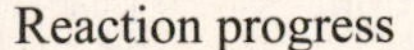

(c) Given a mechanism, determine the rate expression in terms of concentrations of reactants (and possibly products).

This mechanism has a fast initial step. Following the description given in Section 13.7, write the rate expression from the rate-determining step that includes an intermediate as a reactant. Set up an equation showing the steady state for the creation and destruction of the intermediate. Solve that equation for the intermediate concentration in terms of the concentrations of reactants (and possibly products). Use that equation to eliminate the intermediate concentration from the rate expression.

The slow step is step 2, so the rate of the reaction is equal to the rate of this rate-determining step. Because this reaction is elementary, its rate expression is related to the stoichiometry of its reactants:

$$\text{rate of reaction} = \text{rate}_{\text{reaction 2}} = k_2[CH_3OH_2^+][Br^-]$$

$CH_3OH_2^+$ is an intermediate, so we need to eliminate it from the proposed mechanism. Look for all the ways this intermediate is created and destroyed during the mechanism. It is only created in the forward reaction of step 1. It is destroyed in the reverse reaction of step 1 and also in the reaction of step 2. Because all three of these reactions are elementary, use their reactants' stoichiometric coefficients to write rate expressions for each of these three reactions:

$$\text{rate}_{\text{forward reaction 1}} = k_1[CH_3OH][H^+]$$

$$\text{rate}_{\text{reverse reaction 1}} = k_{-1}[CH_3OH_2^+]$$

$$\text{rate}_{\text{reaction 2}} = k_2[CH_3OH_2^+][Br^-]$$

The rate of C creation is equal to the rate of its destruction once the steady state condition is reached:

$$\text{rate}_{\text{forward reaction 1}} = \text{rate}_{\text{reverse reaction 1}} + \text{rate}_{\text{reaction 2}}$$

$$k_1[CH_3OH][H^+] = k_{-1}[CH_3OH_2^+] + k_2[CH_3OH_2^+][Br^-]$$

Because the rate of step 2 is presumed to be much smaller than the rate of step 1, the second term is presumed to be negligibly small compared to the first term:

$$\text{rate}_{\text{forward reaction 1}} \cong \text{rate}_{\text{reverse reaction 1}}$$

$$k_1[CH_3OH][H^+] = k_{-1}[CH_3OH_2^+]$$

$$[CH_3OH_2^+] = \frac{k_1}{k_{-1}}[CH_3OH][H^+]$$

Substitute the equality just derived into the rate expression: $\text{rate} = k_2[CH_3OH_2^+][Br^-]$

$$\text{rate} = k_2\left(\frac{k_1}{k_{-1}}[CH_3OH][H^+]\right)[Br^-]$$

We will define the new group of rate constants by one variable, $k' = k_2\frac{k_1}{k_{-1}}$.

$$\text{rate} = k'[CH_3OH][H^+][Br^-]$$

Catalysts and Reaction Rate (Section 13.8)

97. *Answer:* **(a) True. (b) False. "A catalyst must never be consumed in a reaction." (c) False. "A catalyst need not be the same phase as the reactants." (d) False. "A catalyst can change the course of a reaction, but the same products are always formed."**

Strategy and Explanation:

(a) The concentration of a homogeneous catalyst may appear in the rate law.

(b) A catalyst may be consumed in a reaction, but must always be recreated. A correct statement is: "A catalyst must never be consumed in a reaction."

(c) A homogeneous catalyst is in the same phase as the reactants, but a heterogeneous catalyst is always in a different phase. A correct statement is: "A catalyst need not be the same phase as the reactants."

(d) A catalyst can change the mechanism of a reaction and allow different intermediates to be produced, but it cannot change the products formed. A chemical that changes the products is not called a catalyst. A correct statement is: "A catalyst can change the course of a reaction, but the same products are always formed."

Notice that if a multiple set of reactions occurs simultaneously with the same reactants, a catalyst that speeds only one of these reactions would help produce one product in favor of some others. In such case, the statement might be considered true, but it is likely not something students studying this chapter will think of.

99. *Answer:* **(a) and (c); (a) homogeneous (c) heterogeneous**

Strategy and Explanation: A catalyst may appear in the rate law, though it will not appear in the overall equation representing the reaction. Any rate law with chemicals that are not reactants or products, involve catalysts. If the catalyst has the same phase as the reactants, it is homogeneous. If the catalyst has a different phase from the reactants, it is heterogeneous.

(a) This aqueous reaction involves a homogeneous catalyst, $H_3O^+(aq)$.

(b) This reaction does not appear to involve a catalyst. Both of the concentrations in the rate law are of reactants.

(c) This gas phase reaction involves a heterogeneous catalyst, Pt(s).

(d) This reaction does not appear to involve a catalyst. Both of the concentrations in the rate law are of reactants.

Enzymes: Biological Catalysts (Section 13.9)

101. *Answer:* **38**

Strategy and Explanation: Given the Ea before and after the addition of a catalyst, determine by what factor the rate constant would increase (assuming the frequency factor, A, remains unchanged). The Arrhenius equation given in Section 13.5 on page 616 is $k = Ae^{-E_a/RT}$ Set up a ratio:

$$\frac{k_2}{k_1} = \frac{Ae^{-E_{a2}/RT}}{Ae^{-E_{a1}/RT}} = e^{-E_{a2}/RT + E_{a1}/RT} = e^{(E_{a1} - E_{a2})/RT}$$

$$\frac{k_2}{k_1} = e^{(215 \text{ kJ/mol} - 206 \text{ kJ/mol})/(0.008314 \text{J/mol·K})(25+273.15)\text{K}} = 38$$

102. *Answer/Explanation:*

Active site: That part of an enzyme where the substrate binds to the enzyme in preparation for conversion into products.

Cofactor: A small molecule that interacts with an enzyme to allow it to catalyze a biological reaction.

Enzyme: A protein that catalyzes a biological reaction.

Inhibition: When a molecule that is not the substrate enters and occupies the active site of an enzyme preventing substrate binding and reaction.

Lysozyme: An enzyme that speeds the hydrolysis of certain polysaccharides.

Monomer: The small molecule reactant in the formation of a polymer.

Polypeptides: Molecules made by polymerizing several amino acids.

Polysaccharide: An ether-linked chain of monomer sugars.

Proteins: Large polypeptide molecules that serve as structural and functional molecules in living organisms.

Substrate: The reactant of a reaction catalyzed by an enzyme.

104. *Answer:* **The rate is 30. times faster.**

Strategy and Explanation: We will use information described in Section 13.2, and methods similar to those in Questions 30 - 32.

The reaction rate is proportional to the concentration in a first-order reaction, rate = k[E]. That means the rate increase is related to the ratios:

$$\frac{\text{rate}_2}{\text{rate}_1} = \frac{[E]_2}{[E]_1} = \frac{4.5 \times 10^{-6} \text{ M}}{1.5 \times 10^{-6} \text{ M}} = 30.$$

The reaction rate is increased by a factor of 30. times.

106. *Answer/Explanation:* The succinate dehydrogenase enzyme catalyzes the reaction of substrate succinate ion, $^-OOCCH_2CH_2COO^-$. The active site (where the substrate binds to the enzyme in preparation for conversion into products) is shaped to accommodate the four O atoms, two with negative charges, on the ends of the molecule in the formation of the enzyme-substrate complex. The malonate ion, $^-OOCCH_2COO^-$, and the oxalate ion, $^-OOCCOO^-$, also have four O atoms, two with negative charges, at the ends of their structures. Evidently, these two ions enter and occupy the active site of an enzyme preventing the substrate from binding, inhibiting the reaction, and reducing the rate of the succinate dehydrogenation reaction.

108. *Answer:* **rate** $= \mathbf{k}\dfrac{[X][HA]}{[A^-]}$**; first-order with respect to HA; and doubling the [HA] doubles the rate**

Strategy and Explanation: We will use information described in Section 13.2, and methods similar to those in Questions 30 – 32.

Given a mechanism, determine the rate expression in terms of concentrations of reactants (and possibly products).

This mechanism has fast first and second step. Following the description given in Section 13.7, write the rate expression from the rate-determining step that includes an intermediate as a reactant. Set up equations showing the steady state for the creation and destruction of the intermediates. Solve those equation for the intermediate concentrations in terms of the concentrations of reactants (and possibly products). Use those equations to eliminate the intermediate concentrations from the rate expression.

The slow step is step 3, so the rate of the reaction is equal to the rate of this rate-determining step. Because this reaction is elementary, its rate expression is related to the stoichiometry of its reactants:

$$\text{rate of reaction} = \text{rate}_{\text{reaction 3}} = k_3[XH^+]$$

XH^+ is an intermediate, so we need to eliminate it from the proposed mechanism. Look for all the ways this intermediate is created and destroyed during the mechanism. It is only created in the forward reaction of step 2. It is destroyed in the reverse reaction of step 1 and also in the reaction of step 3. Because all three of these reactions are elementary, use their reactants' stoichiometric coefficients to write rate expressions for each of these three reactions:

$$\text{rate}_{\text{forward reaction 2}} = k_2[X][H_3O^+]$$

$$\text{rate}_{\text{reverse reaction 2}} = k_{-2}[XH^+][H_2O]$$

$$\text{rate}_{\text{reaction 3}} = k_3[XH^+]$$

The rate of E creation is equal to the rate of its destruction once the steady state condition is reached:

$$\text{rate}_{\text{forward reaction 1}} = \text{rate}_{\text{ reverse reaction 1}} + \text{rate}_{\text{reaction 2}}$$

$$k_2[X][H_3O^+] = k_{-2}[XH^+][H_2O] + k_3[XH^+]$$

Because the rate of step 3 is presumed to be much smaller than the rate of step 2, the second term is presumed to be negligibly small compared to the first term:

$$k_2[X][H_3O^+] = k_{-2}[XH^+][H_2O]$$

$$[XH^+] = \frac{k_2}{k_{-2}}[X][H_3O^+]$$

Substitute the equality just derived for [E] into the rate expression: $\text{rate} = k_3[XH^+]$

$$\text{rate} = k_3\left(\frac{k_2}{k_{-2}}[X][H_3O^+]\right)$$

$$\text{rate} = \frac{k_3k_2}{k_{-2}}[X][H_3O^+]$$

H_3O^+ is an intermediate, so we need to eliminate it from the proposed mechanism. Look for all the ways this intermediate is created and destroyed during the mechanism. It is only created in the forward reaction of step 2. It is destroyed in the reverse reaction of step 1 and in the forward reaction of step 2. Because all three of these reactions are elementary, use their reactants' stoichiometric coefficients to write rate expressions for both reactions:

$$\text{rate}_{\text{forward reaction 1}} = k_1[HA][H_2O]$$

$$\text{rate}_{\text{reverse reaction 1}} = k_{-1}[H_3O^+][A^-]$$

$$\text{rate}_{\text{forward reaction 2}} = k_2[X][H_3O^+]$$

The rate of C creation is equal to the rate of its destruction once the steady state condition is reached:

$$\text{rate}_{\text{forward reaction 1}} = \text{rate}_{\text{ reverse reaction 1}} + \text{rate}_{\text{ forward reaction 2}}$$

$$k_1[HA][H_2O] = k_{-1}[H_3O^+][A^-] + k_2[X][H_3O^+]$$

Because the rate of step 3 is presumed to be much smaller than the rate of step 2, the second term is presumed to be negligibly small compared to the first term:

$$k_1[HA][H_2O] = k_{-1}[H_3O^+][A^-]$$

$$[H_3O^+] = \frac{k_1}{k_{-1}} \frac{[HA][H_2O]}{[A^-]}$$

Substitute the equality just derived for [C] in the rate expression derived above: $\text{rate} = \frac{k_3k_2}{k_{-2}}[X][H_3O^+]$

$$\text{rate} = \frac{k_3k_2}{k_{-2}}[X]\left(\frac{k_1}{k_{-1}} \frac{[HA][H_2O]}{[A^-]}\right)$$

We will define the new group of rate constants by one variable, $k'' = \frac{k_3k_2k_1}{k_{-2}k_{-1}}$. The concentration of liquid water in an aqueous solution is also a constant, so incorporate that into the rate constant, too.

$$k' = \frac{k_3k_2k_1}{k_{-2}k_{-1}}[H_2O]$$

$$\text{rate} = k' \frac{[X][HA]}{[A^-]}$$

The reaction is first order with respect to HA. Doubling the concentration of HA doubles the rate of the reaction.

Catalysts in Industry (Section 13.10)

109. *Answer/Explanation:* Catalysts make possible the production of vital products. Without them many necessities and luxuries could not be made efficiently, if at all.

111. *Answer/Explanation:*

(a) The honeycomb arrangement of the ceramic support is mostly likely to maximize the surface area and increase contact with the heterogeneous catalyst.

(b) The catalysis happens only at the surface of the metal. So, using expensive platinum in the form of strips or rods would be inefficient and not cost effective.

General Questions

113. *Answer:* **Exothermic**

Strategy and Explanation: Given the activation energy for the forward reaction of an elementary reaction and the activation energy for the reverse of the same reaction, determine if the forward reaction is exothermic or endothermic.

The difference between the forward and reverse activation energies is the ΔE for the reaction. If ΔE is positive, the forward reaction is endothermic, and if it is negative, the forward reaction is exothermic.

$\Delta E = E_{a,forward} - E_{a,reverse} = 32 \text{ kJ/mol} - 58 \text{ kJ/mol} = -26 \text{ kJ/mol}$

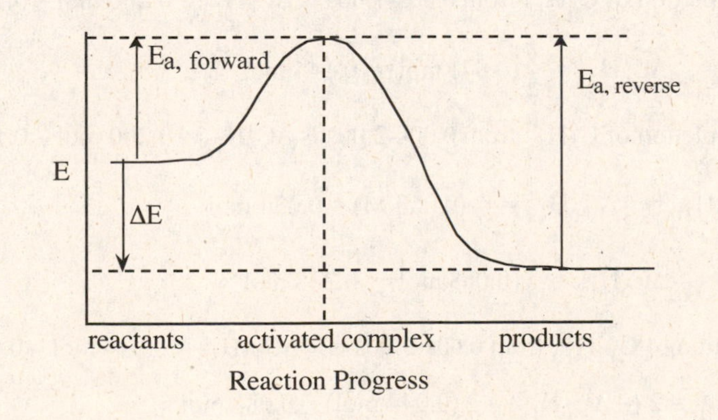

The reaction is exothermic.

✓ *Reasonable Answer Check:* The reverse reaction requires a higher activation energy than the forward reaction, suggesting that the reactants are higher in energy than the products. It makes sense that the reaction is exothermic.

114. *Answer:* **see diagram and energy relationships below**

Strategy and Explanation: An exothermic reaction has the products at a lower-energy state than the reactants.

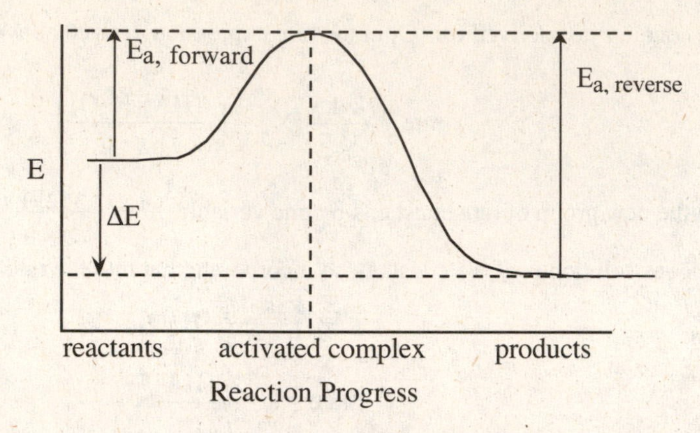

$$\Delta E = E_{a,forward} - E_{a,reverse}$$

116. *Answer:* **(a)**

Time(s)	$[C_6H_{12}]$ ($\frac{mol}{V}$)	$[C_{12}H_{10}]$ ($\frac{mol}{V}$)	$[H_2]$ ($\frac{mol}{V}$)
0.0	0.200	0.000	0.000
1.00	0.159	0.021	**0.144**
2.00	0.132	**0.034**	**0.238**
3.00	**0.088**	0.044	**0.308**

(b) 0.036 mol V^{-1} s^{-1}

Strategy and Explanation: Use stiochiometry to relate reactants to products and use the time-dependence of the concentration changes to determine the reaction rate.

(a) The balanced equation gives the ratio: 2 mol C_6H_{12} reactant: 1 mol $C_{12}H_{10}$ product: 7 mol H_2 product

Time(s)	$[C_6H_{12}]$ ($\frac{mol}{V}$)	$[C_{12}H_{10}]$ ($\frac{mol}{V}$)	$[H_2]$ ($\frac{mol}{V}$)
0.0	0.200	0.000	0.000
1.00	0.159	0.021	(i)
2.00	0.132	(ii)	(ii)
3.00	(iii)	0.044	(iii)

Pick a convenient volume, V and relate the stoichiometry of the reactants and products.

(i) The depletion of C_6H_{12} from 0.00s-1.00s is ΔC_6H_{12} = 0.200 mol – 0.159 mol = 0.041 mol C_6H_{12}

$$-\Delta H_2 = \frac{7}{2}\Delta C_6H_{12} = \frac{7}{2}(0.041 \text{ mol}) = 0.144 \text{ mol}$$

(ii) The depletion of C_6H_{12} from 0.00s-2.00s is ΔC_6H_{12} = 0.200 mol – 0.132 mol = 0.068 mol C_6H_{12}

$$-\Delta C_{12}H_{10} = \frac{1}{2}\Delta C_6H_{12} = \frac{1}{2}(0.068 \text{ M}) = 0.034 \text{ mol}$$

$$-\Delta H_2 = \frac{7}{2}\Delta C_6H_{12} = \frac{7}{2}(0.068 \text{ mol}) = 0.238 \text{ mol}$$

(iii) Formation of $C_{12}H_{10}$ from 0.00s-3.00s is $-\Delta C_{12}H_{10}$ = 0.044 mol – 0.000 mol = 0.044 M $C_{12}H_{10}$

$$\Delta C_{12}H_{10} = 2(-\Delta C_6H_{12}) = 2(0.044 \text{ mol}) = 0.088 \text{ mol}$$

$- \Delta H_2 = 7\,(-\Delta C_6H_{12}) = 7\,(0.044 \text{ mol}) = 0.308 \text{ mol}$

Time(s)	$[C_6H_{12}]\ (\frac{mol}{V})$	$[C_{12}H_{10}]\ (\frac{mol}{V})$	$[H_2]\ (\frac{mol}{V})$
0.0	0.200	0.000	0.000
1.00	0.159	0.021	**0.144**
2.00	0.132	**0.034**	**0.238**
3.00	**0.088**	0.044	**0.308**

(b) The rate of the reaction requires the reaction order. A plot of $[C_6H_{12}]$ vs. time gives a linear graph, so the reaction is zero order.

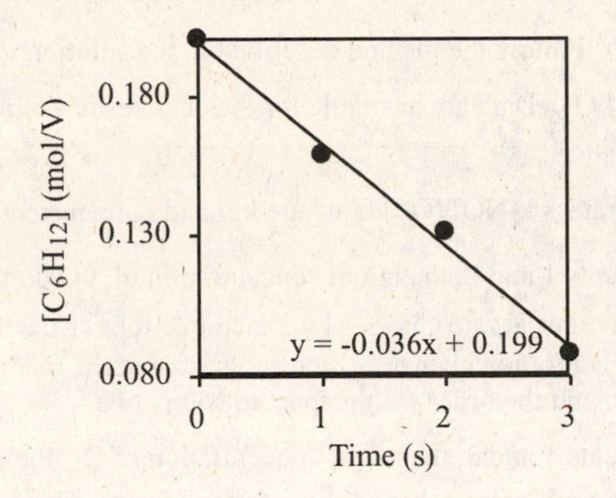

The slope of the graph is 0.036 mol V^{-1} s^{-1}, which also represents the value of k.

Since this is a zero-order reactions, $\frac{-\Delta[C_6H_{12}]}{\Delta t} = k = 0.036 \text{ mol } V^{-1}\ s^{-1}$

The zero-order reaction is independent of concentration, so the rate at 1.5 s is 0.036 mol V^{-1} s^{-1}

118. *Answer:* **(a) True (b) False. "The reaction rate decreases as a first-order reaction proceeds at a constant temperature." (c) True (d) False. "As a second-order reaction proceeds at a constant temperature, the rate constant does not change." (Other corrections for (b) and (d) are also possible.)**

Strategy and Explanation:

(a) True: It is possible to change the rate constant for a reaction by changing the temperature. The Arrehenius equation in Section 13.5 shows the temperature dependence of the rate constant, k. Unless E_a is zero, k would change when T changes.

(b) False. A related true statement is: The reaction rate decreases as a first-order reaction proceeds at a constant temperature.

(c) True: The rate constant for a reaction is independent of reaction concentrations. The rate law describes the rate constant as the proportionality constant between the concentrations and the rate.

(d) False. A related true statement is: As a second-order reaction proceeds at a constant temperature, the rate constant does not change.

120. *Answer:* **rate = $k[H_2]^1[NO]^2$**

Strategy and Explanation: We will use information described in Section 13.2, and methods similar to those in Questions 30 - 33.

The reactants are H_2 and NO, so the rate law looks like this:

rate = $k[H_2]^i[NO]^j$ where k, i and j are currently unknown.

When the initial concentration of H_2 is halved (presuming that the initial concentration of NO stays constant), the initial rate is halved. The rate change is the same as the concentration change, which suggests that the rate is proportional to the concentration of H_2, and the order with respect to H_2 is one.

When the initial concentration of NO increases by a factor of 3 (presuming that the initial concentration of H_2 stays constant), the initial rate changes by a factor of 9. The rate change is the square of the concentration change, which suggests that the rate is proportional to the square of the concentration of NO, and the order with respect to NO is two.

The rate equation, also called the rate law, now looks like this: $\text{rate} = k[H_2]^1[NO]^2$

122. *Answer:* **(a) NO is second order, O_2 is first order. (b) rate = $k[NO]^2[O_2]$ (c) 25 $L^2mol^{-2}s^{-1}$ (d) 7.8×10^{-4} mol $L^{-1}s^{-1}$ (e) rate for NO: 2.0×10^{-4} mol $L^{-1}s^{-1}$; rate for NO_2: 2.0×10^{-4} mol $L^{-1}s^{-1}$**

Strategy and Explanation: Follow the method described in the solution to Question 29.

The reactants are NO and O_2. Data are available for the changes in each of these reactants' concentrations, so the rate law looks like this:

$$\text{rate} = k[NO]^i[O_2]^j \quad \text{where k, i and j are currently unknown.}$$

(a) Looking at Experiments 1 and 2, the initial concentration of NO doubles, the initial concentration of O_2 stays constant, and the initial rate changes by a factor of four ($1.0 \times 10^{-4}/2.5 \times 10^{-5}$). The rate change is the square of the concentration change, which suggests that the rate is proportional to the square of the concentration of NO, and the order with respect to NO is two.

Looking at Experiments 1 and 3, the initial concentration of O_2 doubles, the initial concentration of NO stays constant, and the initial rate changes by a factor of two. The rate change is the same as the concentration change, which suggests that the rate is proportional to the concentration of O_2, and the order with respect to O_2 is one.

(b) The rate law (also called the rate equation) now looks like this: $\text{rate} = k[NO]^2[O_2]^1$

(c) Solve the rate law for k: $$k = \frac{\text{rate}}{[NO]^2[O_2]}$$

Plug in each experiment's data. Here is an example of the calculation for Experiment 1:

$$k_1 = \frac{2.5 \times 10^{-5}\ \text{mol L}^{-1}\text{s}^{-1}}{(0.010\ \text{mol/L})^2(0.010\ \text{mol/L})} = 25\ \text{L}^2\ \text{mol}^{-2}\text{s}^{-1}$$

[NO] (mol/L)	$[O_2]$ (mol/L)	Initial rate (mol $L^{-1}s^{-1}$)	Rate constant ($L^2mol^{-2}s^{-1}$)
0.010	0.010	2.5×10^{-5}	25
0.020	0.010	1.0×10^{-4}	25
0.010	0.020	5.0×10^{-5}	25

The average of these three rate constants is 25 $L^2mol^{-2}s^{-1}$.

(d) $\text{rate} = k[NO]^2[O_2]^1 = (25\ L^2mol^{-2}s^{-1})(0.025\ mol/L)^2(0.050\ mol/L)^1 = 7.8 \times 10^{-4} mol\ L^{-1}s^{-1}$

(e) The relative rates depend on the stoichiometric coefficients (Equation 13.3). Here, the stoichiometric relationship is 2NO : $1O_2$: $2NO_2$

$$-\frac{1}{2}\left(\frac{\Delta[NO]}{\Delta t}\right) = -\frac{1}{1}\left(\frac{\Delta[O_2]}{\Delta t}\right) = \frac{1}{2}\left(\frac{\Delta[NO_2]}{\Delta t}\right)$$

$$-\frac{\Delta[NO]}{\Delta t} = 2\left(-\frac{\Delta[O_2]}{\Delta t}\right) = 2\left(1.0 \times 10^{-4}\ \text{mol L}^{-1}\text{s}^{-1}\right) = 2.0 \times 10^{-4}\ \text{mol L}^{-1}\text{s}^{-1}$$

$$\frac{\Delta[NO_2]}{\Delta t} = 2\left(-\frac{\Delta[O_2]}{\Delta t}\right) = 2\left(1.0\times10^{-4}\ \text{mol L}^{-1}\text{s}^{-1}\right) = 2.0\times10^{-4}\ \text{mol L}^{-1}\text{s}^{-1}$$

124. *Answer:* **(a) first-order in $HCrO_4^-$, first-order in H_2O_2, and first-order in H_3O^+ (b) see equations below (c) second step**

Strategy and Explanation: Follow methods similar to those used the solutions to Questions 85 - 91.

(a) The observed rate law can be used to find the reaction orders of the reactants:

$$\text{rate} = k[HCrO_4^-][H_2O_2][H_3O^+]$$

Since all the concentrations are raised to the same power, all three of them are first-order. That is, the reaction is first-order in $HCrO_4^-$, first-order in H_2O_2, and first-order in H_3O^+.

(b) Cancel intermediates, H_2CrO_4 and $H_2CrO(O_2)_2$, and add the three reactions.

$$HCrO_4^- + H_3O^+ \rightleftharpoons \cancel{H_2CrO_4} + H_2O$$
$$\cancel{H_2CrO_4} + H_2O_2 \longrightarrow \cancel{H_2CrO(O_2)_2} + H_2O$$
$$+\ \cancel{H_2CrO(O_2)_2} + H_2O_2 \longrightarrow CrO(O_2)_2 + 2\,H_2O$$
$$HCrO_4^- + 2\,H_2O_2 + H_3O^+ \longrightarrow CrO(O_2)_2 + 4\,H_2O$$

(c) It is clear, for two reasons, that the first step is not the rate limiting step. First, the double arrow used between the reactants and products indicates that the reaction is fast enough to reach a steady state. Second, the rate law would be: rate = $k[HCrO_4^-][H_3O^+]$, if the first step was slow. That rate law is incompatible with the observed rate law.

If we derive the rate law for the mechanism assuming that the second step were slow, we have a step that looks like this: rate = $k_2[H_2CrO_4][H_2O_2]$

We set up a steady state condition for the intermediate, H_2CrO_4.

$$k_1[HCrO_4^-][H_3O^+] = k_{-1}[H_2CrO_4][H_2O] + k_2[H_2CrO_4][H_2O_2]$$

Assume that the second term is negligibly small since the rate of step 2 is small:

$$k_1[HCrO_4^-][H_3O^+] = k_{-1}[H_2CrO_4][H_2O]$$

Solve for the concentration of the intermediate: $[H_2CrO_4] = \frac{k_1}{k_{-1}}\frac{[HCrO_4^-][H_3O^+]}{[H_2O]}$

Plug into the rate law: $\text{rate} = k_2\frac{k_1}{k_{-1}}\frac{[HCrO_4^-][H_3O^+]}{[H_2O]}[H_2O_2]$

Because the concentration of water in aqueous solutions is essentially constant, the observed rate constant is defined as $k = k_2\frac{k_1}{k_{-1}[H_2O]}$, giving the derived rate law the same functional form as the observed rate law.

$$\text{rate} = k[HCrO_4^-][H_2O_2][H_3O^+]$$

Therefore, the second step is the rate-limiting step.

126. *Answer:* **See derivations below**

Strategy and Explanation: Follow methods similar to those used the solutions to Questions 85 - 91.

It is clear, for two reasons, that the first step is not the rate limiting step. First, the double arrow used between the reactants and products indicates that the reaction is fast enough to reach a steady state. Second, the rate laws would be: rate = $k[NO][O_2]$ or rate = $k[NO]^2$, if the first step was slow. That rate law is incompatible with the observed rate law.

Mechanism 1: If we derive the rate law for the mechanism assuming that the second step were slow, we have a step that looks like this: rate = $k_2[NO_3][NO]$

We set up a steady state condition for the intermediate, NO_3.

$$k_1[NO][O_2] = k_{-1}[NO_3] + k_2[NO_3][NO]$$

Assume that the second term is negligibly small if the rate of step 2 is small:

$$k_1[NO][O_2] = k_{-1}[NO_3]$$

Solve for the concentration of the intermediate: $[NO_3] = \frac{k_1}{k_{-1}}[NO][O_2]$

Plug into the rate law: rate = $k_2[NO_3][NO] = k_2\left(\frac{k_1}{k_{-1}}[NO][O_2]\right)[NO]$

$$\text{rate} = k[NO]^2[O_2]$$

Mechanism 2: If we derive the rate law for the mechanism assuming that the second step were slow, we have a step that looks like this: rate = $k_2[N_2O_2][O_2]$

We set up a steady state condition for the intermediate, N_2O_2.

$$k_1[NO]^2 = k_{-1}[N_2O_2] + k_2[N_2O_2][O_2]$$

Assume that the second term is negligibly small if the rate of step 2 is small:

$$k_1[NO]^2 = k_{-1}[N_2O_2]$$

Solve for the concentration of the intermediate: $[N_2O_2] = \frac{k_1}{k_{-1}}[NO]^2$

Plug into the rate law: rate = $k_2[N_2O_2][O_2] = k_2\left(\frac{k_1}{k_{-1}}[NO]^2\right)[O_2]$

$$\text{rate} = k[NO]^2[O_2]$$

128. *Answer/Explanation:* The catalytic role of the chlorine atom (produced from the light decomposition of chlorofluorocarbons) in the mechanism for the destruction of ozone indicates that even small amounts of CFCs released into the atmosphere pose a serious risk.

Applying Concepts

130. *Answer:* **Curve A represents $[H_2O(g)]$ increase with time, Curve B represents $[O_2(g)]$ increase with time, and Curve C represents $[H_2O_2(g)]$ decrease with time.**

Strategy and Explanation: This question uses methods described in the solutions to Questions 11 - 13.

Cuve A represents the increase in the concentration of the product $H_2O(g)$ with time. Curve B represents the increase in the concentration of product O_2 with time. The O_2 curve is half as steep as the H_2O line because the stoichiometric relationship between them is 1:2. Curve C represents the decrease in the concentration of reactant $H_2O_2(g)$ with time.

132. *Answer:* **Snapshot (b)**

Strategy and Explanation: Products form more quickly at higher temperatures, so (b) the snapshot with fewer HI molecules is the one corresponding to a lower temperature.

134. *Answer:* **rate = k[A]^3[B][C]^2**

Strategy and Explanation: We will use the corrected information described above, the techniques described in Section 13.2, and methods similar to those in the solutions to Questions 31 - 33.

The reactants are A, B and C, so the rate law looks like this:

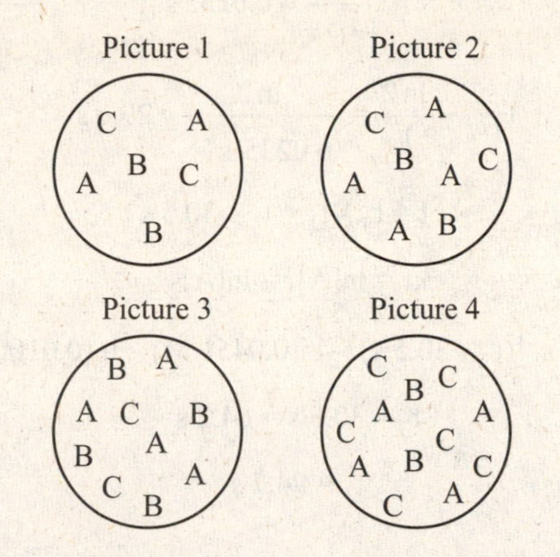

rate = k[A]^i[B]^j[C]^h where k, i, j, and h are currently unknown.

Between pictures 1 and 2, the initial concentration of A doubles (2 A → 4 A), and the initial concentrations of B and C stay constant (2 of each), the initial rate changes by a factor of (12/1.5 =) eight. The rate change is the cube of the concentration change ($8 = 2^3$), which suggests that the order with respect to A is three.

Between pictures 2 and 3, the initial concentration of B doubles (2 B → 4 B), and the initial concentrations of A and C stay constant (4 A and 2 C), the initial rate changes by a factor of (23/12 =) two. The rate change is equal to the concentration change, which suggests that the order with respect to B is one.

Between pictures 2 and 4, the initial concentration of C triples (2 C → 6 C), and the initial concentrations of A and B stay constant (4A and 2B), the initial rate changes by a factor of nine. The rate change is the square of the concentration change($9 = 3^2$), which suggests that the order with respect to C is two.

The rate equation, also called the rate law, now looks like this: rate = k[A]^3[B][C]^2

136. *Answer:* **(a) Three (b) Exothermic**

Strategy and Explanation:

(a) Each hill in the reaction energy diagram represents one elementary reaction. So, this mechanism has three steps.

(b) The products are lower in energy than the reactants, so the reaction causes a release of energy, making it exothermic.

137. *Answer:* **E_a is very, very small—approximately zero.**

Strategy and Explanation: The Arrhenius equation described in Section 13.5 describes the temperature dependence of the rate constant.

$$k = A e^{-E_a/RT}$$

If a radioactive decay reaction is independent of temperature, that says the activation energy E_a is indistinguishable from zero, and $k = A e^{-0/RT} = Ae^0 = A(1) = A$, with no functional dependence on T.

139. *Answer/Explanation:* The lead must bind with the platinum on the surface occupying the actives sites of the catalyst. Deposits of lead would mask the surface of the catalyst reducing or destroying the catalytic properties.

140. *Answer:* **29.6 s, 94.7 s**

Strategy and Explanation: Use the method described in the solution to Question 51.

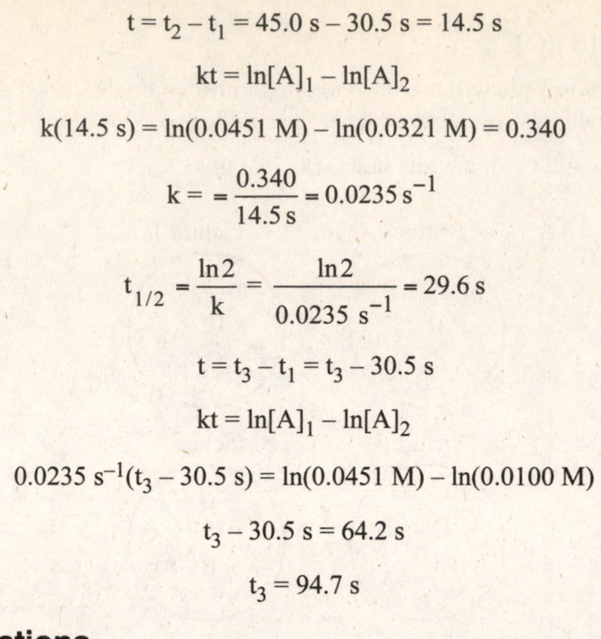

$$t = t_2 - t_1 = 45.0\ \text{s} - 30.5\ \text{s} = 14.5\ \text{s}$$

$$kt = \ln[A]_1 - \ln[A]_2$$

$$k(14.5\ \text{s}) = \ln(0.0451\ \text{M}) - \ln(0.0321\ \text{M}) = 0.340$$

$$k = = \frac{0.340}{14.5\ \text{s}} = 0.0235\ \text{s}^{-1}$$

$$t_{1/2} = \frac{\ln 2}{k} = \frac{\ln 2}{0.0235\ \text{s}^{-1}} = 29.6\ \text{s}$$

$$t = t_3 - t_1 = t_3 - 30.5\ \text{s}$$

$$kt = \ln[A]_1 - \ln[A]_2$$

$$0.0235\ \text{s}^{-1}(t_3 - 30.5\ \text{s}) = \ln(0.0451\ \text{M}) - \ln(0.0100\ \text{M})$$

$$t_3 - 30.5\ \text{s} = 64.2\ \text{s}$$

$$t_3 = 94.7\ \text{s}$$

More Challenging Questions

142. *Answer:* **(a) rate = 2.4×10^{-7}mol $L^{-1}s^{-1}$ + k[BSC][F^-] (b) 0.3 L $mol^{-1}s^{-1}$**

Strategy and Explanation: Given the initial concentrations and initial rates of a reaction at several different experimental conditions, determine the rate law and rate constant for the reaction.

In Section 13.2, the method of finding the rate law from initial rates is described for getting the orders. However, before we compare the experimental rates, we need to subtract the residual rate. Then try comparing pairs of experiments where only one of the concentrations is different and relating that to the changes in the rate. If the result of the pair-wise comparison is ambiguous, make a linear graph. Once the orders are determined, plug the data into the rate law to determine the value of k.

Benzenesulfonyl chloride (abbreviated BSC) has a known effect on the rate (first-order), but it has a constant concentration in this experiment, 2×10^{-4} M. As a result, we can't say how it is involved in the constant term representing the residual rate, so we will seek a rate law looks like this:

rate = 2.4×10^{-7}mol $L^{-1}s^{-1}$ + k[BSC][F^-]^i where k and i are currently unknown.

To study how [F^-] affects the rate of the reaction, subtract the residual rate from the observed rate.

Experiment	[F^-] × 10^2 (mol/L)	Initial rate × 10^7 (mol $L^{-1}s^{-1}$)	Adjusted initial rate × 10^7 (mol $L^{-1}s^{-1}$)
1	0	2.4	2.4 – 2.4 = 0.0
2	0.5	5.4	5.4 – 2.4 = 3.0
3	1.0	7.9	7.9 – 2.4 = 5.5
4	2.0	13.9	13.9 – 2.4 = 11.5
5	3.0	20.2	20.2 – 2.4 = 17.8
6	4.0	25.2	25.2 – 2.4 = 22.8
7	5.0	32.0	32.0 – 2.4 = 29.6

(a) The adjusted rate law has the functional form: adjusted rate = k[BSC][F^-]^i

To get the order of F^-, compare experiments 2 to 3, 3 to 4, and 4 to 6. They each show a concentration increase of a factor of two. The ratio of the adjusted initial rate in each of these instances is:

$$2\text{ to }3:\ \frac{5.5}{3.0} = 1.8,\qquad 3\text{ to }4:\ \frac{11.5}{5.5} = 2.1,\qquad 4\text{ to }6:\ \frac{22.8}{11.5} = 2.0$$

It looks like the adjusted rate change is approximately the same as the concentration change, suggesting that the adjusted rate is proportional to the concentration of fluoride ions. The relationship is only approximate, though, so let's prepare a graph that uses all of the data. By taking the log of the rate law, the order becomes the slope of a linear graph.

$$\log(\text{adjusted rate}) = \log(k[\text{BSC}]) + \log([\text{F}^-]^i)$$

$$\log(\text{adjusted rate}) = i \log[\text{F}^-] + \log(k[\text{BSC}])$$

$\log [\text{F}^-]$	log(adjusted initial rate) ($\text{mol L}^{-1}\text{s}^{-1}$)
–2.3	–6.52
–2.00	–6.26
–1.70	–5.939
–1.52	–5.750
–1.40	–5.642
–1.30	–5.529

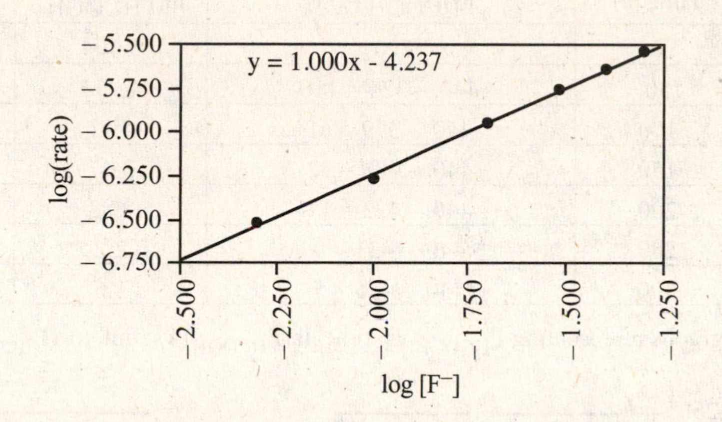

This confirms that the order of the adjusted rate law with respect to fluoride (i) is one, and the most complete rate law we can write for the reaction looks like this:

$$\text{rate} = 2.4 \times 10^{-7}\text{mol L}^{-1}\text{s}^{-1} + k[\text{BSC}][\text{F}^-]$$

(b) Solve the adjusted rate law for k: $k = \dfrac{\text{adjusted rate}}{[\text{BSC}][\text{F}^-]}$

Plug in each set of data. Here is a sample of a calculation for the first experiment with nonzero $[\text{F}^-]$:

$$k_2 = \frac{3.0 \times 10^{-7}\text{ mol L}^{-1}\text{s}^{-1}}{(2 \times 10^{-4}\text{ M})(0.5 \times 10^{-2}\text{ mol/L})} = 0.3\text{ M}^{-1}\text{s}^{-1}$$

Experiment	$[\text{F}^-] \times 10^2$ (mol/L)	Adjusted initial rate × 10^7 ($\text{mol L}^{-1}\text{s}^{-1}$)	Adjusted rate constant ($\text{M}^{-1}\text{s}^{-1}$)
2	0.5	3.0	0.3
3	1.0	5.5	0.3
4	2.0	11.5	0.3
5	3.0	17.8	0.3
6	4.0	22.8	0.3
7	5.0	29.6	0.3

The average of these seven rate constants is $0.3\text{ L mol}^{-1}\text{s}^{-1}$

✓ *Reasonable Answer Check:* It is satisfying to get the same rate constant for each data set.

144. *Answer:* 0.0127 s^{-1}, 54.6 s

Strategy and Explanation: The chemical reaction takes one mole of gas and makes it two, so the partial pressure rises as a result of the formation of products. The total pressure P_{TOT} initially is entirely due to the pressure exerted by HCOOH(g), 220 torr. As the reaction proceeds the reduction in the partial pressure of HCOOH(g) causes an increase in the partial pressure of products CO_2(g) and H_2(g). The stoichiometry is 1:1:1. So, at any given time, the partial pressure of HCOOH(g), P_{HCOOH} = 220. torr – x, The partial pressures of CO_2(g) and H_2(g) increase to $P_{CO_2} = P_{H_2}$ = x. Dalton's law of partial pressures (Chapter 10, page 446) indicates that the sum of the partial pressures is equal to the total pressure.

$$P_{TOT} = P_{HCOOH} + P_{CO_2} + P_{H_2} = (220.\text{ torr} - x) + x + x = 220.\text{ torr} + x$$

$$P_{TOT} - 220.\text{ torr} = x$$

$$P_{HCOOH} = 220.\text{ torr} - (P_{TOT} - 220.\text{ torr}) = 440.\text{ torr} - P_{TOT}$$

Time (s)	P_{HCOOH} (torr)	ln(P_{HCOOH})	1/P_{HCOOH}
0	220.	5.394	0.00455
50	440. – 324 = 116	4.754	0.00862
100	440. – 379 = 61	4.11	0.016
150	440. – 408 = 32	3.47	0.031
200	440. – 423 = 17	2.83	0.059
250	440. – 431 = 9	2.2	0.1
300	440. – 435 = 5	1.6	0.2

Create three graphs representing P_{HCOOH} vs. time, ln(P_{HCOOH}) vs. time, 1/(P_{HCOOH}) vs. time:

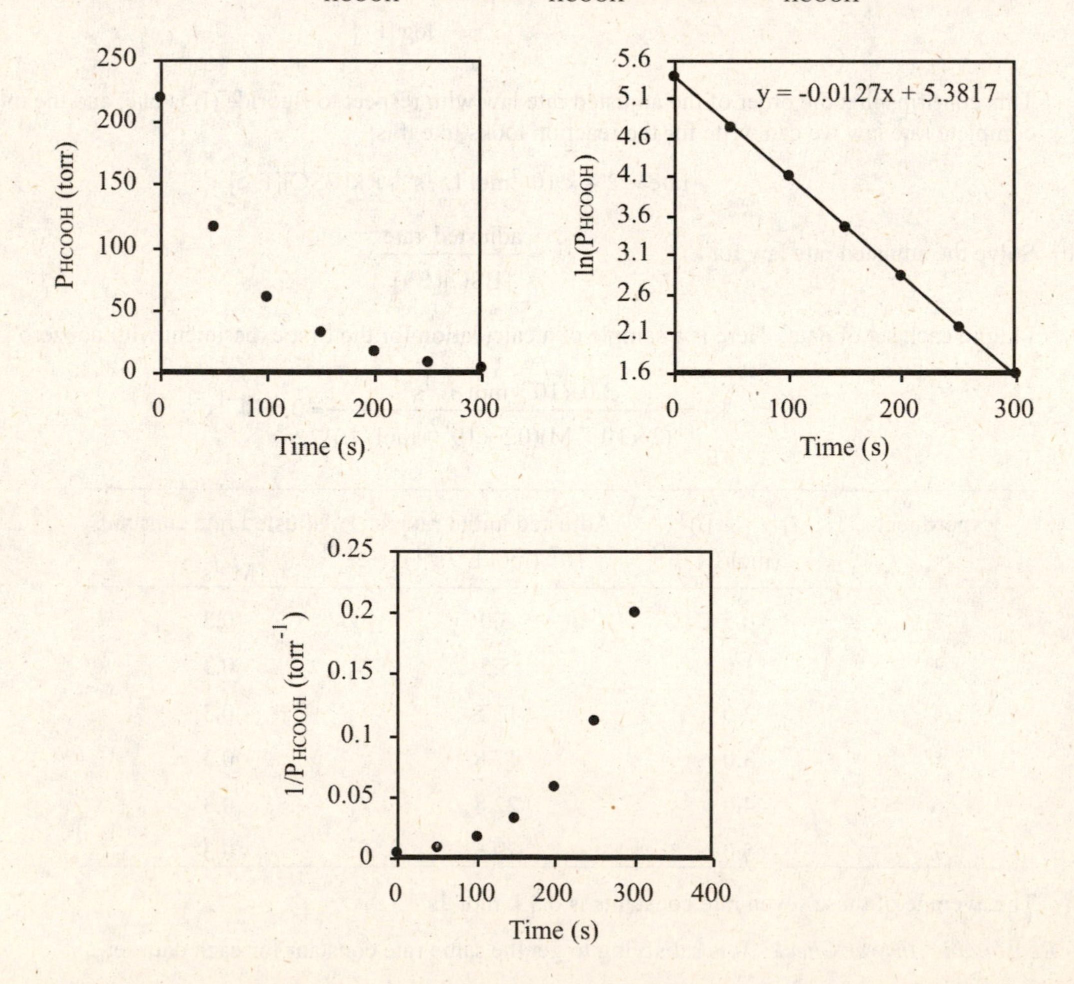

The linear graph is the second one, which represents a first-order graph. The slope of the first-order graph is –k; therefore, $k = 0.0127\ s^{-1}$. The first-order half-life equation looks like this:

$$t_{1/2} = \frac{\ln 2}{k} = \frac{\ln 2}{0.0127\,s^{-1}} = 54.6\ s$$

✓ *Reasonable Answer Check:* It makes sense that the half life is a little more than 50 s since half the initial pressure of HCOOH is 110 torr, and the pressure at t = 50 pressure is 116 torr.

146. *Answer:* **(a) 2.8×10^3 s (b) 1.4×10^4 s (c) 2.0×10^4 s**

Strategy and Explanation: Use methods and equations similar to ones described in the solution to Question 51.

(a)
$$t_{1/2} = \frac{\ln 2}{k} = \frac{\ln 2}{2.5\times 10^{-4}\,s^{-1}} = 2.8\times 10^{3}\ s$$

(b)
$$[A]_t = \frac{1}{32}\times[A]_0$$

Notice, the initial concentration is not known, but the ratio is all we need:

$$kt = \ln[A]_0 - \ln[A]_t = \ln\left(\frac{[A]_0}{[A]_t}\right) = \ln\left(\frac{[A]_0}{\frac{1}{32}[A]_0}\right) = \ln(32) = 3.47$$

$$t = \frac{3.47}{k} = \frac{3.47}{2.5\times 10^{-4}\,s^{-1}} = 1.4 \times 10^4\ s$$

Alternatively, we could also use the results derived in Question 141 to answer this question:

$$[A]_t = [A]_0\left(\frac{1}{2}\right)^x \qquad x = \frac{t}{t_{1/2}}$$

Since $\left(\frac{1}{2}\right)^x = \frac{1}{32}$, when x = 5; therefore, $t = 5 \times (2.8 \times 10^3\ s) = 1.4 \times 10^4\ s$.

(c) $kt = \ln[A]_0 - \ln[A]_t = \ln(3.4 \times 10^{-3}\ mol/L) - \ln(2.3 \times 10^{-5}\ mol/L) = 5.00$

$$t = \frac{5.00}{k} = \frac{5.00}{2.5\times 10^{-4}\,s^{-1}} = 2.0 \times 10^4\ s$$

148. *Answer:* **(a) 3×10^2 kJ (b) 2×10^3 s**

Strategy and Explanation: Given two different temperatures and the rate constants at those temperatures, determine the activation energy of a reaction.

(a) Convert the temperatures to Kelvin, then use the equation derived in Problem-Solving Example 13.9 to calculate the activation energy:

$$\ln\left(\frac{k_1}{k_2}\right) = \frac{E_a}{R}\left(\frac{1}{T_2} - \frac{1}{T_1}\right)$$

470. °C + 273.15 = 743 K   510. °C + 273.15 = 783 K

$$\ln\left(\frac{1.10\times 10^{-4}\ s^{-1}}{1.02\times 10^{-3}\ s^{-1}}\right) = \frac{E_a}{R}\left(\frac{1}{783\ K} - \frac{1}{743\ K}\right)$$

$$\ln(0.108) = \frac{E_a}{(0.008134 kJ/mol\cdot K)}\left(0.00128\ K^{-1} - 0.00135\ K^{-1}\right)$$

$$-2.227 = \frac{E_a}{(0.008134 kJ/mol\cdot K)}\left(0.00007\ K^{-1}\right)$$ *(subtraction leaves only one sig fig)*

$$-2.227 = E_a\,(-.008\ mol/kJ)$$

$$269 \text{ kJ} = E_a$$

$$3\times 10^2 \text{ kJ} = E_a \quad \textit{(with one sig fig)}$$

(b) Use the same equation to determine the rate constant at 500. °C.

$$500.\ °C + 273.15 = 773 \text{ K}$$

$$\ln\left(\frac{1.10\times10^{-4}\ s^{-1}}{k_2}\right) = \frac{E_a}{R}\left(\frac{1}{773 \text{ K}} - \frac{1}{743 \text{ K}}\right)$$

$$\ln\left(\frac{1.10\times10^{-4}\ s^{-1}}{k_2}\right) = \frac{(3\times10^2 \text{ kJ})}{(0.008134 \text{ kJ/mol}\cdot\text{K})}\left(0.00129 \text{ K}^{-1} - 0.00135 \text{ K}^{-1}\right) \quad \textit{(with 1 sig fig in } E_a\textit{)}$$

$$\ln\left(\frac{1.10\times10^{-4}\ s^{-1}}{k_2}\right) = -2 \quad \textit{(with one sig fig in } E_a \textit{ and in subtraction)}$$

$$\frac{1.10\times10^{-4}\ s^{-1}}{e^{-2}} = k_2 \quad \textit{(with one sig fig in exponent of e)}$$

$$8\times10^{-4}\ s^{-1} = k_2 \quad \textit{(round to one sig fig)}$$

The rate constants are first-order units, so use first order integrated rate law to determine the time:

$$kt = \ln[A]_0 - \ln[A]_t = \ln(0.10) - \ln(0.023) = 1.47$$

$$t = \frac{1.47}{k} = \frac{1.47}{8\times10^{-4}\ s^{-1}} \quad \textit{(with one sig fig in k)}$$

$$t = 2\times10^3 \text{ s} \quad \textit{(rounded to one sig fig)}$$

150. *Answer:* **See mechanisms below (There can be other correct answers.)**

Strategy and Explanation: You may come up with correct answers to this question that are NOT the same as the answers given here. In all cases, a proposed mechanism must have three things: First, it should be composed of first or second order elementary reactions. Second, it must recreate the given overall net reaction, including the elimination of all intermediates and the recreation of all the catalysts. Third, it must recreate the observed rate law. If the rate law is simple first or second order, then make the first step the slow step, and use the chemicals in the rate law as reactants. If the rate law is more complicated, start with a first step that is fast, and include some of the chemicals from the rate law in that reaction, then use the product of that reaction as a reactant in the next reaction. In each case, once you have written the mechanism, confirm that its mechanism matches the observed (given) mechanism.

(a)

$$CH_3CO_2CH_3 + H_3O^+ \longrightarrow CH_3COHOCH_3 + H_2O \quad \text{slow}$$

$$CH_3COHOCH_3 + H_2O \longrightarrow CH_3COH(OH_2)OCH_3 \quad \text{fast}$$

$$CH_3COH(OH_2)OCH_3 \longrightarrow CH_3C(OH)_2 + CH_3OH \quad \text{fast}$$

$$CH_3C(OH)_2 + H_2O \longrightarrow CH_3COOH + H_3O^+ \quad \text{fast}$$

Check this mechanism: Each reaction is unimolecular or bimolecular Check the overall reaction:

$$CH_3CO_2CH_3 + \cancel{H_3O^+} \longrightarrow CH_3C\cancel{OHOCH_3} + \cancel{H_2O}$$

$$CH_3\cancel{COHOCH_3} + \cancel{H_2O} \longrightarrow CH_3CO\cancel{H(OH_2)}OCH_3$$

$$CH_3CO\cancel{H(OH_2)}OCH_3 \longrightarrow CH_3\cancel{C(OH)_2} + CH_3OH$$

$$CH_3\cancel{C(OH)_2} + H_2O \longrightarrow CH_3COOH + \cancel{H_3O^+}$$

$$CH_3CO_2CH_3 + H_2O \longrightarrow CH_3COOH + CH_3OH \quad \text{This is the net equation.}$$

The rate law of the slow first elementary reaction is the rate law for the whole mechanism. As described in Section 13.6, the stoichiometric coefficient of a reactant in an elementary reaction is the reaction order for that reactant.

$$CH_3CO_2CH_3 + H_3O^+ \longrightarrow \text{products} \qquad \text{rate} = k_1[CH_3CO_2CH_3][H_3O^+]$$

This matches the reported rate law for the reaction. Therefore, the proposed mechanism is plausible.

(b) $H_2 + I_2 \longrightarrow 2\text{ HI}$ slow

Check this mechanism: The reaction is bimolecular and represents the overall reaction. As described in (a), check the mechanism.

$$H_2 + I_2 \longrightarrow \text{products} \qquad \text{rate} = k[H_2][I_2]$$

This matches the reported rate law for the reaction. Therefore, the proposed mechanism is plausible.

(c) The presence of I_2 and Pt in the rate law suggests that they are catalyst in this reaction:

$H_2 + Pt(s) \longrightarrow PtH_2$ fast

$PtH_2 + I_2 \longrightarrow PtH_2I_2$ slow

$PtH_2I_2 + O_2 \longrightarrow PtI_2O + H_2O$ fast

$PtI_2O + H_2 \longrightarrow Pt(s) + I_2 + H_2O$ fast

Check this mechanism: Each reaction is unimolecular or bimolecular. Overall:

$H_2 + \cancel{Pt} \longrightarrow \cancel{PtH_2}$

$\cancel{PtH_2} + \cancel{I_2} \longrightarrow \cancel{PtH_2I_2}$

$\cancel{PtH_2I_2} + O_2 \longrightarrow \cancel{PtI_2O} + H_2O$

$\cancel{PtI_2O} + H_2 \longrightarrow \cancel{Pt} + \cancel{I_2} + H_2O$

$2\,H_2 + O_2 \longrightarrow 2\,H_2O$ This is the overall equation.

Determine the rate law for this mechanism.

The slow step is step 2, so the rate of the reaction is equal to the rate of this rate-determining step. Because this reaction is elementary, its rate law is related to the stoichiometry of its reactants:

$$\text{rate of reaction} = \text{rate}_{\text{reaction 2}} = k_2[PtH_2][I_2]$$

PtH_2 is an intermediate, so we need to eliminate it from the proposed mechanism. Look for all the ways this intermediate is created and destroyed during the mechanism. It is only created in the forward reaction of step 1. It is destroyed in the reverse reaction of step 1 and also in the reaction of step 2. Because all three of these reactions are elementary, use their reactants' stoichiometric coefficients to write rate laws for each of these three reactions:

$$\text{rate}_{\text{forward reaction 1}} = k_1[H_2](\text{area of Pt surface})$$

$$\text{rate}_{\text{reverse reaction 1}} = k_{-1}[PtH_2]^1$$

$$\text{rate}_{\text{reaction 2}} = k_2[PtH_2][I_2]$$

The rate of PtH_2 creation is equal to the rate of its destruction once the steady state condition is reached:

$$\text{rate}_{\text{forward reaction 1}} = \text{rate}_{\text{reverse reaction 1}} + \text{rate}_{\text{reaction 2}}$$

$$k_1[H_2](\text{area of Pt surface}) = k_{-1}[PtH_2]^1 + k_2[PtH_2][I_2]$$

Because the rate of step 2 is presumed to be much smaller than the rate of step 1, the second term is presumed to be negligibly small compared to the first term:

$$\text{rate}_{\text{forward reaction 1}} \cong \text{rate}_{\text{reverse reaction 1}}$$

$$k_1[H_2](\text{area of Pt surface}) = k_{-1}[PtH_2]^1$$

$$[PtH_2] = \frac{k_1}{k_{-1}}[H_2](\text{area of Pt surface})$$

Substitute the equality just derived into the rate law: $\text{rate} = k[PtH_2][I_2]$

$$\text{rate} = k_2\frac{k_1}{k_{-1}}[H_2](\text{area of Pt surface})[I_2]$$

We will define the new group of rate constants by one variable, $k' = k_2\frac{k_1}{k_{-1}}$.

$$\text{rate} = k'[H_2][I_2](\text{area of Pt surface})$$

This matches the reported rate law for the reaction. Therefore, the proposed mechanism is a plausible one.

(d)

$H_2 \longrightarrow 2\,H$	fast
$H + CO \longrightarrow HCO$	slow
$HCO + H \longrightarrow H_2CO$	fast

Check this mechanism: Each reaction is unimolecular or bimolecular.

$$H_2 \longrightarrow 2\cancel{H}$$
$$\cancel{H} + CO \longrightarrow \cancel{HCO}$$
$$\cancel{HCO} + \cancel{H} \longrightarrow H_2CO$$

$H_2 + CO \longrightarrow H_2CO$ This is the overall equation.

Determine the rate law for this mechanism.

The slow step is step 2, so the rate of the reaction is equal to the rate of this rate-determining step. Because this reaction is elementary, its rate law is related to the stoichiometry of its reactants:

$$\text{rate of reaction} = \text{rate}_{\text{reaction 2}} = k_2[H][CO]$$

PtH_2 is an intermediate, so we need to eliminate it from the proposed mechanism. Look for all the ways this intermediate is created and destroyed during the mechanism. It is only created in the forward reaction of step 1. It is destroyed in the reverse reaction of step 1 and also in the reaction of step 2. Because all three of these reactions are elementary, use their reactants' stoichiometric coefficients to write rate laws for each of these three reactions:

$$\text{rate}_{\text{forward reaction 1}} = k_1[H_2]$$
$$\text{rate}_{\text{reverse reaction 1}} = k_{-1}[H]^2$$
$$\text{rate}_{\text{reaction 2}} = k_2[H][CO]$$

The rate of H creation is equal to the rate of its destruction once the steady state condition is reached:

$$\text{rate}_{\text{forward reaction 1}} = \text{rate}_{\text{reverse reaction 1}} + \text{rate}_{\text{reaction 2}}$$
$$k_1[H_2] = k_{-1}[H]^2 + k_2[H][CO]$$

Because the rate of step 2 is presumed to be much smaller than the rate of step 1, the second term is presumed to be negligibly small compared to the first term:

$$\text{rate}_{\text{forward reaction 1}} \cong \text{rate}_{\text{reverse reaction 1}}$$
$$k_1[H_2] = k_{-1}[H]^2$$
$$[H]^2 = \frac{k_1}{k_{-1}}[H_2]$$

$$[H] = \left(\frac{k_1}{k_{-1}}[H_2]\right)^{1/2}$$

Substitute the equality just derived into the rate law: rate = $k_2[H][CO]$

$$\text{rate} = k_2\left(\frac{k_1}{k_{-1}}\right)^{1/2}[H_2]^{1/2}$$

We will define the new group of rate constants by one variable, $k' = k_2\left(\frac{k_1}{k_{-1}}\right)^{1/2}$.

$$\text{rate} = k'[H_2]^{1/2}[CO]$$

This matches the reported rate law for the reaction. Therefore, the proposed mechanism is plausible.

152. *Answer:* **Approximately 26 times faster**

Strategy and Explanation: The instructions in Question 151 were to use the equation derived in that solution to solve this question. This is the equation derived:

$$\frac{k_1}{k_2} = e^{\frac{E_a' - E_a}{RT}}$$

In Exercise 13.9, we are given E_a = 10. kJ/mol and T = 370 K. If E_a' = 0 kJ/mol, the rate increases because the E_a is lowered, the ratio of the rates equals the ratio of the rate constants. The rate changes by a factor of:

$$\frac{\text{rate}_1}{\text{rate}_2} = e^{(10.\text{ kJ/mol} - 0\text{ kJ/mol})/(0.008314\text{ kJ/mol}\cdot\text{K})(370\text{ K})} = 26$$

The reaction goes approximately 26 times faster.

154. *Answer:* **Estimate 402 × 10^{-21} J/molecule (similar to 435 × 10^{-21} J/molecule)**

Strategy and Explanation: Given the description of a partial reaction and bond enthalpies, estimate the activation energy for the whole reaction and compare to the activation energy for the whole reaction given elsewhere.

The "rotation reaction" looks like this: $CH_3CH{=}CHCH_3 \longrightarrow CH_3CH{-}CHCH_3$

Use Table 8.2 to find the bond enthalpy for the single and double bonds between carbon atoms. The difference between them gives an estimate of the enthalpy required to break the π-bond in kJ/mol. Active Figure 13.7 gives energies in J/molecule, so convert the estimated energy to J/molecule.

$$D_{C=C} - D_{C\text{-}C} = 598\text{ kJ/mol} - 356\text{ kJ/mol} = 242\text{ kJ/mol}$$

$$\frac{242\text{ kJ}}{1\text{ mole}} \times \frac{1\text{ mol}}{6.022\times 10^{23}\text{ molecules}} \times \frac{1000\text{ J}}{1\text{ kJ}} = 402\times 10^{-21}\text{ J/molecule}$$

In Active Figure 13.7, the activation energy is given 435 × 10^{-21} J/molecule. It is comparable in the first significant figure to the estimated energy for breaking a π-bond.

✓ *Reasonable Answer Check:* The calculation of the estimated energy of breaking a π-bond uses bond enthalpies, which are derived from the average of the enthalpies for several molecules containing these kinds of bonds.

156. *Answer/Explanation:* The speed of the reaction is related to the energy of the production of a photon that can break a bond in iodine and make a free radical catalyst.

$$E_{photon} = \frac{hc}{\lambda} = \frac{6.626\times 10^{-34}\text{ J}\cdot\text{s}\times 2.998\times 10^{8}\text{ m/s}}{800\text{ nm}\times\frac{1\times 10^{-9}\text{ m}}{1\text{ nm}}} = 2.5\times 10^{-19}\text{ J}$$

$$\text{energy per mole of bonds} = \frac{2.5\times 10^{-19}\text{ J}\times}{1\text{ photon}} \times \frac{6.022\times 10^{23}\text{ photons}}{1\text{ mol photons}} \times \frac{1\text{ kJ}}{1000\text{ J}} = 150\ \frac{\text{kJ}}{\text{mol}}$$

This is approximately the bond enthalpy of I_2. Wavelengths longer than 800 nm have photons with energy less than the bond enthalpy of I_2.

158. *Answer/Explanation:* Use Section 13.3 and the margin notes early in Section 13.3 to assist in the solution to this question.

(i) Define the reaction rate in terms of [A]: $\text{Rate} = -\dfrac{\Delta[A]}{\Delta t}$

(ii) Write the rate law in terms of [A], k and t: $-\dfrac{\Delta[A]}{\Delta t} = k[A]$

Calculus uses "d" to replace "Δ" to describe very small changes: $-\dfrac{d[A]}{dt} = k[A]$

(iii) Separate the variables by putting everything with [A] on one side and everything with t on the other side:

$$\frac{d[A]}{[A]} = -\,kdt$$

(iv) Integrate each side, using calculus: $\int \dfrac{dx}{x} = \ln x$ and $\int dx = x$

and use the initial and final conditions to resolve the definite integral:

$$\int_0^t \frac{d[A]}{[A]} = -k\int_0^t dt$$

$$\ln[A]_t - \ln[A]_0 = -\,k(t_t - t_0)$$

(v) If the reaction starts at time t=0 and goes to time t, then $t_t - t_0$ is the elapsed time, t.

$$\ln[A]_t - \ln[A]_0 = -\,kt$$

$$\ln[A]_t = -\,kt + \ln[A]_0$$

This equation is in the form of y = mx + b, where variable $y = \ln[A]_t$, m = – k (a constant), variable x = t, and $b = \ln[A]_0$ (also a constant)

(vi) Half life is the elapsed time for $[A]_0$ to drop by half, so:

$$[A]_t = \frac{1}{2}[A]_0\text{, when } t = t_{1/2}$$

$$\ln\left(\frac{1}{2}[A]_0\right) - \ln[A]_0 = -\,kt_{1/2}$$

$$\ln[A]_0 - \ln 2 - \ln[A]_0 = -\,kt_{1/2}$$

$$-\ln 2 = -\,kt_{1/2}$$

$$t_{1/2} = \frac{-\ln 2}{-k} = \frac{\ln 2}{k}$$

Chapter 14: Chemical Equilibrium

Solutions for Blue-Numbered Questions for Review and Thought

Topical Questions

Characteristics of Chemical Equilibrium (Section 14.1)

8. *Answer/Explanation:* There are many answers to this question. One possible answer is given later in Question 139. Prepare a sample of N_2O_4 in which the N atoms are the heavier isotopes ^{15}N. Introduce the heavy isotope of N_2O_4 into an equilibrium mixture of N_2O_4 and NO_2. Use spectroscopic methods, such as infrared spectroscopy to observe the distribution of the radioisotope among the reactants and products.

10. *Answer:* (a) 0 °C (b) Dynamic equilibrium (see explanation below)

    Strategy and Explanation:
    (a) The temperature of an equilibrium mixture of ice and water is the melting point of water: 0 °C.
    (b) This is a dynamic equilibrium. Molecules are not smart enough to stay in a particular phase. Some molecules at the interface between the water and the ice detach from the ice and enter the liquid phase or attach to the solid phase leaving the liquid phase.

The Equilibrium Constant (Section 14.2)

12. *Answer:* **see drawings below**

    Strategy and Explanation: Draw reactant concentration vs. time as a downward curve to level off flat at the equilibrium concentration. Draw product concentration vs. time curve as an upward curve to level off flat at the equilibrium concentration. The slope of [NO] increase is steeper than slope of [N_2] decrease, and the equilibrium concentration of NO is equal to half of the equilibrium concentration of N_2 or O_2.

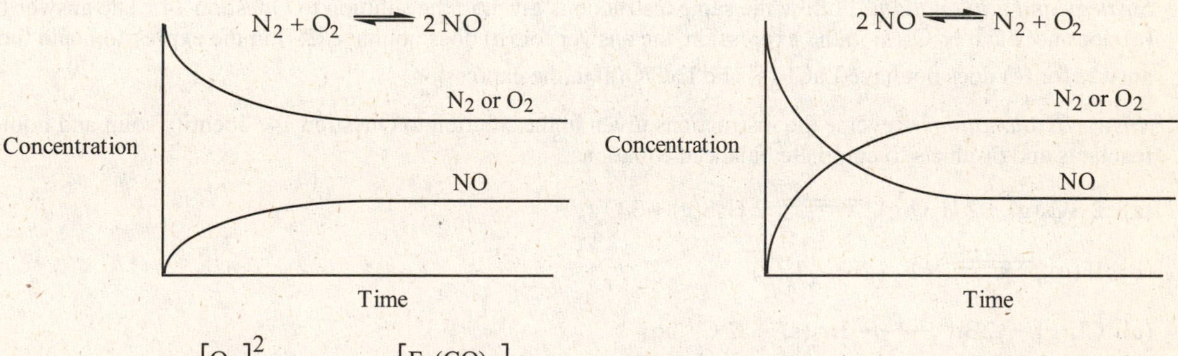

14. *Answer:* (a) $K_c = \frac{[O_3]^2}{[O_2]^3}$ (b) $K_c = \frac{[Fe(CO)_5]}{[CO]^5}$ (c) $K_c = [NH_3]^2[CO_2][H_2O]$ (d) $K_c = [Ag^+]^2[SO_2^{2-}]$

    Strategy and Explanation: For the expression of the equilibrium constant, use this form: $K_c = \frac{[\text{products}]}{[\text{reactants}]}$; if a stoichiometric coefficient precedes a species, that number is used as the mathematical power of the concentration in the expression. Remember that the equilibrium state is not affected by the relative quantity of any solids or liquids, so those materials do not show up in the equilibrium expression. Hence, in particular, the answer for (b) does not have Fe(s) in the expression, the answer for (c) does not have $(NH_4)_2CO_3$ (s) in the expression, and the answer for (d) does not have Ag_2SO_4 (s) in the expression.

16. *Answer:* (a) $K_c = \frac{[CO][H_2O]}{[CO_2][H_2]}$ (b) $K_c = [CO_2]$ (c) $K_c = \frac{[CH_4]^2[Cl_2]^3}{[CHCl_3]^2[H_2]^3}$ (d) $K_c = \frac{[HI]^2}{[H_2][I_2]}$

    (e) $K_c = \frac{[NO]^2[Cl_2]}{[NOCl]^2}$

Strategy and Explanation: Follow the method described in Question 14. The answer for (b) does not have $SrCO_3(s)$ in the expression.

17. *Answer:* (a) $K_c = \frac{[H_2O]^2[O_2]}{[H_2O_2]^2}$ (b) $K_c = \frac{[PCl_5]}{[PCl_3][Cl_2]}$ (c) $K_c = [CO]^2$ (d) $K_c = \frac{[H_2S]}{[H_2]}$

Strategy and Explanation: Follow the method described in Question 14. The answer for (c) does not have $SiO_2(s)$, C(s), or SiC(s) in the expression, and the answer for (d) does not have $S_8(s)$ in the expression.

19. *Answer:* (a) $K_c = \frac{[Ni(CO)_4]}{[CO]^4}$ (b) $K_c = \frac{[H_3O^+][F^-]}{[HF]}$ (c) $K_c = \frac{[Cl^-]^2}{[Cl_2][Br^-]^2}$

Strategy and Explanation: Follow the same instructions given in the solution to Question 14. The answer for (a) does not have Ni(s) in the expression, the answer for (b) does not have $H_2O(\ell)$ in the expression, and the answer for (c) does not have $Br(\ell)$ in the expression.

21. *Answer:* (a) $K_c = \frac{[PCl_3][Cl_2]}{[PCl_5]}$ (b) $K_c = \frac{[CoCl_4^{2-}]}{[Co(H_2O)_6^{2+}][Cl^-]^4}$ (c) $K_c = \frac{[CH_3COO^-][H_3O^+]}{[CH_3COOH]}$

(d) $K_c = \frac{[OF_2][HF]^2}{[F_2]^2[H_2O]}$

Strategy and Explanation: Follow the same instructions given in the solution to Question 14. The answer for (b) and (c) do not have $H_2O(\ell)$ in the expression, however (d) does include the gaseous $H_2O(g)$.

23. *Answer:* (a) $K_c = [H_2O]^2$ (b) $K_c = \frac{[HF]^4}{[SiF_4][H_2O]^2}$ (c) $K_c = \frac{[HCl]^2}{[H_2O]}$

Strategy and Explanation: Follow the same instructions given in the solution to Question 14. The answer for (a) does not have $N_2O_5(s)$ in the expression, the answer for (b) does not have C(s) in the expression, and the answer for (c) does not have $LaCl_3(s)$ and LaClO(s) in the expression.

25. *Answer/Explanation:* Reverse the instructions given in the solution to Question 14. Identify solid and liquid reactants and products to complete balanced equations.

(a) $2\,SO_2(g) + 2\,H_2O(g) \rightleftharpoons 2\,H_2S(g) + 3\,O_2(g)$

(b) $IF(g) \rightleftharpoons \frac{1}{2}F_2(g) + \frac{1}{2}I_2(g)$

(c) $Cl_2(g) + 2Br^- \longrightarrow Br_2(s) + 2\,Cl^-(aq)$

27. *Answer:* **Equation (e)**

Strategy and Explanation: The two equations are related, so we will use the information in Section 14.2 to identify how their equilibrium constants are related.

Multiplying the first equation by a constant factor of 2 and then reversing that equation gives the second equation. The first change means we need to raise the K from the first reaction to the power of 2. The second change means we need to take the reciprocal of the K after the first change. Therefore the second equation's equilibrium constant is related to the first equation's equilibrium constant as represented in (e), $K_{c_2} = 1/K_{c_1}{}^2$.

29. *Answer:* **(a) 0.87 (b) 1.3**

Strategy and Explanation: The two equations are related, so we will use the information in Section 14.2 to identify how their equilibrium constants are related.

(a) The synthesis of 1 mol sulfur trioxide gas is represented by the formation reaction:

$$SO_2(g) + \frac{1}{2}O_2(g) \rightleftharpoons SO_3(g)$$

Multiplying the given equation by a constant factor of $\frac{1}{2}$ gives the second equation. This change means we need to raise the K_c from the first reaction to the power of $\frac{1}{2}$.

$$K_{(a)} = K_c^{1/2} = (0.76)^{1/2} = 0.87$$

(b) The decomposition of 2 mol sulfur trioxide gas is represented by the formation reaction:

$$2\ SO_3(g) \rightleftharpoons 2\ SO_2(g) + O_2(g)$$

This reaction is the reverse of the given reaction, so we need to take the reciprocal of the K_c:

$$K_{(b)} = 1/K_c = 1/(0.76) = 1.3$$

31. *Answer:* **48**

Strategy and Explanation: The two equations are related, so we will use the information in Section 14.2 to identify how their equilibrium constants are related. The reverse of an equation requires us to take the reciprocal of the original K_c: $K_{c,\ reverse} = 1/K_{c,\ given} = 1/(0.0021) = 48$

33. *Answer:* (a) $K_p = \dfrac{P^2_{H_2O} P_{O_2}}{P^2_{H_2O_2}}$ (b) $K_p = \dfrac{P_{PCl_5}}{P_{PCl_3} P_{Cl_2}}$ (c) $K_p = P^2_{CO}$ (d) $K_p = \dfrac{P_{H_2S}}{P_{H_2}}$

Strategy and Explanation: For the expression of the equilibrium constant, use this form: $K_p = \dfrac{P_{products}}{P_{reactants}}$; if a stoichiometric coefficient precedes a species, that number is used as the mathematical power of the pressure in the expression. Remember that the equilibrium is not affected by the quantity of any solids or liquids, so those materials do not show up in the equilibrium expression. Hence, in particular, the answer for (c) does not have $SiO_2(s)$, C(s), or SiC(s) in the expression, and the answer for (d) does not have $S_8(s)$ in the expression.

35. *Answer:* **3.0×10^{20}**

Strategy and Explanation: Given three chemical equations and their equilibrium constants, calculate the K for a reaction described.

Use the naming system described in Chapter 3 for binary compounds. Dinitrogen oxide is N_2O and dinitrogen tetraoxide is N_2O_4. Write equation for the desired equation.

$$N_2O(g) + O_2(g) \rightleftharpoons N_2O_4\ (g) \qquad K = ?$$

Follow the method shown in Problem-Solving Example 14.2, look at the desired equation and ascertain which of the given equations can provide reactants and products, then arrange them in a way that, when added together produce the desired equation as described for Hess' Law in Chapter 6. Use the information in Section 14.2 to identify how the equilibrium constants are affected by manipulating these equations. The reactant $N_2O(g)$ is only found in the first equation, but in that equation it is a product. We need to reverse the first reaction and take the reciprocal of the K value:

$$N_2O(g) \rightleftharpoons 2\ N_2(g) + O_2(g) \qquad K = 1/(1.2 \times 10^{-35})$$

The product $N_2O_4(g)$ is only found in the second equation, but in that equation it is a reactant. We need to reverse the second reaction and take the reciprocal of the K value:

$$2\ NO_2(g) \rightleftharpoons N_2O_4(g) \qquad K = 1/(4.6 \times 10^{-3})$$

The reactant in the equation above uses $NO_2(g)$, but that compound is not in the desired equation, so we need to recreate that compound. $NO_2(g)$ is only found in the third equation, but there's only one equation and we need two. We need to multiply the third reaction by two and square the K value:

$$N_2(g) + 2\ O_2(g) \rightleftharpoons 2\ NO_2(g) \qquad K = (4.1 \times 10^{-9})^2$$

The sum of the three equations, results in the mathematical product of their K values:

$$N_2O(g) \rightleftharpoons 2\ N_2(g) + O_2(g) \qquad K = 1/(1.2 \times 10^{-35})$$

$$2\ NO_2(g) \rightleftharpoons N_2O_4(g) \qquad K = 1/(4.6 \times 10^{-3})$$

$$+ \quad N_2(g) + 2\ O_2(g) \rightleftharpoons 2\ NO_2(g) \qquad K = (4.1 \times 10^{-9})^2$$

$$N_2O(g) + O_2(g) \rightleftharpoons N_2O_4\ (g) \qquad K = \frac{\left(4.1\times10^{-9}\right)^2}{\left(1.2\times10^{-35}\right)\left(4.6\times10^{-3}\right)} = 3.0 \times 10^{20}$$

37. *Answer:* **K_c = 0.0161**

Strategy and Explanation: Given a phase change reaction and the vapor pressure of the gaseous product, find the value of K_c.

Write the K_p expression for the reaction and plug in the known value. Then use the relationship between K_p and K_c (Equation 14.5), with $R = 0.08206\ \frac{L \cdot atm}{mol \cdot K}$, and Δn = change in the number of moles of gas in the reaction, to get K_c.

$$H_2O(\ell) \rightleftharpoons H_2O(g) \quad K_p = P_{H_2O(g)} = 0.467\ atm$$

(Assume temperature is known within ± 1 °C.) T = 80. °C + 273.15 = 353 K

$$\Delta n = 1\ mol\ H_2O\ gas\ product - 0\ mol\ gas\ reactants = 1$$

$$K_p = K_c(RT)^{\Delta n} \qquad K_p(RT)^{-\Delta n} = K_c$$

$$K_c = (0.467\ atm) \times \left\{\left(0.08206\ \frac{L \cdot atm}{mol \cdot K}\right) \times (353\ K)\right\}^{-1} = 0.0161$$

✓ *Reasonable Answer Check:* This $(RT)^{-1}$ factor makes K_c smaller than K_p.

Determining Equilibrium Constants (Section 14.3)

39. *Answer:* **(a) [CO] = 0.0071 mol/L, [$COCl_2$] = 0.00308 mol/L (b) 1.4 × 10^2**

Strategy and Explanation: Given an equation for a reaction, the initial concentration of the gaseous reactants and the equilibrium concentration of a gaseous product, find the value of K_c.

Following the procedure given in Section 14.3, write the equation and construct a reaction table. No products are initially present.

	$CO(g)$	+	$Cl_2(g)$	$\rightleftharpoons$	$COCl_2(g)$
initial conc. (mol/L)	0.0102		0.00609		0
change as reaction occurs (mol/L)	______		______		______
equilibrium conc. (mol/L)	______		0.00301		______

Describe the stoichiometric changes in terms of one variable, x.

	$CO(g)$	+	$Cl_2(g)$	$\rightleftharpoons$	$COCl_2(g)$
initial conc. (mol/L)	0.0102		0.00609		0
change as reaction occurs (mol/L)	– x		– x		+ x
equilibrium conc. (mol/L)	0.0102 – x		0.00609 – x		x

(a) Calculate the concentrations of the gases at equilibrium using equilibrium [$COCl_2$]

$$[Cl_2] = 0.00609 - x = 0.00301\ mol/L$$

$$x = 0.00308 \text{ mol/L}$$

$$[CO] = 0.0102 - x = 0.0102 \text{ mol/L} - 0.00308 \text{ mol/L} = [CO] = 0.0071 \text{ mol/L}$$

$$[COCl_2] = x = 0.00308 \text{ mol/L}$$

(b) Then write the K_c expression and plug the concentrations into the expression to get the value of K_c.

$$K_c = \frac{[COCl_2]}{[CO][Cl_2]} = \frac{(0.00308)}{(0.0071)(0.00301)} = 1.4 \times 10^2 = 0.080$$

✓ *Reasonable Answer Check:* The larger product concentrations suggest this reaction is product-favored, so it makes sense that the K_c is greater than 1.

41. *Answer:* **$K_c = 2.6 \times 10^{-9}$**

Strategy and Explanation: Use a method similar to that described in the solution to Question 39(b).

$$K_c = \frac{[Br_2][F_2]^5}{[BrCl_5]^2} = \frac{(0.0018) \times (0.0090)^5}{(0.0064)^2} = 2.6 \times 10^{-9}$$

43. *Answer:* **$K_p = 1.6$**

Strategy and Explanation: Given an equation for a reaction and the moles of the gaseous reactants and products at equilibrium in a known volume, find the value of K_p.

Write the K_c expression for the reaction, calculate the concentrations of the gases, and plug them into the expression to get the value of K_c. Use the relationship between K_p and K_c (also described in the solution to Question 37) to get K_p:

$$K_p = K_c(RT)^{\Delta n}$$

$$H_2(g) + CO_2(g) \rightleftharpoons H_2O(g) + CO(g)$$

$$K_c = \frac{[H_2O][CO]}{[H_2][CO_2]} = \frac{\left(\frac{0.11 \text{ mol}}{1.0 \text{ L}}\right) \times \left(\frac{0.11 \text{ mol}}{1.0 \text{ L}}\right)}{\left(\frac{0.087 \text{ mol}}{1.0 \text{ L}}\right) \times \left(\frac{0.087 \text{ mol}}{1.0 \text{ L}}\right)} = 1.6$$

$$T = 986 \text{ °C} + 273.15 = 1259 \text{ K}$$

$$\Delta n = 1 \text{ mol } H_2O(g) + 1 \text{ mol } CO(g) - 1 \text{ mol } H_2(g) - 1 \text{ mol } CO_2(g) = 0$$

$$K_p = (1.6) \times \left\{\left(0.08206 \frac{\text{L} \cdot \text{atm}}{\text{mol} \cdot \text{K}}\right) \times (1259 \text{ K})\right\}^0 = 1.6$$

✓ *Reasonable Answer Check:* The slightly larger moles of products suggest this reaction is slightly product-favored, so it makes sense that the K_c is larger than 1. The equal moles of gas-phase reactants and gas-phase products, which is responsible for the power in the $(RT)^0$ factor, makes K_p the same as K_c.

45. *Answer:* **(a) 1.0 (b) 1.0 (c) 1**

Strategy and Explanation: Given the equation describing the interconversion of two isomers and the concentrations of the gaseous reactant and product at equilibrium, find the value of K_c.

Write the K_c expression for the reaction and plug in the known values.

$$A(g) \rightleftharpoons B(g) \qquad K_c = \frac{[B]}{[A]}$$

(a)

$$K_c = \frac{0.74 \text{ mol/L}}{0.74 \text{ mol/L}} = 1.0$$

(b)

$$K_c = \frac{2.0 \text{ mol/L}}{2.0 \text{ mol/L}} = 1.0$$

(c)
$$K_c = \frac{0.01\ \text{mol/L}}{0.01\ \text{mol/L}} = 1$$

✓ *Reasonable Answer Check:* The change in temperature does not affect this chemical reaction, suggesting that the reactants and products have the same energy. That means that the rate of the forward reaction and the rate of the reverse reaction are

47. *Answer:* $\mathbf{K_p = 0.108}$

Strategy and Explanation: Given an equation for a reaction, the initial mass of solid reactant, the partial pressure of both products, and the percentage increase of a product at equilibrium, find the value of K_P.

Construct a reaction table. Put all known pressures in the table:

	$NH_4HS(s)$	⇌	$NH_3(g)$	+ $H_2S(g)$
initial pressure (atm)	solid		0.692	0.0532
change as reaction occurs (atm)	______		______	______
equilibrium pressure (atm)	______		______	______

Describe the stoichiometric changes in terms of one variable, x.

	$NH_4HS(s)$	⇌	$NH_3(g)$	+ $H_2S(g)$
initial conc. (mol/L)	solid		0.692	0.0532
change as reaction occurs (mol/L)	– x		+x	+x
equilibrium conc. (mol/L)	solid		0.692 + x	0.0532 + x

Use equilibrium change in pressure to solve for x. The partial pressure of ammonia increases by 12.4%.

$$\frac{12.4\ \text{atm}}{100\ \text{atm}} \times 0.692\ \text{atm} = 0.0858\ \text{atm} = \text{x}$$

Use the value of x to find the equilibrium concentrations of the other substances:

$$P_{NH_3} = 0.692 + \text{x} = 0.692 + 0.0858 = 0.778\ \text{atm}$$

$$P_{H_2S} = 0.0532 + \text{x} = 0.0532 + 0.0858 = 0.1390\ \text{atm}$$

Then write the K_P expression and plug the concentrations into the expression to get the value of K_P.

$$K_P = P_{NH_3}P_{H_2S} = (0.778)(0.1390) = 0.108$$

✓ *Reasonable Answer Check:* The small pressures suggest this reaction is reactant-favored, so it makes sense that the K_P is smaller than 1.

48. *Answer:* $\mathbf{K_c = 0.075}$

Strategy and Explanation: Given an equation for a reaction, the initial moles of the gaseous reactant in a known volume, and the equilibrium concentration of the gaseous reactant, find the value of K_c.

Following the procedure given in Section 14.3, write the equation and construct a reaction table. Put all known concentrations in the table. There is no $NO_2(g)$ present, initially. The equilibrium $[N_2O_4] = 0.00090$ M

$$(\text{conc. } N_2O_4) = 0.010\ \text{mol } N_2O_4/2.0\ \text{L} = 0.0050\ \text{M}$$

	$N_2O_4(g)$	⇌	$2\ NO_2(g)$
initial conc. (M)	0.0050		0
change as reaction occurs (M)	______		______
equilibrium conc. (M)	0.00090		______

Describe the stoichiometric changes in terms of one variable, x.

	$N_2O_4(g)$	⇌	$2\ NO_2(g)$
initial conc. (M)	0.0050		0
change as reaction occurs (M)	– x		+ 2x
equilibrium conc. (M)	0.00090 = 0.0050 – x		0 + 2x

Calculate the concentrations of the gases at equilibrium using what you know about the equilibrium concentration of the reactant.

Use equilibrium $[N_2O_4]$ to solve for x: 0.00090 M = 0.0050 M – x

$$x = 0.0050\text{ M} - 0.00090\text{ M} = 0.0041\text{ M}$$

Use the value of x to find the equilibrium $[NO_2]$: 2x = 2(0.0041 M) = 0.0082 M

Then write the K_c expression and plug the concentrations into the expression to get the value of K_c.

$$K_c = \frac{[NO_2]^2}{[N_2O_4]} = \frac{(2x)^2}{0.0050 - x} = \frac{(0.0082)^2}{0.00090} = 0.075$$

✓ *Reasonable Answer Check:* The smaller product concentration suggests this reaction is reactant-favored, so it makes sense that the K_c is smaller than 1.

50. *Answer:* **$K_c = 75$**

Strategy and Explanation: Use a method similar to that described in the first part of the solution to Question 43.

$$CO(g) + Cl_2(g) \rightleftharpoons COCl_2(g) \qquad K_c = \frac{[COCl_2]}{[CO][Cl_2]}$$

We will assume the volume is measured with a precision of ± 1 L.

$$K_c = \frac{\left(\frac{9.00\text{ mol}}{50.\text{ L}}\right)}{\left(\frac{3.00\text{ mol}}{50.\text{ L}}\right) \times \left(\frac{2.00\text{ mol}}{50.\text{ L}}\right)} = 75$$

The larger moles of products suggest this reaction is slightly product-favored, so it makes sense that the K_c is larger than 1.

52. *Answer:* **3.9×10^{-4}**

Strategy and Explanation: Given an equation for a reaction, the initial moles of the gaseous reactant and the percent dissociation at equilibrium, find the value of K_c.

Following the procedure given in Section 14.3, write the equation and construct a reaction table. No products are present, initially.

$$(\text{conc. } Br_2) = 0.086\text{ mol } Br_2/1.26\text{ L} = 0.068\text{ mol/L}$$

And describe the stoichiometric changes in terms of one variable, x.

	$Br_2(g)$	$\rightleftharpoons$	2 Br(g)
initial conc. (M)	0.068		0
change as reaction occurs (M)	– x		+ 2x
equilibrium conc. (M)	0.068 – x		2x

Calculate the concentrations of the Br_2 gas at equilibrium using what you know about the percent dissociation. Given that 3.7% Br_2 dissociates, x = 0.037 × (0.068 M) = 0.0025

Write the expression for K_c and plug in the concentrations from part (a)

$$K_c = \frac{[Br]^2}{[Br_2]} = \frac{(2x)^2}{(0.068 - x)} = \frac{(2 \times 0.0025)^2}{(0.068 - 0.0025)} = 3.9 \times 10^{-4}$$

✓ *Reasonable Answer Check:* The small percent ionization causes smaller product concentrations. Both of these are consistent with a reactant-favored reaction. It also makes sense that the K_c is smaller than 1.

54. *Answer:* **1.4×10^3**

Strategy and Explanation: Given an equation for a reaction, the initial concentrations of the gaseous reactants and the concentration of one reactant at equilibrium, find the value of K_c.

Following the procedure given in Section 14.3, write the equation and construct a reaction table. No products are present, initially.

Describe the stoichiometric changes in terms of one variable, x.

	$2\ SO_2(g)$	+	$O_2(g)$	$\rightleftharpoons$	$2\ SO_3(g)$
initial conc. (M)	0.0076		0.0036		0
change as reaction occurs (M)	– 2x		– x		+2x
equilibrium conc. (M)	0.0076 – 2x		0.0036 – x		2x

Use equilibrium $[SO_2]$, to find the value of x, then use x to determine the other equilibrium concentrations.

$$[SO_2] = 0.0032 = 0.0076 - 2x$$

$$2x = 0.0044$$

$$x = 0.0022\ M$$

$$[O_2] = 0.0036\ M - 0.002\ M = 0.0014\ M$$

$$[SO_3] = 0.0044\ M$$

Write the expression for K_c and plug in the concentrations from part (a)

$$K_c = \frac{[SO_3]^2}{[SO_2]^2[O_2]} = \frac{(0.0044)^2}{(0.0032)^2 \times (0.0014)} = 1.4 \times 10^3$$

✓ *Reasonable Answer Check:* The reduction in the concentration of a reactant causes relatively large product concentrations. Both of these are consistent with a product-favored reaction. It also makes sense that the K_c is larger than 1.

The Meaning of the Equilibrium Constant (Section 14.4)

55. *Answer:* **Reactions (b) and (c) are product-favored. Most reactant-favored (a), then (b), then (c).**

Strategy and Explanation: Given a table of equations with values of K_c and K_p, order the members of a set or equations from most reactant-favored to most product-favored.

Look up the given chemical equation or a related chemical equation, determine the size of its K_c, using techniques described at the end of Section 14.2, as needed. K_c values larger than 1 are product-favored. The smaller K_c is, the more reactant-favored the reaction is. The larger K_c is, the more product-favored the reaction is. Order the equations from smallest K_c to largest K_c.

(a) $2\ NH_3(g) \rightleftharpoons N_2(g) + 3\ H_2(g)$

is the reverse of the third reaction in Table 14.1, so $K_{c,(a)} = 1/K_c = 1/(3.5 \times 10^8) = 2.9 \times 10^{-9}$.

K_c is smaller than 1, so the reaction is not product-favored.

(b) $NH_4^+(aq) + OH^-(aq) \rightleftharpoons NH_3(aq) + H_2O(\ell)$

is the reverse of the twelfth reaction in Table 14.1, so $K_{c,(b)} = 1/K_c = 1/(1.8 \times 10^{-5}) = 5.6 \times 10^4$.

K_c is larger than 1, so the reaction is product-favored.

(c) $2\ NO(g) \rightleftharpoons N_2(g) + O_2(g)$ is the reverse of the fourth reaction in Table 14.1, so

$K_{c,(c)} = 1/K_c = 1/(4.5 \times 10^{-31}) = 2.2 \times 10^{30}$. K_c is larger than 1, so the reaction is product-favored.

Therefore, the order is (a) 2.9×10^{-9}, then (b) 5.6×10^4, then (c) 2.2×10^{30}.

✓ *Reasonable Answer Check:* Reversing a reaction takes what were the products and makes them the reactants and vice versa, so the reverse of a product-favored reaction will be a reactant-favored reaction and vice versa. The values of K_c make sense. Comparing the values of K to determine which is the most reactant-favored and product-favored among reaction with different stoichiometric relationships is not always legitimate. Here, while the denominators all have two concentration values, we find four concentration values in the numerator of (a), only two in (c), and only one in (b). These differences will affect the size of K_c dramatically. In general, one should compare the values of K_c only among reactions with the same stoichiometric relationships (i.e., with the same number of reactants and products.)

57. *Answer:* **(a), (b), and (c)**

Strategy and Explanation: Given chemical equations and their associated K value, determine which of them favors reactants.

A reaction favors reactants if the value of K is less than 1, because the concentrations of the reactant would be larger than those of the products, making the fraction [prod]/[react] < 1. The reactions in (a), (b) and (c) all have K < 1, so they favor reactants.

58. *Answer:* **(a) see reactions and equations below (b) Ag_2SO_4 (c) Ag_2S**

Strategy and Explanation: Dissolving ionic compounds produces aqueous solutions of ions as described in Chapters 2 and 3. Water is not explicitly included in the dissolving equation, because its only function is as the solvent to stabilize the aqueous products via intermolecular forces. Write equilibrium expressions as was done in the solution to Question 14. Assess solubility by comparing the size of K_c as was done in the solution to Question 55.

(a) $Ag_2SO_4(s) \rightleftharpoons 2\ Ag^+(aq) + SO_4^{2-}(aq)$ $\qquad K_c = [Ag^+]^2[SO_4^{2-}] = 1.7 \times 10^{-5}$

$Ag_2S(s) \rightleftharpoons 2\ Ag^+(aq) + S^{2-}(aq)$ $\qquad K_c = [Ag^+]^2[S^{2-}] = 6 \times 10^{-30}$

(b) When the dissociation stoichiometry is comparable, as it is here, the compound that is more soluble has the larger K_c, and that is $Ag_2SO_4(s)$.

(c) When the dissociation stoichiometry is comparable, as it is here, the compound that is least soluble has the smaller K_c, and that is $Ag_2S(s)$.

Using Equilibrium Constants (Section 14.5)

60. *Answer:* **(a) see table below (b) $\frac{0.100 + x}{0.100 - x} = 2.5, x = 0.043$ (c) [2-methylpropane] = 0.024 M, [butane] = 0.010 M**

Strategy and Explanation: Given an equation for a reaction, the initial concentration or moles of the gaseous reactant in a known volume, and the value of the equilibrium constant, determine the change in the concentrations and the equilibrium concentrations of the reactants and products.

Follow the procedure given in Section 14.5. When necessary, find Q to determine direction of the reaction. Then write the equation and construct a reaction table. Describe the stoichiometric changes in terms of one variable, x, the change in the concentration of butane. Write the K_c expression for the reaction in terms of x. Solve that equation to get x, and use it to calculate the concentrations of the gases at equilibrium.

$$\text{butane (g)} \rightleftharpoons \text{2-methylpropane (g)}$$

(a) (conc. butane) = (conc. 2-methylpropane) = 0.100 M

$$Q_c = \frac{\text{(conc. 2 - methylpropane)}}{\text{(conc. butane)}} = \frac{0.100\ M}{0.100\ M} = 1.00 < 2.5$$

$Q_c < K_c$, so reaction goes from reactants toward products:

	butane (g) $\rightleftharpoons$	2-methylpropane (g)
initial conc. (M)	0.100	0.100
change as reaction occurs (M)	– x	+ x
equilibrium conc. (M)	0.100 – x	0.100 + x

(b) $$K_c = \frac{[\text{2-methylpropane}]}{[\text{butane}]} = 2.5$$

At equilibrium $$\frac{0.100 + x}{0.100 - x} = 2.5$$

$$0.100 + x = 2.5(0.100 - x) = 0.25 - 2.5x$$

$$x + 2.5x = 0.25 - 0.100$$

$$3.5x = 0.15$$

$$x = 0.043$$

(c) (conc. butane) = 0.017 mol butane/0.50 L = 0.034 M

	butane (g) $\rightleftharpoons$	2-methylpropane (g)
initial conc. (M)	0.034	0
change as reaction occurs (M)	– x	+ x
equilibrium conc. (M)	0.034 – x	x

At equilibrium $$\frac{x}{0.034 - x} = 2.5$$

$$x = 2.5(0.034 - x) = 0.085 - 2.5x$$

$$x + 2.5x = 0.085$$

$$3.5x = 0.085$$

$$x = 0.024 \text{ M} = [\text{2-methylpropane}]$$

$$[\text{butane}] = 0.034 \text{ M} - 0.024 \text{ M} = 0.010 \text{ M}$$

✓ *Reasonable Answer Check:* Substituting the equilibrium concentrations into the equilibrium expression will reproduce K_c:

Part (b), [butane] = 0.100 – 0.043 = 0.057 M, [2-methylpropane] = 0.100 + 0.043 = 0.143 M. $K_c = \frac{0.143}{0.057} = 2.5$.

(c) $K_c = \frac{0.024}{0.010} = 2.4$. These are both right, within the uncertainty of the data.

62. *Answer:* **3.39 g**

Strategy and Explanation: Given an equation for a reaction, the initial mass of the gaseous reactant in a known volume, and the value of the equilibrium constant, determine the mass of the reactant present at equilibrium.

Convert grams to moles using the molar mass. Then follow the method described in the solution to Question 60 to get the equilibrium concentration of the reactant gas. Then use the volume and molar mass to determine the grams:

$$\text{cyclohexane (g)} \rightleftharpoons \text{methylcyclopentane (g)}$$

$$(\text{conc. cyclohexane}) = \frac{3.79 \text{ g cyclohexane}}{2.80 \text{ L}} \times \frac{1 \text{ mol cyclohexane}}{84.15 \text{ g cyclohexane}} = 0.0161 \text{ M}$$

	cyclohexane (g) $\rightleftharpoons$	methylcyclopentane (g)
initial conc. (M)	0.0161	0
change as reaction occurs (M)	– x	+ x
equilibrium conc. (M)	0.0161 – x	x

At equilibrium $$K_c = \frac{[\text{methylcyclopentane}]}{[\text{cyclohexane}]} = 0.12$$

$$\frac{x}{0.0161 - x} = 0.12$$

$$x = 0.12(0.0161 - x) = 0.0019 - 0.12x$$

$$x + 0.12x = 0.0019$$

$$1.12x = 0.0019$$

$$x = 0.0017 \text{ M} = [\text{methylcyclopentane}]$$

$$[\text{cyclohexane}] = 0.0161 \text{ M} - 0.0017 \text{ M} = 0.0144 \text{ M}$$

$$2.80 \text{ L} \times \frac{0.0144 \text{ mol cyclohexane}}{1 \text{ L}} \times \frac{84.15 \text{ g cyclohexane}}{1 \text{ mol cyclohexane}} = 3.39 \text{ g}$$

✓ *Reasonable Answer Check:* The quantity present at equilibrium is less than the initial quantity. The equilibrium concentrations can be used to recreate the value of the equilibrium constant: $K_c = \frac{0.0017}{0.0144} = 0.12$.

64. *Answer:* **[HI] = 4 × 10^{-2} M, [I_2]= 5 × 10^{-3} M, [H_2] = 5 × 10^{-3} M**

Strategy and Explanation: Use a method similar to that described in the solution to Question 62.

(conc. HI) = 0.05 mol/1.0 L = 0.05 M

	H_2 (g)	+ I_2 (g)	⇌ 2 HI (g)
initial conc. (M)	0	0	0.05
change as reaction occurs (M)	+ x	+ x	– 2x
equilibrium conc. (M)	x	x	0.05– 2x

At equilibrium $$K_c = \frac{[HI]^2}{[H_2][I_2]} = \frac{(0.05 - 2x)^2}{(x)(x)} = 76$$

Take the square root of each side: $$\frac{(0.05 - 2x)}{x} = 8.7$$

$$0.05 - 2x = 8.7x$$

$$0.05 = 8.7x + 2x = 10.7x$$

$$x = 5 \times 10^{-3} \text{ M} = [H_2] = [I_2]$$

$$[HI] = 0.05 - 2x = 0.05 \text{ M} - 2 \times (5 \times 10^{-3} \text{ M}) = 4 \times 10^{-2} \text{ M}$$

✓ *Reasonable Answer Check:* The larger concentration of product is consistent with a K greater than 1. Substituting the equilibrium concentrations into the equilibrium expression will reproduce K_c within the precision of the data: $K_c = \frac{(4 \times 10^{-2})^2}{(5 \times 10^{-3}) \times (5 \times 10^{-3})} = 6 \times 10^1$.

66. *Answer:* **[Br_2] = [F_2] =0.047 M, [BrF] = 0.347 M**

Strategy and Explanation: Use a method similar to that described in the solution to Question 63(b).

	Br_2(g)	+ F_2(g)	⇌ 2 BrF(g)
initial conc. (mol/L)	0.220	0.220	0
change as reaction occurs (mol/L)	– x	– x	+ 2x
equilibrium conc. (mol/L)	0.220 – x	0.220 –x	2x

At equilibrium $$K_c = \frac{[BrF]^2}{[Br_2][F_2]} = \frac{(2x)^2}{(0.220 - x)^2} = 55.3$$

Take the square root of each side: $$\frac{(2x)}{(0.220 - x)} = \sqrt{55.3} = 7.44$$

Solve for x: $$7.44(0.220 - x) = 2x$$

$$1.64 = 2x + 7.44x = (2 + 7.44)x = 9.44x$$

$$1.64 = 9.44x$$

$$x = 0.173 \text{ M}$$

$$0.220 - x = 0.220 - 0.173 = 0.047 \text{ M} = [Br_2] = [F_2]$$

$$[BrF] = 2 \times (0.173) = 0.347 \text{ M}$$

✓ *Reasonable Answer Check:* The large concentration of product is consistent with a K greater than 1.

67. *Answer:* **(a) 1.94 mol (b) 1.92 mol (c) 1.98 mol**

Strategy and Explanation: Use a method similar to that described in the solution to Question 62.

(a) $(\text{conc. } I_2) = 1.00 \text{ mol}/10.00 \text{ L} = 0.100 \text{ M}$

$(\text{conc. } H_2) = 3.00 \text{ mol}/10.00 \text{ L} = 0.300 \text{ M}$

	H_2 (g)	+ I_2 (g)	⇌	2 HI (g)
initial conc. (M)	0.300	0.100		0
change as reaction occurs (M)	– x	– x		+ 2x
equilibrium conc. (M)	0.300 – x	0.100 – x		2x

At equilibrium $$K_c = \frac{[HI]^2}{[H_2][I_2]} = \frac{(2x)^2}{(0.300 - x)(0.100 - x)} = 50.0$$

Set up to use the quadratic equation (see Appendix A, Section A.7, page A.13):

$$50.0\{(0.300)(0.100) - (0.300 + 0.100)x + x^2\} = 4x^2$$

$$1.50 - 20.0x + 50.0x^2 = 4x^2$$

$$1.50 - 20.0x + 46.0x^2 = 0$$

Plug into the quadratic equation:

$$x = \frac{-b \pm \sqrt{b^2 - 4ac}}{2a} = \frac{20.0 \pm \sqrt{(-20.0)^2 - 4(46.0)(1.50)}}{2(46.0)}$$

The two roots are:

$$x = \frac{20.0 - 11.1}{2(46.0)} = \frac{8.9}{92.0} = 0.097$$

$$x = = \frac{20.0 + 11.1}{2(46.0)} = \frac{31.1}{92.0} = 0.338$$

Notice: Both of these roots are mathematically sound solutions for the quadratic equation, but the second of these roots must be discarded because it is too large and produces a nonsensical result. In particular, here, the equilibrium concentration of I_2 (0.100 – x) must be a positive number, but using the second root produces a negative answer, 0.100 – 0.338 = – 0.238.

$$x = 0.097 \text{ M}$$

$$2x = 2(0.097 \text{ M}) = 0.194 \text{ M} = [HI]$$

$$10.00 \text{ L} \times \frac{0.194 \text{ mol}}{\text{L}} = 1.94 \text{ mol}$$

$$[H_2] = 0.300 - x = 0.300 \text{ M} - (0.097 \text{ M}) = 0.203 \text{ M}$$

$$[I_2] = 0.100 - x = 0.100 \text{ M} - (0.097 \text{ M}) = 0.003 \text{ M}$$

✓ *Reasonable Answer Check:* (a) The larger concentration of product is consistent with a K greater than 1. Substituting the equilibrium concentrations into the equilibrium expression will reproduce K_c within the precision of the data: $K_c = \frac{(0.194)^2}{(0.203) \times (0.003)} = 6 \times 10^1 \cong 50.0$ within the uncertainty of 1 significant figure.

(b)

$$(\text{conc. } I_2) = 1.00 \text{ mol}/5.00 \text{ L} = 0.200 \text{ M}$$
$$(\text{conc. } H_2) = 3.00 \text{ mol}/5.00 \text{ L} = 0.600 \text{ M}$$

	H_2 (g)	+ I_2 (g)	⇌ 2 HI (g)
initial conc. (M)	0.600	0.200	0
change as reaction occurs (M)	– x	– x	+ 2x
equilibrium conc. (M)	0.600 – x	0.200 – x	2x

At equilibrium $K_c = \dfrac{[HI]^2}{[H_2][I_2]} = \dfrac{(2x)^2}{(0.600-x)(0.200-x)} = 50.0$

Set up to use the quadratic equation (Appendix A, Section A.7, page A.13):

$$50.0\{(0.600)(0.200) - (0.600 + 0.200)x + x^2\} = 4x^2$$
$$6.00 - 40.0x + 46.0x^2 = 0$$

Plug into the quadratic equation:

$$x = \frac{-b \pm \sqrt{b^2 - 4ac}}{2a} = \frac{40.0 \pm \sqrt{(-40.0)^2 - 4(46.0)(6.00)}}{2(46.0)} = \frac{17.7}{92.0} = 0.192$$
$$2x = 0.384 \text{ M} = [HI]$$
$$5.00 \text{ L} \times \frac{0.384 \text{ mol}}{\text{L}} = 1.92 \text{ mol}$$

✓ *Reasonable Answer Check:* (b) The larger concentration of product is consistent with a K greater than 1. Substituting the equilibrium concentrations into the equilibrium expression will reproduce K_c within the precision of the data:

$$[H_2] = 0.600 - x = 0.600 \text{ M} - (0.192 \text{ M}) = 0.408 \text{ M}$$
$$[I_2] = 0.200 - x = 0.200 \text{ M} - (0.192 \text{ M}) = 0.008 \text{ M}$$

$K_c = \dfrac{(0.384)^2}{(0.408) \times (0.008)} = 5 \times 10^1 \cong 50.0$ with only one significant figure. The answers for (a) and (b) should be the same, because the volume dependence cancels due to the fact that the number of products and reactants are the same. Within the cited uncertainty (a) and (b) give the same results.

(c)

$$(\text{conc. } I_2) = 1.00 \text{ mol}/10.00 \text{ L} = 0.100 \text{ M}$$
$$(\text{conc. } H_2) = (3.00 \text{ mol} + 3.00 \text{ mol})/10.00 \text{ L} = 0.600 \text{ M}$$

	H_2 (g)	+ I_2 (g)	⇌ 2 HI (g)
initial conc. (M)	0.600	0.100	0
change as reaction occurs (M)	– x	– x	+ 2x
equilibrium conc. (M)	0.600 – x	0.100 – x	2x

At equilibrium $K_c = \dfrac{[HI]^2}{[H_2][I_2]} = \dfrac{(2x)^2}{(0.600-x)(0.100-x)} = 50.0$

Set up to use the quadratic equation (Appendix A, Section A.7, page A.13):

$$50.0\{(0.600)(0.100) - (0.600 + 0.100)x + x^2\} = 4x^2$$
$$3.00 - 35.0x + 46.0x^2 = 0$$

Plug into the quadratic equation:

$$x = \frac{-b \pm \sqrt{b^2 - 4ac}}{2a} = \frac{35.0 \pm \sqrt{(-35.0)^2 - 4(46.0)(3.00)}}{2(46.0)} = \frac{9.1}{92.0} = 0.099$$
$$2x = 0.198 \text{ M} = [HI]$$

$$10.00\ \text{L} \times \frac{0.198\ \text{mol}}{\text{L}} = 1.98\ \text{mol}$$

✓ *Reasonable Answer Check:* (c) The larger concentration of product is consistent with a K greater than 1. Substituting the equilibrium concentrations into the equilibrium expression will reproduce K_c within the precision of the data:

$$[H_2] = 0.300 - x = 0.600\ \text{M} - (0.099\ \text{M}) = 0.501\ \text{M}$$

$$[I_2] = 0.100 - x = 0.100\ \text{M} - (0.099\ \text{M}) = 0.001\ \text{M}$$

$K_c = \dfrac{(0.198)^2}{(0.501)\times(0.001)} = 7\times10^1 \cong 50.0$. With only one significant figure round off errors have started to dramatically affect the smallest number, $[I_2]$. These results still look reasonable.

69. *Answer:* **(a) $[CO] = [H_2O] = 0.95$ M, $[CO_2] = [H_2] = 0.0489$ M (b) $[CO] = [H_2O] = 1.90$ M, $[CO_2] = [H_2] = 0.0977$ M**

Strategy and Explanation: Use a method similar to that described in the solution to Question 62.

(a) (conc. CO) = 1.00 mol/1.00 L = 1.00 M and (conc. H_2O) = 1.00 mol/1.00 L = 1.00 M

	CO (g)	+ H_2O(g)	⇌ CO_2 (g)	+ H_2 (g)
initial conc. (M)	1.00	1.00	0	0
change as reaction occurs (M)	– x	– x	+ x	+ x
equilibrium conc. (M)	1.00 – x	1.00 – x	x	x

At equilibrium $K_c = \dfrac{[CO_2][H_2]}{[CO][H_2O]} = \dfrac{(x)(x)}{(1.00-x)(1.00-x)} = 2.64\times10^{-3}$

Take the square root of each side: $\dfrac{(x)}{(1.00-x)} = 0.0514$

$$x = 0.0514 - 0.0514x$$

$$x + 0.0514x = 0.0514$$

$$(1.0514)x = 0.0514$$

$$x = 0.0489\ \text{M} = [CO_2] = [H_2]$$

$$[CO] = [H_2O] = 1.00\ \text{M} - x = 1.00\ \text{M} - 0.0489\ \text{M} = 0.95\ \text{M}$$

✓ *Reasonable Answer Check:* The smaller concentration of products is consistent with a K smaller than 1. Substituting the equilibrium concentrations into the equilibrium expression will reproduce K_c within the precision of the data: $K_c = \dfrac{(0.0489)\times(0.0489)}{(0.95)\times(0.95)} = 2.6\times10^{-3}$.

(b) (conc. CO) = (1.00 mol + 1.00 mol)/1.00 L = 2.00 M

(conc. H_2O) = (1.00 mol + 1.00 mol)/1.00 L = 2.00 M

	CO (g)	+ H_2O(g)	⇌ CO_2 (g)	+ H_2 (g)
initial conc. (M)	2.00	2.00	0	0
change as reaction occurs (M)	– x	– x	+ x	+ x
equilibrium conc. (M)	2.00 – x	2.00 – x	x	x

At equilibrium $K_c = \dfrac{[CO_2][H_2]}{[CO][H_2O]} = \dfrac{(x)(x)}{(2.00-x)(2.00-x)} = 2.64\times10^{-3}$

Take the square root of each side: $\dfrac{(x)}{(2.00-x)} = 0.0514$

Solve for x $x = 2.00(0.0514) - 0.0514x$

$$x + 0.0514x = 0.103$$

$$(1.0514)x = 0.103$$

$$x = 0.0977 \text{ M} = [CO_2] = [H_2]$$

$$[CO] = [H_2O] = 2.00 \text{ M} - x = 2.00 \text{ M} - 0.0977 \text{ M} = 1.90 \text{ M}$$

✓ *Reasonable Answer Check:* The smaller concentration of products is consistent with a K smaller than 1. Substituting the equilibrium concentrations into the equilibrium expression will reproduce K_c within the precision of the data: $K_c = \frac{(0.0977)\times(0.0977)}{(1.90)\times(1.90)} = 2.64 \times 10^{-3}$

71. *Answer:* **(a) $[CO] = [Br_2] = 0.143$ M, $[COBr_2] = 0.107$ M (b) 57.1%**

Strategy and Explanation: Given an equation for a reaction, the initial moles of the gaseous reactant in a known volume, and the value of the equilibrium constant, determine the equilibrium concentrations and calculate the percent decomposition.

(a) Use a method similar to that described in the solution to Question 62 to find the equilibrium concentrations. $K_c = 0.190$

$$(\text{conc. } COBr_2) = 0.500 \text{ mol}/2.00 \text{ L} = 0.250 \text{ M}$$

	$COBr_2(g)$ ⇌	$CO(g)$	+ $Br_2(g)$
initial conc. (M)	0.250	0	0
change as reaction occurs (M)	– x	+ x	+ x
equilibrium conc. (M)	0.250 – x	x	x

At equilibrium $K_c = \frac{[CO][Br_2]}{[COBr_2]} = \frac{(x)^2}{(0.250 - x)} = 0.190$

Set up to use the quadratic equation (Appendix A, Section A.7, page A.13):

$$x^2 = (0.250)(0.190) - 0.190x$$

$$x^2 + 0.190x - 0.0475 = 0$$

Plug into the quadratic equation:

$$x = \frac{-b \pm \sqrt{b^2 - 4ac}}{2a} = \frac{-0.190 \pm \sqrt{(0.190)^2 - 4(1)(-0.0475)}}{2(1)} = \frac{-0.190 + 0.4755}{2} = 0.143$$

$$[CO] = [Br_2] = 0.143 \text{ M}$$

$$[COBr_2] = 0.250 \text{ M} - x = 0.250 \text{ M} - 0.143 \text{ M} = 0.107 \text{ M}$$

(b) The percent decomposition is the calculated by dividing the amount of reactant decomposed by the initial quantity and multiplying by 100%:

$$\text{Percentage } COBr_2 \text{ decomposed} = \frac{\text{amount of } COBr_2 \text{ converted}}{\text{initial amount of } COBr_2} \times 100\ \%$$

$$= \frac{0.107 \text{ M}}{0.250 \text{ M}} \times 100\ \% = 57.1\ \%$$

✓ *Reasonable Answer Check:* The large concentrations of products are consistent with a K greater than 1. Substituting the equilibrium concentrations into the equilibrium expression will reproduce K_c within the limits of the significant figures:

$$K_c = = \frac{(0.143)^2}{(0.107)} = 0.191.$$

72. *Answer:* **(a) No (b) proceeds toward products**

Strategy and Explanation: Given an equation for a reaction, the initial moles of the gaseous reactants and products in a known volume, and the value of the equilibrium constant, determine if the system is at equilibrium and, if it is not, determine which direction it must proceed to reach equilibrium.

Follow the procedure given in Section 14.5. Calculate Q and compare it with K. If Q = K, then the system is at equilibrium. If Q < K, the reaction proceeds toward products to reach equilibrium; if Q > K, the reaction proceeds toward reactants to reach equilibrium. $K_c = 3.58 \times 10^{-3}$

$$(\text{conc. } SO_3) = 0.15 \text{ mol}/10.0 \text{ L} = 0.015 \text{ M}$$

$(\text{conc. } SO_2) = 0.015 \text{ mol}/10.0 \text{ L} = 0.0015 \text{ M}$ $(\text{conc. } O_2) = 0.0075 \text{ mol}/10.0 \text{ L} = 0.00075 \text{ M}$

$$Q_c = \frac{(\text{conc. } SO_2)^2(\text{conc. } O_2)}{(\text{conc. } SO_3)^2} = \frac{(0.0015)^2(0.00075)}{(0.015)^2} = 7.5 \times 10^{-6}$$

(a) $Q_c \neq K_c$, so the reaction is not at equilibrium.

(b) $Q_c < K_c$, so the reaction must proceeds toward products to reach equilibrium.

✓ *Reasonable Answer Check:* Considering the powers of 10, the size of Q makes sense

$$\frac{(10^{-3})^2(10^{-4})}{(10^{-2})^2} = \frac{(10^{-6})(10^{-4})}{(10^{-4})} = 10^{-6}$$

73. *Answer:* **No; NO_2 increases**

Strategy and Explanation: Given an equation for a reaction, the initial moles of the gaseous reactants and products in a known volume, and the value of the equilibrium constant, determine if the system is at equilibrium and, if it is not, determine whether the reactant concentration increases or decreases as the system proceeds to equilibrium.

Follow the procedure given in Section 14.5. Calculate Q and compare it with K. If Q = K, then the system is at equilibrium. If Q < K, the reaction proceeds toward products to reach equilibrium; if Q > K, the reaction proceeds toward reactants to reach equilibrium. $K_c = 170$

$$(\text{conc. } NO_2) = 2.0 \times 10^{-3} \text{ mol}/10. \text{ L} = 2.0 \times 10^{-4} \text{ M}$$

$$(\text{conc. } N_2O_4) = 1.5 \times 10^{-3} \text{ mol}/10. \text{ L} = 1.5 \times 10^{-4} \text{ M}$$

$$Q_c = \frac{(\text{conc. } N_2O_4)}{(\text{conc. } NO_2)^2} = \frac{1.5 \times 10^{-4})}{(2.0 \times 10^{-4})^2} = 3.8 \times 10^3$$

$Q_c \neq K_c$, so the reaction is not at equilibrium.

$Q_c > K_c$, so the products must proceed toward reactants to reach equilibrium, and the concentration of the reactant, NO_2, increases.

✓ *Reasonable Answer Check:* Looking at the powers of 10, the size of Q makes sense: $(10^{-4})/(10^{-8}) = 10^4$. The increase in the reactant concentration and decrease of the product concentration lowers the size of the Q fraction bringing it close to the value of K.

76. *Answer:* **(a) No (b) proceed toward reactants (c) $[N_2] = [O_2] = 0.025$ M; [NO] = 0.0010 M**

Strategy and Explanation: Given an equation for a reaction, the initial moles of the gaseous reactant in a known volume, and the value of the equilibrium constant, determine if the system is at equilibrium and, if it is not, determine which direction it must proceed to reach equilibrium, and calculate the equilibrium concentrations.

Start with the procedure given in Section 14.5. Calculate Q and compare it to K to see if the system is at equilibrium, and, as needed, which direction the reaction proceeds in order to reach equilibrium. If the system is not already at equilibrium, use a method similar to that described in the solution to Question 62 to find the equilibrium concentrations. $K_c = 1.7 \times 10^{-3}$

$$(\text{conc. NO}) = 0.015 \text{ mol}/10.0 \text{ L} = 0.0015 \text{ M}$$

$(\text{conc. } N_2) = 0.25 \text{ mol}/10.0 \text{ L} = 0.025 \text{ M}$ $(\text{conc. } O_2) = 0.25 \text{ mol}/10.0 \text{ L} = 0.025 \text{ M}$

$$Q_c = \frac{(\text{conc. NO})^2}{(\text{conc. N}_2)(\text{conc. O}_2)} = \frac{(0.0015)^2}{(0.025)(0.025)} = 3.6\times10^{-3}$$

(a) $Q_c \neq K_c$, so the reaction is not at equilibrium.

(b) $Q_c > K_c$, so the reaction must proceeds toward reactants to reach equilibrium.

(c)

	$N_2(g)$	+ $O_2(g)$	$\rightleftharpoons$ 2 NO(g)
initial conc. (M)	0.025	0.025	0.0015
change as reaction occurs (M)	+ x	+ x	– 2x
equilibrium conc. (M)	0.025 + x	0.025 + x	0.0015 – 2x

At equilibrium $$K_c = \frac{[NO]^2}{[N_2][O_2]} = \frac{(0.0015-2x)^2}{(0.025+x)(0.025+x)} = 1.7\times10^{-3}$$

Take the square root of each side: $$\frac{(0.0015-2x)}{(0.025+x)} = \sqrt{1.7\times10^{-3}} = 0.041$$

Solve for x: $$0.0015 - 2x = 0.041(0.025 + x)$$

$$0.0015 - 2x = 0.001025 + 0.041x$$

$$0.0015 - 0.001025 = (2 + 0.041)x$$

$$x = 2.3 \times 10^{-4}\ \text{M}$$

$$[N_2] = [O_2] = 0.025 - x = 0.025 - 2.3 \times 10^{-4}\ \text{M} = 0.025\ \text{M}$$

$$[NO] = 0.0015 - 2x = 0.0015\ \text{M} - 2 \times (2.3 \times 10^{-4}\ \text{M}) = 0.0010\ \text{M}$$

✓ *Reasonable Answer Check:* The smaller concentrations of products are consistent with a Q > K and a K less than 1.

78. *Answer:* **(a) No (b) proceed toward reactants**

Strategy and Explanation: Given an equation for a reaction, the initial concentrations of the gaseous reactants and products, and the value of the equilibrium constant, determine if the system is at equilibrium and, if it is not, determine which direction it must proceed to reach equilibrium, and calculate the equilibrium concentrations.

Start with the procedure given in Section 14.5. Calculate Q and compare it to K to see if the system is at equilibrium, and, as needed, which direction the reaction proceeds in order to reach equilibrium. If the system is not already at equilibrium, use a method similar to that described in the solution to Question 62 to find the equilibrium concentrations. $K_c = 19.9$

$$Q_c = \frac{(\text{conc. ClF})^2}{(\text{conc. Cl}_2)(\text{conc. F}_2)} = \frac{(7.3)^2}{(0.50)(0.20)} = 5.3\times10^2$$

(a) $Q_c \neq K_c$, so the reaction is not at equilibrium.

(b) $Q_c > K_c$, so the reaction must proceeds toward reactants to reach equilibrium.

Shifting a Chemical Equilibrium: Le Chatelier's Principle (Section 14.6)

80. *Answer:* **(a) left (b) left (c) left (d) right**

Strategy and Explanation: Use the methods described in Section 14.6 to answer this question. Changing concentrations or pressures of reactants and products or changing the available energy in the system will take the system out of equilibrium. The response to that change will be a shift away from the increase. Because this reaction has a positive ΔH, heat energy is absorbed as the reaction goes from reactants to products.

(a) Adding more Br_2 increases the concentration of a product, so the equilibrium responds by shifting to the **left** to reduce the product concentration.

(b) Removing NOBr decreases the concentration of a reactant, so the equilibrium responds by shifting to the **left** to increase the reactant concentration.

(c) Decreasing the temperature takes energy away from the system, so the equilibrium responds by shifting to the **left** to release some energy.

(d) Increasing the volume of the container, lowers the concentrations of every gas-phase reactant. To raise the concentrations, the reaction shifts to the **right** because combining two reactant molecules makes three product molecules.

82. *Answer:* **Choice (b); heat energy is absorbed, increasing temperature drives reaction towards products.**

Strategy and Explanation: As described in Section 14.6, the positive ΔH indicates that heat energy is absorbed in this reaction. Increasing the temperature will drive the reaction toward products and away from N_2O_4. That means the N_2O_4 concentration will decrease.

84. *Answer:* **(a) left (b) right (c) no shift**

Strategy and Explanation: An expansion of the volume of a system decreases all the partial pressures. The system will respond by making more gas. Examine each equation and compare the number of moles of gas on each side of the equation. The equilibrium shifts to the side with greater number of gas molecules.

(a) This reaction has four gaseous reactants, 4 CO(g), and only one gaseous product, $Ni(CO)_4$(g), so the equilibrium **shifts to the left**.

(b) This reaction has one gaseous reactant, ClF_5(g), and two gaseous products, ClF_3 and F_2, so the equilibrium **shifts to the right**.

(c) This reaction has one gaseous reactant, HBr(g), and two halves of gaseous products, $\frac{1}{2}H_2$ and $\frac{1}{2}Br_2$, so the equilibrium will **not shift** as a result of this expansion.

85. *Answer:* **(a) reverse reaction (b) forward reaction (c) forward reaction**

Strategy and Explanation: Changing concentrations or pressures of reactants and products will take the system out of equilibrium. The response to that change will be a shift away from the increase, because the reaction rate will be temporarily faster for the reaction with the newly increased concentrations.

(a) Adding more of the product F_2 causes the **reverse reaction** rate to be faster than the forward reaction immediately after the change.

(b) Decreasing the amount of the product ClF_3 causes the reverse reaction rate to be slower than the forward reaction, so the **forward reaction** rate will be faster immediately after the change.

(c) Doubling the total volume of the system lowers all the concentrations by half. The rate of the forward reaction is first order, rate = k[ClF_5]. The rate of the reverse reaction is second order, rate = k[ClF_3][F_2]. Therefore the reverse reaction will be slowed more than the forward reaction when the concentrations drop and the **forward reaction** rate will be faster immediately after the change.

87. *Answer:* **see chart below**

Strategy and Explanation: Use the methods described in Section 14.6 to answer this question. Changing concentrations or pressures of reactants and products or changing the available energy in the system will take the system out of equilibrium. The response to that change will be a shift away from the increase. Changing the amounts of solids or liquids will not affect the equilibrium position. Only temperature can affect the value of the equilibrium constant. Because this reaction is exothermic, energy is released.

$H_2(g) + Br_2(g) \rightleftharpoons 2\ HBr(g)$ $\Delta H = -103.7$ kJ

Change	[Br_2]	[HBr]	K_c	K_p
Some H_2 *(a reactant)* is added to the container	decrease	increase	no change	no change
The temperature of the gases in the container is increased. *(increase in energy, energy released in the reaction)*	increase	decrease	decrease	decrease
The pressure of HBr (*a product*) is increased.	increase	increase (since extra is added)	no change	no change

89. *Answer:* **(a) no change (b) left (c) left**

Strategy and Explanation: Use the methods described in Section 14.6 to answer this question. Changing concentrations or pressures of reactants and products or changing the available energy in the system will take the system out of equilibrium. The response to that change will be to shift away from an increase (or shift towards a decrease). Changing the amounts of solids or liquids will not affect the equilibrium position. Because the reaction is endothermic, energy is absorbed.

$$2\ NOBr(g) \rightleftharpoons 2\ NO(g) + Br_2(\ell) \qquad \Delta H = 16.1\ kJ$$

(a) The equilibrium is not affected by the addition of a pure liquid, so adding more liquid Br_2 will cause **no change**.

(b) Removing reactant, NOBr, shifts the reaction toward the reactants (**left**) to increase the [NOBr].

(c) Decreasing the temperature removes energy. The reaction shifts toward the reactants (**left**), to evolve more energy.

91. *Answer:* **(a) right (b) left (c) right**

Strategy and Explanation: Use methods described in Section 14.6.

$$2\ NO(g) + O_2(s) \rightleftharpoons 2\ NO_2(g)$$

The reaction is exothermic.

(a) Adding more reactant, $O_2(g)$, shifts the reaction toward the products (**right**) to decrease the $[O_2]$.

(b) Adding more product, $NO_2(g)$, shifts the reaction toward the reactants (**left**) to decrease the $[NO_2]$.

(c) Decreasing the temperature removes energy. The reaction shifts toward the products (**right**), to evolve energy.

93. *Answer:* **(a) (i) (b) (ii) (c) (i) (d) (iii) (e) (iii)**

Strategy and Explanation: Use methods described in Section 14.6.

(a) Adding solid $BaCO_3$ causes no shift, so the answer is **(i)**.

(b) Adding the product, CO_2, causes a shift to the left, so the answer is **(ii)**.

(c) Adding solid BaO causes no shift, so the answer is **(i)**.

(d) Raising the temperature increases the heat energy, which leads to decomposition, so the answer is **(iii)**.

(e) Increasing the volume decreases the pressure, which leads to a lower concentration of the gaseous product CO_2, causing a shift to the right, so the answer is **(iii)**.

95. *Answer:* **(a) (1) increase (2) increase (3) no shift (4) increase (5) increase (b) No change increases K; Change (5) decreases K.**

Strategy and Explanation: Use methods described in the solutions to Section 14.6.

(a) (1) Removing reactant O_2 causes a shift to the **left** to replace some of the missing O_2. The shift toward reactants increases the amount of NH_3.

(2) Adding product N_2 causes a shift to the **left** to remove some of the new O_2. The shift toward reactants increases the amount of NH_3.

(3) Adding liquid water causes **no shift** in the position of equilibrium, so there is no change in the amount of NH_3.

(4) Expanding the volume of the container decreases the concentration of all the gases. The reactants have seven moles of gas and the products have two moles of gas, so the decrease in concentration will be more prominent on the reactant side, causing a shift to the **left** to increase the concentration. The shift toward reactants increases the amount of NH_3.

(5) Increasing the temperature on an exothermic reaction drives the reactions to the **left**. The shift toward reactants increases the amount of NH_3.

(b) Only changes in the temperature affect the value of K. The change described in **change (5)** affects the value of K. None of the changes increase the value of K. Section 4.6 explains, for an exothermic reaction, an increase in temperature always means a decrease in K_c. The reaction will become less product-favored at higher temperatures in Change (5) of part (a).

97. *Answer:* **(a) left (b)** $[PCl_5] = 0.0198$ M, $[PCl_3] = 0.0231$M, $[Cl_2] = 0.0403$ M

Strategy and Explanation: Given an equation for a reaction, the equilibrium mass of the gaseous reactants and products in a known volume and a mass of product subsequently added, determine how the addition of the product will affect the equilibrium and the equilibrium concentrations when equilibrium is re-established.

(a) Adding a product will shift the equilibrium to the **left**.

(b) First, calculate the equilibrium concentrations from the masses and the volume.

$$\frac{3.120 \text{ g } PCl_5}{1.00 \text{ L}} \times \frac{1 \text{ mol } PCl_5}{208.2388 \text{ g } PCl_5} = 0.0150 \text{ M } PCl_5$$

$$\frac{3.845 \text{ g } PCl_3}{1.00 \text{ L}} \times \frac{1 \text{ mol } PCl_3}{137.3328 \text{ g } PCl_3} = 0.0280 \text{ M } PCl_3$$

$$\frac{1.787 \text{ g } Cl_2}{1.00 \text{ L}} \times \frac{1 \text{ mol } Cl_2}{70.906 \text{ g } Cl_2} = 0.0252 \text{ M } Cl_2$$

Use a method similar to that described in the solution to Question 29 and Question 42 to calculate the equilibrium constant.

$$K_c = \frac{[PCl_3][Cl_2]}{[PCl_5]} = \frac{(0.0280 \text{ M}) \times (0.0252 \text{ M})}{(0.0150 \text{ M})} = 0.0471$$

Use a method similar to that described in the solution to Question 62 to find the equilibrium concentrations.

Total mass of $Cl_2 = 1.418 \text{ g} + 1.787 \text{ g} = 3.205 \text{ g } Cl_2$

$$(\text{conc. } Cl_2) = \frac{3.205 \text{ g } Cl_2}{1.00 \text{ L}} \times \frac{1 \text{ mol } Cl_2}{70.906 \text{ g } Cl_2} = 0.0452 \text{ M } Cl_2$$

	$PCl_5(g)$ $\rightleftharpoons$	$PCl_3(g)$ +	$Cl_2(g)$
initial conc. (M)	0.0150	0.0280	0.0452
change as reaction occurs (M)	+ x	– x	– x
equilibrium conc. (M)	0.0150 + x	0.0280 – x	0.0452 – x

At equilibrium $$K_c = \frac{[PCl_3][Cl_2]}{[PCl_5]} = \frac{(0.0280 - x)(0.0452 - x)}{(0.0150 + x)} = 0.0471$$

Set up to use the quadratic equation (Appendix A, Section A.7, page A.13):

$$(0.0280 - x)(0.0452 - x) = 0.0471(0.0150 + x)$$

$$(0.0280)(0.0452) - (0.0280 + 0.0452)x + x^2 = (0.0471)(0.0150) + 0.0471x$$

$$0.00127 - 0.07320x + x^2 = 0.000706 + 0.0471x$$

$$x^2 - 0.1203x - 0.00056 = 0$$

Plug into the quadratic equation:

$$x = \frac{-b \pm \sqrt{b^2 - 4ac}}{2a} = \frac{-(-0.1203) \pm \sqrt{(-0.1203)^2 - 4(1)(-0.00056)}}{2(1)} = \frac{0.1203 - 0.1106}{2} = 0.00485$$

$$[PCl_5] = 0.0150 \text{ M} + x = 0.0150 \text{ M} + 0.00485 \text{ M} = 0.0198 \text{ M}$$

$$[PCl_3] = 0.00280 \text{ M} - x = 0.0280 \text{ M} - 0.00485 \text{ M} = 0.0231\text{M}$$

$$[Cl_2] = 0.0452 \text{ M} - x = 0.0452 \text{ M} - 0.00485 \text{ M} = 0.0403 \text{ M}$$

✓ *Reasonable Answer Check:* Substituting the equilibrium concentrations into the equilibrium expression will reproduce K_c within the uncertainty of the significant figures:

$$K_c = = \frac{(0.0231)\,(0.0403)}{(0.0198)} = 0.0470$$

Equilibrium at the Nanoscale (Section 14.7)

99. *Answer:* **(a) energy effect (b) entropy effect (c) neither**

Strategy and Explanation: We will use methods described in Section 14.7 to answer this question.

The entropy effect is related to randomness and probability. The more random a system is, the more probable it is. In many of the chemical reactions we study, we can get a qualitative idea of the entropy change by looking at the physical phases of the reactants and products. In general, gases have more entropy than liquids, which have more entropy than solids. Aqueous substances have more entropy than solids, but less than gases.

The energy effect is related to the relative stability of the reactants and products. If the products are more stable than the reactants (exothermic, ΔH = negative), then the products are favored energetically. If the reactants are more stable than the products (endothermic, ΔH = positive), then the reactants are favored energetically.

(a) 4 mol gas $\rightleftharpoons$ 2 mol gas Entropy effect favors reactants, not products.

Reaction is exothermic, so energy effect favors products.

Reaction (a) (in the forward direction) is favored only by the energy effect.

(b) 1 mol gas $\rightleftharpoons$ 2 mol gas Entropy effect favors products.

Reaction is endothermic, so energy effect favors reactants, not products.

Reaction (b) (in the forward direction) is favored only by the entropy effect.

(c) 4 mol gas $\rightleftharpoons$ 2 mol gas Entropy effect favors reactants, not products.

Reaction is endothermic, so energy effect favors reactants, not products.

Reaction (c) (in the forward direction) is not favored by either effect.

101. *Answer:* **(a) Insufficient information is available (b) Greater than 1; product-favored; (c) Less than 1; reactant-favored.**

Strategy and Explanation: We will use methods described in the solution to Question 99 and in Section 14.7. If the reaction is favored by both effects, it is product-favored and K will be greater than 1. If the reaction is favored by neither effect, it is reactant-favored and K will be less than 1. If the reaction is only favored by one of the two effects, we have insufficient information available.

(a) 3 mol gas $\rightleftharpoons$ 2 mol gas Entropy effect favors reactants, not products.

Reaction is exothermic, so energy effect favors products.

We have insufficient information to judge the size of K for reaction (a) or which side of the reaction is favored at equilibrium.

(b) 2 mol gas $\rightleftharpoons$ 3 mol gas Entropy effect favors products.

Reaction is exothermic, so energy effect favors products.

Reaction (b) (in the forward direction) is favored by both effects, so it is product-favored and its K will be greater than 1.

(c) 4 mol gas $\rightleftharpoons$ 2 mol gas Entropy effect favors reactants, not products.

Reaction is endothermic, so energy effect favors reactants, not products.

Reaction (c) (in the forward direction) is not favored by either effect, so it is reactant-favored and K will be less than 1.

Controlling Chemical Reactions: The Haber-Bosch Process (Section 14.8)

103. *Answer/Explanation:* A reaction will only go significantly towards products if it is product-favored. As is described in Section 14.7 and 14.8, the change in entropy and the change in enthalpy both affect the whether a reaction is product-favored. An increase in entropy favors products and having products lower in energy favors products. In some instances, these two quantities have the opposite signs, in which cases the reaction is either always product-favored or never product-favored:

Sign of entropy change	Sign of enthalpy change	Product favored?
+	–	Yes
–	+	No

In some instances, these two quantities have the same signs, in which cases the reaction is product-favored only at some temperatures. If the reaction is favored only by an entropy effect, then it needs high temperatures to assist the endothermic process. If the reaction is favored only by an energy effect, then it needs low temperatures to keep the randomness at a minimum:

Sign of entropy change	Sign of enthalpy change	Product favored?
+	+	Only at high temperatures
–	–	Only at low temperatures

105. *Answer:* **(a) Step 1: –296.830 kJ/mol, Step 2: –197.78 kJ/mol, Step 3: –132.44 kJ/mol (b) all three steps are exothermic (c) None have entropy increase, Steps 2 and 3 have entropy decrease, the entropy in Step 1 stays about the same. (d) All three steps have products favored more at low temperatures.**

Strategy and Explanation:

(a) Follow the method described in Chapter 6. Look up the ΔH°_f values in Appendix J.

Step 1: $\Delta H^\circ = (1 \text{ mol}) \times \Delta H^\circ_f(SO_2) - (1 \text{ mol}) \times \Delta H^\circ_f(S) - (1 \text{ mol}) \times \Delta H^\circ_f(O_2)$

$\Delta H^\circ = (1 \text{ mol}) \times (-296.830 \text{ kJ/mol}) - (1 \text{ mol}) \times (0 \text{ kJ/mol}) - (1 \text{ mol}) \times (0 \text{ kJ/mol}) = -296.830 \text{ kJ/mol}$

Step 2: $\Delta H^\circ = (2 \text{ mol}) \times \Delta H^\circ_f(SO_3) - (2 \text{ mol}) \times \Delta H^\circ_f(SO_2) - (1 \text{ mol}) \times \Delta H^\circ_f(O_2)$

$\Delta H^\circ = (2 \text{ mol}) \times (-395.72 \text{ kJ/mol}) - (2 \text{ mol}) \times (-296.830 \text{ kJ/mol}) - (1 \text{ mol}) \times (0 \text{ kJ/mol})$
$= -197.78 \text{ kJ/mol}$

Step 3: $\Delta H^\circ = (1 \text{ mol}) \times \Delta H^\circ_f(H_2SO_4(\ell)) - (1 \text{ mol}) \times \Delta H^\circ_f(SO_3) - (1 \text{ mol}) \times \Delta H^\circ_f(H_2O(\ell))$

$\Delta H^\circ = (1 \text{ mol}) \times (-813.989 \text{ kJ/mol}) - (1 \text{ mol}) \times (-395.72 \text{ kJ/mol})$
$- (1 \text{ mol}) \times (-285.830 \text{ kJ/mol}) = -132.44 \text{ kJ/mol}$

(b) All three reactions are exothermic (ΔH = negative)

(c) Follow the method described in the solution to Question 99.

Step 1: 1 mol gas $\rightleftharpoons$ 1 mol gas — No large entropy change.

Step 2: 3 mol gas $\rightleftharpoons$ 2 mol gas — Entropy decreases.

Step 3: 1 mol gas $\rightleftharpoons$ 0 mol gas — Entropy decreases.

(d) Heat is produced in all three steps. A low temperature in each step would serve to reduce the energy and drive the reaction toward products. All three steps would favor products more at lower temperatures.

General Questions

107. *Answer:* **First Reaction:** $K_c = [H^+][OH^-]$ **Second Reaction:** $K_c = \dfrac{[CH_3COO^-][H^+]}{[CH_3COOH]}$

Third Reaction: $K_c = \dfrac{[NH_3]^2}{[N_2][H_2]^3}$ **Fourth Reaction:** $K_c = \dfrac{[CO_2][H_2]}{[CO][H_2O]}$

For answers to (a), (b) and (c) for each reaction, see below.

Strategy and Explanation:

For each reaction, follow the directions given in the question: Assume that all gases and solutes have initial concentrations of 1.0 mol/L. Then let the *first* reagent in each reaction changes its concentration by –x. *(Notice: This is not always the best approach, but those are the instructions for this question. Additional details are given afterward.)*

First reaction:	$H_2O\ (\ell)$	$\rightleftharpoons$	H^+ (aq)	+	OH^- (aq)
initial conc. (mol/L)	liquid		1.0		1.0
change as reaction occurs (mol/L)	– x		+ x		+ x
equilibrium conc. (mol/L)	liquid		1.0 + x		1.0 + x

(a) $K_c = [H^+][OH^-] = (1.0 + x)(1.0 + x) = 1.0 \times 10^{-14}$

(b) This equation is a quadratic equation.

(c) Not applicable.

Second reaction:	CH_3COOH (aq)	$\rightleftharpoons$	CH_3COO^- (aq)	+ H^+ (aq)
initial conc. (mol/L)	1.0		1.0	1.0
change as reaction occurs (mol/L)	– x		+ x	+ x
equilibrium conc. (mol/L)	1.0 – x		1.0 + x	1.0 + x

(a) $K_c = \dfrac{[CH_3COO^-][H^+]}{[CH_3COOH]} = \dfrac{(1.0 + x)(1.0 + x)}{(1.0 - x)} = 1.8 \times 10^{-5}$

(b) This equation is a quadratic equation.

(c) Not applicable.

Third reaction:	$N_2(g)$	+	$3\ H_2(g)$	$\rightleftharpoons$	$2\ NH_3$ (g)
initial conc. (mol/L)	1.0		1.0		1.0
change as reaction occurs (mol/L)	– x		– 3x		+2x
equilibrium conc. (mol/L)	1.0 – x		1.0 – 3x		1.0 + 2x

(a) $K_c = \dfrac{[NH_3]^2}{[N_2][H_2]^3} = \dfrac{(1.0 + 2x)^2}{(1.0 - x)(1.0 - 3x)^3} = 3.5 \times 10^8$

(b) This equation is not a quadratic equation.

(c) This K is so large that it might be worth starting with a limiting reactant problem first and running the reaction as far to the right as possible using stoichiometry, before defining an appropriate change variable:

	N_2 (g)	+ $3\ H_2$ (g)	$\rightleftharpoons$ $2\ NH_3$
initial conc. (mol/L)	1.0	1.0	1.0
change as forward reaction occurs (mol/L)	– 0.33	– 1.0	+ 0.67
final conc. (mol/L)	0.7	0.0	1.7
change as reverse reaction occurs (mol/L)	+ x	+ 3x	– 2x
equilibrium conc. (mol/L)	0.7 + x	+ 3x	1.7 – 2x

Since this x is assumed to be very small compared to 0.7 or 1.7, we can ignore subtraction of x from 0.7 and addition of 2x from 1.7:

$$K_c = \frac{[NH_3]^2}{[N_2][H_2]^3} = \frac{(1.7-2x)^2}{(0.7+x)(3x)^3} \cong \frac{(1.7)^2}{(0.7)(3x)^3} = 4\times10^8\text{, then solve for x.}$$

x = 0.0008 mol/L<<1.7 or 0.7

Fourth reaction:	CO (g)	+ $H_2O(g)$	$\rightleftharpoons$ CO_2 (g)	+ H_2 (g)
initial conc. (M)	1.0	1.0	1.0	1.0
change as reaction occurs (M)	– x	– x	+ x	+ x
equilibrium conc. (M)	1.0 – x	1.0 – x	1.0 + x	1.0 + x

(a) $K_c = \frac{[CO_2][H_2]}{[CO][H_2O]} = \frac{(1.0+x)(1.0+x)}{(1.00-x)(1.00-x)} = 4.00$

(b) This equation is a quadratic equation.

(c) Not applicable.

IMPORTANT NOTICE FOR THIS QUESTION: If we were actually doing this question to get the answers, it would be a very good idea to calculate Q to determine the direction of reaction before deciding which side of the reaction experiences a decrease in concentration (– x) and which side experiences an increase in concentration (+ x). The side where concentration decreases should get the – x, so that we can be certain that the value of x is a positive quantity. Doing that significantly simplifies the mathematics, especially when dealing with quadratic equations and other higher order polynomials. For the reactions in this question, here are the Q calculations.

First reaction: $Q = (1.0M)(1.0M) = 1.0 > K_c = 1.0 \times 10^{-14}$

Reaction goes toward reactants, not products. The x defined earlier for the first reaction will be negative!

Second reaction: $Q = \frac{(1.0)(1.0)}{(1.0)} = 1.0 > K_c = 1.8\times10^{-5}$

Reaction goes toward reactants, not products. The x defined earlier for the second reaction will be negative!

Third reaction: $Q = \frac{(1.0)^2}{(1.0)(1.0)^3} = 1.0 < K_c = 3.5\times10^8$

Reaction goes toward products. The x defined above will be positive.

Fourth reaction: $Q = \frac{(1.0)(1.0)}{(1.0)(1.0)} = 1.0 < K_c = 4.00$

Reaction goes toward products. The x defined above will be positive.

Notice: The fourth reaction has a large K value and is product-favored, so it might be worth starting with a limiting reactant problem first and running the reaction as far forward as possible using stoichiometry then running the reverse reaction (a reactant-favored reaction) to establish the equilibrium, so that x will be small.

109. *Answer:* **(a) negative (b) 4.0 × 10^2 (c) 0.50%**

Strategy and Explanation:

(a) A decrease in K_c with an increase in temperature indicates the reaction is exothermic, so $\Delta H°$ is **negative**. The reaction will become less product-favored at higher temperatures.

(b) The reaction given here is one-half of the original reaction, so the value of K_c must be raised to the one-half power.

$$K_c = (1.6 \times 10^5)^{1/2} = \mathbf{4.0 \times 10^2}$$

(c) The decomposition of HBr, can be written as the reverse of the original reaction, so we must reciprocal the value of K_c.

$$K_c = 1/1.6 \times 10^5 = 6.3 \times 10^{-6}$$

	$2\,HBr(g)$ ⇌	$H_2(g)$	+ $Br_2(g)$
initial fraction	1.00	0	0
change as reaction occurs	$-2x$	$+x$	$+x$
equilibrium fraction	$1.00 - 2x$	x	x

At equilibrium $$K_c = \frac{[H_2][Br_2]}{[HBr]^2} = \frac{x^2}{(1.00-2x)^2} = 6.3 \times 10^{-6}$$

Take the square root of each side: $$\frac{x}{1.00-2x} = 2.5 \times 10^{-3}$$

$$x = 2.5 \times 10^{-3}(1.00 - 2x) = 2.5 \times 10^{-3} - 5.0 \times 10^{-3}x$$

$$1.005x = 2.5 \times 10^{-3}$$

$$x = 2.5 \times 10^{-3}$$

The fraction of HBr that decomposed is: $2x = 5.0 \times 10^{-3}$

The percentage of HBr that decomposed is: $5.0 \times 10^{-3} \times 100 =$ **0.50%**

✓ *Reasonable Answer Check:* Substituting the equilibrium concentrations into the equilibrium expression will reproduce K_c within the uncertainty of the significant figures:

$$K_c = \frac{(2.5\times10^{-3})(2.5\times10^{-3})}{(1.00)} = 6.3\times10^{-6}$$

110. *Answer:* **(a)**

Species	Br_2	Cl_2	F_2	H_2	N_2	O_2
[E] (mol/L)	**0.28**	**0.057**	**1.44**	$\mathbf{1.76\times10^{-5}}$	$\mathbf{4\times10^{-14}}$	$\mathbf{4.0\times10^{-6}}$

(b) F_2; see explanation below.

Strategy and Explanation: Use a method similar to that described in the solution to Question 62.

(a) (conc. E_2) = 1.00 mol E_2/1.0 L = 1.0 M

	E_2 (g) ⇌	2 E (g)
initial conc. (M)	1.0	0
change as reaction occurs (M)	$-x$	$+2x$
equilibrium conc. (M)	$1.0 - x$	2x

At equilibrium $$K_c = \frac{[E]^2}{[E_2]} = \frac{(2x)^2}{1.0-x}$$

Set up to use the quadratic equation (Appendix A, Section A.7, page A.13):

$$(2x)^2 = K_c(1.0 - x) = K_c - K_cx$$

$$4x^2 + K_cx - K_c = 0$$

Plug into the quadratic equation:

$$x = \frac{-b \pm \sqrt{b^2 - 4ac}}{2a} = \frac{-K_c \pm \sqrt{K_c^2 - 4(K_c)(-K_c)}}{2(4)}$$

$$[E] = 2x$$

Here is an example of the calculation for the first species: E = Br, $K_c = 8.9 \times 10^{-2}$.

$$x = \frac{-(8.9\times10^{-2}) - \sqrt{(8.9\times10^{-2})^2 - 4(8.9\times10^{-2})(-8.9\times10^{-2})}}{2(4)} = \frac{1.1}{8} = 0.14$$

$$[Br] = 2\times(0.14\text{ M}) = 0.28\text{ M}$$

Similar calculations give these results:

E species	K_c	x (M)	[E] (M)
Br	8.9×10^{-2}	0.14	0.28
Cl	3.4×10^{-3}	0.029	0.057
F	7.4	0.72	1.44
H	3.1×10^{-10}	8.8×10^{-6}	1.76×10^{-5}
N	1×10^{-27}	2×10^{-14}	4×10^{-14}
O	1.6×10^{-11}	2.0×10^{-6}	4.0×10^{-6}

✓ *Reasonable Answer Check:* [E] is small when K is small. [E] is large when K is large. The equilibrium concentrations combine to reproduce $K_c = \dfrac{(2x)^2}{1.0 - x}$.

(b) At this temperature, the lowest bond enthalpy is predicted from the reaction that gives the most products, so F_2 is predicted to have the lowest bond dissociation enthalpy.

Comparing these results to the bond enthalpy values in Table 8.2, the product production decreases as the bond enthalpy increases: 158 kJ F_2, 193 kJ Br_2, 242 kJ Cl_2, 436 kJ H_2, 498 kJ O_2, 946 kJ N_2.

Lewis structures of F_2, Br_2, Cl_2, H_2 have a single bonds and more products in these calculations than O_2 with double bond and N_2 with a triple bond.

111. *Answer:* $\mathbf{P_{NO_2} = 0.40\text{ atm}, P_{N_2O_4} = 1.1\text{ atm}}$

Strategy and Explanation: Given the total pressure for a mixture of two gases, the value of K_P and an equation for a reaction, calculate the partial pressure of each gas in the mixture.

Use a combination of Dalton's law of partial pressures (Section 10.7) and equilibrium expression to relate the pressures of the two gases.

$$P_{tot} = 1.5\text{ atm} = P_{NO_2} + P_{N_2O_4}$$

$$K_p = 7.0 = \frac{P_{N_2O_4}}{P_{NO_2}^{\ 2}}$$

Solve the first equation for $P_{N_2O_4}$ and plug it into the second

$$P_{N_2O_4} = 1.5\text{ atm} - P_{NO_2}$$

$$7.0 = \frac{1.5 - P_{NO_2}}{P_{NO_2}^{\ 2}}$$

Set up to use the quadratic equation (Appendix A, Section A.7, page A.13):

$$7.0\, P_{NO_2}^{\ 2} + P_{NO_2} - 1.5 = 0$$

Plug into the quadratic equation:

$$P_{NO_2} = \frac{-b \pm \sqrt{b^2 - 4ac}}{2a} = \frac{-1 \pm \sqrt{(1)^2 - 4(7.0)(-1.5)}}{2(7.0)} = 0.40$$

Using the quadratic equation,

$$P_{NO_2} = 0.40\text{ atm}$$

$$P_{N_2O_4} = 1.5\text{ atm} - P_{NO_2} = 1.1\text{ atm}$$

✓ *Reasonable Answer Check:* Plugging the pressures into the K_P expression gives K_P within uncertainty:

$$K_p = \frac{(1.1)}{(0.40)^2} = 6.9$$

114. *Answer:* **0.0014 atm**

Strategy and Explanation: Given the K_p and equation for a reaction, and the equilibrium partial pressures for one reactant and one product, determine the equilibrium concentation of another reactant. Set up the expression for K_p and solve for the P_{H_2}.

$$K_p = \frac{P_{N_2} P_{H_2}^{\ 3}}{P_{NH_3}^{\ 2}} = \frac{(1.2)P_{H_2}^{\ 3}}{(0.015)^2} = 1.5 \times 10^{-5}$$

$$P_{H_2} = 0.0014 \text{ atm}$$

✓ *Reasonable Answer Check:* Plugging the equilibrium concentrations into the K_p expression produces

$$K_p = \frac{(1.2)(0.0014)^3}{(0.015)^2} 1.5 \times 10^{-5}.$$

116. *Answer:* **(a) 0.927 atm (b) 0.0420 moles**

Strategy and Explanation: Given K_p and the equation for the decomposition of a solid into two gases, the mass of the reactant and the volume of the container, determine the total pressure of gas in the flask at equilibrium and how much of the solid decomposed.

Use a method similar to that described in the solution to Question 62.

Use the molar mass of NH_4I to determine how many moles of the solid are present initially:

$$15.0 \text{ g } NH_4I \times \frac{1 \text{ mol } NH_4I}{145.9507 \text{ g } NH_4I} = 0.1028 \text{ mol } NH_4I$$

Remember, the quantity of solids does not affect the equilibrium state.

	$NH_4I(s) \rightleftharpoons$	NH_3 (g)	+ HI (g)
initial press. (atm)	solid	0	0
change as reaction occurs (atm)	- x	+ x	+ x
equilibrium press. (atm)	solid - x	x	x

At equilibrium $\quad K_p = P_{NH_3}P_{HI} = x^2 = 0.215$

$$x = 0.464$$

$$P_{NH_3} = P_{HI} = x = 0.46\underline{3}7 \text{ atm}$$

(a) Use Dalton's law: $P_{tot} = P_{NH_3} + P_{HI} = 0.46\underline{3}7 \text{ atm} + 0.46\underline{3}7 \text{ atm} = 0.927 \text{ atm}$

(b) First, calculate the moles of NH_3(g) at equilibrium, using the ideal gas law, PV = nRT (Section 10.4):

$$P_{NH_3}V = n_{NH_3}RT$$

$$n_{NH_3} = \frac{P_{NH_3} V}{RT} = \frac{(0.4637 \text{ atm})(5.00 \text{ L})}{\left(0.08206 \frac{\text{atm} \cdot \text{L}}{\text{K} \cdot \text{mol}}\right)(673 \text{ K})} = 0.0420 \text{ mol } NH_3$$

The stoichiometric relationship between production of NH_3 and the decomposition of NH_4I is 1:1, so 0.0420 moles of ammonium iodide decomposes.

✓ *Reasonable Answer Check:* (a) Substituting the equilibrium pressures into the equilibrium expression will reproduce K_p:

$$K_p = P_{NH_3}P_{HI} = 0.464$$

Less than half of the solid decomposes, which makes sense because the K_p is slightly smaller than 1.

118. *Answer:* **Cases c and d will have increased [HI]; Cases a and b will have decreased [HI]**

Strategy and Explanation: Use some of the methods described in the solution to Question 55.

$$2\ HI(g) \rightleftharpoons H_2(g) + I_2(g) \qquad K_c = \frac{[H_2][I_2]}{[HI]^2} = 0.0200$$

Construct Q_c:
$$Q_c = \frac{(\text{conc. } H_2)(\text{conc. } I_2)}{(\text{conc. HI})^2}$$

Then compare to K_c.

Case a:
$$Q_c = \frac{\left(\frac{0.10 \text{ mol } H_2}{10.00 \text{ L}}\right)\left(\frac{0.10 \text{ mol } I_2}{10.00 \text{ L}}\right)}{\left(\frac{1.0 \text{ mol HI}}{10.00 \text{ L}}\right)^2} = 0.010 < 0.0200$$

Reaction goes toward products, and the concentration of HI decreases.

Case b:
$$Q_c = \frac{\left(\frac{1.0 \text{ mol } H_2}{10.00 \text{ L}}\right)\left(\frac{1.0 \text{ mol } I_2}{10.00 \text{ L}}\right)}{\left(\frac{10. \text{ mol HI}}{10.00 \text{ L}}\right)^2} = 0.010 < 0.0200$$

Reaction goes toward products, and the concentration of HI decreases.

Case c:
$$Q_c = \frac{\left(\frac{10. \text{ mol } H_2}{10.00 \text{ L}}\right)\left(\frac{1.0 \text{ mol } I_2}{10.00 \text{ L}}\right)}{\left(\frac{10. \text{ mol HI}}{10.00 \text{ L}}\right)^2} = 0.10 > 0.0200$$

Reaction goes toward reactants, and the concentration of HI increases.

Case d:
$$Q_c = \frac{\left(\frac{0.381 \text{ mol } H_2}{10.00 \text{ L}}\right)\left(\frac{1.75 \text{ mol } I_2}{10.00 \text{ L}}\right)}{\left(\frac{5.62 \text{ mol HI}}{10.00 \text{ L}}\right)^2} = 0.0211 < 0.0200$$

Reaction goes toward reactants a very small amount, and the concentration of HI increases a very small amount.

In conclusion, the HI concentration increases in case c and very slightly in case d. The HI concentration decreases in cases a and b.

120. *Answer:* **(a) 0.18 (b) P_{CO} =1.02 atm, P_{Br_2}= 0.52 atm, P_{COBr_2} =0.10 atm**

Strategy and Explanation: Given the partial pressures of three gases at equilibrium for a given equation, determine the value of K_p. Given the reduced partial pressure of one of the gases (due to partial condensation), determine the equilibrium partial pressures of all the gases after equilibrium is re-established.

(a) Set up the K_p expression and plug in the equilibrium partial pressures:

$$K_p = \frac{P_{COBr_2}}{P_{CO}P_{Br_2}} = \frac{(0.12)}{(1.00)(0.65)} = 0.1846 = 0.18 \text{ (2 sig figs)}$$

(b) LeChatlier's principle tells us that when a reactant is removed, the equilibrium will shift to the left to make more reactant. So, use signs on the change variable, x, that indicate the reaction going to the left. This way we can assure that the value of x we are seeking is positive.

	CO(s)	+ $Br_2(g)$ ⇌	+ $COBr_2(g)$
initial press. (atm)	1.00	0.50	0.12
change as reaction occurs (atm)	+ x	+ x	− x
equilibrium press. (atm)	1.00 + x	0.50 + x	0.12 − x

At equilibrium $$K_p = \frac{P_{COBr_2}}{P_{CO}P_{Br_2}} = \frac{(0.12 - x)}{(1.00 + x)(0.50 + x)} = 0.1846$$

Set up to use the quadratic equation (Appendix A, Section A.7, page A.13):

$$(0.1\underline{8}46)(1.00)(0.50) + (0.1\underline{8}46)(1.00)x + (0.1\underline{8}46)(0.50)x + (0.1\underline{8}46)x^2 = 0.12 - x$$

$$(0.09\underline{2}31) + (0.1\underline{8}46 + 0.09231)x + (0.1\underline{8}46)x^2 = 0.12 - x$$

$$0.1\underline{8}46x^2 + 1.2\underline{7}69x - 0.0\underline{2}769 = 0$$

Plug into the quadratic equation:

$$x = \frac{-b \pm \sqrt{b^2 - 4ac}}{2a} = \frac{-1.2769 \pm \sqrt{(1.2769)^2 - 4(0.1846)(-0.0276)}}{2(0.1846)}$$

$$x = 0.02\underline{1}6$$

$$P_{CO} = 1.00 + x = 1.02 \text{ atm}$$

$$P_{Br_2} = 0.50 + x = 0.52 \text{ atm}$$

$$P_{COBr_2} = 0.12 - x = 0.10 \text{ atm}$$

✓ *Reasonable Answer Check:* The equilibrium has greater pressures of reactants, so it makes sense that the value of K_p is less than 1. Plug the equilibrium pressures into the K_p expression, recreates the value calculated in (a), within the limits of uncertainty:

$$K_p = \frac{P_{COBr_2}}{P_{CO}P_{Br_2}} = \frac{(0.10)}{(1.02)(0.52)} = 0.18$$

Applying Concepts

123. *Answer:* **Yes, the system is at equilibrium at 600. K; no further experiments would be needed.**

Strategy and Explanation: For the system to be at equilibrium, it must contain some quantity of both reactants and products, the concentrations of all the reactants and products must be constant over time, and the same relative concentrations are established regardless of which direction the equilibrium was approached. The 42% *cis* concentration was achieved in two different experiments; therefore, the system is at equilibrium. No further experiments are needed.

125. *Answer/Explanation:* When pressure is applied to a sample of ice, Le Chatelier's principle says the phase change equilibrium will shift to compensate for the change. The pressure causes the H_2O molecules in the ice to be pushed closer together. Because the density of water is higher than the density of ice, some of the molecules that were in the solid phase respond by melting into the liquid phase, thus reducing the pressure.

127. *Answer:* **(a) Reaction (iii) (b) Reaction (i)**

Strategy and Explanation: As described near the end of Section 14.6, an endothermic reaction will have an increased K_c at higher temperatures, and an exothermic reaction will have a decrease in K_c at higher temperatures. In addition, the larger the change, the more exothermic or endothermic the reaction is, respectively.

(a) When R = CH_3, the values of K_c for the *cis-trans* isomerism decrease at higher temperatures. This is the only data set that shows this trend, so (iii) the *cis-trans* isomerism for R = CH_3 is the most exothermic.

(b) When R = F and R = Cl, the values of K_c for the *cis-trans* isomerism increase at higher temperatures. That means these are both endothermic *cis-trans* isomerism reactions. The R = F isomerism has a larger change, so (i) the *cis-trans* isomerism for R = F is the most endothermic.

129. *Answer/Explanation:* In the warmer sample, the molecules would be moving faster, and more NO_2 molecules would be seen. In the cooler sample, the molecules would be moving slower, and fewer NO_2 molecules would be seen. In both samples, the molecules are moving very fast. The average speed of gas molecules is commonly hundreds of miles per hour. In both samples, I would see some N_2O_4 molecules decomposing and some NO_2 molecules reacting with each other, at equal rates.

131. *Answer:* **Diagrams (b), (c), and (d)**

Strategy and Explanation: The equilibrium constant expression for this reaction is: $K_c = \dfrac{[AB]^2}{[A_2][B_2]}$.

We need to find diagrams that fit this range: $10^2 >= \dfrac{[AB]^2}{[A_2][B_2]} > 0.1$

Diagram (a) has only reactants, A_2 and B_2. No AB molecules are found in the mixture. That means the value of $K_c = 0$. Diagram (e) has only product molecules, AB. No A_2 or B_2 is found in the mixture. That means the value of $K_c = \infty$. Neither of these K_c values is within the stated range.

In the other three diagrams, we notice that the number of A_2 molecules and the number of B_2 molecules are equal.

$$K = \frac{[AB]^2}{[A_2][B_2]} = \frac{[(\text{Number AB}) \times N_0 / V]^2}{(\text{Number } A_2) \times N_0 / V \times (\text{Number } B_2) \times N_0 / V}$$

$$= \frac{(\text{Number AB})^2}{(\text{Number } A_2) \times (\text{Number } B_2)} = \frac{(\text{Number AB})^2}{(\text{Number } A_2)^2} = \left(\frac{\text{Number AB}}{\text{Number } A_2}\right)^2$$

That means we can simplify the search range by taking the square-root of the range variables and compare them to the ratio of the numbers:

$$\sqrt{10^2} = 10 >= \frac{\text{Number of AB}}{\text{Number of } A_2} > = \sqrt{0.1} = 0.3$$

Diagram (b) has 4 AB and 2 A_2 = 2 B_2: $\dfrac{\text{Number of AB}}{\text{Number of } A_2} = \dfrac{4}{2} = 2$

Diagram (c) has 6 AB and 1 A_2 = 1 B_2: $\dfrac{\text{Number of AB}}{\text{Number of } A_2} = \dfrac{6}{1} = 6$

Diagram (d) has 2 AB and 3 A_2 = 3 B_2: $\dfrac{\text{Number of AB}}{\text{Number of } A_2} = \dfrac{2}{3} = 0.667$

Therefore, all three of these remaining diagrams, (b), (c), and (d), represent equilibrium mixtures where $10^2 > K_c > 0.1$.

133. *Answer:* **See diagram below (many other answers are possible)**

Strategy and Explanation: The nanoscale-level diagram will have CO, H_2O, CO_2, and H_2 gas molecules in relative proportions dictated by the relationship described by the equilibrium constant expression.

$$K_c = 4.00 = \frac{[CO_2][H_2]}{[CO][H_2O]}$$

Assume the container has a volume of V and contains one CO molecule and one H_2O molecule. We can use the equilibrium expression to calculate how many product molecules to put in the container at equilibrium with these reactants.

The concentrations of each of the reactants is: $[CO] = [H_2O] = 1$ molecule / V = 1/V. If we let x represent the number of CO_2 and H_2 molecule, their concentrations are: $[CO_2] = [H_2] =$ x molecules / V. Substitute the concentrations into the equilibrium expression and solve for x:

$$4.00 = \frac{(\text{Number of } CO_2 / V)(\text{Number of } H_2 / V)}{(\text{Number of CO}/ V)(\text{Number of } H_2O / V)} = \frac{(x / V)^2}{(1 / V)^2}$$

$$4.00 = x^2$$

$$2.00 = x = \text{number of } CO_2 \text{ molecules} = \text{number of } H_2 \text{ molecules}$$

To represent an equilibrium state, we need to put two CO_2 molecules and two H_2 molecules in the box with one CO molecule and one H_2O molecule.

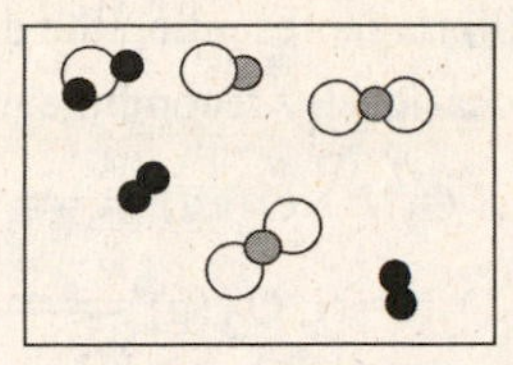

Many other answers are possible as long as K_c is satisfied, such as: four CO_2 molecules and one H_2 molecules in the box with one CO molecule and one H_2O molecule $\frac{(4/V)(1/V)}{(1/V)(1/V)} = 4$, or: four CO_2 molecules and two H_2 molecules in the box with one CO molecule and two H_2O molecule $\frac{(4/V)(2/V)}{(1/V)(2/V)} = 4$, etc.

✓ *Reasonable Answer Check:* The equilibrium constant can be reproduced by plugging in the concentrations of the reactants and products: $\frac{(2/V)(2/V)}{(1/V)(1/V)} = 4$

135. *Answer:* **Box (b)**

Strategy and Explanation: Count the molecules of each kind in each box. Divide the number of molecules by volume, V, to get concentrations, and combine them in an expression matching the K_c:

$$K_c = \frac{[CH_4][CCl_4]}{[CH_2Cl_2]} = \frac{\left(\frac{N_{CH_4}}{V}\right)\left(\frac{N_{CCl_4}}{V}\right)}{\left(\frac{N_{CH_2Cl_2}}{V}\right)^2} = \frac{\left(N_{CH_4}\right)\left(N_{CCl_4}\right)}{N_{CH_2Cl_2}{}^2}$$

The box that produces a K_c closest to 1.05 is the answer.

Box (a) has four CCl_4, four CH_4, and one CH_2Cl_2.

$$K_c = \frac{(4)(4)}{(1)^2} = 16$$

Box (b) has three CCl_4, three CH_4, and three CH_2Cl_2.

$$K_c = \frac{(3)(3)}{(3)^2} = 1.0$$

Box (c) has two CCl_4, two CH_4, and five CH_2Cl_2.

$$K_c = \frac{(2)(2)}{(5)^2} = 0.16$$

Box (b) has an equilibrium constant closest to the given value.

137. *Answer/Explanation:* Dynamic equilibria with small values of K introduce a small amount of D^+ ions in place of H^+ ions in the place of the acidic hydrogen.

$$H_2O(\ell) \rightleftharpoons H^+(aq) + OH^-(aq)$$

$$D_2O(\ell) \rightleftharpoons D^+(aq) + OH^-(aq)$$

$$C_6H_5COOH\ (s) \rightleftharpoons C_6H_5COOH\ (aq) \rightleftharpoons H^+(aq) + C_6H_5COO^-(aq)$$

$$C_6H_5COOD\ (s) \rightleftharpoons C_6H_5COOD\ (aq) \rightleftharpoons D^+(aq) + C_6H_5COO^-(aq)$$

$$D^+(aq) + C_6H_5COO^-(aq) \rightleftharpoons C_6H_5COOD\ (aq) \rightleftharpoons C_6H_5COOD\ (s)$$

139. *Answer/Explanation:* Dynamic equilibria representing the decomposition of the dimer $N_2O_4(g)$ produces NO_2 (g) and N^*O_2 (g), which will occasionally recombine into the mixed dimer, $O_2N^*–NO_2$ (g)

$$O_2N–NO_2\ (g) \rightleftharpoons 2\ NO_2\ (g)$$

$$O_2N^*–N^*O_2\ (g) \rightleftharpoons 2\ N^*O_2\ (g)$$

$$O_2N^*–NO_2\ (g) \rightleftharpoons N^*O_2\ (g) + NO_2\ (g)$$

No $O_2N^*–N^*O_2$ is observed since the chance of two N^*O_2 molecules meeting is vanishingly small.

More Challenging Questions

141. *Answer:* **1.34 atm**

Strategy and Explanation: Given the mass of a mixture of two compounds in equilibrium, the volume and temperature of the container, the equation and equilibrium constant, K_p, for the reaction, determine the total pressure of gas in the container.

There are a few ways to solve this question. Below we use a "moles focused"-approach.

Let x = moles of NO_2 and y = moles of N_2O_4. Now, we'll set up two equations that relate these quantities. One is the equilibrium expression relating concentration, in moles per liter. The second is the mass relationship identified for the mixture.

First, find K_c, adapting the method described in the solution to Question 33. T = 300. K

$$\Delta n = 1\ \text{mol}\ N_2O_4(g) - 2\ \text{mol}\ NO_2(g) = -1$$

$$K_c = K_p(RT)^{-\Delta n}$$

$$K_c = (6.67) \times \left\{\left(0.08206\ \frac{\text{L}\cdot\text{atm}}{\text{mol}\cdot\text{K}}\right) \times (300.\ \text{K})\right\}^{-(-1)} = 164$$

Next, set up the equilibrium constant expression to relate the concentrations of the reactants and products in terms of x and y.

$$K_c = \frac{[N_2O_4]}{[NO_2]^2} = \frac{\left(\frac{y}{15.0\ \text{L}}\right)}{\left(\frac{x}{15.0\ \text{L}}\right)^2} = \frac{y}{x^2} \times 15.0$$

$$164 = = \frac{y}{x^2} \times 15.0$$

$$y = 10.\underline{9}468\ x^2$$

Now, relate the relative masses of the two substances to the total mass, in terms of x and y.

$$\text{Mass total} = \text{mass of}\ NO_2 + \text{mass of}\ N_2O_4$$

$$64.4\ \text{g total} = x\left(\frac{46.0055\ \text{g}\ NO_2}{1\ \text{mol}\ NO_2}\right) + y\left(\frac{92.0110\ \text{g}\ N_2O_4}{1\ \text{mol}\ N_2O_4}\right)$$

Plug in the relationship for y, which was derived in the equilibrium equation:

$$64.4 = 46.0055\ x + 92.0110\ y = 46.0055\ x + 92.0110\ (10.\underline{9}468\ x^2)$$

Set up to use the quadratic equation (Appendix A, Section A.7, page A.13):

$$-64.4 + 46.0055\,x + 10\underline{0}7.226\,x^2 = 0$$

Plug into the quadratic equation:

$$x = \frac{-b \pm \sqrt{b^2 - 4ac}}{2a} = \frac{-46.0055 \pm \sqrt{(46.0055)^2 - 4(1007.226)(-64.4)}}{2(1007.226)}$$

$$x = 0.231 \text{ mol}$$

$$y = 10.\underline{9}468\,x^2 = 10.\underline{9}468\,(0.231)^2 = 0.584 \text{ mol.}$$

Now, use a combination of Dalton's law of partial pressures (Section 10.7) and the ideal gas law, PV = nRT (Section 10.4), to calculate the total pressure.

$$n_{tot} = 0.231 \text{ mol } NO_2 + 0.584 \text{ mol } N_2O_4 = 0.815 \text{ mol total}$$

$$P_{tot}V = n_{tot}RT$$

$$P_{tot} = \frac{n_{tot}RT}{V} = \frac{(0.815 \text{ mol}) \times \left(0.08206 \frac{\text{atm} \cdot \text{L}}{\text{K} \cdot \text{mol}}\right) \times (300.\text{ K})}{(15.0 \text{ L})} = 1.34 \text{ atm}$$

✓ *Reasonable Answer Check:* The values of x and y need to simultaneously satisfy the equilibrium constant equation and the total mass equation. Use the resulting moles and total pressure to determine the partial pressures and recreate the given K_p.

$$P_{N_2O_4} = Ptot\left(\frac{n_{NO_2}}{n_{tot}}\right) = 1.34 \text{ atm} \times \left(\frac{0.231 \text{ mol}}{0.815 \text{ mol}}\right) = 0.379 \text{ atm}$$

$$P_{NO_2} = Ptot\left(\frac{n_{N_2O_4}}{n_{tot}}\right) = 1.34 \text{ atm} \times \left(\frac{0.584 \text{ mol}}{0.815 \text{ mol}}\right) = 0.959 \text{ atm}$$

$$K_p = \frac{P_{N_2O_4}}{P_{NO_2}^{\;2}} = \frac{0.959}{0.379^2} = 6.67$$

Use the calculated moles and the molar masses to recreate the given total mass:

$$0.231 \text{ mol}\left(\frac{46.0055 \text{ g } NO_2}{1 \text{ mol } NO_2}\right) + 0.584 \text{ mol}\left(\frac{92.0110 \text{ g } N_2O_4}{1 \text{ mol } N_2O_4}\right) = 64.4 \text{ g total}$$

143. *Answer:* **(a) $K_p = 1.00 \times 10^{-6}$ (b) 0.089 mol $La(C_2O_4)_3$**

Strategy and Explanation: Use a method similar to that described in the solution to Question 62.

(a) Remember, the quantity of solids does not affect the equilibrium state.

	$La(C_2O_4)_3(s)$ ⇌	$La_2O_3(s)$	$+ 3\,CO(g)$	$+ 3\,CO_2(g)$
initial press. (atm)	solid	0	0	0
change as reaction occurs (atm)	$-\frac{1}{3}x$	$+\frac{1}{3}x$	+ x	+ x
equilibrium press. (atm)	solid – some	solid	x	x

At equilibrium $\quad K_p = P_{CO}^{\;3}\,P_{CO_2}^{\;3} = x^3x^3 = x^6$

$$P_{tot} = P_{CO} + P_{CO_2}$$

$$0.200 \text{ atm} = 2x$$

$$x = 0.100 \text{ atm}$$

$$K_p = x^6 = (0.100)^6 = 1.00 \times 10^{-6}$$

(b) First, calculate the moles of CO(g) at equilibrium, using the ideal gas law, PV = nRT (Section 10.4) and T = 373 K:

$$P_{CO}V = n_{CO}RT$$

$$n_{CO} = \frac{P_{CO}V}{RT} = \frac{(0.100 \text{ atm})(10.0 \text{ L})}{\left(0.08206 \frac{\text{atm} \cdot \text{L}}{\text{K} \cdot \text{mol}}\right)(373 \text{ K})} = 0.03267 \text{ mol CO}$$

The equation gives the stoichiometric relationship between production of CO and the decomposition of $La(C_2O_4)_3$:

$$0.03267 \text{ mol CO} \times \frac{1 \text{ mol } La(C_2O_4)_3}{3 \text{ mol CO}} = 0.0189 \text{ mol } La(C_2O_4)_3 \text{ decomposed}$$

Calculate the amount of $La(C_2O_4)_3$ unreacted:

0.100 mol $La(C_2O_4)_3$ initial – 0.0189 mol $La(C_2O_4)_3$ decomposed = 0.089 mol $La(C_2O_4)_3$ unreacted

✓ *Reasonable Answer Check:* More than 80% of the solid remains unreacted, because the K_p is much smaller than 1 and the reaction should favor the reactants.

145. *Answer:* **(a) $[PCl_5]$ = 0.005 M, $[PCl_5]$ = $[Cl_2]$ = 0.14 M (b) $[Cl_2]$ = 0.14 M, $[PCl_5]$ = 0.01 M, $[PCl_3]$ = 0.29 M**

Strategy and Explanation: Use a method similar to that described in the solution to Question 62.

(a) (conc. PCl_5) = 0.75 mol PCl_5/5.00 L = 0.15 M

	PCl_5 (g)	⇌	PCl_3 (g)	+ Cl_2 (g)
initial conc. (M)	0.15		0	0
change as reaction occurs (M)	– x		+ x	+ x
equilibrium conc. (M)	0.15 – x		x	x

At equilibrium $K_c = \frac{[PCl_3][Cl_2]}{[PCl_5]} = \frac{(x)(x)}{(0.15-x)} = 3.30$

Set up to use the quadratic equation (Appendix A, Section A.7, page A.13):

$$x^2 = 3.30(0.15 - x) = 0.50 - 3.30x$$

$$x^2 + 3.30x - 0.5 = 0$$

Plug into the quadratic equation:

$$x = \frac{-b \pm \sqrt{b^2 - 4ac}}{2a} = \frac{-3.30 \pm \sqrt{(3.30)^2 - 4(1)(-0.15)}}{2(1)} = \frac{0.29}{2} = 0.14$$

$$x = [PCl_5] = [Cl_2] = 0.14 \text{ M}$$

$$[PCl_5] = 0.15 - x = 0.15 \text{ M} - 0.14 \text{ M} = 0.01 \text{ M}$$

The reaction condition in (a) causes more than 90% formation of products, so let's take the reaction completely to the product's side, then bringing it back to equilibrium. This is a common tactic when reactants form products in the reaction with a K larger than 1. Use Chapter 4 stoichiometric "limiting reactant" procedure to accomplish this (Section 4.5).

	PCl_5 (g)	⇌	PCl_3 (g)	+ Cl_2 (g)
initial conc. (M)	0.15		0	0
final conc. (M)	0		0.15	0.15
change as reaction occurs (M)	+ x		– x	– x
equilibrium conc. (M)	x		0.15 – x	0.15 – x

At equilibrium $K_c = \frac{[PCl_3][Cl_2]}{[PCl_5]} = \frac{(0.15-x)(0.15-x)}{(x)} = 3.30$

Set up to use the quadratic equation (Appendix A, Section A.7, page A.13):

$$(0.15)^2 - 2(0.15)x + x^2 = 3.30x$$

$$x^2 - 3.60x - 0.023 = 0$$

Plug into the quadratic equation:

$$x = \frac{-b \pm \sqrt{b^2 - 4ac}}{2a} = \frac{3.60 \pm \sqrt{(-3.60)^2 - 4(1)(0.023)}}{2(1)} = \frac{0.01}{2} = 0.005$$

$$[PCl_5] = 0.005 \text{ M}$$

$$x = [PCl_5] = [Cl_2] = 0.15 - 0.005 \text{ M} = 0.14 \text{ M}$$

This way we get slightly more reliable information about the smallest concentration, $[PCl_5]$.

(b) $(\text{conc. } PCl_5) = 0.75 \text{ mol } PCl_5/5.00 \text{ L} = 0.15 \text{ M}$

$(\text{conc. } PCl_3) = 0.75 \text{ mol } PCl_5/5.00 \text{ L} = 0.15 \text{ M}$

	PCl_5 (g) ⇌	PCl_3 (g)	+ Cl_2 (g)
initial conc. (M)	0.15	0.15	0
change as reaction occurs (M)	– x	+ x	+ x
equilibrium conc. (M)	0.15 – x	0.15 + x	x

At equilibrium $$K_c = \frac{[PCl_3][Cl_2]}{[PCl_5]} = \frac{(0.15 + x)(x)}{(0.15 - x)} = 3.30$$

Set up to use the quadratic equation (Appendix A, Section A.7, page A.13):

$$0.15x + x^2 = 3.30(0.15 - x) = 0.50 - 3.30x$$

$$x^2 + 3.45x - 0.50 = 0$$

Plug into the quadratic equation:

$$x = \frac{-b \pm \sqrt{b^2 - 4ac}}{2a} = \frac{-3.45 \pm \sqrt{(3.45)^2 - 4(1)(0.50)}}{2(1)} = \frac{0.28}{2} = 0.14$$

$$x = [Cl_2] = 0.14 \text{ M}$$

$$[PCl_5] = 0.15 - x = 0.15 \text{ M} - 0.14 \text{ M} = 0.01 \text{ M}$$

$$[PCl_3] = 0.15 + x = 0.15 \text{ M} + 0.14 \text{ M} = 0.29 \text{ M}$$

The reaction condition in (b) also causes more than 90% formation of products, but taking the reaction completely to the products side, as we did in (a), produces the same result, within the given significant figures, so it is not shown here.

✓ *Reasonable Answer Check:* Substituting the equilibrium concentrations into the equilibrium expression will reproduce K_c:

$$K_{c,(a)} = \frac{(0.14)(0.14)}{(0.005)} = 4$$

$$K_{c,(a)} = \frac{(0.29)(0.14)}{(0.01)} = 4$$

Both are approximately equal to 3.30, within the uncertainty of the results (±1).

147. *Answer:* **(a) 56 (b) 5.2 atm before, 5.2 atm after (c) $P_{I_2} = P_{H_2} = 0.55$ atm, $P_{HI} = 4.1$ atm**

Strategy and Explanation: Given the description for a reaction, the initial moles of the gaseous reactant in a known volume, and the value of K_c, determine the value of K_p. Given the moles of the reactants calculate the total pressure of the mixture before and after equilibrium, and calculate the partial pressure of each gas at equilibrium.

First, balance the equation:

$$H_2(g) + I_2(g) \rightleftharpoons 2\text{ HI}(g)$$

(a) Relate the K_c to the K_p:

$$K_p = K_c(RT)^{\Delta n}$$

Δn = 2 mol HI gas product – (1 mol H_2 gas reactant + 1 mol I_2 gas reactant) = 0

$$K_p = K_c(RT)^0 = K_c$$

$$K_p = 56$$

(b) Now, use a combination of the ideal gas law, PV = nRT (Section 10.4) and Dalton's law of partial pressures (Section 10.7) to calculate the total pressure.

$$P_{H_2}V = n_{H_2}RT$$

$$P_{H_2} = \frac{n_{H_2}RT}{V} = \frac{(0.45 \text{ mol}) \times \left(0.08206 \frac{\text{atm} \cdot \text{L}}{\text{K} \cdot \text{mol}}\right) \times (435 + 273.15) \text{ K}}{(10.0 \text{ L})} = 2.6 \text{ atm}$$

$$P_{I_2} = P_{H_2} = 2.6$$

$$P_{tot} = P_{I_2} + P_{H_2} = 2.6 \text{ atm } H_2 + 2.6 \text{ atm } I_2 = 5.2 \text{ atm total}$$

Because the reaction produces two moles of gas product for every two moles of gas reactant, the final pressure must be the same as the initial total pressure. 5.2 atm.

(c) Set up an ICE table for the reaction. Because the K_p is larger than 1, the reaction will be product-favored, so take the reaction all the way to products using stoichiometry, then bring it back to equilibrium. Doing this will assure that x is small and preserves the greatest number of significant figures.

	$H_2(g)$	+	$I_2(g)$	$\rightleftharpoons$	2 HI(g)
initial press. (atm)	2.6		2.6		0
Change as reaction goes all the way to the right (atm)	– 2.6		– 2.6		+ 5.2
"new initial" pressure (atm)	0		0		5.2
change as reaction occurs (atm)	+ x		+ x		– 2x
equilibrium press. (atm)	x		x		5.2 – 2x

At equilibrium $$K_p = \frac{P_{HI}^2}{P_{H_2} P_{I_2}} = \frac{(5.2 - 2x)^2}{(x)^2} = 56$$

Take the square root of each side: $$\frac{5.2 - 2x}{x} = \sqrt{56} = 7.5$$

Solve for x: $$5.2 - 2x = 7.5x$$

$$5.2 = 9.5x$$

$$x = 0.55 \text{ atm}$$

$$P_{I_2} = P_{H_2} = 0.55 \text{ atm}$$

$$P_{HI} = 5.2 - 2x = 5.2 - 2 \times (0.55) = 4.1 \text{ atm}$$

✓ *Reasonable Answer Check:* The sum of the partial pressures calculated in (c) add up to the total pressure identified in (b): 0.55 atm + 0.55 atm + 4.1 atm = 5.2 atm

149. *Answer/Explanation:* Look first at the relationship between the equilibrium constant and the activation energies of the forward and reverse reactions.

$$K_c = \frac{A_f e^{(-E_{a,f}/RT)}}{A_r e^{(-E_{a,r}/RT)}} = \frac{A_f}{A_r} e^{(-E_{a,f}/RT)} e^{(E_{a,r}/RT)} = \frac{A_f}{A_r} e^{[(-E_{a,f}/RT)+(E_{a,r}/RT)]}$$

$$K_c = \frac{A_f}{A_r} e^{[(E_{a,r}-E_{a,f})/RT]}$$

When a catalyst is added, the activation energies of both the forward and reverse reactions get smaller and the equation that describes the K_c for the catalyzed reaction is:

$$K_{c,cat} = \frac{A_{f,cat}}{A_{r,cat}} e^{[(E_{a,r,cat} - E_{a,f,cat})/RT]}$$

We were told to assume that the frequency factors did not change. The activation energies are reduced, and because the reaction has to conserve energy, they must each be reduced by the same amount. Let us call that quantity of energy X, here.

$$A_{f,\ cat} = A_f \qquad E_{a,f,\ cat} = E_{a,f} - X$$

$$A_{r,\ cat} = A_r \qquad E_{a,r,\ cat} = E_{a,r} - X$$

Notice that $E_{a,f,\ cat} - E_{a,r,\ cat} = (E_{a,f} - X) - (E_{a,r} - X) = E_{a,f} - X - E_{a,r} + X) = E_{a,f} - E_{a,r}$

Substituting into the $K_{c,\ cat}$ equation gives: $K_{c,cat} = \frac{A_f}{A_r} e^{[(E_{a,r} - E_{a,f})/RT]} = K_c$

Therefore, the equilibrium state, as described quantitatively by the equilibrium constant, does not change with the addition of a catalyst.

150. *Answer:* **(a) $2\ NO_2(g) + Br_2(g) \longrightarrow 2\ NOBr(g)$ (b) see equations below (c) see derivation below (d) Yes**

Strategy and Explanation:

(a) As in Chapter 6 Hess's law questions, cancel species that end up on both sides and add up the rest.

$$NO(g) + Br_2(g) \rightleftharpoons \cancel{NOBr_2}(g)$$

$$+\ \cancel{NOBr_2}(g) + NO(g) \longrightarrow 2\ NOBr(g)$$

$$2\ NO_2(g) + Br_2(g) \longrightarrow 2\ NOBr(g)$$

(b) At equilibrium, the rate of the forward reaction equals the rate of the reverse reaction. Use techniques from Chapter 13:

Rate forward, step 1 = rate reverse, step 1

$$k_1[NO][Br_2] = k_{-1}[NOBr_2]$$

$$K_c, step1 = \frac{k_1}{k_{-1}} = \frac{[NOBr_2]}{[NO][Br_2]}$$

Rate forward, step 2 = rate reverse, step 2

$$k_2[NOBr_2][NO] = k_{-2}[NOBr]^2$$

$$K_{c,step2} = \frac{k_2}{k_{-2}} = \frac{[NOBr]^2}{[NOBr_2][NO]}$$

(c) $$K_{c,overall} = \frac{[NOBr]^2}{[NO]^2[Br_2]} = \left(\frac{[NOBr_2]}{[NO][Br_2]}\right)\left(\frac{[NOBr]^2}{[NO][NOBr_2]}\right) = \left(\frac{k_1}{k_{-1}}\right)\left(\frac{k_2}{k_{-2}}\right) = \frac{k_1k_2}{k_{-1}k_{-2}}$$

(d) The result in (c) confirms the statement that the equilibrium constant can be obtained by taking the product of the rate constants for all forward steps (here, k_1k_2) and dividing by the product of the rate constants for all reverse steps (here, $k_{-1}k_{-2}$):

$$K_{c,overall} = \frac{k_1k_2}{k_{-1}k_{-2}}$$

151. *Answer:* **(a) No; proceed to the right (b) decrease**

Strategy and Explanation: Given an equation for a reaction, the initial concentrations of the gaseous reactants and products, and the value of the equilibrium constant, determine if the system is at equilibrium and, if it is not, determine which direction it must proceed to reach equilibrium and calculate the equilibrium concentrations.

(a) Calculate Q and compare it to K to see if the system is at equilibrium, and, as needed, which direction the reaction proceeds in order to reach equilibrium. $K_c = 6.0 \times 10^{34}$

$$Q_c = \frac{(\text{conc. } O_2)(\text{conc. } NO_2)^2}{(\text{conc. } O_3)(\text{conc. NO})} = \frac{(8.2\times10^{-6})(2.5\times10^{-4})}{(1.0\times10^{-6})(1.0\times10^{-5})} = 2.1\times10^5$$

$Q_c \neq K_c$, so the reaction is not at equilibrium.

$Q_c < K_c$, so the reaction must proceeds to the **right** (toward products) to reach equilibrium.

(b) To determine how K changes with temperature, calculate the ΔH, using Hess' Law and Appendix J.

$$\Delta H = (1 \text{ mol}) \times \Delta H_f^\circ(O_2) + (1 \text{ mol}) \times \Delta H_f^\circ(NO_2) - [(1 \text{ mol}) \times \Delta H_f^\circ(O_3) + (1 \text{ mol}) \times \Delta H_f^\circ(NO)]$$

$$= (1 \text{ mol})\times(0 \text{ kJ/mol}) + (1 \text{ mol})\times(33.81 \text{ kJ/mol}) - (1 \text{ mol})\times(142.7 \text{ kJ/mol}) - (1 \text{ mol})\times(90.25 \text{ kJ/mol})$$

$= -199.1$ kJ, exothermic reaction

As described near the end of Section 14.6, an exothermic reaction will have a decrease in K_c at higher temperatures. The concentrations of the products decrease.

✓ *Reasonable Answer Check:* As huge as the K_c is, it may not be possible to detect the small concentration of reactants remaining when the reaction reaches equilibrium. So, it makes sense that these measured values do not represent an equilibrium state and that the reaction at equilibrium will have much larger relative concentrations of products. The production of stable O_2 and NO_2 should be favored over less-stable O_3 and NO; hence the reaction has a negative enthalpy change, which indicates a product-favored reaction.

153. *Answer:* **The $PbCl_2$ beaker**

Strategy and Explanation: The larger-K_c reaction favors products more, so the beaker containing the $PbCl_2$ ($1.7 \times 10^{-5} > 3.7 \times 10^{-8}$) has greater concentration of Pb^{2+} ion.

155. *Answer:* **(a) $K_p = 0.03126$; $K_c = 0.0128$ (b) $K_p = 1$ (c) $K_c = 1/(RT_{bp})$**

Strategy and Explanation:

(a) Use the method described in the solution to Question 33.

$$H_2O(\ell) \rightleftharpoons H_2O(g) \qquad K_p = P_{H_2O(g)} = 0.03126 \text{ atm}$$

$$T = 25.\ °C + 273.15 = 298 \text{ K}$$

$$\Delta n = 1 \text{ mol } H_2O \text{ gas product} - 0 \text{ mol gas reactants} = 1$$

$$K_p = K_c(RT)^{\Delta n} \qquad K_p(RT)^{-\Delta n} = K_c$$

$$K_c = (0.03126 \text{ atm}) \times \left\{\left(0.08206 \frac{\text{L}\cdot\text{atm}}{\text{mol}\cdot\text{K}}\right)\times(298 \text{ K})\right\}^{-1} = 0.0128$$

(b) At the normal boiling point, the vapor pressure of the gas above the liquid is exactly 1 atmosphere. Because $K_p = P_{H_2O(g)}$, then $K_p = 1$, under these conditions.

(c) Δn for a pure liquid's vaporization equation will always be 1. Also, when the temperature is the boiling point $T = T_{bp}$ and $K_p = P_{(g)} = 1$. So the equation, $K_c = K_p(RT)^{-\Delta n}$, can be simplified to:

$$K_c = 1/(RT_{bp})$$

157. *Answer:* **1.15%**

Strategy and Explanation: Use a method similar to that described in the solution to Question 62.

$$(\text{conc. } N_2) = (\text{conc. } H_2) = 1.00 \text{ mol}/10.00 \text{ L} = 0.100 \text{ M}$$

	N_2 (g)	+ $3\ H_2$ (g)	$\rightleftharpoons 2\ NH_3$ (g)
initial conc. (M)	0.100	0.100	0
change as reaction occurs (M)	– x	– 3x	+ 2x
equilibrium conc. (M)	0.100 – x	0.100 – 3x	2x

At equilibrium $$K_c = \frac{[NH_3]^2}{[N_2][H_2]^3} = \frac{(2x)^2}{(0.100 - x)(0.100 - 3x)^3} = 5.97 \times 10^{-2}$$

The reactant-favored reaction will have larger reactant concentrations than product concentrations, so let's assume x is small, such that subtraction from the reactant concentrations is negligible: $0.100 - x \cong 0.100$, and $0.100 - 3x \cong 0.100$.

$$\frac{(2x)^2}{(0.100)(0.100)^3} = 5.97 \times 10^{-2}$$

$$x = \sqrt{\frac{5.97 \times 10^{-2}(0.100)(0.100)^3}{4}} = 0.00122$$

However, this value of x is not very small, so the assumption may not be a good one. Using the method described in Appendix A, Section A.7, plug this approximate value of x back into the equation in the places where we ignored it, to obtain a more accurate value. Repeat the procedure until the value of x stops changing:

$$x = \sqrt{\frac{5.97 \times 10^{-2}(0.100 - 0.00122)(0.100 - 3(0.00122))^3}{4}} = 0.00115$$

$$x = \sqrt{\frac{5.97 \times 10^{-2}(0.100 - 0.00115)(0.100 - 3(0.00115))^3}{4}} = 0.00115$$

$$\text{Percentage } N_2 \text{ converted} = \frac{\text{amount of } N_2 \text{ converted}}{\text{initial amount of } N_2} \times 100\ \%$$

$$= \frac{0.00115\ M}{0.100} \times 100\ \% = 1.15\ \%$$

✓ *Reasonable Answer Check:* The equilibrium concentrations should combine to reproduce K_c:

$[N_2] = 0.100 - x = 0.099$ M, $[H_2] = 0.100 - 3x = 0.097$ M, $[NH_3] = 2x = 0.00230$ M

$$K_c = \frac{(0.00230)^2}{(0.099)(0.097)^3} = 5.9 \times 10^{-2}$$

Chapter 15: The Chemistry of Solutes and Solutions

Solutions for Blue-Numbered Questions for Review and Thought

Topical Questions

How Substances Dissolve (Section 15.1-15.5)

25. *Answer/Explanation:* If the solid interacts with the solvent using similar or stronger intermolecular forces, it will dissolve readily. If the solute interacts with the solvent using different intermolecular forces than those experienced in the solvent, it will be almost insoluble. For example, consider dissolving an ionic solid in water and oil. The interactions between the ions in the solid and water are very strong, since ions would be attracted to the highly polar water molecule; hence the solid would have a relatively high solubility. However, the ions in the solid interact with each other much more strongly than the London dispersion forces experienced between the nonpolar hydrocarbons in the oil; hence, the solid would have a low solubility.

27. *Answer:* **Beaker (c)**

Strategy and Explanation: Benzene is a nonpolar organic compound that interacts with other molecules via London dispersion forces. Water molecules are very polar and are capable of experiencing hydrogen bonding interactions. The two are not miscible. Water has a higher density than benzene, so benzene will sit on top of the water. Beaker (c) best represents the results of their mixing.

29. *Answer/Explanation:* The dissolving process was endothermic, so the temperature dropped as more solute was added. The solubility of the solid at the lower temperature is lower, so some of the solid did not dissolve. As the solution warmed up, however, the solubility increased again. What remained of the solid dissolved. The solution was saturated at the lower temperature, but is no longer saturated at the current temperature.

30. *Answer:* **exothermic**

Strategy and Explanation: The relationship between these processes is described in Section 15.3.

Lattice formation: $CaCl_2(s) \longrightarrow Ca^{2+}(g) + 2\ Cl^-(g)$ $\Delta H_{latt} = -2258$ kJ/mol

Hydration: $+\ Ca^{2+}(g) + 2\ Cl^-(g) \longrightarrow Ca^{2+}(aq) + 2\ Cl^-(aq)$ $+\ \Delta H_{hyd} = +2175$ kJ/mol

Hess's law sum: $CaCl_2(s) \longrightarrow Ca^{2+}(aq) + 2\ Cl^-(aq)$ $\Delta H_{sum} = -83$ kJ/mol

Solution equation: $CaCl_2(s) \longrightarrow Ca^{2+}(aq) + 2\ Cl^-(aq)$ $\Delta H_{soln} = -83$ kJ/mol

The dissolving reaction is exothermic.

31. *Answer/Explanation:* When an organic acid has a large (nonpolar) piece, it interacts primarily using London dispersion intermolecular forces. Since water interacts via hydrogen bonding intermolecular forces, it would rather interact with itself than with the acid. Hence, the solubility of the large organic acids drops, and some are completely insoluble.

33. *Answer/Explanation:* The positive side (H side) of the very polar water molecule interacts with the negative ions. The negative side of the very polar water molecule (O side) interacts with the positive ions.

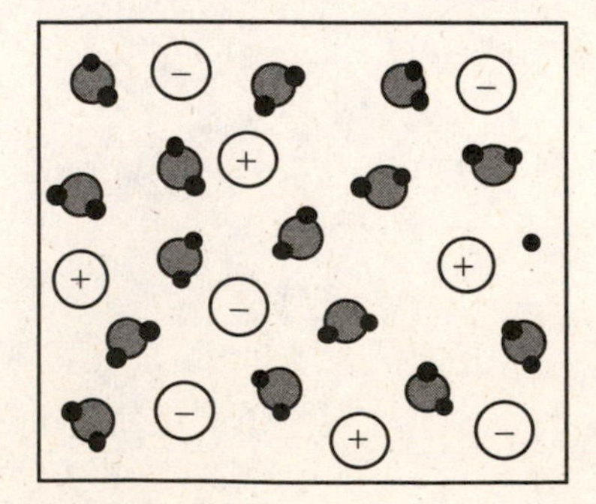

35. *Answer:* **1×10^{-3} M N_2**

Strategy and Explanation: Given the partial pressure of a gas, the temperature, and the Henry's law constant, determine the concentration of the gas dissolved in the water.

Find the partial pressure of the gas in units of mm Hg, then use Henry's law described in Section 15.5: $s_g = k_H P_g$ Where s_g is the solubility of the gas, k_H is the Henry's law constant, and P_g is the partial pressure of a gas.

78% by volume of gaseous N_2 in air is directly proportional to the percentage by mole (Avogadro's law, Section 10.4). The mole percentage is directly proportional to the pressure (Section 10.7), so we get 0.78 atm N_2 in every 1.00 atm air.

$$2.5 \text{ atm air} \times \frac{0.78 \text{ atm } N_2}{1 \text{ atm air}} \times \frac{760 \text{ mmHg } N_2}{1 \text{ atm } N_2} = 1.5 \times 10^3 \text{ mmHg } N_2$$

$$s_g = (8 \times 10^{-7} \text{ mol L}^{-1}\text{mm Hg}^{-1})(1.5 \times 10^3 \text{ mm Hg}) = 1 \times 10^{-3} \text{ mol L}^{-1} = 1 \times 10^{-3} \text{ M}$$

✓ *Reasonable Answer Check:* Always double check to be sure that the gas does not react with the water, since Henry's law does not apply to those gases.

Concentration Units (Section 15.6)

38. *Answer:* **7.32×10^{-3}%**

Strategy and Explanation: The relationship between ppm and percent was derived in the solution to Question 10.13: 1% = 10,000 ppm.

$$73.2 \text{ ppm} \times \frac{1\%}{10000 \text{ ppm}} = 7.32 \times 10^{-3}\%$$

40. *Answer/Explanation:* The definition of parts per billion can be expressed by this ratio: $1 \text{ ppb} = \dfrac{1 \text{ g part}}{10^9 \text{ g whole}}$

$$\text{Therefore, } 1\text{ppb} = \frac{1 \text{ g part}}{10^9 \text{ g whole}} \times \frac{1 \text{ µg part}}{10^{-6} \text{ g part}} \times \frac{1000 \text{ g whole}}{1 \text{ kg whole}} = \frac{1 \text{ µg part}}{1 \text{ kg whole}}$$

42. *Answer:* **90. g ethanol**

Strategy and Explanation: Use standard conversion factors (see Section 15.6 for examples).

$$750 \text{ mL solution} \times \frac{1.00 \text{ g solution}}{1 \text{ mL solution}} \times \frac{12 \text{ g ethanol}}{100 \text{ g solution}} = 90. \text{ g ethanol}$$

45. *Answer:* **1.6×10^{-6} g lead**

Strategy and Explanation: Use standard conversion factors (see Section 15.6 for examples). *(Assume that "1 gallon" of paint is measured with at least two significant figures.)*

$$1.0 \text{ cm}^2 \times \left(\frac{1 \text{ in}}{2.54 \text{ cm}}\right)^2 \times \left(\frac{1 \text{ ft}}{12 \text{ in}}\right)^2 \times \frac{1 \text{ gal paint}}{500. \text{ ft}^2} \times \frac{8.0 \text{ lb paint}}{1 \text{ gal paint}} \times \frac{453.6 \text{ g paint}}{1 \text{ lb pa int}} \times \frac{200. \text{ g Pb}}{10^6 \text{ g paint}} = 1.6 \times 10^{-6} \text{ g Pb}$$

47. *Answer:* **(a) 160. g NH_4Cl (b) 83.9 g KCl (c) 7.46 g Na_2SO_4**

Strategy and Explanation: Use standard conversion factors (see Section 15.6 for examples).

(a) $$750. \text{ mL} \times \frac{1 \text{ L}}{1000 \text{ mL}} \times \frac{4.00 \text{ mol } NH_4Cl}{1 \text{ L}} \times \frac{53.491 \text{ g } NH_4Cl}{1 \text{ mol } NH_4Cl} = 160. \text{ g } NH_4Cl$$

(b) $$1.50 \text{ L} \times \frac{0.750 \text{ mol KCl}}{1 \text{ L}} \times \frac{74.5513 \text{ g KCl}}{1 \text{ mol KCl}} = 83.9 \text{ g KCl}$$

(c) $$150. \text{ mL} \times \frac{1 \text{ L}}{1000 \text{ mL}} \times \frac{0.350 \text{ mol } Na_2SO_4}{1 \text{ L}} \times \frac{142.0422 \text{ g } Na_2SO_4}{1 \text{ mol } Na_2SO_4} = 7.46 \text{ g } Na_2SO_4$$

48. *Answer:* **(a) 0.762 M KCl (b) 0.174 M K_2CrO_4 (c) 0.0126 M $KMnO_4$ (d) 0.167 M $C_6H_{12}O_6$**

Strategy and Explanation: Use standard conversion factors (see Section 15.6 for examples). Calculate quantity of solute in moles then divide by volume of solution in liters.

(a) $$\frac{14.2 \text{ g KCl} \times \dfrac{1 \text{ mol KCl}}{74.5513 \text{ g KCl}}}{250. \text{ mL} \times \dfrac{1 \text{ L}}{1000 \text{ mL}}} = 0.762 \text{ M KCl}$$

The calculation in (a) can be written more concisely as:

$$\frac{14.2 \text{ g KCl}}{250. \text{ mL}} \times \frac{1 \text{ mol KCl}}{74.5513 \text{ g KCl}} \times \frac{1000 \text{ mL}}{1 \text{ L}} = 0.762 \text{ M KCl}$$

(b) $$\frac{5.08 \text{ g } K_2CrO_4}{150. \text{ mL}} \times \frac{1 \text{ mol } K_2CrO_4}{194.1903 \text{ g } K_2CrO_4} \times \frac{1000 \text{ mL}}{1 \text{ L}} = 0.174 \text{ M } K_2CrO_4$$

(c) $$\frac{0.799 \text{ g } KMnO_4}{400. \text{ mL}} \times \frac{1 \text{ mol } KMnO_4}{158.0339 \text{ g } KMnO_4} \times \frac{1000 \text{ mL}}{1 \text{ L}} = 0.0126 \text{ M } KMnO_4$$

(d) $$\frac{15.0 \text{ g } C_6H_{12}O_6}{500. \text{ mL}} \times \frac{1 \text{ mol } C_6H_{12}O_6}{180.1554 \text{ g } C_6H_{12}O_6} \times \frac{1000 \text{ mL}}{1 \text{ L}} = 0.167 \text{ M } C_6H_{12}O_6$$

51. *Answer:* **96% H_2SO_4**

Strategy and Explanation: Use standard conversion factors (see Section 15.6 for examples).

$$\frac{18 \text{ mol } H_2SO_4}{1 \text{ L solution}} \times \frac{1 \text{ L solution}}{1000 \text{ cm}^3 \text{ solution}} \times \frac{1 \text{ cm}^3 \text{ solution}}{1.84 \text{ g solution}} \times \frac{98.078 \text{ g } H_2SO_4}{1 \text{ mol } H_2SO_4} \times 100\ \% = 96\%\ H_2SO_4$$

53. *Answer:* **0.1 M NaCl**

Strategy and Explanation: Calculate quantity of solute in moles then divide by volume of solution in liters.

$$\frac{4. \text{ mg NaCl} \times \dfrac{1 \text{ g}}{1000 \text{ mg}} \times \dfrac{1 \text{ mol NaCl}}{58.4428 \text{ g NaCl}}}{0.6 \text{ mL} \times \dfrac{1 \text{ L}}{1000 \text{ mL}}} = 0.1 \text{ M NaCl}$$

55. *Answer:* **59 g $C_2H_4(OH)_2$**

Strategy and Explanation: Use standard conversion factors (see Section 15.6 for examples)

$$950. \text{ g } H_2O \times \frac{1 \text{ kg } H_2O}{1000 \text{ g } H_2O} \times \frac{1.0 \text{ mol } C_2H_4(OH)_2}{1 \text{ kg } H_2O} \times \frac{62.07 \text{ g } C_2H_4(OH)_2}{1 \text{ mol } C_2H_4(OH)_2} = 59 \text{ g } C_2H_4(OH)_2$$

57. *Answer:* **1.2×10^{-7} M lead**

Strategy and Explanation: Use standard conversion factors (see Section 15.6 for examples).

$$\frac{25 \text{ g lead}}{10^9 \text{ g solution}} \times \frac{1.00 \text{ g solution}}{1 \text{ mL solution}} \times \frac{1000 \text{ mL solution}}{1 \text{ L solution}} \times \frac{1 \text{ mol lead}}{207.2 \text{ g lead}} = 1.2 \times 10^{-7} \text{ M lead}$$

60. *Answer:* **(a) 108 ppm (b) 5.64×10^{-4} M (c) 5.58×10^{-4} mol/kg**

Strategy and Explanation: Given volume of solution and the mass of solute, determine the concentration in units of ppm, molarity and molality of the solute.

(a) Parts per million (ppm) is usually by mass (see Section 15.6, page 722). To get ppm, divide the milligrams of solute by the kilograms of solution.

$$355 \text{ mL solution} \times \frac{1.01 \text{ g solution}}{1 \text{ mL solution}} \frac{1 \text{ kg}}{1000 \text{ g}} = 0.359 \text{ kg solution}$$

$$\frac{38.9 \text{ mg caffeine}}{0.359 \text{ kg water}} = 108 \text{ ppm}$$

(b) Molarity is moles solute per liter solution. The molar mass of caffeine is 194.2 g/mol.

$$38.9 \text{ mg caffeine} \times \frac{1 \text{ g}}{1000 \text{ mg}} \times \frac{1 \text{ mol caffeine}}{194.2 \text{ g caffeine}} = 2.00 \times 10^{-4} \text{ mol caffeine}$$

$$355 \text{ mL solution} \times \frac{1 \text{ L}}{1000 \text{ mL}} = 0.355 \text{ mL solution}$$

$$\frac{2.00 \times 10^{-4} \text{ mol caffeine}}{0.355 \text{ L}} = 5.64 \times 10^{-4} \text{ M}$$

(c) To get molality, m, divide the moles of solute by the mass of solvent in kilograms.

$$m_{\text{solute}} = \frac{2.00 \times 10^{-4} \text{ mol caffeine}}{0.359 \text{ kg water}} = 5.58 \times 10^{-4} \; m$$

Colligative Properties (Section 15.7)

63. *Answer/Explanation*: The addition of a solute to a solvent lowers the vapor pressure of the solvent, so the upper curve is for benzene and the lower curve is for the solution.

65. *Answer:* **100.26 °C**

Strategy and Explanation: Given the moles of solute, the mass of solvent and the normal boiling point of the solvent, determine the boiling point of the solution.

Use method described in Section 15.7.

Calculate the molality of the solute. Use molality to get ΔT_b with the molal boiling-point-elevation constant of the solvent, given in the text as 0.52 °C kg mol^{-1}. Then add ΔT_b to the normal boiling point (exactly 100 °C) to get the boiling point of the solution.

$$m_{\text{solute}} = \frac{15.0 \text{ g } (NH_2)_2CO}{0.500 \text{ kg water}} \times \frac{1 \text{ mol } (NH_2)_2CO}{60.06 \text{ g} (NH_2)_2CO} = 0.500 \text{ mol / kg}$$

$$\Delta T_b = (0.52 \text{ °C kg mol}^{-1}) \times (0.500 \text{ mol/kg}) = 0.26 \text{ °C}$$

$$T_b(\text{solution}) = 100.00 \text{ °C} + 0.26 \text{ °C} = 100.26 \text{ °C}$$

✓ *Reasonable Answer Check:* The boiling point is elevated by the presence of the solute.

67. *Answer:* **boiling point of (a) < boiling point of (d) < boiling point of (b) < boiling point of (c)**

Strategy and Explanation: In Section 15.7, we learn that the freezing point decreases as the molality of solute particles increases. As described near the end of Section 15.7, page 733, solutes that ionize provide a larger number of particles in the solution than solutes that do not ionize. So, we must calculate the concentration of particles in each solution and order the solutions from the smallest concentration to the largest concentration.

(a) Each methanol (a covalent molecule) provides one particle. = 0.10 mol/kg particles

(b) Each KCl ionizes into K^+ and Cl^-, two particles.

$$0.10 \text{ mol/kg KCl} \times (2 \text{ particles/mol KCl}) = 0.20 \text{ mol/kg particles}$$

(c) Each $BaCl_2$ ionizes into Ba^{2+} and 2 Cl^-, three particles.

$$0.080 \text{ mol/kg } BaCl_2 \times (3 \text{ particles/mol } BaCl_2) = 0.24 \text{ mol/kg particles}$$

(d) Each Na_2SO_4 ionizes into 2 Na^+ and SO_4^{2-}, three particles.

$$0.040 \text{ mol/kg } Na_2SO_4 \times (3 \text{ particles/mol } Na_2SO_4) = 0.12 \text{ mol/kg particles}$$

Order of decreasing freezing points: (a) < (d) < (b) < (c)

69. *Answer:* **$T_f = -1.65$ °C, $T_b = 100.46$ °C**

Strategy and Explanation: Follow the method described in the solution to Question 65 for the boiling point calculation and a similar method for the freezing point calculation, as described in Section 15.7:

$$T_f(\text{solution}) = T_f(\text{solvent}) - \Delta T_f$$

$$\Delta T_f = K_f m_{\text{solute}}$$

Molal boiling-point-elevation and molal freezing-point-lowering constants for water are given in the text.

$$m_{solute} = \frac{4.00 \text{ g } CO(NH_2)_2}{75.0 \text{ g water}} \times \frac{1 \text{ mol } CO(NH_2)_2}{60.06 \text{ g } CO(NH_2)_2} \times \frac{1000 \text{ g water}}{1 \text{ kg water}} = 0.888 \text{ mol/kg}$$

$$\Delta T_b = (0.52\ °C \text{ kg mol}^{-1}) \times (0.888 \text{ mol/kg}) = 0.46\ °C$$

$$T_b(\text{solution}) = 100.00\ °C + 0.46\ °C = 100.46\ °C$$

$$\Delta T_f = (1.86\ °C \text{ kg mol}^{-1}) \times (0.888 \text{ mol/kg}) = 1.65\ °C$$

$$T_f(\text{solution}) = 0.00\ °C - 1.65\ °C = -1.65\ °C$$

71. *Answer:* **X_{H_2O} = 0.79999; 712 g sucrose**

Strategy and Explanation: Given the vapor pressure of the solvent at a given temperature, the vapor pressure of a solution at the same temperature and the mass of solvent, determine the mole fraction of the solvent and the mass of solute in the solution. Use Raoult's law, described in Section 15.7.

$$X_{H_2O} = \frac{P_{H_2O}}{P^0_{H_2O}} = \frac{119.55 \text{ mmHg}}{149.44 \text{ mmHg}} = 0.79999$$

$$150. \text{ g } H_2O \times \frac{1 \text{ mol } H_2O}{18.02 \text{ g } H_2O} = 8.32 \text{ mol } H_2O$$

$$X_{H_2O} = \frac{n_{H_2O}}{n_{H_2O} + n_{sucrose}}$$

Solve for $n_{sucrose}$:

$$X_{H_2O}(n_{H_2O} + n_{sucrose}) = X_{H_2O}n_{H_2O} + X_{H_2O}n_{sucrose} = n_{H_2O}$$

$$X_{H_2O}n_{sucrose} = n_{H_2O} - X_{H_2O}n_{H_2O}$$

$$n_{sucrose} = \frac{(1-X_{H_2O})n_{H_2O}}{X_{H_2O}} = \frac{(1-0.79999)(8.32 \text{ mol})}{0.79999} = \frac{(0.20001)(8.32 \text{ mol})}{0.79999} = 2.08 \text{ mol}$$

$$2.08 \text{ mol } C_{12}H_{22}O_{11} \times \frac{342.297 \text{ g } C_{12}H_{22}O_{11}}{1 \text{ mol } C_{12}H_{22}O_{11}} = 712 \text{ g } C_{12}H_{22}O_{11}$$

✓ *Reasonable Answer Check:* The mole fraction of sucrose = 2.08 mol/(8.32 mol + 2.08 mol) = 0.200. The sum of all the mole fractions (0.79999 + 0.200) is one.

73. *Answer:* **190 g/mol**

Strategy and Explanation: Adapt the method described in the solution to Question 65 to find moles of solute. Divide mass by moles to get molar mass.

$$\Delta T_b = 0.65\ °C$$

$$\Delta T_b = K_b m_{unknown}$$

$$m_{unknown} = \frac{\Delta T_b}{K_b} = \frac{0.65\ °C}{2.53\ °C \text{ kg mol}^{-1}} = 0.26 \text{ mol/kg}$$

$$100. \text{ g benzene} \times \frac{1 \text{ kg benzene}}{1000 \text{ g benzene}} \times \frac{0.26 \text{ mol unknown}}{1 \text{ kg benzene}} = 0.026 \text{ mol unknown}$$

$$\text{Molar mass} = \frac{5.0 \text{ g}}{0.026 \text{ mol}} = 190 \text{ g/mol}$$

75. *Answer:* **3.6×10^2 g/mol; $C_{20}H_{16}Fe_2$**

Strategy and Explanation: Adapt the method described in the solution to Question 73 to find moles of solute, then divide mass by moles for molar mass. Use methods described in Chapter 3 to find molecular formula.

$$\Delta T_b = T_b(\text{solution}) - T_b(\text{solvent}) = 80.26\ °C - 80.10\ °C = 0.16\ °C$$

$$\Delta T_b = K_b m_{C_{10}H_8Fe}$$

$$m_{C_{10}H_8Fe} = \frac{\Delta T_b}{K_b} = \frac{0.16\ ^\circ C}{2.53\ ^\circ C\ kg\ mol^{-1}} = 0.063\ mol/kg$$

$$11.12\ g\ benzene \times \frac{1\ kg\ benzene}{1000\ g\ benzene} \times \frac{0.063\ mol\ C_{10}H_8Fe}{1\ kg\ benzene} = 0.00070\ mol\ C_{10}H_8Fe$$

$$\text{Molar mass of the compound} = \frac{0.255\ g}{0.00070\ mol} = 3.6 \times 10^2\ g/mol$$

The molecular formula is a multiple of the empirical formula: $(C_{10}H_8Fe)_n$

The molar mass of the empirical formula $C_{10}H_8Fe$ is 184.01 g/mol

$$n = \frac{3.6 \times 10^2\ g/mol}{184.01\ g/mol} = 2.0 \cong 2$$

The molecular formula is: $C_{20}H_{16}Fe_2$

77. *Answer:* **1.8 × 10² g/mol; $C_{14}H_{10}$**

Strategy and Explanation: Adapt the method described in the solution to Question 73 to find moles of solute, then divide mass by moles for molar mass. Use methods described in Chapter 3 to find the molecular formula.

$$\Delta T_b = T_b(\text{solution}) - T_b(\text{solvent})$$

$$\Delta T_b = 80.34\ ^\circ C - 80.10\ ^\circ C = 0.24\ ^\circ C$$

$$\Delta T_b = K_b m_{C_7H_5}$$

$$m_{C_7H_5} = \frac{\Delta T_b}{K_b} = \frac{0.24\ ^\circ C}{2.53\ ^\circ C\ kg\ mol^{-1}} = 0.095\ mol/kg\ \textit{(two sig figs)}$$

$$30.0\ g\ benzene \times \frac{1\ kg\ benzene}{1000\ g\ benzene} \times \frac{0.095\ mol\ C_7H_5}{1\ kg\ benzene} = 0.0029\ mol\ C_7H_5\ \textit{(two sig figs)}$$

$$\text{Molar mass of the compound} = \frac{0.500\ g}{0.0029\ mol} = 1.8 \times 10^2\ g/mol\ \textit{(two sig figs)}$$

The molecular formula is a multiple of the empirical formula: $(C_7H_5)_n$

The molar mass of the empirical formula C_7H_5 is 89.117 g/mol

$$n = \frac{1.8 \times 10^2\ g/mol}{89.117\ g/mol} = 2.0 \cong 2$$

The molecular formula is: $C_{14}H_{10}$

✓ *Reasonable Answer Check:* Anthracene is described in Question 8.77, and its formula is determined in the solution to Question 8.77 to be $C_{14}H_{10}$.

79. *Answer:* **(a) 2.5 kg $C_2H_6C_2$ (b) 104.2 °C**

Strategy and Explanation: Adapt the method described in the solutions to Questions 60 and 69:

(a)

$$\Delta T_f = T_f(\text{solvent}) - T_f(\text{solution}) = 0.0\ ^\circ C - (-15.0\ ^\circ C) = 15.0\ ^\circ C$$

$$m_{C_2H_6O_2} = \frac{\Delta T_f}{K_f} = \frac{15.0\ ^\circ C}{1.86\ ^\circ C\ kg\ mol^{-1}} = 8.06\ mol/kg$$

$$5.0\ kg\ water \times \frac{8.06\ mol\ C_2H_6O_2}{1\ kg\ water} \times \frac{62.07\ g\ C_2H_6O_2}{1\ mol\ C_2H_6O_2} \times \frac{1\ kg\ C_2H_6O_2}{1000\ g\ C_2H_6O_2} = 2.5\ kg\ C_2H_6O_2$$

You must add 2.5 kg of ethylene glycol to 5.0 kg of water for this much freezing protection.

(b)

$$\Delta T_b = (0.52\ ^\circ C\ kg\ mol^{-1}) \times (8.06\ mol/kg) = 4.2\ ^\circ C$$

$$T_b(\text{solution}) = 100.00\ ^\circ C + 4.2\ ^\circ C = 104.2\ ^\circ C$$

81. *Answer:* **29 atm**

Strategy and Explanation: Given the freezing point of a solution, the temperature, and an assumption that the density is the same as pure water, determine the osmotic pressure.

Adapt the method described in Problem-Solving Example 15.12 to find the molality of the solution. With the molality and the density of the solution, determine the molar concentration. Then use the equation described in Section 15.7 to get the osmotic pressure: $\Pi = cRT$ Π is the osmotic pressure, c is the concentration of the solute in the solution, R is the familiar gas constant with liter and atmosphere units, 0.08206 L·atm/mol·K, and T is the absolute temperature in kelvin units.

$\Delta T_f = T_f(\text{solvent}) - T_f(\text{solution}) = 0.0\ °C - (-2.3\ °C) = 2.3\ °C$

$$m_{\text{solute}} = \frac{\Delta T_f}{K_f} = \frac{2.3\ °C}{1.86\ °C\ \text{kg mol}^{-1}} = 1.2\ \text{mol/kg}$$

Assume the mass of the solvent is essentially equal to the mass of the solution: T = 20.0 °C + 273.15 = 293.2 K

$$\frac{1.2\ \text{mol solute}}{1\ \text{kg solvent}} \times \frac{1\ \text{kg solvent}}{1\ \text{kg solution}} \times \frac{1.00\ \text{kg solution}}{1\ \text{L solution}} = 1.2\ \text{mol/L}$$

$$\Pi = (1.2\ \text{mol/L})(0.08206\ \text{L·atm/mol·K})(293.2\ \text{K}) = 29\ \text{atm}$$

✓ *Reasonable Answer Check:* The solute concentration of seawater is fairly high, so it makes sense that the osmotic pressure is relatively high, also.

Water: Purification and Solutions (Section 15.10)

84. *Answer:* **5×10^{-4} g As**

Strategy and Explanation: This is a standard conversion factor problem.

$$1\ \text{week} \times \frac{7\ \text{days}}{1\ \text{week}} \times \frac{6\ \text{glasses}}{1\ \text{day}} \times \frac{8\ \text{oz}}{1\ \text{glass}} \times \frac{1\ \text{quart}}{32\ \text{oz}} \times \frac{1\ \text{L}}{1.0567\ \text{quart}} \times \frac{1000\ \text{mL}}{1\ \text{L}}$$

$$\times \frac{1.00\ \text{g water}}{1\ \text{mL}} \times \frac{0.050\ \text{g As}}{10^6\ \text{g water}} = 5 \times 10^{-4}\ \text{g As}$$

86. *Answer:* **No**

Strategy and Explanation: The relationship between ppm and ppb was derived in the solution to Question 10.13:

$1\ \text{ppm} = 1{,}000\ \text{ppb}.$ $$5\ \text{ppb chlorodane} \times \frac{1\ \text{ppm chlorodane}}{1000\ \text{ppb chlorodane}} = 0.005\ \text{ppm}$$

This concentration is higher than 0.002 ppm, so it is not within the MCL for chlordane.

88. *Answer/Explanation:* The lime-soda process relies on the precipitation of insoluble compounds to remove the "hard water" ions. The ion exchange process relies on the high charge of the "hard water" ions to attract them to an ion exchange resin, thereby removing them from the water.

90. *Answer/Explanation:* Water is sprayed into the air to oxidize organic substances dissolved in it.

92. *Answer/Explanation:* Fish breathe oxygen by extracting it from the water. Plants "breathe" carbon dioxide by extracting it from the water. The concentrations of these gases in calm water drop, unless they are replenished. The concentration of dissolved gases in the water is replenished by bubbling air through the water in the aquarium.

General Questions

94. *Answer:* **4%**

Strategy and Explanation: This is a standard percent problem: *Assume all masses are known ± 0.1 g.*

Total mass = 5.0 g + 0.2 g + 0.3 g = 5.5 g $$\text{Weight percent of A} = \frac{0.2\ \text{g A}}{5.5\ \text{g total}} \times 100\% = 4\%$$

96. *Answer/Explanation:* Water in the cells of the wood leaked out, since the osmotic pressure inside the cells was less than that of the seawater the wood was sitting in. See the discussion about hypertonic solutions and cells in Section 15.7.

98. *Answer:* **28% NH_3**

Strategy and Explanation: Use standard conversion factors (see Section 15.6 for examples).

$$\frac{14.8 \text{ mol } NH_3}{1 \text{ L solution}} \times \frac{1 \text{ L solution}}{1000 \text{ cm}^3 \text{ solution}} \times \frac{1 \text{ cm}^3 \text{ solution}}{0.90 \text{ g solution}} \times \frac{17.03 \text{ g } NH_3}{1 \text{ mol } NH_3} \times 100\% = 28\% \text{ } NH_3$$

100. *Answer:* **0.982 mol/kg; 10.2% $C_4H_8N_2O_2$**

Strategy and Explanation: Use standard conversion factors (see Section 15.6 for examples)

$$500. \text{ mL } CH_3OH \times \frac{0.7893 \text{ g } CH_3OH}{1 \text{ mL } CH_3OH} = 395. \text{ g } CH_3OH$$

$$\text{molality} = \frac{45.0 \text{ g } C_4H_8N_2O_2}{395 \text{ g ethanol}} \times \frac{1 \text{ mol } C_4H_8N_2O_2}{116.13 \text{ g } C_4H_8N_2O_2} \times \frac{1000 \text{ g}}{1 \text{ kg}} = 0.982 \text{ mol/kg}$$

$$\text{Total mass} = 45.0 \text{ g} + 395. \text{ g} = 440. \text{ g}$$

$$\text{Weight percent of } C_4H_8N_2O_2 = \frac{45.0 \text{ g } C_4H_8N_2O_2}{440. \text{ g total}} \times 100\% = 10.2\% \text{ } C_4H_8N_2O_2$$

102. *Answer:* **freezing point of (a) = freezing point of (c) < freezing point of (d) < freezing point of (b)**

Strategy and Explanation: Use the method described in the solution to Question 67.

(a) Each ethylene glycol (a covalent molecule) provides one particle, so the actual concentration is 0.20 mol/kg particles.

(b) Each Na_2SO_4 ionizes into 2 Na^+ and SO_4^{2-}, three particles.

$$0.12 \text{ mol/kg } Na_2SO_4 \times (3 \text{ particles/mol } Na_2SO_4) = 0.36 \text{ mol/kg particles}$$

(c) Each NaBr ionizes into Na^+ and Br^-, two particles.

$$0.10 \text{ mol/kg NaBr} \times (2 \text{ particles/mol NaBr}) = 0.20 \text{ mol/kg particles}$$

(d) Each KI ionizes into K^+ and I^-, two particles.

$$0.12 \text{ mol/kg KI} \times (2 \text{ particles/mol KI}) = 0.24 \text{ mol/kg particles}$$

Order of decreasing freezing points: (a) = (c) < (d) < (b)

104. *Answer:* **1.77 g $[(C_4H_9)_4N][ClO_4]$**

Strategy and Explanation: Adapt the method described in the solution to Question 78.

$$\Delta T_b = T_b(\text{solution}) - T_b(\text{solvent}) = 63.20\ °C - 61.70\ °C = 1.50\ °C$$

$$\Delta T_b = K_b m_{solute} i$$

One $[(C_4H_9)_4N][ClO_4]$ ionizes into $(C_4H_9)_4N^+$ and ClO_4^-, two particles, so i = 2.

$$m_{solute} = \frac{\Delta T_b}{K_b i} = \frac{1.50\ °C}{(3.63\ °C \text{ kg mol}^{-1})(2)} = 0.207 \text{ mol / kg}$$

$$25.0 \text{ g chloroform} \times \frac{1 \text{ kg chloroform}}{1000 \text{ g chloroform}} \times \frac{0.207 \text{ mol } [(C_4H_9)_4N][ClO_4]}{1 \text{ kg chloroform}} \times \frac{341.90 \text{ g } [(C_4H_9)_4N][ClO_4]}{1 \text{ mol } [(C_4H_9)_4N][ClO_4]} = 1.77 \text{ g } [(C_4H_9)_4N][ClO_4]$$

106. *Answer:* **(a) see equation below (b) 15.0 mL $Ba(NO_3)_2$ (b) 12.0 mL Na_2SO_4**

Strategy and Explanation: Use the methods from Chapter 3, 4 and 5.

(a) The balanced molecular equation has two sodium ions, two nitrate ions, one barium ion and one sulfate ion on each side of the equation:

$$Na_2SO_4(aq) + Ba(NO_3)_2(aq) \longrightarrow 2\ NaNO_3(aq) + BaSO_4(s)$$

(b) $0.700\ g\ BaSO_4 \times \frac{1\ mol\ BaSO_4}{233.390\ g\ BaSO_4} \times \frac{1\ mol\ Ba(NO_3)_2}{1\ mol\ BaSO_4} \times \frac{1\ L\ Ba(NO_3)_2}{0.200\ mol\ Ba(NO_3)_2} \times \frac{1000\ mL}{1\ L} = 15.0\ mL\ Ba(NO_3)_2$

(c) $0.700\ g\ BaSO_4 \times \frac{1\ mol\ BaSO_4}{233.390\ g\ BaSO_4} \times \frac{1\ mol\ Na_2SO_4}{1\ mol\ BaSO_4} \times \frac{1\ L\ Na_2SO_4}{0.250\ mol\ Na_2SO_4} \times \frac{1000\ mL}{1\ L} = 12.0\ mL\ Na_2SO_4$

Applying Concepts

108. *Answer:* **(a)**

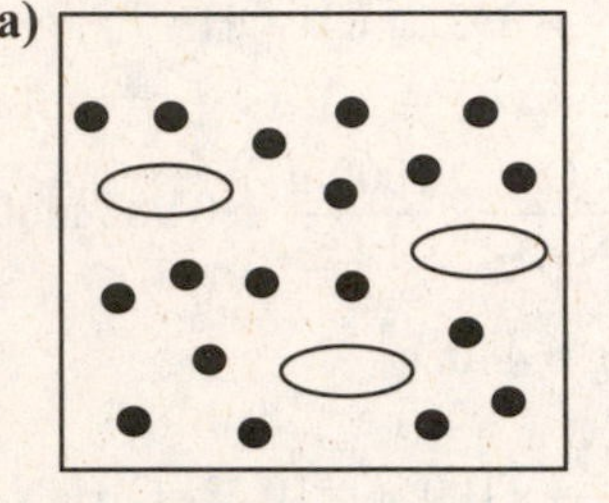

(b)

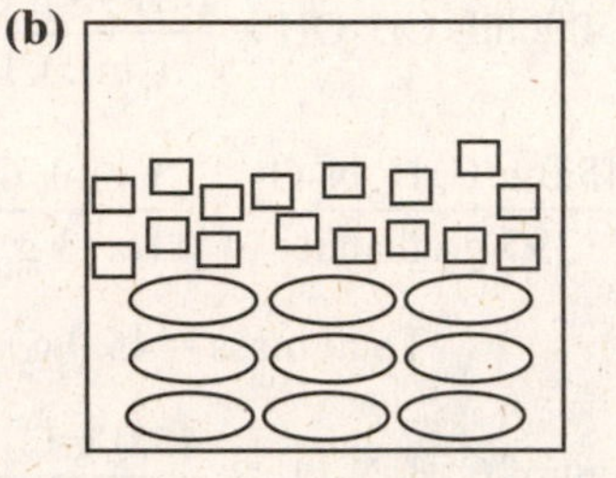

Strategy and Explanation:

(a) Sugar and water interact with the same hydrogen bonding attractive forces, so they will commingle.

(b) Carbon tetrachloride and sugar interact with very different interactive forces, so they will remain separate phases.

110. *Answer:* **(a) unsaturated (b) supersaturated (c) supersaturated (d) unsaturated**

Strategy and Explanation: Figure 15.11 gives curves showing solubility versus temperature. If the concentration is on the curve at the given temperature, the solution is saturated. If the concentration is above the curve at the given temperature, the solution is supersaturated. If the concentration is below the curve at the given temperature, the solution is unsaturated.

(a) 40 g NH_4Cl/100 g H_2O at 80 °C: This point is below the curve, so the solution is unsaturated.

(b) 100 g LiCl/100 g H_2O at 30 °C: This point is above the curve, so the solution is supersaturated.

(c) 120 g $NaNO_3$/100 g H_2O at 40 °C: This point is above the curve, so the solution is supersaturated.

(d) 25 g Li_2SO_4/100 g H_2O at 50 °C: This point is below the curve, so the solution is unsaturated.

112. *Answer:* **see chart below**

Strategy and Explanation: Mass fraction is calculated by dividing the mass of the substance by the total mass of the solution. Calculate weight percent by multiplying mass fraction by 100%. Calculate ppm, using the equality 1% = 10,000 ppm.

Compound	Mass of compound	Mass of water	Mass fraction	Weight percent	ppm of solute
Lye	**75.0 g**	125 g	0.375	**37.5%**	**3.75×10^5**
Glycerol	33 g	200. g	**0.14**	**14%**	**1.4×10^5**
Acetylene	0.0015 g	**2×10^2 g**	**0.000009**	0.0009%	**9**

114. *Answer/Explanation:* Molecules slow down and move less. The reduced motion prevents them from randomly translocating as they had in the liquid state. As a result, the intermolecular forces between one molecule and the next begin to organize them into a crystal form. The presence of a nonvolatile solute disrupts the formation of the crystal. Its size and shape will be different from that of the solute. Intermolecular forces between the solute and solvent are also different from those of solvent molecules with each other. To form the crystalline solid, the solute has to be excluded. If the ice in an iceberg is in regular crystalline form, the water will be pure. Only if the ice is crushed or dirty will other particles be included. So, melting an iceberg will produce relatively pure water.

116. *Answer/Explanation:*

(a) Sea water contains more dissolved solutes than fresh water. The presence of a solute lowers the freezing point. That means a lower temperature is required to freeze the sea water than to freeze fresh water.

(b) Salt added to a mixture of ice and water will lower the freezing point of the water. If the ice cream is mixed at a lower temperature, its temperature will drop faster; hence, it will freeze faster.

More Challenging Questions

119. *Answer:* **(a) Empirical formula is $C_{18}H_{24}Cr$. (b) Molecular formula is $C_{18}H_{24}Cr$.**

Strategy and Explanation: This is a combination of the problem-solving from several different chapters.

(a) From Chapter 3: 100.00 g compound – 73.94 g C – 8.27 g H = 17.79 g Cr

$$73.94 \text{ g C} \times \frac{1 \text{ mol C}}{12.0107 \text{ g C}} = 6.156 \text{ mol C} \qquad 8.27 \text{ g H} \times \frac{1 \text{ mol H}}{1.0079 \text{ g H}} = 8.21 \text{ mol H}$$

$$17.79 \text{ g Cr} \times \frac{1 \text{ mol Cr}}{51.996 \text{ g Cr}} = 0.342 \text{ mol Cr}$$

Mole Ratio: 6.156 mol C : 8.21 mol H : 0.342 mol Cr

Simplify the ratio: 18 C : 24 H : 1 Cr — Empirical formula is $C_{18}H_{24}Cr$.

(b) Adapt the solution to Question 81.

$$T = 25\ °C + 273.15 = 298 \text{ K}$$

$$c = \frac{\Pi}{RT} = \frac{3.17 \text{ mmHg}\left(\frac{1 \text{ atm}}{760 \text{ mmHg}}\right)}{\left(0.08206 \frac{\text{L}\cdot\text{atm}}{\text{mol}\cdot\text{K}}\right)(298 \text{ K})} = 1.71 \times 10^{-4} \text{ mol/L}$$

Assuming that the addition of solute does not change the volume of the solution, so the volume of the solvent is equal to the volume of the solution:

$$100. \text{ mL chloroform} \times \frac{1 \text{ L chloroform}}{1000 \text{ mL chloroform}} \times \frac{1 \text{ L solution}}{1 \text{ L chloroform}}$$

$$\times \frac{1.71 \times 10^{-4} \text{ mol solute}}{1 \text{ L solution}} = 1.71 \times 10^{-5} \text{ mol solute}$$

$$\text{Molar mass of the compound} = \frac{5.00 \text{ mg}}{1.71 \times 10^{-5} \text{ mol}} \times \frac{1 \text{ g}}{1000 \text{ mg}} = 292 \text{ g/mol}$$

The molecular formula is a multiple of the empirical formula: $(C_{18}H_{24}Cr)_n$

The molar mass of the empirical formula is 292.37 g/mol, so the molecular formula is $C_{18}H_{24}Cr$.

121. *Answer:* **(a) No (b) 108.9 °C**

Strategy and Explanation: Adapt the method described in the solutions to Questions 60 and 66:

(a) Calculate the molality from the proof:

$$\frac{100 \text{ proof ethanol}}{2} = 50\% \text{ by volume ethanol}$$

$$\frac{50.0 \text{ mL } C_2H_5OH}{50.0 \text{ mL water}} \times \frac{0.789 \text{ g } C_2H_5OH}{1 \text{ mL } C_2H_5OH} \times \frac{1 \text{ mL } H_2O}{1.00 \text{ g } H_2O} \times \frac{1000 \text{ g}}{1 \text{ kg}}$$

$$\times \frac{1 \text{ mol } C_2H_5OH}{46.068 \text{ g } C_2H_5OH} = \frac{17.13 \text{ mol } C_2H_5OH}{\text{kg } H_2O}$$

$$\Delta T_f = m_{C_2H_6O_2} K_f = (17.13 \text{ mol/kg})(1.86\ °\text{C kg mol}^{-1}) = 31.9\ °C$$

$$T_f(50\% \text{ solution}) = T_f(H_2O) - \Delta T_f = 0.0\ °C - (31.9\ °C) = -31.9\ °C$$

So, the vodka will not freeze at a temperature of –15°C.

(b) $\Delta T_b = m_{C_2H_6O_2}K_b = (17.13 \text{ mol/kg})(0.52\ °C \text{ kg mol}^{-1}) = 8.9\ °C$

$$T_b(50\% \text{ solution}) = T_b(H_2O) + \Delta T_b = 100.00\ °C + 8.9\ °C = 108.9\ °C$$

(Notice: This answer assumes that the dissolved alcohol will not evaporate from the solution as the temperature is raised, which is probably not true. If the alcohol does evaporate, that will lower the concentration of the solute, and the boiling temperature will be closer to that of pure water than that calculated here.)

123. *Answer:* **28 m**

Strategy and Explanation: Given the concentration of the non-electrolyte solute present in tree sap, determine the height of the sap in a tree.

Use methods described in Section 15.7 and in the solution to Question 16 to determine the osmotic pressure (in mm Hg) for a nonelectrolyte related to the difference in the concentration of the sap inside and outside of the tree, then compare the measure of liquid height by relating their densities (see Table 1.1 of the textbook for these densities). Assume the temperature is 25.00°C and the density of a dilute aqueous solution is the same as water.

$$\Pi = cRT$$

$$\Pi = (0.13\ \text{M} - 0.020\ \text{M}) \times \left(0.08206 \frac{\text{L}\cdot\text{atm}}{\text{mol}\cdot\text{K}}\right) \times (298.15\ \text{K}) \times \left(\frac{760\ \text{mmHg}}{1\ \text{atm}}\right) = 2.0 \times 10^3\ \text{mmHg}\ \textit{(two sig figs)}$$

Since pressure is proportional to density of the liquid, 1 mm Hg is directly related to the density of mercury with the same proportionality constant as 1 mm sap is related to the density of sap.

$$2.0 \times 10^3\ \text{mmHg} \times \frac{13.55\ \text{g/mL}}{1\ \text{mmHg}} \times \frac{1\ \text{mm sap}}{0.998\ \text{g/mL}} \times \frac{1\ \text{m}}{1000\ \text{mm}} = 28\ \text{m sap}$$

125. *Answer:* **0.30 mol/L**

Strategy and Explanation: Adapt method in the solution to Question 79. T = 37 °C + 273.15 = 310. K

$$c = \frac{\Pi}{RTi} = \frac{7.7\ \text{atm}}{\left(0.08206 \frac{\text{L}\cdot\text{atm}}{\text{mol}\cdot\text{K}}\right)(310.\ \text{K})(1.00)} = 0.30\ \text{mol/L}$$

127. *Answer:* **12.3 M; 44.1 *m***

Strategy and Explanation:

(a) A graph of mass percent of ethanol (on the x-axis) and boiling point (on the y-axis).

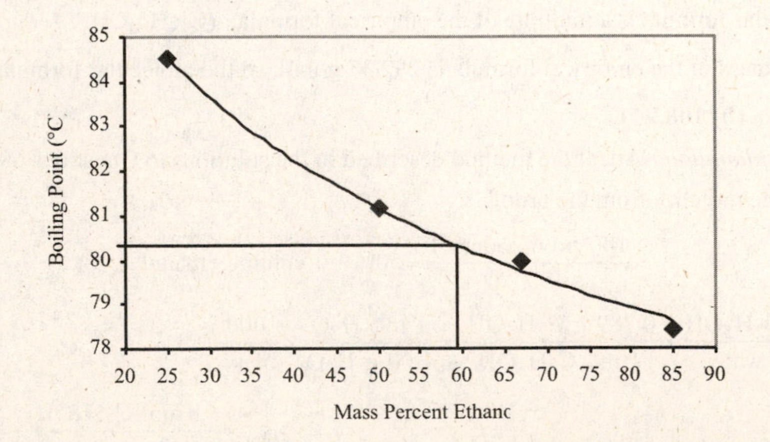

The vertical line drawn up to the best fit non-linear curve from the point representing 59% ethanol by mass produces a horizontal line at 80.4 °C, which therefore represents the estimated boiling point of a solution at that concentration.

(b) In exactly 100 grams of the solution that is 67.0% by mass ethanol, there are 67.0 grams of ethanol and 33.0 grams of water.

Find molarity: We need the density of the solution, which we will assume is a weighted average of the densities of the two liquids:

$$67.0\text{ g ethanol} \times \frac{1\text{ mL}}{0.789\text{ g ethanol}} + 33.0\text{ g water} \times \frac{1\text{ mL}}{0.998\text{ g water}} = 117.9\text{ mL in exactly 100 grams}$$

$$d_{\text{soln}} = \frac{100\text{ g}}{117.9\text{ mL}} = 0.848\text{ g/mL}$$

$$\frac{67.0\text{ g ethanol}}{100\text{ g solution}} \times \frac{0.848\text{ g solution}}{1\text{ mL solution}} \times \frac{1000\text{ mL}}{1\text{ L}} \times \frac{1\text{ mole ethanol}}{46.0682\text{ ethanol}} = 12.3\text{ M}$$

Find molality: We need the mass of solvent = 100 g – 67 g ethanol = 33 g water

$$\frac{67.0\text{ g ethanol}}{33.0\text{ g solvent}} \times \frac{1000\text{ g}}{1\text{ kg}} \times \frac{1\text{ mol ethanol}}{46.0682\text{ g ethanol}} = 44.1\text{ mol/kg} = 44.1\ m$$

129. *NOTE: This question involves an aqueous solution that is not characterized well by the simple equations describing colligative properties given in the textbook. Using those equations in this question does not produce sensible answers. Plan to describe non-ideal solutions before assigning this question.*

Answer: **(a) see graph below; –3.36 °C (b) 39000%, 9400%, 810%, 80%**

Strategy and Explanation:

(a) A graph of molality of ammonium chloride (on the x-axis) and freezing point (on the y-axis).

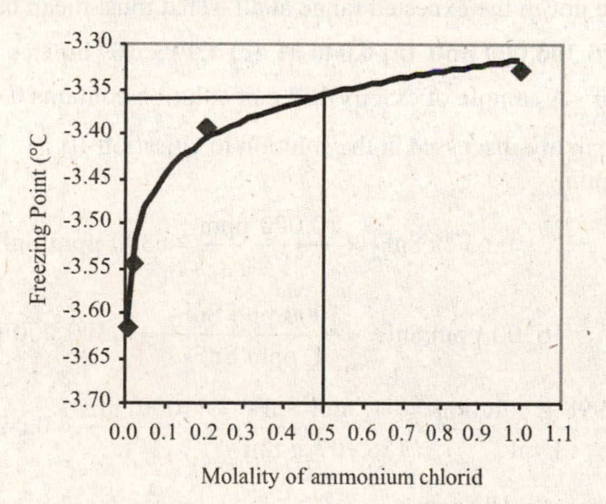

The vertical line drawn down to the best fit non-linear curve from the point representing 0.50 m ammonium chloride produces a horizontal line at –3.36 °C, which therefore represents the estimated freezing point of a solution at that concentration.

(b) If ammonium chloride does not dissociate at all, the value of i would be 1. If it dissociates completely it would form NH_4^+ and Cl^-, and theoretically i would be 2. As a result, the expected range of i values is from 1 (at 0% ionized) to 2 (at 100%).

First solution: $\Delta T_f = T_f(\text{solution}) - T_f(\text{solvent}) = 0.000\text{ °C} - 3.617\text{ °C} = 3.617\text{ °C}$

$$i = \frac{\Delta T_f}{K_f m_{\text{solute}}} = \frac{3.617\ ^\circ\text{C}}{(1.86\ ^\circ\text{C kg mol}^{-1})(0.0050\text{ mol/kg})} = 390 \quad \textit{(two sig figs)}$$

$$\text{percent dissociation} = \frac{i_{\text{expt}} - i_{0\%}}{i_{100\%} - i_{0\%}} \times 100\% = \frac{390 - 1.00}{2.00 - 1.00} \times 100\% = 39{,}000\% \quad \textit{(two sig figs)}$$

Second solution: $\Delta T_f = T_f(\text{solution}) - T_f(\text{solvent}) = 0.000\text{ °C} - 3.544\text{ °C} = 3.544\text{ °C}$

$$i = \frac{3.544\ ^\circ\text{C}}{(1.86\ ^\circ\text{C kg mol}^{-1})(0.020\text{ mol/kg})} = 95$$

$$\text{percent} = \frac{95 - 1.00}{2.00 - 1.00} \times 100\% = 9{,}400\% \quad \textit{(two sig figs)}$$

Third solution: $\Delta T_f = T_f(\text{solution}) - T_f(\text{solvent}) = 0.000\ ^\circ\text{C} - 3.392\ ^\circ\text{C} = 3.392\ ^\circ\text{C}$

$$i = \frac{3.392\ ^\circ\text{C}}{(1.86\ ^\circ\text{C kg mol}^{-1})(0.20\ \text{mol/kg})} = 9.1$$

$$\text{percent} = \frac{9.1 - 1.0}{2.0 - 1.0} \times 100\% = 810\%$$

Fourth solution: $\Delta T_f = T_f(\text{solution}) - T_f(\text{solvent}) = 0.00\ ^\circ\text{C} - 3.33\ ^\circ\text{C} = 3.33\ ^\circ\text{C}$

$$i = \frac{3.33\ ^\circ\text{C}}{(1.86\ ^\circ\text{C kg mol}^{-1})(1.0\ \text{mol/kg})} = 1.8$$

$$\text{percent} = \frac{1.8 - 1.0}{2.0 - 1.0} \times 100\% = 80\%$$

	i	percent dissociation
First Solution	390	39000%
Second Solution	95	9400%
Third Solution	9.1	810%
Fourth Solution	1.8	80%

Except for the fact that the percent ionization is higher at lower concentrations, the first three of these calculated results are not in the expected range at all. That must mean this is not an ideal solution.

131. *Answer:* **(a) 6300 ppm, 6,300,000 ppb (b) 0.040 M (c) 4.99 × 10⁵ bottles**

Strategy and Explanation: A sample of exactly 100 g of solution contains 0.63 g SnF_2 and 99.37 g water.

(a) The units ppm and ppb are discussed in the solution to Question 10.13. In the solution to Question 112 is given 1% = 10,000 ppm.

$$0.63\ \%\ SnF_2 \times \frac{10{,}000\ \text{ppm}}{1\ \%} = 6300\ \text{ppm}\ SnF_2$$

$$6300\ \text{ppm}\ SnF_2 \times \frac{1000\ \text{ppb}\ SnF_2}{1\ \text{ppm}\ SnF_2} = 6{,}300{,}000\ \text{ppb}\ SnF_2$$

(b) $$\frac{0.63\ \text{g}\ SnF_2}{10^2\ \text{g solution}} \times \frac{0.998\ \text{g solution}}{1\ \text{mL}} \times \frac{1\ \text{mol}\ SnF_2}{156.707\ \text{g}\ SnF_2} \times \frac{1000\ \text{mL}}{1\ \text{L}} = 0.040\ \text{M}\ SnF_2$$

(c) In one metric ton, there are 10^6 grams.

$$10^6\ \text{g}\ SnO_2 \times \frac{1\ \text{mol}\ SnO_2}{150.709\ \text{g}\ SnO_2} \times \frac{1\ \text{mol Sn}}{1\ \text{mol}\ SnO_2} \times \frac{118.710\ \text{g Sn}}{1\ \text{mol Sn}} = 7.88 \times 10^5\ \text{g Sn theoretical}$$

$$7.88 \times 10^5\ \text{g Sn theoretical} \times \frac{80\ \text{g Sn actual}}{100\ \text{g Sn theoretical}} = 6.30 \times 10^5\ \text{g Sn actual}$$

$$6.30 \times 10^5\ \text{g Sn} \times \frac{1\ \text{mol Sn}}{118.710\ \text{g Sn}} \times \frac{1\ \text{mol}\ SnF_2}{1\ \text{mol Sn}} \times \frac{156.707\ \text{g}\ SnF_2}{1\ \text{mol}\ SnF_2} = 8.32 \times 10^5\ \text{g}\ SnF_2\ \text{theoretical}$$

$$8.32 \times 10^5\ \text{g}\ SnF_2\ \text{theoretical} \times \frac{94\ \text{g}\ SnF_2\ \text{actual}}{100\ \text{g}\ SnF_2\ \text{theoretical}} = 7.82 \times 10^5\ \text{g}\ SnF_2\ \text{actual}$$

Use the answer from (b), in units of molarity to determine volume.

$$7.82 \times 10^5\ \text{g}\ SnF_2 \times \frac{1\ \text{mol}\ SnF_2}{156.707\ \text{g}\ SnF_2} \times \frac{1\ \text{L}}{0.040\ \text{mol}\ SnF_2} \times \frac{1000\ \text{mL}}{1\ \text{L}} \times \frac{1\ \text{bottle}}{250.\ \text{mL}} = 4.99 \times 10^5\ \text{bottles}$$

Chapter 16: Acids and Bases

Solutions for Blue-Numbered Questions for Review and Thought

Topical Questions

The Brønsted-Lowry Concept of Acids and Bases (Section 16.1)

12. *Answer:* **see equations below**

Strategy and Explanation: An $H^+(aq)$ ion is transferred from the acid to the water molecule to make $H_3O^+(aq)$ and the conjugate base of the acid.

(a) $HCO_3^-(aq) + H_2O(\ell) \rightleftharpoons H_3O^+(aq) + CO_3^{2-}(aq)$

(b) $HCl(aq) + H_2O(\ell) \rightleftharpoons H_3O^+(aq) + Cl^-(aq)$

(c) $CH_3COOH(aq) + H_2O(\ell) \rightleftharpoons H_3O^+(aq) + CH_3COO^-(aq)$

(d) $HCN(aq) + H_2O(\ell) \rightleftharpoons H_3O^+(aq) + CN^-(aq)$

Notice: These reactions will proceed toward products to a variable extent depending on their strength.

14. *Answer:* **see equations below**

Strategy and Explanation: Use the method shown in the solution to Question 12.

(a) $HIO(aq) + H_2O(\ell) \rightleftharpoons H_3O^+(aq) + IO^-(aq)$

(b) $CH_3(CH_2)_4COOH(aq) + H_2O(\ell) \rightleftharpoons H_3O^+(aq) + CH_3(CH_2)_4COO^-(aq)$

(c) $HOOCCOOH(aq) + H_2O(\ell) \rightleftharpoons H_3O^+(aq) + HOOCCOO^-(aq)$

$HOOCCOO^-(aq) + H_2O(\ell) \rightleftharpoons H_3O^+(aq) + {}^-OOCCOO^-(aq)$

(d) $CH_3NH_3^+(aq) + H_2O(\ell) \rightleftharpoons H_3O^+(aq) + CH_3NH_2(aq)$

Notice: These reactions will proceed toward products to a variable extent depending on their strength.

16. *Answer:* **see equations below**

Strategy and Explanation: An $H^+(aq)$ ion is transferred from the water molecule to the base to make the conjugate acid of the base and $OH^-(aq)$.

(a) $HSO_4^-(aq) + H_2O(\ell) \rightleftharpoons H_2SO_4(aq) + OH^-(aq)$

(b) $CH_3NH_2(aq) + H_2O(\ell) \rightleftharpoons CH_3NH_3^+(aq) + OH^-(aq)$

(c) $I^-(aq) + H_2O(\ell) \rightleftharpoons HI(aq) + OH^-(aq)$

(d) $H_2PO_4^-(aq) + H_2O(\ell) \rightleftharpoons H_3PO_4(aq) + OH^-(aq)$

Notice: These reactions will proceed toward products to a variable extent depending on their strength. In particular, the two bases in (a) and (c) I^- are very weak bases. The reaction that is actually observed between HSO_4^- and water is shown in the solution to Question 11(c). I^- ion is not observed to react with water.

20. *Answer:* **(a) H_2SO_4 (b) HNO_3 (c) $HClO_4$ (d) $HClO_3$ (e) H_2SO_4**

Strategy and Explanation: Two relationships between acids can be used to help us decide: If the central atoms are the same, then the acid with more O atoms is stronger. If the numbers of O atoms are the same, then the acid with the more electronegative central atom is stronger. (Electronegativities are found in Figure 8.6.)

(a) H_2SO_4 is stronger than H_2SO_3 because it has more O atoms, and H_2SO_3 is comparable in strength to H_2CO_3 because the electronegativities of C and S are almost identical ($EN_S = 2.3$, $EN_C = 2.4$), therefore H_2SO_4 is stronger than H_2CO_3.

(b) HNO_3 is stronger than HNO_2 because it has more O atoms.

(c) $HClO_4$ is stronger than H_2SO_4 because it has a more electronegative atom ($EN_{Cl} > EN_S$).

(d) $HClO_3$ is stronger than H_3PO_4 because it has a more electronegative atom ($EN_{Cl} > EN_P$).

(e) H_2SO_4 is stronger than H_2SO_3 because it has more O atoms.

22. *Answer:* **(a) conjugate base, I^-, iodide ion (b) conjugate acid, HNO_3, nitric acid (c) conjugate acid, HCO_3^-, hydrogen carbonate ion (d) conjugate base, HCO_3^-, hydrogen carbonate ion (e) conjugate acid of H_2SO_4, sulfuric acid, and conjugate base, SO_4^{2-}, sulfate ion (f) conjugate base of HSO_3^{2-}, hydrogen sulfite ion**

Strategy and Explanation: Conjugate acid-base pairs differ by one H^+ ion.

(a) HI is a Brønsted-Lowry acid. Its conjugate base is I^-, iodide ion.

(b) NO_3^- is a Brønsted-Lowry base. Its conjugate acid is HNO_3, nitric acid.

(c) CO_3^{2-} is a Brønsted-Lowry base. Its conjugate acid is HCO_3^-, hydrogen carbonate ion.

(d) H_2CO_3 is a Brønsted-Lowry acid. Its conjugate base is HCO_3^-, hydrogen carbonate ion.

(e) HSO_4^- as a Brønsted-Lowry base, has a conjugate acid of H_2SO_4, sulfuric acid. HSO_4^- as a Brønsted-Lowry acid, has a conjugate base of SO_4^{2-}, sulfate ion.

(f) SO_3^{2-} is a Brønsted-Lowry base. Its conjugate is HSO_3^-, hydrogen sulfite ion.

24. *Answer:* **pairs (b), (c), and (d)**

Strategy and Explanation: Conjugate acid-base pairs differ by only one H^+ ion.

(a) NH_2^- is the conjugate base of NH_3, not NH_4^+. NH_3 is the conjugate base of NH_4^+.

(b) NH_3 is the conjugate acid of NH_2^-.

(c) H_3O^+ is the conjugate acid of H_2O.

(d) OH^- is the conjugate acid of O^{2-}.

(e) H_3O^+ is the conjugate acid of H_2O, not OH^-. H_2O is the conjugate acid of OH^-.

To conclude: Proper conjugate acid-base pairs are described in (b), (c), and (d).

26. *Answer:* **see identifications below**

Strategy and Explanation: The conjugate base of an acid has one less H^+ than its acid partner; the conjugate acid of a base has one more H^+ than its base partner.

(a) $HS^-(aq)$	+	$H_2O(\ell)$	$\rightleftharpoons$	$H_2S(aq)$	+	$OH^-(aq)$
reactant base		**reactant acid**		**conj. acid of HS^-**		**conj. base of H_2O**

H_2O/OH^- and H_2S/HS^- are the two acid-base conjugate pairs.

(b) $S^{2-}(aq)$	+	$NH_4^+(aq)$	$\rightleftharpoons$	$NH_3(aq)$	+	$HS^-(aq)$
reactant base		**reactant acid**		**conj. acid of NH_4^+**		**conj. base of S^{2-}**

NH_4^+/NH_3 and H_2S/HS^- are the two acid-base conjugate pairs.

(c) $HCO_3^-(aq)$	+	$HSO_4^-(aq)$	$\rightleftharpoons$	$H_2CO_3(aq)$	+	$SO_4^{2-}(aq)$
reactant base		**reactant acid**		**conj. acid of HCO_3^-**		**conj. base of HSO_4^-**

HSO_4^-/SO_4^{2-} and H_2CO_3/HCO_3^- are the two acid-base conjugate pairs.

(d) $NH_3(aq)$	+	$NH_2^-(aq)$	$\rightleftharpoons$	$NH_2^-(aq)$	+	$NH_3(aq)$
reactant acid		**reactant base**		**conj. base of NH_3**		**conj. acid of NH_2^-**

NH_3/NH_2^- and NH_3/NH_2^- are the two acid-base conjugate pairs.

28. *Answer:* **see identifications below**

Strategy and Explanation: Use the method described in the solution to Question 26.

(a) $CN^-(aq)$ + $CH_3COOH(aq)$ $\rightleftharpoons$ $CH_3COO^-(aq)$ + $HCN(aq)$

reactant base	reactant acid	conj. base of CH_3COOH	conj. acid of CN^-

CH_3COOH/CH_3COO^- and HCN/CN^- are the two acid-base conjugate pairs.

(b) $O^{2-}(aq)$ + $H_2O(\ell)$ $\rightleftharpoons$ $OH^-(aq)$ + $OH^-(aq)$

reactant base	reactant acid	conj. acid of O^{2-}	conj. base of H_2O

H_2O/OH^- and OH^-/O^{2-} are the two acid-base conjugate pairs.

(c) $HCO_2^-(aq)$ + $H_2O(\ell)$ $\rightleftharpoons$ $HCOOH(aq)$ + $OH^-(aq)$

reactant base	reactant acid	conj. acid of HCO_2^-	conj. base of H_2O

H_2O/OH^- and $HCOOH/HCO_2^-$ are the two acid-base conjugate pairs.

(Notice: The reactant base is more often written with the formula $HCOO^-$.)

31. *Answer:* **see equations below**

Strategy and Explanation: For acids, sequentially transfer H^+ from the acid to water. For bases, sequentially transfer H^+ from water to the base.

(a) CO_3^{2-} is the anion of a diprotic acid, H_2CO_3, so write two protonation equations.

$CO_3^{2-}(aq) + H_2O(\ell) \rightleftharpoons HCO_3^-(aq) + OH^-(aq)$

$HCO_3^-(aq) + H_2O(\ell) \rightleftharpoons H_2CO_3(aq) + OH^-(aq)$

(b) $H_3AsO_4(aq)$ is a triprotic acid, write three deprotonation equations.

$H_3AsO_4(aq) + H_2O(\ell) \rightleftharpoons H_3O^+(aq) + H_2AsO_4^-(aq)$

$H_2AsO_4^-(aq) + H_2O(\ell) \rightleftharpoons H_3O^+(aq) + HAsO_4^{2-}(aq)$

$HAsO_4^{2-}(aq) + H_2O(\ell) \rightleftharpoons H_3O^+(aq) + AsO_4^{3-}(aq)$

(c) $NH_2CH_3COO^-(aq)$ can be protonated in two places, at the N atom and at the O^- atom, so write two protonation equations.

$NH_2CH_2COO^-(aq) + H_2O(\ell) \rightleftharpoons {}^+NH_3CH_2COO^-(aq) + OH^-(aq)$

${}^+NH_3CH_2COO^-(aq) + H_2O(\ell) \rightleftharpoons {}^+NH_3CH_2COOH + OH^-(aq)$

pH Calculations (Section 16.3)

33-43. The following is a summary of relationships between H_3O^+ concentration, OH^-concentration, pH, and pOH.

$$pH = -\log[H_3O^+] \qquad pOH = -\log[OH^-] \qquad 14 = pH + pOH$$

To get $[H_3O^+]$ or $[OH^-]$ concentrations from pH or pOH, use the following relationships:

$$[H_3O^+] = 10^{-pH} \quad \text{and} \quad [OH^-] = 10^{-pOH}$$

When $pH < 7$ and $pOH > 7$, the solution is acidic; when $pH = 7 = pOH$, the solution is neutral; when $pH < 7$ and $pOH > 7$, the solution is acidic.

33. *Answer:* **3×10^{-11} M; basic**

Strategy and Explanation: Use the appropriate relationship from among those provided <u>before</u> the solution to Question 33. When pH = 10.5, $[H_3O^+] = 10^{-10.5} = 3 \times 10^{-11}$ M. The solution is basic.

The high pH gives an $[H_3O^+]$ that is relatively small ($< 10^{-7}$). $Mg(OH)_2$, found in some antacids, is an ionic compound that puts hydroxide ions in the saturated solution.

35. *Answer:* **3.6×10^{-3} M, acidic**

Strategy and Explanation: Use the appropriate relationship from among those provided before the solution to Question 33. When pH = 2.44,

$$[H_3O^+] = 10^{-2.44} = 3.6 \times 10^{-3} \text{ M}$$

37. *Answer:* **pH = 12.40, pOH = 1.60**

Strategy and Explanation: Use the appropriate relationship from among those provided before the solution to Question 33.

NaOH is a soluble hydroxide and a strong base. It completely ionizes to form Na^+ and OH^-:

$[OH^-] = 0.025$ M, pOH = – log$[OH^-]$ = 1.60 pH = 14.00 – pOH = 14.00 – 1.60 = 12.40

39. *Answer:* **pOH = 12.51**

Strategy and Explanation: Use the appropriate relationship from among those provided before the solution to Question 33. $[H_3O^+] = 0.032$ M pH = – log$[H_3O^+]$ = 1.49 pOH = 14.00 – pH = 14.00 – 1.49 = 12.51

41. *Answer:* **5×10^{-2} M; 2 g HCl**

Strategy and Explanation: Use the appropriate relationship from among those provided before the solution to Question 33 and Chapter 5. When pH = 1.3, $[H_3O^+] = 10^{-1.3} = 5 \times 10^{-2}$ M.

HCl is a strong acid. One mole of HCl completely ionizes to one mole of H_3O^+ and one mole Cl^- ions.

$$1000.\text{ mL solution} \times \frac{1\text{ L}}{1000\text{ mL}} \times \frac{5 \times 10^{-2}\text{ mol } H_3O^+}{1\text{ L solution}} \times \frac{1\text{ mol HCl}}{1\text{ mol } H_3O^+} \times \frac{36.4609\text{ g HCl}}{1\text{ mol HCl}} = 2\text{ g HCl}$$

43. *Answer:*

	pH	$[H_3O^+]$	$[OH^-]$	acidic or basic
(a)	**6.21**	6.1×10^{-7}	**1.6×10^{-8}**	**acidic**
(b)	**5.34**	**4.5×10^{-6}**	2.2×10^{-9}	**acidic**
(c)	4.67	**2.1×10^{-5}**	**4.7×10^{-10}**	**acidic**
(d)	**1.60**	2.5×10^{-2}	**4.0×10^{-13}**	**acidic**
(e)	9.12	**7.6×10^{-10}**	**1.3×10^{-5}**	**basic**

Strategy and Explanation: Use the appropriate relationship from among those provided before the solution to Question 33.

(a) $[H_3O^+] = 6.1 \times 10^{-7}$ M pH = $-\log[H_3O^+]$ = 6.21 acidic solution (pH < 10^{-7})

pOH = 14 – pH = 7.79 $[OH^-] = 10^{-pOH} = 1.6 \times 10^{-8}$ M

(b) $[OH^-] = 2.2 \times 10^{-9}$ M pOH = $-\log[OH^-]$ = 8.66

pH = 14 – pOH = 5.34 $[H_3O^+] = 10^{-pH} = 4.5 \times 10^{-6}$ M acidic solution (pH < 10^{-7})

(c) pH = 4.67 $[H_3O^+] = 10^{-pH} = 2.1 \times 10^{-5}$ M acidic solution (pH < 10^{-7})

pOH = 14 – pH = 9.33 $[OH^-] = 10^{-pOH} = 4.7 \times 10^{-10}$ M

(d) $[H_3O^+] = 2.5 \times 10^{-2}$ M pH = $-\log[H_3O^+]$ = 1.60 acidic solution (pH < 10^{-7})

pOH = 14 – pH = 12.40 $[OH^-] = 10^{-pOH} = 4.0 \times 10^{-13}$ M

(e) pH = 9.12 $[H_3O^+] = 10^{-pH} = 7.6 \times 10^{-10}$ M basic solution (pH > 10^{-7})

pOH = 14 – pH = 4.88 $[OH^-] = 10^{-pOH} = 1.3 \times 10^{-5}$ M

45. *Answer:* **(a) 100 times more acidic (b) 10,000 times more basic (c) 16 times more basic (d) 20,000 times more acidic**

Strategy and Explanation: As we see from Figure 16.3, the pH scale is logarithmic. Each time the pH unit changes by one, the concentrations of H_3O^+ and OH^- change by a factor of ten. $[H_3O^+] = 10^{-pH}$

(a) Black coffee has a pH about 5.0. $[H_3O^+] = 1 \times 10^{-5}$ M, about 100 times more acidic than neutral.

(b) Household ammonia (assuming that is aqueous ammonia on the table) has a pH about 11.0. $[H_3O^+] = 1 \times 10^{-11}$ M, about 10,000 times more basic than neutral.

(c) Baking soda (also known as sodium bicarbonate) has a pH about 8.2. $[H_3O^+] = 7 \times 10^{-9}$ M, about 16 times more basic than neutral.

(d) Vinegar has a pH about 2.7. $[H_3O^+] = 2 \times 10^{-3}$ M, about 20,000 times more acidic than neutral.

47. *Answer:* **(a) 5.0×10^{-9} M (b) basic**

Strategy and Explanation: Use the appropriate relationship from among those provided before the solution to Question 33.

(a) When pH = 8.30, $[H_3O^+] = 10^{-8.30} = 5.0 \times 10^{-9}$ M

(b) For pH > 7, the solution is basic.

Acid-Base Strengths (Section 16.6)

49. *Answer:* **see chemical equations and equilibrium expressions below**

Strategy and Explanation: In acid ionization reactions, the acid donates H^+ to a water molecule to make H_3O^+ and the conjugate base. In base ionization reactions, the base receives H^+ from a water molecule to make OH^- and the conjugate acid.

(a) $F^-(aq) + H_2O(\ell) \rightleftharpoons HF(aq) + OH^-(aq)$ $K = \frac{[HF][OH^-]}{[F^-]}$

(b) $NH_3(aq) + H_2O(\ell) \rightleftharpoons NH_4^+(aq) + OH^-(aq)$ $K = \frac{[NH_4^+][OH^-]}{[NH_3]}$

(c) $H_2CO_3(aq) + H_2O(\ell) \rightleftharpoons HCO_3^-(aq) + H_3O^+(aq)$ $K = \frac{[HCO_3^-][H_3O^+]}{[H_2CO_3]}$

(d) $H_3PO_4(aq) + H_2O(\ell) \rightleftharpoons H_2PO_4^-(aq) + H_3O^+(aq)$ $K = \frac{[H_2PO_4^-][H_3O^+]}{[H_3PO_4]}$

(e) $CH_3COO^-(aq) + H_2O(\ell) \rightleftharpoons CH_3COOH(aq) + OH^-(aq)$ $K = \frac{[CH_3COOH][OH^-]}{[CH_3COO^-]}$

(f) $S^{2-}(aq) + H_2O(\ell) \rightleftharpoons HS^-(aq) + OH^-(aq)$ $K = \frac{[HS^-][OH^-]}{[S^{2-}]}$

51. *Answer:* **(a) 0.10 M NH_3 (b) 0.10 M K_2S (c) 0.10 M $NaCH_3COO$ (d) 0.10 M KCN**

Strategy and Explanation: In all these comparisons, the concentrations are the same, so we can use the ionization constants from Appendix F or Table 16.2 to compare their strengths. The larger the K_b the more basic the solution. Ionic compounds provide cations or anions that may be bases, so watch for them.

(a) NH_3 is a base. ($K_b = 1.8 \times 10^{-5}$) $NaF(s) \longrightarrow Na^+(aq) + F^-(aq)$, and F^- is a base. ($K_b = 1.5 \times 10^{-11}$)

Given equal concentrations, a solution of NH_3 is more basic than a solution of NaF.

(b) $K_2S(s) \longrightarrow 2\,K^+(aq) + S^{2-}(aq)$, and S^{2-} is a base. ($K_b = 1 \times 10^5$)

$K_3PO_4(s) \longrightarrow 3\,K^+(aq) + PO_4^{3-}(aq)$, and PO_4^{3-} is a base. ($K_b = 2.2 \times 10^{-2}$)

Given equal concentrations, a solution of K_2S is more basic than a solution of K_3PO_4.

(c) $NaNO_3(s) \longrightarrow Na^+(aq) + NO_3^-(aq)$, NO_3^- is base. ($K_b = 5 \times 10^{-16}$)

$NaCH_3COO(s) \longrightarrow Na^+(aq) + CH_3COO^-(aq)$, and CH_3COO^- is a base. ($K_b = 5.6 \times 10^{-10}$)

Given equal concentrations, a solution of CH_3COONa is more basic than a solution of $NaNO_3$.

(d) $KCN(s) \longrightarrow K^+(aq) + CN^-(aq)$, and CN^- is a base. ($K_b = 3.0 \times 10^{-5}$) NH_3 is a base. ($K_b = 1.8 \times 10^{-5}$)

Given equal concentrations, a solution of KCN is more basic than a solution of NH_3.

Using K_a and K_b (Section 16.7)

54. *Answer:* **(a) 2.19 is between 2 and 6 (b) 5.13 is between 2 and 6 (c) 8.09 is between 8 and 12 (d) 9.02 is between 8 and 12 (e) 13.30 pH 12 (or higher) (f) 1.55 is close to being between 2 and 6 (g) 9.68 is between 8 and 12 (h) 7.00 is between 6 and 8**

Strategy and Explanation: Follow the procedure described in Problem-Solving Example 16.8 in Section 16.7, adapting the methods described in the solution to Question 14.60. In all parts, the initial concentration of the solute is 0.10 M.

(a) The equation for the equilibrium and the equilibrium expression are:

$$HNO_2(aq) + H_2O(\ell) \rightleftharpoons H_3O^+(aq) + NO_2^-(aq) \qquad K_a = \frac{[H_3O^+][NO_2^-]}{[HNO_2]}$$

As the reactants decompose, the concentrations of the products increase stoichiometrically, until they reach equilibrium concentrations.

	$HNO_2(aq)$	$H_3O^+(aq)$	$NO_2^-(aq)$
initial conc. (M)	0.10	$1.0 \times 10^{-7*}$	0
change as reaction occurs (M)	– x	+ x	+ x
equilibrium conc. (M)	0.10 – x	x	x

* from the dissociation of pure water. This number is small compared to the acid added.

At equilibrium
$$K_a = \frac{(x)(x)}{(0.10 - x)} = 4.5 \times 10^{-4}$$

$$x^2 = (4.5 \times 10^{-4})(0.10 - x)$$

$$x^2 + 4.5 \times 10^{-4}x - 4.5 \times 10^{-5} = 0$$

Use the quadratic equation: (see Appendix A, Section A.7, page A.13)

$$x = 6.5 \times 10^{-3}\ M = [H_3O^+] \qquad pH = -\log[H_3O^+] = -\log(6.5 \times 10^{-3}) = 2.19$$

If you did Question 52, compare this pH to your qualitative prediction. You should have predicted between 2 and 6.

(b) The equation for the equilibrium and the equilibrium expression are:

$$NH_4^+(aq) + H_2O(\ell) \rightleftharpoons H_3O^+(aq) + NH_3(aq) \qquad K_a = \frac{[H_3O^+][NH_3]}{[NH_4^+]}$$

As the reactants decompose, the concentrations of the products increase stoichiometrically, until they reach equilibrium concentrations.

	$NH_4^+(aq)$	$H_3O^+(aq)$	$NH_3(aq)$
initial conc. (M)	0.10	$1.0 \times 10^{-7*}$	0
change as reaction occurs (M)	– x	+ x	+ x
equilibrium conc. (M)	0.10 – x	x	x

* from the dissociation of pure water. This number is small compared to the acid added.

At equilibrium
$$K_a = \frac{(x)(x)}{(0.10 - x)} = 5.6 \times 10^{-10}$$

The small-K, reactant-favored reaction will have a larger reactant concentration than product concentrations, so assume x is very small, such that subtraction from the reactant concentration is negligible: $0.10 - x \cong 0.10$.

$$x^2 = (5.6 \times 10^{-10})(0.10)$$

$$x = 7.5 \times 10^{-6}\ M = [H_3O^+]$$

$$pH = -\log[H_3O^+] = -\log(7.5 \times 10^{-6}) = 5.13$$

If you did Question 52, compare this pH to your qualitative prediction. You should have predicted between 2 and 6.

(c) The equation for the equilibrium and the equilibrium expression are:

$$F^-(aq) + H_2O(\ell) \rightleftharpoons HF(aq) + OH^-(aq) \qquad K_b = \frac{[HF][OH^-]}{[F^-]}$$

As the reactants decompose, the concentrations of the products increase stoichiometrically, until they reach equilibrium concentrations.

	$F^-(aq)$	$HF(aq)$	$OH^-(aq)$
initial conc. (M)	0.10	0	1.0×10^{-7}*
change as reaction occurs (M)	$-x$	$+x$	$+x$
equilibrium conc. (M)	$0.10 - x$	x	x

* from the dissociation of pure water. This number is small compared to the base added.

At equilibrium $\qquad K_b = \dfrac{(x)(x)}{(0.10 - x)} = 1.5 \times 10^{-11}$

For the same reasons as in (b), assume x is very small and: $0.10 - x \cong 0.10$.

$$1.5 \times 10^{-11} = \frac{x^2}{(0.10)}$$

$$x^2 = (1.5 \times 10^{-11})(0.10)$$

$$x = 1.2 \times 10^{-6}\ M = [OH^-]$$

$$pOH = -\log[OH^-] = -\log(1.2 \times 10^{-6}) = 5.91$$

$$pH = 14.00 - pOH = 8.09$$

If you did Question 52, compare this pH to your qualitative prediction. You should have predicted between pH 8 and 12.

(d) $(\text{conc. } CH_3COO^-) = 0.10\ M\ Ba(CH_3COO)_2 \times \dfrac{2 \text{ mol } CH_3COO^-}{1 \text{ mol } Ba(CH_3COO)_2} = 0.20\ M$

The equation for the equilibrium and the equilibrium expression are:

$$CH_3COO^-(aq) + H_2O(\ell) \rightleftharpoons CH_3COOH(aq) + OH^-(aq) \qquad K_b = \frac{[CH_3COOH][OH^-]}{[CH_3COO^-]}$$

As the reactants decompose, the concentrations of the products increase stoichiometrically, until they reach equilibrium concentrations.

	$CH_3COO^-(aq)$	$CH_3COOH(aq)$	$OH^-(aq)$
initial conc. (M)	0.20	0	1.0×10^{-7}*
change as reaction occurs (M)	$-x$	$+x$	$+x$
equilibrium conc. (M)	$0.20 - x$	x	x

* from the dissociation of pure water. This number is small compared to the base added.

At equilibrium $\qquad K_b = \dfrac{(x)(x)}{(0.20 - x)} = 5.6 \times 10^{-10}$

For the same reasons as in (b), assume x is very small and: $0.20 - x \cong 0.20$.

$$5.6 \times 10^{-10} = \frac{x^2}{(0.20)}$$

$$x^2 = (5.6 \times 10^{-10})(0.20)$$

$$x = 1.1 \times 10^{-5}\ M = [OH^-]$$

$$pOH = -\log[OH^-] = -\log(1.1 \times 10^{-5}) = 4.98$$

$$pH = 14.00 - pOH = 14.00 - 4.98 = 9.02$$

If you did Question 52, compare this pH to your qualitative prediction. You should have predicted between pH 8 and 12.

(e) $O^{2-}(aq) + H_2O(\ell) \longrightarrow 2\ OH^-(aq)$

This must have a very large K. 100% reaction.

$$[OH^-] = 0.10\ M\ O^{2-} \times \frac{2\ mol\ OH^-}{1\ mol\ O^{2-}} = 0.20\ M$$

$$pOH = -\log[OH^-] = -\log(0.20) = 0.70$$

$$pH = 14.00 - pOH = 14.00 - 0.70 = 13.30$$

If you did Question 52, compare this pH to your qualitative prediction. You should have predicted pH 12 (or higher).

(f) $HSO_4^-(aq) + H_2O(aq) \rightleftharpoons H_3O^+(aq) + SO_4^{2-}(aq)$ $\quad K_a = \dfrac{[H_3O^+][SO_4^{2-}]}{[HSO_4^-]} = 1.1 \times 10^{-2}$

$HSO_4^-(aq) + H_2O(aq) \rightleftharpoons H_2SO_4(aq) + OH^-(aq)$ $\quad K_b = \dfrac{[H_2SO_4][OH^-]}{[HSO_4^-]} = \text{very small}$

We will use the first reaction since it has the largest K value, and produces more products.

As the reactants decompose, the concentrations of the products increase stoichiometrically, until they reach equilibrium concentrations.

	$HSO_4^-(aq)$	$H_3O^+(aq)$	$SO_4^{2-}(aq)$
initial conc. (M)	0.10	1.0×10^{-7}*	0
change as reaction occurs (M)	– x	+ x	+ x
equilibrium conc. (M)	0.10 – x	x	x

* from the dissociation of pure water. This number is small compared to the base added.

At equilibrium $\quad K_a = \dfrac{(x)(x)}{(0.10 - x)} = 1.1 \times 10^{-2}$

$$x^2 = (1.1 \times 10^{-2})(0.10 - x)$$

$$x^2 + 1.1 \times 10^{-2}x - 1.1 \times 10^{-3} = 0$$

Use the quadratic equation: (see Appendix A, Section A.7, page A.13)

$$2.8 \times 10^{-2}\ M = [H_3O^+] \qquad pH = -\log[H_3O^+] = -\log(2.8 \times 10^{-2}) = 1.55$$

If you did Question 52, compare this pH to your qualitative prediction. You should have predicted between 2 and 6. While this calculated pH doesn't fall in that range, it is still greater than the pH of 1 we would predict for a 0.10 M strong acid.

(g) $HCO_3^-(aq) + H_2O(aq) \rightleftharpoons H_3O^+(aq) + CO_3^{2-}(aq)$ $\quad K_a = \dfrac{[H_3O^+][CO_3^{2-}]}{[HCO_3^-]} = 4.7 \times 10^{-11}$

$HCO_3^-(aq) + H_2O(aq) \rightleftharpoons H_2CO_3(aq) + OH^-(aq)$ $\quad K_b = \dfrac{[H_2CO_3][OH^-]}{[HCO_3^-]} = 2.3 \times 10^{-8}$

We will use the second reaction since it has the largest K value, and produces more products.

As the reactants decompose, the concentrations of the products increase stoichiometrically, until they reach equilibrium concentrations.

	$HCO_3^-(aq)$	$HCO_3^-(aq)$	$OH^-(aq)$
initial conc. (M)	0.10	0	1.0×10^{-7}*
change as reaction occurs (M)	– x	+ x	+ x
equilibrium conc. (M)	0.10 – x	x	x

* from the dissociation of pure water. This number is small compared to the base added.

At equilibrium $$K_b = \frac{(x)(x)}{(0.10 - x)} = 2.3 \times 10^{-8}$$

For the same reasons as in (b), assume x is very small and: $0.10 - x \cong 0.10$.

$$2.3 \times 10^{-8} = \frac{x^2}{(0.10)} \qquad x^2 = (2.3 \times 10^{-8})(0.10) \qquad x = 4.8 \times 10^{-5}\text{ M} = [OH^-]$$

$$pOH = -\log[OH^-] = -\log(4.8 \times 10^{-5}) = 4.32$$

$$pH = 14.00 - pOH = 14.00 - 4.32 = 9.68$$

If you did Question 52, compare this pH to your qualitative prediction. You should have predicted between pH 8 and 12.

(h) $BaCl_2 \longrightarrow Ba^{2+} + 2\,Cl^-$ Neither of these ions affect the pH of a water solution, because Ba^{2+} is the cation of a soluble ionic hydroxide compound, $Ba(OH)_2$ and Cl^- is a weak base (K_b = very small), so pH 7.00 and no calculation is needed. If you did Question 52, you should have predicted pH 7, which is between 6 and 8.

56. *Answer:* **1.6×10^{-5}**

Strategy and Explanation: Combine the methods described before Question 33 and in Question 54 with those described in Section 16.7 and in the solution to Question 14.60.

We do not know the formula of butyric acid, so we will assume it is monoprotic and give it the symbol HBu. The equation for the equilibrium and the equilibrium expression are:

$$HBu(aq) + H_2O(\ell) \rightleftharpoons H_3O^+(aq) + Bu^-(aq) \qquad K_a = \frac{[H_3O^+][Bu^-]}{[HBu]}$$

As the reactants decompose, the concentrations of the products increase stoichiometrically, until they reach equilibrium concentrations.

At equilibrium, pH = 3.21, so $[H_3O^+] = 10^{-pH} = 10^{-3.21} = 6.2 \times 10^{-4}$. Since the H_3O^+ ions are produced from the decomposition of HBu, an equal quantity of Bu^- is formed, and the concentrations change in the following way:

	$HBu(aq)$	$H_3O^+(aq)$	$Bu^-(aq)$
initial conc. (M)	0.015	1.0×10^{-7}*	0
change as reaction occurs (M)	-6.2×10^{-4}	$+6.2 \times 10^{-4}$	$+6.2 \times 10^{-4}$
equilibrium conc. (M)	$0.015 - 6.2 \times 10^{-4}$	6.2×10^{-4}	6.2×10^{-4}

* from the dissociation of pure water. This number is small compared to the acid added.

$$K_a = \frac{(6.2 \times 10^{-4})(6.2 \times 10^{-4})}{(0.025 - 6.2 \times 10^{-4})} = 1.6 \times 10^{-5}$$

Assume x is very small and: $0.20 - x \cong 0.20$.

$$1.8 \times 10^{-5} = \frac{x^2}{(0.20)}$$

$$x^2 = (1.8 \times 10^{-5})(0.20)$$

$$x = 1.9 \times 10^{-3}\text{ M} = [H_3O^+] = [CH_3COO^-]$$

$$[CH_3COOH] = 0.20\text{ M} - x = 0.20\text{ M} - 1.9 \times 10^{-3}\text{ M} = 0.20\text{ M (as assumed)}$$

58. *Answer:* **$[H_3O^+] = 1.3 \times 10^{-5}$; $[A^-] = 1.3 \times 10^{-5}$; [HA] = 0.040**

Strategy and Explanation: Combine the methods described in Questions 33 and 54 with those described in Section 16.7 and in the solution to Question 14.64. The equation for the equilibrium and the equilibrium expression are:

$$HA(aq) + H_2O(\ell) \rightleftharpoons H_3O^+(aq) + A^-(aq) \qquad K_a = \frac{[H_3O^+][A^-]}{[HA]}$$

As the reactants decompose, the concentrations of the products increase stoichiometrically, until they reach equilibrium concentrations.

	HA(aq)	$H_3O^+(aq)$	$A^-(aq)$
initial conc. (M)	0.040	1.0×10^{-7}*	0
change as reaction occurs (M)	– x	+ x	+ x
equilibrium conc. (M)	0.040 – x	x	x

* from dissociation of pure water. Assume this concentration will be small compared to the acid added.

At equilibrium $$K_a = \frac{(x)(x)}{(0.040 - x)} = 4.0 \times 10^{-9}$$

Assume x is very small and: $0.040 - x \cong 0.040$.

$$4.0 \times 10^{-9} = \frac{x^2}{(0.040)}$$

$$x^2 = (4.0 \times 10^{-9})(0.040)$$

$$x = 1.3 \times 10^{-5} \text{ M} = [H_3O^+] = [A^-]$$

$$[HA] = 0.040 \text{ M} - x = 0.040 \text{ M} - 1.3 \times 10^{-5} \text{ M} = 0.040 \text{ M (as assumed)}$$

60. *Answer:* **1.4×10^{-5}**

Strategy and Explanation: Use the method described in the solution to Question 56. The equation for the equilibrium and the equilibrium expression are:

$$CH_3CH_2COOH(aq) + H_2O(\ell) \rightleftharpoons H_3O^+(aq) + CH_3CH_2COO^-(aq) \qquad K_a = \frac{[H_3O^+][CH_3CH_2COO^-]}{[CH_3CH_2COOH]}$$

At equilibrium, pH = 2.93, so $[H_3O^+] = 10^{-pH} = 10^{-2.93} = 1.2 \times 10^{-3}$ M. Since the H_3O^+ ions are produced from the decomposition of C_6H_5COOH, an equal quantity of $C_6H_5COO^-$ is formed, and the concentrations change in the following way:

	$C_6H_5COOH(aq)$	$H_3O^+(aq)$	$C_6H_5COO^-(aq)$
initial conc. (M)	0.15	1.0×10^{-7}*	0
change as reaction occurs (M)	$- 1.2 \times 10^{-3}$	$+ 1.2 \times 10^{-3}$	$+ 1.2 \times 10^{-3}$
equilibrium conc. (M)	$0.015 - 1.2 \times 10^{-3}$	1.2×10^{-3}	1.2×10^{-3}

* from dissociation of pure water. Assume this concentration will be small compared to the acid added.

$$K_a = \frac{(1.2 \times 10^{-3})(1.2 \times 10^{-3})}{(0.015 - 1.2 \times 10^{-3})} = 1.4 \times 10^{-5}$$

62. *Answer:* **8.84**

Strategy and Explanation: Adapt the methods described in Questions 33 and 54 with those described in Section 16.7 and in the solution to Question 14.60. The equation for the equilibrium and the equilibrium expression are:

$$C_6H_5NH_2(aq) + H_2O(\ell) \rightleftharpoons C_6H_5NH_3^+(aq) + OH^-(aq) \qquad K_b = \frac{[C_6H_5NH_3^+][OH^-]}{[C_6H_5NH_2]}$$

As the reactants decompose, the concentrations of the products increase stoichiometrically, until they reach equilibrium concentrations.

	$C_6H_5NH_2(aq)$	$C_6H_5NH_3^+(aq)$	$OH^-(aq)$
initial conc. (M)	0.12	0	1.0×10^{-7}*
change as reaction occurs (M)	– x	+ x	+ x
equilibrium conc. (M)	0.12 – x	x	x

* from dissociation of pure water. Assume this concentration will be small compared to the base added.

At equilibrium $K_b = \dfrac{(x)(x)}{(0.12 - x)} = 3.9 \times 10^{-10}$

Assume x is very small and: $0.12 - x \cong 0.12$.

$$x^2 = (3.9 \times 10^{-10})(0.12)$$

$$x = 6.8 \times 10^{-6}\ M = [\ OH^-\ ]$$

$$pOH = -\log[OH^-\ ] = -\log(6.8 \times 10^{-6}) = 5.16$$

$$pH = 14.00 - pOH = 14.00 - 5.16 = 8.84$$

64. *Answer:* 1.3×10^{-3}

Strategy and Explanation: Modify the method described in Question 56.

Piperidine, $C_5H_{11}N$, is a base. The equation for the equilibrium and the equilibrium expression are:

$$C_5H_{11}N(aq) + H_2O(\ell) \rightleftharpoons C_5H_{11}NH^+(aq) + OH^-(aq) \qquad K_a = \frac{[C_5H_{11}NH^+][OH^-]}{[C_5H_{11}N]}$$

As the reactants decompose, the concentrations of the products increase stoichiometrically, until they reach equilibrium concentrations.

At equilibrium, pH = 11.72, so $[OH^-] = 10^{-pOH} = 10^{-(14.00-pH)} = 10^{(pH-14.00)} = 10^{(11.72-14.00)} = 10^{-2.30} = 5.0 \times 10^{-3}$. Since the OH^- ions are produced from the hydrolysis of $C_5H_{11}N(aq)$, an equal quantity of $C_5H_{11}NH^+$ is formed, and the concentrations change in the following way:

	$C_5H_{11}N(aq)$	$C_5H_{11}NH^+(aq)$	$OH^-(aq)$
initial conc. (M)	0.025	0	1.0×10^{-7}*
change as reaction occurs (M)	-5.0×10^{-3}	$+5.0 \times 10^{-3}$	$+5.0 \times 10^{-3}$
equilibrium conc. (M)	$0.025 - 5.0 \times 10^{-3}$	5.0×10^{-3}	5.0×10^{-3}

* from dissociation of pure water. This concentration is very small compared to the base added.

$$K_a = \frac{(5.0 \times 10^{-3})(5.0 \times 10^{-3})}{(0.015 - 5.0 \times 10^{-3})} = 1.3 \times 10^{-3}$$

66. *Answer:* **(a) $C_{10}H_{15}NH_2(aq) + H_2O(\ell) \rightleftharpoons C_{10}H_{15}NH_3^+(aq) + OH^-(aq)$ (b) 10.47**

Strategy and Explanation:

(a) Amantadine reacts with water, undergoing hydrolysis to form a basic solution. The equation for the equilibrium is:

$$C_{10}H_{15}NH_2(aq) + H_2O(\ell) \rightleftharpoons C_{10}H_{15}NH_3^+(aq) + OH^-(aq)$$

(b) The K_a given is for the dissociation of $C_{10}H_{15}NH^+(aq)$. We need to find the K_b from the K_a:

$$K_b = \frac{K_w}{K_a} = \frac{1.00 \times 10^{-14}}{7.9 \times 10^{-11}} = 1.3 \times 10^{-4}$$

The reactants decompose, the concentrations of the products increase stoichiometrically, until they reach equilibrium concentrations.

	$C_{10}H_{15}N(aq)$	$C_{10}H_{15}NH^+(aq)$	$OH^-(aq)$
initial conc. (M)	0.0010	0	1.0×10^{-7}*
change as reaction occurs (M)	– x	+ x	+ x
equilibrium conc. (M)	0.0010 – x	x	x

* from dissociation of pure water. Assume this concentration will be small compared to the base added.

At equilibrium $K_b = \frac{[C_{10}H_{15}NH_3^+][OH^-]}{[C_{10}H_{15}NH_2]} = \frac{(x)(x)}{(0.0010 - x)} = 1.3 \times 10^{-4}$

$$x^2 = (1.3 \times 10^{-4})(0.0010 - x)$$

$$x^2 + 1.3 \times 10^{-4}x - 1.5 \times 10^{-7} = 0$$

Use the quadratic equation: (see Appendix A, Section A.7, page A.13)

$$x = 3.0 \times 10^{-4}\ M = [\ OH^-]$$

$$pOH = -\log[OH^-] = -\log(3.0 \times 10^{-4}) = 3.53$$

$$pH = 14.00 - pOH = 14.00 - 3.53 = 10.47$$

68. *Answer:* **3.28**

Strategy and Explanation: First use methods from Chapters 3 and 4 to find the initial concentration of the $C_3H_6O_3$. Then combine the methods described in Questions 33 and 54 with those described in Section 16.7 and in the solution to Question 14.60.

$$\frac{56\ mg\ C_3H_6O_3}{250\ mL\ soln} \times \frac{1\ g}{1000\ mg} \times \frac{1\ mol\ C_3H_6O_3}{90.08\ g\ C_3H_6O_3} \times \frac{1000\ mL}{1\ L} = 0.0025\ M$$

According to the structure shown in Section 16.2, lactic acid is a monoprotic acid, so we'll write the formula as: $HC_3H_5O_3$. The equation for the equilibrium and the equilibrium expression are:

$$HC_3H_5O_3(aq) + H_2O(\ell) \rightleftharpoons H_3O^+(aq) + C_3H_5O_3^-(aq) \qquad K_a = \frac{[H_3O^+][C_3H_5O_3^-]}{[HC_3H_5O_3]}$$

As the reactants decompose, the concentrations of the products increase stoichiometrically, until they reach equilibrium concentrations.

	$HC_3H_5O_3(aq)$	$H_3O^+(aq)$	$C_3H_5O_3^-(aq)$
initial conc. (M)	0.0025	1.0×10^{-7}*	0
change as reaction occurs (M)	– x	+ x	+ x
equilibrium conc. (M)	0.0025 – x	x	x

* from dissociation of pure water. Assume this concentration will be small compared to the acid added.

At equilibrium $K_a = \frac{(x)(x)}{(0.0025 - x)} = 1.4 \times 10^{-4}$

$$x^2 = (1.4 \times 10^{-4})(0.0025 - x)$$

$$x^2 + 1.4 \times 10^{-4}x - 3.5 \times 10^{-7} = 0$$

Use the quadratic equation: (see Appendix A, Section A.7, page A.13)

$x = 5.2 \times 10^{-4}\ M = [H_3O^+]$ $\qquad pH = -\log[H_3O^+] = -\log(5.2 \times 10^{-4}) = 3.28$

Acid-Base Reactions of Salts (Section 16.8)

72. *Answer:* **(a) CN^-; product-favored (b) HS^-; reactant-favored (c) $H_2(g)$; product-favored**

Strategy and Explanation: Use the methods described in the solutions to Questions 22, 26, and 49. Compare the reactant acid to the product acid and identify which is stronger and which is weaker. Do the same with the bases. Equilibrium favors the weaker species in the reaction.

(a) $\mathbf{CN^-(aq)}$ + $HSO_4^-(aq)$ $\rightleftharpoons$ $HCN(aq)$ + $SO_4^{2-}(aq)$
stronger base — stronger acid — weaker acid — weaker base
The reaction is product-favored.

(b) $H_2S(aq)$ + $H_2O(\ell)$ $\rightleftharpoons$ $H_3O^+(aq)$ + $\mathbf{HS^-(aq)}$
weaker acid — weaker base — stronger acid — stronger base
The reaction is reactant-favored.

(c) $H^-(aq)$ + $H_2O(\ell)$ $\rightleftharpoons$ $OH^-(aq)$ + $\mathbf{H_2(g)}$
stronger base — stronger acid — weaker base — weaker acid
The reaction is product-favored.

74. *Answer:* **Reactions (b) and (c) are product-favored; reactions (a) and (d) are reactant-favored; see equations below.**

Strategy and Explanation: Use the method described in the solution to Question 72.

(a) $NH_4^+(aq)$ + $HPO_4^{2-}(aq)$ $\rightleftharpoons$ $\mathbf{NH_3(aq)}$ + $\mathbf{H_2PO_4^-(aq)}$
weaker acid — weaker base — stronger base — stronger acid
The reaction is reactant-favored.

(b) $CH_3COOH(aq)$ + $OH^-(aq)$ $\rightleftharpoons$ $\mathbf{CH_3COO^-(aq)}$ + $\mathbf{H_2O(\ell)}$
stronger acid — stronger base — weaker base — weaker acid
The reaction is product-favored.

(c) We choose to use HSO_4^- as the reactant acid, since it is a stronger acid than $H_2PO_4^-$

$HSO_4^-(aq)$ + $H_2PO_4^-(aq)$ $\rightleftharpoons$ $\mathbf{H_3PO_4(aq)}$ + $\mathbf{SO_4^{2-}(aq)}$
stronger acid — stronger base — weaker acid — weaker base
The reaction is product-favored.

(d) $CH_3COOH(aq)$ + $F^-(aq)$ $\rightleftharpoons$ $\mathbf{CH_3COO^-(aq)}$ + $\mathbf{HF(aq)}$
weaker acid — weaker base — stronger base — stronger acid
The reaction is reactant-favored.

76. *Answer:* **(a) less than 7 (b) greater than 7 (c) equal to 7; see explanations below**

Strategy and Explanation: Adapt the method described in the solutions to Question 50-52.

(a) $AlCl_3(s) \longrightarrow Al^{3+}(aq) + 3\ Cl^-(aq)$ The Cl^- ions do not affect the pH of a water solution, because HCl is a strong acid. However, the hydrated aqueous aluminum ion is formed:

$$Al^{3+} + 6\ H_2O(aq) \longrightarrow [Al(H_2O)_6]^{3+}$$

It is a weak acid, so we predict pH less than 7.

(b) $Na_2S(s) \longrightarrow 2\ Na^+(aq) + S^{2-}(aq)$ The Na^+ ions do not affect the pH of a water solution. S^{2-} is a strong base, so we predict pH greater than 7.

(c) $NaNO_3(s) \longrightarrow Na^+(aq) + NO_3^-(aq)$ Neither of these ions affects the pH of a water solution, because NaOH is a strong base and HNO_3 is a strong acid, so we predict pH equal to 7.

78. *Answer:* **(a) greater than 7 (b) greater than 7 (c) greater than 7; see explanations below**

Strategy and Explanation: Adapt the method described in the solution to Question 76.

(a) $Na_2HPO_4(s) \longrightarrow 2\ Na^+(aq) + HPO_4^{2-}(aq)$ The Na^+ ions do not affect the pH of a water solution. The HPO_4^{2-} ion is a weak acid and a weak base, so we must compare the size of K_a and K_b:

$HPO_4^{2-}(aq) + H_2O(aq) \rightleftharpoons H_3O^+(aq) + PO_4^{3-}(aq)$ $K_a = 4.6 \times 10^{-13}$

$HPO_4^{2-}(aq) + H_2O(aq) \rightleftharpoons H_2PO_4^-(aq) + OH^-(aq)$ $K_b = 1.6 \times 10^{-7}$

$H_2PO_4^-$ is a stronger base than it is an acid, so we predict pH greater than 7.

(b) $(NH_4)_2S(s) \longrightarrow 2\ NH_4^+(aq) + S^{2-}(aq)$ The NH_4^+ ion is a weak acid, but the S^{2-} ion is a strong base, so we predict pH greater than 7.

(c) $KCH_3COO(s) \longrightarrow K^+(aq) + CH_3COO^-(aq)$ The K^+ ions do not affect the pH of a water solution.

80. *Answer:* **(a), (b), (c), (d), and (e) are more soluble at pH 2 than at pH 7**

Strategy and Explanation: All of these solids have the same cation, whose aqueous form, $Cu(H_2O)_6^{2+}$, is an acid with $K_a = 1.6 \times 10^{-7}$. The presence of extra H_3O^+ will not affect the cation concentration from one solution to the next. Therefore, we will compare the strength of the anionic bases to judge variations in the solubility in acid solutions. The solid will be more soluble at pH 2 than pH 7 if its anion reacts with H_3O^+ to form a weak conjugate acid.

(a) $Cu(OH)_2(s) \longrightarrow Cu^{2+}(aq) + 2\ OH^-(aq)$ The conjugate of OH^- is H_2O, a very weak acid. This salt would be more soluble at pH 2.

(b) $CuSO_4(s) \longrightarrow Cu^{2+}(aq) + SO_4^{2-}(aq)$ The conjugate of SO_4^{2-} is HSO_4^-, a slightly weak acid. This salt would be slightly more soluble at pH 2.

(c) $CuCO_3(s) \longrightarrow Cu^{2+}(aq) + CO_3^{2-}(aq)$ The conjugate of CO_3^{2-} is HCO_3^-, a weak acid. This salt would be more soluble at pH 2.

(d) $CuS(s) \longrightarrow Cu^{2+}(aq) + S^{2-}(aq)$ The conjugate of S^{2-} is HS^-, a weak acid. This salt would be more soluble at pH 2.

(e) $CuS(s) \longrightarrow 3\ Cu^{2+}(aq) + 2\ PO_4^{3-}(aq)$ The conjugate of PO_4^{3-} is HPO_4^{2-}, a weak acid. This salt would be more soluble at pH 2.

In conclusion, all of these salts would be more soluble at pH 2 than at pH 7.

Lewis Acids and Bases (Section 16.9)

82. *Answer:* **Lewis acid: molecule (b); Lewis bases: molecules (a), (b), and (c)**

Strategy and Explanation: The Lewis model focuses on what electron pairs are doing. The substance capable of donating the electron pair to form a new bond is called a Lewis base. The substance capable of accepting an electron pair is a Lewis acid.

(a) O^{2-} has a lone pair of electrons that can form a new bond, so O^{2-} can be a Lewis base. It cannot accept any more electrons, so it is not a Lewis acid.

:Ö:

(b) CO_2 has a lone pair of electrons on the O atoms that can form a new bond, so CO_2 can be a Lewis base. Its central C atom can interact with lone pairs on other Lewis bases, so CO_2 can also be a Lewis acid.

:Ö═C═Ö:

(c) H^- has a lone pair of electrons that can form a new bond, so H^- can be a Lewis base. It cannot accept any more electrons, so it is not a Lewis acid.

H:

84. *Answer:* **Lewis acids: molecules (a), (b) and (c); Lewis bases: molecules (b) and (c)**

Strategy and Explanation: Follow the method described in the solution to Question 82.

(a) Cr^{3+} can interact with lone pairs on Lewis bases, so Cr^{3+} can also be a Lewis acid. It has no valence electrons that can form a new bond, so it cannot be a Lewis base.

(b) The S atoms in SO_3 could interact with lone pairs on other Lewis bases, so SO_3 can also be a Lewis acid.

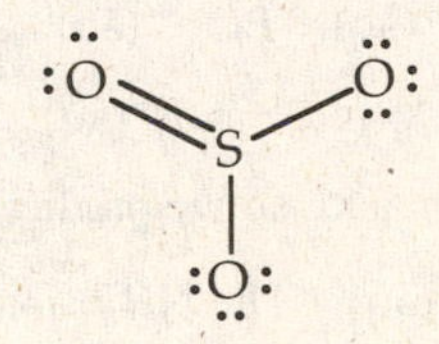

(c) CH_3NH_2 has a lone pair of electrons on N that can form a new bond, so CH_3NH_2 can be a Lewis base. The H atoms polar-covalently bonded to N could interact with other Lewis bases and be removed, so it could function as a Lewis acid; however, that reaction requires a very strong Lewis base.

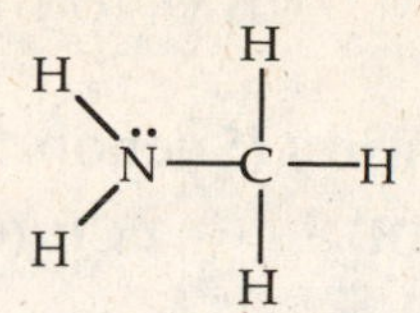

86. *Answer:* **(a) Lewis acid: I_2; Lewis base: I^- (b) Lewis acid: BF_3; Lewis base: SO_2 (c) Lewis acid: Au^+; Lewis base: CN^- (d) Lewis acid: CO_2; Lewis base: H_2O**

Strategy and Explanation: Identify which reactant is donating electrons and which is accepting them.

(a) The curved arrow shows how the lone pair on I^- becomes a new bond on the right I of I_2.

$$:\ddot{I}-\ddot{I}: + [:\ddot{I}:]^- \longrightarrow [:\ddot{I}-\ddot{I}-\ddot{I}:]^-$$

Therefore, I_2 is the Lewis acid and I^- is the Lewis base.

(b) The curved arrow shows how the lone pair on S becomes a new bond between the S atom and B atom.

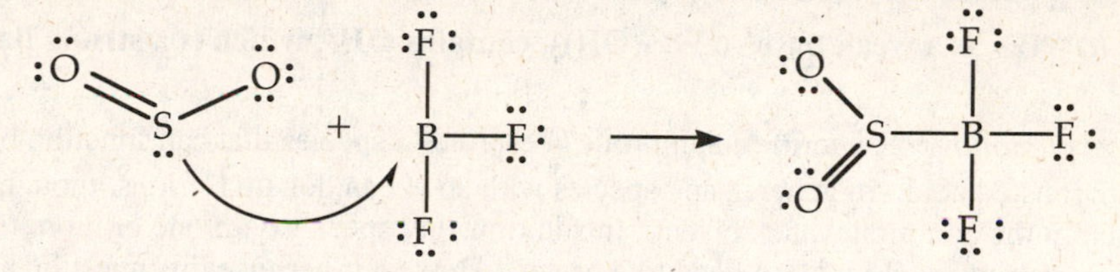

Therefore, BF_3 is the Lewis acid and SO_2 is the Lewis base.

(c) Au^+ metal ion has no bonds to start with and only two low-energy valence electrons. Electrons on CN^- make new bonds with Au^+, so Au^+ is Lewis acid and CN^- is Lewis base.

(d) A similar reaction is shown at the end of Section 16.9 on page 789. The curved arrow below shows how the lone pair on O becomes a new bond between C atom and O atom. (Notice that after the initial Lewis acid-base reaction, one of the H atoms migrates to a nearby O atom to balance the positive and negative charges.)

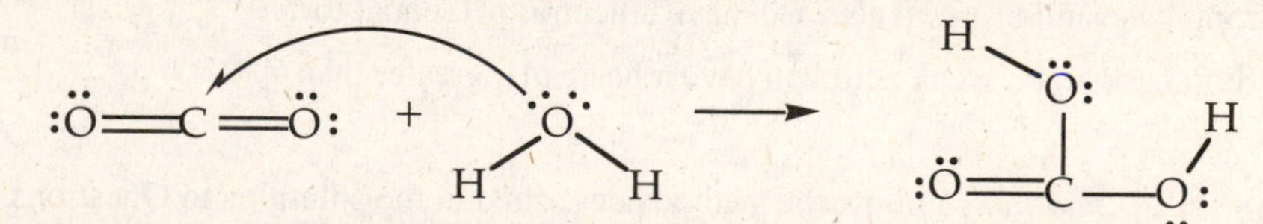

Therefore, CO_2 is the Lewis acid and H_2O is the Lewis base.

88. *Answer:* **See Lewis structure below; ICl_3 is T-shaped; ICl_3 functions as the Lewis acid to form ICl_4^-; see below for Lewis structure of ICl_4^-; it has a square planar geometry.**

Strategy and Explanation: See the solution to Question 9.92 for the shape determinations. The ICl_3 molecule has AB_3E_2 form, so the shape is T-shaped. The curved arrow shows how the lone pair on Cl^- becomes a new bond between the Cl atom and I atom. Therefore, ICl_3 is the Lewis acid in the reaction with chloride ion. The ICl_4^- ion has AB_4E_2 form, so the shape is square planar.

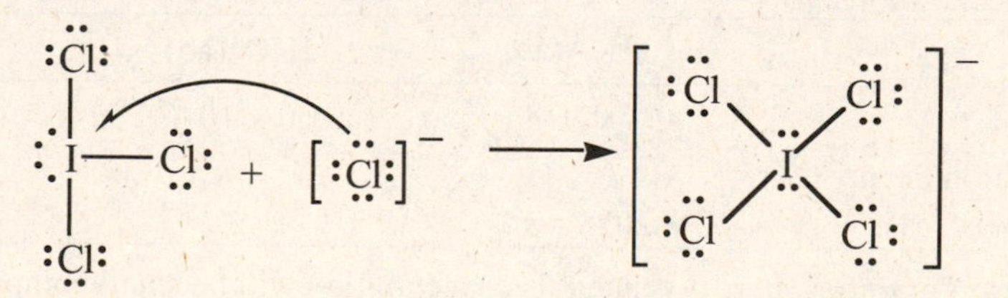

90. *Answer/ Explanation:* Conjugates in an acid-base pair must differ by only one H^+ ion. The conjugate acid of the $[Zn(H_2O)_3(OH)]^+$ ion will have one less H^+ ion, removed from one of the H_2O oxygen atoms:

$$Zn(H_2O)_2(OH)_2$$

Additional Applied Acid-Base Chemistry (Section 16.10)

92. *Answer:* **$Na_2CO_3 + 2\ CH_3(CH_2)_{16}COOH \longrightarrow 2\ CH_3(CH_2)_{16}COONa + H_2O + CO_2$**

Strategy and Explanation: Na_2CO_3 dissolves to form Na^+ and CO_3^{2-}. Proton transfer occurs between stearic acid, $CH_3(CH_2)_{16}COOH$, and the anionic base, CO_3^{2-}, resulting in the production of weak acid H_2CO_3, which decomposes into $CO_2(g)$ and water.

94. *Answer/Explanation:* Dishwasher detergent is very basic, and should not be used to wash anything by hand, including a car. If it gets into the engine area, it can also dissolve automobile grease and oil, which could prevent the engine from running correctly.

96. *Answer/Explanation:* Lemon juice contains citric acid. The acid protonates the basic amines. The conjugate acids formed from the neutralized bases are ions and not volatile.

General Questions

98. *Answer:* **(a) CH_3COOH is a weak acid. (b) Na_2O, contains O^{2-}, which is a strong base. (c) H_2SO_4 is a strong acid. (d) NH_3 is a weak base. (e) $Ba(OH)_2$, contains OH^-, which is a strong base. (f) $H_2PO_4^-$ is amphiprotic.**

Strategy and Explanation: The term "amphiprotic" describes a species that can function both as a Bronsted base and as a Bronsted acid. In general, any species with an H can donate H^+ ions, though sometimes that will not happen in the common water solvent. In addition, any species with one or more lone pairs can be a base, since the electrons could make a bond to a proton. However, because this question asks for a judgment of "weak" and "strong," we will restrict our designations to the reactions of these species in water and use Table 16.2 to assist us. Notice that Na_2O, contains O^{2-}.

100. *Answer:* **(a) less than 7 (b) equal to 7 (c) greater than 7**

Strategy and Explanation: If equal molar amounts of an acid and a base are combined, the stronger of them will dictate the pH of the solution.

(a) A weak base and a strong acid will have an acidic pH, less than 7.

(b) A strong base and a strong acid will have a neutral pH, equal to 7.

(c) A strong base and a weak acid will have a basic pH, greater than 7.

102. *Answer:* **2.85**

Strategy and Explanation: Follow the methods described in the solutions to Questions 54 - 60.

$$\frac{5.0\text{ mg } C_6H_8O_6}{1\text{ mL soln}} \times \frac{1\text{ g}}{1000\text{ mg}} \times \frac{1\text{ mol } C_6H_8O_6}{176\text{ g } C_6H_8O_6} \times \frac{1000\text{ mL}}{1\text{ L}} = 0.028\text{ M}$$

Ascorbic acid, $C_6H_8O_6$, is diprotic acid, so give it a formula of H_2A. The equation for the equilibrium and the equilibrium expression are:

$$H_2A(aq) + H_2O(\ell) \rightleftharpoons H_3O^+(aq) + HA^-(aq) \qquad K_a = \frac{[H_3O^+][HA^-]}{[H_2A]}$$

As the reactants decompose, the concentrations of the products increase stoichiometrically, until they reach equilibrium concentrations.

	$H_2A(aq)$	$H_3O^+(aq)$	$HA^-(aq)$
initial conc. (M)	0.028	1.0×10^{-7}*	0
change as reaction occurs (M)	– x	+ x	+ x
equilibrium conc. (M)	0.028 – x	x	x

* from dissociation of pure water. Assume this concentration will be small compared to the acid added.

At equilibrium $$K_a = \frac{(x)(x)}{(0.028 - x)} = 7.9 \times 10^{-5}$$

$$x^2 = (7.9 \times 10^{-5})(0.028 - x)$$

If we assume that x is small, and $0.028 - x \cong 0.028$

$$x^2 = (7.9 \times 10^{-5})(0.028)$$

$$x = 1.5 \times 10^{-3}\ M = [H_3O^+]$$

$$pH = -\log[H_3O^+] = -\log(1.5 \times 10^{-3}) = 2.82$$

If we use the quadratic equation: (see Appendix A, Section A.7, page A.13)

$$x^2 + 7.9 \times 10^{-5}x - 2.2 \times 10^{-6} = 0$$

$$x = 1.4 \times 10^{-3}\ M = [H_3O^+]$$

$$pH = -\log[H_3O^+] = -\log(1.4 \times 10^{-3}) = 2.85$$

104. *Answer:* **(a) increases (b) stays the same (c) stays the same**

Strategy and Explanation: Adding acids to bases decreases the pH. Adding bases to acids increases the pH. Adding weak acids or bases to strong acids or bases, respectively, does not change the pH significantly.

(a) $Na_2C_2O_4$ contains $C_2O_4^{2-}$, a weak base. The base is added to an acid, so the pH increases.

(b) NH_4Cl contains NH_4^+, a weak acid. The weak acid is added to a strong acid, so no change in pH will be noticed. pH stays the same.

(c) Adding a neutral salt, NaCl, will not change the pH, so the pH stays the same.

106. *Answer:* **9.59**

Strategy and Explanation: Use the method similar to that given in the solution to Question 62. The equation for the equilibrium and the equilibrium expression are:

$$ClO^-(aq) + H_2O(\ell) \rightleftharpoons HClO(aq) + OH^-(aq) \qquad K_b = \frac{[HClO][OH^-]}{[ClO^-]}$$

As the reactants decompose, the concentrations of the products increase stoichiometrically, until they reach equilibrium concentrations.

	$ClO^-(aq)$	$HClO(aq)$	$OH^-(aq)$
initial conc. (M)	0.010	0	1.0×10^{-7}*
change as reaction occurs (M)	$-x$	$+x$	$+x$
equilibrium conc. (M)	$0.010 - x$	x	x

* from dissociation of pure water. Assume this concentration will be small compared to the base added.

At equilibrium $$K_b = \frac{(x)(x)}{(0.010 - x)} = 1.5 \times 10^{-7}$$

Assume x is very small and: $0.010 - x \cong 0.010$.

$$x^2 = (1.5 \times 10^{-7})(0.010)$$

$$x = 3.9 \times 10^{-5}\ M = [OH^-]$$

$$pOH = -\log[OH^-] = -\log(3.9 \times 10^{-5}) = 4.41$$

$$pH = 14.00 - pOH = 14.00 - 4.41 = 9.59$$

Applying Concepts

108. *Answer:* **7.27, 6.998, 6.631; all three solutions are neutral.**

Strategy and Explanation: Table 16.1 provides the ionization constant for water at various temperatures. The equation representing the autoionization of water and the expression for its equilibrium constant are:

$$2\ H_2O(\ell) \rightleftharpoons H_3O^+(aq) + OH^-(aq) \qquad K_w = [H_3O^+][OH^-]$$

The concentration of products increase stoichiometrically, until they reach equilibrium concentrations.

	$H_3O^+(aq)$	$OH^-(aq)$
initial conc. (M)	0	0
change as reaction occurs (M)	+ x	+ x
equilibrium conc. (M)	x	x

So, at equilibrium $K_w = x^2$

At 10 °C $x = \sqrt{0.29 \times 10^{-14}} = 5.4 \times 10^{-8}\ M = [H_3O^+]$

$pH = -\log[H_3O^+] = -\log(5.4 \times 10^{-8}) = 7.27$

At 25 °C $x = \sqrt{1.01 \times 10^{-14}} = 1.00 \times 10^{-7}\ M = [H_3O^+]$

$pH = -\log[H_3O^+] = -\log(1.00 \times 10^{-7}) = 6.998$

At 50 °C $x = \sqrt{5.48 \times 10^{-14}} = 2.34 \times 10^{-7}\ M = [H_3O^+]$

$pH = -\log[H_3O^+] = -\log(2.34 \times 10^{-7}) = 6.631$

The solutions are neutral, since $[H_3O^+] = [OH^-]$ in each of them.

110. *Answer:* **(a) $H_2O > H_3O^+ = Cl^- >> OH^-$ (b) $H_2O > Na^+ = ClO_4^- >> H_3O^+ = OH^-$ (c) $H_2O > HNO_2 > H_3O^+ = NO_2^- >> OH^-$ (d) $H_2O > Na^+ \cong ClO^- > OH^- = HClO >> H_3O^+$ (e) $H_2O > NH_4^+ \cong Cl^- > H_3O^+ = NH_3 >> OH^-$ (f) $H_2O > Na^+ = OH^- >> H_3O^+$**

Strategy and Explanation: Adapt the methods described in the solution to Question 52.

(a) HCl is a strong acid, resulting in a solution of H_3O^+ and Cl^-. The solution contains the following molecules and ions in order of decreasing concentrations:

$$H_2O > H_3O^+ = Cl^- >> OH^-$$

(b) $NaClO_4$ is a soluble ionic compound, resulting in a solution of Na^+ and ClO_4^-. Neither of these ions reacts with water. The solution contains the following molecules and ions in order of decreasing concentrations:

$$H_2O > Na^+ = ClO_4^- >> H_3O^+ = OH^-$$

(c) HNO_2 is a weak acid, resulting in a solution of a large proportion of HNO_2 and a small proportion of H_3O^+ and NO_2^- The solution contains the following molecules and ions in order of decreasing concentrations:

$$H_2O > HNO_2 > H_3O^+ = NO_2^- >> OH^-$$

(d) NaClO is a soluble ionic compound, resulting in a solution of Na^+ and ClO^-. The anion is a weak base and reacts with water to a small extent to make a small proportion of HClO and OH^-. The solution contains the following molecules and ions in order of decreasing concentrations:

$$H_2O > Na^+ \cong ClO^- > OH^- = HClO >> H_3O^+$$

(e) NH_4Cl is a soluble ionic compound, resulting in a solution of NH_4^+ and Cl^-. The cation is a weak acid and reacts with water to a small extent to make a small proportion of NH_3 and H_3O^+. The solution contains the following molecules and ions in order of decreasing concentrations:

$$H_2O > NH_4^+ \cong Cl^- > H_3O^+ = NH_3 >> OH^-$$

(f) NaOH is a strong base, resulting in a solution of Na^+ and OH^-. The solution contains the following molecules and ions in order of decreasing concentrations:

$$H_2O > Na^+ = OH^- >> H_3O^+$$

112. *Answer/Explanation:* Conjugates in an acid-base pair must differ by only one H^+ ion. The acids and bases were identified correctly, but the conjugates were not. HCO_3 is the conjugate of H_2CO_3 and HSO_4 is the conjugate of SO_4^{2-}.

114. *Answer/Explanation:* The other two acid-base theories are described by the Lewis theory in the following ways:

Arrhenius theory: Electron pairs on the solvent water molecules (Lewis base) form a bond with the hydrogen ion (Lewis acid) producing aqueous H^+ ions. Electron pairs on the OH^- ions (Lewis base) form a bond with the hydrogen ion (Lewis acid) in the solvent water molecule, producing aqueous OH^- ions.

Bronsted Theory: The H^+ ion from the Bronsted acid is bonded to a Bronsted base using an electron pair on the base. The electron-pair acceptor, the H^+ ion, is the Lewis acid and the electron-pair donor is the Lewis base.

116. *Answer:* **(a) 2.01 (b) 23 times**

Strategy and Explanation: Adapt the method shown in Question 54.

(a) The equation for the equilibrium and the equilibrium expression are:

$$CCl_3COOH(aq) + H_2O(\ell) \rightleftharpoons H_3O^+(aq) + CCl_3COO^-(aq) \qquad K_a = \frac{[H_3O^+][CCl_3COO^-]}{[CCl_3COOH]}$$

The concentrations of products increase stoichiometrically, until equilibrium is reached.

	$CCl_3COOH(aq)$	$H_3O^+(aq)$	$CCl_3COO^-(aq)$
initial conc. (M)	0.010	1.0×10^{-7}*	0
change as reaction occurs (M)	– x	+ x	+ x
equilibrium conc. (M)	0.010 – x	x	x

* from dissociation of pure water. Assume this concentration will be small compared to the acid added.

At equilibrium $$K_a = \frac{(x)(x)}{(0.010 - x)} = 3.0 \times 10^{-1}$$

$$x^2 = (3.0 \times 10^{-1})(0.010 - x)$$

$$x^2 + 3.0 \times 10^{-1}x - 3.0 \times 10^{-5} = 0$$

Use the quadratic equation: (see Appendix A, Section A.7, page A.13)

$$x = 9.7 \times 10^{-3}\ M = [H_3O^+]$$

$$pH = -\log[H_3O^+] = -\log(9.7 \times 10^{-3}) = 2.01$$

(b) The equation for the equilibrium and the equilibrium expression are:

$$CH_3COOH(aq) + H_2O(\ell) \rightleftharpoons H_3O^+(aq) + CH_3COO^-(aq) \qquad K_a = \frac{[H_3O^+][CH_3COO^-]}{[CH_3COOH]}$$

The concentrations of products increase stoichiometrically, until equilibrium is reached.

	$CH_3COOH(aq)$	$H_3O^+(aq)$	$CH_3COO^-(aq)$
initial conc. (M)	0.010	1.0×10^{-7}*	0
change as reaction occurs (M)	– x	+ x	+ x
equilibrium conc. (M)	0.010 – x	x	x

* from dissociation of pure water. Assume this concentration will be small compared to the acid added.

At equilibrium $$K_a = \frac{(x)(x)}{(0.010 - x)} = 1.8 \times 10^{-5}$$

Assume x is very small and: $0.20 - x \cong 0.20$.

$$1.8 \times 10^{-5} = \frac{x^2}{(0.010)}$$

$$x^2 = (1.8 \times 10^{-5})(0.010)$$

$$x = 4.2 \times 10^{-4}\ M = [H_3O^+]$$

To determine the number of times more hydronium ions in the trichloroacetic acid versus the acetic acid, divide the hydronium ion concentration in the trichloroacetic acid solution by that in the acetic acid solution.

$$\frac{9.7 \times 10^{-3}\ M}{4.2 \times 10^{-4}\ M} = 23 \text{ times more } H_3O^+ \text{ in } CCl_3COOH(aq)$$

119. *Answer:* **(a) HY; see explanation below (b) Z^-; see explanation below**

Strategy and Explanation:

(a) For each box, count the number of each species in the box and dividing by the box volume, V, to get the concentration. Then calculate the K_a value by putting the concentrations into the form of the reaction quotient. (A = X, Y, or Z)

$$HA(aq) + H_2O(\ell) \rightleftharpoons H_3O^+(aq) + A^-(aq) \qquad K_a = \frac{[H_3O^+][A^-]}{[HA]}$$

In the first box, there are eight HX molecules, four X^-, and four H_3O^+:

$$K_a = \frac{\left(\frac{4}{V}\right)\left(\frac{4}{V}\right)}{\left(\frac{8}{V}\right)} = \frac{2}{V}$$

In the second box, there are six HY molecules, six Y^-, and six H_3O^+:

$$K_a = \frac{\left(\frac{6}{V}\right)\left(\frac{6}{V}\right)}{\left(\frac{6}{V}\right)} = \frac{6}{V}$$

In the third box, there are six HZ molecules, two Z^-, and two H_3O^+:

$$K_a = \frac{\left(\frac{2}{V}\right)\left(\frac{2}{V}\right)}{\left(\frac{6}{V}\right)} = \frac{4}{6V}$$

Assuming the boxes are all the same volume, HY has the largest K_a value.

(b) The weaker the acid, the stronger the conjugate base. The acid with the smallest K_a value, HZ, is the weakest of these three acid, so it's conjugate base, Z^-, is the strongest conjugate base.

122. *Answer:* **(a) Boxes C, F, and G (b) Box I (c) Boxes A and H (d) Boxes D and E (e) Box B (f) Box I (g) Boxes D and E**

Strategy and Explanation:

(a) The pH > 7.0 in basic solutions, so look for solutions with $[OH^-] > 10^{-7}$ M or $[H_3O^+] < 10^{-7}$ M. There are four boxes with concentrations given, Boxes C, F, G, and I; examine each one.

Box C has $[OH^-]\ 10^{-5}\ M > 10^{-7}\ M$

Box F has $[OH^-] = 0.10\ M > 10^{-7}\ M$

Box G has $[H_3O^+] = 10^{-8}\ M < 10^{-7}\ M$

Box I has $[H_3O^+] = 0.01\ M > 10^{-7}\ M$

The boxes that answer this item are **Boxes C, F, and G**.

(b) $[H_3O^+] > 1.0 \times 10^{-7}$ M in acidic solutions, so look for ones with $[OH^-] < 10^{-7}$ M or $[H_3O^+] > 10^{-7}$ M. There are four boxes with concentrations given, Boxes C, F, G and I; examine each one as done in (a).

The only box that answers this item is **Box I**.

(c) A basic salt is a salt that contains an anion that is the conjugate acid of a weak acid. There are two boxes that contain salts, Boxes A and H; examine each one.

Box A contains a salt with the anion, CO_3^{2-}. The acid H_2CO_3 is a weak acid

Box H contains a salt with the anion, NO_2^-. The acid HNO_2 is a weak acid

The boxes that answer this item are **Boxes A and H**.

(d) The item is limited to the species in the grid that can function as a base and as an acid. Those are found in Boxes D and E; examine each one.

Box D, H_2SO_3, is a fairly strong weak acid. The compound does still have lone pairs of electrons on the O atoms, so it can also be a base, but it would be a very weak base. $K_a > K_b$

Box E, $H_2PO_4^-$, contains an acid anion that can be found in Table 16.2. $K_a = 10^{-8}$ and $K_b = 10^{-12}$, so $K_a > K_b$

(Note: Boxes C, F, G, and I also contain acids or bases that could be considered, but because they are given with specific concentration values, they will not be considered in the answer to this question. If they were, the acids in Boxes G and I, H_3O^+ and HNO_2, would also qualify as answers to this question.)

The boxes that answer this item are **Boxes D and E**.

(e) A Lewis acid is an electron-pair acceptor. The metallic cation Cu^{2+} ion can function as a Lewis acid, by accepting electron pairs. The answer is **Box B**.

(f) A solution with a pH = 2.00, has $[H_3O^+] = 10^{-2.00}\ M = 0.01\ M$. Because $HClO_4$ is a strong acid, the solution identified in Box I has the right concentration for pH 2.00. The answer is **Box I**.

(g) A polyprotic acid has more than one acidic hydrogen atom in its formula. There are two acids listed in the grid, in Boxes D and E.

Box D, H_2SO_3, is a diprotic acid.

Box E, $H_2PO_4^-$, is a diprotic acid.

(Note: Boxes G and I also contain acids that could be considered, but because they are given with specific concentration values, they will not be considered in the answer to this question. If they were, the acids in Boxes G, triprotic H_3O^+, would also qualify as an answer to this question.)

The boxes that answer this item are **Boxes D and E**.

More Challenging Questions

124. *Answer:* **lactic acid sample**

Strategy and Explanation: This question involves some interpretation and research. Of the three materials described, two are solutions containing acids: Vinegar contains acetic acid (CH_3COOH, $K_a = 1.8 \times 10^{-5}$) at a concentration of approximately 5% and lemon juice contains triprotic citric acid ($H_3C_5H_5O_7$, Table 12.5) at a concentration of approximately 1%. The third material is lactic acid ($C_3H_6O_3$, $K_a = 1.4 \times 10^{-4}$, from Question 68), a 100% pure molecular acid that is sometimes found in milk.

The ion being neutralized is $HCO_3^-(aq)$.

$$H_3O^+(aq) + HCO_3^-(aq) \rightleftharpoons H_2CO_3(aq) + H_2O(\ell) \qquad K = \frac{1}{K_{a,H_2CO_3}} = \frac{1}{4.3 \times 10^{-7}} = 2.3 \times 10^6$$

$$HA(aq) + H_2O(\ell) \rightleftharpoons H_3O^+(aq) + A^-(aq) \qquad K_{a,HA}$$

Because lactic acid, acetic acid, and citric acid each presumably have $K_a > 10^{-7}$, the three neutralization reactions we are studying here will all be product-favored. ($K > 1$). So, all we need to do is determine which sample provides the largest number of H_3O^+ ions.

Using the molar mass of lactic acid, we can determine the moles of H_3O^+ ions available if one gram of lactic acid reacts:

$$1\ g\ C_3H_6O_3 \times \frac{1\ mol\ C_3H_6O_3}{90.08\ g\ C_3H_6O_3} \times \frac{1\ mol\ H_3O^+}{1\ mol\ C_3H_6O_3} = 0.01\ mol\ H_3O^+$$

Using the percentages of acids in solution and the molar masses, determine the H_3O^+ ions available in the vinegar and lemon juice samples.

$$1\ g\ vinegar \times \frac{0.05\ g\ CH_3COOH}{1\ g\ vinegar} \times \frac{1\ mol\ CH_3COOH}{60.0518\ g\ CH_3COOH} \times \frac{1\ mol\ H_3O^+}{1\ mol\ CH_3COOH} = 0.0008\ mol\ H_3O^+$$

$$1\ g\ lemon\ juice \times \frac{0.01\ g\ H_3C_5H_3O_7}{1\ g\ lemon\ juice} \times \frac{1\ mol\ H_3C_5H_3O_7}{178.097\ g\ H_3C_5H_3O_7} \times \frac{3\ mol\ H_3O^+}{1\ mol\ H_3C_5H_3O_7} = 0.0002\ mol\ H_3O^+$$

With this analysis, we find that the 1gram sample of pure lactic acid will make more CO_2, even though the citric acid in the lemon juice is the strongest acid.

125. *Answer:* **0.76 L; no**

Strategy and Explanation: This problem can be solved by methods described in Chapter 5 and 10. Use the chemical reaction in Table 16.8. The product-favored reaction goes essentially to completion in the presence of sufficient acid.

$$2.5\ g\ NaHCO_3 \times \frac{1\ mol\ NaHCO_3}{84.0066\ g\ NaHCO_3} \times \frac{1\ mol\ CO_2(g)}{1\ mol\ NaHCO_3} = 0.030\ mol\ CO_2(g)$$

$$V_{CO_2} = \frac{n_{CO_2}RT}{P} = \frac{(0.030\ mol) \times \left(0.08206 \frac{L \cdot atm}{mol \cdot K}\right) \times (37 + 273)K}{1\ atm} = 0.76\ L$$

This, by itself, is an insufficient volume of CO_2 to rupture a 1-L stomach, though it would be enough to be uncomfortable. To be able to predict a rupture we'd need to know how much volume the food occupied and whether this added CO_2 would exceed the capacity of his stomach. We don't know what pressure of gas the stomach at maximum volume can take, nor do we know if some other involuntary action such as burping or vomiting would occur to prevent the excess gas from causing damage. One would need to be a gastroenterologist, to accurately answer this part of the question.

126. *Answer:* **Yes, pH increases**

Strategy and Explanation: When the can is initially pressurized with CO_2, the following chemical equilibrium expression is shifted toward the formation of products. The beverage becomes acidic.

$$CO_2(aq) + 2\ H_2O(\ell) \rightleftharpoons H_3O^+(aq) + HCO_3^-(aq)$$

When the can is opened and warmed, carbon dioxide escapes from the carbonated beverage. The loss of $CO_2(aq)$ shifts the above equilibrium to the left. Hydronium ion concentration is decreased and the solution pH increases (i.e. becomes more basic).

128. *Answer/Explanation:* Br has a higher electronegativity than H. The bromine withdraws electron density from the nitrogen atom, reducing the nitrogen's ability to bind the positively charged proton. The K_b value is smaller for $BrNH_2$ than for NH_3. $ClNH_2$ is a weaker base than $BrNH_2$ because Cl is more electronegative than Br.

130. *Answer:* **HM > HQ > HZ ; HZ, $K_{a,HZ} = 1 \times 10^{-5}$; HQ, $K_{a,HQ} = 1 \times 10^{-3}$; HM, $K_{a,HM} = 1 \times 10^{-1}$ or larger**

Strategy and Explanation: Stronger acids result in weaker conjugate bases. The stronger base has the larger pH. So, the salt solution with the highest pH contains the anion of the weakest acid.

Smallest pH: NaM > NaQ > NaZ :Largest pH

Strongest: HM > HQ > HZ :Weakest

The general equation for the equilibrium and the equilibrium expression are:

$$A^-(aq) + H_2O(\ell) \rightleftharpoons HA(aq) + OH^-(aq) \qquad K_b = \frac{[HA][OH^-]}{[A^-]}$$

(where A = M, Q, or Z)

The initial $[OH^-]$ of neutral water is 1×10^{-7} M. The initial concentration of the base is 0.1 M. The concentrations of products increase stoichiometrically, until they reach equilibrium concentrations.

In general:	$A^-(aq)$	$HA(aq)$	$OH^-(aq)$
initial conc. (M)	0.1	0	1×10^{-7}*
change as reaction occurs (M)	– x	+ x	+ x
equilibrium conc. (M)	0.1 – x	x	10^{-7} + x

* from the dissociation of pure water.

At equilibrium $$K_b = \frac{(x)(1\times10^{-7}+x)}{(0.1-x)}$$

$$1 \times 10^{-7} + x = [OH^-] = 10^{-pOH} = 10^{-(14.00-pH)} = 10^{(pH-14.00)}$$

$$x = 10^{(pH-14.00)} - 1 \times 10^{-7}$$

For NaZ, A^- above is Z^- and pH = 9.0, so

$$x = 10^{(9.0-14.00)} - 1 \times 10^{-7} = 1 \times 10^{-5} - 1 \times 10^{-7} = 1 \times 10^{-5}, \text{ and}$$

$$K_b = \frac{(1\times10^{-5})(1\times10^{-7}+1\times10^{-5})}{(0.1-1\times10^{-5})} = 1\times10^{-9}$$

For NaQ, A^- above is Q^- and pH = 8.0, so

$$x = 10^{(8.0-14.00)} - 1 \times 10^{-7} = 1 \times 10^{-6} - 1 \times 10^{-7} = 1 \times 10^{-6}, \text{ and}$$

$$K_b = \frac{(1\times10^{-6})(1\times10^{-7}+1\times10^{-6})}{(0.1-1\times10^{-6})} = 1\times10^{-11}$$

For NaM, A^- above is M^- and pH = 7.0, so $x = 10^{(7.0-14.00)} - 1 \times 10^{-7}$

$x = 1 \times 10^{-7} - 1 \times 10^{-7} = 0$, suggesting that the reaction with water does not go toward products at all. That means K_b of M^- may be so small that the pH of the solution is accounted for exclusively by the ionization of water. If indeed the base does provide all the OH^- ions to make the solution's pH: $[HM] = [OH^-] = 10^{-7.0} = 1 \times 10^{-7}$:

$$K_b = \frac{(1\times10^{-7})(1\times10^{-7})}{(0.1)} = 1\times10^{-13}$$

A relationship between K_a and K_b is described at the end of Section 16.7 on page 780.

For HZ, $K_a = K_w/K_b = (1.0 \times 10^{-14})/(1 \times 10^{-9}) = 1 \times 10^{-5}$

For HQ, $K_a = K_w/K_b = (1.0 \times 10^{-14})/(1 \times 10^{-11}) = 1 \times 10^{-3}$

For HM, $K_a = K_w/K_b = (1.0 \times 10^{-14})/(1 \times 10^{-13}) = 1 \times 10^{-1}$ or larger

132. *Answer:* **(a) weak (b) weak (c) conjugate acid-base pair (d) 6.29**

Strategy and Explanation:

(a) The acid ionization constant for hydrogen is very small. H_2O_2 is a weak acid.

(b) K_b for OOH^- is larger than K_a, but it is still small. OOH^- is a weak base.

(c) H_2O_2 and OOH^- are a conjugate acid and base.

(d) Hydrogen peroxide will ionize in solution according to the following equation with the following equilibrium expression:

$$HOOH(aq) + H_2O(\ell) \rightleftharpoons H_3O^+(aq) + HOO^-(aq) \qquad K_b = \frac{[H_3O^+][HOO^-]}{[HOOH]}$$

	HOOH	$H_3O^+(aq)$	$HOO^-(aq)$
initial conc. (M)	0.100	1.0×10^{-7}	0
change as reaction occurs (M)	– x	+ x	+ x
equilibrium conc. (M)	0.100 – x	$1.0 \times 10^{-7} + x$	x

$$K_b = \frac{(1.0\times10^{-7} + x)(x)}{(0.100 - x)} = 2.1 \times 10^{-12}$$

$$(1.0 \times 10^{-7} + x)x = (2.1 \times 10^{-12})(0.100 - x)$$

If we assume that x is small, and $0.100 - x \cong 0.100$

$$(1.0 \times 10^{-7} + x)x = (2.1 \times 10^{-12})(0.100)$$

$$x^2 + x(1.0 \times 10^{-7}) - 2.1 \times 10^{-13} = 0$$

Use the quadratic equation: (see Appendix A, Section A.7, page A.13)

$$x = 4.1 \times 10^{-7}$$

$$[H_3O^+] = 1.0 \times 10^{-7} + 4.1 \times 10^{-7} = 5.1 \times 10^{-7}\ M$$

$$pH = -\log[H_3O^+] = -\log(5.1 \times 10^{-7}) = 6.29$$

134. *Answer:* **0.1% solution is 9% dissociation; saturated solution is 2% dissociation.**

Strategy and Explanation: Determination of the percent dissociation for adipic acid requires the calculation of the initial concentration of the acid and the concentration of H_3O^+ created. The concentration of adipic acid is determined from the mass of the acid per 100 mL of solution.

Assume the 0.1% solution is a mass percent. At low concentrations, the density of a solution can be assumed equal to the density of water. Convert to molarity as follows:

$$\frac{0.1\ g\ C_5H_9O_2COOH}{100\ g\ solution} \times \frac{0.998\ g\ solution}{1\ mL} \times \frac{1000\ mL}{L} \times \frac{1\ mol\ C_5H_9O_2COOH}{146.1408\ g\ C_5H_9O_2COOH} = 7 \times 10^{-3}\ M$$

Determine the concentration of H_3O^+ from pH = 3.2.

$$[H_3O^+] = 10^{-pH} = 10^{-3.2} = 6 \times 10^{-4}$$

The percent ionization of the 0.1% adipic acid solution is:

$$\frac{[H_3O^+]\text{ at equilibrium}}{\text{initial acid conc.}} \times 100\% = \frac{6\times10^{-4}\ M}{7\times10^{-3}\ M} \times 100\% = 9\%$$

Repeat the calculations for the saturated adipic acid solution.

$$\frac{1.44\ g\ C_5H_9O_2COOH}{100\ g\ solution} \times \frac{0.998\ g\ solution}{1\ mL} \times \frac{1000\ mL}{L} \times \frac{1\ mol\ C_5H_9O_2COOH}{146.1408\ g\ C_5H_9O_2COOH} = 9.9 \times 10^{-2}\ M$$

Determine the concentration of H_3O^+ from pH = 2.7. $[H_3O^+] = 10^{-pH} = 10^{-3.2} = 2 \times 10^{-3}$

The percent ionization of the 0.1% adipic acid solution is: $\dfrac{2\times10^{-3}\ M}{9.84\times10^{-2}\ M} \times 100\% = 2\%$

136. *Answer:* **(a) $H_2NCH_2CH_2CH_2CH_2CH_2NH_2$ (b) $^+H_3NCH_2CH_2CH_2CH_2CH_2NH_3^+$ (c) 9.13**

Strategy and Explanation:

(a) The formula for cadaverine is $H_2NCH_2CH_2CH_2CH_2CH_2NH_2$.

(b) The formula for protonated cadaverine is $^{+}H_3NCH_2CH_2CH_2CH_2CH_2NH_3^{+}$.

(c) The smaller pK_a (i.e. the larger K_a), 9.13, refers to the first deprotonation of cadaverine. The larger pK_a (i.e. the smaller K_a), 10.25, refers to the second deprotonation of cadaverine.

138. *Answer/Explanation:* Lysine is given in Question 136. It has three functional groups, a carboxylic acid and two amine groups. All three of these have acid-base chemistry. The most acidic functional group is the carboxylic acid, so it must account for the pK_a of 2.18. The amine groups are bases, which means that their pK_b's would be less than 7; hence their pK_a's will be greater than 7. The amine group closest to the carboxylic acid accounts for the pK_a of 8.95. The second amine function accounts for the pK_a of 10.53.

Chapter 17: Additional Aqueous Equilibria

Solutions for Blue-Numbered Questions for Review and Thought

Topical Questions

Buffer Solutions (Section 17.1)

16. *Answer:* **combination (c)**

Strategy and Explanation: To determine the pH of a buffer, look up the pK_a (Table 17.1) or look up the K_a (Table 16.2) and calculate the pK_a ($pK_a = -\log K_a$). The pK_a closest to the desired pH is the best buffer, since close to equal quantities of the acid and base would be used, giving the solution approximately equal ability to neutralize added acid or added base.

(a) The $CH_3COOH/NaCH_3COO$ buffer system has $pK_a = 4.74$.

(b) The acid in HCl/NaCl is HCl. It has K_a = very large. This is not a buffer.

(c) The NH_3/NH_4Cl buffer system has $pK_a = 9.25$

The combination that would make the best pH 9 buffer system is (c) NH_3/NH_4Cl.

18. *Answer:* **(a) 2.2 (b) 7.20 (c) 12.33**

Strategy and Explanation: To answer this question quantitatively we need the value of pK_a, since that is equal to the pH in an equimolar buffer solution. In some cases, that value is not in the textbook. Without doing any calculations, we must estimate the pK_a from the K_a. We will look at the size of the K_a value and estimate its power of ten. We do this by determining which two powers of ten the number is between, then we can also use the values provided in Table 17.1 to give us advice about the fractional part of the pK_a.

(a) The acid of the pair is H_3PO_4. It has $K_a = 7.2 \times 10^{-3}$. Here the K_a is between 10^{-2} and 10^{-3}. In Table 17.1, the dihydrogen phosphate buffer K_a has a slightly smaller number multiplying its power of ten, 6.3, and its pK_a has a fractional component of .20, so we will estimate the $HNO_2/NaNO_2$ buffer system will have a pH of about 2.2.

(b) The NaH_2PO_4/Na_2HPO_4 buffer system has $pK_a = 7.20$, so its pH is 7.20.

(c) The acid of the pair is HPO_4^{2-}, with $K_a = 4.6 \times 10^{-13}$. Here the K_a is between 10^{-12} and 10^{-13}. In Table 17.1, the hydrogen carbonate ion K_a has a similar number multiplying its power of ten, 4.7, and its pK_a has a fractional component of .33, so we will estimate the Na_2HPO_4/Na_3PO_4 buffer system will have a pH of about 12.33.

✓ *Reasonable Answer Check:* To check the answers, we will disobey the explicit instructions and do the calculation. In this case, you can also compare the estimates to the calculations you may have already done in Question 15. (a) $pK_a = -\log(7.2 \times 10^{-3}) = 2.14$ (c) $pK_a = -\log(4.6 \times 10^{-13}) = 12.34$

20. *Answer:* **(a) lactic acid/lactate (b) acetic acid/acetate (c) hypochlorous acid/hypochlorite (d) hydrogen carbonate/carbonate; see explanations, below**

Strategy and Explanation: We will compare the pH to the values of pK_a in Table 17.1, since that is equal to the pH in an equimolar buffer solution. The pK_a closest to the desired pH is most suitable.

(a) pH = 3.45, needs a lactic acid/lactate buffer ($pK_a = 3.85$).

(b) pH = 5.48, needs an acetic acid/acetate buffer ($pK_a = 4.74$). Its pH is slightly closer to the desired pH than any of the other buffer systems in Table 17.1.

(c) pH = 8.32, needs a hypochlorous acid/hypochlorite buffer ($pK_a = 7.17$). Its pH is slightly closer to the desired pH than any of the other buffer systems.

(d) pH =10.15, needs a hydrogen carbonate/carbonate buffer ($pK_a = 10.33$).

22. *Answer:* **4.8 g NH_4Cl**

Strategy and Explanation: This question uses methods learned in several previous chapters. First use the Henderson-Hasselbalch equation and Table 17.1 to find the concentration of the NH_4^+ present in the equilibrium solution.

$$pH = pK_a + \log\left(\frac{[\text{conj. base}]}{[\text{conj. acid}]}\right)$$

Table 17.1 gives pK_a of ammonium/ammonia buffer as 9.25 The pH is 9.00. The initial concentration of NH_3, the conjugate base, is 0.10 M. The conjugate acid, NH_4^+, is being added, but its initial concentration is unknown, so use the variable "c" to identify that quantity.

$$9.00 = 9.25 + \log\left(\frac{0.10}{c}\right)$$

Solve for c: $c = 0.10 \text{ M} \times 10^{-(9.00 - 9.25)} = 0.18 \text{ M}$

Then adapt the methods from Chapter 5 to find the mass of the salt.

$$500.\text{ mL} \times \frac{1\text{ L}}{1000\text{ mL}} \times \frac{0.18\text{ mol NH}_4^+}{1\text{ L}} \times \frac{1\text{ mol NH}_4\text{Cl}}{1\text{ mol NH}_4^+} \times \frac{53.49\text{ g NH}_4\text{Cl}}{1\text{ mol NH}_4\text{Cl}} = 4.8\text{ g NH}_4\text{Cl}$$

24. *Answer:* **8.62**

Strategy and Explanation: Adapt the method described in the solution to Question 22.
Calculate the initial concentration of NH_4^+, the conjugate acid, from methods described in Chapter 5:

$$(\text{conc. NH}_4^+) = \frac{5.15\text{ g NH}_4\text{NO}_3}{0.10\text{ L}} \times \frac{1\text{ mol NH}_4\text{NO}_3}{80.0432\text{ g NH}_4\text{NO}_3} \times \frac{1\text{ mol NH}_4^+}{1\text{ mol NH}_4\text{NO}_3} = 0.64\text{ M NH}_4^+$$

Table 17.1 gives pK_a of ammonium/ammonia buffer as 9.43. The initial concentration of NH_3, the conjugate base, is 0.15 M.

$$pH = 9.25 + \log\left(\frac{0.15}{0.64}\right) = 8.62$$

26. *Answer:* **9.55; 9.51**

Strategy and Explanation: Adapt the method described in the solutions to Question 22 and 24.

Calculate the initial concentration of the conjugate acid, NH_4^+:

$$(\text{conc. NH}_4^+) = \frac{0.125\text{ mol NH}_4\text{Cl}}{500.\text{ mL}} \times \frac{1000\text{ mL}}{1\text{ L}} \times \frac{1\text{ mol NH}_4^+}{1\text{ mol NH}_4\text{Cl}} = 0.250\text{ M NH}_4^+$$

Table 17.1 gives pK_a of ammonium/ammonia buffer as 9.25. The initial concentration of NH_3, the conjugate base, is 0.500 M.

$$pH = 9.25 + \log\left(\frac{0.500}{0.250}\right) = 9.55 \text{ before the HCl is added}$$

Assume that all the HCl gas bubbled through the solution is actually dissolved in the solution, find the initial concentration of H_3O^+ ions from the ionization of HCl, after it is dissolved but before it reacts:

$$(\text{conc. H}_3\text{O}^+) = \frac{0.0100\text{ mol HCl}}{500.\text{ mL}} \times \frac{1000\text{ mL}}{1\text{ L}} \times \frac{1\text{ mol H}_3\text{O}^+}{1\text{ mol HCl}} = 0.0200\text{ M H}_3\text{O}^+$$

The acid neutralizes the strongest base in the solution, NH_3. Write the product-favored neutralization equation:

$$NH_3(aq) + H_3O^+(aq) \longrightarrow NH_4^+(aq) + H_2O(\ell)$$

This reaction is product-favored, so we will first make as many products as possible. Using the method of limiting reactants, run the reaction towards products until one of the reactants runs out.

	$NH_3(aq)$	$H_3O^+(aq)$	$NH_4^+(aq)$
initial conc. (M)	0.500	0.0200	0.250
change as reaction occurs (M)	– 0.0200	– 0.0200	+ 0.0200
final conc. (M)	0.480	0	0.270

The solution is a still buffer solution, so, using the same technique, we can find the pH.

$$pH = 9.25 + \log\left(\frac{0.480}{0.270}\right) = 9.51 \text{ after the HCl is added}$$

✓ *Reasonable Answer Check:* The original buffer solution has a pH of 9.55. After adding some quantity of strong acid, it makes sense that the pH is slightly more acidic.

28. *Answer:* **Sample (b); see explanations below**

Strategy and Explanation: A 1-L solution of 0.20 M NaOH provides a strong base. The added solution must provide a source of a weak acid in sufficient quantity for some of it to completely neutralize the strong base and some of it to remain in the solution.

(a) Adding 0.10 mol CH_3COOH to the 1-L solution makes a 0.10 M CH_3COOH solution. The strong base neutralizes the acid in the solution, CH_3COOH.

$$CH_3COOH(aq) + OH^-(aq) \longrightarrow CH_3COO^-(aq) + H_2O(\ell)$$

This reaction is product-favored, so we will first make as many products as possible. Using the method of limiting reactants, run the reaction towards products until one of the reactants runs out.

	$CH_3COOH(aq)$	$OH^-(aq)$	$CH_3COO^-(aq)$
initial conc. (M)	0.10	0.20	0
change as reaction occurs (M)	– 0.10	– 0.10	+ 0.10
final conc. (M)	0.00	0.10	0.10

The solution produced is a not buffer solution. Too much base is present and all the CH_3COOH is neutralized.

(b) Adding 0.30 mol CH_3COOH to the 1-L solution makes a 0.30 M CH_3COOH solution. The strong base neutralizes the acid in the solution, as in (a). This reaction is product-favored, so we will first make as many products as possible. Using the method of limiting reactants, run the reaction towards products until one of the reactants runs out.

	$CH_3COOH(aq)$	$OH^-(aq)$	$CH_3COO^-(aq)$
initial conc. (M)	0.30	0.20	0
change as reaction occurs (M)	– 0.20	– 0.20	+ 0.20
final conc. (M)	0.10	0	0.20

The solution produced is a buffer solution, containing an acid-base conjugate pair.

(c) Adding a strong acid to a strong base will not produce a buffer. A buffer needs a weak acid-base conjugate pair in solution and this solution has neither.

(d) Adding a weak base to a strong base will not produce a buffer. A buffer needs a weak acid-base conjugate pair in solution and this solution has no significant acid concentration.

Therefore, of the four substances added to the base, only the addition of sample (b) produce buffer solutions.

30. *Answer:* **(a) ΔpH = 0.1 (b) ΔpH = 3.8 (c) ΔpH = 7.25**

Strategy and Explanation: Calculate the initial pH by finding the pK_a. Calculate initial OH^- concentration using the methods from Chapter 5. Then adapt the method used in the solution to Question 26.

Equimolar buffer solutions have $pH = pK_a = -\log K_a$. So, $pH = -\log(1.8 \times 10^{-5}) = 4.74$

Calculate the concentration of hydroxide using the total volume of the solution after the addition, but before the reaction.

$$V = 0.100 \text{ L} \times \frac{1000 \text{ mL}}{1 \text{ L}} + 1.0 \text{ mL}$$

$$(\text{conc. } OH^-) = \frac{1.0 \text{ mL}}{101 \text{ mL}} \times \frac{1.0 \text{ mol } OH^-}{1 \text{ L}} = 0.0099 \text{ M } OH^-$$

The strong base neutralizes the acid in the solution, CH_3COOH.

$$CH_3COOH(aq) + OH^-(aq) \longrightarrow CH_3COO^-(aq) + H_2O(\ell)$$

This reaction is product-favored, so we will first make as many products as possible. Using the method of limiting reactants, run the reaction towards products until one of the reactants runs out.

(a)

	$CH_3COOH(aq)$	$OH^-(aq)$	$CH_3COO^-(aq)$
initial conc. (M)	0.10	0.0099	0.10
change as reaction occurs (M)	– 0.0099	– 0.0099	+ 0.0099
final conc. (M)	0.09	0	0.11

The solution is still a buffer solution, containing an acid-base conjugate pair, so we can find the pH using the Henderson-Hasselbalch equation. In Table 17.1, the pK_a is 4.74.

$$pH = 4.74 + \log\left(\frac{0.11}{0.09}\right) = 4.8$$

$$\Delta pH = 4.8 - 4.74 = 0.1$$

(b)

	$CH_3COOH(aq)$	$OH^-(aq)$	$CH_3COO^-(aq)$
initial conc. (M)	0.010	0.0099	0.010
change as reaction occurs (M)	– 0.0099	– 0.0099	+ 0.0099
final conc. (M)	0.000	0.0000	0.020

Within known significant figures (three decimal places), the solution is no longer a buffer solution. It contains only the conjugate base. We must find the pH using K_b of the base, as was done in Chapter 16, such as Question 16.54.

$$CH_3COO^-(aq) + H_2O(\ell) \rightleftharpoons CH_3COOH(aq) + OH^-(aq) \qquad K_b = \frac{[CH_3COOH][OH^-]}{[CH_3COO^-]}$$

As the reactants decompose, the concentrations of the products increase stoichiometrically, until they reach equilibrium concentrations.

	$CH_3COO^-(aq)$	$CH_3COOH(aq)$	$OH^-(aq)$
initial conc. (M)	0.020	0	0
change as reaction occurs (M)	– x	+ x	+ x
equilibrium conc. (M)	0.020 – x	x	x

At equilibrium $K_b = \frac{(x)(x)}{(0.020 - x)} = 5.6 \times 10^{-10}$

Assume x is very small and does not affect the difference.

$$5.6 \times 10^{-10} = \frac{x^2}{(0.020)}$$

$$x^2 = (5.6 \times 10^{-10})(0.20)$$

$$x = 3.3 \times 10^{-6}\ M = [OH^-]$$

$$pOH = -\log[OH^-] = -\log(3.3 \times 10^{-6}) = 5.48$$

$$pH = 14.00 - pOH = 8.52$$

$$\Delta pH = 8.52 - 4.74 = 3.8$$

Notice: Even if we could ignore the limitation of the significant figures, we should not use the Henderson-Hassalbalch equation for such low acid concentrations.

(c) In this situation, the acetic acid is the limiting reactant:

	$CH_3COOH(aq)$	$OH^-(aq)$	$CH_3COO^-(aq)$
initial conc. (M)	0.0010	0.0099	0.0010
change as reaction occurs (M)	– 0.0010	– 0.0010	+ 0.0010
final conc. (M)	0.0000	0.0089	0.0020

The solution is no longer a buffer solution. This time, some OH^- ions are left over. The minor amount produced by the weak base reaction will not change this value, so:

$$pH = 14.00 + \log[OH^-] = 14.00 + \log(0.0089) = 11.95$$

$$\Delta pH = 11.95 - 4.74 = 7.25$$

32. *Answer:* **(a) 5.02 (b) 4.99 (c) 4.06**

Strategy and Explanation: Adapt the method described in the solutions to Question 24, 28, and 30.

Use the abbreviation HProp for the monoprotic propanoic acid. The salt sodium propanoate contains the $Prop^-$ ion.

(a) Find the pH using Henderson-Hassalbalch. The K_a of propanoic acid is given: 1.4×10^{-5}. As described in Section 17.1, calculate the pK_a:

$$pK_a = -\log K_a = -\log(1.4 \times 10^{-5}) = 4.85$$

The concentration of the conjugate acid, HProp, is 0.20 M; the concentration of the conjugate base, $Prop^-$, is 0.30 M.

$$pH = 4.85 + \log\left(\frac{0.30}{0.20}\right) = 4.85 + 0.18 = 5.02$$

(b) Calculate the concentration of hydrogen ion, using the total volume of the solution after the addition, but before the reaction.

$$V = 0.010\ L \times \frac{1000\ mL}{1\ L} + 1.0\ mL = 11\ mL$$

$$(\text{conc. } H_3O^+) = \frac{1.0\ mL}{11\ mL} \times \frac{1.0\ mol\ H_3O^+}{1\ L} = 0.0091\ M\ H_3O^+$$

The strong acid neutralizes the base in the solution, CH_3COO^-.

$$\text{Prop}^-(\text{aq}) + \text{H}_3\text{O}^+(\text{aq}) \longrightarrow \text{HProp} + \text{H}_2\text{O}(\ell)$$

This reaction is product-favored, so we will first make as many products as possible. Using the method of limiting reactants, run the reaction towards products until one of the reactants runs out.

	$Prop^-(aq)$	$H_3O^+(aq)$	HProp(aq)
initial conc. (M)	0.30	0.0091	0.20
change as reaction occurs (M)	– 0.0091	– 0.0091	+ 0.0091
final conc. (M)	0.29	0	0.21

The solution is still a buffer solution, containing an acid-base conjugate pair, so determine the pH as in (a).

$$\text{pH} = 4.85 + \log\left(\frac{0.29}{0.21}\right) = 4.85 + 0.14 = 4.99$$

(c) Calculate the concentration of hydrogen ion, using the total volume of the solution after the addition, but before the reaction.

$$V = 0.010\ \text{L} \times \frac{1000\ \text{mL}}{1\ \text{L}} + 3.0\ \text{mL} = 13\ \text{mL}$$

$$(\text{conc. H}_3\text{O}^+) = \frac{3.0\ \text{mL}}{13\ \text{mL}} \times \frac{1.0\ \text{mol H}_3\text{O}^+}{1\ \text{L}} = 0.23\ \text{M H}_3\text{O}^+$$

The strong acid neutralizes the base in the solution, $Prop^-$.

$$\text{Prop}^-(\text{aq}) + \text{H}_3\text{O}^+(\text{aq}) \longrightarrow \text{HProp} + \text{H}_2\text{O}(\ell)$$

This reaction is product-favored, so we will first make as many products as possible. Using the method of limiting reactants, run the reaction towards products until one of the reactants runs out.

	$Prop^-(aq)$	$H_3O^+(aq)$	HProp(aq)
initial conc. (M)	0.30	0.23	0.20
change as reaction occurs (M)	– 0.23	– 0.23	+ 0.23
final conc. (M)	0.07	0	0.43

The solution is still a buffer solution, containing an acid-base conjugate pair, so we can find pH as in (a).

$$\text{pH} = 4.85 + \log\left(\frac{0.07}{0.43}\right) = 4.85 - 0.79 = 4.06$$

Acid-Base Titrations and Titration Curves (Section 17.2)

33. *Answer:* **(a) Curve 2; see explanation below (b) acid 1: pH ≈ 7; acid 2: pH ≈ 9.5 (c) see explanation below (d) see explanation below (e) the best choice is bromthymol blue; see explanation below**

Strategy and Explanation:

(a) The effects of acid strength on the shape of the titration curve are shown in Figure 17.7. The curve for acid 2 is the curve for the weak acid.

(b) The equivalence point for acid 1 is about pH 7, and the equivalence point for acid 2 is about pH 9.5.

(c) At the equivalence point of a strong acid and a strong base, the solution contains the cation of the strong base and the anion of the strong acid, both of which are too weak to change the pH of the water solution. Thus the solution is neutral, and pH = 7. At the equivalence point of a weak acid and a strong base, the solution contains the cation of the strong base and the anion of the weak acid. The cation is too weak to change the pH of the water solution, but the anion of a weak acid is a weak base. Thus the solution is basic, and pH > 7.

(d) The pH of the weak acid starts at a higher value because of a smaller degree of ionization of the weak acid compared to the ionization of a strong acid.

(e) Bromthymol blue should be used for acid 1 and phenolphthalein should be used for acid 2, because their color changes are near the respective equivalence points.

34. *Answer/Explanation:* At the equivalence point, the solution contains the conjugate base of the weak acid. Weaker acids have stronger conjugate bases. Stronger bases have higher pH, so the titration of a weaker acid will have a more basic equivalence point.

36. *Answer:* **Best choices: (a) bromthymol blue (b) phenolphthalein (c) methyl red (d) bromthymol blue; see explanations below**

Strategy and Explanation: The color change needs to be near the pH of the equivalence point.

(a) The strong base, NaOH, titrated with a strong acid, $HClO_4$, has a neutral equivalence point. It would be best to choose bromthymol blue, which is shown changing color in Figure 17.5 at a pH near 7. In practice, any of them would be suitable, because of the extreme change in the pH of the solution very close to the equivalence point.

(b) The weak acid, CH_3COOH, titrated with a strong base, KOH, has a basic equivalence point, due to the presence of the weak base CH_3COO^- in the solution. It would be best to choose phenolphthalein, which is shown changing color in Figure 17.5 at a pH near 9.

(c) The weak base, NH_3, titrated with a strong acid, HBr, has an acidic equivalence point, due to the presence of the weak acid NH_4^+ in the solution. It would be best to choose methyl red, which is shown changing color in Figure 17.5 at a pH near 5.

(d) The strong base, KOH, titrated with a strong acid, HNO_3, has a neutral equivalence point. It would be best to choose bromthymol blue, which is shown changing color in Figure 17.5 at a pH near 7. .In practice, any of them would be suitable, because of the extreme change in the pH of the solution very close to the equivalence point.

38. *Answer:* **0.0253 M HCl**

Strategy and Explanation: Use the methods described in Chapter 5.

$$\frac{22.6 \text{ mL Ba(OH)}_2}{25.00 \text{ mL HCl}} \times \frac{1 \text{ L Ba(OH)}_2}{1000 \text{ mL Ba(OH)}_2} \times \frac{1000 \text{ mL HCl}}{1 \text{ L HCl}}$$

$$\times \frac{0.0140 \text{ mol Ba(OH)}_2}{1 \text{ L Ba(OH)}_2} \times \frac{2 \text{ mol OH}^-}{1 \text{ mol Ba(OH)}_2} \times \frac{1 \text{ mol HCl}}{1 \text{ mol OH}^-} = 0.0253 \frac{\text{mol HCl}}{\text{L HCl}} = 0.0253 \text{ M HCl}$$

40. *Answer:* **93.6%**

Strategy and Explanation: Use the methods described in Chapter 5.

$$24.4 \text{ mL NaOH} \times \frac{1 \text{ L}}{1000 \text{ mL}} \times \frac{0.110 \text{ mol NaOH}}{1 \text{ L NaOH}} \times \frac{1 \text{ mol C}_6\text{H}_8\text{O}_6}{1 \text{ mol NaOH}} \times \frac{176.1238 \text{ g C}_6\text{H}_8\text{O}_6}{1 \text{ mol C}_6\text{H}_8\text{O}_6} = 0.473 \text{ g C}_6\text{H}_8\text{O}_6$$

$$\frac{0.473 \text{ g C}_6\text{H}_8\text{O}_6}{0.505 \text{ g capsule}} \times 100\% = 93.6\%$$

42. *Answer:* **(a) 29.2 mL HCl (b) 600. mL HCl (c) 1.20 L HCl (d) 2.7 mL HCl**

Strategy and Explanation: Use the methods described in Chapter 5.

(a) $$25.0 \text{ mL KOH} \times \frac{0.175 \text{ mol KOH}}{1 \text{ L KOH}} \times \frac{1 \text{ mol HCl}}{1 \text{ mol KOH}} \times \frac{1 \text{ L HCl}}{0.150 \text{ mol HCl}} = 29.2 \text{ mL HCl}$$

(b) $$15.0 \text{ mL NH}_3 \times \frac{6.00 \text{ mol NH}_3}{1 \text{ L NH}_3} \times \frac{1 \text{ mol HCl}}{1 \text{ mol NH}_3} \times \frac{1 \text{ L HCl}}{0.150 \text{ mol HCl}} = 600. \text{ mL HCl}$$

(c) $$15.0 \text{ mL C}_3\text{H}_7\text{NH}_2 \times \frac{0.712 \text{ g C}_3\text{H}_7\text{NH}_2}{1 \text{ mL C}_3\text{H}_7\text{NH}_2} \times \frac{1 \text{ mol C}_3\text{H}_7\text{NH}_2}{59.1099 \text{ g C}_3\text{H}_7\text{NH}_2}$$

$$\times \frac{1 \text{ mol HCl}}{1 \text{ mol C}_3\text{H}_7\text{NH}_2} \times \frac{1 \text{ L HCl}}{0.150 \text{ mol HCl}} = 1.20 \text{ L HCl}$$

(d) $$40.0\text{ mL Ba(OH)}_2 \times \frac{0.0050\text{ mol Ba(OH)}_2}{1\text{ L Ba(OH)}_2} \times \frac{2\text{ mol OH}^-}{1\text{ mol Ba(OH)}_2} \times \frac{1\text{ mol HCl}}{1\text{ mol OH}^-} \times \frac{1\text{ L HCl}}{0.150\text{ mol HCl}} = 2.7\text{ mL HCl}$$

44. *Answer:* **(a) 3.62 (b) 8.31 (c) 12.15**

Strategy and Explanation: Adapt the method used in Problem-Solving Example 17.7.

The chemical formula for benzoic acid is given in Question 23, C_6H_5COOH. Benzoate anion has the formula $C_6H_5COO^-$.

(a) Calculate the moles of hydroxide and the benzoic acid.

$$\text{mol OH}^- = 10.00\text{ mL} \times \frac{1\text{ L}}{1000\text{ mL}} \times \frac{0.100\text{ mol OH}^-}{1\text{ L}} = 0.00100\text{ mol OH}^-$$

$$\text{mol C}_6\text{H}_5\text{COOH} = 30.00\text{ mL} \times \frac{1\text{ L}}{1000\text{ mL}} \times \frac{0.100\text{ mol C}_6\text{H}_5\text{COOH}}{1\text{ L}} = 0.00300\text{ mol C}_6\text{H}_5\text{COOH}$$

The strong base neutralizes the acid in the solution:

C_6H_5COOH	+ $OH^-(aq)$	$\longrightarrow$	$C_6H_5COO^-(aq)$	+ $H_2O(\ell)$
0.00300 mol in acid soln	0.00100 mol added		0.00100 mol formed	

After neutralization, mol C_6H_5COOH = 0.00300 mol – 0.00100 mol = 0.00200 mol, and the total volume is 30.0 mL + 10.0 mL = 40.0 mL or 0.0400 L.

Use the K_a expression associated with the benzoic acid ionization equilibrium to calculate $[H_3O^+]$, then calculate the pH:

$$K_a = \frac{[H_3O^+][C_6H_5COO^-]}{[C_6H_5COOH]} = \frac{[H_3O^+](0.00100\text{ mol}/0.0400\text{ L})}{(0.00200\text{ mol}/0.0400\text{ L})} = 1.2 \times 10^{-4}$$

$$[H_3O^+] = (1.2 \times 10^{-4}) \times 2.00 = 2.4 \times 10^{-4}\text{ M}$$

$$\text{pH} = -\log[H_3O^+] = -\log(2.4 \times 10^{-4}) = 3.62$$

(b) The initial moles of benzoic acid are unchanged, but we need to calculate the new moles of hydroxide.

$$\text{mol OH}^- = 30.00\text{ mL} \times \frac{1\text{ L}}{1000\text{ mL}} \times \frac{0.100\text{ mol OH}^-}{1\text{ L}} = 0.00300\text{ mol OH}^-$$

C_6H_5COOH	+ $OH^-(aq)$	$\longrightarrow$	$C_6H_5COO^-(aq)$	+ $H_2O(\ell)$
0.00300 mol in acid soln	0.00300 mol added		0.00300 mol formed	

The total volume is 30.0 mL + 30.0 mL = 60.0 mL or 0.0600 L. At this point in the neutralization, all the C_6H_5COOH has been neutralized and the solution's pH is governed by the hydrolysis of conjugate base, $C_6H_5COO^-$. Use K_a for benzoic acid to find the K_b for the conjugate base hydrolysis reaction.

$$K_b = \frac{K_w}{K_a} = \frac{1.00 \times 10^{-14}}{1.2 \times 10^{-4}} = 8.3 \times 10^{-11}$$

Then use the K_b expression for the hydrolysis reaction to calculate OH^-, then calculate the pH.conc.

$$C_6H_5COO^- = \frac{0.00300\text{ mol C}_6\text{H}_5\text{COO}^-}{0.06000\text{ L}} = 0.0500\text{ mol C}_6\text{H}_5\text{COO}^-$$

Substituting into the K_b expression, we let $x = [C_6H_5COOH] = [OH^-]$:

$$K_b = \frac{[C_6H_5COOH][OH^-]}{[C_6H_5COO^-]} = 8.3 \times 10^{-11} = \frac{x^2}{0.0500}$$

$$x^2 = (8.3 \times 10^{-11})(0.0500)$$

$$x = 2.0 \times 10^{-6}\ M = [OH^-]$$

$$pOH = -\log[OH^-] = -\log(2.0 \times 10^{-6}) = 5.69$$

$$pH = 14.00 + pOH = 14.00 + 5.69 = 8.31$$

(c) The initial moles of benzoic acid are unchanged, but we need to calculate the moles of hydroxide.

$$\text{mol } OH^- = 40.00\ \text{mL} \times \frac{1\ \text{L}}{1000\ \text{mL}} \times \frac{0.100\ \text{mol } OH^-}{1\ \text{L}} = 0.00400\ \text{mol } OH^-$$

Neutralization occurs, as in (a), but this time benzoic acid runs out first:

C_6H_5COOH	+ $OH^-(aq)$	⟶	$C_6H_5COO^-(aq)$	+ $H_2O(\ell)$
0.00300 mol in acid soln	0.00400 mol added		0.00300 mol formed	

After neutralization, mol OH^- = 0.00400 mol – 0.00300 mol = 0.00100 mol. The total volume in the solution after the two solutions are mixed is 30.00 mL + 40.00 mL = 70.00 mL or 0.07000 L.

The pH of the solution now is governed by the presence of excess strong base. We calculate the final concentration of OH^- after neutralization, then determine the pH.

$$\text{final } OH^- \text{ conc.} = \frac{0.00100\ \text{mol } OH^-}{0.07000\ \text{L}} = 0.0143\ \text{M } OH^-$$

$$pOH = -\log[OH^-] = -\log(0.0143) = 1.85$$

$$pH = 14.00 - pOH = 14.00 - 1.85 = 12.15$$

46. *Answer:* **(a) 0.824 (b) 1.30 (c) 3.8 (d) 7.000 (e) 10.2 (f) 12.48**

Strategy and Explanation: Adapt the method shown in Problem-Solving Example 17.6. For each of the points, determine the limiting reactant and use the concentration of the excess reactant to calculate the pH.

$$\text{volume of acid in liters} = 50.00\ \text{mL} \times \frac{1\ \text{L}}{1000\ \text{mL}} = 0.05000\ \text{L}$$

$$\text{original mol } H_3O^+ \text{ added} = 0.05000\ \text{L} \times \frac{0.150\ \text{mol } H_3O^+}{1\ \text{L}} = 0.00750\ \text{mol } H_3O^+$$

(a) This solution contains only 0.150 M HCl, which ionizes to make 0.150 M H_3O^+, so

$$pH = \log[H_3O^+] = -\log(0.150) = 0.824$$

(b) After the addition of the titrant, use the equation given a the beginning of Problem-Solving Example 17.6 to determine the equilibrium concentration of H_3O^+

$$[H_3O^+] = \frac{\text{original moles acid} - \text{total moles base added}}{\text{volume of acid (L)} + \text{volume of base (L)}}$$

$$\text{volume of base in liters} = 25.00\ \text{mL} \times \frac{1\ \text{L}}{1000\ \text{mL}} = 0.02500\ \text{L}$$

$$\text{total mol } OH^- \text{ added} = 0.02500\ \text{L} \times \frac{0.150\ \text{mol } OH^-}{1\ \text{L}} = 0.00375\ \text{mol } OH^-$$

$$[H_3O^+] = \frac{0.00750 \text{ mol} - 0.00375 \text{ mol}}{0.05000 \text{ L} + 0.02500 \text{ L}} = 0.0500 \text{ M}$$

$$pH = -\log[H_3O^+] = -\log(0.050) = 1.30$$

(c) Use the equation given in (b), again. volume of base (L) = $49.9 \text{ mL} \times \frac{1 \text{ L}}{1000 \text{ mL}} = 0.0490 \text{ L}$

$$\text{total mol OH}^- \text{ added} = 0.0499 \text{ L} \times \frac{0.150 \text{ mol OH}^-}{1 \text{ L}} = 0.007485 \text{ mol OH}^- \approx 0.00749 \text{ mol OH}^-$$

$$[H_3O^+] = \frac{0.00750 \text{ mol} - 0.007485 \text{ mol}}{0.05000 \text{ L} + 0.0499 \text{ L}} = \frac{0.000015 \text{ mol}}{0.0999 \text{ L}} = 0.00015 \text{ M} \approx 0.0002 \text{ M} \textit{ (1 sig fig)}$$

$$pH = -\log[H_3O^+] = -\log(0.00015) = 3.82 \approx 3.8 \textit{ (1 decimal place)}$$

(d) Calculate volume and total moles of base: volume of base (L) = $50.00 \text{ mL} \times \frac{1 \text{ L}}{1000 \text{ mL}} = 0.05000 \text{ L}$

$$\text{total mol } H_3O^+ \text{ added} = 0.05000 \text{ L} \times \frac{0.150 \text{ mol } H_3O^+}{1 \text{ L}} = 0.00750 \text{ mol } H_3O^+$$

Total moles of H_3O^+ added is equal to the original moles OH^-, so the solution is neutral, and the pH = 7.000.

(e) At this point, the moles of base begin to exceed the moles of acid, so we adapt the expression we used in (b) for a solution with excess base to determine the equilibrium concentration of OH^-:

$$[OH^-] = \frac{\text{total moles base added} - \text{original moles acid}}{\text{volume of acid (L)} + \text{volume of base (L)}}$$

$$\text{volume of base (L)} = 50.1 \text{ mL} \times \frac{1 \text{ L}}{1000 \text{ mL}} = 0.0501 \text{ L}$$

$$\text{total mol OH}^- \text{ added} = 0.0501 \text{ L} \times \frac{0.150 \text{ mol OH}^-}{1 \text{ L}} = 0.007515 \text{ mol OH}^- \approx 0.00752 \text{ mol OH}^-$$

$$[OH^-] = \frac{0.007515 \text{ mol} - 0.00750 \text{ mol}}{0.05000 \text{ L} + 0.0501 \text{ L}} = \frac{0.000015 \text{ mol}}{0.1001 \text{ L}} = 0.00015 \text{ M}$$

$$pH = -\log[H_3O^+] = -\log(0.000015) = 3.8$$

$$pH = 14.00 - pOH = 14.00 - 3.8 = 10.2$$

(f) Use the equation given in (e) again.

$$\text{volume of base (L)} = 75.00 \text{ mL} \times \frac{1 \text{ L}}{1000 \text{ mL}} = 0.0750 \text{ L}$$

$$\text{total mol OH}^- \text{ added} = 0.0750 \text{ L} \times \frac{0.150 \text{ mol } H_3O^+}{1 \text{ L}} = 0.0113 \text{ mol } H_3O^+$$

$$[OH^-] = \frac{0.0113 \text{ mol} - 0.00750 \text{ mol}}{0.05000 \text{ L} + 0.0750 \text{ L}} = \frac{0.00038 \text{ mol}}{0.1250 \text{ L}} = 0.030 \text{ M}$$

$$pH = -\log[H_3O^+] = -\log(0.030) = 1.52$$

$$pH = 14.00 - pOH = 14.00 - 1.52 = 12.48$$

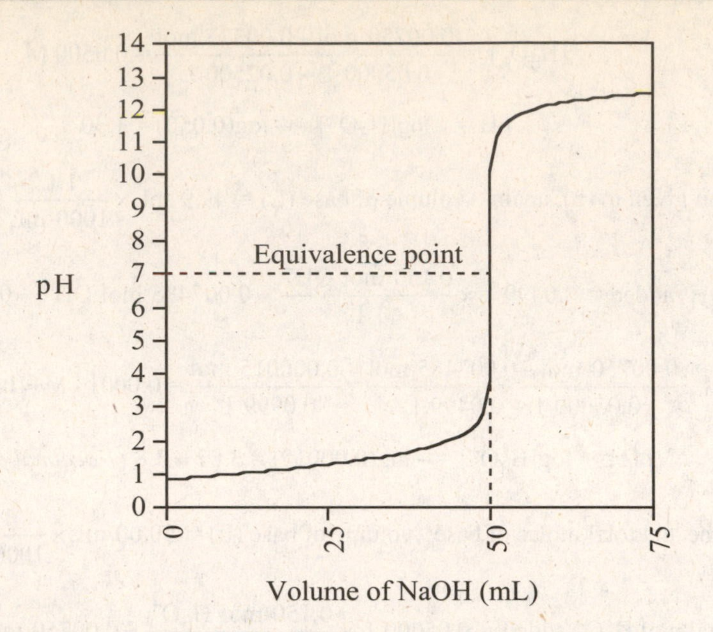

Acid Rain (Section 17.3)

48. *Answer:* **NO_2 and SO_3; see equations below**

Strategy and Explanation: Two oxides that are key producers of acid rain are NO_2 and SO_3. NO_2 reacts with water in the air to make nitric acid and nitrous acid:

$$2\ NO_2(g) + H_2O(\ell) \longrightarrow HNO_3(aq) + HNO_2(aq)$$

SO_2 reacts with oxygen in the air to make SO_3 and SO_3 reacts water in the air to make sulfuric acid:

$$2\ SO_2(g) + O_2(g) \longrightarrow 2\ SO_3(g)$$

$$SO_3(g) + H_2O(\ell) \longrightarrow H_2SO_4(aq)$$

50. *Answer:* **$CaCO_3(s) + 2\ H^+(aq) \longrightarrow Ca^{2+}(aq) + CO_2(g) + H_2O(\ell)$**

Strategy and Explanation: Limestone is made of calcium carbonate. The anion of this compound is a base and reacts with acid in rainwater to neutralize it by a gas-forming exchange reaction (Section 5.2 page 176).

Solubility Product (Section 17.4)

54. *Answer:* **see equations and expressions below**

Strategy and Explanation: Adapt the method developed in Problem-Solving Example 17.8.

(a) $FeCO_3(s) \rightleftharpoons Fe^{2+}(aq) + CO_3^{2-}(aq)$ $\quad K_{sp} = [Fe^{2+}][CO_3^{2-}]$

(b) $Ag_2SO_4(s) \rightleftharpoons 2\ Ag^+(aq) + SO_4^{2-}(aq)$ $\quad K_{sp} = [Ag^+]^2[SO_4^{2-}]$

(c) $Ca_3(PO_4)_2(s) \rightleftharpoons 3\ Ca^{2+}(aq) + 2\ PO_4^{3-}(aq)$ $\quad K_{sp} = [Ca^{2+}]^3[PO_4^{3-}]^2$

(d) $Mn(OH)_2(s) \rightleftharpoons Mn^{2+}(aq) + 2\ OH^-(aq)$ $\quad K_{sp} = [Mn^{2+}][OH^-]^2$

55. *Answer:* **$K_{sp} = 3.0 \times 10^{-18}$**

Strategy and Explanation: Adapt the method developed in Problem-Solving Example 17.10.

Write the chemical equation and the equilibrium expression for the dissociation of the solute:

$$Ag_3AsO_4(s) \rightleftharpoons 3\ Ag^+(aq) + AsO_4^{3-}(aq) \qquad K_{sp} = [Ag^+]^3[AsO_4^{3-}]$$

At equilibrium, the moles of solid that dissolve per liter are:

$$\frac{8.5 \times 10^{-6}\ \text{g}\ Ag_3AsO_4}{1\ \text{mL}} \times \frac{1\ \text{mol}\ Ag_3AsO_4}{462.5238\ \text{g}\ Ag_3AsO_4} \times \frac{1000\ \text{mL}}{1\ \text{L}} = 1.8 \times 10^{-5}\ \text{M}$$

The stoichiometry of the equation shows that the arsenate ion concentration is the identical to the number of moles of solid that dissolve per liter, 1.8×10^{-5} M, and the silver ion concentration is three times that value, $3 \times (1.8 \times 10^{-5}\text{ M}) = 5.4 \times 10^{-5}$ M.

$$K_{sp} = (5.4 \times 10^{-5})^3(1.8 \times 10^{-5}) = 3.0 \times 10^{-18}$$

56. *Answer:* $\mathbf{K_{sp} = 2.22 \times 10^{-4}}$

Strategy and Explanation: Adapt the method developed in the solution to Question 55.

Write the chemical equation and the equilibrium expression for the dissociation of the solute:

$$CaSO_4(s) \rightleftharpoons Ca^{2+}(aq) + SO_4^{3-}(aq) \qquad K_{sp} = [Ca^{2+}][SO_4^{2-}]$$

At equilibrium, the moles of solid that dissolve per liter is:

$$\frac{2.03\text{ g }CaSO_4}{1\text{ L}} \times \frac{1\text{ mol }CaSO_4}{136.14\text{ g }CaSO_4} = 0.0149\text{ M}$$

The stoichiometry of the equation shows that the concentrations of calcium ion and sulfate ion are both the same as the moles of solid that dissolve per liter, 0.0149 M.

$$K_{sp} = [Ca^{2+}][SO_4^{2-}] = (0.0149)(0.0149) = 2.22 \times 10^{-4}$$

58. *Answer:* $\mathbf{K_{sp} = 2.2 \times 10^{-12}}$

Strategy and Explanation: Follow the method developed in the solution to Question 55.

Write the chemical equation and the equilibrium expression for the dissociation of the solute:

$$Ag_2CrO_4(s) \rightleftharpoons 2\,Ag^+(aq) + CrO_4^{2-}(aq) \qquad K_{sp} = [Ag^+]^2[CrO_4^{2-}]$$

At equilibrium, the moles of solid that dissolve per liter is:

$$\frac{2.7 \times 10^{-3}\text{ g }Ag_2CrO_4}{100.\text{ mL}} \times \frac{1\text{ mol }Ag_2CrO_4}{331.7301\text{ g }Ag_2CrO_4} \times \frac{1000\text{ mL}}{1\text{ L}} = 8.1 \times 10^{-5}\text{ M}$$

The stoichiometry of the equation shows that the concentration of chromate ion is the same as the moles of solid that dissolve per liter, 8.1×10^{-5} M, and the silver ion concentration is twice that value, $2 \times (8.1 \times 10^{-5}\text{ M}) = 1.6 \times 10^{-4}$ M.

$$K_{sp} = (1.6 \times 10^{-4})^2(8.1 \times 10^{-5}) = 2.2 \times 10^{-12}$$

60. *Answer:* $\mathbf{K_{sp} = 1.7 \times 10^{-5}}$

Strategy and Explanation: Follow the method developed in the solution to Question 55.

Write the chemical equation and the equilibrium expression for the dissociation of the solute:

$$PbCl_2(s) \rightleftharpoons Pb^{2+} + 2\,Cl^-(aq) \qquad K_{sp} = [Pb^{2+}][Cl^-]^2$$

At equilibrium, the moles of solid that dissolve per liter is given as the solubility 1.62×10^{-2} M. The stoichiometry of the equation shows that the concentration of lead ion is the same as the moles of solid that dissolve per liter, 1.62×10^{-2} M, and the chloride ion concentration is two times that value, $2 \times (1.62 \times 10^{-2}\text{ M}) = 3.24 \times 10^{-2}$ M.

$$K_{sp} = (1.62 \times 10^{-2})(3.24 \times 10^{-2})^2 = 1.70 \times 10^{-5}$$

Common Ion Effect (Section 17.5)

62. *Answer:* **6.2×10^{-11} mol/L**

Strategy and Explanation: We will use the equilibrium expression for K_{sp} and the known values to determine the unknown value. We will get the value of K_{sp} from Appendix H.

The soluble Na_2CO_3 salt produces carbonate ions in the solution.

The anion base hydrolysis reaction should be considered because its K is larger than the K_{sp} above. The equation for that reaction is: $CO_3^{2-} + H_2O \rightleftharpoons HCO_3^- + OH^-$, some of the carbonate (–x) reacts to form hydrogen carbonate (x) and hydroxide (x). $K_b = \frac{x^2}{0.25 - x} = 2.1 \times 10^{-4}$. Solving for x (0.0072 M) and calculating the carbonate concentration, we find that it is slightly but measurably lower:

$$[CO_3^{2-}] = 0.25\ M - x = 0.25\ M - 0.0072\ M = 0.24\ M$$

$$K_{sp} = [Zn^{2+}][CO_3^{2-}] = [Zn^{2+}] \times (0.24) = 1.5 \times 10^{-11},\ [Zn^{2+}] = 6.2 \times 10^{-11}\ mol/L$$

63. *Answer:* **(a) 3.7×10^{-6} mol/L (b) 2.8×10^{-10} mol/L (c) 2.8×10^{-10} mol/L**

Strategy and Explanation: Adapt the methods described in Problem-Solving Examples 17.9 and 17.11. Insert variables describing the concentrations of the ions present from another sources as initial concentrations in the equilibrium calculation.

	$ZnCO_3(s) \rightleftharpoons$	$Zn^{2+}(aq)$	+ $CO_3^{2-}(aq)$
conc. initial (M)	solid	c_{zinc}	c_{carb}
change as reaction occurs (M)	solid	+ S	+ S
equilibrium conc. (M)	solid	c_{zinc} + S	c_{carb} + S

At equilibrium $K_{sp} = [Zn^{2+}][CO_3^{2-}] = (c_{zinc} + S)(c_{carb} + S) = 1.4 \times 10^{-11}$

(a) In water, no other source of Zn^{2+} or CO_3^{2-} is present so, before dissociation, $c_{zinc} = c_{carb} = 0$.

$$(S)(S) = 1.4 \times 10^{-11}$$

$$S = 3.7 \times 10^{-6}\ mol/L$$

(b) In $Zn(NO_3)_2$, an external source of Zn^{2+} is present, so $c_{zinc} = 0.050$ M. No external sources of CO_3^{2-} are present so $c_{carb} = 0$.

$$(0.050 + S)(S) = 1.4 \times 10^{-11}$$

Assuming S is small compared to 0.050, ignore its addition.

$$(0.050)(S) = 1.4 \times 10^{-11}$$

$$S = 2.8 \times 10^{-10}\ mol/L$$

(c) In K_2CO_3, an external source of CO_3^{2-} is present. Neglecting the reaction of carbonate as a base, $c_{carb} = 0.050$ M. No external sources of Zn^{2+} are present so, before dissociation, $c_{zinc} = 0$.

$$(S)(0.050 + S) = 1.4 \times 10^{-11}$$

Assuming S is small compared to 0.050, ignore its addition.

$$(S)(0.050) = 1.4 \times 10^{-11}$$

$$S = 2.8 \times 10^{-10}\ mol/L$$

(*Notice the assumption that S is small is a good one.*)

64. *Answer:* **3.2×10^{-5} mol/L**

Strategy and Explanation: Follow the method developed Problem-Solving Example 17.11.

The soluble Na_2SO_4 salt produces sulfate ions in the solution. Neglecting the reaction of sulfate as a base, (conc. SO_4^{2-}) = 0.010 mol/L.

	$SrSO_4(s) \rightleftharpoons$	$Sr^{2+}(aq)$	+ $SO_4^{2-}(aq)$
conc. initial (M)		0	0.010
change as reaction occurs (M)		+ S	+ S
equilibrium conc. (M)		S	0.010 + S

At equilibrium $K_{sp} = [Sr^{2+}][SO_4^{2-}] = (S)(0.010 + S) = 3.2 \times 10^{-7}$

Assuming S is small compared to 0.010, ignore its addition.

$$3.2 \times 10^{-7} = (S)(0.010) \qquad S = 3.2 \times 10^{-5} \text{ mol/L}$$

The solubility of strontium sulfate in a 0.010 M solution of sodium sulfate is 3.2×10^{-5} mol/L.

66. *Answer:* **(a) 1 × 10^{-11} (b) 8 × 10^{-3} M or higher**

Strategy and Explanation: Follow the method developed in the solutions to Questions 55 and 62.

(a) Write the chemical equation and the equilibrium expression for the dissociation of the solute:

$$Mg(OH)_2(s) \rightleftharpoons Mg^{2+}(aq) + 2\ OH^-(aq) \qquad K_{sp} = [Mg^{2+}][OH^-]^2$$

At equilibrium, the moles of solid that dissolve per liter are:

$$\frac{9 \text{ mg } Mg(OH)_2}{1 \text{ L}} \times \frac{1 \text{ g}}{1000 \text{ mg}} \times \frac{1 \text{ mol } Mg(OH)_2}{58.32 \text{ g } Mg(OH)_2} = 1.5 \times 10^{-4} \text{ M} \cong 2 \times 10^{-4} \text{ M} \ \textit{(1 sig. fig.)}$$

The stoichiometry of the equation shows that the concentration of magnesium ion is the same as the moles of solid that dissolve per liter, 1.5×10^{-4}, and the hydroxide ion concentration is two times that value, $2 \times (1.5 \times 10^{-4} \text{ M}) = 3.0 \times 10^{-4}$ M.

$$K_{sp} = (1.5 \times 10^{-4})(3.0 \times 10^{-4})^2 = 1.47 \times 10^{-11}$$

$$K_{sp} \cong 1 \times 10^{-11} \ \textit{(rounded to 1 sig. fig.)}$$

(b) First we determine the concentration of the iron(II) ion in a 1.0μg Mg^{2+} ion solution:

$$[Mg^{2+}] = \frac{5.0 \ \mu\text{g } Mg^{2+}}{1 \text{ L}} \times \frac{10^{-6} \text{ g}}{1 \ \mu\text{g}} \times \frac{1 \text{ mol } Mg^{2+}}{24.305 \text{ g } Mg^{2+}} = 2.1 \times 10^{-7} \text{ M}$$

Next, we use the calculated K_{sp} from (a) to calculate the hydroxide ion concentration in this solution:

$$K_{sp} = [Mg^{2+}][OH^-]^2$$

$$1.47 \times 10^{-11} = (2.1 \times 10^{-7})[OH^-]^2 \ \textit{(rounded to 1 sig. fig.)}$$

$$[OH^-] = 8 \times 10^{-3} \text{ M}$$

An equilibrium $[OH^-]$ of 8×10^{-3} M or higher will keep the $[Mg^{2+}]$ at or below 1.0 μg/L

Factors Affecting the Solubility of Sparingly Soluble Solutes (Section 17.5)

68. *Answer:* **1.2 × 10^{-9} M or lower**

Strategy and Explanation: Adapt the method developed in the solution to Question 64 and use the relationships provided in Chapter 16 to determine the hydroxide concentration from the pH.
Given the pH, we can determine the concentration of the aqueous ion, OH^-. The precipitation of sparingly soluble $Zn(OH)_2$ will occur above a certain concentration of Zn^{2+}.

$$[OH^-] = 10^{-pOH} = 10^{pH - 14.00} = 10^{pH - 14.00} = 10^{10.00 - 14.00} = 1.0 \times 10^{-4} \text{ M}$$

$$K_{sp} = [Zn^{2+}][OH^-]^2 = 1.2 \times 10^{-17} \text{ (from Appendix H)}$$

$$1.2 \times 10^{-17} = [Zn^{2+}](1.0 \times 10^{-4})^2$$

$$[Zn^{2+}] = 1.2 \times 10^{-9} \text{ M}$$

An equilibrium $[Zn^{2+}]$ of 1.2×10^{-9} M or lower can exist in a solution with pH 10.00. Above that concentration, $Zn(OH)_2$ will precipitate.

70. *Answer:* **9.16**

Strategy and Explanation: Adapt the method developed in the solution to Question 68. We will use the K_{sp} derived for $Mg(OH)_2$ in part (a) of Question 66: $K_{sp} = 1 \times 10^{-11}$.

The acid added affects the concentration of the hydroxide. So, if all the solid is dissolved it must be acidic enough to hold all the magnesium ions without precipitation. So, we will calculate the resulting magnesium ion concentration:

$$[Mg^{2+}] = \frac{5.00 \text{ g } Mg(OH)_2}{1 \text{ L}} \times \frac{1 \text{ mol } Mg(OH)_2}{58.32 \text{ g } Mg(OH)_2} \times \frac{1 \text{ mol } Mg^{2+}}{1 \text{ mol } Mg(OH)_2} = 0.0857 \text{ M}$$

$$K_{sp} = [Mg^{2+}][OH^-]^2$$

$$1.8 \times 10^{-11} = (0.0857)[OH^-]^2$$

$$[OH^-] = 1.4 \times 10^{-5} \text{ M}$$

$$pOH = -\log[OH^-] = -\log(1.4 \times 10^{-5}) = 4.84$$

$$pH = 14.00 - pOH = 14.00 - 4.84 = 9.16$$

Enough acid must be added to drop the pH to 9.16, before all the solid will dissolve.

Complex Ion Formation (Section 17.5)

72. *Answer:* **see equations and expressions below**

Strategy and Explanation: The charge of the reactant metal ion is determined by subtracting the Lewis base's charge(s), if any, from the complex ion charge.

(a) $Ag^+ + 2\,CN^- \rightleftharpoons [Ag(CN)_2]^-$ $\quad K_f = \dfrac{[[Ag(CN)_2]^-]}{[Ag^+][CN^-]^2}$

(b) $Cd^{2+} + 4\,NH_3 \rightleftharpoons [Cd(NH_3)_4]^{2+}$ $\quad K_f = \dfrac{[[Cd(NH_3)_4]^{2+}]}{[Cd^{2+}][NH_3]^4}$

74. *Answer:* **6.3×10^{-3} mol or more**

Strategy and Explanation: Adapt the method described in the solution to Question 70, and Problem-Solving Example 17.13 on pages 836-7. The $Na_2S_2O_3$ salt provides a source of $S_2O_3^{2-}$, a Lewis base capable of forming a complex ion with the silver ion. If all the solid is dissolved it must be have enough $S_2O_3^{2-}$ to complex enough the silver ions to prevent the precipitation of AgBr in the resulting Br^- solution. First, we get the balanced equation for the reaction from Problem-Solving Example 17.13 and determine the value of its equilibrium constant using Tables 17.2 and 17.3:

$$AgBr(s) + 2\,S_2O_3^{2-}(aq) \rightleftharpoons [Ag(S_2O_3)_2]^{3-}(aq) + Br^-(aq) \qquad K = \frac{[[Ag(S_2O_3)_2]^{3-}][Br^-]}{[S_2O_3^{2-}]^2}$$

$$K = K_{sp} \times K_f = [Ag^+][Br^-] \times \frac{[[Ag(S_2O_3)_2]^{3-}]}{[Ag^+][S_2O_3^{2-}]^2} = \frac{[[Ag(S_2O_3)_2]^{3-}][Br^-]}{[S_2O_3^{2-}]^2}$$

$$K = K_{sp} \times K_f = (5.0 \times 10^{-13}) \times (2.0 \times 10^{13}) = 10.$$

Now, calculate the $[Br^-]$ and $[[Ag(S_2O_3)_2]^{3-}]$, once all the AgBr has dissolved:

$$[Br^-] = \frac{0.020 \text{ mol AgBr}}{1.0 \text{ L}} \times \frac{1 \text{ mol } Br^-}{1 \text{ mol AgBr}} = 0.020 \text{ M},$$

Similarly, $[[Ag(S_2O_3)_2]^{3-}] = 0.020$ M

Now, we can calculate the necessary $[S_2O_3^{2-}]$:

$$10. = \frac{(0.020)(0.020)}{[S_2O_3^{2-}]^2}$$

$$[S_2O_3^{2-}] = 6.3 \times 10^{-3}\ M$$

One mole of $S_2O_3^{2-}$ is found in each mole of $Na_2S_2O_3$, so we must add 6.3×10^{-3} mol of $Na_2S_2O_3$ to this 1.0 L solution.

76. *Answer:* **see equations below**

Strategy and Explanation: The hydroxide anion of the solid is neutralized in excess acid, and the aqueous complex ion is formed in excess base.

(a) $$Zn(OH)_2(s) + 2\ H_3O^+(aq) \longrightarrow Zn^{2+}(aq) + 4\ H_2O(\ell)$$

$$Zn(OH)_2(s) + 2\ OH^-(aq) \longrightarrow [Zn(OH)_4]^{2-}(aq)$$

(b) $$Sb(OH)_3(s) + 3\ H_3O^+(aq) \longrightarrow Sb^{3+}(aq) + 6\ H_2O(\ell)$$

$$Sb(OH)_3(s) + OH^-(aq) \longrightarrow [Sb(OH)_4]^-(aq)$$

General Questions

78. *Answer:* **(a) H_2O, CH_3COO^-, Na^+, CH_3COOH, H_3O^+, OH^- (b) 4.95 (c) 5.05 (d) $CH_3COOH(aq) + H_2O(\ell) \rightleftharpoons H_3O^+(aq) + CH_3COO^-(aq)$**

Strategy and Explanation:

(a) The solution contains an abundance of water, more sodium acetate than acetic acid, a small amount of H_3O^+ since the solution is an acidic buffer, and an even smaller amount of OH^-. Therefore, the ions and molecules in solution from the largest concentration to the smallest is: H_2O, CH_3COO^-, Na^+, CH_3COOH, H_3O^+, OH^-.

(b) Adapt the methods used in the solutions to Question 30. The equation for the equilibrium and the equilibrium expression are:

$$CH_3COOH(aq) + H_2O(\ell) \rightleftharpoons H_3O^+(aq) + CH_3COO^-(aq) \qquad K_a = \frac{[H_3O^+][CH_3COO^-]}{[CH_3COOH]}$$

$$\frac{4.95\ g\ NaCH_3COO}{250.\ mL} \times \frac{1\ mol\ NaCH_3COO}{82.0337\ g\ NaCH_3COO} \times \frac{1000\ mL}{1\ L} \times \frac{1\ mol\ CH_3COO^-}{1\ mol\ NaCH_3COO} = 0.241\ M\ CH_3COO^-$$

As the reactants decompose, the concentrations change stoichiometrically, until they reach equilibrium.

	$CH_3COOH(aq)$	$H_3O^+(aq)$	$CH_3COO^-(aq)$
initial conc. (M)	0.150	0	0.241
change as reaction occurs (M)	– x	+ x	+ x
equilibrium conc. (M)	0.150 – x	x	0.241 + x

At equilibrium $$K_a = \frac{(x)(0.241 + x)}{(0.150 - x)} = 1.8 \times 10^{-5}$$

Assume x is very small and does not affect the sum or the difference.

$$1.8 \times 10^{-5} = \frac{(x)(0.241)}{(0.150)}$$

$$x = 1.1 \times 10^{-5}\ M = [H_3O^+]$$

$$pH = -\log[H_3O^+] = -\log(1.1 \times 10^{-5}) = 4.95$$

(c) The strong base neutralizes the acid in the solution, CH_3COOH.

$$CH_3COOH(aq) + OH^-(aq) \longrightarrow CH_3COO^-(aq) + H_2O(\ell)$$

$$(\text{conc. } OH^-) = \frac{80.\text{ mg NaOH}}{100.\text{ mL}} \times \frac{1000\text{ mL}}{1\text{ L}} \times \frac{1\text{ g}}{1000\text{ mg}} \times \frac{1\text{ mol NaOH}}{40.00\text{ g NaOH}} \times \frac{1\text{ mol }OH^-}{1\text{ mol NaOH}} = 0.020\text{ M}$$

We take it as far to the products as possible, using the method of limiting reactants.

	$CH_3COOH(aq)$	$OH^-(aq)$	$CH_3COO^-(aq)$
initial conc. (M)	0.150	0.020	0.241
change as reaction occurs (M)	– 0.020	– 0.020	+ 0.020
conc. final (M)	0.130	0	0.261

The solution is still a buffer solution, containing an acid-base conjugate pair, so we can find the pH using K_a of the acid.

$$CH_3COOH(aq) + H_2O(\ell) \rightleftharpoons H_3O^+(aq) + CH_3COO^-(aq) \qquad K_a = \frac{[H_3O^+][CH_3COO^-]}{[CH_3COOH]}$$

As the reactants decompose, the concentrations change stoichiometrically, until they reach equilibrium.

	$CH_3COOH(aq)$	$H_3O^+(aq)$	$CH_3COO^-(aq)$
initial conc. (M)	0.130	0	0.261
change as reaction occurs (M)	– x	+ x	+ x
equilibrium conc. (M)	0.130 – x	x	0.261 + x

At equilibrium $\quad K_a = \frac{(x)(0.261 + x)}{(0.130 - x)} = 1.8 \times 10^{-5}$

Assume x is very small and does not affect the sum or the difference.

$$1.8 \times 10^{-5} = \frac{(x)(0.261)}{(0.130)} \qquad x = 9.0 \times 10^{-6}\text{ M} = [H_3O^+]$$

$$pH = -\log[H_3O^+] = -\log(9.0 \times 10^{-6}) = 5.05$$

(d) $CH_3COOH(aq) + H_2O(\ell) \rightleftharpoons H_3O^+(aq) + CH_3COO^-(aq)$

80. *Answer:* **ratio = 1.6**

Strategy and Explanation: Adapt the methods used in the solution to Question 22.

$$pH = pK_a + \log\left(\frac{[o\text{-ethylbenzoate}]}{[o\text{-ethylbenzoic acid}]}\right)$$

At equilibrium, pH = 4.0 and $pK_a = 3.79$

$$pH - pK_a = \log\left(\frac{[o\text{-ethylbenzoate}]}{[o\text{-ethylbenzoic acid}]}\right)$$

$$4.0 - 3.79 = 0.2 = \log\left(\frac{[o\text{-ethylbenzoate}]}{[o\text{-ethylbenzoic acid}]}\right)$$

$$\frac{[o\text{-ethylbenzoate}]}{[o\text{-ethylbenzoic acid}]} = 10^{0.2} = 1.6$$

The [potassium *o*-ethylbenzoate] is approximately two times the [*o*-ethylbenzoic acid].

82. *Answer:* **0.020 mol HNO_2**

Strategy and Explanation: Adapt the method used in the solution to Question 22.

$$[NO_2^-] = \frac{7.50\text{ g }KNO_2}{1\text{ L}} \times \frac{1\text{ mol }KNO_2}{85.11\text{ g }KNO_2} \times \frac{1\text{ mol }NO_2^-}{1\text{ mol }KNO_2} = 0.0881\text{ M}$$

$$pH = pK_a + \log\left(\frac{[\text{conj. base}]}{[\text{conj. acid}]}\right)$$

In Table 16.2, the K_a of nitrous acid is given: 4.5×10^{-4}. As described in Section 17.1, calculate the pK_a:

$$pK_a = -\log K_a = -\log(4.5 \times 10^{-4}) = 3.35$$

The pH is 4.00. The initial concentration of NO_2^-, the conjugate base, is 0.0881 M. The conjugate acid, HNO_2, is being added, but its initial concentration is unknown, so use the variable "c" to identify that quantity.

$$4.00 = 3.35 + \log\left(\frac{0.0881}{c}\right)$$

Solve for c: $c = 0.0881\ M \times 10^{-(4.00 - 3.35)} = 0.020\ M$

Add 0.020 mol HNO_2 to one liter of solution to make this buffer.

84. *Answer:* **(a) 2.78 (b) 5.39**

Strategy and Explanation:

(a) Use the method similar to that used in the solution to Question 16.54.

The equation for the equilibrium and the equilibrium expression are:

$$CH_3COOH(aq) + H_2O(\ell) \rightleftharpoons H_3O^+(aq) + CH_3COO^-(aq) \qquad K_a = \frac{[CH_3COO^-][H_3O^+]}{[CH_3COOH]}$$

As the reactants decompose, the concentrations change stoichiometrically, until they reach equilibrium.

	$CH_3COOH(aq)$	$H_3O^+(aq)$	$CH_3COO^-(aq)$
initial conc. (M)	0.15	0	0
change as reaction occurs (M)	– x	+ x	+ x
equilibrium conc. (M)	0.15 – x	x	x

At equilibrium $K_a = \frac{(x)(x)}{(0.15 - x)} = 1.8 \times 10^{-5}$

Assume x is very small and does not affect the difference.

$$x^2 = (1.8 \times 10^{-5})(0.15) \qquad x = 1.6 \times 10^{-3}\ M = [H_3O^+]$$

$$pH = -\log[H_3O^+] = -\log(1.6 \times 10^{-3}) = 2.78$$

(b) Use the methods described in the solutions to Questions 27(a) and 31.

The solution is a buffer solution, containing an acid-base conjugate pair, so we can find the pH using K_a of the acid described in (a).

$$[CH_3COO^-] = \frac{83\ g\ NaCH_3COO}{1.50\ L} \times \frac{1\ mol\ NaCH_3COO}{82.03\ g\ NaCH_3COO} \times \frac{1\ mol\ CH_3COO^-}{1\ mol\ NaCH_3COO} = 0.67\ M\ CH_3COO^-$$

The solution is now a buffer solution, containing an acid-base conjugate pair, so we can find the pH using the Henderson-Hasselbalch equation. In Table 17.1, the pK_a is 4.74.

$$pH = 4.74 + \log\left(\frac{0.67}{0.15}\right) = 4.74 + 0.65 = 5.39$$

86. *Answer:* **3.5×10^{-6}**

Strategy and Explanation: Adapt the methods described in the solution to Question 44. We will use the equivalence point data to find the total moles of the weak acid, HA in the solution.

$$35.00\ mL\ NaOH \times \frac{1\ L}{1000\ mL} \times \frac{0.100\ mol\ NaOH}{1\ L\ NaOH} \times \frac{1\ mol\ HA}{1\ mol\ NaOH} = 0.00350\ mol\ HA$$

Calculate the moles of hydroxide added, after 20.00 mL of titrant have been added.

$$\text{mol OH}^- = 20.00\ \text{mL} \times \frac{1\ \text{L}}{1000\ \text{mL}} \times \frac{0.100\ \text{mol OH}^-}{1\ \text{L}} = 0.00200\ \text{mol OH}^-$$

The strong base neutralizes the acid, HA, in the solution.

$$HA + OH^-(aq) \longrightarrow A^-(aq) + H_2O(\ell)$$

We take it as far to the products as possible, using the method of limiting reactants.

	Initial	After Reaction of HA with NaOH
mol A^-	0	0 + 0.00200 = 0.00200
mol HA	0.00350	0.00300 – 0.00200 = 0.00100

The equation for the equilibrium and the equilibrium expression are:

$$HA(aq) + H_2O(\ell) \rightleftharpoons H_3O^+(aq) + A^-(aq) \qquad K_a = \frac{[H_3O^+][A^-]}{[HA]}$$

We calculate the concentrations using the total volume after 20.0 mL of titrant has been added:

20.0 mL + 40.0 mL = 60.0 mL, which is 0.0600 L.

$$(\text{conc. } A^-) = \frac{0.00200\ \text{mol}}{0.0600\ \text{L}} = 0.0333\ \text{M} \qquad (\text{conc. HA}) = \frac{0.00100\ \text{mol}}{0.0600\ \text{L}} = 0.0167\ \text{M}$$

The solution is now a buffer solution, so use the Hendersen-Hasselbalch equation to calculate pK_a. The concentration of the conjugate acid, HA, is 0.0167 M; the concentration of the conjugate base, A^-, is 0.0333 M. The pH is 5.75.

$$pH = pK_a + \log\left(\frac{[\text{conj. base}]}{[\text{conj. acid}]}\right)$$

$$pK_a = 5.75 - \log\left(\frac{0.0333}{0.0167}\right) = 5.75 - 0.30 = 5.45$$

$$pK_a = -\log K_a, \text{ so } K_a = 10^{-pK_a} = 10^{-5.45} = 3.5 \times 10^{-6}$$

88. *Answer:* **no effect**

Strategy and Explanation: There is no effect on the equilibrium state if more solid is added, since the quantity of that solid present does not affect the equilibrium state.

90. *Answer:* **2.2×10^{-12}**

Strategy and Explanation: Follow the method developed in the solution to Questions 55 - 61.

	$Ag_2CrO_4(s)$ $\rightleftharpoons$	$2\ Ag^+(aq)$	+ $CrO_4^{2-}(aq)$
conc. initial (M)		0	0
change as reaction occurs (M)		+ 2S	+ S
equilibrium conc. (M)		2S	S

At equilibrium, $$\frac{2.7 \times 10^{-3}\ \text{g } Ag_2CrO_4}{100.\ \text{mL}} \times \frac{1\ \text{mol } Ag_2CrO_4}{331.8\ \text{g } Ag_2CrO_4} \times \frac{1000\ \text{mL}}{1\ \text{L}} = 8.1 \times 10^{-5}\ \text{M} = S$$

$$K_{sp} = [Ag^+]^2[CrO_4^{2-}] = (2S)^2(S) = 4S^3 = 4 \times (8.1 \times 10^{-5})^3 = 2.2 \times 10^{-12}$$

Applying Concepts

92. *Answer/Explanation:* The tiny amount of base (CH_3COO^-) present is insufficient to prevent the pH from changing dramatically if a strong acid is introduced into the solution.

94. *Answer:* **2.3×10^{-4}**

Strategy and Explanation: When exactly half of the acid in the original solution has been neutralized, that means that equal quantities of the acid and base are present in the solution. As described in the solution to Question 18, when equimolar quantities are present, $pH = pK_a$.

$$pH = 3.64 = pK_a$$

$$K_a = 10^{-pK_a} = 10^{-3.64} = 2.3 \times 10^{-4}$$

96. *Answer/Explanation:* Blood pH decreases because of an increase in H_2CO_3, which leads to acidosis, acidification of the blood.

98. *Answer/Explanation:* $Ca_5(PO_4)_3OH(s) \rightleftharpoons 5\ Ca^{2+}(aq) + 3\ PO_4^{3-}(aq) + OH^-(aq)$

(a) Apatite is a relatively insoluble compound. Drinking milk containing calcium ion increases the concentration of a product in the above reaction, and, according to LeChatelier's principle, drives the equilibrium toward the formation of apatite.

(b) Lactic acid will react with OH^-. The subsequent decrease in OH^- concentration will cause apatite to dissolve as equilibrium is re-established.

100. *Answer/Explanation:* During a recent television medical drama, a person went into cardiac arrest and stopped breathing. A doctor quickly injected sodium hydrogen carbonate into the heart. This would indicate that cardiac arrest leads to **acidosis** and that the sodium hydrogen carbonate helps to **increase** the pH. Explanation: When the patient stopped breathing, his blood pH decreases because of the increase in H_2CO_3. This leads to acidosis, an acidification of the blood. The pH of the blood drops too low. To counteract the extra acidity, the weak base sodium hydrogen carbonate is introduced to increase the pH back to normal.

More Challenging Questions

102. *Answer:* **(a) Boxes C and D (b) Boxes C and D (c) Box E (d) Box D (e) Box A (f) Box F (g) Box E**

Strategy and Explanation:

(a) The amphiprotic $Zn(OH)_2$ will dissolve in sufficient acid (enough to neutralize the OH^-) and sufficient base, to form the complex ion, $Zn(OH)_4^{2-}$. So, **Boxes C and D** can be correct in answering this prompt.

(b) HPO_4^{2-} can be partially neutralized with NaOH to form a HPO_4^{2-}/PO_4^{3-} buffer. It can also be partially protonated with an increase in the solution's acidity by decreasing the pH to form $H_2PO_4^-/HPO_4^{2-}$. Therefore, **Boxes C and D** both serve as correct answers for this prompt.

(c) At halfway to the equivalence point in a titration of a weak monoprotic acid, the solution has equal concentrations of acid and conjugate base:

$$pH = pK_a + \log\left(\frac{[\text{base}]}{[\text{acid}]}\right)$$

Because, $\log(1) = 0$, $pH = pK_a$. This answer is given in **Box E**.

(d) When SO_2 is present in the atmosphere, precipitation becomes acidic. Decreasing pH is in **Box D**.

(e) A Lewis acid-base reaction occurs between Cr^{3+} ion (a Lewis acid) and the four OH^- ions (Lewis bases) to form $[Cr(OH)_4]^-$. The answer is **Box A**.

(f) The reaction quotient must be greater than the solubility product for a precipitation reaction to be assured of no precipitation. $Q < K_{sp}$ is in **Box F**.

(g) The buffer solution has equal concentrations of acid and conjugate base, so as described in (c) $pH = pK_a$, given in **Box E**.

104. *Answer:* **(a) Box III (b) Box IV (c) Box II (d) Box I**

Strategy and Explanation: The solution starts with an aqueous acidic solution of strong acid, HA, because the acid is strong, the solution will contain an equal quantity (16) of H_3O^+ and A^- ions present.

(a) After a very small amount of base has been added, a very small amount of H_3O^+ ions will be neutralized, leaving slightly fewer H_3O^+ ions (15) than A^- ions (16). This is shown in **Box III**.

(b) When the solution is halfway to the equivalence point, there will be half as many H_3O^+ molecules as there were to begin with. Eight are shown in **Box IV**.

(c) When enough titrant has been added to take the solution to just past the equivalence point, there will be no H_3O^+ ions left and some leftover OH^-. This is shown in **Box II**.

(d) At the equivalence point, there will be only A^- ions. This is shown in **Box I**.

106. *Answer:* **Sample A: $NaHCO_3$; Sample B: NaOH; Sample C: NaOH and $NaHCO_3$; Sample D: Na_2CO_3 (other answers can also be correct)**

Strategy and Explanation: We must assume that any of these substances, if present, are present in "reasonable" concentrations.

Sample A: Phenolphthalein is colorless, suggesting the pH is below 8.3. The most acidic of the choices is $NaHCO_3$, and that is a likely guess for this sample's identity. (Notice: any of these compounds in solution, even NaOH, can have a pH below 8.3, if it is sufficiently dilute, that's why we have to assume reasonable concentrations.)

Sample B: We interpret the evidence to say that the methyl orange changed to its acidic color as soon as it was added. That means the solution became acidic as soon as the phenolphthalein end point was reached. That suggests the titration of a strong base with a very dramatic decline to acidic pH values after the equivalence point. In addition, at very low pH values (below pH = 3.01), bubbles would have been observed as the carbonate reacted to form $CO_2(g)$. All of these interpretations lead us to believe that this sample is probably NaOH.

Sample C: Presumably, both indicators changed color rapidly, suggesting that they were true endpoints, and not just a result of pH changes. Because two different acidic end points were reached with different indicators, this sample must be a mixture of excess strong base, NaOH, and one or more of the salts, either Na_2CO_3 or $NaHCO_3$. It is not possible to distinguish which salt is present, due to the immediate reaction of the strong base with $NaHCO_3$ to make the same product Na_2CO_3.

Sample D: This sample has a two endpoints the second at exactly twice the volume of the first, suggesting that two neutralizations occurred. That means the sample may have been pure Na_2CO_3, which undergoes two sequential neutralization reactions during the titration. Alternative scenarios, equally plausible, would be a mixture of equimolar quantities of NaOH and $NaHCO_3$ or a mixture of NaOH, Na_2CO_3, and $NaHCO_3$ with the following proportions:

$$(\text{conc. NaOH}) = (\text{conc. NaHCO}_3) = \frac{1}{2}(\text{conc. Na}_2\text{CO}_3)$$

Other proportions are also possible. Without some more data, or other limits on the quantities, no more definite answers are possible.

108. *Answer:* **5.64**

Strategy and Explanation: Adapt the method in the solution to Question 25. Find the initial moles of H_3O^+ and C_5H_5N, after the solutions are combined:

$$25.0\ \text{mL} \times \frac{1\ \text{L}}{1000\ \text{mL}} \times \frac{0.085\ \text{mol C}_5\text{H}_5\text{N}}{1\ \text{L}} = 0.00212\ \text{mol C}_5\text{H}_5\text{N}$$

$$5.5\ \text{mL} \times \frac{1\ \text{L}}{1000\ \text{mL}} \times \frac{0.102\ \text{mol HCl}}{1\ \text{L}} \times \frac{1\ \text{mol H}_3\text{O}^+}{1\ \text{mol HCl}} = 0.00056\ \text{mol H}_3\text{O}^+$$

The acid neutralizes the base, via product-favored neutralization reaction.

$$\text{H}_3\text{O}^+(aq) + \text{C}_5\text{H}_5\text{N}(aq) \longrightarrow \text{C}_5\text{H}_5\text{NH}^+(aq) + \text{H}_2\text{O}(\ell)$$

We take it as far to the products as possible, using the method of limiting reactants.

	$H_3O^+(aq)$	$C_5H_5N(aq)$	$C_5H_5NH^+(aq)$
init. mol	0.00056	0.00212	0
change in mol	-0.00056	-0.00056	+0.00056
after rxn (mol)	0	0.00156	0.00056

The solution is a buffer solution, containing an acid-base conjugate pair, so we can find the pH using Henderson-Hasselbalch equation. In the question, we are given K_b for $C_5H_5N = 1.6x10^{-9}$. We will use that to find the K_a for the dissociation of the pyradinium cation, $C_5H_5NH^+(aq)$:

$$K_a = \frac{K_w}{K_b} = \frac{1.00 \times 10^{-14}}{1.6 \times 10^{-9}} = 6.3 \times 10^{-6}$$

Determine the post-neutralization concentrations of $C_5H_5NH^+$ and C_5H_5N, using moles and the combined volumes of solutions: 25.0 mL + 5.5 mL = 30.5 mL, which is 0.0305 L

$$[C_5H_5NH^+] = \frac{0.00056 \text{ mol } C_5H_5NH^+}{0.03050 \text{ L}} = 0.0184 \text{ M } C_5H_5N^+$$

$$[C_5H_5N] = \frac{0.00156 \text{ mol } C_5H_5N}{0.03050 \text{ L}} = 0.0511 \text{ M } C_5H_5N$$

As described in Section 17.1, calculate the pK_a: $pK_a = -\log K_a = -\log(6.3 \times 10^{-6}) = 5.20$

$$pH = 5.20 + \log\left(\frac{0.0511}{0.0184}\right) = 5.20 + 0.44 = 5.64$$

110. *Answer:* **3.22**

Strategy and Explanation: Distilled water with a pH of 5.6, has an H_3O^+ concentration.

$$[H_3O^+] = 10^{-pH} = 10^{-5.6} = 2.5 \times 10^{-6} \text{ M}$$

The acidic pH is a result of the following equlibria:

$$CO_2(aq) + H_2O(\ell) \rightleftharpoons H_2CO_3(aq)$$

$$H_2CO_3(aq) + H_2O(\ell) \rightleftharpoons H_3O^+(aq) + HCO_3^-(aq)$$

HCl is strong acid. It ionizes completely to produce H_3O^+. As a result, the equilibria reactions above shift the left in order to re-establish a new equilibrium state. That means the $H_3O^+(aq)$ concentration *resulting from the ionization of the weak acid* will be even less than that calculated, above.

Determine the $H_3O^+(aq)$ concentration resulting from the ionization of the HCl after one drop of concentrated HCl solution is added to the 1.0 L of water. Notice that the drop adds negligible volume to the 1.0 L solution.

$$\frac{1 \text{ drop}}{1.0 \text{ L}} \times \frac{1 \text{ mL}}{20 \text{ drop}} \times \frac{1 \text{ L}}{1000 \text{ mL}} \times \frac{12 \text{ mol HCl}}{1 \text{ L}} = 6.0 \times 10^{-4} \text{ M HCl}$$

It is clear that the addition of HCl produces so much H_3O^+ that the presence of aqueous CO_2 as an acid source is negligible. Ultimately, the pH depends only on the molarity of the HCl.

$$pH = -\log[H_3O^+] = -\log(6.0 \times 10^{-4}) = 3.22$$

112. *Answer:* **3.2 g glacial acetic acid**

Strategy and Explanation: Glacial acetic acid is pure liquid CH_3COOH. It contains no water.

Assuming that all 300. mL of 0.100 M sodium acetate is used to make the buffer, calculate the moles of CH_3COO^-.

$$300.\text{ mL} \times \frac{1\text{ L}}{1000\text{ mL}} \times \frac{0.100\text{ mol CH}_3\text{COONa}}{1\text{ L}} \times \frac{1\text{ mol CH}_3\text{COO}^-}{1\text{ mol CH}_3\text{COONa}} = 0.0300\text{ mol}$$

The mole ratio between two solutes in a single solution is identical to their concentration ratio. Use this fact to adapt the Henderson-Hasselbalch equation (Section 17.1) to determine the moles of glacial acetic acid needed.

$$\text{pH} = \text{pK}_a + \log\left(\frac{[\text{conj. base}]}{[\text{conj. acid}]}\right) \qquad \frac{[\text{conj. base}]}{\text{conj. }[\text{acid}]} = \frac{\text{mol conj. base}}{\text{mol conj. acid}}$$

Here $$\text{pH} = \text{pK}_a + \log\left(\frac{\text{mol CH}_3\text{COO}^-}{\text{mol CH}_3\text{COOH}}\right)$$

Solve the above equation for the moles of glacial acetic acid, x.

$$4.50 = 4.74 + \log\left(\frac{0.0300\text{ mol}}{\text{x}}\right)$$

$$-0.24 = \log\left(\frac{0.0300\text{ mol}}{\text{x}}\right)$$

$$\frac{0.0300\text{ mol}}{\text{x}} = 10^{-0.24} = 0.57$$

$$\text{x} = \frac{0.0300\text{ mol}}{0.57} = 0.053\text{ mol CH}_3\text{COOH}$$

Use molar mass to determine the mass of glacial acetic acid:

$$0.053\text{ mol CH}_3\text{COOH} \times \frac{60.0519\text{ g}}{1\text{ mol}} = 3.2\text{ g CH}_3\text{COOH}$$

Notice that, while a small volume change happens: $3.2\text{ g} \times \frac{1\text{ mL}}{1.05\text{ g}} = 3.05\text{ mL}$, the volume change affects both the numerator and the denominator of the concentration ratio. So, it is not necessary to use the given density of the glacial acetic acid.

114. *Answer:* **(a) 1.1×10^{-9} M (b) 0.010 M**

Strategy and Explanation: Determine the barium ion concentration in the presence of each these anions at equilibrium.

(a) $BaSO_4$ equilibrium: $BaSO_4(s) \rightleftharpoons Ba^{2+}(aq) + SO_4^{2-}(aq)$

$$K_{sp} = [Ba^{2+}][SO_4^{2-}] = [Ba^{2+}](0.010) = 1.1 \times 10^{-10}$$

$$[Ba^{2+}] = 1.1 \times 10^{-9}\text{ M}$$

The $[Ba^{2+}]$ will be 1.1×10^{-9} M when the $BaSO_4$ begins to precipitate.

(b) BaF_2 equilibrium: $BaF_2(s) \rightleftharpoons Ba^{2+}(aq) + 2\,F^-(aq)$

$$K_{sp} = [Ba^{2+}][F^-]^2 = [Ba^{2+}](0.010)^2 = 1.0 \times 10^{-6}$$

$$[Ba^{2+}] = 0.010\text{ M}$$

The $[Ba^{2+}]$ will be 0.010 M when the BaF_2 begins to precipitate.

Chapter 18: Thermodynamics: Directionality of Chemical Reactions

Solutions for Blue-Numbered Questions for Review and Thought

Topical Questions

Reactant-Favored and Product-Favored Processes (Section 18.1)

15. *Answer:* **(a) $2\ H_2O(\ell) \longrightarrow 2\ H_2(g) + O_2(g)$; reactant-favored (b) $C_8H_{18}(\ell) \longrightarrow C_8H_{18}(g)$; product-favored (c) $C_{12}H_{22}O_{11}(s) \longrightarrow C_{12}H_{22}O_{11}(aq)$; product-favored**

Strategy and Explanation: Think about the process and whether you have observed it happening spontaneously. If it has, it will be classified as product-favored. If it has not, it will be classified as reactant-favored.

(a) Water decomposing into its elements. $2\ H_2O(\ell) \longrightarrow 2\ H_2(g) + O_2(g)$.

The decomposition of water has not been observed to occur spontaneously at normal temperature and pressure, thus this reaction is classified as reactant-favored.

(b) Gasoline spills and evaporates, changing from liquid to gas phase: $C_8H_{18}(\ell) \longrightarrow C_8H_{18}(g)$

The process has been observed to occur spontaneously at normal temperature and pressure, thus this reaction is classified as product-favored.

(c) Dissolving sugar at room temperature, changing it from solid to aqueous phase:

$$C_{12}H_{22}O_{11}(s) \longrightarrow C_{12}H_{22}O_{11}(aq)$$

The process has been observed to occur spontaneously at normal temperature and pressure, thus this reaction is classified as product-favored.

Chemical Reactions and Dispersal of Energy (Section 18.2)

17. *Answer:* **(a) $\frac{1}{2}$ (b) $\frac{1}{2}$ (c) 50 "heads" and 50 "tails"**

Strategy and Explanation: Determine the total number of possible outcomes and use that to build a fraction describing the probability of obtaining one outcome.

(a) "Heads" is one of two possible outcomes, so the probability is $\frac{1}{2}$.

(b) "Tails" is one of two possible outcomes, so the probability is $\frac{1}{2}$.

(c) If you flip a coin 100 times, you get 100 outcomes—each outcome has an equal probability, $\frac{1}{2}$, of being "heads" or "tails". The most likely number of "heads" or "tails" would be half of the total number of flips, or 50 "heads" and 50 "tails".

19. *Answer:* **(a) $\frac{1}{2}$ in A ; $\frac{1}{2}$ in B (b) 50 molecules in flask A and 50 molecules in flask B is most probably arrangement and has the highest entropy.**

Strategy and Explanation: Determine the total number of possible outcomes and use that to build a fraction describing the probability of obtaining one outcome. We will assume that the connecting tube between the two flasks is so short than there is no probability that the molecules will actually be found there.

(a) The probability of the molecule being in flask A is one of two possible outcomes, so the probability is $\frac{1}{2}$.

The probability of the molecule being in flask B is one of two possible outcomes, so the probability is $\frac{1}{2}$.

(b) If you put 100 molecules in the two-flask system, the most likely arrangement of the molecules in flask A and flask B would be half of the molecules in each, or 50 in flask A and 50 in flask B. This 50-50 arrangement has the highest entropy, because the molecules have a higher disorder than any other state that would crowd more than 50 atoms into one flask or the other.

Measuring Dispersal of Energy: Entropy (Section 18.3)

21. *Answer:* **(a) negative (b) positive (c) positive**

Strategy and Explanation: Use the qualitative guidelines for entropy changes described in Section 18.3.

(a) $H_2O(g) \longrightarrow H_2O(s)$

The solid product has lower entropy than the gas-phase reactant, so the entropy change is negative.

(b) $CO_2(aq) \longrightarrow CO_2(g)$

The gas-phase product has higher entropy than the aqueous reactant, so the entropy change is positive.

(c) $\text{glass(s)} \longrightarrow \text{glass}(\ell)$

The liquid product has higher entropy than the solid reactant, so the entropy change is positive.

23. *Answer:* **(a) positive (b) positive (c) positive**

Strategy and Explanation: Use the qualitative guidelines for entropy changes described in Section 18.3.

(a) $2\ H_2O(\ell) \longrightarrow 2\ H_2(g) + O_2(g)$

The gas-phase products have higher entropy than the liquid reactant, so the entropy change is positive.

(b) $C_8H_{18}(\ell) \longrightarrow C_8H_{18}(g)$

The gas-phase product has higher entropy than the liquid reactant, so the entropy change is positive.

(c) $C_{12}H_{22}O_{11}(s) \longrightarrow C_{12}H_{22}O_{11}(aq)$

The aqueous product has higher entropy than the solid reactant, so the entropy change is positive.

25. *Answer:* **(a) Item 2 (b) Item 2 (c) Item 2**

Strategy and Explanation: Use the qualitative guidelines for entropy changes described in Section 18.3.

(a) Item 2 has higher entropy since it is identical to item 1 except that its temperature is higher. Molecules at higher temperature have higher entropy.

(b) Item 2, dissolved sugar, has higher entropy than item 1, solid sugar, because solute molecules are more random than those in a solid crystal.

(c) Item 2, the mixture of water and alcohol together has higher entropy than item 1, water and alcohol separate. Mixing makes the molecules more random.

27. *Answer:* **(a) NaCl (b) P_4 (c) $NH_4NO_3(aq)$**

Strategy and Explanation: Use the methods described in Problem-Solving Example 18.2 and the qualitative guidelines for entropy changes described in Section 18.3.

(a) Comparing NaCl and CaO, we find that the biggest difference between these two ionic solids is the attractions due to the charges on the ions. According to Coulomb's law, the Ca^{2+} and O^{2-} ions have greater interaction than the Na^+ and Cl^- ions, so NaCl has a larger entropy/mol than CaO.

(b) P_4 molecules have more atoms than Cl_2 molecules, so P_4 has a larger entropy per mol than Cl_2.

(c) The solid NH_4NO_3 crystal is more ordered than the aqueous NH_4^+ and NO_3^- ions, so the aqueous NH_4NO_3 has a larger entropy per mol than the solid NH_4NO_3.

29. *Answer:* **(a) $Ga(\ell)$ (b) AsH_3 (c) NaF**

Strategy and Explanation: Use the methods described in Question 27.

(a) Solid-state Ga atoms are more ordered than liquid Ga atoms, so Ga(ℓ) has more entropy than Ga(s).

(b) AsH_3 is a molecular gas and Kr is an atomic gas. Because they have similar mass, we look at the relative complexity. The molecule has more complexity. That means AsH_3 has a larger standard molar entropy than Kr.

(c) Comparing NaF and MgO, we find that the biggest difference between these two ionic solids is the attractions due to the charges on the ions. According to Coulomb's law, the Mg^{2+} and O^{2-} ions have greater interaction than the Na^+ and F^- ions, so NaF has a larger standard molar entropy than MgO.

31. *Answer:* **(a) negative (b) positive (c) negative (d) positive**

Strategy and Explanation: Use the qualitative guidelines for entropy changes described in Section 18.3.

(a) The reaction has more gas-phase reactants (2 mol) than gas-phase products (1 mol), so the entropy change will be negative.

(b) The reaction has fewer gas-phase reactants ($\frac{3}{2}$ mol) than gas-phase products (3 mol), so the entropy change will be positive.

(c) The reaction has more gas-phase reactants (4 mol) than gas-phase products (2 mol), so the entropy change will be negative.

(d) The reaction has fewer gas-phase reactants (0 mol) than gas-phase products (1 mol), so the entropy change will be positive.

33. *Answer:* **(a) negative (b) negative (c) positive**

Strategy and Explanation: Use the qualitative guidelines for entropy changes described in Section 18.3.

(a) The reaction has more gas-phase reactants (3 mol) than gas-phase products (2 mol), so the entropy change will be negative.

(b) The reaction has more gas-phase reactants (3 mol) than gas-phase products (0 mol), so the entropy change will be negative.

(c) The reaction has fewer gas-phase reactants (2 mol) than gas-phase products (3 mol), so the entropy change will be positive.

Calculating Entropy Changes (Section 18.4)

35. *Answer:* **112 J K^{-1}mol^{-1}**

Strategy and Explanation: Rearrange Equation 18.5: $T = \frac{\Delta H^\circ}{\Delta S^\circ}$ (when ΔG =0) to solve for ΔS°.

Use the enthalpy of vaporization and the normal boiling point of ethanol:

$$\Delta S^\circ = \frac{\Delta H^\circ}{T} = \frac{\left(39.3\ \frac{\text{kJ}}{\text{mol}}\right)\times\left(\frac{1000\ \text{J}}{1\ \text{kJ}}\right)}{(78.3+273.15)\text{K}} = 112\ \text{J K}^{-1}\text{mol}^{-1}$$

37. *Answer:* **(a) 2.63 J/K (b) 1000 J/K (c) 2.45 J/K**

Strategy and Explanation: Adapt the method described in Problem-Solving Example 18.1. Use Equation 18.1: $\Delta S = \frac{q_{rev}}{T}$. For example, the calculation for (a) looks like this:

$$\Delta S^\circ = \frac{q_{rev}}{T} = \frac{(0.775\ \text{kJ})\times\left(\frac{1000\ \text{J}}{1\ \text{kJ}}\right)}{295\ \text{K}} = 2.63\ \text{J K}^{-1}$$

	q°_{rev} (kJ)	T (K)	ΔS° (J K^{-1})
(a)	0.775	295	2.63
(b)	500	500.	1000
(c)	2.45	1000.	2.45

39. *Answer:* **(a) 113.0 J/K (b) 38.17 kJ**

Strategy and Explanation:

(a) Adapt the method described in Problem-Solving Example 18.3. Write the balanced chemical equation and use the equation from the beginning of Section 18.4:

$$\Delta S° = \sum\left[(\text{moles of product}) \times S°(\text{product})\right] - \sum\left[(\text{moles of reactant}) \times S°(\text{reactant})\right]$$

Vaporization equation: $CH_3OH(\ell) \longrightarrow CH_3OH(g)$

$$\Delta S° = (1\text{ mole}) \times S°(CH_3OH(g)) - (1\text{ mole}) \times S°(CH_3OH(\ell))$$

$$\Delta S° = (1\text{ mole}) \times (239.81\text{ J K}^{-1}\text{mol}^{-1}) - (1\text{ mole}) \times (126.8\text{ J K}^{-1}\text{mol}^{-1}) = 113.0\text{ J K}^{-1}$$

(b) Adapt the method described in the solution to Question 35.

$$\Delta H° = T\Delta S° = (64.6 + 273.15)\text{ K} \times (113.0\text{ J K}^{-1}) \times \frac{1\text{ kJ}}{1000\text{ J}} = 38.17\text{ kJ}$$

41. *Answer:* **(a) –120.64 J/K; negative ΔS prediction confirmed (b) 156.9 J/K; positive ΔS prediction confirmed (c) –198.76 J/K; negative ΔS prediction confirmed (d) 160.6 J/K; positive ΔS prediction confirmed**

Strategy and Explanation: Adapt the method described in the solution to Question 39(a).

(a) $\Delta S° = (1\text{ mol}) \times S°(C_2H_6) - (1\text{ mol}) \times S°(C_2H_4) - (1\text{ mol}) \times S°(H_2)$

$= (1\text{ mol}) \times (229.60\text{ J K}^{-1}\text{mol}^{-1}) - (1\text{ mol}) \times (219.56\text{ J K}^{-1}\text{mol}^{-1})$

$- (1\text{ mol}) \times (130.684\text{ J K}^{-1}\text{mol}^{-1}) = -120.64\text{ J K}^{-1}$

This confirms the negative ΔS prediction from Question 31.

(b) $\Delta S° = (1\text{ mol}) \times S°(CO_2) + (2\text{ mol}) \times S°(H_2O(g)) - (1\text{ mol}) \times S°(CH_3OH(\ell)) - (\frac{3}{2}\text{ mol}) \times S°(O_2)$

$= (1\text{ mol}) \times (213.74\text{ J K}^{-1}\text{mol}^{-1}) + (2\text{ mol}) \times (188.825\text{ J K}^{-1}\text{mol}^{-1})$

$- (1\text{ mol}) \times (126.8\text{ J K}^{-1}\text{mol}^{-1}) - (\frac{3}{2}\text{ mol}) \times (205.138\text{ J K}^{-1}\text{mol}^{-1}) = 156.9\text{ J K}^{-1}$

This confirms the positive ΔS prediction from Question 31.

(c) $\Delta S° = (2\text{ mol}) \times S°(NH_3) - (1\text{ mol}) \times S°(N_2) - (3\text{ mol}) \times S°(H_2)$

$= (2\text{ mol}) \times (192.45\text{ J K}^{-1}\text{mol}^{-1}) - (1\text{ mol}) \times (191.61\text{ J K}^{-1}\text{mol}^{-1})$

$- (3\text{ mol}) \times (130.684\text{ J K}^{-1}\text{mol}^{-1}) = -198.76\text{ J K}^{-1}$

This confirms the negative ΔS prediction from Question 31.

(d) $\Delta S° = (1\text{ mol}) \times S°(CaO) + (1\text{ mol}) \times S°(CO_2) - (1\text{ mol}) \times S°(CaCO_3)$

$= (1\text{ mol}) \times (39.75\text{ J K}^{-1}\text{mol}^{-1}) + (1\text{ mol}) \times (213.74\text{ J K}^{-1}\text{mol}^{-1})$

$- (1\text{ mol}) \times (92.9\text{ J K}^{-1}\text{mol}^{-1}) = 160.6\text{ J K}^{-1}$

This confirms the positive ΔS prediction from Question 31.

43. *Answer:* **(a) –173.01 J/K; negative ΔS prediction confirmed (b) –326.69 J/K; negative ΔS prediction confirmed (c) 137.55 J/K; positive ΔS prediction confirmed**

Strategy and Explanation: Adapt the method described in the solution to Question 39(a).

(a) $\Delta S° = (2\text{ mol}) \times S°(CO_2) - (2\text{ mol}) \times S°(CO) - (1\text{ mol}) \times S°(O_2)$

$= (2\text{ mol}) \times (213.74\text{ J K}^{-1}\text{mol}^{-1}) - (2\text{ mol}) \times (197.674\text{ J K}^{-1}\text{mol}^{-1})$

$- (1\text{ mol}) \times (205.138\text{ J K}^{-1}\text{mol}^{-1}) = -173.01\text{ J K}^{-1}$

This confirms the negative ΔS prediction from Question 33.

(b) $\Delta S° = (2\text{ mol}) \times S°(H_2O(\ell)) - (2\text{ mol}) \times S°(H_2) - (1\text{ mol}) \times S°(O_2)$

$= (2\text{ mol}) \times (69.91\text{ J K}^{-1}\text{mol}^{-1}) - (2\text{ mol}) \times (130.684\text{ J K}^{-1}\text{mol}^{-1})$

$- (1\text{ mol}) \times (205.138\text{ J K}^{-1}\text{mol}^{-1}) = -326.69\text{ J K}^{-1}$

This confirms the negative ΔS prediction from Question 33.

(c) $\Delta S° = (3\text{ mol}) \times S°(O_2) - (2\text{ mol}) \times S°(O_3)$

$= (3\text{ mol}) \times (205.138\text{ J K}^{-1}\text{mol}^{-1}) - (2\text{ mol}) \times (238.93\text{ J K}^{-1}\text{mol}^{-1}) = 137.55\text{ J K}^{-1}$

This confirms the positive ΔS prediction from Question 33.

Entropy and the Second Law of Thermodynamics (Section 18.5)

45. *Answer:* **No, we need ΔH° also; the reaction is product-favored.**

Strategy and Explanation: Adapt the method described in the solution to Question 39(a).

$\Delta S° = (1\text{ mol}) \times S°(C_2H_5OH) - (1\text{ mol}) \times S°(C_2H_4) - (1\text{ mol}) \times S°(H_2O(g))$

$= (1\text{ mol}) \times (160.7\text{ J K}^{-1}\text{mol}^{-1}) - (1\text{ mol}) \times (219.56\text{ J K}^{-1}\text{mol}^{-1})$

$- (1\text{ mol}) \times (188.825\text{ J K}^{-1}\text{mol}^{-1}) = -247.7\text{ J K}^{-1}$

We cannot tell from the results of this calculation whether this reaction is product-favored. We need the ΔH°, also. Use Equation 6.11:

$$\Delta H° = \sum\left[(\text{moles of product}) \times \Delta H_f°(\text{product})\right] - \sum\left[(\text{moles of reactant}) \times \Delta H_f°(\text{reactant})\right]$$

$\Delta H° = (1\text{ mol}) \times \Delta H_f°(C_2H_5OH) - (1\text{ mol}) \times \Delta H_f°(C_2H_4) - (1\text{ mol}) \times \Delta H_f°(H_2O(g))$

$= (1\text{ mol}) \times (-277.69\text{ kJ/mol}) - (1\text{ mol}) \times (52.26\text{ kJ/mol}) - (1\text{ mol}) \times (-241.818\text{ kJ/mol}) = -88.13\text{ kJ}$

Since, both ΔH° and ΔS° are negative, the product-favored nature of the reaction depends on temperature.

$$T = \frac{(-88.13\text{ kJ}) \times \left(\frac{1000\text{ J}}{1\text{ kJ}}\right)}{-247.7\text{ J K}^{-1}} = 355.8\text{ K}$$

When 355.8 K or lower, the reaction is product-favored. The temperature here is 25 °C + 273.15 = 298 K, so the reaction is product-favored.

47. *Answer:* **low temperatures**

Strategy and Explanation: Adapt the method described in the solution to Question 32, then consult Table 18.2.

The reaction has more gas-phase reactants (1 mol) than gas-phase products (0 mol), so the entropy change will be negative. When ΔH° and ΔS° are both negative, the reaction is product-favored only at low temperatures. The exothermicity is sufficient to favor products, if the temperature is low enough to overcome the decrease in entropy.

49. *Answer/Explanation:* Exothermic reactions with an increase in disorder, exhibited by a larger number of gas-phase products (7 mol) than gas-phase reactants (5 mol), never need help from the surroundings to favor products.

51. *Answer:* **(a) ΔS° is positive; ΔH° is negative. (b) ΔH° = –184.28 kJ, which is negative as predicted; ΔS° = –7.7 J/K which is not positive as predicted, but it is very small.**

Strategy and Explanation:

(a) The products of this reaction have a larger number of gas-phase products (1 mol) than gas-phase reactants (0 mol), so we predict the entropy change, ΔS°, to be positive. Sodium reacts violently, suggesting that a great amount of heat is produced in the reaction. We will predict the enthalpy change, ΔH°, is negative.

(b) Adapt the method described in the solution to Question 41.

$$\Delta H° = (1 \text{ mol}) \times \Delta H°_f(\text{NaOH}) + (\tfrac{1}{2} \text{ mol}) \times \Delta H°_f(\text{H}_2) - (1 \text{ mol}) \times \Delta H°_f(\text{Na}) - (1 \text{ mol}) \times \Delta H°_f(\text{H}_2\text{O}(\ell))$$

$$= (1 \text{ mol}) \times (-470.114 \text{ kJ/mol}) + (\tfrac{1}{2} \text{ mol}) \times (0 \text{ kJ/mol}) - (1 \text{ mol}) \times (0 \text{ kJ/mol}) - (1 \text{ mol}) \times (-285.83 \text{ kJ/mol}) = -184.28 \text{ kJ}$$

$$\Delta S° = (1 \text{ mol}) \times S°(\text{NaOH}) + (\tfrac{1}{2} \text{ mol}) \times S°(\text{H}_2) - (1 \text{ mol}) \times \Delta S°(\text{Na}) - (1 \text{ mol}) \times \Delta S°(\text{H}_2\text{O}(\ell))$$

$$= (1 \text{ mol}) \times (48.1 \text{ J K}^{-1}\text{mol}^{-1}) + (\tfrac{1}{2} \text{ mol}) \times (130.684 \text{ J K}^{-1}\text{mol}^{-1}) - (1 \text{ mol}) \times (51.21 \text{ J K}^{-1}\text{mol}^{-1}) - (1 \text{ mol}) \times (69.91 \text{ J K}^{-1}\text{mol}^{-1}) = -7.7 \text{ J K}^{-1}$$

The enthalpy change is negative, as predicted in (a); however, the entropy change is negative, not as predicted in (a). Looking at the values of S°, the aqueous solute has lower entropy than pure water, providing sufficient order to compensate for the higher disorder of the gas. The value of -7.7 J K^{-1} is pretty small.

53. *Answer:* **ΔS° is positive; ΔH° is positive. The reaction is reactant-favored. See explanation below.**

Strategy and Explanation: The equation for the chemical reaction is: $\text{H}_2\text{O}(\ell) \longrightarrow \text{H}_2(\text{g}) + \tfrac{1}{2}\,\text{O}_2(\text{g})$

The reaction written here is the reverse of the highly exothermic reaction described in the first sentence of the Question, so this reaction is highly endothermic, with a positive enthalpy change (ΔH°). With more gas products ($1\tfrac{1}{2}$ mol) than gas reactants (0 mol), the entropy change (ΔS°) is positive. Because we do not see water spontaneously decomposing, we will conclude that the reaction is reactant-favored. The entropy increase is insufficient to drive this highly endothermic reaction to form products without assistance from the surroundings at the temperature of 25 °C.

55. *Answer:* **(a) –851.5 kJ; –37.52 J/K; product-favored at low temperatures (b) 66.36 kJ; – 21.77 J/K; never product-favored**

Strategy and Explanation: Adapt the method described in the solution to Question 45 and use Table 18.2.

(a) $$\Delta H° = (2 \text{ mol}) \times \Delta H°_f(\text{Fe}) + (1 \text{ mol}) \times \Delta H°_f(\text{Al}_2\text{O}_3) - (1 \text{ mol}) \times \Delta H°_f(\text{Fe}_2\text{O}_3) - (2 \text{ mol}) \times \Delta H°_f(\text{Al})$$

$$= (2 \text{ mol}) \times (0 \text{ kJ/mol}) + (1 \text{ mol}) \times (-1675.7 \text{ kJ/mol}) - (1 \text{ mol}) \times (-824.2 \text{ kJ/mol}) - (2 \text{ mol}) \times (0 \text{ kJ/mol}) = -851.5 \text{ kJ}$$

$$\Delta S° = (2 \text{ mol}) \times S°(\text{Fe}) + (1 \text{ mol}) \times S°(\text{Al}_2\text{O}_3) - (1 \text{ mol}) \times S°(\text{Fe}_2\text{O}_3) - (2 \text{ mol}) \times S°(\text{Al})$$

$$= (2 \text{ mol}) \times (27.78 \text{ J K}^{-1}\text{mol}^{-1}) + (1 \text{ mol}) \times (50.92 \text{ J K}^{-1}\text{mol}^{-1}) - (1 \text{ mol}) \times (87.40 \text{ J K}^{-1}\text{mol}^{-1}) - (2 \text{ mol}) \times (28.3 \text{ J K}^{-1}\text{mol}^{-1}) = -37.52 \text{ J K}^{-1}$$

The reaction is exothermic, but its entropy change is negative, so it would stop being product-favored above a specific temperature. (Calculated from Equation 18.5, this temperature is 22,700 K.)

(b) $$\Delta H° = (2 \text{ mol}) \times \Delta H°_f(\text{NO}_2) - (1 \text{ mol}) \times \Delta H°_f(\text{N}_2) - (2 \text{ mol}) \times \Delta H°_f(\text{O}_2)$$

$$= (2 \text{ mol}) \times (33.18 \text{ kJ/mol}) - (1 \text{ mol}) \times (0 \text{ kJ/mol}) - (2 \text{ mol}) \times (0 \text{ kJ/mol}) = 66.36 \text{ kJ}$$

$$\Delta S° = (2 \text{ mol}) \times S°(\text{NO}_2) - (1 \text{ mol}) \times S°(\text{N}_2) - (2 \text{ mol}) \times S°(\text{O}_2)$$

$$= (2 \text{ mol}) \times (240.06 \text{ J K}^{-1}\text{mol}^{-1}) - (1 \text{ mol}) \times (191.61 \text{ J K}^{-1}\text{mol}^{-1}) - (2 \text{ mol}) \times (205.138 \text{ J K}^{-1}\text{mol}^{-1}) = -21.77 \text{ J K}^{-1}$$

The reaction is endothermic; the entropy change is negative, so it will never be product-favored.

Gibbs Free Energy (Section 18.6)

57. *Answer:* **(a) $\Delta S_{universe} = 4.92 \times 10^3$ J/K (b) $\Delta G^\circ_{system} = -1.47 \times 10^3$ kJ and $\Delta G^\circ_{system} = -T\Delta S_{universe}$ (c) Yes, since ethane is used as a fuel.**

Strategy and Explanation: Adapt the method described in Problem-Solving Example 18.4, using three equations:

Equation 18.2 with the Kelvin temperature: $\Delta S_{universe} = -\dfrac{\Delta H^\circ_{system}}{T} + \Delta S^\circ_{system}$

$$T(K) = T(°C) + 273.15 = 25\ °C + 273.15 = 298\ K$$

Equation 18.3 $\quad \Delta G^\circ_{system} = \Delta H^\circ_{system} - T\Delta S^\circ_{system}$

and Equation 18.4, as described in Problem-Solving Example 18.5.

$$\Delta G^\circ = \sum\Big[(\text{moles of product}) \times \Delta G^\circ_f(\text{product})\Big] - \sum\Big[(\text{moles of reactant}) \times \Delta G^\circ_f(\text{reactant})\Big]$$

(a) $\Delta H^\circ_{system} = (2\ \text{mol}) \times \Delta H^\circ_f(CO_2) + (3\ \text{mol}) \times \Delta H^\circ_f(H_2O(\ell)) - (1\ \text{mol}) \times \Delta H^\circ_f(C_2H_6) - (\frac{7}{2}\ \text{mol}) \times \Delta H^\circ_f(O_2)$

$= (2\ \text{mol}) \times (-393.509\ \text{kJ/mol}) + (3\ \text{mol}) \times (-285.83\ \text{kJ/mol})$

$- (1\ \text{mol}) \times (-84.68\ \text{kJ/mol}) - (\frac{7}{2}\ \text{mol}) \times (0\ \text{kJ/mol}) = -1559.83\ \text{kJ}$

$\Delta S^\circ_{system} = (2\ \text{mol}) \times S^\circ(CO_2) + (3\ \text{mol}) \times S^\circ(H_2O(\ell)) - (1\ \text{mol}) \times S^\circ(C_2H_6) - (\frac{7}{2}\ \text{mol}) \times S^\circ(O_2)$

$= (2\ \text{mol}) \times (213.74\ \text{J K}^{-1}\text{mol}^{-1}) + (3\ \text{mol}) \times (69.91\ \text{J K}^{-1}\text{mol}^{-1})$

$- (1\ \text{mol}) \times (229.60\ \text{J K}^{-1}\text{mol}^{-1}) - (\frac{7}{2}\ \text{mol}) \times (205.138\ \text{J K}^{-1}\text{mol}^{-1}) = -310.37\ \text{J K}^{-1}$

$$\Delta S_{universe} = -\frac{(-1559.83\ \text{kJ})}{(298\ \text{K})} \times \frac{1000\ \text{J}}{1\ \text{kJ}} + (-310.37\ \text{J K}^{-1}) = 4.92 \times 10^3\ \text{J K}^{-1}$$

(b) $\Delta G^\circ_{system} = \Delta H^\circ_{system} - T\Delta S^\circ_{system} = (-1559.83\ \text{kJ}) - (298\ \text{K})(-310.37\ \text{J K}^{-1}) \times \dfrac{1\ \text{kJ}}{1000\ \text{J}} = -1467.3\ \text{kJ}$

Independently, we can get ΔG°_{system} using Equation 18.4:

$\Delta G^\circ_{system} = (2\ \text{mol}) \times \Delta G^\circ_f(CO_2) + (3\ \text{mol}) \times \Delta G^\circ_f(H_2O(\ell)) - (1\ \text{mol}) \times \Delta G^\circ_f(C_2H_6) - (\frac{7}{2}\ \text{mol}) \times \Delta G^\circ_f(O_2)$

$= (2\ \text{mol}) \times (-394.359\ \text{kJ/mol}) + (3\ \text{mol}) \times (-237.129\ \text{kJ/mol})$

$- (1\ \text{mol}) \times (-32.82\ \text{kJ/mol}) - (\frac{7}{2}\ \text{mol}) \times (0\ \text{kJ/mol}) = -1467.28\ \text{kJ}$

A negative ΔG°_{system} is consistent with a positive $\Delta S_{universe}$. In addition:

$$-T\Delta S_{universe} = (298\ \text{K})(4.92 \times 10^3\ \text{J K}^{-1}) \times \frac{1\ \text{kJ}}{1000\ \text{J}} = -1.47 \times 10^3\ \text{kJ} = \Delta G^\circ_{system}.$$

(c) Yes. Ethane is used as a fuel; hence, we would expect its combustion reaction to be product–favored.

59. *Answer:* **First line: sign of ΔG_{system} = negative; last line: sign of ΔG_{system} = positive; see table below**

Strategy and Explanation: Adapt the method described in the solution to Question 55 using a generalized form of Equation 18.3:

$$\Delta G_{system} = \Delta H_{system} - T\Delta S_{system}$$

$$(\text{sign of } \Delta G_{system}) = (\text{sign of } \Delta H_{system}) - (\text{positive Kelvin temperature}) \times (\text{sign of } \Delta S^\circ)$$

For the first line of the table: (sign of ΔG°) = (negative) – (positive) × (positive) = negative

For the last line of the table: (sign of ΔG°) = (positive) – (positive) × (negative) = positive

Sign of ΔH_{system}	Sign of ΔS_{system}	Product-favored?	Sign of ΔG_{system}
Negative (exothermic)	Positive	Yes	Negative
Negative (exothermic)	Negative	Yes at low T; no at high T	*(see Question 60)*
Positive (endothermic)	Positive	No at low T; yes at high T	*(see Question 60)*
Positive (endothermic)	Negative	No	Positive

61. *Answer/Explanation:* Adapt the method described in the solution to Question 57 using Equation 18.3:

$$\Delta G° = \Delta H° - T\Delta S°$$

Here: (sign of $\Delta H°$) = negative, and (sign of $\Delta S°$) = positive. The Kelvin temperature is always positive.

$$\Delta G° = \text{(negative } \Delta H°) - \text{(positive Kelvin temperature)} \times \text{(positive } \Delta S°)$$

Take the absolute values of the each term and then include the resulting signs explicitly:

$$\Delta G° = -|\Delta H°| - |T\Delta S°| = -(|\Delta H°| + |T\Delta S°|)$$

$$\text{(sign of } \Delta G°) = -\text{(positive)} = \text{negative}$$

So, $\Delta G° < 0$ for all temperature values.

63. *Answer:* **$\Delta G° = 28.63$ kJ, so reaction is reactant-favored.**

Strategy and Explanation: Adapt the method described in the solution to Question 57(b).

$$\Delta G° = \Delta H° - T\Delta S° = (41.17 \text{ kJ}) - (298 \text{ K}) \times (42.08 \text{ J/K}) \times \frac{1 \text{ kJ}}{1000 \text{ J}} = 28.63 \text{ kJ}$$

The sign of $\Delta G°$ is positive, so the reaction is reactant-favored.

65. *Answer:* **$\Delta G° = 462.28$ kJ, so reaction is not a good way to make pure Si.**

Strategy and Explanation: Adapt the method described in the solution to Question 57(b).

$$\Delta G° = (1 \text{ mol}) \times \Delta G_f°(Si) + (1 \text{ mol}) \times \Delta G_f°(CO_2) - (1 \text{ mol}) \times \Delta G_f°(SiO_2) - (1 \text{ mol}) \times \Delta G_f°(C)$$
$$= (1 \text{ mol}) \times (0 \text{ kJ/mol}) + (1 \text{ mol}) \times (-394.359 \text{ kJ/mol})$$
$$- (1 \text{ mol}) \times (-856.64 \text{ kJ/mol}) - (1 \text{ mol}) \times (0 \text{ kJ/mol}) = 462.28 \text{ kJ}$$

The reaction is not product-favored, so this would not be a good way to make pure silicon.

67. *Answer:* **(a) $\Delta G° = -141.05$ kJ (b) $\Delta G° = 141.73$ kJ (c) $\Delta G° = -959.43$ kJ; reactions (a) and (c) are product-favored.**

Strategy and Explanation: Adapt the method described in the solution to Question 57(b).

(a) $\Delta G° = (1 \text{ mol}) \times \Delta G_f°(C_2H_4) - (1 \text{ mol}) \times \Delta G_f°(C_2H_2) - (1 \text{ mol}) \times \Delta G_f°(H_2)$

$= (1 \text{ mol}) \times (68.15 \text{ kJ/mol}) - (1 \text{ mol}) \times (209.20 \text{ kJ/mol}) - (1 \text{ mol}) \times (0 \text{ kJ/mol}) = -141.05 \text{ kJ}$

(b) $\Delta G° = (2 \text{ mol}) \times \Delta G_f°(SO_2) + (1 \text{ mol}) \times \Delta G_f°(O_2) - (2 \text{ mol}) \times \Delta G_f°(SO_3)$

$= (2 \text{ mol}) \times (-300.194 \text{ kJ/mol}) + (1 \text{ mol}) \times (0 \text{ kJ/mol}) - (2 \text{ mol}) \times (-371.06 \text{ kJ/mol}) = 141.73 \text{ kJ}$

(c) $\Delta G° = (4 \text{ mol}) \times \Delta G_f°(NO) + (6 \text{ mol}) \times \Delta G_f°(H_2O(g)) - (4 \text{ mol}) \times \Delta G_f°(NH_3) - (5 \text{ mol}) \times \Delta G_f°(O_2)$

$= (4 \text{ mol}) \times (86.55 \text{ kJ/mol}) + (6 \text{ mol}) \times (-228.572 \text{ kJ/mol})$

$- (4 \text{ mol}) \times (-16.45 \text{ kJ/mol}) - (5 \text{ mol}) \times (0 \text{ kJ/mol}) = -959.43 \text{ kJ}$

69. *Answer:* **(a) 385.7 K (b) 835.1 K**

Strategy and Explanation: Adapt the method described in the solution to Question 39(a).

(a) $\Delta H° = (1 \text{ mol}) \times \Delta H_f°(CH_3OH) + (1 \text{ mol}) \times \Delta H_f°(CO) - (2 \text{ mol}) \times \Delta H_f°(H_2)$

$= (1 \text{ mol}) \times (-238.66 \text{ kJ/mol}) + (1 \text{ mol}) \times (-110.525 \text{ kJ/mol}) - (2 \text{ mol}) \times (0 \text{ kJ/mol}) = -128.14 \text{ kJ}$

$\Delta S° = (1 \text{ mol}) \times S°(CH_3OH) + (1 \text{ mol}) \times S°(CO) - (2 \text{ mol}) \times S°(H_2)$

$= (1 \text{ mol}) \times (126.8 \text{ J K}^{-1}\text{mol}^{-1}) + (1 \text{ mol}) \times (197.674 \text{ J K}^{-1}\text{mol}^{-1})$

$- (2 \text{ mol}) \times (130.684 \text{ J K}^{-1}\text{mol}^{-1}) = -332.2 \text{ J K}^{-1}$

$$T = \frac{(-128.14 \text{ kJ}) \times \left(\frac{1000 \text{ J}}{1 \text{ kJ}}\right)}{-332.2 \text{ J K}^{-1}} = 385.7 \text{ K}$$

(b) $\Delta H° = (4 \text{ mol}) \times \Delta H_f°(Fe) + (3 \text{ mol}) \times \Delta H_f°(CO_2) - (2 \text{ mol}) \times \Delta H_f°(Fe_2O_3) - (3 \text{ mol}) \times \Delta H_f°(C)$

$= (4 \text{ mol}) \times (0 \text{ kJ/mol}) + (3 \text{ mol}) \times (-393.509 \text{ kJ/mol})$

$- (2 \text{ mol}) \times (-824.2 \text{ kJ/mol}) - (3 \text{ mol}) \times (0 \text{ kJ/mol}) = 467.9 \text{ kJ}$

$\Delta S° = (4 \text{ mol}) \times S°(Fe) + (3 \text{ mol}) \times S°(CO_2) - (2 \text{ mol}) \times S°(Fe_2O_3) - (3 \text{ mol}) \times S°(C)$

$= (4 \text{ mol}) \times (27.78 \text{ J K}^{-1}\text{mol}^{-1}) + (3 \text{ mol}) \times (213.74 \text{ J K}^{-1}\text{mol}^{-1})$

$- (2 \text{ mol}) \times (87.40 \text{ J K}^{-1}\text{mol}^{-1}) - (3 \text{ mol}) \times (5.740 \text{ J K}^{-1}\text{mol}^{-1}) = 560.32 \text{ J K}^{-1}$

$$T = \frac{(467.9 \text{ kJ}) \times \left(\frac{1000 \text{ J}}{1 \text{ kJ}}\right)}{560.32 \text{ J K}^{-1}} = 835.1 \text{ K}$$

71. *Answer:* **(a) 49.7 kJ (b) –178.2 kJ (c) –1267.5 kJ**

Strategy and Explanation: Adapt the method described in the solution to Question 58(b).

The values of $\Delta H°$ and $\Delta S°$ were calculated for the reactions in Question 67 in the solution to Question 68. We will assume $\Delta H°$ and $\Delta S°$ are approximately independent of temperature:

(a) $\Delta H° = -174.47$ kJ and $\Delta S° = -112.1 \text{ J K}^{-1}$

$$\Delta G° = \Delta H° - T\Delta S° = (-174.47 \text{ kJ}) - (2000.) \times (-112.1 \text{ J K}^{-1}) \times \frac{1 \text{ kJ}}{1000 \text{ J}} = 49.7 \text{ kJ}$$

(b) $\Delta H = 197.78$ kJ and $\Delta S° = 188.0 \text{ J K}^{-1}$

$$\Delta G° = \Delta H° - T\Delta S° = (197.78 \text{ kJ}) - (2000.) \times (188.0 \text{ J K}^{-1}) \times \frac{1 \text{ kJ}}{1000 \text{ J}} = -178.2 \text{ kJ}$$

(c) $\Delta H° = -905.47$ kJ and $\Delta S° = 181.0 \text{ J K}^{-1}$

$$\Delta G° = \Delta H° - T\Delta S° = (-905.47 \text{ kJ}) - (2000.) \times (181.0 \text{ J K}^{-1}) \times \frac{1 \text{ kJ}}{1000 \text{ J}} = -1267.5 \text{ kJ}$$

73. *Answer:* **(a) $\Delta H° = 178.32$ kJ; $\Delta S° = 160.6 \text{ J K}^{-1}$; $\Delta G° = 130.5$ kJ (b) reactant-favored (c) It is not product-favored at all temperatures. (d) 1110. K**

Strategy and Explanation: Adapt methods described in the solution to Questions 45 and 55.

(a) $\Delta H° = (1 \text{ mol}) \times \Delta H_f°(CaO) + (1 \text{ mol}) \times \Delta H_f°(CO_2) - (1 \text{ mol}) \times \Delta H_f°(CaCO_3)$

$= (1 \text{ mol}) \times (-635.09 \text{ kJ/mol}) + (1 \text{ mol}) \times (-393.509 \text{ kJ/mol})$

$- (1 \text{ mol}) \times (-1206.92 \text{ kJ/mol}) = 178.32 \text{ kJ}$

$\Delta S° = (1 \text{ mol}) \times S°(CaO) + (1 \text{ mol}) \times S°(CO_2) - (1 \text{ mol}) \times S°(CaCO_3)$

$= (1 \text{ mol}) \times (39.75 \text{ J K}^{-1}\text{mol}^{-1}) + (1 \text{ mol}) \times (213.74 \text{ J K}^{-1}\text{mol}^{-1})$

$- (1 \text{ mol}) \times (92.9 \text{ J K}^{-1}\text{mol}^{-1}) = 160.6 \text{ J K}^{-1}$

$$\Delta G° = \Delta H° - T\Delta S° = (178.32 \text{ kJ}) - (298 \text{ K})(160.6 \text{ J K}^{-1}) \times \frac{1 \text{ kJ}}{1000 \text{ J}} = 130.5 \text{ kJ}$$

(b) The change in Gibbs free energy is positive, so the reaction is reactant-favored.

(c) Both $\Delta H°$ and $\Delta S°$ are positive, so the reaction is only product-favored at high temperatures.

(d) $$T = \frac{(178.32\ \text{kJ}) \times \left(\frac{1000\ \text{J}}{1\ \text{kJ}}\right)}{160.6\ \text{J K}^{-1}} = 1110.\ \text{K}$$

75. *Answer:* **$\Delta G^\circ_f(Ca(OH)_2) = -867.8$ kJ/mol**

Strategy and Explanation: Use Equation 18.4 to determine the value of ΔG°_f sought.

$$\Delta G^\circ = (1\ \text{mol}) \times \Delta G^\circ_f(C_2H_2) + (1\ \text{mol}) \times \Delta G^\circ_f(Ca(OH)_2) - (1\ \text{mol}) \times \Delta G^\circ_f(CaC_2) - (2\ \text{mol}) \times \Delta G^\circ_f(H_2O(\ell))$$

$$\Delta G^\circ_f(Ca(OH)_2) = -\frac{\Delta G^\circ}{(1\ \text{mol})} - \Delta G^\circ_f(C_2H_2) + \Delta G^\circ_f(CaC_2) + 2 \times \Delta G^\circ_f(H_2O(\ell))$$

$$= (-119.282\ \text{kJ/mol}) - (209.20\ \text{kJ/mol}) - (-64.9\ \text{kJ/mol}) - 2 \times (-237.129\ \text{kJ/mol}) = -867.8\ \text{kJ/mol}$$

Gibbs Free Energy Changes and Equilibrium Constants (Section 18.7)

77. *Answer:* **(a) $K_p = 4 \times 10^{-34}$ (b) $K_p = 5 \times 10^{-31}$**

Strategy and Explanation: Use the method shown in the solution to Question 57(b), then apply Equation 18.8 to calculate the equilibrium constant: K°, also called, K_p, for the gas phase reaction, here.

$$\Delta G^\circ = -RT\ln K_p$$

(a) $$\Delta G^\circ = (1\ \text{mol}) \times \Delta G^\circ_f(H_2) + (1\ \text{mol}) \times \Delta G^\circ_f(Cl_2) - (2\ \text{mol}) \times \Delta G^\circ_f(HCl)$$

$$= (1\ \text{mol}) \times (0\ \text{kJ/mol}) + (1\ \text{mol}) \times (0\ \text{kJ/mol}) - (2\ \text{mol}) \times (-95.299\ \text{kJ/mol}) = 190.598\ \text{kJ}$$

$$K_p = e^{(-\Delta G^\circ/RT)} = e^{-\left(\frac{190.598\ \text{kJ}}{(0.008314\ \text{kJ/mol·K})\times(298\ \text{K})}\right)} = e^{-76.9} = 4 \times 10^{-34}$$

(b) $$\Delta G^\circ = (1\ \text{mol}) \times \Delta G^\circ_f(NO) - (1\ \text{mol}) \times \Delta G^\circ_f(N_2) - (2\ \text{mol}) \times \Delta G^\circ_f(O_2)$$

$$= (2\ \text{mol}) \times (86.55\ \text{kJ/mol}) - (1\ \text{mol}) \times (0\ \text{kJ/mol}) - (1\ \text{mol}) \times (0\ \text{kJ/mol}) = 173.10\ \text{kJ}$$

$$K_p = e^{(-\Delta G^\circ/RT)} = e^{-\left(\frac{173.10\ \text{kJ}}{(0.008314\ \text{kJ/mol·K})\times(298\ \text{K})}\right)} = e^{-69.9} = 5 \times 10^{-31}$$

79. *Answer:* **(a) –100.97 kJ; product-favored (b) 5×10^{17}; since ΔG° is negative, K is greater than 1.**

Strategy and Explanation: Use methods described in the solution to Question 77:

(a) $$\Delta G^\circ = (1\ \text{mol}) \times \Delta G^\circ_f(H_3C\text{–}CH_3) - (1\ \text{mol}) \times \Delta G^\circ_f(H_2C{=}CH_2) - (1\ \text{mol}) \times \Delta G^\circ_f(H_2)$$

$$= (1\ \text{mol}) \times (-32.82\ \text{kJ/mol}) - (1\ \text{mol}) \times (-68.15\ \text{kJ/mol}) - (1\ \text{mol}) \times (0\ \text{kJ/mol}) = -100.97\ \text{kJ}$$

The negative sign of ΔG° indicates that the reaction is product-favored.

(b) $$K_p = e^{(-\Delta G^\circ/RT)} = e^{-\left(\frac{-100.97\ \text{kJ}}{(0.008314\ \text{kJ/mol·K})\times(298\ \text{K})}\right)} = e^{+40.8} = 5 \times 10^{17}$$

When ΔG° is positive, K is less than 1. When ΔG° is negative, K is greater than 1. The latter is the case.

81. *Answer:* **(a) $K^\circ_{298} = 2 \times 10^{12}$, product-favored; $K^\circ_{1000} = 2.0 \times 10^{-2}$, reactant-favored (b) $K^\circ_{298} = 10^{135}$, product-favored; $K^\circ_{1000} = 3 \times 10^{33}$, product-favored**

Strategy and Explanation: Adapt the methods described in the solutions to Questions 45 and 77.

(a) $$\Delta H^\circ = (2\ \text{mol}) \times \Delta H^\circ_f(NO_2) - (2\ \text{mol}) \times \Delta H^\circ_f(NO) - (1\ \text{mol}) \times \Delta H^\circ_f(O_2)$$

$$= (2\ \text{mol}) \times (33.18\ \text{kJ/mol}) - (2\ \text{mol}) \times (90.25\ \text{kJ/mol}) - (1\ \text{mol}) \times (0\ \text{kJ/mol}) = -114.14\ \text{kJ}$$

$$\Delta S^\circ = (2\ \text{mol}) \times S^\circ(NO_2) - (2\ \text{mol}) \times S^\circ(NO) - (1\ \text{mol}) \times S^\circ(O_2)$$

$$= (2\ \text{mol}) \times (240.06\ \text{J K}^{-1}\text{mol}^{-1}) - (2\ \text{mol}) \times (210.76\ \text{J K}^{-1}\text{mol}^{-1})$$

$$- (1\ \text{mol}) \times (205.138\ \text{J K}^{-1}\text{mol}^{-1}) = -146.54\ \text{J K}^{-1}$$

$$\Delta G^\circ_{298} = \Delta H^\circ - T\Delta S^\circ = (-114.14\text{ kJ}) - (298\text{ K}) \times (-146.54\text{ J K}^{-1}) \times \frac{1\text{ kJ}}{1000\text{ J}} = -70.5\text{ kJ}$$

$$K^\circ_{298} = e^{(-\Delta G^\circ/RT)} = e^{-\left(\frac{-70.5\text{ kJ}}{(0.008314\text{ kJ/mol}\cdot\text{K})\times(298\text{ K})}\right)} = e^{+28.4} = 2 \times 10^{12} \quad \text{K>1, product-favored}$$

$$\Delta G^\circ_{1000} = \Delta H^\circ - T\Delta S^\circ = (-114.14\text{ kJ}) - (1000.\text{ K}) \times (-146.54\text{ J K}^{-1}) \times \frac{1\text{ kJ}}{1000\text{ J}} = 32.4\text{ kJ}$$

$$K^\circ_{1000} = e^{(-\Delta G^\circ/RT)} = e^{-\left(\frac{32.4\text{ kJ}}{(0.008314\text{ kJ/mol}\cdot\text{K})\times(1000.\text{ K})}\right)} = e^{-3.89} = 2.0 \times 10^{-2} \quad \text{K<1, reactant-favored}$$

(b) $\Delta H^\circ = (2\text{ mol}) \times \Delta H^\circ_f(\text{NaCl}) - (2\text{ mol}) \times \Delta H^\circ_f(\text{Na}) - (1\text{ mol}) \times \Delta H^\circ_f(\text{Cl}_2)$

$= (2\text{ mol}) \times (-411.153\text{ kJ/mol}) - (2\text{ mol}) \times (0\text{ kJ/mol}) - (1\text{ mol}) \times (0\text{ kJ/mol}) = -822.306\text{ kJ}$

$\Delta S^\circ = (2\text{ mol}) \times S^\circ(\text{NaCl}) - (2\text{ mol}) \times S^\circ(\text{Na}) - (1\text{ mol}) \times S^\circ(\text{Cl}_2)$

$= (2\text{ mol}) \times (72.13\text{ J K}^{-1}\text{mol}^{-1}) - (2\text{ mol}) \times (51.21\text{ J K}^{-1}\text{mol}^{-1})$

$- (1\text{ mol}) \times (223.066\text{ J K}^{-1}\text{mol}^{-1}) = -181.23\text{ J K}^{-1}$

$$\Delta G^\circ_{298} = \Delta H^\circ - T\Delta S^\circ = (-822.306\text{ kJ}) - (298\text{ K}) \times (-181.23\text{ J K}^{-1}) \times \frac{1\text{ kJ}}{1000\text{ J}} = -768.3\text{ kJ}$$

$$K^\circ_{298} = e^{(-\Delta G^\circ/RT)} = e^{-\left(\frac{-768.3\text{ kJ}}{(0.008314\text{ kJ/mol}\cdot\text{K})\times(298\text{ K})}\right)} = e^{+310.} = 10^{135} \quad \text{K>1, product-favored}$$

$$\Delta G^\circ_{1000} = \Delta H^\circ - T\Delta S^\circ = (-822.306\text{ kJ}) - (1000.\text{ K}) \times (-181.23\text{ J K}^{-1}) \times \frac{1\text{ kJ}}{1000\text{ J}} = -641.08\text{ kJ}$$

$$K^\circ_{1000} = e^{(-\Delta G^\circ/RT)} = e^{-\left(\frac{-641.1\text{ kJ}}{(0.008314\text{ kJ/mol}\cdot\text{K})\times(1000.\text{ K})}\right)} = e^{77.1} = 3 \times 10^{33} \quad \text{K>1, product-favored}$$

83. *Answer:* **(a) $K^\circ_{800} = 8.7 \times 10^{26}$, product-favored (b) $K^\circ_{500} = 7 \times 10^{10}$, product-favored (c) $K^\circ_{2000} = 1.3 \times 10^{-15}$, reactant-favored**

Strategy and Explanation: Adapt the methods described in the solution to Question 81.

(a) $\Delta H^\circ = (2\text{ mol}) \times \Delta H^\circ_f(\text{H}_2\text{O(g)}) - (2\text{ mol}) \times \Delta H^\circ_f(\text{H}_2) - (1\text{ mol}) \times \Delta H^\circ_f(\text{O}_2)$

$= (2\text{ mol}) \times (-241.818\text{ kJ/mol}) - (2\text{ mol}) \times (0\text{ kJ/mol}) - (1\text{ mol}) \times (0\text{ kJ/mol}) = -483.636\text{ kJ}$

$\Delta S^\circ = (2\text{ mol}) \times S^\circ(\text{H}_2\text{O(g)}) - (2\text{ mol}) \times S^\circ(\text{H}_2) - (1\text{ mol}) \times S^\circ(\text{O}_2)$

$= (2\text{ mol}) \times (188.825\text{ J K}^{-1}\text{mol}^{-1}) - (2\text{ mol}) \times (130.684\text{ J K}^{-1}\text{mol}^{-1})$

$- (1\text{ mol}) \times (205.138\text{ J K}^{-1}\text{mol}^{-1}) = -88.856\text{ J K}^{-1}$

$$\Delta G^\circ_{800} = \Delta H^\circ - T\Delta S^\circ = (-483.636\text{ kJ}) - (800.\text{ K}) \times (-88.856\text{ J K}^{-1}) \times \frac{1\text{ kJ}}{1000\text{ J}} = -412.6\text{ kJ}$$

$$K^\circ_{800} = e^{(-\Delta G^\circ/RT)} = e^{-\left(\frac{-412.6\text{ kJ}}{(0.008314\text{ kJ/mol}\cdot\text{K})\times(800.\text{ K})}\right)} = e^{62.0} = 8.7 \times 10^{26} \quad \text{K>1, product-favored}$$

(b) $\Delta H^\circ = (2\text{ mol}) \times \Delta H^\circ_f(\text{SO}_3) - (2\text{ mol}) \times \Delta H^\circ_f(\text{SO}_2) - (1\text{ mol}) \times \Delta H^\circ_f(\text{O}_2)$

$= (2\text{ mol}) \times (-395.72\text{ kJ/mol}) - (2\text{ mol}) \times (-296.830\text{ kJ/mol}) - (1\text{ mol}) \times (0\text{ kJ/mol}) = -197.78\text{ kJ}$

$\Delta S^\circ = (2\text{ mol}) \times S^\circ(\text{SO}_3) - (2\text{ mol}) \times S^\circ(\text{SO}_2) - (1\text{ mol}) \times S^\circ(\text{O}_2)$

$= (2\text{ mol}) \times (256.76\text{ J K}^{-1}\text{mol}^{-1}) - (2\text{ mol}) \times (248.22\text{ J K}^{-1}\text{mol}^{-1})$

$- (1\text{ mol}) \times (205.138\text{ J K}^{-1}\text{mol}^{-1}) = -188.06\text{ J K}^{-1}$

$$\Delta G^\circ_{500} = \Delta H^\circ - T\Delta S^\circ = (-197.78\text{ kJ}) - (500.\text{ K}) \times (-188.06\text{ J K}^{-1}) \times \frac{1\text{ kJ}}{1000\text{ J}} = -103.8\text{ kJ}$$

$$K^\circ_{500} = e^{(-\Delta G^\circ/RT)} = e^{-\left(\frac{-103.8\ \text{kJ}}{(0.008314\ \text{kJ/mol·K})\times(500.\ \text{K})}\right)} = e^{25.0} = 7 \times 10^{10} \quad \text{K>1, product-favored}$$

(c) $\Delta H^\circ = (1\ \text{mol}) \times \Delta H^\circ_f(H_2) + (1\ \text{mol}) \times \Delta H^\circ_f(F_2) - (2\ \text{mol}) \times \Delta H^\circ_f(HF)$

$= (1\ \text{mol}) \times (0\ \text{kJ/mol}) + (1\ \text{mol}) \times (0\ \text{kJ/mol}) - (2\ \text{mol}) \times (-271.1\ \text{kJ/mol}) = 542.2\ \text{kJ}$

$\Delta S^\circ = (1\ \text{mol}) \times S^\circ(H_2) + (1\ \text{mol}) \times S^\circ(F_2) - (2\ \text{mol}) \times S^\circ(HF)$

$= (1\ \text{mol}) \times (130.684\ \text{J K}^{-1}\text{mol}^{-1}) + (1\ \text{mol}) \times (202.78\ \text{J K}^{-1}\text{mol}^{-1})$
$- (2\ \text{mol}) \times (173.779\ \text{J K}^{-1}\text{mol}^{-1}) = -14.09\ \text{J K}^{-1}$

$$\Delta G^\circ_{2000} = \Delta H^\circ - T\Delta S^\circ = (542.2\ \text{kJ}) - (2000.\ \text{K}) \times (-14.09\ \text{J K}^{-1}) \times \frac{1\ \text{kJ}}{1000\ \text{J}} = 570.4\ \text{kJ}$$

$$K^\circ_{2000} = e^{(-\Delta G^\circ/RT)} = e^{-\left(\frac{570.4\ \text{kJ}}{(0.008314\ \text{kJ/mol·K})\times(2000.\ \text{K})}\right)} = e^{-34.30} = 1.3 \times 10^{-15} \quad \text{K<1, reactant-favored}$$

85. *Answer:* **(a) –106 kJ/mol (b) 8.55 kJ/mol (c) –33.8 kJ/mol use conversion between bar not atm LOG**

Strategy and Explanation: Adapt the strategy described in the solution to Question 77. In part (c), we will also need to use Equation 14.5 as in the solution to Question 14.147 to get K_p from K_c. Use the appropriate value of R in each equation. To convert K_p to K_c, keep in mind that standard pressure is 1 bar. 1 atm is 1.01325 bar, so R = (0.08206 L·atm/mol·K)(1atm / 1.01325 bar) = 0.083147 L·bar/mol·K

(a) $\Delta G^\circ = -RT\ln K_p = -(0.008314\ \text{kJ/mol·K}) \times (298\ \text{K}) \times \ln(4.4 \times 10^{18}) = -106\ \text{kJ/mol}$

(b) $\Delta G^\circ = -RT\ln K_p = -(0.008314\ \text{kJ/mol·K}) \times (298\ \text{K}) \times \ln(3.17 \times 10^{-2}) = 8.55\ \text{kJ/mol}$

(c) Get K_p from K_c: $K_p = K_c(RT)^{\Delta n}$ In this equation use R = 0.083147 L·bar/mol·K

Δn = 2 mol product gas – 4 mol reactant gas = –2

$\Delta G^\circ = -RT\ln K_p = -RT\ln[K_c(RT)^{-2}] = -(0.008314\ \text{kJ/mol·K}) \times (298\ \text{K}) \times$
$\ln[(0.083147\ \text{L·bar/mol·K} \times 298\ \text{K})^{-2} \times (3.5 \times 10^{8})] = -33.8\ \text{kJ/mol}$

Gibbs Free Energy, Maximum Work, and Energy Resources (Section 18.8)

87. *Answer:* **Reaction (a) can be used to do useful work; reactions (b) and (c) require work to be done.**

Strategy and Explanation: In Section 10.2 (page 428), we were told that for a gaseous substance the standard thermodynamic properties are given for a gas pressure of 1 bar. If we also assume that 25 °C is actually 25.00 °C, or 298.15 K, then we can use the standard thermodynamic table in Appendix J. We will calculate the ΔG°, the Gibbs free energy change, which is a measure of maximum useful work. If it is negative, the reaction can be harnessed to do useful work. If it is positive, the reaction cannot be harnessed to do useful work.

(a) $\Delta G^\circ = (12\ \text{mol}) \times \Delta G^\circ_f(CO_2) + (6\ \text{mol}) \times \Delta G^\circ_f(H_2O(g)) - (2\ \text{mol}) \times \Delta G^\circ_f(C_6H_6(\ell)) - (15\ \text{mol}) \times \Delta G^\circ_f(O_2)$

$= (12\ \text{mol}) \times (-394.359\ \text{kJ/mol}) + (6\ \text{mol}) \times (-228.572\ \text{kJ/mol})$
$- (2\ \text{mol}) \times (124.5\ \text{kJ/mol}) - (15\ \text{mol}) \times (0\ \text{kJ/mol}) = -6352.7\ \text{kJ}$

The reaction can be harnessed to do useful work

(b) $\Delta G^\circ = (1\ \text{mol}) \times \Delta G^\circ_f(N_2) + (3\ \text{mol}) \times \Delta G^\circ_f(F_2) - (2\ \text{mol}) \times \Delta G^\circ_f(NF_3)$

$= (1\ \text{mol}) \times (0\ \text{kJ/mol}) + (3\ \text{mol}) \times (0\ \text{kJ/mol}) - (2\ \text{mol}) \times (-83.2\ \text{kJ/mol}) = 166.4\ \text{kJ}$

The reaction requires work to be done.

(c) $\Delta G^\circ = (1\ \text{mol}) \times \Delta G^\circ_f(Ti) + (3\ \text{mol}) \times \Delta G^\circ_f(O_2) - (2\ \text{mol}) \times \Delta G^\circ_f(TiO_2)$

$= (1\ \text{mol}) \times (0\ \text{kJ/mol}) + (3\ \text{mol}) \times (0\ \text{kJ/mol}) - (1\ \text{mol}) \times (-884.5\ \text{kJ/mol}) = 884.5\ \text{kJ}$

The reaction requires work to be done.

89. *Answer:* **Reaction (b) needs 5.068 g; reactions (c) needs 26.94 g.**

Strategy and Explanation: Adapt the method described in the solution to Question 6.102. Find the Gibbs free energy of the graphite reaction and use that as a thermochemical conversion factor to supply the Gibbs free energy of the reactions that require work be done in Question 87 items (b) and (c).

$$\text{C(graphite)} + \text{O}_2\text{(g)} \longrightarrow \text{CO}_2\text{(g)}$$

This equation is identical to the equation describing the formation reaction for CO_2, so $\Delta G° = -394.359$ kJ/mol

The calculation for 87(b) looks like this:

$$166.4 \text{ kJ endergonic reaction} \times \frac{1 \text{ mol C}}{394.359 \text{ kJ provided}} \times \frac{12.011 \text{ g C}}{1 \text{ mol C}} = 5.068 \text{ g C}$$

The rest of the results are described in the table below.

Endergonic Reaction Reference	Calculated ΔG° (X) (kJ)	Mass of graphite needed (g)
87(b)	166.4 kJ	5.068
87(c)	884.5 kJ	26.94

91. *Answer:* **(a) $2\text{ C(s)} + 2\text{ Cl}_2\text{(g)} + \text{TiO}_2\text{(s)} \longrightarrow \text{TiCl}_4\text{(g)} + 2\text{ CO(g)}$ (b) $\Delta H° = -44.6$ kJ; $\Delta S° = 242.7 \text{ J K}^{-1}$; $\Delta G° = -116.5$ kJ (c) product-favored (d) more reactant-favored**

Strategy and Explanation: Balance the equation, then use the method described in the solution to Questions 51 and 57.

(a) $2\text{ C(s)} + 2\text{ Cl}_2\text{(g)} + \text{TiO}_2\text{(s)} \longrightarrow \text{TiCl}_4\text{(g)} + 2\text{ CO(g)}$ Check: 2 C, 4 Cl, 1 Ti, 2 O

(b) $\Delta X° = (1 \text{ mol}) \times X°(\text{TiCl}_4) + (2 \text{ mol}) \times X°(\text{CO}) - (2 \text{ mol}) \times X°(\text{C}) - (2 \text{ mol}) \times X°(\text{Cl}_2) - (1 \text{ mol}) \times X°(\text{TiO}_2)$ $(X = \Delta H_f, \text{S, or } \Delta G_f)$

$\Delta H° = (1 \text{ mol}) \times (-763.2 \text{ kJ/mol}) + (2 \text{ mol}) \times (-110.525 \text{ kJ/mol}) - (2 \text{ mol}) \times (0 \text{ kJ/mol}) - (2 \text{ mol}) \times (0 \text{ kJ/mol}) - (1 \text{ mol}) \times (-939.7 \text{ kJ/mol}) = -44.6 \text{ kJ}$

$\Delta S° = (1 \text{ mol}) \times (354.9 \text{ J K}^{-1}\text{mol}^{-1}) + (2 \text{ mol}) \times (197.674 \text{ J K}^{-1}\text{mol}^{-1}) - (2 \text{ mol}) \times (5.740 \text{ J K}^{-1}\text{mol}^{-1}) - (2 \text{ mol}) \times (223.066 \text{ K}^{-1}\text{mol}^{-1}) - (1 \text{ mol}) \times (49.92 \text{ J K}^{-1}\text{mol}^{-1}) = 242.7 \text{ J K}^{-1}$

$\Delta G° = (1 \text{ mol}) \times (-726.7 \text{ kJ/mol}) + (2 \text{ mol}) \times (-137.168 \text{ kJ/mol}) - (2 \text{ mol}) \times (0 \text{ kJ/mol}) - (2 \text{ mol}) \times (0 \text{ kJ/mol}) - (1 \text{ mol}) \times (-884.5 \text{ kJ/mol}) = -116.5 \text{ kJ}$

(c) $\Delta G°$ is negative, so the reaction is product-favored.

(d) As the reaction temperature increases, heat energy is added to the system. The reaction is exothermic, evolving heat, so the reaction becomes more reactant-favored,

93. *Answer:* **CuO, Ag_2O, HgO, and PbO**

Strategy and Explanation: Adapt the method used in the solution to Question 81. $(X = \Delta H_f \text{ or S})$

Kelvin temperature = 800 + 273.15 = 1.1×10^3 K

(a) $\Delta X° = (1 \text{ mol}) \times X°(\text{Cu}) + (1 \text{ mol}) \times X°(\text{CO}) - (1 \text{ mol}) \times X°(\text{CuO}) - (1 \text{ mol}) \times X°(\text{C})$

$\Delta H° = (1 \text{ mol}) \times (0 \text{ kJ/mol}) + (1 \text{ mol}) \times (-110.525 \text{ kJ/mol}) - (1 \text{ mol}) \times (-157.3 \text{ kJ/mol}) - (1 \text{ mol}) \times (0 \text{ kJ/mol}) = 46.8 \text{ kJ}$

$\Delta S° = (1 \text{ mol}) \times (33.15 \text{ J K}^{-1}\text{mol}^{-1}) + (1 \text{ mol}) \times (197.674 \text{ J K}^{-1}\text{mol}^{-1}) - (1 \text{ mol}) \times (42.63 \text{ J K}^{-1}\text{mol}^{-1}) - (1 \text{ mol}) \times (5.740 \text{ J K}^{-1}\text{mol}^{-1}) = 182.45 \text{ J K}^{-1}$

$$\Delta G°_{800} = \Delta H° - T\Delta S° = (46.8 \text{ kJ}) - (1.1 \times 10^3 \text{ K}) \times (182.45 \text{ J K}^{-1}) \times \frac{1 \text{ kJ}}{1000 \text{ J}}$$

$$= 46.8 \text{ kJ} - 2.0 \times 10^2 \text{ kJ} = -1.5 \times 10^2 \text{ kJ}$$

(b) $\Delta X° = (2\text{ mol}) \times X°(Ag) + (1\text{ mol}) \times X°(CO) - (1\text{ mol}) \times X°(Ag_2O) - (1\text{ mol}) \times X°(C)$

$\Delta H° = (2\text{ mol}) \times (0\text{ kJ/mol}) + (1\text{ mol}) \times (-110.525\text{ kJ/mol})$
$- (1\text{ mol}) \times (-31.05\text{ kJ/mol}) - (1\text{ mol}) \times (0\text{ kJ/mol}) = -79.47\text{ kJ}$

$\Delta S° = (2\text{ mol}) \times (42.55\text{ J K}^{-1}\text{mol}^{-1}) + (1\text{ mol}) \times (197.674\text{ J K}^{-1}\text{mol}^{-1})$
$- (1\text{ mol}) \times (121.3\text{ J K}^{-1}\text{mol}^{-1}) - (1\text{ mol}) \times (5.740\text{ J K}^{-1}\text{mol}^{-1}) = 155.7\text{ J K}^{-1}$

This is an exothermic reaction ($\Delta H°$ = negative) with a positive entropy change, so it will be spontaneous at any temperature, including 800. °C.

(c) $\Delta X° = (1\text{ mol}) \times X°(Hg) + (1\text{ mol}) \times X°(CO) - (1\text{ mol}) \times X°(HgO) - (1\text{ mol}) \times X°(C)$ $(X = \Delta H_f \text{ or } S)$

$\Delta H° = (1\text{ mol}) \times (0\text{ kJ/mol}) + (1\text{ mol}) \times (-110.525\text{ kJ/mol})$
$- (1\text{ mol}) \times (-90.83\text{ kJ/mol}) - (1\text{ mol}) \times (0\text{ kJ/mol}) = -19.70\text{ kJ}$

$\Delta S° = (1\text{ mol}) \times (29.87\text{ J K}^{-1}\text{mol}^{-1}) + (1\text{ mol}) \times (197.674\text{ J K}^{-1}\text{mol}^{-1})$
$- (1\text{ mol}) \times (70.29\text{ J K}^{-1}\text{mol}^{-1}) - (1\text{ mol}) \times (5.740\text{ J K}^{-1}\text{mol}^{-1}) = 151.51\text{ J K}^{-1}$

This is an exothermic reaction ($\Delta H°$ = negative) with a positive entropy change, so it will be spontaneous at any temperature, including 800. °C.

(d) $\Delta X° = (1\text{ mol}) \times X°(Mg) + (1\text{ mol}) \times X°(CO) - (1\text{ mol}) \times X°(MgO) - (1\text{ mol}) \times X°(C)$

$\Delta H° = (1\text{ mol}) \times (0\text{ kJ/mol}) + (1\text{ mol}) \times (-110.525\text{ kJ/mol})$
$- (1\text{ mol}) \times (-601.70\text{ kJ/mol}) - (1\text{ mol}) \times (0\text{ kJ/mol}) = 491.18\text{ kJ}$

$\Delta S° = (1\text{ mol}) \times (32.68\text{ J K}^{-1}\text{mol}^{-1}) + (1\text{ mol}) \times (197.674\text{ J K}^{-1}\text{mol}^{-1})$
$- (1\text{ mol}) \times (26.94\text{ J K}^{-1}\text{mol}^{-1}) - (1\text{ mol}) \times (5.740\text{ J K}^{-1}\text{mol}^{-1}) = 197.67\text{ J K}^{-1}$

$$\Delta G°_{800} = \Delta H° - T\Delta S° = (491.18\text{ kJ}) - (1.1 \times 10^3\text{ K}) \times (197.67\text{ J K}^{-1}) \times \frac{1\text{ kJ}}{1000\text{ J}}$$
$$= 491.18\text{ kJ} - 2.1 \times 10^2\text{ kJ} = 2.8 \times 10^2\text{ kJ}$$

(e) $\Delta X° = (1\text{ mol}) \times X°(Pb) + (1\text{ mol}) \times X°(CO) - (1\text{ mol}) \times X°(PbO) - (1\text{ mol}) \times X°(C)$

$\Delta H° = (1\text{ mol}) \times (0\text{ kJ/mol}) + (1\text{ mol}) \times (-110.525\text{ kJ/mol})$
$- (1\text{ mol}) \times (-217.32\text{ kJ/mol}) - (1\text{ mol}) \times (0\text{ kJ/mol}) = 106.80\text{ kJ}$

$\Delta S° = (1\text{ mol}) \times (64.81\text{ J K}^{-1}\text{mol}^{-1}) + (1\text{ mol}) \times (197.674\text{ J K}^{-1}\text{mol}^{-1})$
$- (1\text{ mol}) \times (68.7\text{ J K}^{-1}\text{mol}^{-1}) - (1\text{ mol}) \times (5.740\text{ J K}^{-1}\text{mol}^{-1}) = 188.0\text{ J K}^{-1}$

$$\Delta G°_{800} = \Delta H° - T\Delta S° = (106.80\text{ kJ}) - (1.1 \times 10^3\text{ K}) \times (188.0\text{ J K}^{-1}) \times \frac{1\text{ kJ}}{1000\text{ J}}$$
$$= 106.80\text{ kJ} - 2.0 \times 10^2\text{ kJ} = -1 \times 10^2\text{ kJ}$$

The coupled reactions that have negative $\Delta G°$ can be used to produce the respective metals, so CuO, Ag_2O, HgO, and PbO can be used to obtain Cu, Ag, Hg, and Pb, respectively, by this method at 800 °C.

Gibbs Free Energy in Biological Systems (Section 18.9)

95. *Answer:* **(a) $\Delta H° \cong -2873$ kJ; Broken: 5 O–H, 7 C–O, 7 C–H, 5 C–C, 6 O=O; formed: 12 C=O, 12 O–H (b) It is close; intermolecular forces in solid glucose and liquid water are being neglected.**

Strategy and Explanation: Adapt the method described in Section 8.6.

(a) Looking at the given ball-and-stick model structure we see that we must break five mol of O–H bonds, seven mol of C–O bonds, seven mol of C–H bonds, and five mol of C–C bonds in glucose. We must also break six mol of O=O bonds. Two C=O bonds in each of six mol of CO_2 and two O–H bonds in each of six mol of H_2O are formed.

The enthalpy of a reaction is approximated by adding the energy required to break one mole of bonds of each type of bond broken, described by the bond energies (D), to the energy required to form one mole of bonds of each type of bonds produced in the reactants, described by the negative of the bond energies (–D).

$\Delta H° \cong$ (bonds broken in glucose) + (bonds broken in O_2)
+ (bonds formed in CO_2) + (bonds formed in H_2O)

$= (5\text{ mol} \times D_{O\text{–}H} + 7\text{ mol} \times D_{C\text{–}O} + 7\text{ mol} \times D_{C\text{–}H} + 5\text{ mol} \times D_{C\text{–}C})$

$+ 6\text{ mol} \times D_{O=O} - 6 \times (2\text{ mol} \times D_{C=O}) - 6 \times (2\text{ mol} \times D_{O\text{–}H})$

Notice that the number of moles multiplied by the energy per mol (in kJ/mol) gives the result for each term in kJ.

$\Delta H° \cong 5\,D_{O\text{–}H} + 7\,D_{C\text{–}O} + 7\,D_{C\text{–}H} + 5\,D_{C\text{–}C} + 6\,D_{O=O} - 12\,D_{C=O} - 12\,D_{O\text{–}H}$

$= 7\,D_{C\text{–}O} + 7\,D_{C\text{–}H} + 5\,D_{C\text{–}C} + 6\,D_{O=O} - 12\,D_{C=O} - 7\,D_{O\text{–}H}$

Now use the data in Table 8.2.

$= 7 \times (336\text{ kJ}) + 7 \times (416\text{ kJ}) + 5 \times (356\text{ kJ}) + 6 \times (498\text{ kJ})$

$- 12 \times (803\text{kJ}) - 7 \times (467\text{ kJ}) = -2873\text{ kJ} \cong \Delta H°$

(b) The actual ΔH° (–2816 kJ) is close to the estimated value in (a). Intermolecular forces in condensed phases (solid glucose and liquid water) are being neglected in this calculation, which could explain the discrepancy. Also, the bond enthalpies given in Table 8.2 are average values; the actual enthalpy of a particular bond in a particular molecule may be slightly greater or less depending upon other factors.

97. *Answer:* **(a) 6.46 mol ATP/mol glucose (b) –106 kJ (c) product-favored**

Strategy and Explanation:

(a) This is a conversion factor question:

$$\frac{197\text{ kJ produced}}{1\text{ mol glucose}} \times \frac{1\text{ mol ATP}}{30.5\text{ kJ needed}} = 6.46\text{ mol ATP per mol of glucose}$$

(b) This is a conversion factor question:

$$\frac{3\text{ mol ATP produced}}{1\text{ mol glucose}} \times \frac{30.5\text{ kJ needed}}{1\text{ mol ATP}} = 91.5\text{ kJ per mol of glucose}$$

The actual reaction must be less exergonic than the given reaction to produce fewer ATP.

$$\Delta G°_{\text{overall reaction}} = \Delta G°_{\text{conversion}} + \Delta G°_{\text{ATP}}$$

$$\Delta G°_{\text{overall reaction}} = -197\text{ kJ} + 91.5\text{ kJ} = -106\text{ kJ}$$

(c) The overall reaction in part (b) has a negative Gibbs free energy change, so it is product-favored.

Conservation of Gibbs Free Energy (Section 18.10)

99. *Answer/Explanation:* Food we eat provides us with a supply of Gibbs free energy. Coal, petroleum, and natural gas are the most common fuel sources used to supply Gibbs free energy by combustion. We also use solar and nuclear energy, as well as the kinetic energy of wind and water.

Thermodynamic and Kinetic Stability (Section 18.11)

101. *Answer:* **(a) ΔG° = –86.5 kJ so the reaction is product-favored (b) ΔG° = –873.1 kJ (c) No (d) Yes**

Strategy and Explanation: Balance the equation, when not provided, and adapt the method in Question 57(b).

(a) $\Delta G° = (1\text{ mol}) \times \Delta G_f°(CH_3COOH) - (1\text{ mol}) \times \Delta G_f°(CH_3OH) - (1\text{ mol}) \times \Delta G_f°(CO)$

$= (1\text{ mol}) \times (-389.9\text{ kJ/mol})$

$- (1\text{ mol}) \times (-166.27\text{ kJ/mol}) - (1\text{ mol}) \times (-137.168\text{ kJ/mol}) = -86.5\text{ kJ}$

Because ΔG° is negative, the reaction is product-favored.

(b) $$CH_3COOH(\ell) + 2\,O_2(g) \longrightarrow 2\,CO_2(g) + 2\,H_2O(\ell)$$

$\Delta G° = (2\text{ mol}) \times \Delta G_f°(CO_2) + (2\text{ mol}) \times \Delta G_f°(H_2O(\ell))$

$- (1\text{ mol}) \times \Delta G_f°(CH_3COOH) - (2\text{ mol}) \times \Delta G_f°(O_2)$

$= (2\text{ mol}) \times (-394.359\text{ kJ/mol}) + (2\text{ mol}) \times (-237.129\text{ kJ/mol})$

$- (1\text{ mol}) \times (-389.9\text{ kJ/mol}) - (2\text{ mol}) \times (0\text{ kJ/mol}) = -873.1\text{ kJ}$

(c) Based on the answer to (b), the products of oxidation are more stable, so acetic acid is not thermodynamically stable.

(d) Acetic acid can be kept both in liquid form and in solution form if stored properly. In the presence of air, it does not explode, so we will classify it as kinetically stable.

103. *Answer:* **$CH_4(g) + 2\ O_2(g) \longrightarrow CO_2(g) + 2\ H_2O(\ell)$; $\Delta G° = -817.90$ kJ, therefore product-favored; $C_6H_6(g) + \frac{15}{2}\ O_2(g) \longrightarrow 6\ CO_2(g) + 3\ H_2O(\ell)$; $\Delta G° = -3202.0$ kJ, therefore product-favored; $CH_3OH(\ell) + \frac{3}{2}\ O_2(g) \longrightarrow CO_2(g) + 2\ H_2O(\ell)$; $\Delta G° = -702.34$ kJ, therefore product-favored; complex molecules require significant rearrangement of atoms, indicating kinetic stability is likely.**

Strategy and Explanation: Balance the equations, and adapt the methods from the solution to Question 57(b).

$$CH_4(g) + 2\ O_2(g) \longrightarrow CO_2(g) + 2\ H_2O(\ell)$$

$$\Delta G° = (1\text{ mol}) \times \Delta G°_f(CO_2) + (2\text{ mol}) \times \Delta G°_f(H_2O(\ell)) - (1\text{ mol}) \times \Delta G°_f(CH_4) - (2\text{ mol}) \times \Delta G°_f(O_2)$$

$$= (1\text{ mol}) \times (-394.359\text{ kJ/mol}) + (2\text{ mol}) \times (-237.129\text{ kJ/mol})$$

$$- (1\text{ mol}) \times (-50.72\text{ kJ/mol}) - (2\text{ mol}) \times (0\text{ kJ/mol}) = -817.90\text{ kJ for } CH_4(g)$$

Therefore the $CH_4(g)$ combustion reaction is product-favored.

$$C_6H_6(g) + \frac{15}{2}\ O_2(g) \longrightarrow 6\ CO_2(g) + 3\ H_2O(\ell)$$

$$\Delta G° = (6\text{ mol}) \times \Delta G°_f(CO_2) + (3\text{ mol}) \times \Delta G°_f(H_2O(\ell)) - (1\text{ mol}) \times \Delta G°_f(CH_4) - (\tfrac{15}{2}\text{ mol}) \times \Delta G°_f(O_2)$$

$$= (6\text{ mol}) \times (-394.359\text{ kJ/mol}) + (3\text{ mol}) \times (-237.129\text{ kJ/mol})$$

$$- (1\text{ mol}) \times (124.5\text{ kJ/mol}) - (\tfrac{15}{2}\text{ mol}) \times (0\text{ kJ/mol}) = -3202.0\text{ kJ for } C_6H_6(g)$$

Therefore the $C_6H_6(g)$ combustion reaction is product-favored.

$$CH_3OH(\ell) + \frac{3}{2}\ O_2(g) \longrightarrow CO_2(g) + 2\ H_2O(\ell)$$

$$\Delta G° = (1\text{ mol}) \times \Delta G°_f(CO_2) + (2\text{ mol}) \times \Delta G°_f(H_2O(\ell)) - (1\text{ mol}) \times \Delta G°_f(CH_3OH) - (\tfrac{3}{2}\text{ mol}) \times \Delta G°_f(O_2)$$

$$= (1\text{ mol}) \times (-394.359\text{ kJ/mol}) + (2\text{ mol}) \times (-237.129\text{ kJ/mol})$$

$$- (1\text{ mol}) \times (-166.27\text{ kJ/mol}) - (\tfrac{3}{2}\text{ mol}) \times (0\text{ kJ/mol}) = -702.34\text{ kJ for } CH_3OH(\ell)$$

Therefore the $CH_3OH(\ell)$ combustion reaction is product-favored.

Organic compounds are complex molecular systems that require significant rearrangement of atoms and bonds to undergo combustion. This makes them likely candidates for being kinetically stable.

General Questions

105. *Answer:* **(a) 5.5×10^{18} J/yr (b) 1.5×10^{16} J/day (c) 1.7×10^{11} J/s (d) 1.7×10^{11} W (e) 6×10^2 W/person**

Strategy and Explanation: Use standard conversion factors to solve this question.

Agriculture, mining and construction industries are represented on the graph with approximately 5.2 quadrillion BTUs per year. One quadrillion is a thousand times more than a trillion, or 10^{15}. One BTU (British thermal unit) is defined in Appendix B in Table B.4. as 1055.06 J.

(a) $$\frac{5.2 \times 10^{15}\text{ BTU}}{\text{yr}} \times \frac{1055.06\text{ J}}{1\text{ BTU}} = \frac{5.5 \times 10^{18}\text{ J}}{\text{yr}}$$

(b) $$\frac{5.5 \times 10^{18}\text{ J}}{\text{yr}} \times \frac{1\text{ year}}{365\text{ day}} = \frac{1.5 \times 10^{16}\text{ J}}{\text{day}}$$

(c) $$\frac{1.5 \times 10^{16}\text{ J}}{\text{day}} \times \frac{1\text{ day}}{24\text{ hr}} \times \frac{1\text{ hr}}{3600\text{ s}} = \frac{1.7 \times 10^{11}\text{ J}}{\text{s}}$$

(d) $$\frac{1.7\times10^{11}\ \text{J}}{\text{s}}\times\frac{1\ \text{W}}{1\ \text{J/s}}=1.7\times10^{11}\ \text{W}$$

(e) $$\frac{1.7\times10^{11}\ \text{W}}{307\times10^{6}\ \text{persons}}=6\times10^{2}\ \frac{\text{W}}{\text{person}}$$

107. *Answer:* **(a) Reaction 2 (b) Reactions 1 & 5 (c) Reaction 2 (d) Reactions 2 & 3 (e) None**

Strategy and Explanation: Adapt the method described in the solution to Question 85(c) and concepts described in Sections 14.2 and 14.5.

(a) Because of the relationship: $K_p = K_c(RT)^{\Delta n}$, all we need to do is check Δn for each reaction. K_p is larger than K_c when Δn is positive, as long as RT >1(RT <1 only below T = 12 K!)

Reaction	Δn	$K_p > K_c$
1	2 mol product gas – 2 mol reactant gas = 0	No
2	1 mol product gas – 0 mol reactant gas = 1	Yes
3	2 mol product gas – 2 mol reactant gas = 0	No
4	1 mol product gas – 3 mol reactant gas = –2	No
5	2 mol product gas – 2 mol reactant gas = 0	No

Only reaction 2 has K_p larger than K_c.

(b) Now we actually need the value of K_p, because $K_p = K° > 1$ for a gas-phase reaction that is product-favored.

$$(RT) = (0.08206\ \text{L atm K}^{-1}\text{mol}^{-1})(298\ \text{K}) = 24.5$$

$$K_p = K_c(RT)^{\Delta n}$$

Reaction	$(RT)^{\Delta n}$	K_c	K_p	Product-favored
1	1	3.6×10^{20}	3.6×10^{20}	Yes
2	$(24.5)^1$	1.24×10^{-5}	3.03×10^{-4}	No
3	1	9.5×10^{-13}	9.5×10^{-13}	No
4	$(24.5)^{-2}$	3.76	6.29×10^{-3}	No
5	1	1.9×10^{24}	1.9×10^{24}	Yes

Only reactions 1 and 5 are product-favored.

(c) Every gas contributes a concentration to the K_c expression, as described in Section 14.2. Only one reaction given has just one gas-phase component: Reaction 2.

(d) The product concentrations get higher when the value of K_c increases. In Section 14.6, we learned that the K_c increases with an increase in temperature when the reaction is endothermic. So, reactions 2 and 3 will have an increase in the product concentrations when the temperature increases.

(e) Only two reactions (1 and 2) have a product that is H_2O. The difference between $\Delta G_f^\circ(H_2O(\ell))$ and $\Delta G_f^\circ(H_2O(g))$ is 8.557 kJ/mol, so the switch would make the reaction ΔG° less negative by 8.557 kJ. If these reactions have positive ΔG° between 8.557 kJ and zero, then the sign would switch. If these reactions have negative ΔG° values with gas-phase water, then the ΔG° will still be negative with liquid water.

Reaction	K_p	ΔG° = – RTlnK°	ΔG° sign affected by (ℓ) to (g)?
1	3.6×10^{20}	– 117	No
2	3.03×10^{-4}	20.1	No

None of these has a ΔG° whose sign is affected by a switch from $H_2O(g)$ to $H_2O(\ell)$.

109. *Answer:* **(a) 141.73 kJ (b) no (c) yes (d) $K_p = 1 \times 10^4$ (d) $K_c = 7 \times 10^1$**

Strategy and Explanation: $2\ SO_3(g) \longrightarrow 2\ SO_2(g) + O_2(g)$

(a) $\Delta G° = (2\text{ mol}) \times \Delta G°_f(SO_2) + (1\text{ mol}) \times \Delta G°_f(O_2) - (2\text{ mol}) \times \Delta G°_f(SO_3)$

$= (2\text{ mol}) \times (-300.194\text{ kJ/mol}) + (1\text{ mol}) \times (0\text{ kJ/mol}) - (2\text{ mol}) \times (-371.06\text{ kJ/mol}) = 141.73\text{ kJ}$

(b) The positive Gibbs free energy change indicates that the reaction is not product-favored at 25 °C.

(c) $\Delta H° = (2\text{ mol}) \times \Delta H°_f(SO_2) + (1\text{ mol}) \times \Delta H°_f(O_2) - (2\text{ mol}) \times \Delta H°_f(SO_3)$

$= (2\text{ mol}) \times (-296.830\text{ kJ/mol}) + (1\text{ mol}) \times (0\text{ kJ/mol}) - (2\text{ mol}) \times (-395.72\text{ kJ/mol}) = 197.78\text{ kJ}$

$\Delta S° = (2\text{ mol}) \times S°(SO_2) + (1\text{ mol}) \times S°(O_2) - (2\text{ mol}) \times S°(SO_3)$

$= (2\text{ mol}) \times (248.22\text{ J K}^{-1}\text{mol}^{-1}) + (1\text{ mol}) \times (205.138\text{ J K}^{-1}\text{mol}^{-1})$

$- (2\text{ mol}) \times (256.76\text{ J K}^{-1}\text{mol}^{-1}) = 188.06\text{ J K}^{-1}$

With both $\Delta H°$ and $\Delta S°$ positive, the reaction will be product-favored at some high temperature, one higher than 25 °C.

(d) $T = 1500\ °C + 273.15 = 1773\text{ K} = 1.8 \times 10^3\text{ K}$ *(round to hundreds place)*

$$\Delta G°_{1773} = \Delta H° - T\Delta S° = (197.78\text{ kJ}) - (1.8 \times 10^3\text{ K}) \times (188.06\text{ J K}^{-1}) \times \frac{1\text{ kJ}}{1000\text{ J}} = -1.4 \times 10^2\text{ kJ}$$

$$K°_{1773} = e^{(-\Delta G°/RT)} = e^{-\left(\frac{-1.4\times10^2\text{ kJ}}{(0.008314\text{ kJ/mol}\cdot\text{K})\times(1.8\times10^3\text{ K})}\right)} = e^{+9.2} = 1 \times 10^4$$

(e) $K_c = K_p(RT)^{-\Delta n}$ $\quad\Delta n = 3\text{ mol} - 2\text{ mol} = 1\text{ mol}$

$$K_c = (1 \times 10^4) \times [(0.08206\text{ L atm K}^{-1}\text{mol}^{-1}) \times (1.8 \times 10^3\text{ K})]^{-1} = 7 \times 10^1$$

111. *Answer:* **(a) $\Delta G° = 31.8$ kJ (b) $K_p = P_{Hg(g)}$ (c) $K° = 2.7 \times 10^{-6}$ (d) 2.7×10^{-6} atm (e) 450 K**

Strategy and Explanation: Use the method described in the solution to Question 77. Then use the equation derived in the solution to Question 138.

(a) $\Delta G° = (1\text{ mol}) \times \Delta G°_f(Hg(g)) - (1\text{ mol}) \times \Delta G°_f(Hg(\ell))$

$= (1\text{ mol}) \times (31.8\text{ kJ/mol}) - (1\text{ mol}) \times (0\text{ kJ/mol}) = 31.8\text{ kJ}$

(b) $K° = K_p = P_{Hg(\ell)}$

(c) $$K° = e^{(-\Delta G°/RT)} = e^{-\left(\frac{31.8\text{ kJ/mol}}{(0.008314\text{ kJ/mol}\cdot\text{K})\times(298\text{ K})}\right)} = e^{-12.8} = 10^{-5.56} = 2.7 \times 10^{-6}$$

(d) $K° = K_p = P_{Hg(\ell)} = 2.7 \times 10^{-6}$ atm

(e) $K°_1 = 2.7 \times 10^{-6}$ atm at $T_1 = 298$ K, $K°_2 = P_{Hg(\ell)} = 10$ mm Hg at T_2

$\Delta H° = (1\text{ mol}) \times \Delta H°_f(Hg(g)) - (1\text{ mol}) \times \Delta H°_f(Hg(\ell))$

$= (1\text{ mol}) \times (61.4\text{ kJ/mol}) - (1\text{ mol}) \times (0\text{ kJ/mol}) = 61.4\text{ kJ}$

$$\ln\left(\frac{K°_1}{K°_2}\right) = \frac{\Delta H°}{R}\left(\frac{1}{T_2} - \frac{1}{T_1}\right)$$

$$\ln\left(\frac{2.7\times10^{-6}\text{ atm}}{10\text{ mmHg}\times\frac{1\text{ atm}}{760\text{ mmHg}}}\right) = \frac{61.4\text{kJ/mol}}{0.008314\text{kJ/mol}\cdot\text{K}}\left(\frac{1}{T_2} - \frac{1}{298\text{ K}}\right)$$

Solve for T_2:

$$T_2 = 450\text{ K}$$

Applying Concepts

113. *Answer/Explanation:* If you don't know the Humpty Dumpty poem, it should be possible to look it up at http://en.wikipedia.org/wiki/Humpty_Dumpty. Humpty Dumpty is a fictional character who was also an egg. A scrambled egg is a very disordered state for an egg. The second law of thermodynamics says that the more disordered state is the more probable state. Putting the delicate tissues and fluids back where they were before the scrambling occurred would take a great deal of energy. Humpty Dumpty fell off a wall. A very probable result of that fall is for the egg to become scrambled. The story goes on to tell that all the energy of the king's horses and men was not sufficient to put Humpty together again.

114. *Answer/Explanation:* It is possible to define conditions for which the entropy of a substance has its lowest possible value, namely zero at T = 0 K, so absolute entropies can be determined. It is not possible to define conditions for a specific minimum value for internal energy, enthalpy, or Gibbs free energy of a substance, so relative quantities must be used.

116. *Answer:* **exothermic; product-favored**

Strategy and Explanation: Many of the oxides have negative enthalpies of reaction, which means their oxidations are exothermic. These are probably product-favored reactions.

117. *Answer/Explanation:* NaCl, in an orderly crystal structure, and pure water, with only O–H hydrogen bonding interactions in the liquid state, are far more ordered than the dispersed hydrated sodium and chloride ions interacting with the water molecules.

119. *Answer:* **(a) false (b) false (c) true (d) true (e) true**

Strategy and Explanation: Use Equation 18.3 ($\Delta G° = \Delta H° - T\Delta S°$) and Equation 18.8 ($\Delta G° = -RT\ln K°$) to relate equilibrium constant to thermodynamic values. When K = 1.0, then $\ln K = \ln(1.0) = 0.0$, so $\Delta G° = 0.0$ and $\Delta H° - T\Delta S° = 0.0$

(a) It is false to say that a chemical reaction with K = 1.0 has $\Delta H° = 0$, unless $\Delta S° = 0$, also (this is unlikely to be true, except in the trivial cases where the reaction has the same reactants as products).

(b) It is false to say that a chemical reaction with K = 1.0 has $\Delta S° = 0$, unless $\Delta H° = 0$, also (this is unlikely to be true, except in the trivial cases where the reaction has the same reactants as products).

(c) It is true to say that a chemical reaction with K = 1.0 has $\Delta G° = 0$.

(d) It is true to say that $\Delta H°$ and $\Delta S°$ have equal sign, since $\Delta H°$ must have the same sign as $T\Delta S°$ and T is always positive.

(e) It is true to say that $\Delta H°/T = \Delta S°$ at the temperature T. The relationship is obtained when we solve this equation: $\Delta H° - T\Delta S° = 0$ for $\Delta S°$.

121. *Answer/Explanation:* Living organisms have a high degree of order, which results in their having relatively low entropy. As a result, they must contain a large amount of free energy to maintain this low-probability state. During metabolism, the energy obtained from nutrients is stored as ATP. Conversion of ATP to ADP is an exergonic reaction. The energy from this reaction is used to drive the endergonic reactions needed to maintain the organized state of the organism. The original source of the energy needed to synthesize the sugars was sunlight used by plants to produce the sugars and other carbohydrates.

123. *Answer/Explanation:* $\Delta G < 0$ means products are favored; however, the equilibrium state will always have some reactants present, too. To get all the reactants to go away requires the removal of the products from the reactants, so that the reaction continues forward.

125. *Answer:* **(a) Case (c) (b) Case (a)**

Strategy and Explanation: We are given the same molar mass and the same number of molecules (five).

(a) The most complex of the molecules (with the most atoms bonded together in one molecule) will have the greatest opportunity for complex motion, so **Case (c)** has the highest entropy.

(b) The least complex of the species shown are the diatomic molecules. Therefore, **Case (a)** has the lowest entropy.

More Challenging Questions

127. *Answer:* **(a) $\Delta S°$ =58.78 J/K (b) $\Delta S°$ = –53.29 J/K (c) $\Delta S°$ = –173.93 J/K; Adding more H atoms decreases the ΔS (i.e., makes it more negative).**

Strategy and Explanation: Adapt the method described in the solution to Question 39. Formation is defined as producing one mol of a substance from standard state elements.

(a) $2\ C(graphite) + H_2(g) \longrightarrow C_2H_2(g)$

$\Delta S° = (1\ mol) \times S°(C_2H_2) - (2\ mol) \times S°(C) - (1\ mol) \times S°(H_2)$

$= (1\ mol) \times (200.94\ J\ K^{-1}mol^{-1}) - (2\ mol) \times (5.740\ J\ K^{-1}mol^{-1})$

$- (1\ mol) \times (130.684\ J\ K^{-1}mol^{-1}) = 58.78\ J\ K^{-1}$

(b) $2\ C(graphite) + 2\ H_2(g) \longrightarrow C_2H_4(g)$

$\Delta S° = (1\ mol) \times S°(C_2H_4) - (2\ mol) \times S°(C) - (2\ mol) \times S°(H_2)$

$= (1\ mol) \times (219.56\ J\ K^{-1}mol^{-1}) - (2\ mol) \times (5.740\ J\ K^{-1}mol^{-1})$

$- (2\ mol) \times (130.684\ J\ K^{-1}mol^{-1}) = -53.29\ J\ K^{-1}$

(c) $2\ C(graphite) + 3\ H_2(g) \longrightarrow C_2H_6(g)$

$\Delta S° = (1\ mol) \times S°(C_2H_6) - (2\ mol) \times S°(C) - (3\ mol) \times S°(H_2)$

$= (1\ mol) \times (229.60\ J\ K^{-1}mol^{-1}) - (2\ mol) \times (5.740\ J\ K^{-1}mol^{-1})$

$- (3\ mol) \times (130.684\ J\ K^{-1}mol^{-1}) = -173.93\ J\ K^{-1}$

Adding more H atoms decreases the ΔS (i.e., makes it more negative).

129. *Answer:* **product-favored; see explanation below**

Strategy and Explanation: Adapt the method described in the solution to Questions 39 and 35.

$\Delta H° = (1\ mol) \times \Delta H°_f(C_8H_{18}) - (1\ mol) \times \Delta H°_f(C_8H_{16}) - (1\ mol) \times \Delta H°_f(H_2)$

$= (1\ mol) \times (-208.45\ kJ/mol) - (1\ mol) \times (-82.93\ kJ/mol) - (1\ mol) \times (0\ kJ/mol) = -125.52\ kJ$

We do not have $S°(C_8H_{16})$, but we can assume that it is very close to that of $S°(C_8H_{18})$. The flexibility and freedom of motion of the atoms in these large eight-carbon-chain molecules is probably about the same except near the one double bond in 1-octene. In support of this presumption, we compare $S°(C_2H_6)$ and $S°(C_2H_4)$, which are both larger than 200 $J\ K^{-1}mol^{-1}$ and differ by only about 10 $J\ K^{-1}mol^{-1}$.

$\Delta S° = (1\ mol) \times S°(C_8H_{18}) - (1\ mol) \times S°(C_8H_{16}) - (1\ mol) \times S°(H_2)$

$\cong -(1\ mol) \times S°(H_2) = -(1\ mol) \times (130.684\ J\ K^{-1}mol^{-1}) = -130.684\ J\ K^{-1}$

We can determine the temperature at which this reaction becomes reactant-favored.

$$T \cong \frac{(-125.52\ kJ) \times \left(\frac{1000\ J}{1\ kJ}\right)}{-130.684\ J\ K^{-1}} = 960.48\ K$$

The given temperature of 25 °C (298K) is much lower, so the reaction is confirmed to be product-favored.

(We could also have estimated the $\Delta G°$ for this reaction and shown that it is negative.)

131. *Answer:* **(a) 331.51 K (b) 371 K**

Strategy and Explanation: Adapt methods described in the solution to Question 39.

(a) $Br_2(\ell) \longrightarrow Br_2(g)$

$\Delta H° = (1\ mol) \times \Delta H°_f(Br_2(g)) - (1\ mol) \times \Delta H°_f(Br_2(\ell))$

$= (1\ mol) \times (30.907\ kJ/mol) - (1\ mol) \times (0\ kJ/mol) = 30.907\ kJ$

$\Delta S° = (1\ mol) \times S°(Br_2(g)) - (1\ mol) \times S°(Br_2(\ell))$

$= (1 \text{ mol}) \times (245.463 \text{ J K}^{-1}\text{mol}^{-1}) - (1 \text{ mol}) \times (152.231 \text{ K}^{-1}\text{mol}^{-1}) = 93.232 \text{ J K}^{-1}$

$$T = \frac{(30.907 \text{ kJ}) \times \left(\frac{1000 \text{ J}}{1 \text{ kJ}}\right)}{93.232 \text{ J K}^{-1}} = 331.51 \text{ K}$$

(b) $SnCl_4(\ell) \longrightarrow SnCl_4(g)$

$\Delta H° = (1 \text{ mol}) \times \Delta H°_f(SnCl_4(g)) - (1 \text{ mol}) \times \Delta H°_f(SnCl_4(\ell))$

$= (1 \text{ mol}) \times (-471.5 \text{ kJ/mol}) - (1 \text{ mol}) \times (-511.3 \text{ kJ/mol}) = 39.8 \text{ kJ}$

$\Delta S° = (1 \text{ mol}) \times S°(SnCl_4(g)) - (1 \text{ mol}) \times S°(SnCl_4(\ell))$

$= (1 \text{ mol}) \times (365.8 \text{ J K}^{-1}\text{mol}^{-1}) - (1 \text{ mol}) \times (258.6 \text{ K}^{-1}\text{mol}^{-1}) = 107.2 \text{ J K}^{-1}$

$$T = \frac{(39.8 \text{ kJ}) \times \left(\frac{1000 \text{ J}}{1 \text{ kJ}}\right)}{107.2 \text{ J K}^{-1}} = 371 \text{ K}$$

133. *Answer:* **(a) $K_p = 1.5 \times 10^7$ (b) product-favored (c) $K_c = 3.7 \times 10^8$**

Strategy and Explanation: Adapt methods described in the solution to Questions 77 and 85.

(a) $\Delta G° = (2 \text{ mol}) \times \Delta G°_f(NOCl) + (2 \text{ mol}) \times \Delta G°_f(NO) - (1 \text{ mol}) \times \Delta G°_f(Cl_2)$

$= (2 \text{ mol}) \times (66.08 \text{ kJ/mol}) + (2 \text{ mol}) \times (86.55 \text{ kJ/mol}) - (1 \text{ mol}) \times (0 \text{ kJ/mol}) = -40.94 \text{ kJ}$

$$K_p = e^{(-\Delta G°/RT)} = e^{-\left(\frac{-40.94 \text{ kJ}}{(0.008314 \text{ kJ/mol·K}) \times (298 \text{ K})}\right)} = e^{16.5} = 10^{7.16} = 1.5 \times 10^7$$

(b) The negative Gibbs free energy change indicates that the reaction is product-favored.

(c) Get K_c from K_p: $K_p = K_c(RT)^{\Delta n}$

$\Delta n = 2 \text{ mol product gas} - 3 \text{ mol reactant gas} = -1$

$K_c = K_p(RT)^{-\Delta n} = (1.5 \times 10^7) \times (0.08206 \text{ L·atm/mol·K} \times 298 \text{ K})^{-(-1)} = 3.7 \times 10^8$

135. *Answer:* **(a) See equations below (b) C_2H_6 reaction: $\Delta H° = 101.02$ kJ; $\Delta S° = -72.71$ J/K; $\Delta G° = 122.72$ kJ; C_3H_8 reaction: $\Delta H° = 76.5$ kJ; $\Delta S° = -86.6$ J/K; $\Delta G° = 102.08$ kJ; CH_3OH reaction: $\Delta H° = 96.44$ kJ; $\Delta S° = -305.2$ J/K; $\Delta G° = 187.39$ kJ; None are feasible.**

Strategy and Explanation: Balance the equations, then use the method described in Question 57.

(a) $7\,C(s) + 6\,H_2O(g) \longrightarrow 2\,C_2H_6(g) + 3\,CO_2(g)$

$5\,C(s) + 4\,H_2O(g) \longrightarrow C_3H_8(g) + 2\,CO_2(g)$

$3\,C(s) + 4\,H_2O(g) \longrightarrow 2\,CH_3OH(\ell) + CO_2(g)$

(b) C_2H_6 reaction:

$\Delta X° = (2 \text{ mol}) \times X°(C_2H_6) + (3 \text{ mol}) \times X°(CO_2)$
$- (7 \text{ mol}) \times X°(C) - (6 \text{ mol}) \times X°(H_2O(g))$ $(X = \Delta H_f, S, \text{ or } \Delta G_f)$

$\Delta H° = (2 \text{ mol}) \times (-84.68 \text{ kJ/mol}) + (3 \text{ mol}) \times (-393.509 \text{ kJ/mol})$
$- (7 \text{ mol}) \times (0 \text{ kJ/mol}) - (6 \text{ mol}) \times (-241.818 \text{ kJ/mol}) = 101.02 \text{ kJ}$

$\Delta S° = (2 \text{ mol}) \times (229.60 \text{ J K}^{-1}\text{mol}^{-1}) + (3 \text{ mol}) \times (213.74 \text{ J K}^{-1}\text{mol}^{-1})$
$- (7 \text{ mol}) \times (5.740 \text{ J K}^{-1}\text{mol}^{-1}) - (6 \text{ mol}) \times (188.825 \text{ J K}^{-1}\text{mol}^{-1}) = -72.71 \text{ J K}^{-1}$

$\Delta G° = (2 \text{ mol}) \times (-32.82 \text{ kJ/mol}) + (3 \text{ mol}) \times (-394.359 \text{ kJ/mol})$
$- (7 \text{ mol}) \times (0 \text{ kJ/mol}) - (6 \text{ mol}) \times (-228.572 \text{ kJ/mol}) = 122.72 \text{ kJ}$

C_3H_8 reaction:

$\Delta X° = (1 \text{ mol}) \times X°(C_3H_8) + (2 \text{ mol}) \times X°(CO_2)$
$- (5 \text{ mol}) \times X°(C) - (4 \text{ mol}) \times X°(H_2O(g))$ $(X = \Delta H_f, S, \text{ or } \Delta G_f)$

$\Delta H° = (1 \text{ mol}) \times (-103.8 \text{ kJ/mol}) + (2 \text{ mol}) \times (-393.509 \text{ kJ/mol})$

$- (5 \text{ mol}) \times (0 \text{ kJ/mol}) - (4 \text{ mol}) \times (-241.818 \text{ kJ/mol}) = 76.5 \text{ kJ}$

$\Delta S° = (1 \text{ mol}) \times (269.9 \text{ J K}^{-1}\text{mol}^{-1}) + (2 \text{ mol}) \times (213.74 \text{ J K}^{-1}\text{mol}^{-1})$
$- (5 \text{ mol}) \times (5.740 \text{ J K}^{-1}\text{mol}^{-1}) - (4 \text{ mol}) \times (188.825 \text{ J K}^{-1}\text{mol}^{-1}) = -86.6 \text{ J K}^{-1}$

$\Delta G° = (1 \text{ mol}) \times (-23.49 \text{ kJ/mol}) + (2 \text{ mol}) \times (-394.359 \text{ kJ/mol})$
$- (5 \text{ mol}) \times (0 \text{ kJ/mol}) - (4 \text{ mol}) \times (-228.572 \text{ kJ/mol}) = 102.08 \text{ kJ}$

CH_3OH reaction

$\Delta X° = (2 \text{ mol}) \times X°(CH_3OH) + (1 \text{ mol}) \times X°(CO_2)$
$- (3 \text{ mol}) \times X°(C) - (4 \text{ mol}) \times X°(H_2O(g))$ $(X = \Delta H_f, S, \text{ or } \Delta G_f)$

$\Delta H° = (2 \text{ mol}) \times (-238.66 \text{ kJ/mol}) + (1 \text{ mol}) \times (-393.509 \text{ kJ/mol})$
$- (3 \text{ mol}) \times (0 \text{ kJ/mol}) - (4 \text{ mol}) \times (-241.818 \text{ kJ/mol}) = 96.44 \text{ kJ}$

$\Delta S° = (2 \text{ mol}) \times (126.8 \text{ J K}^{-1}\text{mol}^{-1}) + (1 \text{ mol}) \times (213.74 \text{ J K}^{-1}\text{mol}^{-1})$
$- (3 \text{ mol}) \times (5.740 \text{ J K}^{-1}\text{mol}^{-1}) - (4 \text{ mol}) \times (188.825 \text{ J K}^{-1}\text{mol}^{-1}) = -305.2 \text{ J K}^{-1}$

$\Delta G° = (2 \text{ mol}) \times (-166.27 \text{ kJ/mol}) + (1 \text{ mol}) \times (-394.359 \text{ kJ/mol})$
$- (3 \text{ mol}) \times (0 \text{ kJ/mol}) - (4 \text{ mol}) \times (-228.572 \text{ kJ/mol}) = 187.39 \text{ kJ}$

None of these is feasible. $\Delta G°$ is positive. In addition, $\Delta H°$ is positive and $\Delta S°$ is negative, suggesting that there is no temperature at which the products would be favored.

137. *Answer/Explanation:* Use Equation 18.8 and Equation 18.3: $\Delta G° = -RT\ln K = \Delta H° - T\Delta S°$

Divide everything by – RT: $$\ln K = -\frac{\Delta H°}{RT} + \frac{\Delta S°}{R}$$

If lnK is plotted against 1/T, the slope would be $-\Delta H°/R$, and the y-intercept would be $\Delta S°/R$. A straight line on this graph proves that the quantities of $\Delta H°$ and $\Delta S°$ are independent of temperature.

139. *Answer:* **(i) Reaction (b) (ii) Reactions (a) & (c) (iii) None of them (iv) None of them**

Strategy and Explanation: Determine the qualitative changes in $\Delta H°$ and $\Delta S°$, then refer to Table 18.3.

(a) Using the hint at the end of the question, we look up bond enthalpies in Chapter 6 Section 6.7. We find that forming a bond is exothermic ($\Delta H°$ = negative). Two mol of gas-phase reactants form one mol of gas-phase products, so the entropy decreases ($\Delta S°$ = negative). That means this reaction is (ii) product-favored at low temperatures but not at high temperatures.

(b) A combustion reaction for a hydrocarbon is exothermic ($\Delta H°$ = negative). Nine mol of gas-phase reactants form eleven mol of gas-phase products, so the entropy increases ($\Delta S°$ = positive). That means this reaction is (i) always product-favored.

(c) Going farther with the hint at the end of the question, we can look up actual bond energies in Chapter 8 (Table 8.2). We conclude that forming very strong P–F bonds (490 kJ/mol) produces more energy than is used breaking the weak P–P (209 kJ/mol) and F–F (158 kJ/mol) bonds. Hence, we will predict that the reaction is exothermic ($\Delta H°$ = negative). Eleven mol of gas-phase reactants form four mol of gas-phase products, so the entropy decreases ($\Delta S°$ = negative). That means this reaction is (ii) product-favored at low temperatures but not at high temperatures.

141. *Answer:* **205.2 J K^{-1}mol^{-1}**

Strategy and Explanation: Calculate the $\Delta H°$ for the reaction, using $\Delta H_f°$ given and Hess' Law. Determine $\Delta S°$, from $\Delta G°$ and $\Delta H°$. Use Hess' Law to calculate $S°(O_2)$.

$$\Delta H° = (2 \text{ mol}) \times \Delta H_f°(H_2O(\ell)) + (1 \text{ mol}) \times \Delta H_f°(CO_2) - (1 \text{ mol}) \times \Delta H_f°(CH_3OH(\ell)) - (\tfrac{3}{2} \text{ mol}) \times \Delta H_f°(O_2)$$
$$= (2 \text{ mol}) \times (-285.83 \text{ kJ/mol}) + (1 \text{ mol}) \times (-393.509 \text{ kJ/mol})$$
$$- (1 \text{ mol}) \times (-238.66 \text{ kJ/mol}) - (\tfrac{3}{2} \text{ mol}) \times (0 \text{ kJ/mol}) = -726.51 \text{ kJ}$$

$$\Delta S° = \frac{\Delta H° - \Delta G°}{T} = \frac{-726.51 \text{ kJ} - (-702.35 \text{ kJ})}{298.15 \text{ K}} \times \frac{1000 \text{ J}}{1 \text{ kJ}} = -81.17 \text{ J K}^{-1}$$

$$\Delta S° = (2\text{ mol}) \times S°(H_2O(\ell)) + (1\text{ mol}) \times S°(CO_2) - (1\text{ mol}) \times S°(CH_3OH(\ell)) - (\frac{3}{2}\text{ mol}) \times S°(O_2)$$

$$-81.17\text{ J K}^{-1} = (2\text{ mol}) \times (69.91\text{ J K}^{-1}\text{mol}^{-1}) + (1\text{ mol}) \times (213.74\text{ J K}^{-1}\text{mol}^{-1})$$

$$- (1\text{ mol}) \times (126.8\text{ J K}^{-1}\text{mol}^{-1}) - (\frac{3}{2}\text{ mol}) \times S°(O_2)$$

$$-307.8 = -\frac{3}{2}\, S°(O_2)$$

$$S°(O_2) = 205.2\text{ J K}^{-1}\text{mol}^{-1}$$

Chapter 19: Electrochemistry and Its Applications

Solutions for Blue-Numbered Questions for Review and Thought

Topical Questions

Using Half-Reactions to Understand Redox Reactions (Section 19.2)

10. *Answer:* **(a) $Zn(s) \longrightarrow Zn^{2+} + 2\ e^-$ (b) $2\ H_3O^+ + 2\ e^- \longrightarrow 2\ H_2O + H_2$ (c) $Sn^{4+} + 2\ e^- \longrightarrow Sn^{2+}$ (d) $Cl_2 + 2\ e^- \longrightarrow 2\ Cl^-$ (e) $6\ H_2O + SO_2 \longrightarrow SO_4^{2-} + 4\ H_3O^+ + 2\ e^-$**

Strategy and Explanation: Follow the methods described in Problem-Solving Example 19.3.

(a) Balance Zn atoms, then balance charge with electrons:

$Zn(s) \longrightarrow Zn^{2+}(aq) + 2\ e^-$ (Check: 1 Zn, zero net charge)

(b) Balance H atoms, then balance O atoms with H_2O, then balance charge with electrons:

$2\ H_3O^+(aq) + 2\ e^- \longrightarrow 2\ H_2O(\ell) + H_2(g)$ (Check: 6 H, 2 O, zero net charge)

(c) Balance Sn atoms, then balance charge with electrons:

$Sn^{4+}(aq) + 2\ e^- \longrightarrow Sn^{2+}(aq)$ (Check: 1 Sn, +2 net charge)

(d) Balance Cl atoms, then balance charge with electrons:

$Cl_2(g) + 2\ e^- \longrightarrow 2\ Cl^-(aq)$ (Check: 2 Cl, –2 net charge)

(e) Balance S atoms, then balance O atoms with H_2O, then balance H atoms with H^+, then balance charge with electrons:

$2\ H_2O(\ell) + SO_2(g) \longrightarrow SO_4^{2-}(aq) + 4\ H^+(aq) + 2\ e^-$

Then add four water molecules to each side to convert the H^+ ions into H_3O^+ ions:

$6\ H_2O(\ell) + SO_2(g) \longrightarrow SO_4^{2-}(aq) + 4\ H_3O^+(aq) + 2\ e^-$ (Check: 12 H, 8 O, 1 S, zero net charge)

12. *Answer:* **(a) $Al(s) \longrightarrow Al^{3+} + 3\ e^-$; $Cl_2 + 2\ e^- \longrightarrow 2\ Cl^-$ (b) $Fe^{2+} \longrightarrow Fe^{3+} + e^-$; $MnO_4^- + 8\ H_3O^+ + 5\ e^- \longrightarrow Mn^{2+} + 12\ H_2O$ (c) $FeS + 12\ H_2O \longrightarrow Fe^{3+} + SO_4^{2-} + 8\ H_3O^+ + 9\ e^-$: $NO_3^- + 4\ H_3O^+ + 3\ e^- \longrightarrow NO + 6\ H_2O$**

Strategy and Explanation: Follow methods described in the solution to Question 10 and Problem-Solving Example 19.3.

The chemicals for oxidation all go in one half-reaction and the chemicals for reduction all go in the other half-reaction. We do not have to determine where the H_3O^+ ions and H_2O molecules go; they will show up where they belong as we balance the half-reactions.

(a) Put Al and Al^{3+} in one half-reaction. Put Cl_2 and Cl^- in the other. Balance as in Question 10.

$Al(s) \longrightarrow Al^{3+}(aq) + 3\ e^-$

$Cl_2(g) + 2\ e^- \longrightarrow 2\ Cl^-(aq)$

(b) Put Fe^{2+} and Fe^{3+} in one half-reaction. Put MnO_4^- and Mn^{2+} in the other. Balance.

$Fe^{2+}(aq) \longrightarrow Fe^{3+}(aq) + e^-$

$MnO_4^-(aq) + 8\ H_3O^+(aq) + 5\ e^- \longrightarrow Mn^{2+}(aq) + 12\ H_2O(\ell)$ (Check: 1 Mn, 12 O, 24 H, +2)

(c) Put FeS, $Fe^{3+}(aq)$ and SO_4^{2-} in one half-reaction. Put NO_3^- and NO in the other. Balance.

$FeS(s) + 12\ H_2O(\ell) \longrightarrow Fe^{3+}(aq) + SO_4^{2-}(aq) + 8\ H_3O^+(aq) + 9\ e^-$

(Check: 1 Fe, 1 S, 12 O, 24 H, zero net charge)

$NO_3^-(aq) + 4\ H_3O^+(aq) + 3\ e^- \longrightarrow NO(g) + 6\ H_2O(\ell)$ (Check: 1 N, 7 O, 12 H, zero net charge)

14. *Answer:* $\mathbf{4\ Zn(s) + 7\ OH^-(aq) + NO_3^-(aq) + 6\ H_2O(\ell) \longrightarrow 4\ Zn(OH)_4^{2-} + NH_3(aq)}$

Strategy and Explanation: Follow methods described in the solution to Question 10 and Problem-Solving Example 19.4.

The chemicals for oxidation all go in one half-reaction and the chemicals for reduction all go in the other half-reaction. Balance the half-reactions, neutralize the acid, then combine the half-reactions.

Put Zn and $Zn(OH)_4^{2-}$ in one half-reaction. Put NO_3^- and NH_3 in the other. Balance as in Question 10.

Note that hydroxide ion is combining with the zinc ion, and we eventually want the reaction in basic solution, so it's simpler to just add hydroxide, than to balance in acid.

$$Zn(s) + 4\ OH^-(aq) \longrightarrow Zn(OH)_4^{2-} + 2\ e^- \qquad \text{(Check: 1 Zn, 4 O, 4 H, –4 net charge)}$$

Balanced the second half-reaction in acid then neutralized the acid.

$$NO_3^-(aq) + 9\ H^+(aq) + 8\ e^- \longrightarrow NH_3(aq) + 3\ H_2O(\ell) \qquad \text{(Check: 1 N, 9 H, 3 O, zero net charge)}$$

$$NO_3^-(aq) + 9\ H^+(aq) + 9\ OH^- + 8\ e^- \longrightarrow NH_3(aq) + 3\ H_2O(\ell) + 9\ OH^-$$

$$NO_3^-(aq) + 9\ H_2O(\ell) + 8\ e^- \longrightarrow NH_3(aq) + 3\ H_2O(\ell) + 9\ OH^-$$

$$NO_3^-(aq) + 6\ H_2O(\ell) + 8\ e^- \longrightarrow NH_3(aq) + 9\ OH^- \qquad \text{(Check: 1 N, 6 O, 12 H, –9 net charge)}$$

Multiply the first half-reaction by 4 so that the electrons balance, then add the half-reactions.

$$4\ Zn(s) + 16\ OH^-(aq) \longrightarrow 4\ Zn(OH)_4^{2-} + 8\ e^-$$
$$NO_3^-(aq) + 6\ H_2O(\ell) + 8\ e^- \longrightarrow NH_3(aq) + 9\ OH^-$$
$$\overline{4\ Zn(s) + 7\ OH^-(aq) + NO_3^-(aq) + 6\ H_2O(\ell) \longrightarrow 4\ Zn(OH)_4^{2-} + NH_3(aq)}$$

(Check: 4 Zn, 16 O, 19 H, 1 N, –8 net charge)

16. *Answer:* **(a)** $\mathbf{3\ CO + O_3 \longrightarrow 3\ CO_2}$**; oxidizing agent:** $\mathbf{O_3}$**; reducing agent: CO (b)** $\mathbf{H_2 + Cl_2 \longrightarrow 2\ HCl}$**; oxidizing agent:** $\mathbf{Cl_2}$**; reducing agent:** $\mathbf{H_2}$ **(c)** $\mathbf{2\ H_3O^+ + H_2O_2 + Ti^{2+} \longrightarrow 4\ H_2O + Ti^{4+}}$**; oxidizing agent:** $\mathbf{H_2O_2}$**; reducing agent:** $\mathbf{Ti^{2+}}$ **(d)** $\mathbf{2\ MnO_4^- + 6\ Cl^- + 8\ H_3O^+ \longrightarrow 2\ MnO_2 + 3\ Cl_2 + 12\ H_2O}$**; oxidizing agent:** $\mathbf{MnO_4^-}$**; reducing agent:** $\mathbf{Cl^-}$ **(e)** $\mathbf{4\ FeS_2 + 11\ O_2 \longrightarrow 2\ Fe_2O_3 + 8\ SO_2}$**; oxidizing agent:** $\mathbf{O_2}$**; reducing agent:** $\mathbf{FeS_2}$ **(f)** $\mathbf{O_3 + NO \longrightarrow O_2 + NO_2}$**; oxidizing agent:** $\mathbf{O_3}$**; reducing agent: NO (g) Zn(Hg) + HgO** $\mathbf{\longrightarrow}$ **ZnO + 2 Hg; oxidizing agent: HgO; reducing agent: Zn(Hg)**

Strategy and Explanation: Follow the methods described in the solution to Questions 10, 12 and Problem-Solving Examples 19.3 and 19.4. After the half-reactions are separated and balanced, equalize the electrons with appropriate multipliers and add the two half-reactions. The conversion of the H^+ ions into H_3O^+ ions can wait until the redox reaction is balanced.

(a) Put CO and CO_2 in one half-reaction. Put O_3 in the other. Balance each, then add.

$$3\ H_2O(\ell) + 3\ CO(g) \longrightarrow 3\ CO_2(g) + 6\ H^+(aq) + 6\ e^-$$
$$6\ e^- + 6\ H^+(aq) + O_3(g) \longrightarrow 3\ H_2O(\ell)$$
$$\overline{3\ CO(g) + O_3(g) \longrightarrow 3\ CO_2(g)}$$

O_3 is the oxidizing agent. CO is the reducing agent.

(b) This reaction is trivial to balance the old-fashioned way: $H_2(g) + Cl_2(g) \longrightarrow 2\ HCl(g)$

Cl_2 is the oxidizing agent. H_2 is the reducing agent.

(c) Put H_2O_2 in one half-reaction. Put Ti^{2+} and Ti^{4+} in the other. Balance each, then add.

$$2\ e^- + 2\ H^+(aq) + H_2O_2(aq) \longrightarrow 2\ H_2O(\ell)$$
$$Ti^{2+}(aq) \longrightarrow Ti^{4+}(aq) + 2\ e^-$$
$$\overline{2\ H^+(aq) + H_2O_2(aq) + Ti^{2+}(aq) \longrightarrow 2\ H_2O(\ell) + Ti^{4+}(aq)}$$

Then add two water molecules to each side to convert the H^+ ions into H_3O^+ ions:

$2\ H_3O^+(aq) + H_2O_2(aq) + Ti^{2+}(aq) \longrightarrow 4\ H_2O(\ell) + Ti^{4+}(aq)$ (Check: 8 H, 4 O, 1 Ti, +4 net charge)

H_2O_2 is the oxidizing agent. Ti^{2+} is the reducing agent.

(d) Put MnO_4^- and MnO_2 in one half-reaction. Put Cl^- and Cl_2 in the other.

$$3\ e^- + MnO_4^-(aq) + 4\ H^+(aq) \longrightarrow MnO_2(s) + 2\ H_2O(\ell)$$

$$2\ Cl^-(aq) \longrightarrow Cl_2(g) + 2\ e^-$$

Multiply each half-reaction by a constant to get the same number of electrons:

$$2 \times [3\ e^- + MnO_4^-(aq) + 4\ H^+(aq) \longrightarrow MnO_2(s) + 2\ H_2O(\ell)]$$

$$3 \times [2\ Cl^-(aq) \longrightarrow Cl_2(g) + 2\ e^-]$$

Now add them:

$$6\ e^- + 2\ MnO_4^-(aq) + 8\ H^+(aq) \longrightarrow 2\ MnO_2(s) + 4\ H_2O(\ell)$$

$$6\ Cl^-(aq) \longrightarrow 3\ Cl_2(g) + 6\ e^-$$

$$2\ MnO_4^-(aq) + 6\ Cl^-(aq) + 8\ H^+(aq) \longrightarrow 2\ MnO_2(s) + 3\ Cl_2(g) + 4\ H_2O(\ell)$$

Then add two water molecules to each side to convert the H^+ ions into H_3O^+ ions:

$2\ MnO_4^-(aq) + 6\ Cl^-(aq) + 8\ H_3O^+(aq) \longrightarrow 2\ MnO_2(s) + 3\ Cl_2(g) + 12\ H_2O(\ell)$

(Check: 2 Mn, 16 O, 24 H, zero net charge)

MnO_4^- is the oxidizing agent. Cl^- is the reducing agent.

(e) Put FeS_2, Fe_2O_3 and SO_2 in one half-reaction. Put O_2 in the other.

$$11\ H_2O(\ell) + 2\ FeS_2(s) \longrightarrow Fe_2O_3(s) + 4\ SO_2(g) + 22\ H^+(aq) + 22\ e^-$$

$$4\ e^- + 4\ H^+(aq) + O_2(g) \longrightarrow 2\ H_2O(\ell)$$

Multiply each half-reaction by a constant to get the same number of electrons:

$$2 \times [11\ H_2O(\ell) + 2\ FeS_2(s) \longrightarrow Fe_2O_3(s) + 4\ SO_2(g) + 22\ H^+(aq) + 22\ e^-]$$

$$11 \times [4\ e^- + 4\ H^+(aq) + O_2(g) \longrightarrow 2\ H_2O(\ell)]$$

Now add them:

$$22\ H_2O(\ell) + 4\ FeS_2(s) \longrightarrow 2\ Fe_2O_3(s) + 8\ SO_2(g) + 44\ H^+(aq) + 44\ e^-]$$

$$44\ e^- + 44\ H^+(aq) + 11\ O_2(g) \longrightarrow 22\ H_2O(\ell)$$

$$4\ FeS_2(s) + 11\ O_2(g) \longrightarrow 2\ Fe_2O_3(s) + 8\ SO_2(g)$$

(Check: 4 Fe, 8 S, 22 O, zero net charge)

O_2 is the oxidizing agent. FeS_2 is the reducing agent.

(f) This reaction is already balanced: $O_3(g) + NO(g) \longrightarrow O_2(g) + NO_2(g)$

O_3 is the oxidizing agent. NO is the reducing agent.

(g) This reaction is already balanced: $Zn(Hg)(amalgam) + HgO(s) \longrightarrow ZnO(s) + 2\ Hg(\ell)$

HgO is the oxidizing agent. Zn(Hg) is the reducing agent.

Electrochemical Cells (Section 19.3)

18. *Answer/Explanation:* The generation of electricity occurs when electrons are transmitted through a wire from the metal to the cation. Here, the transfer of electrons would occur directly from the metal to the cation and the electrons would not flow through any wire.

20. *Answer/Explanation:* Conventionally, in chemistry, they are written as reduction half-reactions. Engineers, material scientists, and physicists may still use an oxidation half-reaction convention.

22. *Answer:* **(a) $Zn + Pb^{2+} \longrightarrow Zn^{2+} + Pb$ (b) Anode: $Zn(s) \longrightarrow Zn^{2+} + 2\ e^-$ Cathode: $2\ e^- + Pb^{2+} \longrightarrow Pb(s)$ (c) see diagram below**

Strategy and Explanation: (a) $Zn(s) + Pb^{2+}(aq) \longrightarrow Zn^{2+}(aq) + Pb(s)$

(b) Oxidation of zinc atoms occurs at the anode, which is metallic zinc:

$$Zn(s) \longrightarrow Zn^{2+}(aq) + 2\ e^-$$

The reduction of lead(II) ions occurs at the cathode, which is metallic lead:

$$2\ e^- + Pb^{2+}(aq) \longrightarrow Pb(s)$$

(c)

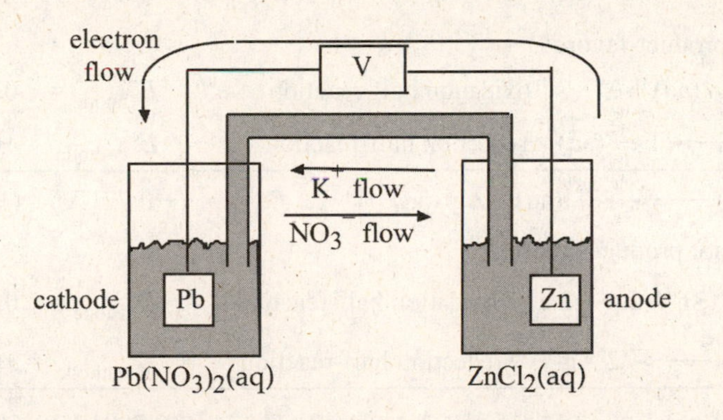

Electrochemical Cells and Voltage (Section 19.4)

24. *Answer:* **7500 C**

Strategy and Explanation: This is a standard conversion factor problem: divide the time by the voltage and use the definition of "watt" given in the question to show that 25 J are transferred per second. 1 J = 1C × 1V

$$\frac{1.0\ \text{hr}}{12\ \text{V}} \times \frac{60\ \text{min}}{1\ \text{hr}} \times \frac{60\ \text{s}}{1\ \text{min}} \times \frac{25\ \text{J}}{1\ \text{s}} \times \frac{1\ \text{C} \times 1\ \text{V}}{1\ \text{J}} = 7500\ \text{C}$$

26. *Answer:* **(a) $Cu \longrightarrow Cu^{2+} + 2\ e^-$; $Ag^+ + e^- \longrightarrow Ag$ (b) Oxidation: Cu half-reaction; Reduction: Ag half-reaction; Anode compartment: Cu half-reaction; Cathode compartment: Ag half-reaction**

Strategy and Explanation:

(a) Copper is converted to copper(II) ion and silver ion is converted to silver metal.

$$Cu(s) \longrightarrow Cu^{2+}(aq) + 2\ e^-$$

$$Ag^+(aq) + e^- \longrightarrow Ag(s)$$

(b) The copper half-reaction is oxidation (since Cu is losing electrons) and it occurs in the anode compartment. The silver half-reaction is reduction (since Ag^+ is gaining electrons) and it occurs in the cathode compartment.

Using Standard Cell Potentials (Section 19.5)

28. *Answer:* **Strongest reducing agent: Li Weakest oxidizing agent: Li^+ Strongest oxidizing agent: F_2 Weakest reducing agent: F^-**

Strategy and Explanation: On a chart where the standard reduction potentials are listed from most positive to most negative (Table 19.1), the reactant of the first reaction is the strongest oxidizing agent, because it has the largest reduction potential. The reactant of the last reaction is the weakest oxidizing agent, because that reaction has the smallest reduction potential. The product of the first reaction is the weakest reducing agent, because that reaction, in reverse, has the smallest oxidation potential. The product of the last reaction is the strongest reducing agent, because that reaction, in reverse, has the largest oxidation potential. The Li is the strongest reducing agent and Li^+ is the weakest oxidizing agent. F_2 is the strongest oxidizing agent and F^- is the weakest reducing agent.

30. *Answer:* **(d) < (c) < (a) < (b)**

Strategy and Explanation: Look up the reduction potential for the half-reactions that have the given species as reactants. The largest positive reduction potential is the best at reducing and represents the best oxidizing agent.

$$H_2O\ (-0.8277\ V) < PbSO_4\ (-0.3505V) < O_2\ (+1.229\ V) < H_2O_2\ (+1.763\ V)$$

32. *Answer:* **(a) 2.891 V (b) –0.028 V (c) 0.65 V (d) 1.164 V; (a), (c) and (d) are product-favored.**

Strategy and Explanation: Use the method described in Problem-Solving Example 19.7.

(a) $Mg(s) \longrightarrow Mg^{2+}(aq) + 2\ e^-$ oxidation half-reaction $E°_{anode} = -2.356\ V$

$I_2(s) + 2\ e^- \longrightarrow 2\ I^-(aq)$ reduction half-reaction $E°_{cathode} = +0.535\ V$

$I_2(s) + Mg(s) \longrightarrow Mg^{2+}(aq) + 2\ I^-(aq)$ $E°_{cell} = (+0.535\ V) - (-2.356\ V) = 2.891\ V$

This reaction is product-favored.

(b) $Ag(s) \longrightarrow Ag^+(aq) + e^-$ oxidation half-reaction $E°_{anode} = +0.7991\ V$

$Fe^{3+}(aq) + e^- \longrightarrow Fe^{2+}(aq)$ reduction half-reaction $E°_{cathode} = +0.771\ V$

$Ag(s) + Fe^{3+}(aq) \longrightarrow Fe^{2+}(aq) + Ag^+(aq)$ $E°_{cell} = (+0.771\ V) - (+0.7991\ V) = -0.028\ V$

This reaction is not product-favored.

(c) $Sn^{2+}(aq) \longrightarrow Sn^{4+}(aq) + 2\ e^-$ oxidation half-reaction $E°_{anode} = +0.15\ V$

$2\ Ag^+(aq) + 2\ e^- \longrightarrow 2\ Ag(s)$ reduction half-reaction $E°_{cathode} = +0.7991\ V$

$Sn^{2+}(aq) + 2\ Ag^+(aq) \longrightarrow Sn^{4+}(aq) + 2\ Ag(s$ $E°_{cell} = (+0.7991\ V) - (+0.15\ V) = 0.65\ V$

This reaction is product-favored.

(d) $2\ Zn(s) \longrightarrow 2\ Zn^{2+}(aq) + 4\ e^-$ oxidation half-reaction $E°_{anode} = -0.763\ V$

$O_2(s) + 2\ H_2O(\ell) + 4\ e^- \longrightarrow 4\ OH^-(aq)$ reduction half-reaction $E°_{cathode} = +0.401\ V$

$2\ Zn(s) + O_2(s) + 2\ H_2O(\ell) \longrightarrow Zn^{2+}(aq) + 4\ OH^-(aq)$ $E°_{cell} = (+0.401\ V) - (-0.763\ V) = 1.164\ V$

This reaction is product-favored.

34. *Answer:* **(a) Al^{3+} (b) Ce^{4+} (c) Al (d) Ce^{3+} (e) Yes (f) No (g) Ce^{4+}, Ag^+, and Hg_2^{2+} (h) Hg, Sn, Ni, and Al**

Strategy and Explanation:

(a) The reactant of the last reaction, Al^{3+}, is the weakest oxidizing agent.

(b) The reactant of the first reaction, Ce^{4+}, is the strongest oxidizing agent.

(c) The product of the last reaction, Al, is the strongest reducing agent.

(d) The product of the first reaction, Ce^{3+}, is the weakest reducing agent

(e) $Sn(aq) \longrightarrow Sn^{2+}(aq) + 2\ e^-$ oxidation half-reaction $E°_{anode} = -0.14\ V$

$2\ Ag^+(aq) + 2\ e^- \longrightarrow 2\ Ag(s)$ reduction half-reaction $E°_{cathode} = +0.80\ V$

$Sn^{2+}(aq) + 2\ Ag^+(aq) \longrightarrow Sn^{4+}(aq) + 2\ Ag(s)$ $E°_{cell} = (+0.80\ V) - (-0.14\ V) = 0.94\ V$

Yes, Sn(s) will reduce Ag^+ to Ag(s).

(f) $2\ Hg(\ell) \longrightarrow Hg_2^{2+}(aq) + 2\ e^-$ oxidation half-reaction $E°_{anode} = +0.80$ V

$Sn^{2+}(aq) + 2\ e^- \longrightarrow Sn(s)$ reduction half-reaction $E°_{cathode} = -0.14$ V

$2\ Hg(\ell) + Sn^{2+}(aq) \longrightarrow Hg_2^{2+}(aq) + Sn(s)$ $E°_{cell} = (-0.14\text{ V}) - (+0.80\text{ V}) = -0.94$ V

No, $Hg(\ell)$ will not reduce Sn^{2+} to Sn(s).

(g) Sn(s) can reduce any ion whose reduction potential is more positive than –0.14 V. In the table provided, Ce^{4+}, Ag^+, and Hg_2^{2+} can reduce Sn(s).

(h) $Ag^+(aq)$ can be oxidized by any metal whose reduction potential is smaller than 0.7994 V. In the table provided, Hg, Sn, Ni, and Al can oxidize $Ag^+(aq)$.

E° and Gibbs Free Energy (Section 19.6)

36. *Answer:* **greater; less**

Strategy and Explanation: In a product-favored chemical reaction, the standard cell potential, $E°$, is **greater** than zero and the Gibbs free energy, $\Delta G°$, is **less** than zero.

38. *Answer:* **(a) 1.55 V (b) –1196 kJ, 1.55 V**

Strategy and Explanation:

(a) Adapt the method described in Problem-Solving Example 19.8: $\Delta G° = -nFE°_{cell}$

Two O atoms are going from Ox. # = 0 to Ox. # = –2. So, n = 4 mol.

$$E°_{cell} = \frac{-\Delta G°}{nF} = \frac{-(-598\text{ kJ}) \times \left(\frac{1000\text{ J}}{1\text{ kJ}}\right) \times \left(\frac{1\text{ C} \times 1\text{ V}}{1\text{ J}}\right)}{(4\text{ mol})(96485\text{ C/mol})} = 1.55\text{ V}$$

(b) When the reaction is written with all the coefficients doubled:

$$\Delta G°_{double} = 2 \times \Delta G°_{original} = 2 \times (-598\text{ kJ}) = -1196\text{ kJ}$$

$E°_{cell}$ is independent of scale, so $E°_{cell} = 1.55$ V.

40. *Answer:* **–409 kJ**

Strategy and Explanation: Adapt the method described in the solution to Question 38 and Problem-Solving Example 19.8.

$$Zn(s) + Cl_2(g) \longrightarrow Zn^{2+}(aq) + 2\ Cl^-(aq)$$

Zn is going from Ox. # = 0 to Ox. # = +2. So, n = 2 mol.

$$\Delta G° = -(2\text{ mol}) \times (96485\text{ C/mol}) \times (2.12\text{ V}) \times \frac{1\text{ J}}{1\text{ C} \times 1\text{ V}} \times \frac{1\text{ kJ}}{1000} = -409\text{ kJ}$$

42. *Answer:* **$K° = 1 \times 10^{-18}$, $\Delta G° = 102$ kJ**

Strategy and Explanation: Adapt the method described in the solution to Question 38 and Problem-Solving Example 19.9.

$2\ Br^-(aq) \longrightarrow Br_2(\ell) + 2\ e^-$ $E°_{anode} = +1.066$ V

$I_2(s) + 2\ e^- \longrightarrow 2\ I^-(aq)$ $E°_{cathode} = +0.535$ V

$I_2(s) + 2\ Br^-(aq) \longrightarrow 2\ I^-(aq) + Br_2(\ell)$ $E°_{cell} = (+0.535\text{ V}) - (+1.066\text{ V}) = -0.531$ V

$$\log K° = \frac{nE°_{cell}}{0.0592\text{ V}} = \frac{(2\text{ mol}) \times (-0.531\text{ V})}{0.0592\text{ V}} = -17.9$$

$$K° = 1 \times 10^{-18}$$

$$\Delta G° = -(2\text{ mol}) \times (96485\text{ C/mol}) \times (-0.531\text{ V}) \times \frac{1\text{ J}}{1\text{ C} \times 1\text{ V}} \times \frac{1\text{ kJ}}{1000} = 102\text{ kJ}$$

46. *Answer:* **$K° = 4 \times 10^1$**

Strategy and Explanation: Adapt the method described in the solution to Question 38 and Problem-Solving Example 19.9.

$$\log K° = \frac{nE°_{cell}}{0.0592\ V}$$

$E°_{cell}$ is given to be 0.046 V and the number of moles of electrons passed to take 2+ ions to neutral is 2.

$$\log K° = \frac{nE°_{cell}}{0.0592\ V} = \frac{(2\ mol)\times(0.046\ V)}{0.0592\ V} = 1.5$$

$$K° = 10^{1.5} = 36 \cong 4 \times 10^1 \quad \textit{(round to 1 sig fig)}$$

Effect of Concentration on Cell Potential (Section 19.7)

48. *Answer:* **(a) $E°_{cell} = 1.202$ V (b) $E_{cell} = 1.16$ V (c) (conc Ag^+) = 3 M**

Strategy and Explanation: Adapt the method described in the solution to Question 32 and Problem-Solving Example 19.7, then use the Nernst equation at T = 298 K, by adapting the method described in the Problem-Solving Example 19.10.

$$E_{cell} = E°_{cell} - \frac{0.0592\ V}{n}\log Q$$

(a) $Cd(s) \longrightarrow Cd^{2+}(aq) + 2\ e^-$ $\qquad E°_{anode} = -0.403$ V

$2\ Ag^+(aq) + 2\ e^- \longrightarrow 2\ Ag(s)$ $\qquad E°_{cathode} = +0.7991$ V

$Cd(s) + 2\ Ag^+(aq) \longrightarrow Cd^{2+}(aq) + 2\ Ag(s)$ $\qquad E°_{cell} = (+0.7991\ V) - (-\ 0.403\ V) = 1.202$ V

For this reaction, $\quad Q = \dfrac{(\text{conc } Cd^{2+})}{(\text{conc } Ag^+)^2}$ and n = 2

(b) $$E_{cell} = E°_{cell} - \frac{0.0592\ V}{2}\log\left(\frac{(\text{conc } Cd^{2+})}{(\text{conc } Ag^+)^2}\right) = 1.202\ V - \frac{0.0592\ V}{2}\log\left(\frac{2.0\ M}{(0.25\ M)^2}\right)$$

$$= 1.202\ V - \frac{0.0592\ V}{2}\log(32) = 1.202\ V - \frac{0.0592\ V}{2} \times 1.5 = 1.202\ V - 0.04\ V$$

$$E_{cell} = 1.16\ V$$

(c) $$\log\left(\frac{(\text{conc } Cd^{2+})}{(\text{conc } Ag^+)^2}\right) = \frac{n}{0.0592\ V}(E°_{cell} - E_{cell}) = \frac{2}{0.0592\ V} \times (1.202\ V - 1.25\ V)$$

$$= \frac{2}{0.0592\ V} \times (-\ 0.05\ V) = -\ 1.62 \cong -\ 2$$

(subtraction of voltages leaves only 1 sig fig, so the answer must be round to 1 sig.fig.)

$$\log\left(\frac{0.100\ M}{(\text{conc } Ag^+)^2}\right) = -\ 2$$

$$10^{-2} = \frac{0.100}{(\text{conc } Ag^+)^2}$$

$$(\text{conc } Ag^+)^2 = 0.100(10^{+2})$$

$$(\text{conc } Ag^+) = 3\ M$$

50. *Answer:* $\mathbf{E_{cell} = -0.378\ V}$

Strategy and Explanation: Adapt the method described in the solution to Question 48 and the methods described in Chapter 16 for determining the concentration of H_3O^+.

$$2\ H_2O(\ell) + H_2(g) \longrightarrow 2\ H_3O^+(aq) + 2\ e^- \qquad E°_{anode} = 0.00\ V$$

$$2\ H_3O^+(aq) + 2\ e^- \longrightarrow 2\ H_2O(\ell) + H_2(g) \qquad E°_{cathode} = 0.00\ V$$

$$H_2(g)_{anode} + 2\ H_3O^+(aq)_{cathode} \longrightarrow 2\ H_3O^+(aq)_{anode} + H_2(g)_{cathode}$$

$$E°_{cell} = (0.00\ V) - (0.00\ V) = 0.00\ V$$

For this reaction, $$Q = \frac{(\text{press } H_2)_{cathode}(\text{conc } H_3O^+)^2_{anode}}{(\text{press } H_2)_{anode}(\text{conc } H_3O^+)^2_{cathode}} \quad \text{and} \quad n = 2$$

The cathode has pH = 7.8, so

$$(\text{conc } H_3O^+)_{cathode} = 10^{-pH} = 10^{-7.8} = 2 \times 10^{-8}\ M \quad \textit{(must round to 1 sig fig)}$$

$$E_{cell} = 0.00\ V - \frac{0.0592\ V}{2}\log\frac{(1.00\ bar)(0.05\ M)^2}{(1.00\ bar)(2\times10^{-8}\ M)^2}$$

$$= 0.00\ V - \frac{0.0592\ V}{2}\log(1 \times 10^{13})$$

$$= 0.00\ V - \frac{0.0592\ V}{2} \times (13.0) = 0.00\ V - (0.378\ V) = -0.378\ V$$

52. *Answer:* $\mathbf{\dfrac{conc\ Pb^{2+}}{conc\ Sn^{2+}} = 0.38}$

Strategy and Explanation: Adapt the method described in the solution to Question 48.

$$Pb(s) \longrightarrow Pb^{2+} + 2e^- \qquad E°_{anode} = -0.125\ V$$

$$+ \quad Sn^{2+} + 2e^- \longrightarrow Sn(s) \qquad E°_{cathode} = -0.1375\ V$$

$$Pb(s) + Sn^{2+} \longrightarrow Pb^{2+} + Sn(s) \qquad E°_{cell} = -0.0125\ V$$

$$= -0.013\ V \quad \textit{(must round to three decimal places)}$$

$$n = 2$$

$$E_{cell} = E°_{cell} - \frac{0.0592\ V}{n}\log\left(\frac{\text{conc } Pb^{2+}}{\text{conc } Sn^{2+}}\right)$$

$$0\ V = -0.013\ V - \frac{0.0592\ V}{2}\log\left(\frac{\text{conc } Pb^{2+}}{\text{conc } Sn^{2+}}\right)$$

$$-0.42 = \log\left(\frac{\text{conc } Pb^{2+}}{\text{conc } Sn^{2+}}\right)$$

$$0.38 = \frac{\text{conc } Pb^{2+}}{\text{conc } Sn^{2+}} \quad \text{(= 0.3 with proper sig figs)}$$

Common Batteries (Section 19.9)

54. *Answer:* **(a) $Ni^{2+} + Cd \longrightarrow Ni + Cd^{2+}$ (b) oxidized: Cd; reduced: Ni^{2+}; oxidizing agent: Ni^{2+}; reducing agent: Cd (c) anode: Cd; cathode: Ni (d) $E°_{cell} = 0.15$ V (e) from the Cd electrode to the Ni electrode (f) toward the anode compartment**

Strategy and Explanation: Apply the methods described in the solutions to Questions 22, 26, and 34.

(a) Nickel ion reacts with cadmium metal to form Nickel metal and cadmium ion.

$$Ni^{2+}(aq) + Cd(s) \longrightarrow Ni(s) + Cd^{2+}(aq)$$

(b)

Substance oxidized	Substance reduced	Oxidizing agent	Reducing agent
Cd	Ni^{2+}	Ni^{2+}	Cd

(c) Metals in the half-reaction will serve as the electrodes. Metallic Cd is the anode and metallic Ni is the cathode.

(d) $Cd(s) \longrightarrow Cd^{2+}(aq) + 2\ e^-$ $\quad E°_{anode} = -\ 0.403$ V

$Ni^{2+}(aq) + 2\ e^- \longrightarrow Ni(s)$ $\quad E°_{cathode} = -\ 0.25$ V

$Cd(s) + Ni^{2+}(aq) \longrightarrow Cd^{2+}(aq) + Ni(s)$ $\quad E°_{cell} = (-\ 0.25\ V) - (-\ 0.403\ V) = 0.15$ V

(e) The half-reactions above show that electrons flow spontaneously from the Cd electrode to the Ni electrode.

(f) The NO_3^- anions in the salt bridge flow toward the anode compartment to replenish the negative charges to neutralize the Cd^{2+} ions being formed.

Fuel Cells (Section 19.10)

56. *Answer/Explanation:* A fuel cell has a continuous supply of reactants, and will be useable for as long as the reactants are supplied. A battery contains all the reactants of the reaction. Once the reactants are gone, the battery is no longer useable.

58. *Answer:* **(a) anode: N_2H_4 oxidation; cathode: O_2 reduction (b) $N_2H_4(g) + O_2(g) \longrightarrow N_2(g) + 2H_2O(\ell)$ (c) 7.5 g N_2H_4 (d) 7.5 g O_2**

Strategy and Explanation: Adapt the methods described in the solutions to Question 48, Questions in Chapter 4, and Problem-Solving Example 19.13.

(a) The N_2H_4 oxidation occurs at the anode. The O_2 reduction occurs at the cathode.

(b) Adding the two half-reactions gives: $N_2H_4(g) + O_2(g) \longrightarrow N_2(g) + 2H_2O(\ell)$

(c) $$50.0\ hr \times 0.50\ A \times \frac{3600\ s}{1\ hr} \times \frac{1\ C}{1\ A \cdot 1\ s} \times \frac{1\ mol\ e^-}{96485\ C} \times \frac{1\ mol\ N_2H_4}{4\ mol\ e^-} \times \frac{32.05\ g\ N_2H_4}{1\ mol\ N_2H_4} = 7.5\ g\ N_2H_4$$

(d) $$7.5\ g\ N_2H_4 \times \frac{1\ mol\ N_2H_4}{32.05\ g\ N_2H_4} \times \frac{1\ mol\ O_2}{1\ mol\ N_2H_4} \times \frac{32.00\ g\ O_2}{1\ mol\ O_2} = 7.5\ g\ O_2$$

Electrolysis: Reactant-Favored Redox Reactions (Section 19.11)

59. *Answer:* **anode O_2; cathode H_2; 2 moles of H_2 produced per mole O_2**

Strategy and Explanation: Electrolysis of water in sulfuric acid involves the oxidation of water to form oxygen gas and the reduction of the strong acid to form hydrogen gas.

Anode - oxidation: $6\ H_2O(\ell) \longrightarrow O_2(g) + 4\ H_3O^+(aq) + 4\ e^-$

Cathode - reduction: $4\ H_3O^+(aq) + 4\ e^- \longrightarrow 2\ H_2(g) + 4\ H_2O(\ell)$

$$6\ H_2O(\ell) + 4\ H_3O^+(aq) \longrightarrow O_2(g) + 2\ H_2(g) + 4\ H_3O^+(aq) + 4\ H_2O(\ell)$$

$$2\ H_2O(\ell) + 4\ H_3O^+(aq) \longrightarrow O_2(g) + 2\ H_2(g) + 4\ H_3O^+(aq)$$

Gaseous O_2 is produced is at the anode, gaseous H_2 is produced at the cathode, and 2 moles of $H_2(g)$ are produced for each mole of $O_2(g)$.

61. *Answer:* **Au^{3+}, Hg^{2+}, Ag^{+}, Hg_2^{2+}, Fe^{3+}, Cu^{2+}, Sn^{4+}, Sn^{2+}, Ni^{2+}, Cd^{2+}, Fe^{2+}, Zn^{2+}**

Strategy and Explanation: All the metals with a reduction potential more positive than the half-reaction with water as a reactant (– 0.8277 V) can be electrolyzed from their aqueous ions to the corresponding metals. The metals in Table 19.1 that qualify are listed above.

63. *Answer:* **$H_2(g)$, $Br_2(\ell)$, and 2 $OH^-(aq)$ are formed; H_2O, Na^+, OH^- (and small amounts of dissolved Br_2 and H_3O^+) are in the solution; H_2 is formed at the cathode. Br_2 is formed at the anode.**

Strategy and Explanation: Electrolysis of NaBr involves the oxidation of the anion Br^- to Br_2 and the reduction of the water.

Anode - oxidation: $2\,Br^-(aq) \longrightarrow Br_2(\ell) + 2\,e^-$

Cathode - reduction: $2\,H_2O(\ell) + 2\,e^- \longrightarrow H_2(g) + 2\,OH^-(aq)$

$$2\,H_2O(\ell) + 2\,Br^-(\ell) \longrightarrow H_2(g) + Br_2(\ell) + 2\,OH^-(aq)$$

H_2 and Br_2 are produced in a basic solution. After the reaction is complete, Br^- is used up and the solution contains Na^+, OH^-, a small amount of dissolved Br_2 (though it has low solubility in water), and a very small amount of H_3O^+. H_2 is formed in the reduction reaction at the cathode. Br_2 is formed in the oxidation reaction at the anode.

Counting Electrons (Section 19.12)

65. *Answer:* **0.16 g Ag**

Strategy and Explanation: Follow the method described in the solutions to Question 58 and Problem-Solving Example 19.13.

$$Ag^+(aq) + e^- \longrightarrow Ag(s)$$

$$155\text{ min} \times 0.015\text{ A} \times \frac{60\text{ s}}{1\text{ min}} \times \frac{1\text{ C}}{1\text{ A}\cdot 1\text{ s}} \times \frac{1\text{ mol e}^-}{96485\text{ C}} \times \frac{1\text{ mol Ag}}{1\text{ mol e}^-} \times \frac{107.9\text{ g Ag}}{1\text{ mol Ag}} = 0.16\text{ g Ag}$$

67. *Answer:* **5.93 g Cu**

Strategy and Explanation: Follow the method described in the solution to Question 65.

$$Cu^{2+}(aq) + 2\,e^- \longrightarrow Cu(s)$$

$$2.00\text{ hr} \times 2.50\text{ A} \times \frac{3600\text{ s}}{1\text{ hr}} \times \frac{1\text{ C}}{1\text{ A}\cdot 1\text{ s}} \times \frac{1\text{ mol e}^-}{96485\text{ C}} \times \frac{1\text{ mol Cu}}{2\text{ mol e}^-} \times \frac{63.55\text{ g Cu}}{1\text{ mol Cu}} = 5.93\text{ g Cu}$$

69. *Answer:* **2.7×10^5 g Al**

Strategy and Explanation: Follow the method described in the solution to Question 65.

$$8.0\text{ hr} \times (1.0 \times 10^5\text{ A}) \times \frac{3600\text{ s}}{1\text{ hr}} \times \frac{1\text{ C}}{1\text{ A}\cdot 1\text{ s}} \times \frac{1\text{ mol e}^-}{96485\text{ C}} \times \frac{1\text{ mol Al}}{3\text{ mol e}^-} \times \frac{26.98\text{ g Al}}{1\text{ mol Al}} = 2.7 \times 10^5\text{ g Al}$$

71. *Answer:* **1.9×10^2 g Pb**

Strategy and Explanation: Follow the method described in the solution to Question 65. Equations for chemical reactions are found in Section 19.9 (page 935).

$$50.\text{ hr} \times 1.0\text{ A} \times \frac{3600\text{ s}}{1\text{ hr}} \times \frac{1\text{ C}}{1\text{ A}\cdot 1\text{ s}} \times \frac{1\text{ mol e}^-}{96485\text{ C}} \times \frac{1\text{ mol Pb}}{2\text{ mol e}^-} \times \frac{207.2\text{ g Pb}}{1\text{ mol Pb}} = 1.9 \times 10^2\text{ g Pb}$$

73. *Answer:* **0.10 g Zn**

Strategy and Explanation: Follow the method described in the solution to Question 65. Equations for chemical reactions are found in Section 19.9. *Assume time is precise to the minute.*

$$20.\ \text{min} \times 250.\ \text{mA} \times \frac{60\ \text{s}}{1\ \text{min}} \times \frac{10^{-3}\ \text{A}}{1\ \text{mA}} \times \frac{1\ \text{C}}{1\ \text{A}\cdot 1\ \text{s}} \times \frac{1\ \text{mol e}^-}{96485\ \text{C}} \times \frac{1\ \text{mol Zn}}{2\ \text{mol e}^-} \times \frac{65.39\ \text{g Zn}}{1\ \text{mol Zn}} = 0.10\ \text{g Zn}$$

75. *Answer:* **The lithium battery uses 0.043 g Li, while the lead battery uses 0.64 g Pb.**

Strategy and Explanation: Follow the method described in the solutions to Question 65.

$$10.\ \text{min} \times 1.0\ \text{A} \times \frac{60\ \text{s}}{1\ \text{min}} \times \frac{1\ \text{C}}{1\ \text{A}\cdot 1\ \text{s}} \times \frac{1\ \text{mol e}^-}{96485\ \text{C}} = 6.2 \times 10^{-3}\ \text{mol e}^-$$

$$6.2 \times 10^{-3}\ \text{mol e}^- \times \frac{1\ \text{mol Li}}{1\ \text{mol e}^-} \times \frac{6.941\ \text{g Li}}{1\ \text{mol Li}} = 0.043\ \text{g Li}$$

$$6.2 \times 10^{-3}\ \text{mol e}^- \times \frac{1\ \text{mol Pb}}{2\ \text{mol e}^-} \times \frac{207.2\ \text{g Pb}}{1\ \text{mol Pb}} = 0.64\ \text{g Pb}$$

The lithium battery consumes only 0.043 g Li, while the lead-acid storage battery consumes 0.64 g Pb.

77. *Answer:* **6.85 min**

Strategy and Explanation: Adapt the method described in the solution to Question 65.

$$\frac{0.500\ \text{g Ni}}{4.00\ \text{A}} \times \frac{1\ \text{mol Ni}}{58.6934\ \text{g Ni}} \times \frac{2\ \text{mol e}^-}{1\ \text{mol Ni}} \times \frac{96485\ \text{C}}{1\ \text{mol e}^-} \times \frac{1\text{A}\cdot 1\text{s}}{1\text{C}} \times \frac{1\ \text{min}}{60\ \text{s}} = 6.85\ \text{min}$$

Corrosion: Product-Favored Redox Reactions (Section 19.13)

80. *Answer/Explanation:* As described in Section 19.13, one requirement for corrosion is an electrolyte in contact with both the anode and the cathode. The presence of sodium and chloride ions in salt water increases the electrolytic capacity of the solution.

82. *Answer/Explanation:* Chromium is highly resistant to corrosion and protects the more active iron metal in the steel from oxidizing.

General Questions

85. *Answer:* **0.00689 g Cu; 0.00195 g Al**

Strategy and Explanation: Adapt the method described in the solution to Question 65.

$Ag^+(aq) + e^- \longrightarrow Ag(s)$, $\quad 0.0234\ \text{g Ag} \times \frac{1\ \text{mol Ag}}{107.9\ \text{g Ag}} \times \frac{1\ \text{mol e}^-}{1\ \text{mol Ag}} = 2.17 \times 10^{-4}\ \text{mol e}^-$

$Cu^{2+}(aq) + 2\,e^- \longrightarrow Cu(s)$, $\quad 2.17 \times 10^{-4}\ \text{mol e}^- \times \frac{1\ \text{mol Cu}}{2\ \text{mol e}^-} \times \frac{63.55\ \text{g Cu}}{1\ \text{mol Cu}} = 0.00689\ \text{g Cu}$

$Al^{3+}(aq) + 3\,e^- \longrightarrow Al(s)$, $\quad 2.17 \times 10^{-4}\ \text{mol e}^- \times \frac{1\ \text{mol Al}}{3\ \text{mol e}^-} \times \frac{26.98\ \text{g Al}}{1\ \text{mol Al}} = 0.00195\ \text{g Al}$

Applying Concepts

87. *Answer:* **B < D < A < C**

Strategy and Explanation: The strongest reducing agent is the most reactive metal. From the information in (b), we find that metal C reacts with all the other metals' ions, so it will be last on the list. From the information in (a), we find that metals A and C are more reactive than the other metals, so A will precede C in the list. From the information in (c), we find that metal D reacts with the ions of metal B, so B will be first on the list.

89. *Answer:* **(a) oxidized: B(s); reduced: A^{2+} (b) oxidizing agent: A^{2+}; reducing agent: B(s) (c) anode: B(s); cathode: A(s) (d) see equations below (e) A(s) (f) from B(s) to A(s) (g) towards the A^{2+} solution**

Strategy and Explanation: The solution labeled "A^{2+}" is getting lighter and the solution labeled "B^{2+}" is getting darker. If we assume that means A^{2+} is getting less concentrated and that B^{2+} is getting more concentrated we can make the following conclusions:

(a) B(s) is being oxidized to B^{2+} and A^{2+} is being reduced to A(s).

(b) A^{2+} is the oxidizing agent since it is being reduced, and B(s) is the reducing agent since it is being oxidized.

(c) B(s) is the anode, the solid at the site of oxidation, and A(s) is the cathode, the solid at the site of reduction.

(d) $A^{2+} + 2\,e^- \longrightarrow A(s)$ and $B(s) \longrightarrow B^{2+} + 2\,e^-$

(e) The A(s) metal gains mass.

(f) Electrons flow from the B(s) electrode to the A(s) electrode.

(g) K^+ ions in the salt bridge will migrate towards the A^{2+} solution to replace the cations that plated out as A(s).

91. *Answer:* **(a) The reaction of Co^{3+} with H_2O is spontaneous. (b) The reaction of Fe^{2+} with O_2 is spontaneous.**

Strategy and Explanation: The spontaneous direction of electron flow has a positive value of $E°_{cell}$. Determine if $E°_{cell}$ is positive.

(a)

$$6\,H_2O \longrightarrow O_2 + 4\,H_3O^+ + 4\,e^- \qquad E°_{anode} = 1.229\text{ V}$$
$$+\ 4\,Co^{3+} + 4\,e^- \longrightarrow 4\,Co^{2+} \qquad E°_{cathode} = 1.92\text{ V}$$
$$4\,Co^{3+} + 6\,H_2O \longrightarrow O_2 + 4\,H_3O^+ + 4\,Co^{2+} \qquad E°_{cell} = 0.69\text{ V}$$

Since the cell potential is positive, the reaction is spontaneous, and Co^{3+} is not stable in water.

(b)

$$O_2 + 4\,H_3O^+ + 4\,e^- \longrightarrow 6\,H_2O \qquad E°_{cathode} = 1.229\text{ V}$$
$$+\ 4\,Fe^{2+} \longrightarrow 4\,Fe^{3+} + 4\,e^- \qquad E°_{anode} = 0.771\text{ V}$$
$$O_2 + 4\,H_3O^+ + 4\,Fe^{2+} \longrightarrow 4\,Fe^{3+} + 6\,H_2O \qquad E°_{cell} = 0.458\text{ V}$$

Since the cell potential is positive, the reaction is spontaneous, and Fe^{2+} is not stable in air.

93. *Answer:* **0.379 V**

Strategy and Explanation: Adapt the method described in the solution to Question 50.

$$E_{cell} = E°_{cell} - \frac{0.0592\text{ V}}{n}\log Q$$

$$2\,H_3O^+ + 2\,e^- \longrightarrow H_2 + 2\,H_2O \qquad E°_{cathode} = 0.0000\text{ V}$$
$$+\ H_2 + 2\,H_2O \longrightarrow 2\,H_3O^+ + 2\,e^- \qquad E°_{anode} = 0.0000\text{ V}$$
$$H_{2,\,anode} + 2\,H_2O + 2\,H_3O^+{}_{cathode} \longrightarrow H_{2,\,cathode} + 2\,H_2O + 2\,H_3O^+{}_{anode}$$
$$E°_{cell} = 0.0000\text{ V} \qquad n = 2$$

$$Q = \frac{P_{H_2,cathode}\left(\text{conc } H_3O^+\right)^2_{anode}}{P_{H_2,anode}\left(\text{conc } H_3O^+\right)^2_{cathode}} = \frac{(1\text{ atm}) \times (1.0 \times 10^{-8}\text{ M})^2}{(1\text{ atm}) \times (0.025\text{ M})^2} = 1.6 \times 10^{-13}$$

$$E°_{cell} = 0.0000\text{ V} - \frac{0.0592\text{ V}}{2}\log(1.6 \times 10^{-13}) = -\frac{0.0592\text{ V}}{2} \times (-12.796) = 0.379\text{ V}$$

More Challenging Questions

95. *Answer:* **(a) 9.5 × 10⁶ g HF (b) 1.7 × 10³ kWh**

Strategy and Explanation: Perform a standard stoichiometry problem (Chapter 4) and then adapt the method described in the solution to Question 86(d) and at the end of Section 19.12, page 945.

(a) $$9.0 \text{ metric tons} \times \frac{1000 \text{ kg}}{1 \text{ metric ton}} \times \frac{1000 \text{ g}}{1 \text{ kg}} \times \frac{1 \text{ mol } F_2}{38.00 \text{ g } F_2} \times \frac{2 \text{ mol HF}}{1 \text{ mol } F_2} \times \frac{20.01 \text{ g HF}}{1 \text{ mol HF}} = 9.5 \times 10^6 \text{ g HF}$$

(b) $$24. \text{ hr} \times (6.0 \times 10^3 \text{ A}) \times \frac{3600 \text{ s}}{1 \text{ hr}} \times \frac{1 \text{ C}}{1 \text{ A} \cdot 1 \text{ s}} \times 12 \text{ V} \times \frac{1 \text{ J}}{1 \text{ C} \cdot 1 \text{ V}} \times \frac{1 \text{ kWh}}{3.60 \times 10^6 \text{ J}} = 1.7 \times 10^3 \text{ kWh}$$

97. *Answer:* $Cl_2 + 2\,Br^- \longrightarrow Br_2 + 2\,Cl^-$

Strategy and Explanation:

$$Br_2 + 2\,e^- \longrightarrow 2\,Br^- \qquad +1.066 \text{ V}$$

$$Cl_2 + 2\,e^- \longrightarrow 2\,Cl^- \qquad +1.358 \text{ V}$$

Reverse the first half-reaction to make a spontaneous reaction (with a positive cell potential).

$$2\,Br^- \longrightarrow Br_2 + 2\,e^- \qquad E°_{anode} = +1.066 \text{ V}$$

$$+ \quad Cl_2 + 2\,e^- \longrightarrow 2\,Cl^- \qquad E°_{cathode} = +1.358 \text{ V}$$

$$Cl_2 + 2\,Br^- \longrightarrow Br_2 + 2\,Cl^- \qquad E°_{cell} = +0.292 \text{ V}$$

99. *Answer:* **4+**

Strategy and Explanation: Adapt methods described in the solution to Question 65 and in Section 19.12.

$$3.00\text{hr} \times 2.00 \text{ A} \times \frac{1\text{C}}{1\text{A} \cdot 1\text{s}} \times \frac{3600 \text{ s}}{1 \text{ hr}} \times \frac{1 \text{ mol } e^-}{96485 \text{ C}} = 0.224 \text{mol } e^-$$

$$10.9 \text{ g Pt} \times \frac{1 \text{ mol Pt}}{195.08 \text{ g Pt}} = 0.0559 \text{ mol Pt}$$

$$\frac{0.224 \text{ mole } e^-}{0.0559 \text{ mol Pt}} = 4.01$$

The platinum ion is a 4+ ion, Pt^{4+}.

101. *Answer:* **75 s**

Strategy and Explanation: Use the methods described in the solution to Question 77 and in Section 19.12.

$$V = 1200 \text{ mm}^2 \times 1.0 \text{ μm} \times \left(\frac{1 \text{ m}}{1000 \text{ mm}} \times \frac{100 \text{ cm}}{1 \text{ m}}\right)^2 \times \left(\frac{10^{-6} \text{ m}}{1 \text{ μm}} \times \frac{100 \text{ cm}}{1 \text{ m}}\right) = 0.0012 \text{ cm}^3$$

$$0.0012 \text{ cm}^3 \times \frac{10.5 \text{ g}}{1 \text{ cm}^3} \times \frac{1 \text{ mol}}{107.8682 \text{ g}} = 1.2 \times 10^{-4} \text{ mol Ag}$$

$$1.2 \times 10^{-4} \text{ mol Ag} \times \frac{1 \text{ mol } Ag^+}{1 \text{ mol Ag}} \times \frac{1 \text{ mol } e^-}{1 \text{ mol } Ag^+} \times \frac{96485 \text{ C}}{1 \text{ mol } e^-} = 11 \text{ C}$$

$$\frac{11 \text{ C}}{150.0 \text{ mA}} \times \frac{1000 \text{ mA}}{1 \text{ A}} \times \frac{1\text{A} \cdot 1\text{s}}{1 \text{ C}} = 75 \text{ s}$$

103. *Answer:* **conc $Cu^{2+} = 4 \times 10^{-6}$ M**

Strategy and Explanation: Use the method shown in the solution to Question 48 involving the Nernst equation.

$$
\begin{array}{llll}
 & Cu(s) \longrightarrow Cu^{2+} + 2e^- & E^\circ_{anode} = 0.340\text{ V} & \\
+ & 2(Ag^+ + e^- \longrightarrow Ag(s)) & E^\circ_{cathode} = 0.7991\text{ V} & \\
\hline
 & Cu(s) + 2\,Ag^+ \longrightarrow Cu^{2+} + 2\,Ag(s) & E^\circ_{cell} = 0.459\text{ V} & n = 2
\end{array}
$$

$$E_{cell} = E^\circ_{cell} - \frac{0.0592\text{ V}}{n}\log\left(\frac{\text{conc } Cu^{2+}}{(\text{conc } Ag^+)^2}\right)$$

$$0.62\text{ V} = 0.459\text{ V} - \frac{0.0592\text{ V}}{2}\log\left(\frac{\text{conc } Cu^{2+}}{(1.00\text{ M})^2}\right)$$

$$\log(\text{conc } Cu^{2+}) = -5.4$$

$$\text{conc } Cu^{2+} = 10^{-5.4} = 4 \times 10^{-6}\text{ M} \quad \textit{(round to 1 sig fig)}$$

Chapter 20: Nuclear Chemistry

Solutions for Blue-Numbered Questions for Review and Thought

Topical Questions

Nuclear Reactions (Section 20.2)

11. *Answer:* **(a) alpha emission (b) beta emission (c) electron capture or positron emission (d) beta emission**

Strategy and Explanation: The mass numbers and atomic numbers must balance. Use that to determine the mass number and the atomic number of the decay particle. Use the periodic table and the atomic number to get the symbol of an element. Use the identity of the decay particle to identify what transformation that occurs.

(a) $^{230}_{90}\text{Th} \longrightarrow ^{226}_{88}\text{Ra} + \underline{^{4}_{2}\text{He}}$ $\quad 230 - 226 = 4,\ 90 - 88 = 2$

Thorium-230 decays by alpha emission to form radium-226.

(b) $^{137}_{55}\text{Cs} \longrightarrow ^{137}_{56}\text{Ba} + \underline{^{0}_{-1}\text{e}}$ $\quad 137 - 137 = 0,\ 55 - 56 = -1$

Cesium-137 decays by beta emission to form barium-137.

(c) $^{38}_{19}\text{K} + \underline{^{0}_{-1}\text{e}} \longrightarrow ^{38}_{18}\text{Ar}$ $\quad 38 - 38 = 0,\ 19 + x = 18,\ x = -1$

$^{38}_{19}\text{K} \longrightarrow ^{38}_{18}\text{Ar} + \underline{^{0}_{+1}\text{e}}$ $\quad 38 - 38 = 0,\ 19 - 18 = 1$

Potassium-38 decays by electron capture or positron emission to form argon-38.

(d) $^{97}_{40}\text{Zr} \longrightarrow ^{97}_{41}\text{Nb} + \underline{^{0}_{-1}\text{e}}$ $\quad 97 - 97 = 0,\ 41 - 40 = -1$

Zirconium-97 decays by beta emission to form niobiom-137.

✓ *Reasonable Answer Check:* Each reaction matches a known radioactive decay process, atomic numbers and mass numbers are conserved.

13. *Answer:* **(a) $^{238}_{92}\text{U}$ (b) $^{32}_{15}\text{P}$ (c) $^{10}_{5}\text{B}$ (d) $^{0}_{-1}\text{e}$ (e) $^{15}_{7}\text{N}$**

Strategy and Explanation: The mass numbers and atomic numbers must balance. Use that to determine the mass number and the atomic number of the missing entry. Use the periodic table and the atomic number to get the symbol of an element.

(a) $^{242}_{94}\text{Pu} \longrightarrow ^{4}_{2}\text{He} + \underline{^{238}_{92}\text{U}}$ $\quad 242 - 4 = 238,\ 94 - 2 = 92$, Element 92 is U.

(b) $\underline{^{32}_{15}\text{P}} \longrightarrow ^{32}_{16}\text{S} + ^{0}_{-1}\text{e}$ $\quad 32 + 0 = 32,\ 16 + (-1) = 15$, Element 15 is P.

(c) $^{252}_{98}\text{Cf} + \underline{^{10}_{5}\text{B}} \longrightarrow 3\ ^{1}_{0}\text{n} + ^{259}_{103}\text{Lr}$ $\quad 3 \times (1) - 259 - 252 = 10$

$3 \times (0) - 103 - 98 = 5$, Element 5 is B.

(d) $^{55}_{26}\text{Fe} + \underline{^{0}_{-1}\text{e}} \longrightarrow ^{55}_{25}\text{Mn}$ $\quad 55 - 55 = 0,\ 25 - 26 = -1$, electron captured $^{0}_{-1}\text{e}$.

(e) $^{15}_{8}\text{O} \longrightarrow \underline{^{15}_{7}\text{N}} + ^{0}_{+1}\text{e}$ $\quad 15 - 0 = 15,\ 8 - 1 = 7$, Element 7 is N.

✓ *Reasonable Answer Check:* Atomic numbers and mass numbers are conserved.

15. *Answer:* **(a) $^{28}_{12}\text{Mg} \longrightarrow ^{28}_{13}\text{Al} + ^{0}_{-1}\text{e}$ (b) $^{238}_{92}\text{U} + ^{12}_{6}\text{C} \longrightarrow 4\ ^{1}_{0}\text{n} + ^{246}_{98}\text{Cf}$ (c) $^{2}_{1}\text{H} + ^{3}_{2}\text{He} \longrightarrow ^{4}_{2}\text{He} + ^{1}_{1}\text{H}$ (d) $^{38}_{19}\text{K} \longrightarrow ^{38}_{18}\text{Ar} + ^{0}_{+1}\text{e}$ (e) $^{175}_{78}\text{Pt} \longrightarrow ^{4}_{2}\text{He} + ^{171}_{76}\text{Os}$**

Strategy and Explanation: Interpret the statement by identifying the nuclear symbol(s) for the given reactant and/or product isotope(s) and identifying the details of the radioactive decay process. Then follow the balancing method described in the solution to Question 11.

(a) Magnesium-28 is ${}^{28}_{12}Mg$ and β emission is the production of ${}^{0}_{-1}e$.

$$ {}^{28}_{12}Mg \longrightarrow {}^{28}_{13}Al + {}^{0}_{-1}e $$

(b) Uranium-238 is ${}^{238}_{92}U$, carbon-12 is ${}^{12}_{6}C$, and the neutron symbol is ${}^{1}_{0}n$.

$$ {}^{238}_{92}U + {}^{12}_{6}C \longrightarrow 4\ {}^{1}_{0}n + {}^{246}_{98}Cf $$

(c) Hydrogen-2 is ${}^{2}_{1}H$, helium-3 is ${}^{3}_{2}He$, and helium-4 is ${}^{4}_{2}He$.

$$ {}^{2}_{1}H + {}^{3}_{2}He \longrightarrow {}^{4}_{2}He + {}^{1}_{1}H $$

(d) Argon-38 is ${}^{38}_{18}Ar$ and positron emission is the production of ${}^{0}_{+1}e$.

$$ {}^{38}_{19}K \longrightarrow {}^{38}_{18}Ar + {}^{0}_{+1}e $$

(e) Platinum-175 is ${}^{175}_{78}Pt$, and osmium-171 is ${}^{171}_{76}Os$.

$$ {}^{175}_{78}Pt \longrightarrow {}^{4}_{2}He + {}^{171}_{76}Os $$

✓ *Reasonable Answer Check:* Atomic numbers and mass numbers are conserved.

17. *Answer:* $\mathbf{{}^{231}_{90}Th,\ {}^{231}_{91}Pa,\ {}^{227}_{89}Ac,\ {}^{227}_{90}Th}$, **and** $\mathbf{{}^{223}_{88}Ra}$

Strategy and Explanation: The first five steps of the decay series are: α, β, α, β, α. Therefore, start with uranium-235 undergoing an α decay reaction. Then take the radioisotope produced and make it the reactant of the second β decay reaction. Repeat this process for the remaining three steps undergoing α, then β, then α.

$$ {}^{235}_{92}U \longrightarrow {}^{4}_{2}He + {}^{231}_{90}Th $$

$$ {}^{231}_{90}Th \longrightarrow {}^{0}_{-1}e + {}^{231}_{91}Pa $$

$$ {}^{231}_{91}Pa \longrightarrow {}^{4}_{2}He + {}^{227}_{89}Ac $$

$$ {}^{227}_{89}Ac \longrightarrow {}^{0}_{-1}e + {}^{227}_{90}Th $$

$$ {}^{227}_{90}Th \longrightarrow {}^{4}_{2}He + {}^{223}_{88}Ra $$

So, the radioisotopes produced in the first five steps are: ${}^{231}_{90}Th$, ${}^{231}_{91}Pa$, ${}^{227}_{89}Ac$, ${}^{227}_{90}Th$, and ${}^{223}_{88}Ra$.

✓ *Reasonable Answer Check:* Atomic numbers and mass numbers are conserved.

19. *Answer:* $\mathbf{{}^{222}_{86}Rn \longrightarrow {}^{4}_{2}He + {}^{218}_{84}Po,\ {}^{218}_{84}Po \longrightarrow {}^{4}_{2}He + {}^{214}_{82}Pb,\ {}^{214}_{82}Pb \longrightarrow {}^{0}_{-1}e + {}^{214}_{83}Bi,}$
$\mathbf{{}^{214}_{83}Bi \longrightarrow {}^{0}_{-1}e + {}^{214}_{84}Po,\ {}^{214}_{84}Po \longrightarrow {}^{4}_{2}He + {}^{210}_{82}Pb,\ {}^{210}_{82}Pb \longrightarrow {}^{0}_{-1}e + {}^{210}_{83}Bi,}$
$\mathbf{{}^{210}_{83}Bi \longrightarrow {}^{0}_{-1}e + {}^{210}_{84}Po,\ {}^{210}_{84}Po \longrightarrow {}^{4}_{2}He + {}^{206}_{82}Pb}$

Strategy and Explanation: The decay series of radon-222 has eight steps: α, α, β, β, α, β, β, α. Start with radon-222 undergoing an α decay reaction. Then take the radioisotope produced and make it the reactant of the second α decay reaction. Repeat this process for the remaining six steps.

$$ {}^{222}_{86}Rn \longrightarrow {}^{4}_{2}He + {}^{218}_{84}Po $$

$$ {}^{218}_{84}Po \longrightarrow {}^{4}_{2}He + {}^{214}_{82}Pb $$

$$^{214}_{82}Pb \longrightarrow {}^{0}_{-1}e + {}^{214}_{83}Bi$$

$$^{214}_{83}Bi \longrightarrow {}^{0}_{-1}e + {}^{214}_{84}Po$$

$$^{214}_{84}Po \longrightarrow {}^{4}_{2}He + {}^{210}_{82}Pb$$

$$^{210}_{82}Pb \longrightarrow {}^{0}_{-1}e + {}^{210}_{83}Bi$$

$$^{210}_{83}Bi \longrightarrow {}^{0}_{-1}e + {}^{210}_{84}Po$$

$$^{210}_{84}Po \longrightarrow {}^{4}_{2}He + {}^{206}_{82}Pb \quad \text{Lead-206 is stable.}$$

✓ *Reasonable Answer Check:* Atomic numbers and mass numbers are conserved.

Nuclear Stability (Section 20.3)

20. *Answer:* **(a) $^{19}_{10}Ne \longrightarrow {}^{19}_{9}F + {}^{0}_{+1}e$ (b) $^{230}_{90}Th \longrightarrow {}^{0}_{-1}e + {}^{230}_{91}Pa$ (c) $^{82}_{35}Br \longrightarrow {}^{0}_{-1}e + {}^{82}_{36}Kr$**

(d) $^{212}_{84}Po \longrightarrow {}^{4}_{2}He + {}^{208}_{82}Pb$

Strategy and Explanation: Identify which type of radioactive decay is most likely for the isotope, by identifying the N/Z ratio and seeing where it falls on Figure 20.2.

(a) The neon-19 isotope has 9 neutrons and 10 protons, giving an N/Z ratio of 0.90. That means it has too few neutrons and undergoes positron emission or electron capture. Neon is relatively small, so we will predict that it undergoes positron emission:

$$^{19}_{10}Ne \longrightarrow {}^{19}_{9}F + {}^{0}_{+1}e$$

(b) The thorium-230 isotope has 140 neutrons and 90 protons, giving an N/Z ratio of 1.56. Stable isotopes in this range of the graph have an N/Z ratio of about 1.52. That means it has too many neutrons, so it undergoes β emission:

$$^{230}_{90}Th \longrightarrow {}^{0}_{-1}e + {}^{230}_{91}Pa$$

(c) The bromine-82 isotope has 47 neutrons and 35 protons, giving an N/Z ratio of 1.34. Stable isotopes in this range of the graph have an N/Z ratio of about 1.20. That means it has too many neutrons, so it undergoes β emission:

$$^{82}_{35}Br \longrightarrow {}^{0}_{-1}e + {}^{82}_{36}Kr$$

(d) The lead-212 isotope has 128 neutrons and 84 protons, more than the threshold limit of 126 neutrons and 83 protons above which all isotopes undergo α decay:

$$^{212}_{84}Po \longrightarrow {}^{4}_{2}He + {}^{208}_{82}Pb$$

✓ *Reasonable Answer Check:* Each reaction is a well-known radioactive decay process. Atomic numbers and mass numbers are conserved.

22. *Answer:* **6.252×10^8 kJ/mol nucleons ^{10}B; 6.688×10^8 kJ/ mol nucleon ^{11}B; ^{11}B is more stable than ^{10}B**

Strategy and Explanation: Use the method described in Section 20.3. We want to compare the binding energy per nucleon. Calculate the change in mass (Δm) when the nucleus is formed from the protons and neutrons. Use Einstein's equation: $E = (\Delta m)c^2$ to calculate the total energy generated, then divide that number by the number of nucleons to determine the binding energy per nucleon.

$\Delta m = 10.01294 \text{ g } {}^{10}B - [(5 \text{ mol } {}^{1}_{1}H \times 1.00783 \text{ g/mol } {}^{1}_{1}H) + (5 \text{ mol } {}^{1}_{0}n \times 1.00867 \text{ g/mol } {}^{1}_{0}n)] = -0.06956 \text{ g}$

The nuclear binding energy $= E_b = (\Delta m)c^2$

$$E_b = (\Delta m)c^2 = \left(-0.06956\ \text{g} \times \frac{1\ \text{kg}}{1000\ \text{g}}\right) \times (2.99792 \times 10^8\ \text{m/s})^2 \times \frac{1\ \text{J}}{1\ \text{kg}\,\text{m}^2\,\text{s}^{-2}} \times \frac{1\ \text{kJ}}{1000\ \text{J}} = -6.252 \times 10^9\ \text{kJ}$$

As described in Section 20.3 on page 967, "Binding energy per nucleon" is really E_b per mol nucleons.

One mol of boron-10 nuclei has 5 mol of protons and 5 mol of neutrons, or a total of 10 mol of nucleons:

$$E_b \text{ per mol nucleon} = \frac{6.252 \times 10^9\ \text{kJ}}{10\ \text{mol nucleons}} = 6.252 \times 10^8\ \text{kJ/mol nucleons}$$

$$\Delta m = 11.00931\ \text{g}\ ^{11}\text{B} - [(5\ \text{mol}\ ^1_1\text{H} \times 1.00783\ \text{g/mol}\ ^1_1\text{H}) + (6\ \text{mol}\ ^1_0\text{n} \times 1.00867\ \text{g/mol}\ ^1_0\text{n})] = -0.08186\ \text{g}$$

$$E_b = (\Delta m)c^2 = \left(-0.08186\ \text{g} \times \frac{1\ \text{kg}}{1000\ \text{g}}\right) \times (2.99792 \times 10^8\ \text{m/s})^2 \times \frac{1\ \text{J}}{1\ \text{kg}\,\text{m}^2\,\text{s}^{-2}} \times \frac{1\ \text{kJ}}{1000\ \text{J}} = -7.357 \times 10^9\ \text{kJ}$$

One mol of boron-11 nuclei has 5 mol of protons and 6 mol of neutrons, or a total of 11 mol of nucleons:

$$E_b \text{ per nucleon} = \frac{7.357 \times 10^9\ \text{kJ}}{11\ \text{mol nucleons}} = 6.688 \times 10^8\ \text{kJ/ mol nucleon}$$

Boron-11 has a larger E_b per mol nucleon than boron-10, so ^{11}B is more stable than ^{10}B.

✓ *Reasonable Answer Check:* The average atomic mass of boron (10.811) is closer to 11 than 10, consistent with the calculation that boron-11 is more stable than boron-10.

24. *Answer:* **– 1.7394 × 10^{11} kJ/mol; 7.3086 × 10^{8} kJ/mol nucleon ^{238}U**

Strategy and Explanation: Use the method described in Question 22. One mol of uranium-238 nuclei has 92 mol of protons and 146 mol of neutrons, or a total of 238 mol of nucleons:

$$\Delta m = 238.0508\ \text{g}\ ^{238}\text{U} - [(92\ \text{mol}\ ^1_1\text{H} \times 1.00783\ \text{g/mol}\ ^1_1\text{H}) + (146\ \text{mol}\ ^1_0\text{n} \times 1.00867\ \text{g/mol}\ ^1_0\text{n})] = -1.9354\ \text{g}$$

$$E_b = (\Delta m)c^2 = \left(-1.9354\ \text{g} \times \frac{1\ \text{kg}}{1000\ \text{g}}\right) \times (2.99792 \times 10^8\ \text{m/s})^2 \times \frac{1\ \text{J}}{1\ \text{kg}\,\text{m}^2\,\text{s}^{-2}} \times \frac{1\ \text{kJ}}{1000\ \text{J}} = -1.7394 \times 10^{11}\ \text{kJ}$$

"Binding energy per nucleon" (as described in Section 20.3) is E_b per mol nucleon.

$$E_b \text{ per mol nucleon} = \frac{1.7394 \times 10^{11}\ \text{kJ}}{238\ \text{mol nucleons}} = 7.3086 \times 10^8\ \text{kJ/mol nucleon}$$

Rates of Disintegration Reactions (Section 20.4)

27. *Answer:* **5 mg**

Strategy and Explanation: Because of the special circumstances where we are given an amount of time that is a multiple of the half-life, we can use the method shown in Problem-Solving Example 20.4:

$$t = 1\ \text{d} \times \left(\frac{24\ \text{h}}{1\ \text{d}}\right) + 6\ \text{h} = 30.\ \text{h}$$ *(Assume in this context that 1 d is exact.)*

$$30.\ \text{h} \times \frac{1\ \text{half-life}}{15\ \text{h}} = 2.0\ \text{half-lives}$$

30. hours is 2.0 times the half-life of 15 hours, so the sample mass is reduced by half twice:

$$\frac{1}{2} \times \frac{1}{2} \times 20\ \text{mg} = 5\ \text{mg}.$$

It is only legitimate to use the quick method described above when the elapsed time is a whole-number multiple of the half-life.

If the elapsed time was not a whole-number multiple of the half-life, it would be necessary to adapt the methods described in other Problem-Solving Examples in Section 20.4 and use these equations:

$$A = kN \qquad \ln\left(\frac{N}{N_0}\right) = -kt \qquad \ln\left(\frac{A}{A_0}\right) = -kt \qquad t_{1/2} = \frac{\ln 2}{k}$$

(Notice: ln2 is used in the last equation instead of 0.693 for two reasons: it is faster to type into a calculator—requiring only two buttons be pushed instead of four, and it is more precise than its 3 sig. fig. approximation.)

Use the last equation to determine the value of k:

$$k = \frac{\ln 2}{t_{1/2}} = \frac{\ln 2}{15\ \text{h}} = 4.6 \times 10^{-3}\ \text{h}^{-1}$$

The mass (m) of a sample of a pure isotopic substance is directly proportional to the number of atoms (N) because its atomic mass is a constant value and the number of atoms in a mole is also a constant. Therefore, the atom ratio is the same as the mass ratio:

$$\frac{N}{N_0} = \frac{m}{m_0}$$

Now, use the second equation to determine the new mass:

$$\ln\left(\frac{N}{N_0}\right) = \ln\left(\frac{m}{m_0}\right) = -kt = -(4.6 \times 10^{-3}\ \text{h}^{-1}) \times (30.\ \text{h}) = -1.4$$

$$m = m_0 e^{-1.4} = (20\ \text{mg}) \times e^{-1.4} = (20\ \text{mg}) \times (0.25) = 5\ \text{mg}$$

✓ *Reasonable Answer Check:* The quick method and the actual calculation produce the same answer, which is a small fraction of the initial mass.

29. *Answer:* **(a)** $\mathbf{{}^{131}_{53}I \longrightarrow {}^{0}_{-1}e + {}^{131}_{54}Xe}$ **(b) 1.56 mg**

Strategy and Explanation: Adapt the methods described in the solutions to Questions 15 and 27.

(a) The reaction's equation is: $\quad {}^{131}_{53}I \longrightarrow {}^{0}_{-1}e + {}^{131}_{54}Xe \qquad 131 = 0 + 131;\ 53 = -1 + 54.$

(b) $$32.2\ \text{d} \times \frac{1\ \text{half-life}}{8.05\ \text{d}} = 4.00\ \text{half-lives}$$

32.2 days is four half-lives, and the sample mass is reduced by half four times:

$$\frac{1}{2} \times \frac{1}{2} \times \frac{1}{2} \times \frac{1}{2} \times 25.0\ \text{mg} = 1.56\ \text{mg}$$

If the elapsed time was not a whole-number multiple of the half-life, it would be necessary to do the actual calculations:

$$k = \frac{\ln 2}{t_{1/2}} = \frac{\ln 2}{8.05\ \text{d}} = 8.61 \times 10^{-2}\ \text{d}^{-1}$$

The mass (m) of a sample of a compound is directly proportional to the number of radioactive atoms (N) because there is a fixed percentage of that atom in the compound, so the ratios are interchangeable, as described in the solution to Question 27:

$$\ln\left(\frac{m}{m_0}\right) = -kt = -(8.61 \times 10^{-2}\ \text{d}^{-1}) \times (32.2\ \text{d}) = -2.77$$

$$m = m_0 e^{-1.4} = (25.0\ \text{mg}) \times e^{-2.77} = (25.0\ \text{mg}) \times (0.063) = 1.6\ \text{mg}$$

✓ *Reasonable Answer Check:* The quick method and the actual calculation produce the same answer with what is known about the significant figures. The mass remaining is a small fraction of the initial mass.

31. *Answer:* **4.0 y**

Strategy and Explanation: Adapt the methods described in the solution to Question 27.

We can adapt the quick method by applying the definition of half-life. Initially, the radioactivity measured is defined as 100%. After one half-life, the radioactivity will drop by half to 50.0% of the original. After the second half-life, the radioactivity will drop by half again to 25.0% of the original. After the third half-life, the radioactivity will drop by half again to 12.5% of the original (i.e., $\frac{1}{2} \times \frac{1}{2} \times \frac{1}{2} \times 100\% = 12.5\%$). We are told that it takes 12 years to get to 12.5% of the original radioactivity; therefore, 12 years must represent three half-lives:

$$t_{1/2} = \frac{1}{3} \times (12\ y) = 4.0\ y$$

If the percentage given did not correspond exactly to one of the subsequent reductions of 100% by half, it would be necessary to do the actual calculations: A_0 = 100.0% and A = 12.5%

$$\ln\left(\frac{A}{A_0}\right) = -kt \qquad \ln\left(\frac{12.5\%}{100.0\%}\right) = -2.079 = -k \times (12\ y) \qquad k = 0.17\ y^{-1}$$

$$t_{1/2} = \frac{\ln 2}{k} = \frac{\ln 2}{0.17\ y^{-1}} = 4.0\ y$$

✓ *Reasonable Answer Check:* The quick method and the actual calculation produce the same answer. The half life must be less than the elapsed time, since the remaining radioactivity is less than half of the original radioactivity.

33. *Answer:* **34.8 d**

Strategy and Explanation: Adapt the methods described in the solutions to Questions 27, 29 and 31.

The percentage decrease given (5.0%) does not correspond exactly to one of the sequential reductions of 100% by half, so we can't use the quick method. Set up the time calculation using the equation given in Question 27.

In the solution to Question 29, we calculated $k = 8.61 \times 10^{-2}\ d^{-1}$. A_0 = 100.0% and A = 5.0%

$$\ln\left(\frac{A}{A_0}\right) = -kt \qquad \ln\left(\frac{5.0\%}{100.0\%}\right) = -3.00 = -(8.61 \times 10^{-2}\ d^{-1})t \qquad t = 34.8\ d$$

✓ *Reasonable Answer Check:* The elapsed time (34.8 d) is much longer than the half-life (8.05 d) and is consistent with the small remaining percent (5.0%).

35. *Answer:* **2.58×10^3 y**

Strategy and Explanation: Adapt the methods described in the solution to Question 33 and Problem-Solving Example 20.5.

$$k = \frac{\ln 2}{5.73 \times 10^3\ y} = 1.21 \times 10^{-4}\ y^{-1}$$

$A_0 = 15.3\ d\ min^{-1}g^{-1}$ and $A = 11.2\ d\ min^{-1}g^{-1}$

$$\ln\left(\frac{11.2\ d\ min^{-1}\ g^{-1}}{15.3\ d\ min^{-1}\ g^{-1}}\right) = -0.312 = -(1.21 \times 10^{-4}\ y^{-1})t \qquad t = 2.58 \times 10^3\ y$$

✓ *Reasonable Answer Check:* The elapsed time is shorter than the half-life and is consistent with the small reduction in the measured activity.

37. *Answer:* **1.6×10^5 y**

Strategy and Explanation: Adapt the methods described in the solution to Question 33.

The percentage decrease (10.0%) does not correspond exactly to one of the sequential reductions of 100% by half, so we can't use the quick method. Set up the calculations for k and then t: N_0 = 100.% and N = 1%.

$$k = \frac{\ln 2}{2.4 \times 10^4\ y} = 2.9 \times 10^{-5}\ y^{-1}$$

$$\ln\left(\frac{1\%}{100\%}\right) = -4.605 = -(2.9 \times 10^{-5}\ y^{-1})t \qquad t = 1.6 \times 10^5\ y$$

✓ *Reasonable Answer Check:* The elapsed time is much longer than the half-life and is consistent with the small remaining percent.

Artificial Transmutation (Section 20.5)

39. *Answer:* $^{239}_{94}\text{Pu} + 2\ ^{1}_{0}\text{n} \longrightarrow\ ^{0}_{-1}\text{e} + ^{241}_{95}\text{Am}$

Strategy and Explanation: Use the method described in the solution to Question 15.

$$^{239}_{94}\text{Pu} + 2\ ^{1}_{0}\text{n} \longrightarrow\ ^{0}_{-1}\text{e} + ^{241}_{95}\text{Am} \qquad 239 + 2\times(1) = 0 + 241;\ \ 94 + 2\times(0) = -1 + 95$$

41. *Answer:* $^{12}_{6}\text{C}$

Strategy and Explanation: Use the method described in the solution to Question 13.

$$^{238}_{92}\text{U} + ^{x}_{y}? \longrightarrow 4\ ^{1}_{0}\text{n} + ^{246}_{98}\text{Cf}$$

$$238 + x = 4\times(1) + 246 = 250;\ \ 92 + y = 0 + 98$$

$$x = 12, \text{ and } y = 6, \text{ so the isotope is: } ^{12}_{6}\text{C}$$

Nuclear Fission and Fusion (Sections 20.6 & 20.7)

43. *Answer/Explanation:* Three components represent the fundamental parts of a nuclear fission reactor. Cadmium rods are used as a neutron absorber to control the rate of the fission reaction. Uranium rods are the source of fuel, since uranium is a reactant in the nuclear equation. Water is used for cooling by removing excess heat energy. It is also used in the form of steam in the steam/water cycle to produce the turning torque for the generator.

45. *Answer:* **(a)** $^{140}_{54}\text{Xe}$ **(b)** $^{104}_{41}\text{Nb}$ **(c)** $^{92}_{36}\text{Kr}$

Strategy and Explanation: Use the method described in Question 41.

(a)

$$^{235}_{92}\text{U} + ^{1}_{0}\text{n} \longrightarrow ^{x}_{y}? + ^{93}_{38}\text{Sr} + 3\ ^{1}_{0}\text{n}$$

$$235 + 1 = x + 93 + 3\times(1) = 236;\ \ 92 + 0 = y + 38 + 3\times(0)$$

$$x = 140, \text{ and } y = 54, \text{ so the isotope is: } ^{140}_{54}\text{Xe}$$

(b)

$$^{235}_{92}\text{U} + ^{1}_{0}\text{n} \longrightarrow ^{x}_{y}? + ^{132}_{51}\text{Sb} + 3\ ^{1}_{0}\text{n}$$

$$235 + 1 = x + 132 + 3\times(1) = 236;\ 92 + 0 = y + 51 + 3\times(0)$$

$$x = 101, \text{ and } y = 41, \text{ so the isotope is: } ^{101}_{41}\text{Nb}$$

(c)

$$^{235}_{92}\text{U} + ^{1}_{0}\text{n} \longrightarrow ^{x}_{y}? + ^{141}_{56}\text{Ba} + 3\ ^{1}_{0}\text{n}$$

$$235 + 1 = x + 141 + 3\times(1) = 236;\ 92 + 0 = y + 56 + 3\times(0)$$

$$x = 92, \text{ and } y = 36, \text{ so the isotope is: } ^{92}_{36}\text{Kr}$$

47. *Answer:* **6.9×10^3 barrels**

Strategy and Explanation: This is a typical conversion factor question.

$$1.0 \text{ lb } ^{235}\text{U} \times \frac{453.6 \text{ g}}{1 \text{ lb}} \times \frac{1 \text{ mol } ^{235}\text{U}}{235 \text{ g U}} \times \frac{2.1\times10^{10} \text{ kJ}}{1 \text{ mol } ^{235}\text{U}} \times \frac{1 \text{ barrel oil}}{5.9\times10^{6} \text{ kJ}} = 6.9 \times 10^3 \text{ tons of coal}$$

Effects of Nuclear Radiation (Section 20.8)

49. *Answer/Explanation:* The unit "rad" is the measure of the amount of radiation absorbed. The unit "rem" includes a quality factor that better describes the biological impact of a radiation dose. The unit rem would be more appropriate when talking about the effects of an atomic bomb on humans. The unit gray (Gy) is 100 rad.

51. *Answer/Explanation:* Since most elements have some proportion of unstable isotopes that decay and we are composed of these elements (e.g., ^{14}C), our bodies emit radiation particles.

Uses of Radioisotopes (Section 20.9)

53. *Answer/Explanation:* The gamma ray is a high-energy photon. Its interaction with matter is most likely just going to be imparting large quantities of energy. The alpha and beta particles are charged particles of matter, which could interact and possibly react with and alter the matter composing the food. Gamma rays penetrate most matter readily, passing through unaffected. Beta particles are slower and penetration is stopped by dense tissues, such as bones. Alpha particles are stopped by skin.

55. *Answer:* **0.13 L**

Strategy and Explanation: Adapt the methods described in the solution to Questions 27 and Chapter 5.

$$A = kN \propto \text{mol}$$

$$\textit{Molarity}(\text{conc}) = \text{mol/L} \propto \text{A/L} \propto \text{A/mL}$$

$$\textit{Molarity}(\text{conc}) \times V(\text{conc}) = \textit{Molarity}(\text{dil}) \times V(\text{dil})$$

$$\text{A(conc)/mL} \times V(\text{conc}) = \text{A(dil)/mL} \times V(\text{dil})$$

$$V(\text{dil}) = \frac{\text{A / mL(conc)} \times V(\text{conc})}{\text{A / mL}} = \frac{\left(2.0 \times 10^6 \text{ dps / mL conc}\right) \times (1.0 \text{ mL conc}) \times \frac{1 \text{ L}}{1000 \text{ mL}}}{(1.5 \times 10^4 \text{ dps / mL dil})} = 0.13 \text{ L}$$

General Questions

57. *Answer:* (a) ${}^{0}_{-1}\text{e}$ (b) ${}^{4}_{2}\text{He}$ (c) ${}^{87}_{35}\text{Br}$ (d) ${}^{216}_{84}\text{Po}$ (e) ${}^{68}_{31}\text{Ga}$

Strategy and Explanation: Follow the method described in the solutions to Questions 13 and 15.

(a) ${}^{214}_{83}\text{Bi} \longrightarrow \underline{{}^{0}_{-1}\text{e}} + {}^{214}_{84}\text{Po}$ $\quad 214 = \underline{0} + 214;\ 83 = \underline{-1} + 84$

(b) $4\ {}^{1}_{1}\text{H} \longrightarrow \underline{{}^{4}_{2}\text{He}} + 2\ {}^{0}_{+1}\text{e}$ $\quad 4\times(1) = \underline{4} + 2\times(0);\ 4\times(1) = \underline{2} + 2\times(1)$

(c) ${}^{249}_{99}\text{Es} + {}^{1}_{0}\text{n} \longrightarrow 2\ {}^{1}_{0}\text{n} + \underline{{}^{87}_{35}\text{Br}} + {}^{161}_{64}\text{Gd}$ $\quad 249 + 1 = 2\times(1) + \underline{87} + 161;\ 99 + 0 = 2\times(0) + \underline{35} + 64$

(d) ${}^{220}_{86}\text{Rn} \longrightarrow \underline{{}^{216}_{84}\text{Po}} + {}^{4}_{2}\text{He}$ $\quad 220 = \underline{216} + 4;\ 86 = \underline{84} + 2$

(e) ${}^{68}_{32}\text{Ge} + {}^{0}_{-1}\text{e} \longrightarrow \underline{{}^{68}_{31}\text{Ga}}$ $\quad 68 + 0 = \underline{68};\ 32 + (-1) = \underline{31}$

59. *Answer:* **2 mg**

Strategy and Explanation: Use the methods described in the solution to Question 27.

$$t = (1 \text{ h}) \times \left(\frac{60 \text{ min}}{1 \text{ h}}\right) = 60 \text{ min}$$

The elapsed time (60 min) is six times the half-life (10 min), so the quantity remaining will be reduced by half six times:

$$\frac{1}{2} \times \frac{1}{2} \times \frac{1}{2} \times \frac{1}{2} \times \frac{1}{2} \times \frac{1}{2} \times 96 \text{ mg} = 2 \text{ mg}$$

If the time elapsed were not a whole-number multiple of the half-life, it would be necessary to do the calculations:

$$k = \frac{\ln 2}{10 \text{ min}} = 7 \times 10^{-2} \text{ min}^{-1} \quad \textit{(one sig fig)}$$

$$\ln\left(\frac{m}{m_0}\right) = -kt = -(7 \times 10^{-2} \text{ min}^{-1}) \times (60 \text{ min}) = -4 \quad \textit{(one sig fig)}$$

$$m = m_0 e^{-4} = (96 \text{ mg}) \times e^{-4} = (96 \text{ mg}) \times (0.02) \cong 2 \text{ mg} \quad \textit{(one sig fig)}$$

61. *Answer:* **3.6 × 10⁹ y**

Strategy and Explanation: Adapt the methods described in the solution to Question 35.

Because the quantity ratio is not a multiple of $\frac{1}{2}$, we cannot use the quick method. Therefore, we will set up the calculations for k and t:

$$\frac{N}{N_0} = 0.951, \quad \ln\left(\frac{N}{N_0}\right) = -kt$$

$$k = \frac{\ln 2}{4.9 \times 10^{10}\ y} = 1.4 \times 10^{-11}\ y^{-1}$$

$$\ln(0.951) = -0.0502 = -(1.4 \times 10^{-11}\ y^{-1})t$$

$$t = 3.6 \times 10^{9}\ y$$

63. *Answer:* **(a) $^{247}_{99}$Es (b) $^{16}_{8}$O (c) $^{4}_{2}$He (d) $^{12}_{6}$C (e) $^{10}_{5}$B**

Strategy and Explanation: Follow the method described in the solution to Question 57.

(a) $^{238}_{92}U + ^{14}_{7}N \longrightarrow \underline{^{247}_{99}Es} + 5\ ^{1}_{0}n$ $\quad 238 + 14 = \underline{247} + 5\times(1) = 252;\ 92 + 7 = \underline{99} + 5\times(0)$

(b) $^{238}_{92}U + \underline{^{16}_{8}O} \longrightarrow ^{249}_{100}Fm + 5\ ^{1}_{0}n$ $\quad 238 + \underline{16} = 249 + 5\times(1) = 254;\ 92 + \underline{8} = 100 + 5\times(0)$

(c) $^{253}_{99}Es + \underline{^{4}_{2}He} \longrightarrow ^{256}_{101}Md + ^{1}_{0}n$ $\quad 253 + \underline{4} = 256 + 1 = 257;\ 99 + \underline{2} = 101 + 0$

(d) $^{246}_{96}Cm + \underline{^{12}_{6}C} \longrightarrow ^{254}_{102}No + 4\ ^{1}_{0}n$ $\quad 246 + \underline{12} = 254 + 4\times(1) = 258;\ 96 + \underline{6} = 102 + 4\times(0)$

(e) $^{252}_{98}Cf + \underline{^{10}_{5}B} \longrightarrow ^{257}_{103}Lr + 5\ ^{1}_{0}n$ $\quad 252 + \underline{10} = 257 + 5\times(1) = 262;\ 98 + \underline{5} = 103 + 5\times(0)$

Applying Concepts

65. *Answer/Explanation:* Alpha and beta radiation decay particles are charged ($^{4}_{2}He^{2+}$ and $^{0}_{-1}e^{-}$), so they are better able to interact with and ionize tissues, disrupting the function of the cancer cells. Gamma radiation, like X-rays, goes through soft tissue without much being absorbed. This is less likely to interfere with the cancerous cells.

67. *Answer/Explanation:* The ^{20}Ne isotope is stable. The ^{17}Ne isotope is likely to decay by positron emission, to increase the ratio of neutrons to protons. The ^{23}Ne isotope is likely to decay by beta emission to decrease the ratio of neutrons to protons. For more details, refer to the discussion in Section 20.3 and examine Figure 20.2.

69. *Answer/Explanation:* A nuclear reaction occurred, making products. Therefore, some of the lost mass is found in the decay particles, if the decay is alpha or beta decay, and almost all the rest is found in the element produced by the reaction.

More Challenging Questions

71. *Answer/Explanation:* All radioactive decays are first order because, to occur, there needs to be one and only one reactant. The reaction involves only the nucleus of an atom.

73. *Answer:* 2.8×10^4 kg

Strategy and Explanation: Use the methods described in Section 20.3. Calculate the amount of energy that must be produced in an hour to provide 7.0×10^{14} kJ/s, then use $E = (\Delta m)c^2$ to determine the mass of solar material.

$$1\ \text{hour} \times \frac{3600\ \text{s}}{1\ \text{hr}} \times \frac{7.0 \times 10^{14}\ \text{kJ}}{1\ \text{s}} = 2.5 \times 10^{18}\ \text{kJ}$$

$$\Delta m = \frac{E}{c^2} = \frac{2.5 \times 10^{14}\ \text{kJ} \times \dfrac{1000\ \text{J}}{1\ \text{kJ}}}{(2.99792 \times 10^{8}\ \text{m/s})^2 \times \dfrac{1\ \text{J}}{1\ \text{kg m}^2\ \text{s}^{-2}}} = 2.8 \times 10^4 \text{ kg of solar material}$$

75. *Answer:* **75.1 y**

Strategy and Explanation: Adapt the methods described in the solution to Question 33.

$$N_0 = 100\%, N = 1.45\%$$

The ratio of percentage activity is not a multiple of $\frac{1}{2}$, so we can't use the quick method. Therefore, set up the calculation for k and t.

$$k = \frac{\ln 2}{12.3\ \text{y}} = 0.0564\ \text{y}^{-1}$$

$$\ln\left(\frac{1.45\ \%}{100\ \%}\right) = -4.234 = -(0.0564\ \text{y}^{-1})t \qquad t = 75.1\ \text{y}$$

77. *Answer:* **3.92 × 10³ y**

Strategy and Explanation: Use methods described in Chapter 3 to determine the mass of carbon, then calculate the disintegrations per minute per gram (d min^{-1} g^{-1}) of carbon, and use the method described in the solution to Question 35.

$$1.14\ \text{g}\ CaCO_3 \times \frac{1\ \text{mole}\ CaCO_3}{100.087\ \text{g}\ CaCO_3} \times \frac{1\ \text{mole C}}{1\ \text{mole}\ CaCO_3} \times \frac{12.0107\ \text{g C}}{1\ \text{mole C}} = 0.137\ \text{g C}$$

$$k = \frac{\ln 2}{5730\ \text{y}} = 1.21 \times 10^{-4}\ \text{y}^{-1}$$

$$A = \frac{\dfrac{2.17 \times 10^{-2}\ \text{dis}}{\text{s}} \times \dfrac{60\ \text{s}}{1\ \text{min}}}{0.137\ \text{g C}} = 9.52\ \text{d min}^{-1}\ \text{g}^{-1} \quad \text{and} \quad A_0 = 15.3\ \text{d min}^{-1}\ \text{g}^{-1}$$

$$\ln\left(\frac{9.52\ \text{d min}^{-1}\ \text{g}^{-1}}{15.3\ \text{d min}^{-1}\ \text{g}^{-1}}\right) = -0.474 = -(1.21 \times 10^{-4}\ \text{y}^{-1})t$$

$$t = 3.92 \times 10^3\ \text{y}$$

Chapter 21: The Chemistry of the Main Group Elements

Solutions for Blue-Numbered Questions for Review and Thought

Topical Questions

Electrolytic Methods (Section 21.4)

20. *Answer:* **(a) cathode (b) $Cl_2(g)$ (c) 2.027×10^9 C (d) 1.6×10^3 kJ/mol**

Strategy and Explanation: This electrolysis process is described in Section 21.4

(a) The magnesium ion is reduced to magnesium metal. This occurs at the cathode.

(b) Chlorine gas, $Cl_2(g)$, is formed at the other electrode.

(c) For each mol of Mg and Cl_2 produced, two mol of electrons are transferred. One mol of electrons is one Faraday. One Faraday is equal to the charge of 96485 C. These are discussed in Chapter 19.

$$1000.\text{ kg MgCl}_2 \times \frac{1000\text{ g}}{1\text{ kg}} \times \frac{1\text{ mol MgCl}_2}{95.211\text{ g MgCl}_2} \times \frac{2\text{ mol e}^-}{1\text{ mol MgCl}_2} \times \frac{1\text{ Faraday}}{1\text{ mol e}^-} = 2.101 \times 10^4\text{ Faradays}$$

$$2.101 \times 10^4\text{ Faradays} \times \frac{96485\text{ C}}{1\text{ Faraday}} = 2.027 \times 10^9\text{ C}$$

(d) The conversion factor 3.60×10^6 J/kWh is obtained from the relationship given in Section 19.12, in the margin note on page 944. The conversion factor is also given in the Estimation Box on pgae 945 and in Question 19.95.

$$\frac{8.4\text{ kwh}}{\text{lb Mg}} \times \frac{3.60 \times 10^6\text{ J}}{1\text{ kwh}} \times \frac{1\text{ kJ}}{1000\text{ J}} \times \frac{1\text{ lb Mg}}{453.6\text{ g Mg}} \times \frac{24.3050\text{ g Mg}}{1\text{ mol Mg}} = 1.6 \times 10^3\ \frac{\text{kJ}}{\text{mol Mg}}$$

22. *Answer:* **7×10^3 kWh**

Strategy and Explanation: Adapt the methods described in the solution to Question 19.95 where the conversion factor 3.60×10^6 J/kWh was given. Notice: it is not necessary to use the amperes given in the question.

$$1\text{ ton Na} \times \frac{2000\text{ lb}}{1\text{ ton}} \times \frac{453.6\text{ g}}{1\text{ lb}} \times \frac{1\text{ mol Na}}{22.9898\text{ g Na}} \times \frac{2\text{ mol e}^-}{2\text{ mol Na}} \times \frac{96485\text{ C}}{1\text{ mol e}^-}$$

$$\times\ 7.0\text{ V} \times \frac{1\text{ J}}{1\text{ C}\cdot 1\text{ V}} \times \frac{1\text{ kWh}}{3.60 \times 10^6\text{ J}} = 7 \times 10^3\text{ kWh}$$

24. *Answer:* **67.1 g Al**

Strategy and Explanation: Use the method described in the solution to Question 22.

$$2.00\text{ hr} \times 100.\text{ A} \times \frac{3600\text{ s}}{1\text{ hr}} \times \frac{1\text{ C}}{1\text{ A}\cdot 1\text{ s}} \times \frac{1\text{ mol e}^-}{96485\text{ C}} \times \frac{1\text{ mol Al}}{3\text{ mol e}^-} \times \frac{26.9815\text{ g Al}}{1\text{ mol Al}} = 67.1\text{ g Al}$$

General Questions

26. *Answer:*

Formula	Name	Oxidation state of phosphorus
P_4	Phosphorus	**0**
$(NH_4)_2HPO_4$	**Ammonium hydrogen phosphate**	**+5**
H_3PO_4	Phosphoric acid	**+5**
P_4O_{10}	Tetraphosphorus decaoxide	**+5**
$Ca_3(PO_4)_2$	**Calcium phosphate**	**+5**
$Ca(H_2PO_4)_2$	Calcium dihydrogen phosphate	**+5**

Strategy and Explanation: Use the methods described in Chapter 3 for names and formulas: Section 3.2 and 3.5, including Table 3.7 and Chapter 5, Table 5.2. For oxidation states refer to Chapter 5, Section 5.4. Here, we use Ox. # H = +1, Ox. # O = –2, Ox. # Ca = +2, and the sum of the oxidation numbers in a compound add up to the net charge.

Formula, name	Calculation of oxidation state	Oxidation state of phosphorus
P_4, phosphorus	4(Ox. # P) = 0	Ox. # P = 0
$(NH_4)_2HPO_4$, ammonium hydrogen phosphate	Ammonium, NH_4^+ hydrogen phosphate, HPO_4^{2-} (+1) + Ox. # P + 4(–2) = –1	Ox. # P = +5
H_3PO_4, phosphoric acid	3(+1) + Ox. # P + 4(–2) = 0	Ox. # P = +5
P_4O_{10}, tetraphosphorus decaoxide	4(Ox. # P) + 10(–2) = 0	Ox. # P = +5
$Ca_3(PO_4)_2$, calcium phosphate	3(+2) + 2(Ox. # P) + 8(–2) = 0	Ox. # P = +5
$Ca(H_2PO_4)_2$, calcium dihydrogen phosphate	(+2) + 4(+1) + 2(Ox. # P) + 8(–2) = 0	Ox. # P = +5

28. *Answer:* **1.7 × 10⁷ tons**

Strategy and Explanation: This is a stoichiometry and conversion factor question as described in Chapter 4:

$$5.0 \times 10^6 \text{ tons Al} \times \frac{2000 \text{ lb}}{\text{ton}} \times \frac{453.6 \text{ g}}{1 \text{ lb}} \times \frac{1 \text{ mol Al}}{26.9815 \text{ g Al}} \times \frac{1 \text{ mol Al}_2\text{O}_3}{2 \text{ mol Al}}$$

$$\times \frac{101.961 \text{ g Al}_2\text{O}_3}{1 \text{ mol Al}_2\text{O}_3} \times \frac{100 \text{ g bauxite}}{55 \text{ g Al}_2\text{O}_3} \times \frac{1 \text{ lb}}{453.6 \text{ g}} \times \frac{1 \text{ ton}}{2000 \text{ lb}} = 1.7 \times 10^7 \text{ tons bauxite}$$

30. *Answer:* **see structure below**

Strategy and Explanation: Use the methods described in Chapter 8. In the solution to Question 8.14(d), we found that phosphate, PO_4^{3-}, has 4 O atoms single-bonded to the P atom. Use that pattern to develop a three-dimensional version of the P_4O_{10} molecule: Each P atom forms four single bonds to O atoms. Three of the four O atoms have single bonds to two different P atoms.

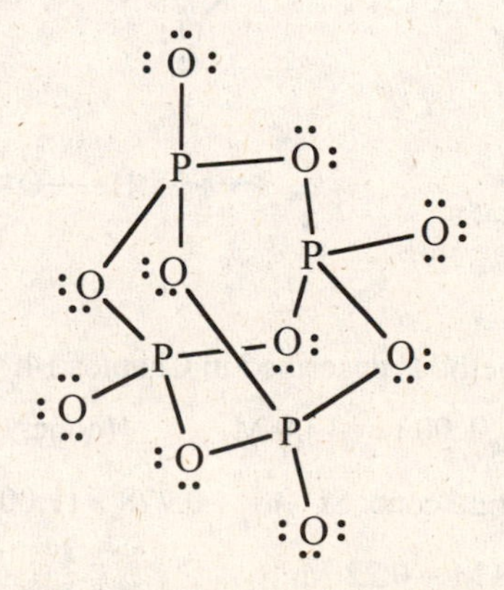

Other Lewis structures are possible, but the structure above is the most plausible one that follows the octet rule. If we look to reduce formal charges, the structure that allows the P atoms to expand their octet would be considered better:

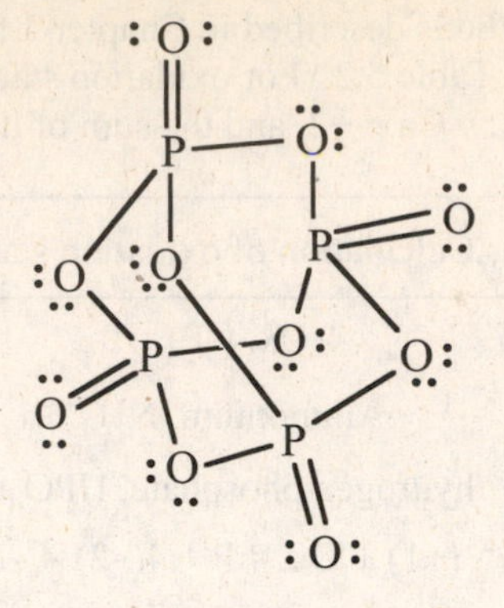

32. *Answer:* **962 K**

Strategy and Explanation: Use the methods described in the solution to Question 18.55. *NOTE: The units of $\Delta S°$ should be $J\,K^{-1}mol^{-1}$.*

$$T = \frac{(-17.6\ \text{kJ/mol}) \times \left(\frac{1000\ \text{J}}{1\ \text{kJ}}\right)}{-18.3\ \text{J K}^{-1}\text{mol}^{-1}} = 962\ \text{K}$$

34. *Answer/Explanation:* The raw materials used in the synthesis of sulfuric acid are sulfur, water, oxygen, and catalyst (Pt or VO_5)

$$S_8(s) + 8\,O_2(g) \longrightarrow 8\,SO_2(g)$$
$$2\,SO_2(g) + O_2(g) \longrightarrow 2\,SO_3(g)$$
$$SO_3(g) + H_2SO_4(\ell) \longrightarrow H_2S_2O_7(\ell)$$
$$H_2S_2O_7(\ell) + H_2O(\ell) \longrightarrow 2\,H_2SO_4(aq)$$

36. *Answer:* **see resonance structures below**

Strategy and Explanation: Use the methods described in Chapter 8. The N atom in nitric acid has one double bond, so we can draw three resonance forms for this molecule. The third of these has the highest formal charges; hence, it is considered to be the worst structure of the three.

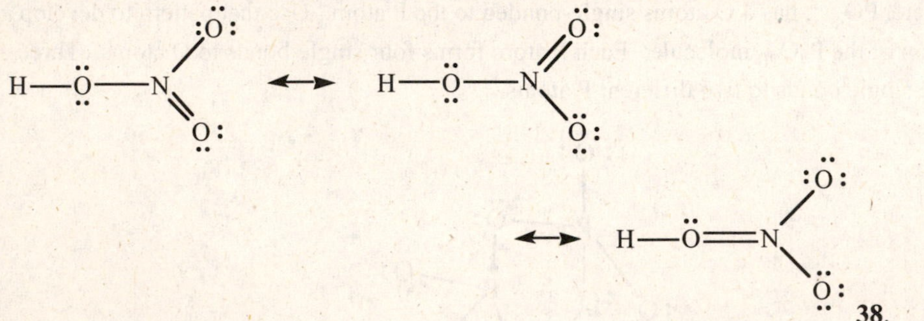

38. *Answer:* $\mathbf{K_c = 2.7}$

Strategy and Explanation: Use the methods described in Chapter 14.

Initially: (conc. SO_2) = 1.00 mol SO_2/1.00 L = 1.00 M (conc. O_2) = 5.00 mol O_2/1.00 L = 5.00 M

Change: 77.8% SO_2 reacted, so (change conc. SO_2) = – 0.778 × (1.00 M) = – 0.778 M

Equilibrium: $[SO_2]$ = 1.00 M – 0.778 M = 0.22 M

	$2\,SO_2(g)$	+	$O_2(g)$	$\rightleftharpoons$	$2\,SO_3(g)$
initial conc. (M)	1.00		5.00		0
change as reaction occurs (M)	– 0.778 = – 2x		– x		+ 2x
equilibrium conc. (M)	0.22		5.00 – x		2x

We use SO_2 concentration change to find x: 0.778 M = 2x = $[SO_3]$

$$x = 0.389\text{ M}$$

$$5.00 - x = 5.00\text{ M} - 0.389\text{ M} = 4.61\text{ M} = [O_2]$$

$$K_c = \frac{[SO_3]^2}{[SO_2]^2[O_2]} = \frac{(2x)^2}{(1.00-2x)^2(5.00-x)} = \frac{(0.778)^2}{(0.22)^2(4.61)} = 2.7$$

40. *Answer:* **3 L**

Strategy and Explanation: Begin by writing a balanced equation for the decomposition of aqueous hydrogen peroxide (H_2O_2).

$$2\ H_2O_2(aq) \longrightarrow O_2(g) + 2\ H_2O(\ell)$$

To determine the amount of oxygen gas produced the amount of hydrogen peroxide must first be determined. We assume the dilute aqueous solution has a density the same as water, and that the percent provided in the question is a mass percent, the mass of hydrogen peroxide decomposed can be determined:

$$250.\text{ mL soln} \times \frac{0.998\text{ g soln}}{1\text{ mL soln}} \times \frac{3\text{ g } H_2O_2}{100\text{ g soln}} = 7.5\text{ g } H_2O_2 \approx 8\text{ g } H_2O_2 \text{ decomposed } \textit{(keep 1 sig fig)}$$

The mass of hydrogen peroxide can then be converted to moles of oxygen using relevant molar masses and the stoichiometry of the balanced equation.

$$7.5\text{ g } H_2O_2 \times \frac{1\text{ mol } H_2O_2}{34.0146\text{ g } H_2O_2} \times \frac{1\text{ mol } O_2}{2\text{ mol } H_2O_2} = 0.11\text{ mol } O_2 \approx 0.1\text{ mol } O_2\ \textit{(keep 1 sig fig)}$$

The moles of oxygen produced are then inserted into the ideal gas law, PV=nRT; recall that the temperature must be converted to Kelvin and the pressure to atmospheres to use the common value of R.

$$V = \frac{(0.11\text{ mol } O_2)\left(0.08206\frac{\text{L}\cdot\text{atm}}{\text{mol}\cdot\text{K}}\right)(22 + 273.15)\text{K}}{750\text{ mmHg} \times \frac{1\text{ atm}}{760\text{ mmHg}}} = 2.70\text{ L} \approx 3\text{ L } \textit{(round to 1 sig fig)}$$

Applying Concepts

42. *Answer:* **K = 3.1 × 10^5; product-favored reaction produces solid Mg(OH)$_2$. The solid can be isolated after it settles.**

Strategy and Explanation: Use the method described in Problem-Solving Example 14.2.

$Ca(OH)_2(s) \rightleftharpoons Ca^{2+}(aq) + 2\ OH^-(aq)$	$K_{sp,1} = 5.5 \times 10^{-6}$
$Mg^{2+}(aq) + 2\ OH^-(aq) \rightleftharpoons Mg(OH)_2(s)$	$K = 1/K_{sp,2} = 1/(1.8 \times 10^{-11})$
$Ca(OH)_2(s) + Mg^{2+}(aq) \rightleftharpoons Ca^{2+}(aq) + Mg(OH)_2(s)$	$K_{net} = K_{sp,1} \times (1/K_{sp,2})$
	$K_{net} = (5.5 \times 10^{-6}) \times [1/(1.8 \times 10^{-11})] = 3.1 \times 10^5$

Putting sea water in the presence of $Ca(OH)_2$ will cause the product-favored precipitation of $Mg(OH)_2$. The solid can be isolated after it settles.

44. *Answer:* $:\ddot{N}{=}N{=}\ddot{N}{-}\ddot{N}{=}\ddot{O}:$ **(Other plausible structures are possible.)**

Strategy and Explanation: Use method described in the solution to Question 8.14. Using 26 electrons, follow the octet rule on each atom. Other plausible resonance structures are also valid. The name helps confirm this structure: "Azide" is formed when three N atoms are bonded together, seen in this structure between the left-most three N atoms. "Nitrosyl" is the –NO functional group seen on the right side of the structure (see above).

46. *Answer:* **(a) see structures below (b) +140. kJ with single and triple bond; +410. kJ with two double**

bonds

Strategy and Explanation:

(a) Use methods described in Chapter 8.

$$\mathrm{H{-}\ddot{N}{=}N{=}\ddot{N}{:}} \longleftrightarrow \mathrm{H{-}\underset{\cdot\cdot}{\ddot{N}}{-}N{\equiv}N{:}}$$

(b) Adapt the method described in the solution to Question 8.42.

Formation is described in Section 6.10, page 244, as the production of one mol of a compound from its standard state elements. That means we do not need to use the given ΔH_f for N and O.

Formation equation for the first structure:

$$\frac{3}{2}(\mathrm{:N{\equiv}N:}) + \frac{1}{2}(\mathrm{H{-}H}) \longrightarrow \mathrm{H{-}\ddot{N}{=}N{=}\ddot{N}{:}}$$

$$\Delta H_f \cong \frac{3}{2}\text{ mol} \times D_{N\equiv N} + \frac{1}{2}\text{ mol} \times D_{H\text{–}H} - 1\text{ mol} \times D_{N\text{–}H} - 2\text{ mol} \times D_{N=N}$$

Use the bond energies (D) in Table 8.2:

$$\Delta H_f \cong \frac{3}{2} \times (946\text{ kJ}) + \frac{1}{2} \times (436\text{ kJ}) - (391\text{ kJ}) - 2 \times (418\text{ kJ}) = 410.\text{ kJ}$$

Formation equation for the second structure:

$$\frac{3}{2}(\mathrm{:N{\equiv}N:}) + \frac{1}{2}(\mathrm{H{-}H}) \longrightarrow \mathrm{H{-}\underset{\cdot\cdot}{\ddot{N}}{-}N{\equiv}N{:}}$$

$$\Delta H_f \cong \frac{3}{2}\text{ mol} \times D_{N\equiv N} + \frac{1}{2}\text{ mol} \times D_{H\text{–}H} - 1\text{ mol} \times D_{N\text{–}H} - 1\text{ mol} \times D_{N\text{–}N} - 1\text{ mol} \times D_{N\equiv N}$$

Use the bond energies (D) in Table 8.2:

$$\Delta H_f \cong \frac{3}{2} \times (946\text{ kJ}) + \frac{1}{2} \times (436\text{ kJ}) - (391\text{ kJ}) - (160\text{ kJ}) - (946\text{ kJ}) = 140\text{ kJ}$$

48. *Answer:* **(a) $NO_2(g) + NO(g) \longrightarrow N_2O_3(g)$ (b) see resonance structures, below (c) O–N–O angle is 120°; N–N–O angle is slightly less than 120°.**

Strategy and Explanation:

(a) $NO_2(g) + NO(g) \longrightarrow N_2O_3(g)$ 2 N and 3 O, on each side.

(b) We use the methods described in Chapter 8.

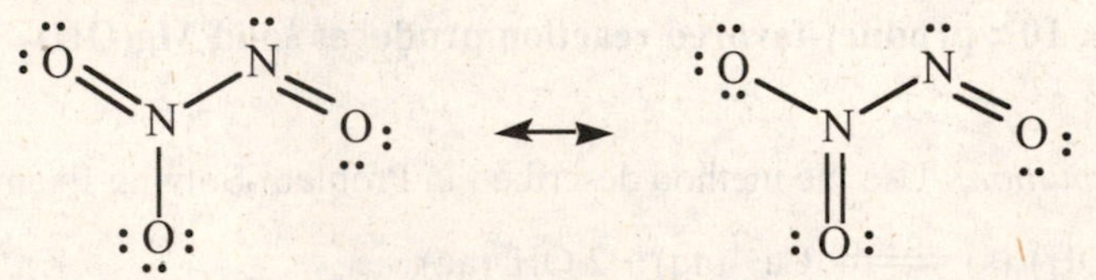

(Notice: The third resonance structure has two adjacent atoms have +1 formal charge, which eliminates that structure from being plausible.)

(c) Use the methods described in Chapter 9. The O–N–O bond is on an N atom that fits the class AX_3. That means the bond angle is 120°. The N–N–O angles involving that same N atom as a central atom are also 120°. The N–N–O angle on the second N atom is slightly less than 120°, due to the repulsion of the lone pair on the central N atom.

51. *Answer:* **(a) 7×10^2 kJ (b) 3 V must be used to overcome the cell potential.**

Strategy and Explanation: Adapt the methods described in Chapter 18 and Chapter 19.

(a) $T = 600\ °C + 273.15 = 9 \times 10^2\ K$

$$\Delta G° = \Delta H° - T\Delta S° = (820\text{ kJ}) - (9 \times 10^2\text{ K}) \times (180\text{ J K}^{-1}) \times \frac{1\text{ kJ}}{1000\text{ J}} = 7 \times 10^2\text{ kJ}$$

(b) $$E°_{cell} = -\frac{\Delta G}{nF} = -\frac{7 \times 10^2\text{ kJ}}{(2)(96485\text{ C})} \times \frac{1000\text{ J}}{1\text{ kJ}} \times \frac{1\text{ C} \times 1\text{ V}}{1\text{ J}} = -3\text{ V}$$

3 volts must be used to overcome this cell potential.

53. *Answer:* **(a) one; 10^{-4} atm and 120°C (b) (i) Solid (rhombic) (ii) liquid (iii) solid (monoclinic) (iv) gaseous**

Strategy and Explanation:

(a) The triple point of a phase diagram is a point at which all phases (solid, liquid, gas) are in equilibrium. Only one point satisfies this definition: point E on the diagram, where the monoclinic form of sulfur exists in equilibrium with liquid and gaseous sulfur. The approximate pressure and temperatures are 10^{-4} atm and 120°C.

Notice that the terms monoclinic and rhombic are specific unit cell configurations used to describe how the sulfur atoms pack in a three dimensional array in a solid. We learned about the cubic unit cells (primitive cubic, fcc, and bcc) that are adapted by many crystalline solids in Section 11.6. Rhombic and monoclinic are two possible variations.

(b) To determine the phase described by the given pressure and temperature, we simply draw a horizontal line intersecting the given pressure and a vertical line intersecting the given temperature. The area in which these two lines intersect identifies the phase of the material.

(i) Solid (rhombic) sulfur

(ii) Liquid sulfur

(iii) Solid (monoclinic) sulfur

(iv) Gaseous sulfur

55. *Answer:* **(a) $4\ H_3O^+ + MnO_2 + 2\ Br^- \longrightarrow Mn^{2+} + 6\ H_2O + Br_2$ (b) 0.0813 mol (c) 3.54 g MnO_2**

Strategy and Explanation:

(a) Since (several) elements in the reaction change oxidation numbers, this is a redox reaction. The half-reaction method of balancing redox reactions introduced in Section 19.2 provides the most straightforward route to determining the balanced overall reaction.

$$2\ e^- + 4\ H_3O^+(aq) + MnO_2(s) \longrightarrow Mn^{2+}(aq) + 6\ H_2O(\ell)$$

$$2\ Br^-(aq) \longrightarrow Br_2(\ell) + 2\ e^-$$

$$4\ H_3O^+(aq) + MnO_2(s) + 2\ Br^-(aq) \longrightarrow Mn^{2+}(aq) + 6\ H_2O(\ell) + Br_2(\ell)$$

(b) Use the stoichiometry of the balanced equation to determine the amount of bromide ions consumed.

$$6.50\text{ g } Br_2 \times \frac{1\text{ mol } Br_2}{159.808\text{ g } Br_2} \times \frac{2\text{ mol } Br^-}{1\text{ mol } Br_2} = 0.0813\text{ mol } Br^-$$

(c) Use the stoichiometry to determine the mass of manganese(II) oxide needed to produce 6.50 g of bromine.

$$6.50\text{ g } Br_2 \times \frac{1\text{ mol } Br_2}{159.808\text{ g } Br_2} \times \frac{1\text{ mol } MnO_2}{1\text{ mol } Br_2} \times \frac{86.9368\text{ g } MnO_2}{1\text{ mol } MnO_2} = 3.54\text{ g } MnO_2$$

More Challenging Questions

63. *Answer:* **6.3×10^5 J/mol**

Strategy and Explanation: The Clausius-Clapeyron equation is discussed in Section 11.2. Two pressure-temperature points can be roughly estimated from the phase diagram vapor-pressure curve (line BE) provided in Question 53. *(Assume that the temperature is known to ±1°C.)*

$$P_1 = 0.0254\text{ mm Hg} \qquad T_1 = 20.°C + 273.15 = 293\text{ K}$$

$$P_2 = 0.133\text{ mm Hg} \qquad T_2 = 40.°C + 273.15 = 313\text{ K}$$

Inserting these values into the two-set form of the Clausius-Clapeyron equation provides the final answer as shown in the Problem-Solving Example 11.1:

$$\ln\left(\frac{0.133 \text{ mmHg}}{0.0254 \text{ mmHg}}\right) = \frac{-\Delta H}{8.31 \text{ JK}^{-1}\text{mol}^{-1}}\left(\frac{1}{313 \text{ K}} - \frac{1}{293 \text{ K}}\right)$$

$$\ln(5.24) = \frac{-\Delta H}{8.31 \text{ JK}^{-1}\text{mol}^{-1}}\left(-0.00022 \text{ K}^{-1}\right)$$

$$\Delta H = \frac{\left(8.31 \text{ J mol}^{-1}\right)\ln(5.24)}{0.00022} = 6.3 \times 10^5 \text{ J/mol}$$

66. *Answer:* **(a) see equations, below (b) 2.25 (c) 25.0 g $Ca_3(PO_4)_2$ (d) 5.42 L**

Strategy and Explanation:

(a) When a nonmetal burns in excess oxygen it typically attains its highest possible oxidation state.

$$P_4(s) + 5\,O_2(g) \longrightarrow 2\,P_2O_5(s) \quad \text{or}$$

$$P_4(s) + 5\,O_2(g) \longrightarrow P_4O_{10}(s)$$

(b) All nonmetal oxides are acidic. In many cases this means they will react with water to form a common aqueous acid.

$$P_2O_5(s) + 3\,H_2O(\ell) \longrightarrow 2\,H_3PO_4(aq) \quad \text{or}$$

$$P_4O_{10}(s) + 6\,H_2O(\ell) \longrightarrow 4\,H_3PO_4(aq)$$

First determine the molarity of the acid produced using the balanced equations provide in (a).

$$\frac{5.00 \text{ g } P_4 \times \frac{1 \text{ mol } P_4}{123.8952 \text{ g } P_4} \times \frac{2 \text{ mol } P_2O_5}{1 \text{ mol } P_4} \times \frac{2 \text{ mol } H_3PO_4}{1 \text{ mol } P_2O_5}}{0.250 \text{ L}} = 0.646 \text{ M } H_3PO_4$$

$$\frac{5.00 \text{ g } P_4 \times \frac{1 \text{ mol } P_4}{123.8952 \text{ g } P_4} \times \frac{1 \text{ mol } P_4O_{10}}{1 \text{ mol } P_4} \times \frac{4 \text{ mol } H_3PO_4}{1 \text{ mol } P_4O_{10}}}{0.250 \text{ L}} = 0.646 \text{ M } H_3PO_4$$

Since phosphoric acid is a weak acid, the pH will be affected by the first ionization of H_3PO_4 to form $H_2PO_4^-$ (other equilibria can be ignored).

The equation for the equilibrium and the equilibrium expression are:

$$H_3PO_4(aq) + H_2O(\ell) \rightleftharpoons H_3O^+(aq) + H_2PO_4^-(aq) \qquad K_{a1} = \frac{[H_3O^+][H_2PO_4^-]}{[H_3PO_4]}$$

As the reactants decompose, the concentrations of the products increase stoichiometrically, until they reach equilibrium concentrations.

	$H_3PO_4(aq)$	$H_3O^+(aq)$	$H_2PO_4^-(aq)$
initial conc. (M)	0.646	0	0
change as the reaction occurs (M)	– x	+ x	+ x
equilibrium conc. (M)	0.646 – x	x	x

K_{a1}, provided in the text (Appendix F) is 7.2×10^{-3}.

At equilibrium: $$K_{a1} = \frac{(x)(x)}{(0.010 - x)} = 7.2 \times 10^{-3}$$

$$x^2 = (7.2 \times 10^{-3})(0.010 - x)$$

$$x^2 + (7.2 \times 10^{-3})x - 7.2 \times 10^{-5} = 0$$

Using the quadratic equation, determine the value of x, then use that to determine the pH:

$$x = 0.0056\ M = [H_3O^+]$$

$$pH = -\log[H_3O^+] = -\log(0.0056) = 2.25$$

(b) Precipitation of calcium phosphate occurs in this reaction, as shown in balanced form below. The limiting reagent, as indicated in the question, is phosphoric acid.

$$3\ Ca(NO_3)_2(aq) + 2\ H_3PO_4(aq) \longrightarrow Ca_3(PO_4)_2(s) + 6\ HNO_3(aq)$$

$$0.1614\ \text{mol}\ H_3PO_4 \times \frac{1\ \text{mol}\ Ca_3(PO_4)_2}{2\ \text{mol}\ H_3PO_4} \times \frac{310.1768\ \text{g}\ Ca_3(PO_4)_2}{1\ \text{mol}\ Ca_3(PO_4)_2} = 25.0\ \text{g}\ Ca_3(PO_4)_2$$

(c) Zinc, an active metal, reacts with acid to form hydrogen gas and a metal salt (ionic compound). Notice that all (active) metals react in a similar manner.

$$Zn(s) + 2\ HNO_3(aq) \longrightarrow Zn(NO_3)_2(aq) + 6\ H_2(g)$$

$$0.1614\ \text{mol}\ H_3PO_4 \times \frac{6\ \text{mol}\ HNO_3}{2\ \text{mol}\ H_3PO_4} \times \frac{1\ \text{mol}\ H_2}{2\ \text{mol}\ HNO_3} = 0.242\ \text{mol}\ H_2\ \text{generated}$$

Use the molar volume of a gas at STP (1 atm and 0°C) to determine the volume of hydrogen gas.

$$0.242\ \text{mol}\ H_2 \times \frac{22.414\ \text{L}}{1\ \text{mol}\ H_2} = 5.42\ \text{L}$$

68. *Answer:* **(a) Al and N are oxidized; Cl is reduced (b) –2674 kJ/mol**

Strategy and Explanation:

(a) Aluminum metal is oxidized (from Ox. # = zero to Al^{+3}).

Nitrogen is oxidized (from Ox. # = +3, in the ammonium ion of NH_4ClO_4 to Ox. # = +2, in NO).

Chlorine is reduced (from Ox. # = +7 in the perchlorate ion of NH_4ClO_4 to Ox. # = –1 in chloride Cl^-).

(a) The enthalpy change is the difference between the sum of the enthalpies of formation of the products and the sum of the enthalpies of formation of the reactants. Enthalpies of formation are found in Appendix J.

$$\Delta H° = \sum\left[(\text{moles of product}) \times \Delta H_f°(\text{product})\right] - \sum\left[(\text{moles of reactant}) \times \Delta H_f°(\text{reactant})\right]$$

$$= [\Delta H_f°(Al_2O_3) + \Delta H_f°(AlCl_3) + 6\ \Delta H_f°(H_2O) + 3\ \Delta H_f°(NO)] - [3\ \Delta H_f°(NH_4ClO_4) + 3\ \Delta H_f°(Al)]$$

$$= [(1\ \text{mol}) \times (-1657.7\ \text{kJ/mol}) + (1\ \text{mol}) \times (-704.2\ \text{kJ/mol}) + (6\ \text{mol}) \times (-241.818\ \text{kJ/mol})$$

$$+ (3\ \text{mol}) \times (90.25\ \text{kJ/mol})] - [(3\ \text{mol}) \times (-295\ \text{kJ/mol}) + (3\ \text{mol}) \times (0\ \text{kJ/mol})] = -2674\ \text{kJ/mol}$$

70. *Answer:* **–2708 kJ/mol**

Strategy and Explanation: Adapt the methods used in the solution to Question 68(b). We will assume that the form of carbon in the reaction is graphite.

First, we will use the enthalpy of sublimation of phosphorus to determine the enthalpy of formation of $P_4(g)$. The standard enthalpy of formation of white phosphorus, $P_4(g) = 0$, according to Appendix J.

$$\Delta H° = \sum\left[(\text{moles of product}) \times \Delta H_f°(\text{product})\right] - \sum\left[(\text{moles of reactant}) \times \Delta H_f°(\text{reactant})\right]$$

Sublimation: $P_4(s) \longrightarrow P_4(g)$ $\quad \Delta H°_{rxn} = 13.06\ \text{kJ} = \Delta H_f°(P_4(g)) - \Delta H_f°(P_4(s))$

$$\Delta H°_{rxn} = [6\ \Delta H_f°(CaSiO_3) + 10\ \Delta H_f°(CO) + \Delta H_f°(P_4(g))]$$

$$- [2\ \Delta H_f°(Ca_3(PO_4)_2 + 6\ \Delta H_f°(SiO_2) + 3\ \Delta H_f°(C(s))]$$

$$\Delta H°_{rxn} = [(6\ \text{mol}) \times \Delta H_f°(CaSiO_3) + (10\ \text{mol}) \times (-110.525\ \text{kJ/mol}) + (1\ \text{mol}) \times (13.06\ \text{kJ/mol})]$$

$- [(2 \text{ mol}) \times (-4138 \text{ kJ/mol}) + (6 \text{ mol}) \times (-910.94 \text{ kJ/mol}) + (10 \text{ mol}) \times (0 \text{ kJ/mol})] = -3060 \text{ kJ/mol}$

Solve for $\Delta H^\circ_f(CaSiO_3)$:

$$(6 \text{ mol}) \times \Delta H^\circ_f(CaSiO_3) = -3060 \text{ kJ} + (1105.25 \text{ kJ}) - (13.06 \text{ kJ}) - (8276 \text{ kJ}) - (5465.64 \text{ kJ})$$

$$\Delta H^\circ_f(CaSiO_3) = -2708 \text{ kJ/mol}$$

72. *Answer:* $\mathbf{K_p = 1.985}$

Strategy and Explanation: The equilibrium constant requested in the question is for the reverse of the equation provided. On reversing a reaction's equation the sign of the Gibbs free energy changes. 25°C + 273.15 = 289.15 K. Use the methods described in Section 18.7 and Equation 18.8 to solve for the equilibrium constant.

$$\Delta G^\circ = -RT\ln K_p$$

$$-1700 \text{J/mol} = -(8.314 \text{ J mol}^{-1}\text{K}^{-1})(298.15\text{K}) \ln K$$

$$K_p = 1.985$$

74. *Answer:* **$\Delta H_{Lattice}$ = 2961 kJ; Mg^{2+} is smaller and has a greater charge density than Sr^{2+}, so the Mg^{2+} ions and the F^- ions are closer together and the ionic bond is much stronger.**

Strategy and Explanation: Lattice enthalpy is defined as the energy necessary to convert a solid salt to its constituent ions in the gas phase, as illustrated for strontium fluoride below.

$$SrF_2(s) \longrightarrow Sr^{2+} + 2\,F^- \qquad \Delta H_{Lattice} = 2496 \text{ kJ}$$

An application of Hess' Law, called a Born-Haber cycle, can be used to determine the lattice enthalpy of an ionic compound. The equations needed are provided in the question. As always, pay close attention to the reaction stoichiometry and sign. For the enthalpies listed the abbreviations are as follows:

sub = sublimation (conversion of a solid to the gas phase);

IE = ionization energy (the energy required to remove an electron)—notice in this case the magnesium metal requires two separate ionizations;

BE = bond enthalpy—the energy required to break a bond between two elements;

EA = electron affinity;

f = formation

Notice in the cycle below, the equation used is the opposite of a formation reaction, thus the sign of the enthalpy must be reversed.

$$Mg(s) \longrightarrow Mg(g) \qquad \Delta H_{sub} = 146 \text{ kJ}$$

$$Mg(g) \longrightarrow Mg^+(g) + e^- \qquad \Delta H_{IE1} = 738 \text{ kJ}$$

$$Mg^+(g) \longrightarrow Mg^{2+}(g) + e^- \qquad \Delta H_{IE2} = 1451 \text{ kJ}$$

$$F_2(g) \longrightarrow 2\,F(g) \qquad \Delta H_{BE} = 158 \text{ kJ}$$

$$2\,F(g) + 2\,e^- \longrightarrow 2\,F^- \qquad 2 \times \Delta H_{EA} = -656 \text{ kJ}$$

$$+\ MgF_2(s) \longrightarrow Mg(s) + F_2(g) \qquad -1 \times \Delta H_f = 1124 \text{ kJ}$$

$$MgF_2(s) \longrightarrow Mg^{2+} + 2\,F^- \qquad \Delta H_{Lattice} = 2961 \text{ kJ}$$

Comparison of strontium fluoride and magnesium fluoride lattice enthalpies: Ionic interactions dominate the bonding in most salts (ionic compounds). This effect is further aided by large differences in electronegativity between the cation and anions that form a given salt. In general, the greater the electronegativity difference between the elements in the salt, the greater the strength of the ionic bond.

For cations of the same charge, as is the case for Mg^{2+} and Sr^{2+}, electrostatic attractions to F^- are affected by the distance between the anion and the cation (electrostatic attractions fall off as $1/r^2$ where r = the radius of the charged particle). The ionic radii for Mg^{2+} and Sr^{2+} are 0.72 and 1.16 angstroms respectively. Because Mg^{2+} is significantly smaller, it can more closely approach the F^- resulting in a stronger interaction. In addition, because of its smaller size (and fewer electrons) the charge of the magnesium ion is distributed over a smaller volume in comparison to the strontium ion, resulting in a greater charge density. The combination of these two factors, both dependent on the ion size, result in a stronger ionic bond and stronger lattice interactions for Mg^{2+} relative to Sr^{2+}.

76. *Answer:* $\mathbf{B_6H_{12}}$

Strategy and Explanation: The mass of boron and hydrogen in a 1 mol of sample (76.7 g) can be determined by multiplying percent by the molecular formula. Converting these masses to moles give rations for the molecular formula.

$$\frac{76.7\text{ g compound}}{1\text{ mol compound}} \times \frac{84.2\text{ g B}}{100\text{ g compound}} \times \frac{1\text{ mol B}}{10.811\text{ g B}} = 6.00\text{ mol B per mol}$$

$$\frac{76.7\text{ g compound}}{1\text{ mol compound}} \times \frac{15.7\text{ g H}}{100\text{ g compound}} \times \frac{1\text{ mol H}}{1.0079\text{ g H}} = 12.0\text{ mol H per mol}$$

This gives a whole number ratio per mol of compound: 6.00 mol B : 12.0 mol H

Molecular Formula = B_6H_{12}

Chapter 22: Chemistry of Selected Transition Elements and Coordination Compounds Solutions for Blue-Numbered Questions for Review and Thought

Topical Questions

Transition Metals (Section 22.1)

14. *Answer:* **(a) Ag [Kr]$4d^{10}5s^1$ and Ag^+ [Kr]$4d^{10}$ (b) Au [Xe]$4f^{14}5d^{10}6s^1$, Au^+ [Xe]$4f^{14}5d^{10}$, and Au^{3+} [Xe]$4f^{14}5d^8$**

Strategy and Explanation: Follow the method described in Sections 7.7, 7.8 and 22.1.

(a) The most common oxidation states of silver are zero and +1.

The $_{47}Ag$ electron configuration: [Kr]$4d^{10}5s^1$ or $1s^22s^22p^63s^23p^63d^{10}4s^24p^64d^{10}5s^1$

The $_{47}Ag^+$ electron configuration: [Kr]$4d^{10}$

(b) The most common oxidation state of gold is zero. The $_{79}Au$ electron configuration:

[Xe]$4f^{14}5d^{10}6s^1$ or $1s^22s^22p^63s^23p^63d^{10}4s^24p^64d^{10}4f^{14}5s^25p^65d^{10}6s^1$

Other oxidation states of gold are +1 and +3.

The $_{79}Au^+$ electron configuration: [Xe]$4f^{14}5d^{10}$

The $_{79}Au^{3+}$ electron configuration: [Xe]$4f^{14}5d^8$

16. *Answer:* **Cr^{2+} and Cr^{3+}**

Strategy and Explanation: The neutral Cr atom has the following orbital diagram, with six unpaired electrons:

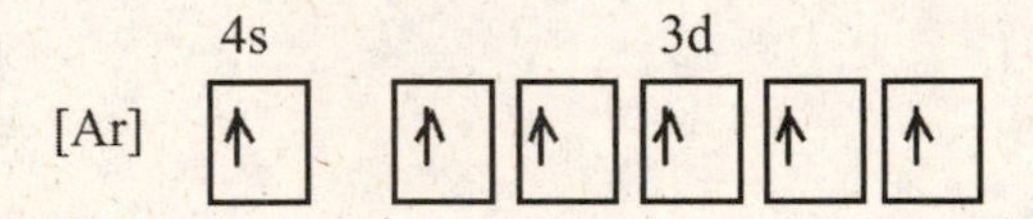

When Cr has lost all six electrons to gain an oxidation state of +6 it will have no unpaired electrons. All oxidation states between zero and +6 will apparently have unpaired electrons. Because the higher oxidation states of Cr^{4+} and Cr^{5+} usually have the chromium atom combined with other elements, which may change the number of unpaired electrons, we will predict that the question is asking for Cr^{2+} and Cr^{3+}, though all four of these oxidation states could be legitimate answers with the information we have.

18. *Answer:* **$Cr_2O_7^{2-}$ (in acid) > Cr^{3+} > Cr^{2+}; the species with a more positive oxidation state of Cr has the greater tendency to be reduced.**

Strategy and Explanation: The species with the more positive oxidation state of Cr has the greater tendency to be reduced, hence acting as a stronger oxidizing agent. $Cr_2O_7^{2-}$ has Ox. # Cr = +6, Cr^{3+} has Ox. # Cr = +3, and Cr^{2+} has Ox. # Cr = +2 so $Cr_2O_7^{2-}$ is the best oxidizing agent, then Cr^{3+}, then Cr^{2+}.

20. *Answer:* **(a) $Fe_2O_3 + 3\ CO \longrightarrow 2\ Fe + 3\ CO_2$ (b) $Fe(s) + 2\ H_3O^+(aq) \longrightarrow Fe^{2+}(aq) + 2\ H_2O(\ell) + H_2(g)$ or $2\ Fe + 6\ H_3O^+ \longrightarrow 2\ Fe^{3+} + 6\ H_2O + 3\ H_2$**

Strategy and Explanation: Roasting is described in Section 22.3:

(a) $$Fe_2O_3(s) + 3\ CO(g) \longrightarrow 2\ Fe(s) + 3\ CO_2(g)$$

(b) Use standard redox balancing methods shown in the solution to Question 19.16.

If Fe^{2+} is produced: $Fe(s) + 2\ H_3O^+(aq) \longrightarrow Fe^{2+}(aq) + 2\ H_2O(\ell) + H_2(g)$

If Fe^{3+} is produced: $2\ Fe(s) + 6\ H_3O^+(aq) \longrightarrow 2\ Fe^{3+}(aq) + 6\ H_2O(\ell) + 3\ H_2(g)$

Fe^{3+} is the more likely product.

22. *Answer:* **$3\ NO_3^-(aq) + 6\ H_3O^+(aq) + Fe(aq) \longrightarrow Fe^{3+}(aq) + 3\ NO_2(g) + 9\ H_2O(\ell)$**

Strategy and Explanation: Use standard redox balancing methods shown in the solution to Question 19.16.

$$3\ NO_3^-(aq) + 6\ H^+(aq) + 3\ e^- \longrightarrow 3\ NO_2(g) + 3\ H_2O(\ell)$$

$$Fe(aq) \longrightarrow Fe^{3+}(aq) + 3\ e^-$$

$$3\ NO_3^-(aq) + 6\ H^+(aq) + Fe(aq) \longrightarrow Fe^{3+}(aq) + 3\ NO_2(g) + 3\ H_2O(\ell)$$

Then add two water molecules to each side to convert the H^+ ions into H_3O^+ ions:

$$3\ NO_3^-(aq) + 6\ H_3O^+(aq) + Fe(aq) \longrightarrow Fe^{3+}(aq) + 3\ NO_2(g) + 9\ H_2O(\ell)$$

(Check: 3 N, 15 O, 18 H, 1 Fe, +3 net charge)

Coordination Compounds (Section 22.6)

26. *Answer:* **(a) $[Cr(NH_3)_2(H_2O)_3(OH)]^{2+}$; +2 (b) The counter ion would be an anion, such as Cl^-.**

Strategy and Explanation: Cr^{3+} is bonded to two NH_3 ligands, three H_2O ligands, and one OH^-.

(a) The complex ion has a formula that looks like this: $[Cr(NH_3)_2(H_2O)_3(OH)]^{2+}$. The net charge is +2.

(b) The complex ion formed is a cation, so it will need a counter ion that is an anion, such as Cl^-.

28. *Answer:* **(a) $C_2O_4^{2-}$ ligands with –2 charge; Cl^- ligands with –1 charge (b) +3 charge (c) $[Co(NH_3)_4Cl_2]^+$ with +1 charge**

Strategy and Explanation:

(a) The complex ion $[Co(C_2O_4)_2Cl_2]^{3-}$ has two $C_2O_4^{2-}$ ligands, each with –2 charge and two Cl^- ions, each with –1 charge.

(b) The net charge is –3. The sum of the individual charges must add up to this number. So:

$$-3 = (\text{cobalt charge}) + 2\times(-2) + 2\times(-1)$$

The cobalt is in the form of Co^{3+} with a charge of +3.

(c) $C_2O_4^{2-}$ is a bidentate ligand and NH_3 is a monodentate, so two $C_2O_4^{2-}$ must be replaced by four NH_3. Replacing two –2 ions with four molecules, reduces the net charge of the ion by 4 also. Therefore, the new complex ion has a formula that looks like this: $[Co(NH_3)_4Cl_2]^+$. It has a +1 charge.

30. *Answer:* **For $Na_3[IrCl_6]$: (a) Six Cl^- (b) Ir^{3+} with +3 charge (c) $[IrCl_6]^{3-}$ with –3 charge (d) Na^+ For $[Mo(CO)_4]Br_2$: (a) Four CO (b) Mo^{2+}; +2 charge (c) $[Mo(CO)_4]^{2+}$ with 2+ charge (d) Br^-**

Strategy and Explanation: Consider $Na_3[IrCl_6]$

(a) The complex ion has six Cl^- ions as ligands.

(b) $0 = +3\times(+1) + (\text{iridium charge}) + 6\times(-1)$ The iridium is in the form of Ir^{3+} with a 3+ charge.

(c) The formula of the complex ion is $[IrCl_6]^{3-}$ with a 3– charge.

(d) The ions that are not in the complex ion are the three Na^+ cations.

Consider $[Mo(CO)_4]Br_2$

(a) The complex ion has four CO molecules as ligands.

(b) 0 = (molybdenum charge) + 4×(0) + 2×(–1) The molybdenum is in the form of Mo^{2+} with a +2 charge.

(c) The formula of the complex ion is $[Mo(CO)_4]^{2+}$ with a 2+ charge.

(d) The ions that are not in the complex ion are the two Br^- anions.

32. *Answer:* **(a) 4 (b) 4**

Strategy and Explanation: Monodentates need one coordination site. Bidentates need two coordination sites. Thus, we need to add the number of monodentate ligands to 2×(number of bidentate ligands) to find the coordination number.

(a) In $[Pt(en)_2]^{2+}$, we have two bidentate ligands (2 en), so the coordination number is 2×(2) = 4.

(b) In $[Cu(ox)_2]^{2-}$, we have two bidentate ligands (2 ox), so the coordination number is 2×(2) = 4.

34. *Answer:* **(a) $[Pt(NH_3)_2Br_2]$ (b) $[Pt(en)(NO_2)_2]$ (c) $[Pt(NH_3)_2BrCl]$**

Strategy and Explanation: Use the method described in the solutions to Questions 25 to 33.

(a) Br^- and NH_3 are monodentate ligands, making $[Pt(NH_3)_2Br_2]$.

(b) The ligand en is a bidentate ligand. NO_2^- is a monodentate ligand. The complex is $[Pt(en)(NO_2)_2]$.

(c) Br^-, Cl^-, and NH_3 are monodentate ligands, making $[Pt(NH_3)_2BrCl]$.

36. *Answer:* **(a) 4 (b) 4 (c) 6 (d) 6**

Strategy and Explanation: Use the method described in the solution to Question 32.

(a) In $[FeCl_4]^-$, we have four monodentate ligands (4 Cl^-), so the coordination number is 4×(1) = 4.

(b) In $[PtBr_4]^{2-}$, we have four monodentate ligands (4 Br^-), so the coordination number is 4×(1) = 4.

(c) In $[Mn(en)_3]^{2+}$, we have three bidentate ligands (3 en), so the coordination number is 3×(2) = 6.

(d) In $[Cr(NH_3)_5H_2O]^{3+}$, we have six monodentate ligands (5 NH_3 and 1 H_2O), so the coordination number is 6×(1) = 6.

38. *Answer:* **(a) monodentate (b) tetradentate (c) tridentate (d) monodentate**

Strategy and Explanation:

(a) $(CH_3)_3P$ has only one electron pair on the P atom, so it is a monodentate ligand.

(b) $(^-OOC)–CH_2–N(COO^-)_2$ has lone pairs of electrons on every O atom as well as the N atom, so in theory it could be heptadentate. In practice, only one of the two O atoms on the carboxyl anion groups and the N atom probably have simultaneous access to the metal atom, due to geometric constraints, so this ligand will usually be tetradentate.

(c) $H_2N–CH_2–CH_2–NH–CH_2–CH_2–NH_2$ has lone pairs of electrons on every N atom, so is a tridentate ligand.

(d) H_2O has electron pairs on only one atom, so is a monodentate ligand.

40. *Answer:* **$FeCl_3$ (Other correct answers are possible.)**

Strategy and Explanation: We need a compound that is composed of a 3+ cation, analogous to the $[Rh(en)]^{3+}$ ion, combined with three simple –1 anions. One example is $FeCl_3$ made with Fe^{3+} and three Cl^- ions.

Naming Complex Ions and Coordination Compounds (Section 22.6)

42. *Answer:* **(a) tetrachloromanganate(II) (b) potassium trioxalatoferrate(III)
(c) diamminedicyanoplatinum(II) (d) pentaquahydroxoiron(III)
(e) diethylenediaminedichoromanganese(II)**

Strategy and Explanation:

(a) Four chloride ligands coordinate-covalently bonded to a Mn^{2+} ion to make a complex anion with a –2 charge, called tetrachloromanganate(II) ion.

(b) Three potassium counter ions combine with a complex anion composed of three oxalate anion ligands coordinate-covalently bonded to Fe^{3+} to form a neutral salt called potassium trioxalatoferrate(III) ion.

(c) Two ammonia ligands (NH_3) and two cyanide ligands (CN^-) are coordinate-covalently bonded to a Pt^{2+} metal ion forming a neutral compound called diamminedicyanoplatinum(II).

(d) Five water ligands and one hydroxide ligand are coordinate-covalently bonded to Fe^{3+} to form a complex cation with a +2 charge called pentaquahydroxoiron(III) ion.

(e) Hexacoordinated manganese(II) ion has two bidentate en ligands and two chloride ligands to form a neutral compound called diethylenediaminedichoromanganese(II). Sometimes this compound is also called bis(ethylenediamine)dichloro manganese (II).

Geometry of Coordination Complexes (Section 22.6)

44. *Answer:* **see sketches, below**

Strategy and Explanation:

(a) *cis*-$[Pt(H_2O)_2Cl_2]^{2-}$ has the H_2O molecules adjacent to each other in the square planar coordination arrangement:

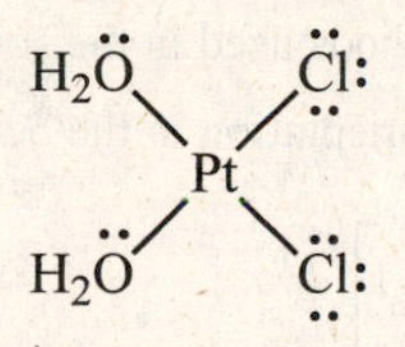

(b) *trans*-$[Cr(H_2O)_4Cl_2]^+$ has the Cl^- ligands across from each other in the octahedral coordination arrangement:

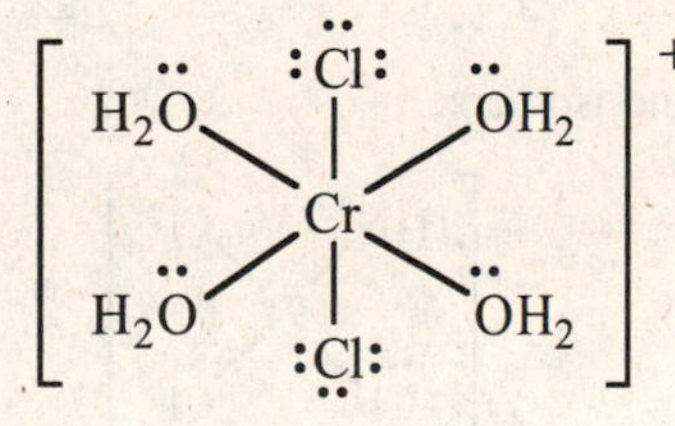

46. *Answer:* **see sketch, below**

Strategy and Explanation: $Co(acac)_3$ has the acetylacetonate ions in an octahedral coordination arrangement:

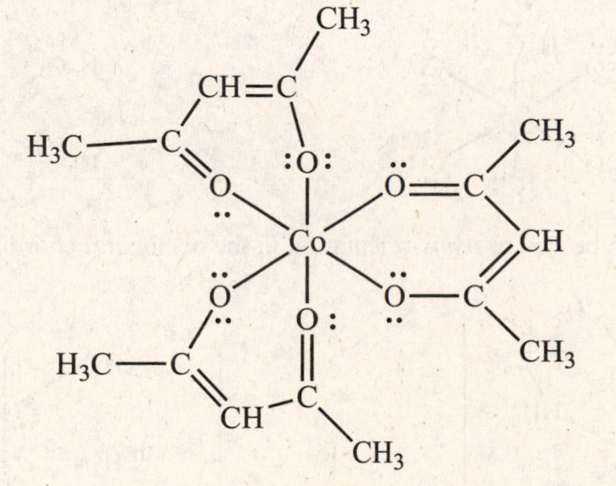

48. *Answer:* **Both (a) and (b)**

Strategy and Explanation:

(a) The triples could have an all-*cis*-orientation, or a pair of each triple could be *trans* while both are *cis* to the third:

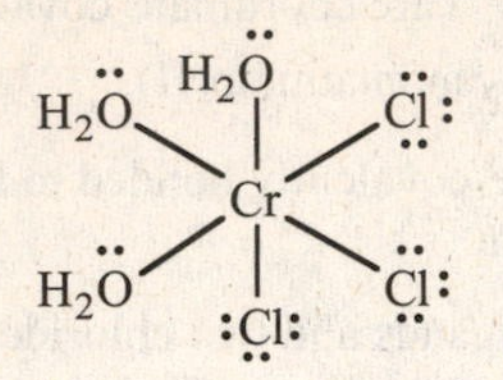

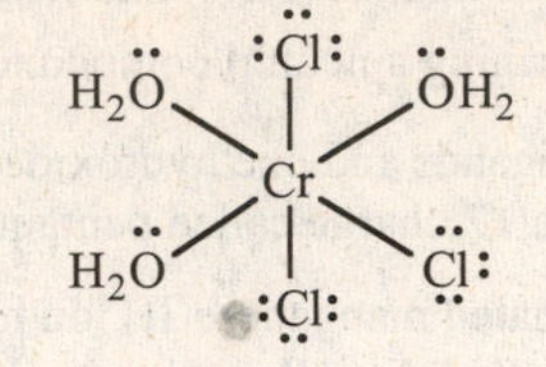

(b) The Cl^- ligands can be *cis*- or *trans*-orientation:

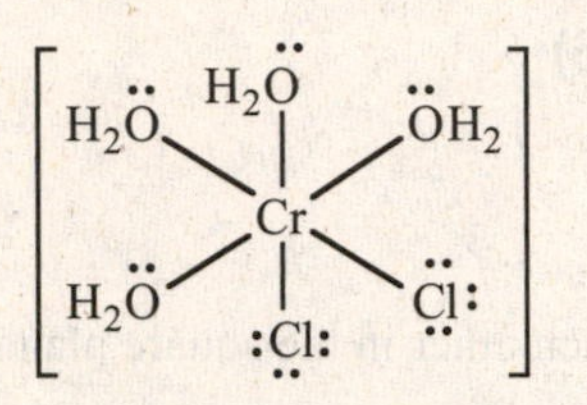

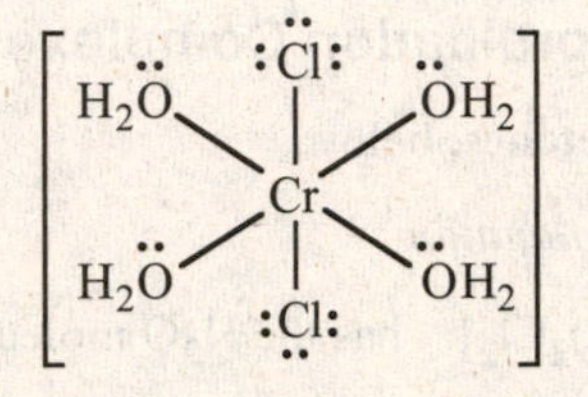

50. *Answer:* **(a), (c), and (d)**

Strategy and Explanation: Adapt the methods used in the solutions to Questions 43 to 49.

(a) The Cl^- ligands can be *cis*- or *trans*-orientation in the octahedral coordination arrangement:

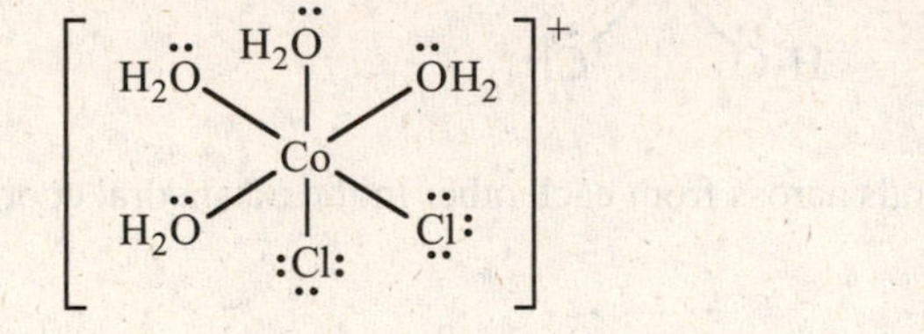

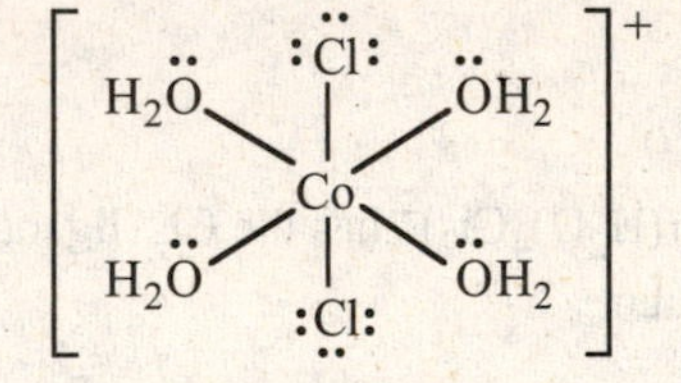

(b) This square planar structure has no isomers:

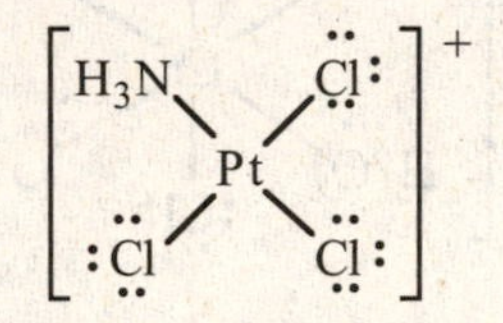

(c) The triples could have an all-*cis*-orientation, or a pair of each triple could be *trans* while both *cis* to the third in the octahedral coordination arrangement:

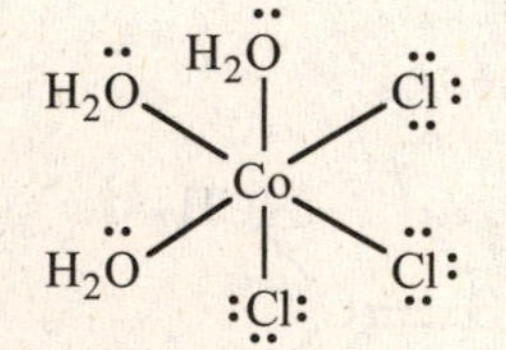

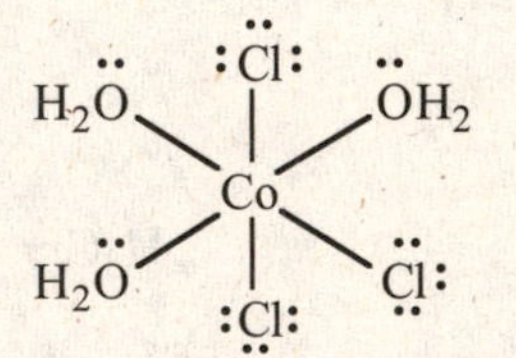

(d) The NH_3 ligands can be *cis*- or *trans*-orientation in the octahedral coordination arrangement:

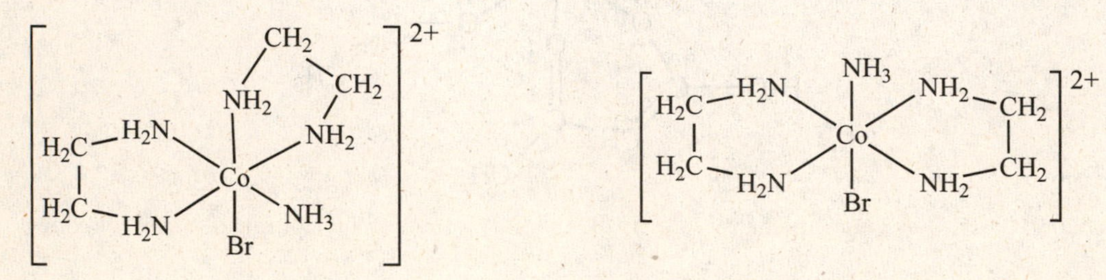

Crystal-Field Theory and Magnetic Properties of Complex Ions (Section 22.7)

52. *Answer:* **(a) 4 (b) 5 (c) 5 (d) 3**

Strategy and Explanation: A truncated version of the spectrochemical series is located in Section 22.7, on page 1094.

(a) Water (H_2O) is a weak field (high spin) ligand. Unpaired electrons = 4

(b) Water (H_2O) is a weak field (high spin) ligand. Unpaired electrons = 5

(c) Fluoride ion (F^-) is a weak field (high spin) ligand. Unpaired electrons = 5

(d) Only one configuration is possible (3 unpaired electrons)

54. *Answer/Explanation:* Only on the addition of a fourth electron will there be a choice between electron pairing or electron promotion to an empty higher-energy orbital. The choice between electron pairing or electron promotion in addition depends on the difference in energy (called Δ_o) between the lower-energy (triply degenerate set) orbitals and the higher-energy (doubly degenerate set) orbitals. If the energy required to pair electrons in the same orbital is greater than the energy required to promote an electron to a higher orbital, a high spin configuration will result.

In this case Cr^{+2} is a d^4 ion and has four d-electrons, while Cr^{+3} is a d^3 ion with three d-electrons. Since there are three low-energy t_2 orbitals, the first three electrons can fill one into each t_2 orbital. The fourth electron (only found in Cr2+) is required to pair up with one of the other electrons in a t_2 orbital (low spin) or to span the Δ_o gap to go into and e orbital (high spin). Only the d^4 ion has a choice of a low spin or high spin configuration.

56. *Answer:* **If Δ_o is large, electrons fill the t_2 orbitals and end up all paired. If Δ_o is small, all five orbitals are half-filled before any pairing begins, resulting in four unpaired electrons, so high-spin Co^{3+} complexes are paramagnetic and low-spin Co^{3+} complexes are diamagnetic.**

Strategy and Explanation: A complex with no unpaired electrons is diamagnetic. A complex with one or more unpaired electrons is paramagnetic. Co^{3+} is a d^6 ion. In a high spin configuration four electrons are unpaired. In a low spin configuration all electrons are paired.

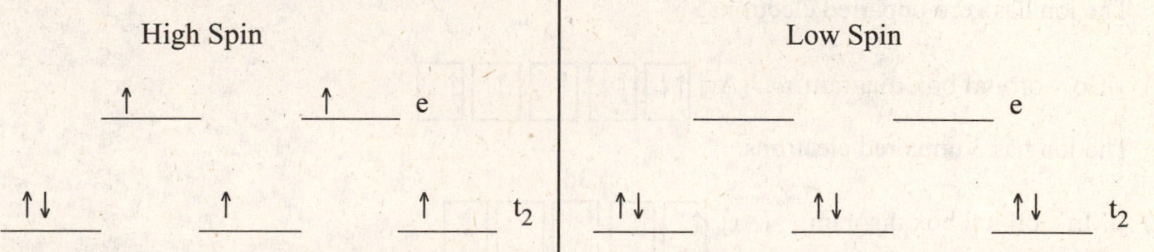

58. *Answer/Explanation:* Cu^{2+} is a d^9 ion. There are so many electrons that three electrons must always go in the high-energy e-orbitals.

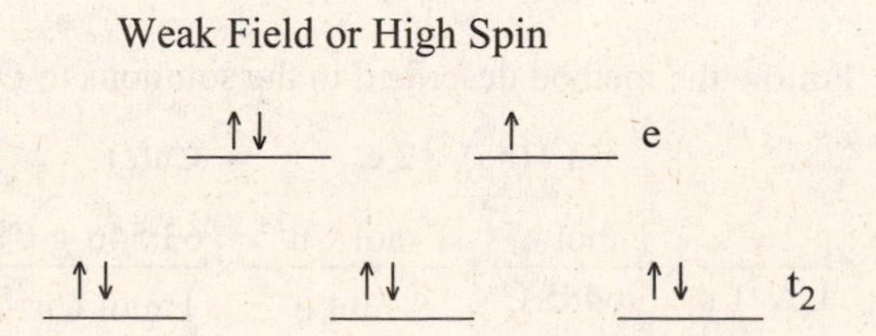

Crystal-Field Theory and Color in Complex Ions (Section 22.7)

60. *Answer:* **violet light (~400 nm)**

Strategy and Explanation: Refer to the color wheel provided in Figure 22.22. Draw a line from the color your eyes detect (assuming you are not color blind) *through the center of the wheel* to the opposite color (the complimentary color). The complimentary color is the color absorbed by the complex. A yellow complex would absorb violet light (~ 400 nm).

62. *Answer:* **With cyanide ligand, Δ_o should increase, color changes from purple to yellow-orange; with chloride ligand, Δ_o should decrease, color changes purple to a blue.**

Strategy and Explanation: Cyanide ion is a strong field ligand, thus Δ_o should increase. The color of the solution is expected to change to a yellow-orange shade (assuming approximately 450 nm or blue light is absorbed). Chloride ion is a weak field ligand (weaker than water), thus Δ_o should decrease. The color of the solution is expected to change to a blue shade (assuming approximately 600 nm or yellow light is absorbed).

General Questions

65. *Answer:* **(a) $[Ar]3d^4$ (b) $[Ar]3d^{10}$ (c) $[Ar]3d^7$ (d) $[Ar]3d^3$**

Strategy and Explanation: Follow the method described in Sections 7.8 and 22.1 and in the solution to Question 14.

(a) ${}_{24}Cr^{2+}$ electron configuration: $[Ar]3d^4$ or $1s^22s^22p^63s^23p^63d^4$

(b) ${}_{30}Zn^{2+}$ electron configuration: $[Ar]3d^{10}$ or $1s^22s^22p^63s^23p^63d^{10}$

(c) ${}_{27}Co^{2+}$ electron configuration: $[Ar]3d^7$ or $1s^22s^22p^63s^23p^63d^7$

(d) ${}_{25}Mn^{4+}$ electron configuration: $[Ar]3d^3$ or $1s^22s^22p^63s^23p^63d^3$

67. *Answer:* **see orbital box dieagrams, below (a) 4 (b) 0 (c) 3 (d) 3**

Strategy and Explanation: Follow the method described in the solution to Question 7.76 using the solution from Question 65.

(a) ${}_{24}Cr^{2+}$ orbital box diagram: [Ar] 3d: [↑] [↑] [↑] [↑] [ ]

The ion has 4 unpaired electrons.

(b) ${}_{30}Zn^{2+}$ orbital box diagram: [Ar] 3d: [↑↓] [↑↓] [↑↓] [↑↓] [↑↓]

The ion has zero unpaired electrons.

(c) ${}_{27}Co^{2+}$ orbital box diagram: [Ar] 3d: [↑↓] [↑↓] [↑] [↑] [↑]

The ion has 3 unpaired electrons.

(d) ${}_{25}Mn^{4+}$ orbital box diagram: [Ar] 3d: [↑] [↑] [↑] [ ] [ ]

The ion has 3 unpaired electrons.

69. *Answer:* **14.8 g Cu**

Strategy and Explanation: Follow the method described in the solutions to Question 19.65.

$$Cu^{2+}(aq) + 2\,e^- \longrightarrow Cu(s)$$

$$5.00\text{ hr} \times 2.50\text{ A} \times \frac{3600\text{ s}}{1\text{ hr}} \times \frac{1\text{ C}}{1\text{ A}\cdot 1\text{ s}} \times \frac{1\text{ mol }e^-}{96485\text{ C}} \times \frac{1\text{ mol }Cu^{2+}}{2\text{ mol }e^-} \times \frac{63.546\text{ g Cu}}{1\text{ mol }Cu^{2+}} = 14.8\text{ g Cu}$$

71. *Answer:* **0.40 ton SO_2**

Strategy and Explanation:

$$2\,Cu_2S(s) + 3\,O_2(g) \longrightarrow 2\,Cu_2O(s) + 2\,SO_2(g)$$

$$1.0\text{ ton CuS} \times \frac{2000\text{ lb}}{1\text{ ton}} \times \frac{453.6\text{ g}}{1\text{ lb}} \times \frac{1\text{ mol }Cu_2S}{159.157\text{ g }Cu_2S} \times \frac{2\text{ mol }SO_2}{2\text{ mol }Cu_2S}$$

$$\times \frac{64.064\text{ g }SO_2}{1\text{ mol }SO_2} \times \frac{1\text{ lb}}{453.6\text{ g}} \times \frac{1\text{ ton}}{2000\text{ lb}} = 0.40\text{ ton }SO_2$$

73. *Answer:* **(a) 4 (b) 6 (c) 4 (d) 6**

Strategy and Explanation: Use the methods described in the solutions to Question 32.

(a) In $[Ni(en)Cl_2]^{2+}$, we have one bidentate ligands (en) and two monodentate ligands (2 Cl^-), so the coordination number is $2 + 2\times(1) = 4$.

(b) In $[Mo(CO)_4Br_2]$, we have six monodentate ligands (4 CO and 2 Br^-), so the coordination number is 6.

(c) In $[Cd(CN)_4]^{2+}$, we have four monodentate ligands (4 CN^-), so the coordination number is 4.

(d) In $[Co(CN)_4(OH)]^{3-}$, we have six monodentate ligands (5 CN^- and OH^-), so the coordination number is 6.

75. *Answer:* **see sketches below**

Strategy and Explanation: Be systematic and remember that the complex ion has six ligands, so one of the ions is an uncomplexed counter ion. (Orientations and locations of ligands may vary, but four distinct combinations can be found: two chlorides in trans orientation with a bromide counter ion, two chlorides in cis orientation with a bromide counter ion, bromide and chloride in trans orientation with a chloride counter ion, and bromide and chloride in cis orientation with a chloride counter ion.)

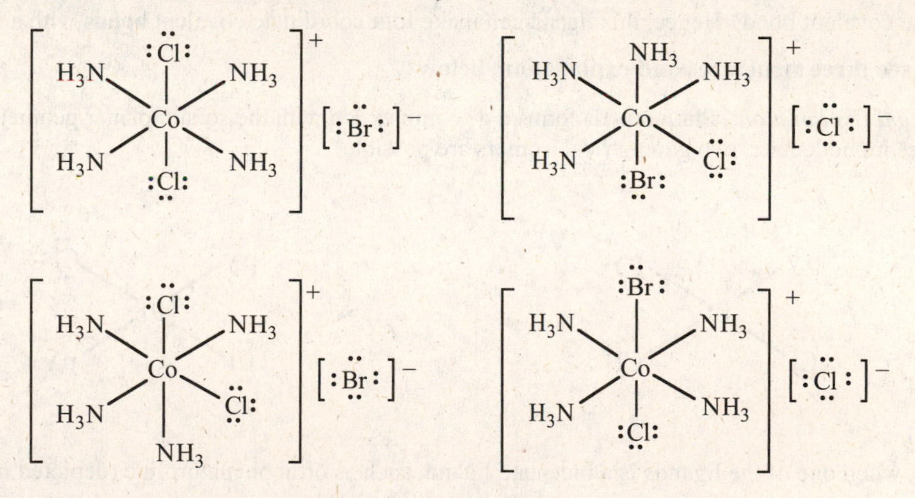

77. *Answer:* **(a) false (b) false (c) false; see corrections, below**

Strategy and Explanation: Use the methods described in the solutions to Questions 25 - 36.

(a) The statement is false. The coordination number of the Fe^{3+} ion in $[Fe(H_2O)_4(C_2O_4)]^+$ is six.

(b) The statement is false. Cu^+ has no unpaired electrons.

(c) The statement is false. The net charge of a coordination complex of Cr^{3+} with two NH_3 and one en is +3.

78. *Answer:* **(a) false; see correction, below (b) true**

Strategy and Explanation: Use the methods described in the solutions to Questions 25 - 36.

(a) The given statement is false. It is most likely that the coordination compound has a complex cation $[Pt(NH_3)_4]^{4+}$ and four Cl^- counter ions. If that is the case, the coordination number of the Pt^{4+} ion in $[Pt(NH_3)_4]Cl_4$ is four. In the unlikely event that the chloride ions are also part of the complex ion, as indicated by the explicit placement of the brackets in the given formula: $[Pt(NH_3)_4Cl_4]$, the coordination number would then have to be eight. The statement can be corrected by changing it to read "In $[Pt(NH_3)_4Cl_4]$ the Pt has a 4+ charge and a coordination number of four or eight, not six."

(b) The given statement is true. In general, Cu^{2+} is more stable than Cu^+ in aqueous solutions.

79. *Answer:* **(a) +2 (b) see structural formula below**

Strategy and Explanation:

(a) The net charge = 0 = (Ox. # Pt) + 2×(0) + (–2). So, Ox. # Pt is +2.

(b) The structural formula for $[Pt(NH_3)_2(C_2O_4)]$:

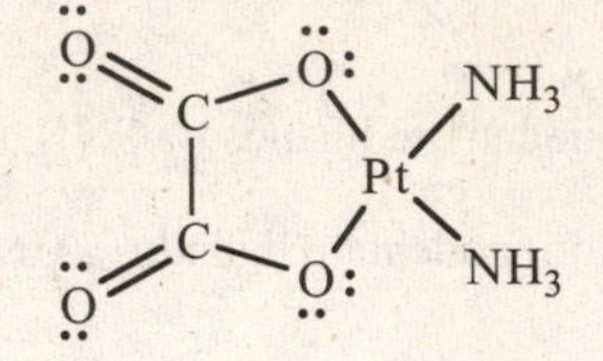

Applying Concepts

81. *Answer/Explanation:* $Fe(s) + 2\ Fe^{3+}(aq) \longrightarrow 3\ Fe^{2+}(aq)$

83. *Answer:* **four**

Strategy and Explanation: Each N atom in the structure has a lone pair of electrons that can be used to make a coordinate covalent bond. Hence, this ligand can make four coordinate covalent bonds with a metal atom.

85. *Answer:* **see three structures and explanation, below**

Strategy and Explanation: Platinum(II) forms a d^8 complex ion with the square planar geometry. When all the ligands are monodentate, two geometrical isomers are possible.

However, when one of the ligands is a bidentate ligand, such as orthophenathroline (depicted in Question 84 as 1,10-phenanthroline), the trans-isomer cannot exist.

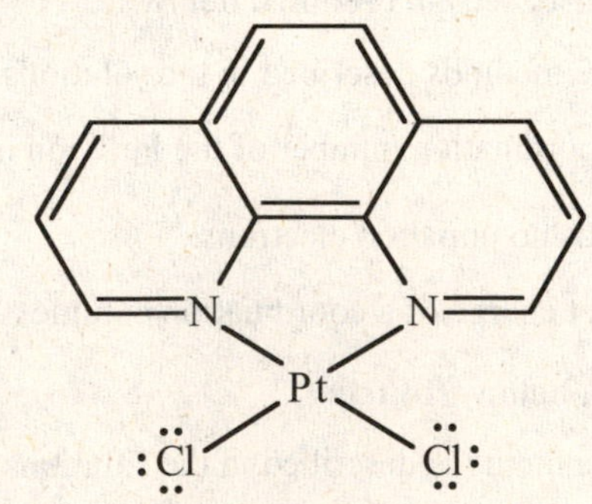

87. *Answer:* **(a) $2\ Ag^+(aq) + Ni(s) \longrightarrow Ni^{2+}(aq) + 2\ Ag(s)$ (b) + 1.05 V (c) see sketch below**

Strategy and Explanation: Use Appendix I or Table 19.1 and methods described in the solution to Question 19.21.

(a) Find a positive $E°_{cell}$ for the product-favored reaction in a cell containing Ni/Ni^{2+} and Ag/Ag^+.

$Ni(s) \longrightarrow Ni^{2+}(aq) + 2\ e^-$	$E°_{anode} = +\ 0.25$ V
$2\ Ag^+(aq) + 2\ e^- \longrightarrow 2\ Ag(s)$	$E°_{cathode} = +\ 0.7991$ V
$2\ Ag^+(aq) + Ni(s) \longrightarrow Ni^{2+}(aq) + 2\ Ag(s)$	$E°_{cell} = +\ 1.05$ V

(b) The cell potential is + 1.05 V.

(c)

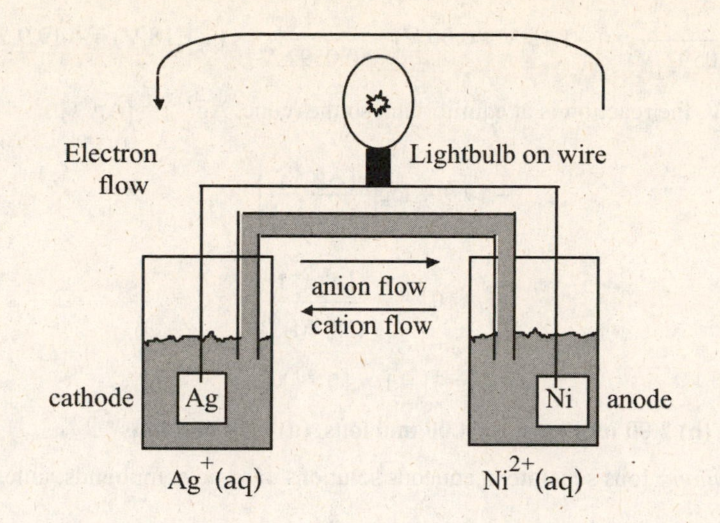

89. *Answer:* **17.4% Fe**

Strategy and Explanation: Balance the equation for the redox reaction as described in the solution to Question 19.16, then use conversion factor methods from Chapters 4 and 5 to determine the percent iron in the ore.

$$1 \times [5\ e^- + 8\ H^+(aq) + MnO_4^-(aq) \longrightarrow Mn^{2+}(aq) + 4\ H_2O(\ell)]$$

$$5 \times [Fe^{2+}(aq) + 1\ e^- \longrightarrow Fe^{3+}(aq)]$$

$$5\ Fe^{2+}(aq) + MnO_4^-(aq) + 8\ H^+(aq) \longrightarrow Mn^{2+}(s) + 5\ Fe^{3+}(aq) + 4\ H_2O(\ell)$$

$$\frac{18.6\ \text{mL}}{1.500\ \text{g ore}} \times \frac{1\ \text{L}}{1000\ \text{mL}} \times \frac{0.05012\ \text{mol MnO}_4^-}{1\ \text{L}} \times \frac{5\ \text{mol Fe}^{2+}}{1\ \text{mol MnO}_4^-}$$

$$\times \frac{1\ \text{mol Fe}}{1\ \text{mol Fe}^{2+}} \times \frac{55.847\ \text{g Fe}}{1\ \text{mol Fe}} \times 100\% = 17.4\%$$

91. *Answer:* **(a) – 1.38 V (b) 1 × 10^{-20} M Ag^{2+}**

Strategy and Explanation: Use the methods described in the solution to Question 19.47.

For this reaction, $Q = \dfrac{(\text{conc Ag}^{2+})}{(\text{conc Ag}^+)^2}$ and n = 1

(a) $$E_{cell} = E°_{cell} - \frac{0.0592\ \text{V}}{n} \log\left(\frac{(\text{conc Ag}^{2+})}{(\text{conc Ag}^+)^2}\right)$$

$$= -1.18\ \text{V} - \frac{0.0592\ \text{V}}{1} \log\left(\frac{\frac{1}{5}(1\times10^{-4}\ \text{M})}{(1\times10^{-4}\ \text{M})^2}\right)$$

$$= -1.18\ \text{V} - 0.0592\ \text{V} \times \log(2 \times 10^3) = -1.18\ \text{V} - 0.0592\ \text{V} \times 3.3$$

$$E = -1.18\ \text{V} - 0.20\ \text{V} = -1.38\ \text{V}$$

(b)
$$\frac{n}{0.0592\text{ V}}(E^\circ_{cell} - E_{cell}) = \log\left(\frac{(\text{conc Ag}^{2+})}{(\text{conc Ag}^{+})^2}\right)$$

$$\frac{1}{0.0592\text{ V}} \times (-1.18\text{ V} - 0.00\text{ V}) = \frac{1}{0.0592\text{ V}} \times (-1.18\text{ V}) = -19.9$$

(When $E_{cell} = 0$ V, the reaction is at equilibrium, so the (conc. Ag^{2+}) = $[Ag^{2+}]$.)

$$-19.9 = \log\left(\frac{[Ag^{2+}]}{(1.0\text{ M})^2}\right)$$

$$1 \times 10^{-20} = \frac{[Ag^{2+}]}{(1.0\text{ M})^2}$$

$$[Ag^{2+}] = 1 \times 10^{-20}\text{ M}$$

93. *Answer:* **(a) no ions (b) 2.00 mol ions (c) 4.00 mol ions (d) 3.00 mol ions**

Strategy and Explanation: Ions separate in aqueous solutions of ionic compounds, unless they are part of stable complex ions.

(a) $[Pt(en)Cl_2]$ is a tetracoordinated coordination complex. When it is dissolved in water, the coordination complex becomes aqueous, but no ions are formed:

$$[Pt(en)Cl_2](s) \longrightarrow [Pt(en)Cl_2](aq)$$

(b) $Na[Cr(en)_2(SO_4)_2]$ is a hexacoordinated complex anion with a sodium counter ion. When it is dissolved in water, it ionizes:

$$Na[Cr(en)_2(SO_4)_2](s) \longrightarrow Na^+(aq) + [Cr(en)_2(SO_4)_2]^-(aq)$$

According to this equation, when 1.00 mol of the solid dissolves, 2.00 mol of ions form.

(c) $K_3[Au(CN)_4]$ is a tetracoordinated complex anion with a potassium counter ion. When it is dissolved in water, it ionizes:

$$K_3[Au(CN)_4](s) \longrightarrow 3\,K^+(aq) + [Au(CN)_4]^{3-}(aq)$$

According to this equation, when 1.00 mol of the solid dissolves, 4.00 mol of ions form.

(d) $[Ni(H_2O)_2(NH_3)_4]Cl_2$ is a hexacoordinated complex cation with a chloride counter ion. When it is dissolved in water, it ionizes:

$$[Ni(H_2O)_2(NH_3)_4]Cl_2(s) \longrightarrow [Ni(H_2O)_2(NH_3)_4]^{2+}(aq) + 2\,Cl^-(aq)$$

According to this equation, when 1.00 mol of the solid dissolves, 3.00 mol of ions form.

96. *Answer:* **see structural formula, below**

Strategy and Explanation: A coordination compound contains one Co^{3+}, one SO_4^{2-}, one Cl^-, and four NH_3. It is clear that Co^{3+} and the four NH_3 molecules are part of the complex ion. We can deduce which of the ions is in the complex ion by observing predicted precipitation reactions. If the ions are in the solution they are free to react; however, they can be protected from participating in a predicted precipitation reaction if they are part of the complex ion. If you need a refresher on solubility rules, refer to Table 5.1. The $BaCl_2$ was added to see if SO_4^{2-} would precipitate, since $BaSO_4$ is insoluble. The $AgNO_3$ was added to see if Cl^- would precipitate, since AgCl is insoluble. Because there was not a precipitate when the $BaCl_2$ was added, the SO_4^{2-} is a part of the complex ion. Because there was a precipitate when the $AgNO_3$ was added, the Cl^- is not part of the complex ion.

$$[Co(NH_3)_4SO_4]Cl$$

Cobalt complexes are hexacoordinated, so the sulfate must be a bidentate, to take up the other two bonding positions that the four ammonia molecules are not occupying. The chloride ion is ionically bonded to the complex cation, shown in structural formula as a separate charged anion, nearby the cation:

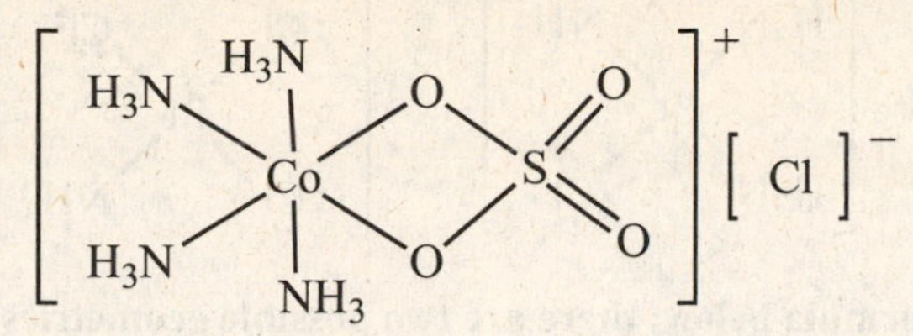

More Challenging Questions

98. *Answer:* **(a) $[Co(NH_3)_6]Cl_3$ (b) $[Co(NH_3)_6]Cl_3(s) \longrightarrow [Co(NH_3)_6]^{3+}(aq) + 3\ Cl^-(aq)$**

Strategy and Explanation: Use the percent by mass to determine the formula, using methods from Chapter 3. Arrange the atoms to form a complex ion with the cobalt such that there are four ions when one formula unit is dissolved in water. For this step, adapt the method described in the solution to Question 80.

(a) In 100.00 g sample of compound, we have 22.0 g Co, 31.4 g N, 6.78 g H, and 39.8 g Cl.

$$22.0\text{ g Co} \times \frac{1\text{ mol Co}}{58.9332\text{ g Co}} = 0.373\text{ mol Co} \qquad 31.4\text{ g N} \times \frac{1\text{ mol N}}{14.0067\text{ g N}} = 2.24\text{ mol N}$$

$$6.78\text{ g H} \times \frac{1\text{ mol H}}{1.0079\text{ g H}} = 6.73\text{ mol H} \qquad 39.8\text{ g Cl} \times \frac{1\text{ mol Cl}}{35.453\text{ g Cl}} = 1.12\text{ mol Cl}$$

$$0.373\text{ mol Co} : 2.24\text{ mol N} : 6.73\text{ mol H} : 1.12\text{ mol Cl}$$

$$1\text{ Co} : 6\text{ N} : 18\text{ H}: 3\text{ Cl}$$

$$CoN_6H_{18}Cl_3$$

A typical ligand in complex ions is NH_3, so let's use the N atoms and H atoms to build NH_3 molecules. $Co(NH_3)_6Cl_3$. We can now arrange the components so that there is one hexacoordinated complex cation, $[Co(NH_3)_6]^{3+}$, with three Cl^- counter ions to make four ions. The formula is $[Co(NH_3)_6]Cl_3$.

(b) The dissociation equation to make an aqueous solution looks like this:

$$[Co(NH_3)_6]Cl_3(s) \longrightarrow [Co(NH_3)_6]^{3+}(aq) + 3\ Cl^-(aq)$$

100. *Answer:* **See structural formula below**

Strategy and Explanation: Adapt the methods described in the solutions to Questions 80 and Chapter 3. The simplest formula, $PtN_2H_6Cl_2$, has a formula mass of 300.05 g/mol. The molar mass is twice that number, so the molecular formula must be $Pt_2N_4H_{12}Cl_4$. A typical ligand in complex ions is NH_3, so let's use the N atoms and H atoms to build NH_3 molecules. $Pt_2(NH_3)_4Cl_4$. We can now arrange the components so that there is one tetracoordinated platinum(II) complex cation and one tetracoordinated platinum(II) complex anion. We do this by putting the neutral NH_3 ligands on Pt^{2+} to form the cation, and the negative Cl^- ligands on Pt^{2+} to form the anion. Therefore, a logical formula for the compound is $[Pt(NH_3)_4]^{2+}[PtCl_4]^{2-}$. Its structural formula looks like this:

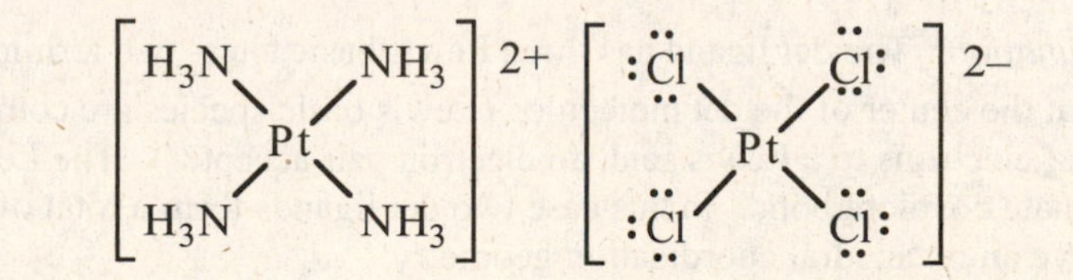

One other formula would also fit the given information: $[Pt(NH_3)_3Cl]^+[Pt(NH_3)Cl_4]^-$.

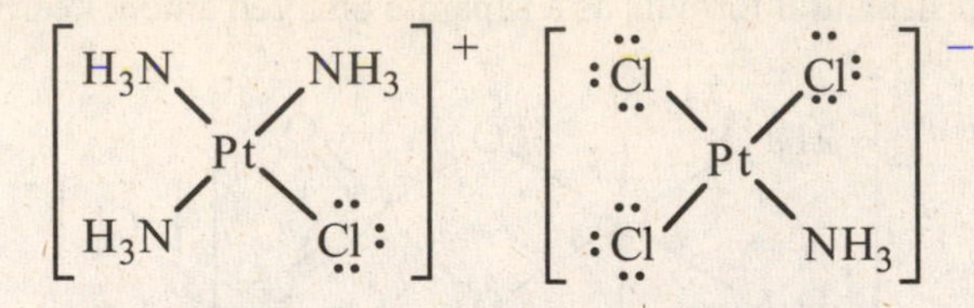

102. *Answer:* **See structural formula below; there are two possible geometries: the copper complex is shown in a trigonal bipyramidal geometry while the nickel complex is shown in a square pyramidal geometry**

Strategy and Explanation: Five-coordinated complexes can adapt two possible geometries, either square pyramidal or trigonal bipyramidal. The geometry a given complex adapts depends on a variety of factors which influence the electronic structure of the complex, such as the types of ligands, the central transition metal oxidation number, the complex ion's surrounding environment (such as counter ion), and crystal packing. In solution both geometries are often in dynamic equilibrium (fluxional) because the energy difference between the two geometries is very small. In the solid state one often sees a structure that is a mixture of the two ideal geometries (i.e., it is neither square pyramidal nor trigonal bypyramidal but rather somewhere in between). Only in cases where there is a clear energetic preference for one structure over another will one geometry be favored. In this case the copper complex is shown in a trigonal bipyramidal geometry while the nickel complex is shown in a square pyramidal geometry.

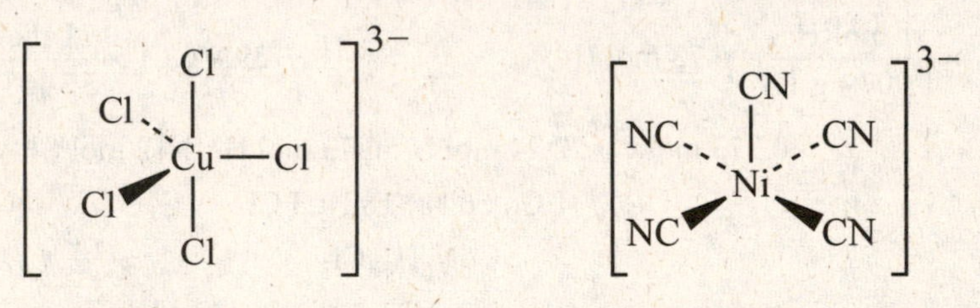

104. *Answer/Explanation:* Assuming the Fe^{+2} is the only oxidation state possible and all complexes are octahedral limits the material to a mixture of two possible geometrical isomers (the cis and trans cyano complexes pictured below). A mixture of high spin and low spin complexes, in which the high spin species is favored by a 2:1 ratio, would give the observed 2.67 unpaired electrons per iron ion. High spin iron(II) has four unpaired electrons, while low spin iron(II) has zero unpaired electrons. This translates to 8 unpaired electrons for every three iron ions or 8/3 = 2.67 unpaired electrons per iron ion.

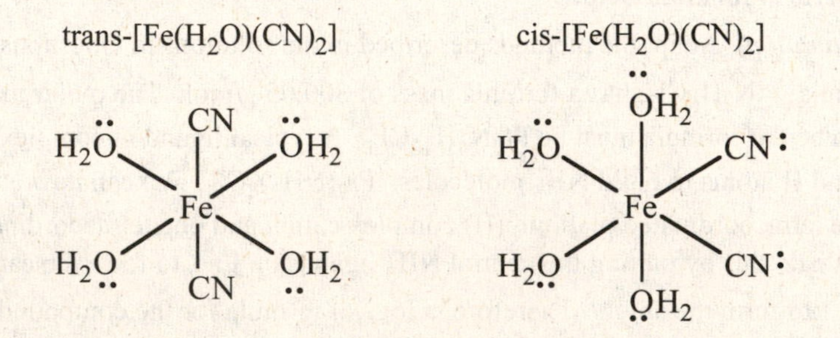

106. *Answer:* **see structural formula below**

Strategy and Explanation: The det ligand has three Lewis basic sites- two terminal amine nitrogens (NH_2), and an amide NH at the center of the det molecule. (Lewis basic species are compounds that can donate one or more lone pair of electrons to a Lewis acid, an electron pair acceptor). The Lewis acid-Lewis base bond is known as a coordinate covalent bond. In this case two det ligands form a total of 6 coordinate covalent bonds with iron(III) to give an octahedral coordination geometry.

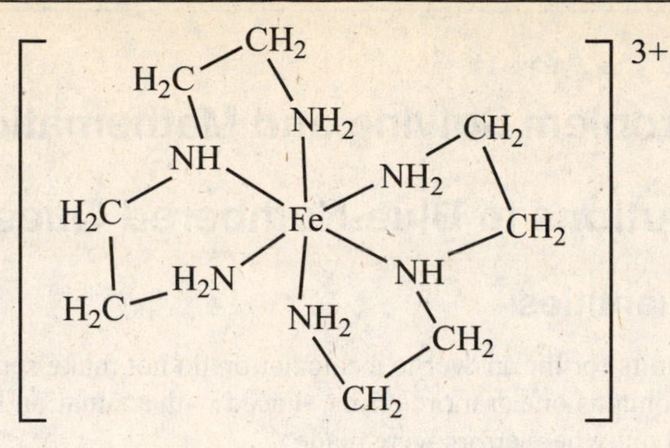

It is also possible for the terminal N atoms in each det ligand to be trans to each other.

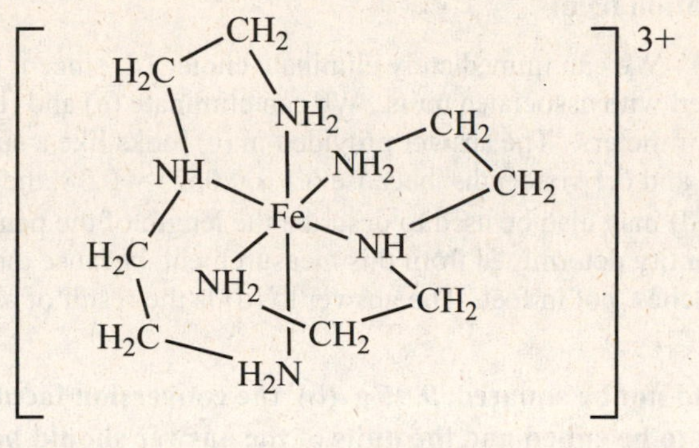

Appendix A: Problem Solving and Mathematical Operations

Solutions to Blue-Numbered Questions

Numbers, Units, and Quantities

4. *Answer/Explanation:* If the units for the answer to a calculation do not make sense, then we must conclude that the solution to the problem contains one or more errors. Faced with a situation like this, we must review the plan and its execution to find out where errors were made.

6. *Answer:* **(c); see explanation below.**

   Strategy and Explanation: We can immediately eliminate choice (d) since it has no units at all and the quantity measured must be reported with associated units. We can eliminate (a) and (b), since the units of this measured quantity will not be feet or meters. The answer provided in (c) looks like a suitable way to record the measurement of 8 inches and 6.1 sixteenths, because $6.1 \times 0.0625 = 0.38$, the decimal part of the reported quantity. The answer in (d) may also be used to describe the length of the pencil, but it should not be used to describe the physical quantity determined from this measurement, because the measuring device is calibrated in inches and fractions of inches, not in feet. The answer to (d) is the result of a conversion calculation, not the observed measurement.

8. *Answer:* **(a) units should not be squared; 9.95 g (b) the conversion factor is upside-down; 16.4 g (c) the conversion factor needs to be cubed and the units of the answer should be cm^3; 9.26 × 10^6 cm**

   Strategy and Explanation: Given flawed calculations, determine what is wrong and show the correct calculation and result for each.

   Review the calculation and determine what is wrong. Fix what is wrong and redo the calculation to get the answer.

   (a) The sum of two quantities with the same units will also have those units. The calculation shown indicates that two masses added together have units of mass squared. That is incorrect. The correct calculation looks like this: $4.32\text{ g} + 5.63\text{ g} = 9.95\text{ g}$

   (b) The conversion shown looks like it was intended to calculate the volume of a sample from the mass. If that is the case, then the conversion factor used is upside-down and the resulting units should be mL. The correct calculation looks like this:

   $$5.23\text{ g} \times \frac{1.00\text{ mL}}{4.87\text{ g}} = 1.07\text{ mL}$$

   (c) The conversion shown looks like it was intended to convert cubic centimeters to cubic meters. If that is the case, then the conversion factor needs to be cubed. The correct calculation looks like this:

   $$3.57\text{ cm}^3 \times \left(\frac{1\text{ m}}{100\text{ cm}}\right)^3 = 3.57 \times 10^{-6}\text{ m}^3$$

   ✓ *Reasonable Answer Check:* The units cancel properly and the size of the result of each calculation makes sense. In (a), when mass is removed, the resulting mass will be smaller and in units of mass. In (b), when the density is grater than 1, the mass quantity will have a larger numerical value than the volume quantity. In (c), the number of cubic centimeters will be much larger than the number of cubic meters, since a centimeter is much shorter than a meter.

Precision, Accuracy, and Significant Figures

10. *Answer:* **(a) 95.9 ±0.59 in (b) three (c) It is accurate.**

    Strategy and Explanation: Given several measured quantities for the length of a pole, determine what should be reported as the length, determine the number of significant figures to be reported, and, given the actual length, determine whether the result is accurate.

    (a) Add up the five individual measurements and divide by five to get the average:

$$\text{Average length of the pole} = \frac{95.31\text{ in} + 96.44\text{ in} + 96.02\text{ in} + 95.78\text{ in} + 95.94\text{ in}}{5} = \frac{479.49\text{ in}}{5} = 95.898\text{ in}$$

The scatter includes measurements lower than the average by as much as (95.898 – 95.31 =) 0.59 inches and higher than the average by as much as (96.44 – 95.898 =) 0.54 inches. So, the pole is reported to be 95.9 ± 0.59 inches long.

(b) The result should be reported with three significant figures, since the uncertainty is found in the first decimal place.

(c) 95.9 ± 0.59 inches. The pole's actual height is given as exactly 8 feet, which is 96 inches, since there are exactly 12 inches per foot. The actual height is within the range of the uncertainty of the calculated height, so the result is accurate, though not very precise.

12. *Answer:* **(a) Four (b) Two (c) Two (d) Four**

Strategy and Explanation: Given several measured quantities, determine the number of significant figures.

Use rules given in Sections 2.4 and A.3, summarized here: All non-zeros are significant. Zeros that precede (sit to the left of) non-zeros are never significant (e.g., 0.003). Zeros trapped between non-zeros are always significant (e.g., 3.003). Zeros that follow (sit to the right of non-zeros may be significant or may not be significant. They are significant if a decimal point is explicitly given (e.g., 3300.) and not significant, if a decimal point is not specified (e.g., 3300).

(a) 3.274 has **four** significant figures (The 3, 2, 7, and 4 digits are each significant.)

(b) 0.0034 L has **two** significant figures (The 3 and the 4 digits are each significant. The zeros are all before the first non-zero-digit, 3, and therefore they are not significant.)

(c) 43,000 m has **two** significant figures (The 4 and the 3 digits are significant. The zeros after the 3 are not significant because the number does not have an explicit decimal point specified.)

(d) 6200. mL has **four** significant figures (The 6 and 2 digits are each significant. The zeros after the 2 are also significant because the number has an explicit decimal point specified.)

✓ *Reasonable Answer Check:* The significant figures rules have been properly applied.

14. *Answer:* **(a) 43.32 (b) 43.32 (c) 43.32 (d) 43.32 (e) 43.32 (f) 43.32**

Strategy and Explanation: Given several measured quantities, round them to four significant figures. Use rules for rounding given in Sections 2.4 and A.3, as summarized here. If the last digit is below 5, then rounding does not change the digit before it. If the last digit is above a five, the digit before it is made one larger. If the last digit is exactly five, round the digit before it to an even number, up if odd and down if even.

(a) 43.3250 has six significant figures. To round it to four significant figures, we need to examine the fifth significant figure, the 5. Since the digit we're rounding is a 5, we look at the number before the 5, which is 2. Since 2 is even, we will round down and leave the 2 unchanged: **43.32**.

(b) 43.3165 has six significant figures. To round it to four significant figures, we need to look at the fifth significant figures, the 6. The removal of the fifth digit 6 rounds up the 1 next to it to a 2. The result is **43.32**.

(c) 43.3237 has six significant figures. To round it to four significant figures, we need to look at the fifth significant figures, the 3. The removal of 3 does not change the 2 next to it. The result is **43.32**.

(d) 43.32499 has seven significant figures. To round it to four significant figures, we need to look at the fifth significant figures, the 4. The removal of 4 doesn't change the 2 next to it. The result is **43.32**.

(e) 43.3150 has six significant figures. To round it to four significant figures, we need to look at the fifth significant figures, the 5. Since the digit we're rounding is a 5, we look at the number before the 5, which is 1. Since 1 is odd, we will round up and change the 1 to a 2: **43.32**.

(f) 43.32501 has seven significant figures. To round it to four significant figures, we need to examine the fifth significant figure, the 5. Since the digit we're rounding is a 5, we look at the number before the 5, which is 2. Since 2 is even, we will round down and leave the 2 unchanged: **43.32**.

✓ *Reasonable Answer Check:* Each answer has four significant figures.

16. *Answer:* **(a) 13.7 (b) 0.247 (c) 12.0**

Strategy and Explanation: Given some numbers combined using calculations, determine the result with proper significant figures.

Perform the mathematical steps according to order of operations, applying the proper significant figures (addition and subtraction retains the least number of decimal places in the result; multiplication and division retain the least number of significant figures in the result). *Notice: if operations are combined that use different rules, it is important to stop and determine the intermediate result any time the rule switches.*

(a) $$\frac{4.47}{0.3260}$$

The ratio uses the division rule. The numerator has three significant figures and the denominator has four significant figures, so the answer will have three significant figures. Therefore, we get **13.7**.

(b) $$\frac{4.03+3.325}{29.75}$$

The numerator uses the addition rule. The first number has two decimal places (the 0 and the 3 are both decimal places -- digit that follow the decimal point to the right) and the second number has three decimal places (the 3, the 2, and the 5 are all decimal places), so the result of the addition has two decimal places.

$$\frac{7.355}{29.75} \cong \frac{7.36}{29.75}$$

The ratio uses the division rule. The numerator has three significant figures and the denominator has four significant figures, so the answer will have three significant figures. Therefore, we get **0.247**.

(c) $$\frac{8.234}{5.673-4.987}$$

The denominator uses the subtraction rule. The first number has three decimal places (the 6, the 7, and the 3 are all decimal places) and the second number has three decimal places (the 9, the 8, and the 7 are all decimal places), so the result of the subtraction has three decimal places.

$$\frac{8.234}{0.686}$$

The ratio uses the division rule. The numerator has four significant figures and the denominator has three significant figures, so the answer will have three significant figures. Therefore, we get **12.0**.

✓ *Reasonable Answer Check:* The proper significant figures rules were used. The size and units of the answers are appropriate.

Exponential or Scientific Notation

18. *Answer:* **(a) 7.6003×10^4 (b) 3.7×10^{-4} (c) 3.4×10^4**

Strategy and Explanation: Adjust the appearance of the number to scientific notation (i.e., write it as a number between 1 and 9.999..., that is multiplied by ten to a whole-number power).

(a) 76,003 is $7.6003 \times 10{,}000$ or 7.6003×10^4. (b) 0.00037 is 3.7×0.0001 or 3.7×10^{-4}.

(c) 34,000 is $3.4 \times 10{,}000$ or 3.4×10^4.

20. *Answer:* **(a) 2.415×10^{-3} (b) 2.70×10^8 (c) 3.236 (d) 116 cm^3**

Strategy and Explanation: Given some numbers combined using calculations, determine the result with proper significant figures.

Perform the mathematical steps according to order of operations, applying the proper significant figures (addition and subtraction retains the least number of decimal places in the result, or rounds the result to the position of greatest uncertainty; multiplication and division retain the least number of significant figures in the result).

(a) $$\frac{0.7346}{304.2}$$

The ratio uses the division rule. The numerator has four significant figures and the denominator has four significant figures, so the answer will have four significant figures. Therefore, we get **2.415×10^{-3}**.

(b) $$\frac{(3.45\times10^{-3})(1.83\times10^{12})}{23.4}$$

The calculation uses the division and multiplication rules. The numbers in the numerator have three significant figures and the denominator has three significant figures, so the answer will have three significant figures. Therefore, we get **2.70×10^{8}**.

(c) $$3.240 - 4.33 \times 10^{-3}$$

The calculation uses the subtraction rule. The first number has three decimal places (the 2, the 4, and the 0 are all decimal places) and the second number, 4.33×10^{-3} is also 0.00433, so it has five decimal places (the zeros after the decimal point, the 4, the 3, and the 3 are all decimal places), so the result of the subtraction has three decimal places.

$$3.240 - 0.00433 = \mathbf{3.236}$$

(d) $$(4.87\text{cm})^3 = 4.87\text{cm} \times 4.87\text{cm} \times 4.87\text{cm}$$

The calculation uses the multiplication rule. The numbers have three significant figures, so the answer will have three significant figures. Therefore, we get **116 cm^3**.

✓ *Reasonable Answer Check:* The proper significant figures rules were used. Size and units are appropriate.

Logarithms

22. *Answer:* **(a) –0.1351 (b) 3.541 (c) 23.7797 (d) 54.7549 (e) –7.455**

Strategy and Explanation: Given some numbers and calculations using logarithms, determine the answers with appropriate significant figures.

Use the information in Section A.6 describing the operations of logarithms and their significant figures. When taking the log of a number, the mantissa (the digits to the right of the decimal point) should have as many significant figures as the numbers whose log was found. So, the number of significant figures in the number we start with gives the number of decimal places in the result.

(a) The log of 4 significant figures, gives an answer with 4 decimal places: log(0.7327) = –0.1351

(b) The ln of 3 significant figures, gives an answer with 3 decimal places: ln(34.5) = 3.541

(c) The log of 4 significant figures, gives an answer with 4 decimal places: $\log(6.022 \times 10^{23}) = 23.7797$

(d) The ln of 4 significant figures, gives an answer with 4 decimal places: $\ln(6.022 \times 10^{23}) = 54.7549$

(e) The ratio of 3 significant figures gives a result with 3 significant figures. The log of 3 significant figures, gives an answer with 3 decimal places:

$$\log\left(\frac{8.34\times10^{-5}}{2.38\times10^{3}}\right)=\log(3.50 \times 10^{-8}) = -7.455$$

✓ *Reasonable Answer Check:* For log calculations, the power of ten is approximately reflected in the ordinate (the number to the right of the decimal point), as in (e), where we see –7.xx when the number was a little larger than 10^{-8}. For ln calculations, the power of ten will be approximately reflected as the ordinate times 2.303. The answer in (d) should be 2.303 times the answer in (c). The ratio is 2.3034 times.

24. *Answer:* **(a) 5.404 (b) 1×10^{15} (c) 110. (d) 1.000320 (e) 3.755**

Strategy and Explanation: Given some numbers and calculations using antilogarithms, determine the answers with appropriate significant figures.

Use the information in Section A.6 describing the operations of antilogarithms and their significant figures.

When taking the antilog of a number, the number of decimal places in the number whose antilog was taken should be the same as the number of significant figures in the result of the antilogarithm operation.

(a) The antilog of 4 decimal places, gives an answer with 4 significant figures: antilog(0.7327) = 5.404

(b) The antiln of 1 decimal place, gives an answer with 1 significant figure: ln(34.5) = 1×10^{15}

(c) The 10^x of 3 decimal places, gives an answer with 3 significant figures: $10^{2.043} = 110$.

(d) The number 3.20×10^{-4} is also 0.000320. Therefore, it has 6 decimal places. The e^x of 6 decimal place, gives an answer with 6 significant figure: : $e^{0.000320} = 1.000320$

(e) The ratio of 4 significant figures gives a result with 4 significant figures. The exp of 4 decimal places, gives an answer with 4 significant figures:

$$\exp\left(\frac{4.333}{3.275}\right) = \exp(1.323) = 3.755$$

✓ *Reasonable Answer Check:* Antilogarithms of small numbers (i.e., numbers near zero) give answers near one. Antilog on larger numbers, give answers with large powers of ten.

Quadratic Equations

26. *Answer:* **(a) 0.480 and –1.80 (b) 3.23 and 1.94**

Strategy and Explanation: Given quadratic equations, find the roots.

Set up the quadratic equation in the form of: $ax^2 + bx + c = 0$, then plug into the quadratic equation:

$$x = \frac{-b \pm \sqrt{b^2 - 4ac}}{2a}$$

The two roots are generated when we use the "+" or the "–" of the "±" in this equation.

(a) $3.27x^2 + 4.32x - 2.83 = 0$, so, a = 3.27, b = 4.32, and c = –2.83:

$$x = \frac{-4.32 \pm \sqrt{4.32^2 - 4(3.27)(-2.83)}}{2(3.27)}$$

$$x = \frac{-4.32 + 7.46}{6.54} = \frac{3.14}{6.54} = 0.480$$

$$x = \frac{-4.32 - 7.46}{6.54} = \frac{-11.78}{6.54} = -1.80$$

(b) Rearrange: $x^2 + 4.32 = 4.57x$ *(Note: The answer for (b) in Appendix M is incorrect.)*

$x^2 - 4.57x + 4.32 = 0$, so, a = 1, b = –4.57, and c = 4.32:

$$x = \frac{-(-4.57) \pm \sqrt{(-4.57)^2 - 4(1)(4.32)}}{2(1)}$$

$$x = \frac{4.57 + 1.90}{2} = \frac{6.47}{2} = 3.23$$

$$x = \frac{4.57 - 1.90}{2} = \frac{2.67}{2} = 1.34$$

✓ *Reasonable Answer Check:* Each root should solve the equation:

(a) $3.27(0.480)^2 + 4.32(0.480) - 2.83 = 0.753 + 2.07 - 2.83 = 0.00$

$3.27(-1.80)^2 + 4.32(-1.80) - 2.83 = 10.6 - 7.78 - 2.83 = 0.0$

(b) $(3.23)^2 + 4.32 = 10.4 + 4.32 = 14.8$ matches $4.57(3.23) = 14.8$

$(1.34)^2 + 4.32 = 1.80 + 4.32 = 6.12$ matches $4.57(1.34) = 6.12$

Graphing

28. *Answer:* **see graph below; yes**

Strategy and Explanation: Given the masses of several volumes of a sample, draw the graph and determine if mass is directly proportional to volume.

Plot the graph and see if the data fits a linear equation.

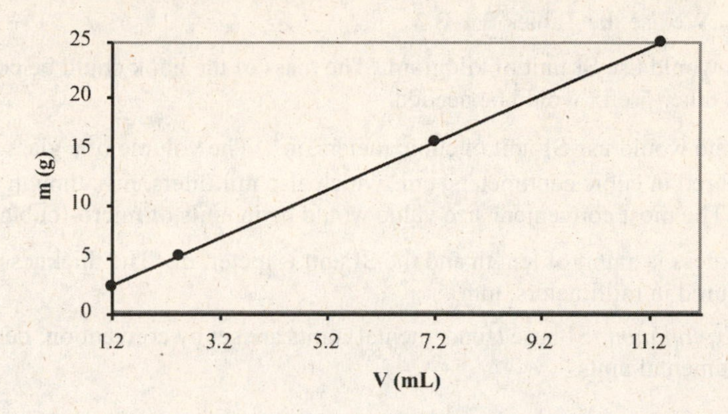

The graph is linear so the mass is directly proportional to volume.

✓ *Reasonable Answer Check:* Mass and volume are extensive properties that are related to each other through a substances fixed density, so it makes sense that the graph is linear, since we expect the mass is directly proportional to volume.

29. *Answer:* **see graph below**

Strategy and Explanation: Given the masses of several volumes of a sample, draw the graph and determine if heat evolved is directly proportional to the amount of reactant.

Plot the graph and see if the data fits a linear equation.

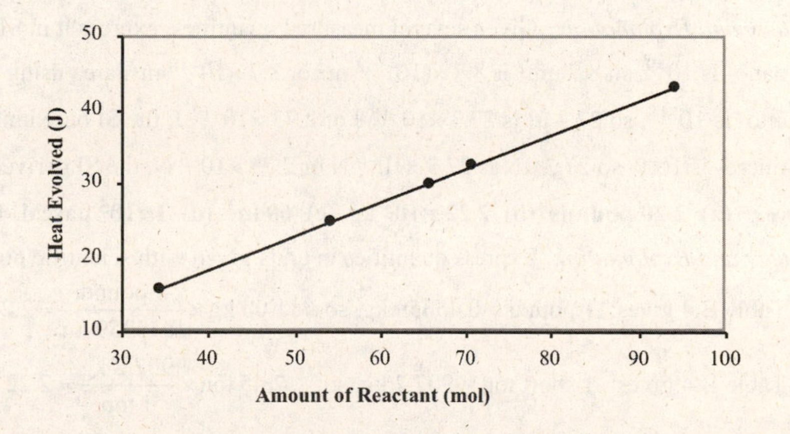

The graph is linear so the mass is directly proportional to volume.

✓ *Reasonable Answer Check:* Mass and volume are extensive properties that are related to each other through a substances fixed density, so it makes sense that the graph is linear, since we expect the mass is directly proportional to volume.

Appendix B: Units, Equivalences, and Conversion Factors

Solutions to Blue-Numbered Questions

Units of the International System

1. *Answer/Explanation:* Determine which SI units and prefix would be used to measure a mass, a volume, and a thickness. We use the Tables B.1-B.3.

   (a) Mass would use SI unit of kilogram. The mass of the book could be conveniently measured in kilograms, so no other prefix would be needed.

   (b) Volume would use SI unit of cubic meters, m^3. The volume of a glass of water could be conveniently measured in cubic centimeters, cm^3 which also milliliters, mL, though that volume will still be hundreds of mL. The most convenient size value would be in units of micro-(cubic meters) or nano-(cubic meters).

   (c) Thickness is a unit of length and the SI unit is meter, m. The thickness of this page could be conveniently measured in millimeters, mm.

3. *Answer/Explanation:* SI base (fundamental) units are set by convention; derived units are derived from the SI base fundamental units.

Conversion of Units for Physical Quantities

5. *Answer:* **(a) 4.75×10^{-10} m (b) 5.6×10^{7} kg (c) 4.28×10^{-6} A**

   Strategy and Explanation: Given several measured quantities, express it in SI base units.

   (a) pico- is 10^{-12}, so 475 pm is 475 $\times 10^{-12}$ m or 4.75 $\times 10^{-10}$ m, the SI base unit of length.

   (b) giga- is 10^{9}, so 56 Gg is 56 $\times 10^{9}$ g or 5.6 $\times 10^{10}$ g, which is 5.6 $\times 10^{7}$ kg, the SI base unit of mass.

   (c) micro- is 10^{-6}, so 4.28 μA is 4.28 $\times 10^{-6}$ A, the SI base unit of electric charge.

7. *Answer:* **(a) 8.7×10^{-18} m^2 (b) 2.73×10^{-17} J (c) 2.73×10^{-5} N**

   Strategy and Explanation: Given several measured quantities, express it in SI base units.

   (a) nano- is 10^{-9}, so 8.7 nm^2 is 8.7 $\times (10^{-9})^2$ m^2 or 8.7 $\times 10^{-18}$ m^2, area using the SI base unit of length.

   (b) atto- is 10^{-18}, so 27.3 aJ is 27.3 $\times 10^{-18}$ J or 2.73 $\times 10^{-17}$ J, the SI base unit of energy.

   (c) micro- is 10^{-6}, so 27.3 μN is 27.3 $\times 10^{-6}$ N or 2.73 $\times 10^{-5}$ N, the SI derived unit for force.

9. *Answer:* **(a) 2.20 pounds (b) 2.22×10^{3} kg (c) 60 in^3 (d) 1×10^{5} pascal, 1 bar**

   Strategy and Explanation: Express quantities in units given with scientific notation.

   (a) Table B.4 gives: 1 pound = 0.45359 kg, so: $1.00\ \text{kg} \times \dfrac{1\ \text{pound}}{0.45359\ \text{kg}} = 2.20\ \text{pounds}$

   (b) Table B.4 gives: 1 short ton = 907.2 kg, so: $2.45\ \text{ton} \times \dfrac{907.2\ \text{kg}}{1\ \text{ton}} = 2.22 \times 10^{3}\ \text{kg}$

   (c) Table B.4 gives: 1 L = 1000 cm^3 and 1 inch = 2.54 cm so: $1\ \text{L} \times \dfrac{1000\ \text{cm}^3}{1\ \text{L}} \times \left(\dfrac{1\ \text{in}}{2.54\ \text{cm}}\right)^3 = 6 \times 10^{1}\ \text{in}^3$

   (d) Table B.4 gives: 1 atm = 1.01325 $\times 10^{5}$ pascals and 1 bar = 10^{5} pascals, so:

   $$1\ \text{atm} \times \frac{1.01325 \times 10^{5}\ \text{pascals}}{1\ \text{atm}} = 1 \times 10^{5}\ \text{pascals} \quad \text{and} \quad 1 \times 10^{5}\ \text{pascals} \times \frac{1\ \text{bar}}{10^{5}\ \text{pascals}} = 1\ \text{bar}$$

11. *Answer:* **(a) 99° F (b) – 10.5 °F (c) – 40.0 °F**

   Strategy and Explanation: Use temperature conversion given in Table B.4: $t_F = \frac{9}{5} t_C + 32$

   (a) $t_F = \frac{9}{5}(37\ ^\circ\text{C}) + 32 = 99\ ^\circ\text{F}$ (b) $t_F = \frac{9}{5}(-23.6\ ^\circ\text{C}) + 32 = -10.5\ ^\circ\text{F}$ (c) $t_F = \frac{9}{5}(-40.0\ ^\circ\text{C}) + 32 = -40.0\ ^\circ\text{F}$